Lecture Notes in Artificial Intelligence 9227

Subseries of Lecture Notes in Computer Science

More information about this series at http://www.springer.com/series/1244

Editors
De-Shuang Huang
Tongji University
Shanghai
China

Kyungsook Han
Incheon
Korea, Republic of (South Korea)

ISSN 0302-9743   ISSN 1611-3349   (electronic)
Lecture Notes in Artificial Intelligence
ISBN 978-3-319-22052-9   ISBN 978-3-319-22053-6   (eBook)
DOI 10.1007/978-3-319-22053-6

Library of Congress Control Number: 2015945139

LNCS Sublibrary: SL7 – Artificial Intelligence

Springer Cham Heidelberg New York Dordrecht London

Printed on acid-free paper

Springer International Publishing AG Switzerland is part of Springer Science+Business Media
(www.springer.com)

De-Shuang Huang · Kyungsook Han (Eds.)

Advanced Intelligent Computing Theories and Applications

11th International Conference, ICIC 2015
Fuzhou, China, August 20–23, 2015
Proceedings, Part III

 Springer

Preface

The International Conference on Intelligent Computing (ICIC) was started to provide an annual forum dedicated to the emerging and challenging topics in artificial intelligence, machine learning, pattern recognition, bioinformatics, and computational biology. It aims to bring together researchers and practitioners from both academia and industry to share ideas, problems, and solutions related to the multifaceted aspects of intelligent computing.

ICIC 2015, held in Fuzhou, China, August 20–23, 2015, constituted the 11th International Conference on Intelligent Computing. It built upon the success of ICIC 2014, ICIC 2013, ICIC 2012, ICIC 2011, ICIC 2010, ICIC 2009, ICIC 2008, ICIC 2007, ICIC 2006, and ICIC 2005 that were held in Taiyuan, Nanning, Huangshan, Zhengzhou, Changsha, China, Ulsan, Korea, Shanghai, Qingdao, Kunming, and Hefei, China, respectively.

This year, the conference concentrated mainly on the theories and methodologies as well as the emerging applications of intelligent computing. Its aim was to unify the picture of contemporary intelligent computing techniques as an integral concept that highlights the trends in advanced computational intelligence and bridges theoretical research with applications. Therefore, the theme for this conference was "Advanced Intelligent Computing Theories and Applications." Papers focusing on this theme were solicited, addressing theories, methodologies, and applications in science and technology.

ICIC 2015 received 671 submissions from 25 countries and regions. All papers went through a rigorous peer-review procedure and each paper received at least three review reports. On the basis of the review reports, the Program Committee finally selected 233 high-quality papers for presentation at ICIC 2015, included in three volumes of proceedings published by Springer: two volumes of *Lecture Notes in Computer Science* (LNCS) and one volume of *Lecture Notes in Artificial Intelligence* (LNAI).

This volume of *Lecture Notes in Artificial Intelligence* (LNAI) 9227 includes 84 papers.

The organizers of ICIC 2015, including Tongji University and Fujian Normal University, China, made an enormous effort to ensure the success of the conference. We hereby would like to thank the members of the Program Committee and the reviewers for their collective effort in reviewing and soliciting the papers. We would like to thank Alfred Hofmann, executive editor at Springer, for his frank and helpful advice and guidance throughout and for his continuous support in publishing the proceedings. In particular, we would like to thank all the authors for contributing their papers. Without the high-quality submissions from the authors, the success of the conference would not have been possible. Finally, we are especially grateful to the International Neural Network Society and the National Science Foundation of China for their sponsorship.

June 2015

De-Shuang Huang
Kyungsook Han

ICIC 2015 Organization

General Co-chairs

De-Shuang Huang   China
Changping Wang   China

Program Committee Co-chairs

Kang-Hyun Jo   Korea
Abir Hussain   UK

Organizing Committee Co-chairs

Hui Li   China
Yi Wu   China

Award Committee Chair

Laurent Heutte   France

Publication Co-chairs

Valeriya Gribova   Russia
Zhi-Gang Zeng   China

Special Session Co-chairs

Phalguni Gupta   India
Henry Han   USA

Special Issue Co-chairs

Vitoantonio Bevilacqua   Italy
Mike Gashler   USA

Tutorial Co-chairs

Kyungsook Han   Korea
M. Michael Gromiha   India

International Liaison

Prashan Premaratne   Australia

Publicity Co-chairs

Juan Carlos Figueroa   Colombia
Ling Wang   China
Evi Syukur   Australia
Chun-Hou Zheng   China

Exhibition Chair

Bing Wang   China

Organizing Committee

Jianyong Cai   China
Qingxiang Wu   China
Suping Deng   China
Lin Zhu   China

Program Committee

Andrea Francesco Abate, Italy
Waqas Haider Khan Bangyal, Pakistan
Shuhui Bi, China
Qiao Cai, USA
Jair Cervantes, Mexico
Chin-Chih Chang, Taiwan, China
Chen Chen, USA
Huanhuan Chen, China
Shih-Hsin Chen, Taiwan, China
Weidong Chen, China
Wen-Sheng Chen, China
Xiyuan Chen, China
Yang Chen, China
Cheng Cheng, China
Ho-Jin Choi, Korea
Angelo Ciaramella, Italy
Salvatore Distefano, Italy
Jianbo Fan, China
Minrui Fei, China
Juan Carlos Figueroa, Colombia

Huijun Gao, China
Shan Gao, China
Dunwei Gong, China
M. Michael Gromiha, India
Zhi-Hong Guan, China
Kayhan Gulez, Turkey
Phalguni Gupta, India
Fei Han, China
Kyungsook Han, Korea
Laurent Heutte, France
Wei-Chiang Hong, Taiwan, China
Yuexian Hou, China
Peter Hung, Ireland
Saiful Islam, India
Li Jia, China
Zhenran Jiang, China
Xin Jin, USA
Joaquín Torres-Sospedra, Spain
Dah-Jing Jwo, Taiwan, China
Vandana Dixit Kaushik, India

Additional Reviewers

Haiqing Li
JinLing Niu
Li Wei
Peipei Xiao
Dong Xianguang
Shasha Tian
Li Liu
Xuan Wang
Sen Xia
Jun Li
Xiao Wang
Anqi Bi
Min Jiang
Wu Yang
Yu Sun
Bo Wang
Wujian Fang
Zehao Chen
Chen Zheng
Hong Zhang
Huihui Wang
Liu Si
Sheng Zou
Xiaoming Liu
Meiyue Song
Bing Jiang
Min-Ru Zhao
Xiaoyong Bian
Chuang Ma
Yujin Wang
Wenlong Hang
Chong He
Rui Wang
Zhu Linhe
Xin Tian
Mou Chen
F.F. Zhang
Gabriela Ramírez
Ahmed Aljaaf
Neng Wan
Abhineet Anand
Antonio Celesti
Aditya Nigam
Aftab Yaseen

Weiyuan Zhang
Shamsul Zulkifli
Alfredo Liverani
Siti Amely Jumaat
Muhammad Amjad
Angelo Ciaramella
Aniza Mohamed Din
En-Shiun Annie Lee
Anoosha Paruchuri
Antonino Staiano
Antony Lam
Alfonso Panarello
Alfredo Pulvirenti
Asdrubal Lopez-Chau
Chun Kit Au
Tao Li
Mohd Ayyub Khan
Azizi Ab Aziz
Azizul Azhar Ramli
Baoyuan Wu
Oscar Belmonte
Biao Li
Jian Gao
Lin Yong
Boyu Zhang
Edwin C. Shi
Caleb Rascon
Yu Zhou
Fuqiang Chen
Fanshu Chen
Chenguang Zhu
Chin-Chih Chang
Chang-Chih Chen
Changhui Lin
Cheng-Hsiung Chiang
Linchuan Chen
Bo Chen
Chen Chen
Chengbin Peng
Cheng Cheng
Jian Chen
Qiaoling Chen
Zengqiang Chen
Ching-Hua Shih

Chua King Lee
Quistina Cuci
Yanfeng Chen
Meirong Chen
Cristian Rodriguez Rivero
Aiguo Chen
Aquil Mirza Mohammed
Sam Kwong
Yang Liu
Yizhang Jiang
Bingbo Cui
Yu Wu
Chuan Wang
Wenbin Chen
Cheng Zhang
Dhiya Al-Jumeily
Hang Dai
Yinglong Dai
Danish Jasnaik
Kui Liu
Dapeng Li
Dan Yang
Davide Nardone
Dawen Xu
Dongbo Bu
Liya Ding
Shifei Ding
Donal O'Regan
Dong Li
Li Kuang
Zaynab Ahmed
David Shultis
Zhihua Du
Chang-Chih Chen
Erkan İMal
Ekram Khan
Eric Wei
Gang Wang
Fadzilah Siraj
Shaojing Fan
Mohammad Farhad
    Bulbul
Mohamad Farhan
    Mohamad Mohsin

Fengfeng Zhou
Feng Jiqiang
Liangbing Feng
Farid García-Lamont
Chien-Yuan Lai
Filipe de O. Saraiva
Francesco Longo
Fabio Narducci
Francesca Nardone
Francesco Camastra
Nhat Linh Bui
Hashim Abdellah
Hashim Moham
Gao Wang
Jungwon Yu
Ge Dingfei
Geethan Mendiz
Na Geng
Fangda Guo
Gülsüm Gezer
Guanghua Sun
Guanghui Wang
Rosalba Giugno
Giovanni Merlino
Xiaoqiang Zhang
Guangchun Cheng
Guanglan Zhang
Tiantai Guo
Weili Guo
Yanhui Guo
Xiaoqing Gu
Hafizul Fahri Hanafi
Haifeng Wang
Mohamad Hairol Jabbar
H.K. Lam
Khalid Isa
Hang Su
Guangjie Han
Hao Chu
Ben Ma
Hao Men
Yonggang Chen
Haza Nuzly
   Abdull Hamed
Mohd Helmy Abd Wahab
   Pang
Hei Man Herbert

Hironobu Fujiyoshi
Guo-Sheng Hao
Huajuan Huang
Hiram Calvo
Hongjie Wu
Hongjun Su
Hitesh Kumar Sharma
Haiguang Li
Hongkai Chen
Qiang Huang
Tengfei Zhang
Zheng Huai
Joe Huang
Jin Huang
Wan Hussain Wan Ishak
Xu Huang
Ying Hu
Ho Yin Sze-To
Haitao Zhu
Ibrahim Venkat
Jooyoung Lee
Josef Moudřk
Li Xu
Jakub Smid
Jianhung Chen
Le Li
Jianbo Lu
Jair Cervantes
Junfeng Xia
Jinhai Li
Hongmei Jiang
Jiaan Zeng
Jian Lu
Jian Wang
Jian Zhang
Jianhua Zhang
Jie Wu
Jim Jing-Yan Wang
Jing Sun
Jingbin Wang
Wu Qi
Jose Sergio Ruiz Castilla
Joaquín Torres
Gustavo Eduardo Juarez
Jun Chen
Junjiang Lin
Junlin Chang

Juntao Liu
Justin Liu
Jianzhong Guo
Jiayin Zhou
Abd Kadir Mahamad
K. Steinhofel
Ka-Chun Wong
Ke Li
Kazuhiro Fukui
Sungshin Kim
Klara Peskova
Kunikazu Kobayashi
Konstantinos Tsirigos
Seeja K.R.
Kwong Sak Leung
Kamlesh Tiwari
Li Kuang
K.V. Arya
Zhenjiang Lan
Ke Liao
Liang-Tsung Huang
Le Yang
Erchao Li
Haitao Li
Wei Li
Meng Lei
Guoqi Luo
Huihui Li
Jing Liang
Liangliang Zhang
Liang Liang
Bo Li
Bing Li
Dingshi LI
Lijiao Liu
Leida Li
Lvzhou Li
Min Li
Kui Lin
Ping Li
Liqi Yi
Lijun Quan
Bingwen Liu
Haizhou Liu
Jing Liu
Li Liu
Ying Liu

Zhaoqi Liu
Zhe Liu
Yang Li
Lin Zhu
Jungang Lou
Lenka Kovářová
Shaoke Lou
Li Qingfeng
Qinghua Li
Lu Huang
Liu Liangxu
Lili Ayu Wulandhari
Xudong Lu
Yiping Liu
Yutong Li
Junming Zhang
Mohammed Khalaf
Maria Musumeci
Shingo Mabu
Yasushi Mae
Manzoor Lone
Liang Ma
Manabu Hashimoto
Md. Abdul Mannan
Qi Ma
Martin Pilat
Asad Khan
Maurizio Fiasché
Max Talanov
Mohd Razali MD Tomari
Mengxing Cheng
Meng Xu
Tianyu Cao
Minfeng Wang
Muhammad Fahad
Michele Fiorentino
Michele Scarpiniti
Zhenmin Zhang
Ming Liu
Miguel Mora-Gonzalez
Aul Montoliu
Yuanbin Mo
Marzio Pennisi
Binh P. Nguyen
Mingqiang Zhang
Muhammad Rashid
Shenglin Mu

Musheer Ahmad
Monika Verma
Naeem Radi
Aditya Nigam
Aditya Nigam
Nagarajan Raju
Mohd Najib Mohd Salleh
Yinan Guo
Zhu Nanli
Mohammad Naved
    Qureshi
Su Rina
Sanders Liu
Chang Liu
Patricio Nebot
Nistor Grozavu
Zhixuan Wei
Nobuyuki Nezu
Nooraini Yusoff
Nureize Arbaiy
Kazunori Onoguchi
R.B. Pachori
Yulei Pang
Xian Pan
Binbin Pan
Peng Chen
Klara Peskova
Petra Vidnerová
Qi Liu
Prashan Premaratne
Puneet Gupta
Prabhat Verma
Peng Zhang
Haoqian Huang
Qiang Fu
Qiao Cai
Haiyan Qiao
Qingnan Zhou
Rabiah Ahmad
Rabiah Abdul Kadir
R. Rakkiyappan
Ramakrishnan
Chandrasekaran
Yunpeng Wang
Hao Zheng
Radhakrishnan Delhibabu
Rohit Katiyar

Wei Cui
Rozaida Ghazali
Raghuraj Singh
Rey-Sern Lin
S.M. Zakariya
Sabooh Ajaz
Alexander Tchitchigin
Kuo-Feng Huang
Shao-Lun Lee
Wei-Chiang Hong
Toshikazu Samura
Sandhya Pundhir
Jin-Xing Liu
Shahzad Alam
Mohd Shamrie Sainin
Shanye Yin
Shasha Tian
Hao Shen
Chong Shen
Jingsong Shi
Nobutaka Shimada
Lu Xingjia
Shun Chen
Silvio Barra
Simone Scardapane
Salvador Juarez Lopez
Shenshen Liang
Sheng Liu
Somnath Dey
Rui Song
Seongpyo Cheon
Sergio Trilles-Oliver
Subir Kumar Nandy
Hung-Chi Su
Sun Jie
QiYan Sun
Shiying Sun
Sushil Kumar
Yu Su
Hu Zhang
Yan Qi
Hotaka Takizawa
Tanggis Bohnuud
Yang Tang
Lirong Tan
Yao Tuozhong
Tian Tian

Tianyi Wang

Toshiaki Kondo

Tofik Ali

Tomáš Ken

Peng Xia

Gurkan Tuna

Tutut Herawan

Zhongneng Xu

Danny Wu

Umarani Jayaraman

Zhichen Gong

Vibha Patel

Vikash Yadav

Esau Villatoro-Tello

Prashan Premaratne

Vishnu Priya Kanakaveti

Vivek Srivastava

Victor Manuel
  Landassuri-Moreno

Hang Su

Yi Wang

Chao Wang

Cheng Wang

Jiahai Wang

Jing Wang

Junxiao Wang

Linshan Wang

Waqas Haider Bangyal

Herdawatie Abdul Kadir

Yi Wang

Wen Zhang

Widodo Budiharto

Wenjun Deng

Wen Wei

Wenbin Chen

Wenzhe Jiao

Yufeng Wang

Haifeng Wang

Deng Weilin

Wei Jiang

Weimin Huang

Wufeng Tian

Wu-Chen Su

Daiyong Wu

Yang Wu

Zhonghua Wu

Jun Zhang

Xiao Wang

Wei Xiong

Weixiang Liu

Wenxi Zhang

Wenye Li

Wenlin Zhang

Li-Xin He

Ming Tan

Yi-Gang Zhang

Xiangliang Zhang

Nan Xiang

Xiaobo Zhang

Xiaohu Wang

Xiaolei Wang

Xiaomo Liu

Lei Wang

Yan Cui

Xiaoyong Zhang

Xiaozhen Xue

Jianming Xie

Xin Gao

Xinhua Xiao

Xin Lu

Xinyu Zhang

Liguang Xu

Hao Wu

Xun Li

Jin Xu

Xin Xu

Yuan Xu

Xiaoyin Xu

Yuan Xu

Shiping Chen

Xiaoyan Sun

Xiaopin Zhong

Atsushi Yamashita

Yanen Guo

Xiaozhan Yang

Zhanlei Yang

Yaqiang Yao

Bei Ye

Yan Fu

Yehu Shen

Yu-Yen Ou

Yingyou Wen

Ying Yang

Yingtao Zhao

Yanjun Zhao

Yong Zhang

Yoshinori Kobayashi

Yesu Feng

Yuan Lin

Lin Yuan

Yugandhar Kumar

Yujia Li

Yupeng Li

Yuting Yang

Yuyan Han

Yong Wang

Yingying Wei

Yingxin Guo

Xiangjuan Yao

Guodong Zhao

Huan Zhang

Bo Zhang

Gongjie Zhang

Yu Zhang

Zhao Yan

Wenrui Zhao

Zhao Yan

Tian Zheng

Zhengxing Hang

Xibei Yang

Zhenxin Zhan

Zhipeng Cai

Lingyun Zhu

Yanbang Zhang

Zheng-Ling Yang

Juan Li

Zongxiao He

Zhengyu Ouyang

Will Zhu

Xuebing Zhang

Zhile Yang

Yi Zhang

Contents

Machine Learning Theory and Methods

A Review of Parameter Learning Methods in Bayesian Network.   3
   Zhiwei Ji, Qibiao Xia, and Guanmin Meng

The Hard-Cut EM Algorithm for Mixture of Sparse Gaussian Processes   13
   Ziyi Chen and Jinwen Ma

Maximum Principle in the Unbounded Domain of Heisenberg Type Group. . .   25
   Zhenhua Wang and Xuemei Yang

An Epidemic Propagation Model with Saturated Infection Rate on a Small
World Network.   34
   Qiao-ling Chen, Liang Chen, Zhi-Qiang Sun, and Zhi-juan Jia

Granular Twin Support Vector Machines Based on Mixture
Kernel Function   43
   Xiuxi Wei and Huajuan Huang

Detecting Multiple Influential Observations in High Dimensional
Linear Regression   55
   Junlong Zhao, Ying Zhang, and Lu Niu

Drift Operator for States of Matter Search Algorithm.   65
   Yuxiang Zhou, Yongquan Zhou, Qifang Luo, Shilei Qiao, and Rui Wang

Classification on Imbalanced Data Sets, Taking Advantage of Errors
to Improve Performance.   72
   Asdrúbal López-Chau, Farid García-Lamont, and Jair Cervantes

PSO-Based Method for SVM Classification on Skewed Data-Sets.   79
   Jair Cervantes, Farid García-Lamont, Asdrúbal López,
   Lisbeth Rodriguez, José S. Ruiz Castilla, and Adrián Trueba

Set-Based Many-Objective Optimization Guided by Preferred Regions   87
   Dunwei Gong, Fenglin Sun, Jing Sun, and Xiaoyan Sun

An Improved Incremental Error Minimized Extreme Learning Machine
for Regression Problem Based on Particle Swarm Optimization   94
   Fei Han, Min-Ru Zhao, and Jian-Ming Zhang

A Systematic Comparison and Evaluation of Supervised Machine Learning
Classifiers Using Headache Dataset...........................   101
  Ahmed J. Aljaaf, Dhiya Al-Jumeily, Abir J. Hussain, Paul Fergus,
  Mohammed Al-Jumaily, and Naeem Radi

A Generative Model with Ensemble Manifold Regularization
for Multi-view Clustering.....................................   109
  Shaokai Wang, Yunming Ye, and Raymond Y.K. Lau

On Classifying Diabetic Patients' with Proliferative Retinopathies
via a Radial Basis Probabilistic Neural Network   115
  Leonarda Carnimeo and Rosamaria Nitti

Passivity Analysis of BAM NNs with Mixed Time Delays.............   127
  Weiyuan Zhang and Yajie Wang

Soft Computing

A Fuzzy Logic Controller for Indirect Matrix Converter Under Abnormal
Input Voltage Conditions.....................................   139
  Quoc-Hoan Tran and Hong-Hee Lee

On Denjoy-McShane-Stieltjes Integral of Fuzzy-Number-Valued Functions...   151
  Wenkai Shao, Jiechang Ruan, and Shu Gong

A Brief Survey on Fuzzy Cognitive Maps Research   159
  Yajie Wang and Weiyuan Zhang

Sensor Data Driven Modeling and Control of Personalized Thermal
Comfort Using Interval Type-2 Fuzzy Sets.......................   167
  Chengdong Li, Weina Ren, Huidong Wang, and Jianqiang Yi

Knowledge Evaluation with Rough Sets..........................   179
  Sylvia Encheva and Torleiv Ese

On Solving CCR-DEA Problems Involving Type-2 Fuzzy Uncertainty
Using Centroid-Based Optimization............................   187
  Juan Carlos Figueroa-García and Carlos Eduardo Castro-Cabrera

Image Processing and Computer Vision

Adaptive Quantization of Local Directional Responses for Infrared Face
Recognition   199
  Zhihua Xie, Zhengzi Wang, and Guodong Liu

Preliminary Study of Tongue Image Classification Based on Multi-label
Learning   208
  XinFeng Zhang, Jing Zhang, GuangQin Hu, and YaZhen Wang

Online Kernel-Based Multimodal Similarity Learning with Application
to Image Retrieval.   221
  Wenping Zhang and Hong Zhang

Reverse Training for Leaf Image Set Classification   233
  Yu-Hui Zhang, Ji-Xiang Du, Jing Wang, and Chuan-Min Zhai

Visual Saliency Detection Based on Color Contrast and Distribution   243
  Yanbang Zhang and Guolong Fan

Moving Vehicle Detection Based on Visual Processing Mechanism
with Multiple Pathways   251
  YanFeng Chen, QingXiang Wu, HaiHui Xie, SanLiang Hong, and Xue Li

A Computer Vision Method for the Italian Finger Spelling Recognition.   264
  Vitoantonio Bevilacqua, Luigi Biasi, Antonio Pepe,
  Giuseppe Mastronardi, and Nicholas Caporusso

A Spiking Neural Network for Extraction of Multi-features in Visual
Processing Pathways   275
  QiYan Sun, QingXiang Wu, Xuan Wang, and Lei Hou

Image Set Classification Based on Synthetic Examples and Reverse
Training.   282
  Qingjun Liang, Lin Zhang, Hongyu Li, and Jianwei Lu

Carried Baggage Detection and Classification Using Part-Based Model   289
  Wahyono and Kang-Hyun Jo

Image Splicing Detection Based on Markov Features in QDCT Domain   297
  Ce Li, Qiang Ma, Limei Xiao, Ming Li, and Aihua Zhang

Facial Expression Recognition Based on Hybrid Approach.   304
  Md. Abdul Mannan, Antony Lam, Yoshinori Kobayashi,
  and Yoshinori Kuno

A Local Feature Descriptor Based on Energy Information for Human
Activity Recognition   311
  Yubo Shi and Yongxiong Wang

An Accurate Online Non-rigid Structure from Motion Algorithm   318
  Ya-Ping Wang, Zhan-Li Sun, Yang Qian, Yun Jing, and De-Xiang Zhang

A Multi-feature Fusion Method for Automatic Multi-label Image
Annotation with Weighted Histogram Integral and Closure
Regions Counting   323
   Sen Xia, Peng Chen, Jun Zhang, Xiao-Ping Li, and Bing Wang

Diffusion-Based Hybrid Level Set Method for Complex Image
Segmentation   331
   Xiao-Feng Wang, Le Zou, and Gang Lv

Knowledge Discovery and Data Mining

The Optimization of Resource Allocation Based on Process Mining   341
   Weidong Zhao, Liu Yang, Haitao Liu, and Ran Wu

Research on Optimum Weighted Combination GM(1,1) Model
with Different Initial Value   354
   Qiumei Chen and Jin Li

Spectral Clustering of High-Dimensional Data via k-Nearest Neighbor
Based Sparse Representation Coefficients.   363
   Fang Chen, Shulin Wang, and Jianwen Fang

Friend Recommendation by User Similarity Graph Based on Interest
in Social Tagging Systems.   375
   Bu-Xiao Wu, Jing Xiao, and Jie-Min Chen

Orchestrating Real-Valued Negative Selection Algorithm with
Computational Efficiency for Crude Oil Price.   387
   Ayodele Lasisi, Rozaida Ghazali, Tutut Herawan, and Haruna Chiroma

A New Algorithm of Automatic Grading in Computer Paperless
Test System   397
   Tian-Lan Liu, Wen-Sheng Tang, Sheng-Chun Wang, and Jun Qin

MLRF: Multi-label Classification Through Random Forest
with Label-Set Partition   407
   Feng Liu, Xiaofeng Zhang, Yunming Ye, Yahong Zhao, and Yan Li

Prediction of Oil and Water Layer by Kernel Local Fisher
Discriminant Analysis   419
   Zehao Chen

Assessment of the Pillar 3 Financial and Risk Information Disclosures
Usefulness to the Commercial Banks Users   429
   Anna Pilkova, Michal Munk, Peter Svec, and Michal Medo

CKNNI: An Improved KNN-Based Missing Value Handling Technique   441
   Chao Jiang and Zijiang Yang

An Item Based Collaborative Filtering System Combined with Genetic
Algorithms Using Rating Behavior 453
   Jing Xiao, Ming Luo, Jie-Min Chen, and Jing-Jing Li

Natural Language Processing and Computational Linguistics

Short Text Classification Based on Semantics....................... 463
   Chenglong Ma, Xin Wan, Zhen Zhang, Taisong Li, and Yan Zhang

Name Disambiguation Using Semi-supervised Topic Model 471
   JinLan Fu, Jie Qiu, Jing Wang, and Li Li

Automatic Evaluation of Machine Translation Through the Residual
Analysis... 481
   Daša Munková and Michal Munk

Is the Most Frequent Sense of a Word Better Connected
in a Semantic Network?... 491
   Hiram Calvo and Alexander Gelbukh

Language Processing and Human Cognition......................... 500
   Daša Munková, Eva Stranovská, and Michal Munk

An Ontology-Based Approach for Measuring Semantic Similarity
Between Words 510
   Ruiling Zhang, Shengwu Xiong, and Zhong Chen

Intelligent Control and Automation

Knowledge Bases' Control for Intelligent Professional Activity
Automatization 519
   Alexander Kleschev and Elena Shalfeeva

Maximum Class Separability-Based Discriminant Feature Selection
Using a GA for Reliable Fault Diagnosis of Induction Motors 526
   Md. Rashedul Islam, Sheraz A. Khan, and Jong-Myon Kim

Multi-fault Diagnosis of Roller Bearings Using Support Vector Machines
with an Improved Decision Strategy 538
   M.M. Manjurul Islam, Sheraz A. Khan, and Jong-Myon Kim

A Local Neural Networks Approximation Control of Uncertain Robot
Manipulators 551
   Minh-Duc Tran and Hee-Jun Kang

Intelligent Communication Networks and Web Applications

A Quantum-Inspired Immune Clonal Algorithm Based Handover Decision
Mechanism with ABC Supported   561
  Tingting Liu, Xingwei Wang, Fuliang Li, and Min Huang

An IEEE 802.21 Based Heterogeneous Access Network Selection
Mechanism.   574
  Renzheng Wang, Xingwei Wang, Fuliang Li, and Min Huang

A Dijkstra Algorithm Based Multi-layer Satellite Network Routing
Mechanism.   586
  Yinchu Sun, Xingwei Wang, Fuliang Li, and Min Huang

The Research on Optimizing Deployment Strategy for Aviation SWIM
Application Servers   598
  Haitao Zhang and Zhijun Wu

Modeling Fault Tolerated Mobile Agents by Colored Petri Nets   607
  Shao-zhen Zhang, Zuo-hua Ding, and Jue-liang Hu

A Self-Adaptive Context-Aware Model for Mobile Commerce   618
  Munir Naveed

Research and Implementation on Autonomic Integration Technology
of Smart Devices Based on DPWS   626
  Yan-qin Mao, Lu Jin, and Su-bin Shen

Experimental Verification of the Dependence Between the Expected
and Observed Visit Rate of Web Pages   637
  Jozef Kapusta, Michal Munk, and Martin Drlik

Bioinformatics Theory and Methods

Kernel Independent Component Analysis-Based Prediction on the Protein
O-Glycosylation Sites Using Support Vectors Machine
and Ensemble Classifiers   651
  Zehao Chen

Semantic Role Labeling for Biomedical Corpus Using Maximum
Entropy Classifier   662
  Lei Han, Dong-hong Ji, and Han Ren

Automatic Detection of Yeast and Pseudohyphal Form Cells in the Human
Pathogen Candida Glabrata   669
  *Luis Frazao, Rui Santos, Miguel Cacho Teixeira, Nipon Theera-Umpon,
  and Sansanee Auephanwiriyakul*

Semi-supervised Feature Extraction for RNA-Seq Data Analysis   679
   Jin-Xing Liu, Yong Xu, Ying-Lian Gao, Dong Wang, Chun-Hou Zheng,
   and Jun-Liang Shang

Compound Identification Using Random Projection for Gas
Chromatography-Mass Spectrometry Data   686
   Li-Li Cao, Zhi-Shui Zhang, Peng Chen, and Jun Zhang

A Random Projection Ensemble Approach to Drug-Target Interaction
Prediction.   693
   Peng Chen, ShanShan Hu, Bing Wang, and Jun Zhang

Kernel Local Fisher Discriminant Analysis-Based Prediction on Protein
O-Glycosylation Sites Using SVM   700
   Xuemei Yang and Shiliang Sun

Identification of Colorectal Cancer Candidate Genes Based on Subnetwork
Extraction Algorithm.   706
   Ran Wei, Hai-Tao Li, Yanjun Wang, Chun-Hou Zheng, and Junfeng Xia

Detection of Protein-Protein Interactions from Amino Acid Sequences
Using a Rotation Forest Model with a Novel PR-LPQ Descriptor   713
   Leon Wong, Zhu-Hong You, Shuai Li, Yu-An Huang, and Gang Liu

Healthcare and Medical Methods

Age-Related Alterations in the Sign Series Entropy of Short-Term Pulse
Rate Variability   723
   Yongxin Chou and Aihua Zhang

An Assessment Method of Tongue Image Quality Based on Random Forest
in Traditional Chinese Medicine   730
   Xinfeng Zhang, Yazhen Wang, Guangqin Hu, and Jing Zhang

Detection of Epileptic Seizures in EEG Signals with Rule-Based
Interpretation by Random Forest Approach.   738
   Guanjin Wang, Zhaohong Deng, and Kup-Sze Choi

Sparse-View X-ray Computed Tomography Reconstruction
via Mumford-Shah Total Variation Regularization.   745
   Bo Chen, Chen Zhang, Zhao-Ying Bian, Wen-Sheng Chen,
   Jian-Hua Ma, Qing-Hua Zou, and Xiao-Hui Zhou

The Utilisation of Dynamic Neural Networks for Medical Data
Classifications- Survey with Case Study.   752
   Abir Jaafar Hussain, Paul Fergus, Dhiya Al-Jumeily, Haya Alaskar,
   and Naeem Radi

Neural Network Classification of Blood Vessels and Tubules Based on
Haralick Features Evaluated in Histological Images of Kidney Biopsy   759
Vitoantonio Bevilacqua, Nicola Pietroleonardo, Vito Triggiani,
Loreto Gesualdo, Anna Maria Di Palma, Michele Rossini,
Giuseppe Dalfino, and Nico Mastrofilippo

Dependable Healthcare Service Automation: A Holistic Approach   766
Kaiyu Wan and Vangalur Alagar

Information Security

Improved Concurrent Signature on Conic Curve   781
Zhanhu Li and Xuemei Yang

Double Compression Detection in MPEG-4 Videos Based on Block
Artifact Measurement with Variation of Prediction Footprint   787
Peisong He, Tanfeng Sun, Xinghao Jiang, and Shilin Wang

Author Index   795

Machine Learning Theory and Methods

A Review of Parameter Learning Methods in Bayesian Network

Zhiwei Ji[1,2], Qibiao Xia[2(✉)], and Guanmin Meng[3]

[1] School of Electronics and Information Engineering, Tongji University,
Shanghai, China
[2] School of Information Engineering, Zhejiang A&F University, Linan, China
xqb8050@163.com
[3] Department of Clinical Laboratory, Tongde Hospital of Zhejiang Province,
Hangzhou, China

Abstract. Bayesian network (BN) is one of the most classical probabilistic graphical models. It has been widely used in many areas, such as artificial intelligence, pattern recognition, and image processing. Parameter learning in Bayesian network is a very important topic. In this study, six typical parameter learning algorithms were investigated. For the completeness of dataset, there are mainly two categories of methods for parameter estimation in BN: one is suitable to deal with the complete data, and another is for incomplete data. We mainly focused on two algorithms in the first category: *maximum likelihood estimate,* and *Bayesian method; Expectation-Maximization algorithm*, *Robust Bayesian estimate*, *Monte-Carlo method*, and *Gaussian approximation method* were discussed for the second category. In the experiment, all these algorithms were applied on a classic example to implement the inference of parameters. The simulating results reveal the inherent differences of these six methods and the effects of the inferred parameters of network on further probability calculation. This study provides insight into the parameter inference strategies of Bayesian network and their applications in different kinds of situations.

Keywords: Bayesian network · Parameter learning · Conditional probability table

1 Introduction

A Bayesian Network (BN) is a graphical representation of knowledge with intuitive structures and parameters [1–3], which was first proposed by J. Pearl. In the past few decades, Bayesian networks has been combined with expert systems and decision theory, and gained rapid development [4, 5]. F.V. Jensen, R.G. Cowell, and R. Neapoliton, et al. conducted a detailed discussion for Bayesian network [6, 7]. More recently, researchers have developed methods for learning Bayesian networks from data.

Bayesian network encodes the conditional dependencies between a set of random variables using a directed acyclic graph (DAG). It can be described as a pair (G, θ), where G is the topological structure of the network, and θ is the set of parameters of the

© Springer International Publishing Switzerland 2015
D.-S. Huang and K. Han (Eds.): ICIC 2015, Part III, LNAI 9227, pp. 3–12, 2015.
DOI: 10.1007/978-3-319-22053-6_1

conditional probability functions of each variable (a node in the graph) given its parents in the network. Both the graph of conditional dependencies and the model parameters can be estimated from a set of observed examples of a domain of interest or can be provided by means of domain-expert knowledge. When only a dataset is provided, a method that learns Bayesian network models usually implements two stages: the extraction of the network structure (structure learning) and the estimation of the model parameters (parameter learning) [8]. In real learning problems, structure learning is looking for relationships among a large number of variables; while the parameter learning is estimating the conditional probability tables (CPT) which involve in the casual relationships among variables. Although some structure learning approaches can infer the candidate network topologies from the observed data, adding sufficient domain knowledge can effectively guide the searching process and quickly obtain an optimal solution. Given a network topology, estimation of all the parameters in the network is obviously much more challenging. Only structure and parameters are both known, the computation of a probability of interest on this model can be expected. In this study, we investigated six typical parameter learning approaches in Bayesian network to provide insight into the differences and scopes of these algorithms. We focused on two categories of parameter learning approaches: one category is for complete data: such as Maximum Likelihood Estimate (MLE) [9], and Bayesian method [9] ; another category is for incomplete data: such as Expectation-maximization (EM) [10], Robust Bayesian Estimate (RBE) [8], Monte-Carlo Method [11, 12], and Gaussian approximation [13]. In the simulation part, we selected a classical example [14] to represent the characteristics of each method. We believe that this study will promote the application of Bayesian network in much more real problems.

2   Parameter Learning in Bayesian Network

A Bayesian network is a graphical model for probabilistic relationships among a set of variables. This graphical model is represented by a directed acyclic graph (DAG). It can be denoted as $G(V, E)$, in which V is the set of all the nodes (with index 1, 2, ..., n) and E is the set of all the edges. The states of j-th node in V can be denoted by discrete variable x_j, in which $j \in V$, $x_j \in X$ and $X = \{x_1, x_2, ..., x_n\}$. For the node $j \in V$, $pa(j)$ is the parents of node j. Therefore, we have $pa(j) \subset V$ and the corresponding variable set of $pa(j)$ is PAX_j ($PAX_j \subset X$). We assumed that there is a conditional probability $P(x_j \mid PAX_j)$ for each node j, then the joint probability distribution of the model can be represented as following:

$$P(X) = \prod_{j \in V} P(x_j \mid PAX_j) \tag{1}$$

where $P(x_j|PAX_j)$ is the local conditional probability of node j. If the states of all the nodes in the network are discrete, we usually use conditional probability table (CPT) to represent the causal relationship between a child node and its parent nodes. Parameter learning is the task to estimate all the CPTs from the observed sample data if the network structure is known.

Figure 1 is a classic example to elaborate the conception of Bayesian network. There are four nodes included in the BN shown in Fig. 1: Cloudy (C), Sprinkler (S), Rain (R), WetGrass (W). In a Bayesian network, each directed edge indicates a type of conditional probability, which can be expressed with a CPT. The nodes are independent if there are no any edges connecting them. Form Fig. 1, we can see C is independent to W; the conditional probability $P(S|C)$ indicates the probability of S under the condition C. So, $P(R|C)$ and $P(W|S, R)$ also have the similar meaning. Therefore, the joint probability distribution in Fig. 1 can be represented as:

$$P(C, S, R, W) = P(C)P(S|C)P(R|C)P(W|S, R) \qquad (2)$$

Bayesian networks provide decision results for expert system, which is based on the network structure and conditional probability tables (CPT). If the conditional probability table is unknown, we only learn from the observed data to obtain the conditional probability parameters. Bayesian network parameter learning is an important part of learning Bayesian networks: Giving a Bayesian network structure and a number of known observation data set, estimate the conditional probability parameters for all the casual relationships in the network. Parameter learning algorithm can be divided into two categories: one is based on a complete data; the other is based on incomplete data. We will discuss several typical algorithms for these two categories.

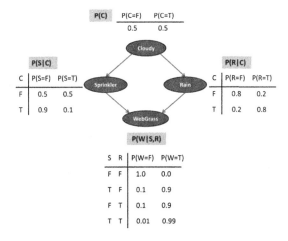

Fig. 1. A simple Bayesian network, adapted from [13].

2.1 The Parameter Learning for Complete Data

We assumed that a Bayesian network included n nodes, which states can be represented by $X = \{x_1, x_2, \ldots, x_j, \ldots, x_n\}$, in which x_j is the state of j-th node. If each node is observed m times, a sample dataset $OSS = \{D_1, D_2, \ldots, D_m\}$ will be obtained. The sample $D_i = \{x_{1i}, x_{2i}, \ldots, x_{ni}\}$ indicates the observed values of all the nodes in the i-th sampling. If there is no missing values occurred in dataset OSS, it was considered as complete data; otherwise, incomplete data. In this section, we firstly introduced maximum likelihood estimate (MLE), and Bayesian methods for complete data.

2.1.1 Maximum Likelihood Estimate

Maximum Likelihood Estimate (MLE), is a common strategy of parameter learning for complete data. The basic idea of MLE is: a random test for an event (C) may induce several possible outcomes $C^1, C^2, \ldots, C^n$. If the result of this test is C^k $(1 \le k \le n)$, it appeared in the maximum likelihood for this event. Hence, the estimated value of $\hat{C}$ will be set as parameter θ if it can maximize the value of likelihood function $P(C|\theta)$. Likelihood is a standard to judge if θ is good or bad. Based on the possibility of θ to sample, the bigger value of likelihood function $L(\theta : D)$ is the better, where $L(\theta : D) = P(D|\theta) = \prod_m P(D_m|\theta)$.

Extended to a general Bayesian network with n nodes, the likelihood function is denoted as:

$$L(\theta : D) = \prod_m P(x_{1m}, \ldots, x_{nm}|\theta) = \prod_m \prod_i P\left(x_{im} \mid pa(i)_m, \theta_i\right)$$
$$= \prod_i \prod_m P\left(x_{im} \mid pa(i)_m, \theta_i\right) = \prod_i L_i(\theta_i : D) \tag{3}$$

If the probability $P(x_i|pa(i))$ satisfies a polynomial distribution, the local likelihood function can be further decomposed as following:

$$L_i(\theta_i : D) = \prod_m P\left(x_{im}|Pa(i)_m, \theta_i\right)$$
$$= \prod_m \prod_{Pa(i)^j} \prod_{x_i'} P\left(x_{im}^k|pa(i)_m^j, \theta_i\right) = \prod_{pa(i)^j} \prod_{x_i'} \theta_{x_i', pa(i)}^{N\left(x_i', pa(i)^j\right)} \tag{4}$$

Considering that the dataset is complete, for each possible value $(pa(i)^j)$ of the parent nodes $pa(i)$, the distribution of probability of $P(x_i|pa(i)^j)$ is the independent polynomial distribution which is not related to all other values $(pa(i)^l(l \ne j))$ of $pa(i)$.

Therefore, MLE method can obtain the estimated parameter θ as following:

$$\theta_{x_i^k} = \frac{N(x_i^k, pa(i)^j)}{N(pa(i)^j)} \tag{5}$$

Based on Eq. (5), we can easily obtain the conditional probability tables on a known network topology. Actually, MLE method uses frequency to instead of probability. It converges slowly due to no prior knowledge was used in the process of parameter estimation.

2.1.2 Bayesian Method

The basic idea of Bayesian method for parameter learning is: given a distribution with unknown parameters and a complete set (C) of instance data (observed data), θ is a random variables with a prior distribution $p(\theta)$; the changes of parameter θ, namely $p(\theta|C)$, can be estimated according to the previous knowledge or the assumption of $p(\theta)$ as the uniform distribution. $p(\theta|C)$ is therefore called as the posterior probability of parameter θ. Here, the aim of this method is to calculate this posterior probability, which will be considered as the basis of parameter learning.

Here, we assumed that prior distribution $p(\theta)$ is Dirichlet distribution:

$$p(\theta) = Dir(\theta|\alpha_1, \ldots, \alpha_r) = \frac{\Gamma(\alpha)}{\prod\limits_{k=1}^{r} \Gamma(\alpha_k)} \prod_{k=1}^{r} \theta_k^{\alpha_k^{-1}} \tag{6}$$

Where $\alpha = \sum\limits_{k=1}^{r} \alpha_k$, $\alpha_k > 0$, $k = 1, \ldots, r$, $\alpha_1, \ldots, \alpha_r$ are hyper parameters; $\Gamma(\cdot)$ is Gama function, $\Gamma(x) = \int\limits_{0}^{\infty} t^{x-1} e^{-t} dt$ satisfies $\begin{cases} \Gamma(x+1) = x\Gamma(x) \\ \quad\quad \Gamma(1) = 1 \end{cases}$. Hence, the probability of observed samples is:

$$p(D) = \int p(\theta)p(D|\theta)d\theta = \int \frac{\Gamma(\alpha)}{\prod\limits_{k=1}^{r} \Gamma(\alpha_k)} \prod_{k=1}^{r} \theta_k^{\alpha_k^{-1}} \times \prod_{k=1}^{r} \theta_k^{N_k} d\theta = \frac{\Gamma(\alpha)}{\Gamma(\alpha+N)} \prod_{k=1}^{r} \frac{\Gamma(\alpha_k + N_k)}{\Gamma(\alpha_k)}$$

In Bayesian method, it assumes that the no prior distribution is uniform distribution, which is consistent with the principle of maximum entropy in information theory, it maximizes the entropy of random variables. Thus, if there is no information used for determination of prior distribution, we set $\alpha_1 = \alpha_2 = \ldots = \alpha_r$.

Here, the Dirichlet distribution was applied as the prior distribution, and we illustrate the parameter learning process in Bayesian network from three situations:

(1) The calculation of non-conditional probability $p(x_i|\theta)$ (nodes which have no parents)

$$p(\theta|D) = \frac{p(\theta)p(D|\theta)}{p(D)} = \frac{\Gamma(\alpha+n)}{\prod\limits_{i} \Gamma(\alpha_i + n_i)} \prod_{k} \theta_k^{\alpha_i - n_i} = Dir(\alpha_1 + n_1, \ldots, \alpha_N + n_N) \tag{7}$$

Therefore,

$$p(\theta_i|D) = \frac{\alpha_i + n_i}{\alpha + N} \tag{8}$$

where n_i is the number of occurrences of i-th possible values for variable x_i in the whole dataset, and N is the number of occurrences of all the possible values for variable x_i in the dataset.

(2) The calculation of conditional probability for each node which has unique parent node.

We assume that the relationship between nodes X and Y can be denoted as $X \rightarrow Y$, and they both are discrete variables. Hence,

$$p(y|x_i, \theta) = Dir(\alpha_{i1} + n_{i1}, \alpha_{i2} + n_{i2}, \ldots, \alpha_{ik} + n_{ik}) \tag{9}$$

(3) The calculation of conditional probability for each node which has multiple parent nodes.

Firstly, an assumption of parameter independence was given: the parameters, which might have different distribution, are mutually independent. Here, θ_{ijk} is represented as the conditional probability when $pa(a) = j$ and $x_i = k$. r_i denotes the number of possible values for variable x_i. $q_i = \prod_{x l \in pa(i)} r_i$ indicates all the situations for its parent nodes. With the assumption of parameter independence, each variable x_i and its parent $pa(i) = j$ obey Dirichlet distribution:

$$p\left(\theta_{ij1}, \theta_{ij2}, \ldots, \theta_{ijl} | \varsigma \right) = c \prod_k \theta_{ijk}^{\alpha_{ijk}-1} \tag{10}$$

Hence, the value of θ_{ijk} can be estimated by following Eq. (11):

$$\theta_{ijk} = \frac{\alpha_{ijk} + n_{ijk}}{\alpha_{ij} + n_{ij}} \quad \left(\alpha_{ij} = \sum \alpha_{ijk}, n_{ij} = \sum n_{ijk} \right) \tag{11}$$

2.2   The Parameter Learning for Incomplete Data

2.2.1   Expectation-Maximization

The basic idea of Expectation Maximization (EM) is: when the observed data is incomplete, we can use the inference algorithms of Bayesian network to estimate the missing value of the dataset so that the dataset will be complete. Little and Rubin confirmed that EM will be an excellent learning approach when the statistical model can be constructed. EM algorithm includes two steps: (1) initialization: θ_s is assigned with a random value; (2) calculation of the exception. We will calculate the statistical expectation factor of each event "$X_i = k$, $pa(i) = j$" under the condition θ_s; For discrete variables, we have $E_{p(X|D, \theta_s, S^h)}(N_{ijk}) = \sum_{i=1}^{N} p(x_i = k, pa(i) = j | y_i, \theta_s, S)$, where N_{ijk} is the sufficient statistical factor of the event "$X_i = k$, $pa(i) = j$"; (3) maximize the exception. The exception sufficient factor was applied to convert the incomplete data D to complete data samples; (4) if the accuracy is reached, stop the process; otherwise, return to (2). Hence, we can calculate the conditional probability via Eq. (12):

$$\theta_{ijk} = \frac{E_{p\left(x|D_i\theta_j S_k\right)}(N_{ijk})}{\sum_{k=1}^{r_i} E_{p\left(x|D_i\theta_j S_k\right)}(N_{ijk})} \tag{12}$$

EM approach uses a Bayesian network inference algorithm to calculate the exception of variables which are un-observed. The selection of inference algorithm is the key for

EM approach. Junction Tree algorithm is widely used in EM approach for parameter learning.

2.2.2 Robust Bayesian Estimate

Robust Bayesian estimate (RBE) [8], is a parameter learning method on incomplete data. The basic idea of RBE is: probability interval-based strategy was applied to estimate the minimal and maximal values of conditional probabilities for each node; and use probability interval to represent the ranges of conditional probabilities. With difference from EM approach, RBE does not need to make any assumptions about the missing data when implementing parameter learning, but directly represent the conditional probability with probability interval. Furthermore, the width of the interval indicates the reliability of estimation [8].

For the complete data, similar as Eq. (5), the conditional probability for "$X_i = k$, $pa(i) = j$" is

$$\theta_{ijk} = \frac{n(x_i = k, pa(i) = j)}{n(pa(i) = j)} \tag{13}$$

Where $n(x_i = k, pa(i) = j)$ is the number of occurrence of event "$x_i = k$, and $pa(i) = j$" in the observed dataset, $n(pa(i) = j)$ is the number of occurrence of j-th situation of $pa(i)$ $(pa(i) = j)$.

For the incomplete data, the statistics in Eq. (13) could not be implemented so that we have to use RBE method to estimate the conditional probability. Please see the steps of RBE as following.

Firstly, we clarify that $n(x_i = k, pa(i) = j)$ and $n(pa(i) = j)$ denote the situations of non-missing values in the observed dataset. For incomplete data, there are only three kinds of situations:

(1). $n(?|pa(i) = j)$: samples fit the event "$pa(i) = j$", but the value of x_i is missing.
(2). $n(x_i = k|?)$: samples fit the event "$x_i = k$", but the values of $pa(i)$ is missing.
(3). $n(?|?)$: samples are missing on both x_i and $pa(i)$.

If the samples fit one of above three situations, the data might be fit the condition "$x_i = k$ and $pa(i) = j$" or not before the data were lost.

Therefore, we can calculate the ranges of parameters. We set:

$$\bar{n}(x_i = k|pa(i) = j) = n(?|pa(i) = j) + n(x_i = k|?) + n(?|?) \tag{14}$$

$$\underline{n}(x_i = k|pa(i) = j) = n(?|pa(i) = j) + \sum_{h \neq k} n(x_i = h|pa(i) = j) + n(?|?) \tag{15}$$

Equation (14) denotes the numbers of all the situations occurred event "$x_i = k$ and $pa(i) = j$" in the observed data, which at least fit one of above three cases. Equation (15) indicates the numbers of all the situations did not occur event "$x_i = k$ and $pa(i) = j$" in the observed data, which at least fit one of above three cases. Equations (14–15) restricts the ranges of conditional probability. For incomplete dataset, the value of θ_{ijk} can be restricted by both Eqs. (16–17):

$$\underline{p}(x_i = k|pa(i) = j) = \frac{\alpha_{ijk} + n(x_i = k, pa(i) = j)}{\alpha_{ij} + n(pa(i) = j) + \bar{n}(x_i = k|pa(i) = j)} \tag{16}$$

$$\bar{p}(x_i = k|pa(i) = j) = \frac{\alpha_{ijk} + n(x_i = k, pa(i) = j) + \bar{n}(x_i = k|pa(i) = j)}{\alpha_{ij} + n(pa(i) = j) + \bar{n}(x_i = k|pa(i) = j)} \tag{17}$$

In here, α_{ijk} is hyper parameter, which represents prior knowledge. Therefore, we have

$$\theta_{ijk} \in [\underline{P}(x_i = k|pa(i) = j), \bar{p}(x_i = k|pa(i) = j)]$$

We can clearly see that RBE applied probability interval to represent conditional probability.

So the probability $P(X_v) = \prod_{v \in V} P(X_{v|U_v})$ can be represented as $P(X_v) \in [\underline{P}(X_v), \bar{P}(X_v))]$, where $\underline{P}(X_v) = \prod_{v \in V} \underline{P}(X_{v|U_v}), \bar{P}(X_v) = \prod_{v \in V} \bar{P}(X_{v|U_v})$.

2.2.3   Monte-Carlo Method

The Gibbs sampling strategy proposed by Geman is belong to the well-known Monte-Carlo method. The basic idea is: the expectation $f(X)$ of joint probability distribution $P(X)$ on variable set X was estimated with Gibbs sampling. This method includes following steps: (1) initialization. Randomly initialization is implemented on incomplete dataset D and obtain a complete dataset D_c. (2) An un-observed variable X_{im} in dataset D was selected out (the m-th sample instance of i-th variable), and it was randomly assigned with the following probability distribution:

$$p\left(X'_{im}|D_c \backslash X_{im}, S\right) = \frac{p\left(X'_{im}|D_c \backslash X_{im}, S\right)}{\sum_{X''_{im}} p(X''_{im}|D_c \backslash X_{im}, S)} \tag{18}$$

Where $D_c \backslash X_{im}$ indicates the situation moving the observation X_{im} from X_{im}; (3) Base on step 2, each un-observed variable in D was assigned, so that a new randomized complete dataset D'_c was finally obtained. (4) calculate $p(\theta_s|D, S)$; (5) repeat step 1 to step 4 so many times, and get the estimated value as the mean value of all the $p(\theta_s|D, S)$.

Monte-Carlo methods are very flexible when other methods are not applicable. The larger the samples, the more accurate results; however, the computational complexity is exponential power of the number of instances of the sample. Gibbs sampling is the most typical one in Monte-Carlo methods.

2.2.4   The Gaussian Approximation

The Gaussian approximation method also can handle large samples and get accurate results; and it has lower complexity than Monte Carlo. The basic idea is: for the large scale data, we can use multivariate Gaussian distribution to approximately simulate $p(\theta_s|D, S) \propto p(D|\theta_s, S)$, $g(\theta_s) = \log (p(D|\theta_s, S)p(\theta_s|S))$. $\tilde{\theta}_s = \arg\max_{\theta_s}\{g(\theta_s)\}$, so $\tilde{\theta}_s$

make $p(\theta_s|D, S)$ reach to maximum; the second-order Taylor expansion of $g(\theta_s)$ in point $\widetilde{\theta}_s$ was used to calculate $g(\theta_s) \approx g(\widetilde{\theta}_s) - 1/2(\theta_s - \widetilde{\theta}_s)A(\theta_s - \widetilde{\theta}_s)^T$, A is the Hess determinant of $g(\theta_s)$ in the point $\widetilde{\theta}_s$. Therefore,

$$p(\theta_s|D, S) \propto p(D|\theta_s, S)p(\theta_s|s) \approx P\left(D|\widetilde{\theta}_s, S\right)p\left(\widetilde{\theta}_s|S\right)$$
$$= \exp\{-\frac{1}{2}\left(\theta_s - \widetilde{\theta}_s\right)A\left(\theta_s - \widetilde{\theta}_s\right)^T\}$$

As can be seen from the above equation, it is necessary to find $\widetilde{\theta}_s$ to complete the Gaussian approximation, and calculate the Hess determinant $\exp\{-\frac{1}{2}(\theta_s - \widetilde{\theta}_s)A(\theta_s - \widetilde{\theta}_s)^T\}$ of $g(\theta_s)$ at the point $\widetilde{\theta}_s$.

3 Discussion and Conclusion

This paper firstly described the structure and representation of Bayesian network; the mainly focus is the parameter learning strategy in Bayesian network. We introduced six typical parameter learning methods of Bayesian network, which were suitable to complete and incomplete dataset, respectively. In the simulation experiment, the detailed comparison of these algorithms was carried out. The results suggest us how to use a reasonable one for different applications in real questions. There are several areas that require further research. First, some methods or approaches can be proposed to process the continuous data for parameters learning. Second, the fast algorithms for incremental learning should be established.

Acknowledgement. This work was supported by the Zhejiang Provincial Education Department Foundation of China (Grant No. Y201432242) and Talent Start Foundation of Zhejiang A&F University (Grant No. 2009FR061). This work was also partially supported by NSFC No. 61133010.

References

1. Shafer, G.: Probabilistic reasoning in intelligent systems - networks of plausible inference - Pearl. J. Synthese **104**(1), 161–176 (1995)
2. Liao, W.H., Ji, Q.: Learning Bayesian network parameters under incomplete data with domain knowledge. Pattern Recogn. **42**(11), 3046–3056 (2009)
3. Pernkopf, F., Wohlmayr, M.: Stochastic margin-based structure learning of Bayesian network classifiers. Pattern Recogn. **46**(2), 464–471 (2013)
4. Liu, Z.K., Liu, Y.H., Cai, B.P., Zheng, C.: An approach for developing diagnostic Bayesian network based on operation procedures. Expert Syst. Appl. **42**(4), 1917–1926 (2015)
5. Hanninen, M., Banda, O.A.V., Kujala, P.: Bayesian network model of maritime safety management. Expert Syst. Appl. **41**(17), 7837–7846 (2014)
6. De Campos, L.M., Fernandez-Luna, J.M., Huete, J.F.: Bayesian networks and information retrieval: an introduction to the special issue. Inform. Process. Manag. **40**(5), 727–733 (2004)

7. Suojanen, M., Olesen, K.G., Andreassen, S.: A method for diagnosing in large medical expert systems based on causal probabilistic networks. In: Keravnou, T., Baud, R., Garbay, C., Wyatt, J. (eds.) AIME 1997. LNCS, vol. 1211. Springer, Heidelberg (1997)

8. Ramoni, M., Sebastiani, P.: Robust learning with missing data. Mach. Learn. **45**(2), 147–170 (2001)

9. Furlotte, N.A., Heckerman, D., Lippert, C.: Quantifying the uncertainty in heritability. J. Hum. Genet. **59**(5), 269–275 (2014)

10. Lauritzen, S.L.: The Em algorithm for graphical association models with missing data. Comput. Stat. Data An. **19**(2), 191–201 (1995)

11. Niu, D.X., Shi, H.F., Wu, D.D.: Short-term load forecasting using bayesian neural networks learned by Hybrid Monte Carlo algorithm. Appl. Soft Comput. **12**(6), 1822–1827 (2012)

12. Titterington, D.M.: Bayesian methods for neural networks and related models. Stat. Sci. **19**(1), 128–139 (2004)

13. Russell, S., Norvig, P.: Articial Intelligence: A Modern Approach, p. 139 (1995)

14. Russell, S., Norvig, P.: Articial Intelligence: A Modern Approach (1995)

The Hard-Cut EM Algorithm for Mixture of Sparse Gaussian Processes

Ziyi Chen and Jinwen Ma[(✉)]

Department of Information Science, School of Mathematical Sciences
and LMAM, Peking University, Beijing 100871, China
jwma@math.pku.edu.cn

Abstract. The mixture of Gaussian Processes (MGP) is a powerful and fast developed machine learning framework. In order to make its learning more efficient, certain sparsity constraints have been adopted to form the mixture of sparse Gaussian Processes (MSGP). However, the existing MGP and MSGP models are rather complicated and their learning algorithms involve various approximation schemes. In this paper, we refine the MSGP model and develop the hard-cut EM algorithm for MSGP from its original version for MGP. It is demonstrated by the experiments on both synthetic and real datasets that our refined MSGP model and the hard-cut EM algorithm are feasible and can outperform some typical regression algorithms on prediction. Moreover, with sparse technique, the parameter learning of our proposed MSGP model is much more efficient than that of the MGP model.

Keywords: Mixture of Gaussian Processes · Sparsity · Hard-cut EM algorithm · Big data

1 Introduction

Gaussian process (GP) is a powerful model for a wide range of applications in machine learning, including regression [1], classification [2], dimensionality reduction [3], reinforcement learning [4], etc. Nevertheless, GP model fails to describe multimodality dataset and the training of GP consumes $O(N^3)$ time for N training samples [5, 6]. In order to solve these problems, Tresp [7] proposed the Mixture of Gaussian Processes (MGP) in 2000, which was adjusted from Mixture of Experts. Since then, various MGP models have been proposed, and the most of them are special cases of mixture of experts where each expert is a GP.

Another useful way of reducing the time cost of training a GP is to adopt the model of sparse Gaussian Process (SGP), which computes the GP likelihood with a pseudo dataset being smaller than the training dataset in size [8]. To combine the advantages of MGP and SGP, the Mixture of Sparse Gaussian Processes (MSGP) has been also proposed, in which each component is a SGP instead of a GP to accelerate the parameter learning [5, 6, 9–11].

For these MGP and MSGP models, there are three main training algorithms: Monte Carlo Markov Chain (MCMC), variational Bayesian inference (VB), and EM algorithm. Actually, MCMC approximated the posterior of parameters by sampling several

© Springer International Publishing Switzerland 2015
D.-S. Huang and K. Han (Eds.): ICIC 2015, Part III, LNAI 9227, pp. 13–24, 2015.
DOI: 10.1007/978-3-319-22053-6_2

long sequences of Markov Chains [12–19]. However, it was quite time consuming as pointed out in [9]. VB tried to approximate the posterior of parameters with factorized forms, which actually may deviate a lot from the true posterior [20–24].

Generally, EM algorithm is an effective approach to the learning of mixture models. However, since the exact posterior of latent variables and the Q function are intractable for MGP and MSGP models, several approximation strategies have been adopted. For example, variational EM algorithm approximated the posterior in E step with variational inference [5, 9–11, 25], which had the same drawbacks as VB. In [26, 27], the Q function was decomposed via leave-one-out cross-validation (LOOCV), which was a complicated form involving much computation. Stachniss et al. [6] and Tresp [7] attempted to simplify the learning process with the help of heuristic estimation in M step. However, kernel parameters were predetermined without learning in [7], whereas the parameters were sampled from data-irrelevant distributions in [6]. Therefore, both learning processes in fact needed more guidance by datasets.

Recently, some hard-cut EM algorithms have also been adopted to improve the learning efficiency, which partition each sample into the component with the maximum posterior in E-step and learn the parameters of each component independently in M-step [9, 28]. Moreover, in the same way, Yang and Ma [26] has adjusted its proposed soft EM algorithm for MGP with the LOOCV probability decomposition into a hard-cut EM algorithm for MGP, which is here referred to as the LOOCV hard-cut EM algorithm for convenience.

After all, these algorithms resorted to some approximation strategies since MGP and MSGP models are usually very complex. Recently, Chen et al. [29] refined the MGP model to a more simplified form, and strictly derived a precise hard-cut EM algorithm for it. In this paper, we develop this approach by adding sparsity constraints and adjusting the refined MGP model into the MSGP model. As a result, the parameter learning becomes more efficient as the MSGP model is refined and the hard-cut EM algorithm is still strictly derived. The experimental results demonstrates that our proposed model and algorithm are feasible and can outperform some typical regression algorithms.

2 The Mixture of Sparse Gaussian Processes

2.1 Gaussian Process (GP)

A GP model for regression is mathematically defined by

$$
\begin{cases}
F = \begin{bmatrix} f_1 \\ f_2 \\ \vdots \\ f_N \end{bmatrix} \sim N \left(\begin{bmatrix} m(x_1) \\ m(x_2) \\ \vdots \\ m(x_N) \end{bmatrix}, \begin{bmatrix} K(x_1,x_1) & K(x_1,x_2) & \cdots & K(x_1,x_N) \\ K(x_2,x_1) & K(x_2,x_2) & \cdots & K(x_2,x_N) \\ \vdots & \vdots & \ddots & \vdots \\ K(x_N,x_1) & K(x_N,x_2) & \cdots & K(x_N,x_N) \end{bmatrix} \right), \\
y_t \sim N(f_t, \sigma^2)
\end{cases} \tag{1}
$$

where x_t, f_t and y_t denote the input, latent response and output of a training sample, respectively, $K(u, v)$ is a mercer kernel function, and σ^2 denotes the noise intensity. As in most cases, we adopt zero mean function ($m \equiv 0$) and the most popular kernel function—the squared exponential (SE) kernel [30]:

$$K(u, v) = l^2 \exp\left[-\frac{1}{2}\sum_{k=1}^{d} b_k^2(u_k - v_k)^2\right], \tag{2}$$

where d is the dimensionality of inputs and each dimension has a different weight b_k to realize automatic feature selection.

There are several ways for learning the hyper-parameters of a GP model, including the approaches of maximum likelihood estimation (MLE), maximizing a posteriori (MAP), surrogate predictive probability (GPP), cross validation (CV), etc. [31].

With the estimated hyper-parameters, the predictive distribution of the output at a test input x^* can be obtained as follows:

$$p(y^*|x^*) \sim N\left[K(x^*, X)K(X, X)^{-1}y, K(x^*, x^*) - K(x^*, X)K(X, X)^{-1}K(X, x^*)\right], \tag{3}$$

where $K(x^*, X) = [K(x^*, x_j)]_{1 \times N}$, $K(X, x^*) = [K(x_i, x^*)]_{N \times 1}$, $K(X, X) = [K(x_i, x_j)]_{N \times N}$ are kernel matrices [30].

2.2 Mixture of Gaussian Processes (MGP)

MGP is a special mixture model in which each component is just a Gaussian process, and these components are independent in most existing MGP models. Denote $z_t = c$ iff (x_t, y_t) belongs to the c-th GP, and denote

$$X_c = \left[x_{t_c(1)}, x_{t_c(2)}, \cdots, x_{t_c(N_c)}\right]^T \quad Y_c = \left[y_{t_c(1)}, y_{t_c(2)}, \cdots, y_{t_c(N_c)}\right]^T$$

as the inputs and outputs from the c-th GP, respectively, where N_c is the number of samples in the c-th GP. Therefore, the output likelihood is mathematically given by

$$Y_c \sim N\left[0, K(X_c, X_c|\theta_c) + \sigma_c^2 I\right]; c = 1, 2, \cdots, C, \tag{4}$$

where the SE kernel given by Eq. (2) is parameterized by $\theta_c = \{l_c, \sigma_c\} \cup \{b_{ck} : k = 1, 2, \cdots, d\}$ for the c-th GP.

MGP has two main advantages over a single GP. Firstly, MGP can capture the heterogeneity among different input locations by fitting them with different GPs. For example, the toy dataset used by some MGP models [12, 17, 26, 29] was generated by 4 continuous functions with Gaussian noise, as is shown in Fig. 1. Therefore, it is better to fit the Toy dataset by MGP with four GPs than only one GP. Secondly, the computational cost can be reduced by dividing the large kernel matrix of one GP into small matrices of components.

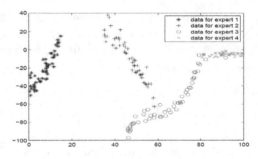

Fig. 1. Toy dataset

2.3 Sparse Gaussian Process (SGP)

Another good scaling technique for GP is to use the model of Sparse Gaussian Process. The main idea of SGP is to approximate N training samples with M pseudo samples (M<<N) [32]. Mathematically, we denote inputs of the training data, test data and pseudo data $X = [x_1, x_2, \cdots, x_N]^T$, $X^* = [x_1^*, x_2^*, \cdots, x_L^*]^T$, $U = [u_1, u_2, \cdots, u_M]^T$ respectively, and their corresponding latent responses are denoted as $F = [f_1, f_2, \cdots, f_N]^T$, $F^* = [f_1^*, f_2^*, \cdots, f_L^*]^T$ and $V = [v_1, v_2, \cdots, v_M]^T$, respectively. Almost, the sparse GP models can be mathematically defined by the following equations [32]:

$$V \sim N[0, K(U,U)], \tag{5}$$

$$p(F, F^*|V) = q(F|V)q(F^*|V), \tag{6}$$

$$y_t|f_t \sim N(f_t, \sigma^2) \ i.i.d., \tag{7}$$

$$y_t^*|f_t^* \sim N(f_t^*, \sigma^2) \ i.i.d., \tag{8}$$

where Eq. (5) means the latent pseudo outputs V fulfill a GP, as in Eqs. (1), and (6) gives the crucial assumption of SGP, i.e., the latent responses for training dataset and test dataset are conditionally independent given V. In Eq. (6), q(F|V) and q(F*|V) can be approximations of the following GP predictive distributions

$$p(F^*|V) \sim N\left[K(X^*,U)^T K(U,U)^{-1}V, K(X^*,X^*) - K(X^*,U)K(U,U)^{-1}K(U,X^*)\right]$$
$$p(F|V) \sim N\left[K(X,U)^T K(U,U)^{-1}V, K(X,X) - K(X,U)K(U,U)^{-1}K(U,X)\right], \tag{9}$$

respectively, and the forms of q(F|V) and q(F*|V) determine the kind of SGP models, including SoR, DTC, FITC and PITC models [32]. Among these models, the Fully Independent Training Conditional (FITC) model proposed in [8] was highly recommended in [32], due to its rich covariance. Experimental results have shown the great advantage of FITC over the other sparse GPs on predictive accuracy, whereas the increase on time consumption is trivial [8]. Therefore, we will use FITC as each component for our proposed mixture of sparse Gaussian Processes model (MSGP).

In FITC model, the conditional distribution of the latent test outputs is the exact GP predictive distribution:

$$q(F^*|V) = p(F^*|V)$$
$$\sim N\left[K(X^*, U)^T K(U, U)^{-1} V, K(X^*, X^*) - K(X^*, U)K(U, U)^{-1}K(U, X^*)\right], \tag{10}$$

whereas the latent training outputs are assumed to be conditionally independent given the latent pseudo output V, as suggested by the name "Fully Independent Training Conditional", i.e.

$$q(F|V) \sim N\left[K(X, U)K(U, U)^{-1}V, \Lambda\right], \tag{11}$$

where $\Lambda = diag\left[K(X, X) - K(X, U)K(U, U)^{-1}K(U, X)\right]$

It can be inferred that the marginal likelihood for the outputs is

$$p(Y) \sim N\left[0, K(X, U)K(U, U)^{-1}K(U, X) + \Lambda + \sigma^2 I\right], \tag{12}$$

and the predictive distribution can be derived as follows.

$$p(Y^*|Y) \sim N\Big\{K(X^*, U)Q^{-1}K(U, X^*)(\Lambda + \sigma^2 I)^{-1}Y,$$
$$K(X^*, X^*) - K(X^*, U)[K(U, U)^{-1} - Q^{-1}]K(U, X^*) + \sigma^2 I\Big\}, \tag{13}$$

where $Q = K(U, U) + K(U, X)(\Lambda + \sigma^2 I)^{-1}K(X, U)$.

The crucial issue on training a SGP is to find appropriate pseudo inputs. One method is to greedily select a fixed number of training inputs one by one with a certain criterion [33–36]. Snelson and Ghahramani [32] have extended the range of pseudo inputs from the training sets to the whole Euclidean space, and learnt pseudo inputs and the hyper parameters together by maximizing the likelihood, which is more flexible than greedy selection and has better performance in experiments.

2.4   The FITC Mixture Model

As the MGP model proposed in [29] is refined and can be trained by the precise learning algorithm, we can inherit its general framework for our MSGP model. The only difference is that SGP is adopted for each component for higher speed in the training process.

As in [29], we adopt the following gating function and Gaussian inputs.

$$\Pr(z_t = c) = \pi_c; c = 1 \sim C \ i.i.d \ for \ t = 1 \sim N, \tag{14}$$

$$p(x_t | z_t = c) \sim N(\mu_c, S_c); c = 1 \sim C, \ i.i.d \ for \ t = 1 \sim N. \tag{15}$$

Furthermore, for each component, we use the FITC model due to its advantages mentioned in Sect. 2.3 to describe the SGP input-output relation

$$p(Y_c | X_c, U_c, \theta_c) \sim N\left[0, K(X_c, U_c | \theta_c) K(U_c, U_c | \theta_c)^{-1} K(U_c, X_c | \theta_c) + \Lambda_c + \sigma_c^2 I\right], \tag{16}$$

where U_c denotes the pseudo inputs in the c-th component,

$$\Lambda_c = diag\left[K(X_c, X_c | \theta_c) - K(X_c, U_c | \theta_c) K(U_c, U_c | \theta_c)^{-1} K(U_c, X_c | \theta_c)\right], \tag{17}$$

and X_c, Y_c, θ_c follow the meanings in Sect. 2.2.

Our MSGP model can be fully described by Eqs. (14)–(16), and it has fewer parameters than previous MSGP models [5, 6, 9–11]. However, it still retains the advantages of combining sparse GPs with the mixture model. For this refined model, we strictly derive its training algorithm, called the precise hard-cut EM algorithm.

3   The Hard-Cut EM Algorithm for the FITC Mixtures

Since the outputs from both MGP and MSGP models are dependent, the computational complexity of Q function is exponential with summation over multiple labels. In order to avoid such an expensive computation, a hard-cut EM algorithm can be adopted by partitioning the training samples into the corresponding component with the maximum posterior in E-step, so that the parameters of each component can be learnt separately via the MLE in M-step. Because our model is developed from the MGP model in [29], where the precise hard-cut EM algorithm was strictly derived and performed quite well on both the prediction accuracy and the learning efficiency, we adjust its general learning framework to train the FITC mixture.

It can be easily derived that the marginal output likelihood for each sample in fact remains exactly the same as [29], i.e.

$$p(y_t | x_t, z_t = c) = N\left[y_t | 0, K(x_t, x_t | \theta_c) + \sigma_c^2\right] = N(y_t | 0, l_c^2 + \sigma_c^2), \tag{18}$$

so, the same posterior in E-step can be derived using the Bayesian formula:

$$\Pr(z_t = c | x_t, y_t) = \frac{\pi_c N(x_t | \mu_c, S_c) N(y_t | l_c^2 + \sigma_c^2)}{\sum\limits_{k=1}^{C} \pi_k N(x_t | \mu_k, S_k) N(y_t | l_k^2 + \sigma_k^2)}. \tag{19}$$

However, in M-step, the GP likelihood in [29] can be replaced by the FITC likelihood given by Eq. (16) with less computation, thus pseudo inputs are added during the learning of the hyper-parameters. With such an adjustment from [29], we obtain the procedure of our proposed precise hard-cut EM algorithm as follows.

- Step1 (Initialize the partition of the training data). Divide the vectors $\{[x_t^T, y_t]\}_{t=1}^N$ into C clusters, and set $z_t \leftarrow$ the indicator of the t-th sample to the cluster.
- Step2 (M-step). Learn the parameters with MLE for each component:

$$\pi_c \leftarrow \frac{N_c}{N}, \quad \mu_c \leftarrow \frac{1}{N_c} \sum_{t=1}^N I(z_t = c)x_t, \quad S_c \leftarrow \frac{1}{N_c} \sum_{t=1}^N I(z_t = c)(x_t - \mu_c)(x_t - \mu_c)^T, \tag{20}$$

$$(\hat{U}_c, \hat{\theta}_c) = \arg\max_{U_c, \theta_c} \ln p(Y_c | X_c, U_c, \theta_c), \tag{21}$$

where N_c is the number of samples in the c-th component, Eq. (20) is the same as learning the parameters of the Gaussian mixture model and Eq. (21) follows from [8].
- Step3 (E-step). Repartition the samples into the corresponding component based on the maximizing a posterior criterion:

$$z_t \leftarrow \arg\max_c \ \Pr(z_t = c | x_t, y_t) = \arg\max_c [\pi_c N(x_t | \mu_c, S_c) N(y_t | l_c^2 + \sigma_c^2)] \tag{22}$$

- Step4. Stop if the number of changed labels falls below η % of the number of training samples ($0 \le \eta \le 10$ is a threshold). Otherwise, return to Step 2.

After the learning process above, we have obtained the estimated parameters and the partition of the training data. Then, for a test input x*, we can classify it into the z-th component of the MSGP by maximizing the posterior as in [29]:

$$z^* \leftarrow \arg\max_c \ \Pr(z^* = c | D, x^*) = \arg\max_c [\pi_c N(x^* | \mu_c, S_c)] \tag{23}$$

According to this classification, we can predict the output of the test input via the z-th component using Eq. (13).

4   Experimental Results

4.1   On a Small Synthetic Dataset from the MGP Model

In order to evaluate the validity and feasibility of the FITC mixture model and its hard-cut EM algorithm, we generate a typical synthetic dataset from MGP model with 3 components. Actually, there are 500 training samples and 100 test samples in each component and each input has 3 features. Then we compare our algorithms with the following typical baselines for regression:

1. The FITC model and its MLE algorithm [8].
2. The MGP model and its precise hard-cut EM algorithm [29].
3. The MGP model and its LOOCV hard-cut EM algorithm [26].
4. The mixture of sparse GPs model and its variational hard-cut EM algorithm [9].
5. Linear regression [37].
6. SVM regression with Gaussian kernel [38].

Each of these algorithms is implemented on this synthetic dataset five times, with an Intel (R) Core (TM) i5 CPU and 4.00 GB of RAM running Matlab R2014b source codes on Secure CRT. For each of these algorithms, the mean and standard deviation of the root mean squared errors (RMSEs) for prediction and the training times are listed in Table 1. In addition, to make the prediction of FITC and its mixture model more accurate, we initialize the kernel parameters by training a GP model on 20 randomly selected training samples before the MLE learning process, as did in [8].

Table 1. The mean and standard deviation of the predictive RMSEs and the training times for each algorithm on the typical synthetic dataset from MGP model

	RMSE	Training time (s)
Our proposed precise hard-cut EM algorithm for FITC mixture ($C = 3$ components, $M = 20$ pseudo inputs, threshold $\eta \% = 5 \%$)	0.5139 ± 0.0222	27.0198 ± 3.9183
MLE for FITC model ($M = 20$)	0.6495 ± 0.0994	22.2497 ± 1.5894
Precise hard-cut EM algorithm for MGP ($C = 3$)	0.4963 ± 0	190.4125 ± 42.4939
LOOCV hard-cut EM algorithm for MGP ($C = 3$)	0.4977 ± 0	2552.0 ± 559.5477
Variational hard-cut EM algorithm ($C = 3$, $M = 20$)	0.6086 ± 0.1634	93.4882 ± 11.3112
Linear regression	0.8514 ± 0	0.0571 ± 0.1273
SVM	0.7109 ± 0	221.1791 ± 5.5132

It can be seen from Table 1 that on the synthetic dataset, our proposed algorithm is much more accurate than the single FITC since the dataset is multimodal. With sparse technique, our algorithm is much more efficient than the two EM algorithms of the MGP models, whereas the loss of accuracy is trivial compared with the increase on speed. The variational hard-cut EM algorithm makes a coarse conditionally independent assumption to the posterior and thus it is not as accurate as our proposed algorithm. Besides, the variational hard-cut EM algorithm also takes longer to train than our proposed algorithm. Traditional regression algorithms like the linear regression and the SVM are rough for such a non-linear synthetic dataset.

Furthermore, our proposed algorithm correctly partitions all the training samples and test samples into their components each time, which, along with the results in Table 1, means that our algorithm is feasible and effective.

4.2   On a Large Synthetic Dataset from the FITC Mixture Model

Moreover, to test the advantage of FITC mixture model and its hard-cut EM algorithm, we generate a typical synthetic dataset from FITC mixture model with 100 components. In each component, there are 200 training samples and 200 test samples, with 1 dimensional inputs. Then we compare our algorithm with the baselines above. The computational environment and experimental details remain the same as above.

On such a big dataset (20000 training samples), some algorithms are prohibitively slow and fail to converge, such as the SVM, the precise hard-cut EM algorithm and the LOOCV hard-cut EM algorithm of MGP, so we omit them in the experiment. It also indicates that our proposed hard-cut EM algorithm is more efficient than the precise hard-cut EM algorithm of the MGP model due to sparsity mechanism.

All the other algorithms are implemented on this synthetic dataset five times and the mean and standard deviation of their RMSEs for prediction and training times are listed in Table 2. Similarly, for FITC model and our proposed FITC mixture model, we initialize the kernel parameters by training a GP model on 10 randomly selected training samples before the MLE learning process, as in [8].

Table 2. The mean and standard deviation of the predictive RMSEs and the training times for each algorithm on the typical synthetic dataset from FITC mixture model

	RMSE	Training time (s)
Our proposed precise hard-cut EM algorithm for FITC mixture (C = 100, M = 10, η % = 5 %)	0.1115 ± 0.0106	86.9691 ± 2.8167
MLE for FITC model (M = 10)	1.0149 ± 0.0190	25.3531 ± 12.8932
Variational hard-cut EM algorithm (C = 100, M = 10)	0.8649 ± 0.0339	890.1411 ± 159.3509
Linear regression	1.0492 ± 0	0.0006 ± 0.0001

Table 2 indicates that for the synthetic dataset with much more components, our proposed FITC mixture model has greater advantage than the FITC model, on both predictive precision and learning efficiency. The conditionally independent assumption of the variational hard-cut EM algorithm, as well as the linear assumption of the linear regression, is not suitable for such a highly non-linear dataset with dependent samples. Therefore, the two algorithms have big prediction errors on this dataset.

4.3   On Kin40k Dataset

Finally, we compare these algorithms on a popular real dataset called kin40k, which is generated by a robot arm simulator, with 10000 training samples, 30000 test samples and 9 attributes [34]. The computational environment and implementation details remain the same as above. The mean and standard deviation of the predictive RMSEs as well as the training times for each algorithm are listed in Table 3. Similarly, for FITC and FITC mixture model, we initialize the kernel parameters by training a GP model on 500 randomly selected training samples before the MLE learning process.

Table 3. The mean and standard deviation of the predictive RMSEs and the training times for each algorithm on kin40k dataset

	RMSE	Training time (s)
Our proposed precise hard-cut EM algorithm for FITC mixture (C = 4, M = 650, η % = 5 %)	0.2007 ± 0.0094	14358.4108 ± 5321.6807
MLE for FITC model (M = 650)	0.2823 ± 0.0033	1547.4120 ± 66.1614
Variational hard-cut EM algorithm for MGP (C = 4, M = 650)	0.2475 ± 0.0269	2627.1 ± 84.1696
Linear regression	1.0492 ± 0	0.0006 ± 0.0001

Still, on the large dataset, the SVM, the precise hard-cut EM algorithm and the LOOCV hard-cut EM algorithm of MGP are prohibitively slow and fail to converge.

From Table 3, we can observe that on the kin40 k dataset, our proposed algorithm is more precise than the other algorithms due to fewer approximations. A possible reason why our algorithm consumes so much time is that on a real dataset that doesn't come from a GP or MGP model, our algorithm needs much more iterations. Therefore, a further improvement can be made by preprocessing a real dataset so that it is more like a dataset simulated from a MGP. Linear regression is still rather coarse since the kin40k dataset is highly non-linear [34].

In general, our algorithm is practical on the real dataset, and its computational cost is acceptable with 10000 training samples.

5   Conclusion

We have refined the MSGP model and developed the hard-cut EM algorithm for its parameter learning. The general framework is adapted from [29] whereas the learning efficiency significantly improves with the sparsity mechanism. The experimental results on both the synthetic dataset and the real dataset demonstrate that the developed hard-cut EM algorithm for MSGP model is precise and efficient, and generally outperforms the conventional linear regression, SVM and some other learning algorithms for GP, MGP and MSGP models.

Acknowledgement. This work was supported by the Natural Science Foundation of China for Grant 61171138. The authors would like to thank Dr. E. Snelson and Dr. Z. Ghahramani for their valuable advice about FITC model.

References

1. Rasmussen, C.E.: Evaluation of Gaussian processes and other methods for non-linear regression. The University of Toronto (1996)
2. Williiams, C.K.I., Barber, D.: Bayesian classification with Gaussian processes. IEEE Trans. Pattern Anal. Mach. Intell. **20**(12), 1342–1351 (1998)

3. Gao, X.B., Wang, X.M., Tao, D.C.: Supervised Gaussian process latent variable model for dimensionality reduction. IEEE Trans. Syst. Man Cybern. B Cyberne. **41**(2), 425–434 (2011)
4. Rasmussen, C.E., Kuss, M.: Gaussian processes in reinforcement learning. In: Thrun, S., Saul, L., Schölkopf, B. (eds.) Advances in Neural Information Processing Systems, vol. 16, pp. 751–759. MIT Press, Cambridge (2003)
5. Yuan, C., Neubauer, C.: Variational mixture of Gaussian process experts. In: Advances in Neural Information Processing Systems, vol. 21, pp. 1897–1904 (2008)
6. Stachniss, C., Plagemann, C., Lilienthal, A.J., et al.: Gas distribution modeling using sparse Gaussian process mixture models. In: Proceedings of Robotics: Science and Systems, pp. 310–317 (2008)
7. Tresp, V.: Mixtures of Gaussian processes. In: Advances in Neural Information Processing Systems, vol. 13, pp. 654–660 (2000)
8. Snelson, E., Ghahramani, Z.: Sparse Gaussian processes using pseudo-inputs. In: Advances in Neural Information Processing Systems, vol. 18, pp. 1257–1264 (2005)
9. Nguyen, T., Bonilla, E.: Fast allocation of Gaussian process experts. In: Proceedings of the 31st International Conference on Machine Learning, pp. 145–153 (2014)
10. Wang, Y., Khardon, R.: Sparse Gaussian processes for multi-task learning. In: The European Conference on Machine Learning and Principles and Practice of Knowledge Discovery in Databases, pp. 711–727 (2012)
11. Sun, S., Xu, X.: Variational inference for infinite mixtures of Gaussian processes with applications to traffic flow prediction. IEEE Trans. Intell. Transp. Syst. **12**(2), 466–475 (2011)
12. Meeds, E., Osindero, S.: An alternative infinite mixture of Gaussian process experts. In: Advances in Neural Information Processing Systems, vol. 18, pp. 883–890 (2005)
13. Gramacy, R.B., Lee, H.K.H.: Bayesian treed Gaussian process models with an application to computer modeling. J. Am. Stat. Assoc. **103**(483), 1119–1130 (2008)
14. Shi, J.Q., Murray-Smith, R., Titterington, D.M.: Bayesian regression and classification using mixtures of Gaussian processes. Int. J. Adapt. Control Sig. Process. **17**(2), 149–161 (2003)
15. Shi, J.Q., Murray-Smith, R., Titterington, D.M.: Hierarchical Gaussian process mixtures for regression. Stat. Comput. **15**(1), 31–41 (2005)
16. Rasmussen, C.E., Ghahramani, Z.: Infinite mixtures of Gaussian process experts. In: Advances in Neural Information Processing Systems, vol. 14, pp. 881–888 (2001)
17. Fergie, M.P.: Discriminative Pose Estimation Using Mixtures of Gaussian Processes. The University of Manchester (2013)
18. Sun, S.: Infinite mixtures of multivariate Gaussian processes. In: Proceedings of the International Conference on Machine Learning and Cybernetics, pp. 1011–1016 (2013)
19. Tayal, A., Poupart, P., Li, Y.: Hierarchical double Dirichlet process mixture of Gaussian processes. In: Proceedings of the 26th Association for the Advancement of Artificial Intelligence, pp. 1126–1133 (2012)
20. Ross, J., Dy, J.: Nonparametric mixture of Gaussian processes with constraints. In: Proceedings of the 30th International Conference on Machine Learning, pp. 1346–1354 (2013)
21. Chatzis, S.P., Demiris, Y.: Nonparametric mixtures of Gaussian processes with power-law behavior. IEEE Trans. Neural Netw. Learn. Syst. **23**(12), 1862–1871 (2012)
22. Platanios, E.A., Chatzis, S.P.: Mixture Gaussian process conditional heteroscedasticity. IEEE Trans. Pattern Anal. Mach. Intell. **36**(5), 888–900 (2014)
23. Kapoor, A., Ahn, H., Picard, R.W.: Mixture of Gaussian processes for combining multiple modalities. In: Oza, N.C., Polikar, R., Kittler, J., Roli, F. (eds.) MCS 2005. LNCS, vol. 3541, pp. 86–96. Springer, Heidelberg (2005)

24. Reece, S., Mann, R., Rezek, I., et al.: Gaussian process segmentation of co-moving animals. Proc. Am. Inst. Phys. **1305**(1), 430–437 (2011)
25. Lázaro-Gredilla, M., Van, V.S., Lawrence, N.D.: Overlapping mixtures of Gaussian processes for the data association problem. Pattern Recogn. **45**(4), 1386–1395 (2012)
26. Yang, Y., Ma, J.: An efficient EM approach to parameter learning of the mixture of Gaussian processes. In: Liu, D., Zhang, H., Polycarpou, M., Alippi, C., He, H. (eds.) ISNN 2011, Part II. LNCS, vol. 6676, pp. 165–174. Springer, Heidelberg (2011)
27. Schiegg, M., Neumann, M., Kersting, K.: Markov logic mixtures of Gaussian processes: towards machines reading regression data. In: Proceedings of the 15th International Conference on Artificial Intelligence and Statistics, JMLR:W&CP, vol. 22, pp. 1002–1011 (2012)
28. Yu, J., Chen, K., Rashid, M.M.: A Bayesian model averaging based multi-kernel Gaussian process regression framework for nonlinear state estimation and quality prediction of multiphase batch processes with transient dynamics and uncertainty. Chem. Eng. Sci. **93** (19), 96–109 (2013)
29. Chen, Z., Ma, J., Zhou, Y.: A precise hard-cut EM algorithm for mixtures of Gaussian processes. In: Huang, D.-S., Jo, K.-H., Wang, L. (eds.) ICIC 2014. LNCS, vol. 8589, pp. 68–75. Springer, Heidelberg (2014)
30. Rasmussen, C.E., Williams, C.K.I.: Gaussian Processes for Machine Learning. MIT Press, Cambridge (2006)
31. Sundararajan, S., Keerthi, S.: Predictive approaches for choosing hyperparameters in Gaussian processes. Neural Comput. **13**(5), 1103–1118 (2001)
32. Quiñonero-Candela, J., Rasmussen, C.E.: A unifying view of sparse approximate Gaussian process regression. J. Mach. Learn. Res. **6**, 1935–1959 (2005)
33. Csató, L., Opper, M.: Sparse online Gaussian processes. Neural Comput. **14**(3), 641–669 (2002)
34. Seeger, M., Williams, C.K.I., Lawrence, N.D.: Fast forward selection to speed up sparse Gaussian process regression. In: Bishop, C.M., Frey, B.J. (eds.) Proceedings of the 9th International Workshop on Artificial Intelligence and Statistics (2003)
35. Keerthi, S.S., Chu, W.: A matching pursuit approach to sparse Gaussian process regression. In: Advances in Neural Information Processing Systems, vol. 18 (2005)
36. Smola, A., Bartlett, P.: Sparse greedy Gaussian process regression. Adv. Neural Inf. Process. Syst. **13**, 619–625 (2000)
37. Murphy, K.P.: Machine Learning: A Probabilistic Perspective. MIT Press, Cambridge (2012)
38. Smola, A.J., Schölkopf, B.: A tutorial on support vector regression. Stat. Comput. **14**(3), 199–222 (2004)

Maximum Principle in the Unbounded Domain of Heisenberg Type Group

Zhenhua Wang[(✉)] and Xuemei Yang

School of Mathematics and Information Science,
Xianyang Normal University, Xianyang, China
mail_wangzhenhua@126.com, yangxuemei691226@163.com

Abstract. In this paper, through polar coordinates and constructing an auxiliary function we prove the maximum principle in the unbounded domain Ω as following: suppose that there exists $u(\xi) \in C(\overline{\Omega})$ bounded above, solution of

$$\begin{cases} Lu(\xi) + c(\xi)u(\xi) \geq 0 \, \xi \in \Omega, \ c(\xi) \leq 0, \\ u(\xi) \leq 0 \, \xi \in \partial\Omega, \end{cases}$$

then $u(\xi) \leq 0$ in Ω. Here Ω is an open connected subset of Heisenberg type group G such that the following condition holds: there exists $\xi_0 \in G$ such that $\overline{\xi_0 \circ \Omega}$ lies on one side of an hyperplane parallel to y_1 axis.

Keywords: Heisenberg type group · Polar coordinates · Sub-Laplace operator

1 Introduction

In 1995, Birindelli [1] defined a compact operator in the bounded domain of Euclidean space, And proved that it had positive eigenvalue and eigenvector. In 1997, H. Berestycki and L.N irenberg [2] proved the maximum principle in the unbounded domain of R^n: let $M \subset R^n$ be unbounded, $\overline{M}$ doesn't intersect with the infinite open connected cone $\Sigma \subset R^n$, if z is bounded on $C(\overline{M})$ and meets

$$\begin{cases} \Delta z + c(x)z \geq 0, \ x \in M, c(x) \leq 0 \\ z \leq 0, \ x \in \partial M \end{cases}$$

then $z \leq 0$.

In 2001, Birindelli (see [3]) generalized the principle to Heisenberg group by compact operator in [1] and the polar coordinates on groups of Heisenberg, precisely as following: let $N \subset H^n$ be unbounded, there exists $v' \in R^m$ such that

$$\overline{\xi_0 \circ N} \subset \{\xi \in H^n | v' \cdot x > 0\},$$

suppose that there exists $u(\xi) \in C(\overline{N})$ bounded above, solution of

$$\begin{cases} \Delta_H u(\xi) + cu(\xi) \geq 0, \ \xi \in N, \ c \leq 0 \\ u(\xi) \leq 0, \ \xi \in \partial N \end{cases}$$

© Springer International Publishing Switzerland 2015
D.-S. Huang and K. Han (Eds.): ICIC 2015, Part III, LNAI 9227, pp. 25–33, 2015.
DOI: 10.1007/978-3-319-22053-6_3

then $u(\xi) \leq 0$ in N. Now we will further generalize the maximum principle to Heisenberg type group G as following:

Theorem 1. Let Ω be an open connected subset of G such that the following conditions hold: there exists $\xi_0 \in G$ such that $\overline{\xi_0 \circ \Omega}$ lies on one side of an hyperplane parallel to y_1 axis, suppose that there exists $u(\xi) \in C(\overline{\Omega})$ bounded above, solution of

$$\begin{cases} Lu(\xi) + cu(\xi) \geq 0, \ \xi \in \Omega, \ c \leq 0, \\ u(\xi) \leq 0, \ \xi \in \partial\Omega, \end{cases} \tag{1}$$

then $u(\xi) \leq 0$ in Ω.

The following definitions are introduced in [4] by Kaplan.

Definition 1. Let G be a Carnot group of step 2, with Lie algebra $\tilde{g} = V_1 \oplus V_2$. the map $J : V_2 \to \text{End}(V_1)$ meets

$$\langle J(\xi_2)\xi_1', \xi_1'' \rangle = \langle \xi_2, [\xi_1', \xi_1''] \rangle, \xi_1', \xi_1'' \in V_1, \xi_2 \in V_2$$

for every $\xi_2 \in V_2$ and $|\xi_2| = 1$, the mapping $J(\xi_2) : V_1 \to V_1$ is orthogonal, then G is called Heisenberg type group. There exists a set of left invariant vector fields

$$X_i = \frac{\partial}{\partial x_i} + \frac{1}{2}\sum_{j=1}^{n} <[\xi, X_i], Y_j > \frac{\partial}{\partial y_j}, i = 1, \ldots, m.$$

If we let

$$\xi = (x, y), \ x = (x_1, \ldots, x_m), y = (y_1, \ldots, y_n),$$
$$\tilde{\xi} = (\tilde{x}, \tilde{y}), \ x = (\tilde{x}_1, \ldots, \tilde{x}_m), y = (\tilde{y}_1, \ldots, \tilde{y}_n),$$

the calculation law in G as following

$$\xi \circ \tilde{\xi} = (x_1 + \tilde{x}_1, \ldots, x_m + \tilde{x}_m, y_1 + \tilde{y}_1 + \frac{1}{2}\sum_{k,l=1}^{m} b_{kl}^1 x_k \tilde{x}_l, \ldots, y_n + \tilde{y}_n + \frac{1}{2}\sum_{k,l=1}^{m} b_{kl}^n x_k \tilde{x}_l).$$

The sub-Laplace operator L on G written as

$$L = -\sum_{j=1}^{m} X_j^* X_j = -\sum_{j=1}^{m} X_j^2$$

or

$$L = \text{div}(\sigma^T \sigma \nabla), \quad X = \sigma\nabla,$$

here ∇ stands for ordinary gradient, and

$$\sigma = \begin{pmatrix} 1 & \cdots & 0 & \frac{1}{2}\langle[\xi, X_1], Y_1\rangle & \cdots & \frac{1}{2}\langle[\xi, X_1], Y_n\rangle \\ \vdots & & \vdots & \vdots & & \vdots \\ 0 & \cdots & 1 & \frac{1}{2}\langle[\xi, X_m], Y_1\rangle & \cdots & \frac{1}{2}\langle[\xi, X_m], Y_n\rangle \end{pmatrix}$$

Measurable function on G is

$$\rho(\xi) = \left(|x(\xi)|^4 + 16|y(\xi)|^2 \right)^{\frac{1}{4}}, \quad \xi \in G,$$

Koranyi unit ball on G is

$$B_G(e, 1) = \left\{ \xi \in G \,\big|\, \|\xi^{-1} \circ e\|_G \leq 1 \right\}, \ e = (0, \ldots, 0).$$

The above contents can be seen in [4–6]. Now we introduce the concept of polar coordinates on G (see [7, 8]).

Let $\rho = \rho(\xi, e) = \|\xi\|_G$, $\theta \in \partial B_G(0, 1)$, then $(\rho, \theta) \in R_+ \times \Omega'$ are polar coordinates on G, here Ω' is a subset in the Koranyi unit sphere $S_G^1 = \partial B_G(e, 1)$, let $\rho = \|\xi\|_G$ and $u : \partial B_G(e, 1) \to R$ be smooth functions, then

$$X_i(u(\theta)\rho^\alpha) = \left(R_i^\alpha u(\theta)\right)\rho^{\alpha-1} \ (i = 1, \ldots, m), \tag{2}$$

$$Y_j(u(\theta)\rho^\alpha) = \left(S_j^\alpha u(\theta)\right)\rho^{\alpha-2} \ (j = 1, \ldots, n), \tag{3}$$

and

$$R_i^\alpha = \widehat{R}_i + \alpha a_i, \tag{4}$$

$$S_j^\alpha = \widehat{S}_j + \alpha\rho b_j, \tag{5}$$

here $a_i = X_i(\rho)$, $b_j = \rho Y_j(\rho)$, $\widehat{R}_i, \widehat{S}_j$ are all tangent vectors of X_i and Y_j on S_G^1, they are equal to zero when acting on constants, and comply with $\left[\widehat{R}_k, \widehat{R}_i\right] = \sum_{j=1}^{n} b_{ki}^j \widehat{S}_j$. A simple computation with (2) and (4) shows that

$$L(u(\theta)\rho^\alpha) = -\sum_{i=1}^{m} X_i^2(u(\theta)\rho^\alpha) = -\sum_{i=1}^{m} X_i\left[\left(R_i^\alpha u(\theta)\right)\rho^{\alpha-1}\right] = -\sum_{i=1}^{m} R_i^{\alpha-1}\left(R_i^\alpha u(\theta)\right)\rho^{\alpha-2}$$

$$= -\rho^{\alpha-2}\sum_{i=1}^{m}\left(R_i^{\alpha-1}R_i^\alpha\right)u(\theta) = -\rho^{\alpha-2}\sum_{i=1}^{m}\left[\left(\widehat{R}_i + (\alpha-1)a_i\right)\left(\widehat{R}_i + \alpha a_i\right)\right]u(\theta) \tag{6}$$

$$= \rho^{\alpha-2}\sum_{i=1}^{m}\left[-\widehat{R}_i^2 - \alpha\widehat{R}_i a_i + (1-\alpha)a_i\widehat{R}_i + (1-\alpha)\alpha a_i^2\right]u(\theta).$$

Let $h = \sum_{i=1}^{m} a_i^2$, we also introduce the following operator

$$A^\alpha = \sum_{i=1}^{m} \left(-\widehat{R}_i^2 - \alpha \widehat{R}_i a_i + (1-\alpha) a_i \widehat{R}_i \right) + (1-\alpha)\alpha h, \tag{7}$$

$$D_1 = \sum_{i=1}^{m} \left(-\widehat{R}_i^2 - \widehat{R}_i a_i \right). \tag{8}$$

Definition 2. If $\lambda > 0$ meets

$$\begin{cases} D_1 u - \lambda h u = 0, \ \xi \in \Omega', \\ u = 0, \ \xi \in \partial\Omega', \end{cases}$$

we call λ the principal weighted eigenvalue of operator D_1.

Lemma 1. (see [9]) (Young inequality with ε) For every $\varepsilon > 0$ and $a, b \geq 0$, $1 < p, q < \infty$, $\frac{1}{p} + \frac{1}{q} = 1$, the following inequality holds

$$ab \leq \frac{\varepsilon a^p}{p} + \frac{\varepsilon^{-q/p} b^q}{q}.$$

Proof. By the Young inequality, we can get

$$ab \leq \frac{a^p}{p} + \frac{b^q}{q},$$

hence

$$ab = (\varepsilon^{1/p} a)\left(\frac{b}{\varepsilon^{1/p}}\right) \leq \frac{(\varepsilon^{1/p} a)^p}{p} + \frac{(\varepsilon^{-1/p} b)^q}{q} = \frac{\varepsilon a^p}{p} + \frac{\varepsilon^{-q/p} b^q}{q}.$$

2   The Proof of Theorem 1

When Ω is bounded there is nothing to prove, we focus on the condition that Ω is unbounded. Now we will give the proof in the unbounded case of Theorem 1.

Proof. Firstly we construct an auxiliary function g to satisfy the following equation

$$\begin{cases} Lg(\xi) + cg(\xi) \leq 0, \ \xi \in \Omega \\ g(\xi) > 0, \ \xi \in \Omega \end{cases}$$

and $\lim\limits_{|\xi|_G \to \infty} g(\xi) = +\infty$, $\xi \in \Omega$. Without loss of generality, we will suppose that the hyperplane parallel to y_1 is $\{x_1 = 0\}$, in addition, $\overline{\Omega} \subset \{\xi | x_1 > 0\} = \pi$. Also let $\Sigma_0 = \pi \cap S_G^1$, for $Lx_1 = 0$, we define the function $u = \frac{x_1}{\rho}$ on S_G^1 to satisfy

$$Lx_1 = L(\rho u) = \rho^{-1} L^1(u) = 0.$$

Due to

$$L^1(u) = \sum_{i=1}^{m} [\widehat{R}_i^2 u + a_i \widehat{R}_i u] + (Q-1)hu = 0, \xi \in \Sigma_0,$$

we get

$$D_1 u(\xi) + (Q-1)hu(\xi) = 0, \ \xi \in \Sigma_0$$

$$u(\xi) = 0, \ \xi \in \partial\Sigma_0$$
$$u(\xi) > 0, \ \xi \in \Sigma_0$$

i.e. $(Q-1)$ is the principal weighted eigenvalue of $-D_1$ in Σ_0. Let $\Sigma_\varepsilon \supset \Sigma_0$ close enough to Σ_0 that $\lambda_1 = \lambda_1(\varepsilon)$ the principal weighted eigenvalue of $-D_1$ in Σ_ε satisfies $Q - 1 - \varepsilon = \lambda_1 < Q - 1$. We can choose Σ_ε such that it has no characteristic points on the boundary. Therefore there exists $\psi_\varepsilon > 0$ in Σ_ε such that

$$\begin{cases} D_1\psi_\varepsilon(\xi) + \lambda_1 h\psi_\varepsilon(\xi) = 0, \ \xi \in \Sigma_\varepsilon \\ \quad\quad \psi_\varepsilon(\xi) = 0, \ \xi \in \partial\Sigma_\varepsilon. \end{cases} \quad\quad (9)$$

If ε meets $\frac{1}{2}(\frac{1}{2} + Q - 2)h < Q - 1 - \varepsilon$, the operator $-(D_1 + \frac{1}{2}(\frac{1}{2} + Q - 2)h)$ will have a positive principal weighted eigenvalue $\frac{1}{2}(\frac{1}{2} + Q - 2)$ in Σ_ε. In addition, it is known that

$$A^\alpha = -\sum_{i=1}^{m} R_i^{\alpha-1} R_i^\alpha = -\sum_{i=1}^{m} \widehat{R}_i^2 - (2\alpha - 1)a_i \widehat{R}_i - \alpha(\alpha-1)a_i^2 - \alpha \widehat{R}_i a_i$$

$$= -\sum_{i=1}^{m} \widehat{R}_i^2 - (2\alpha - 1)a_i \widehat{R}_i - \alpha(\alpha-1)h - \alpha(Q-1)h$$

$$= -\sum_{i=1}^{m} \widehat{R}_i^2 - (2\alpha - 1)a_i \widehat{R}_i - \alpha(Q - 2 + \alpha)h,$$

so that

$$A^{\frac{1}{2}} = -\sum_{i=1}^{m} \widehat{R}_i^2 - \frac{1}{2}\frac{1}{2}(\frac{1}{2} + Q - 2)h.$$

Let $D_1 = \sum_{i=1}^{m} (-\widehat{R}_i^2 - \widehat{R}_i a_i)$, this leads to the following:

Claim 1. there exists $\varepsilon > 0$ such that there exists a function $\gamma(\xi) > 0$ and a constant $\mu > 0$ such that

$$\begin{cases} \sqrt{A}\gamma(\xi) - \mu\gamma(\xi) \geq 0, \ \xi \in \Sigma_\varepsilon, \\ \gamma(\xi) = 0, \ \xi \in \partial\Sigma_\varepsilon. \end{cases}$$

Proof. Firstly let us compute $A^{\frac{1}{2}}(\psi_\varepsilon^\beta)$, $0 < \beta < 1$. For the following equations hold

$$\widehat{R}_i(\psi_\varepsilon^\beta) = \beta\psi_\varepsilon^{\beta-1}\widehat{R}_i\psi_\varepsilon,$$

$$\widehat{R}_i^2(\psi_\varepsilon^\beta) = \beta(\beta-1)\psi_\varepsilon^{\beta-2}(\widehat{R}_i\psi_\varepsilon)^2 + \beta\psi_\varepsilon^{\beta-1}\widehat{R}_i^2\psi_\varepsilon.$$

Let $\gamma = \psi_\varepsilon^\beta$, we get

$$A^{\frac{1}{2}}(\gamma) = \sum_{i=1}^m \widehat{R}_i^2\gamma + \frac{1}{2}\frac{1}{2}(\frac{1}{2} + Q - 2)h\gamma$$

$$= \sum_{i=1}^m [\beta\psi_\varepsilon^{\beta-1}\widehat{R}_i^2\psi_\varepsilon + \beta(\beta-1)\psi_\varepsilon^{\beta-2}(\widehat{R}_i\psi_\varepsilon)^2] + \frac{1}{2}\frac{1}{2}(\frac{1}{2} + Q - 2)h\gamma.$$

Using (9) we can get $D_1\gamma + \lambda_1 h\gamma = 0$, due to $D_1 = \sum_{i=1}^m (-\widehat{R}_i^2 - \widehat{R}_i a_i)$. Hence

$$\lambda_1 h\gamma = -D_1\gamma = \sum_{i=1}^m (\widehat{R}_i^2 + \widehat{R}_i a_i)\gamma = \sum_{i=1}^m \widehat{R}_i^2\psi_\varepsilon^\beta + \sum_{i=1}^m a_i\widehat{R}_i\psi_\varepsilon^\beta$$

$$= \sum_{i=1}^m \beta(\beta-1)\psi_\varepsilon^{\beta-2}(\widehat{R}_i\psi_\varepsilon)^2 + \sum_{i=1}^m \beta\psi_\varepsilon^{\beta-1}\widehat{R}_i^2\psi_\varepsilon + \sum_{i=1}^m \beta\psi_\varepsilon^{\beta-1}a_i\widehat{R}_i\psi_\varepsilon,$$

therefore

$$\sum_{i=1}^m \beta(\beta-1)\psi_\varepsilon^{\beta-2}(\widehat{R}_i\psi_\varepsilon)^2 + \sum_{i=1}^m \beta\psi_\varepsilon^{\beta-1}\widehat{R}_i^2\psi_\varepsilon = \lambda_1 h\gamma - \sum_{i=1}^m \beta\psi_\varepsilon^{\beta-1}a_i\widehat{R}_i\psi_\varepsilon.$$

From (1) we know

$$\sum_{i=1}^m \beta\psi_\varepsilon^{\beta-1}a_i\widehat{R}_i\psi_\varepsilon = \sum_{i=1}^m \beta\psi_\varepsilon^{\frac{\beta}{2}-1}\widehat{R}_i\psi_\varepsilon \cdot a_i\psi_\varepsilon^{\frac{\beta}{2}} \leq \beta\sum_{i=1}^m [2(1-\beta)\frac{(\widehat{R}_i\psi_\varepsilon)^2\psi_\varepsilon^{\beta-2}}{2}$$

$$+ (2(1-\beta))^{-1}\frac{a_i^2\psi_\varepsilon^\beta}{2}] = \sum_{i=1}^m \beta(1-\beta)(\widehat{R}_i\psi_\varepsilon)^2\psi_\varepsilon^{\beta-2} + \sum_{i=1}^m \beta\psi_\varepsilon^\beta\frac{a_i^2}{4(1-\beta)}$$

$$= \sum_{i=1}^m \beta(1-\beta)(\widehat{R}_i\psi_\varepsilon)^2\psi_\varepsilon^{\beta-2} + \frac{\beta}{4(1-\beta)}h\gamma,$$

here $\varepsilon = 2(1 - \beta)$. Therefore

$$A^{\frac{1}{2}}(\gamma) = \sum_{i=1}^{m} [\beta \psi_\varepsilon^{\beta-1} \widehat{R}_i^2 \psi_\varepsilon + \beta(\beta - 1)\psi_\varepsilon^{\beta-2}(\widehat{R}_i \psi_\varepsilon)^2] + \frac{1}{2}\frac{1}{2}(\frac{1}{2} + Q - 2)h\gamma$$

$$= \lambda_1 h\gamma - \sum_{i=1}^{m} \beta \psi_\varepsilon^{\beta-1} a_i \widehat{R}_i \psi_\varepsilon + \frac{1}{2}\frac{1}{2}(\frac{1}{2} + Q - 2)h\gamma$$

$$\geq \sum_{i=1}^{m} \beta(\beta - 1)(\widehat{R}_i \psi_\varepsilon)^2 \psi_\varepsilon^{\beta-2} + \lambda_1 h\gamma - \sum_{i=1}^{m} \beta \psi_\varepsilon^{\beta-1} a_i \widehat{R}_i \psi_\varepsilon + \frac{1}{2}\frac{1}{2}(\frac{1}{2} + Q - 2)h\gamma$$

$$\geq \lambda_1 h\gamma + \frac{\beta}{4(1 - \beta)} h\gamma + \frac{1}{2}\frac{1}{2}(\frac{1}{2} + Q - 2)h\gamma$$

$$= (\lambda_1 - \frac{\beta}{4(1 - \beta)} + \frac{1}{2}\frac{1}{2}(\frac{1}{2} + Q - 2))h\gamma.$$

Let $k(\beta) = -\lambda_1 + \frac{\beta}{4(1-\beta)} - \frac{1}{2}(\frac{1}{2} + Q - 2)$, if we prove $k(\beta) > 0$ for ε sufficiently small, then Lemma 1 is proved by choosing $\mu = k(\beta)$ and $\gamma = \psi_\varepsilon^\beta$. Now we use the analysis method to prove $k(\beta) > 0$.

In fact, if $k(\beta) = -\lambda_1 + \frac{\beta}{4(1-\beta)} - \frac{1}{2}(\frac{1}{2} + Q - 2) > 0$, we can get

$$6(Q - 1)\beta - 6Q + 7 + 4\varepsilon - 4\beta\varepsilon > 0,$$

as $\varepsilon \to 0$

$$6(Q - 1)\beta - 6Q + 7 > 0,$$
$$\beta > \frac{6Q - 7}{6(Q - 1)} = 1 - \frac{1}{6(Q - 1)}.$$

It is easy to see that $k(\beta) > 0$ for $0 < \beta < 1$ while ε sufficiently small. Let $\beta_0 \in (1 - \frac{1}{6(Q-1)}, 1)$, so $\mu = k(\beta_0) > 0$ and $\gamma = \psi_\varepsilon^\beta > 0$.

This completes the proof of claim 1, Next we continue to prove Theorem 1.

Step 1. To choose the auxiliary function $g(\xi) = \rho^{\frac{1}{2}}\gamma(\xi) > 0$. from Lemma 1, we know there exists $c < 0$ such that $A^{\frac{1}{2}}\gamma + c\rho^2\gamma \geq 0$, then

$$\rho^{-\frac{3}{2}}A^{\frac{1}{2}}\gamma + c\rho^{\frac{1}{2}}\gamma \geq 0,$$

$$Lg + cg = L(\rho^{\frac{1}{2}}\gamma) + cg = \rho^{-\frac{3}{2}}A^{\frac{1}{2}}\gamma + c\rho^{\frac{1}{2}}\gamma \geq 0,$$

i.e.

$$\begin{cases} Lg(\xi) + cg(\xi) \geq 0, \ \xi \in \Omega, \\ g(\xi) > 0, \ \xi \in \Omega. \end{cases}$$

Step 2. To construct the function $\sigma = \frac{u}{g}$ defined in Ω, which satisfies the following equation

$$\begin{cases} L\sigma + 2g^{-1}\nabla_L\sigma \cdot \nabla_L g + (Lg + cg)g^{-1}\sigma \geq 0, \ \xi \in \Omega, \\ \qquad\qquad \sigma \leq 0, \ \xi \in \partial\Omega. \end{cases}$$

putting $u = \sigma g$ into (1) we get

$$X_i(\sigma g) = gX_i\sigma + \sigma X_i g,$$

$$X_i^2(\sigma g) = X_i(gX_i\sigma + \sigma X_i g) = gX_i^2\sigma + 2X_i\sigma \cdot X_i g + \sigma X_i^2 g.$$

while

$$\nabla_L\sigma = (X_1\sigma \quad \cdots \quad X_m\sigma), \nabla_L g = (X_1 g \quad \cdots \quad X_m g),$$

then

$$\nabla_L\sigma \cdot \nabla_L g = \sum_{i=1}^{m} X_i\sigma \cdot X_i g.$$

therefore

$$L(\sigma g) = -g\sum_{i=1}^{m} X_i^2\sigma - 2\sum_{i=1}^{m} X_i\sigma \cdot X_i g - \sigma\sum_{i=1}^{m} X_i^2 g$$

$$= gL\sigma - 2\sum_{i=1}^{m} X_i\sigma \cdot X_i g + \sigma Lg$$

$$= gL\sigma - 2\nabla_L\sigma \cdot \nabla_L g + \sigma Lg.$$

Using (1) we can get

$$L(\sigma g) + cu = gL\sigma - 2\nabla_L\sigma \cdot \nabla_L g + \sigma Lg + cu \geq 0,$$

i.e.

$$\begin{cases} L\sigma - 2g^{-1}\nabla_L\sigma \cdot \nabla_L g + (\sigma Lg + cu)g^{-1} \geq 0, \ \xi \in \Omega, \\ \qquad\qquad \sigma \leq 0, \ \xi \in \partial\Omega \end{cases}$$

then

$$\lim_{\rho\to\infty} \sigma = \lim_{\rho\to\infty} \frac{u}{g} \leq 0.$$

By applying the standard maximum principle we obtain that $\sigma \leq 0$ in Ω i.e. $u(\xi) \leq 0$, $\xi \in \Omega$. This completes the proof of Theorem 1.

Acknowledgment. This paper is partially supported by the special research project of education department in Shaanxi province(2013JK1125), and the nature science fund project of Shaanxi province (No. 2014JM1032).

References

1. Birindelli, I.: Hopf's lemma and anti-maximum principle in general domains. J. Diff. Equ. **119**, 450–472 (1995)
2. Berestycki, H., Caffarelli, L., Nirenberg, L.: Monotonicity for elliptic equations in unbounded domains. Commun. Pure Appl. Math. **50**, 1088–1111 (1997)
3. Birindelli, I., Prajapat, J.: One dimensional symmetry in the heisenberg group. Annali della Scuola Normale Superiore di Pisa **3**, 1–17 (2001)
4. Kaplan, A.: Fundamental solutions for a class of hypoelliptic PDE generated by composition of quadratic forms. Trans. Am. Math. Soc. **258**, 147–153 (1980)
5. Garofalo, N., Vassilev, D.N.: Regularity near the characteristic set in the nonlinear dirichlet problem and conformal geometry of sub-laplacians. Math. Ann. **318**, 453–516 (2000)
6. Garofalo, N., Vassilev, D.: Symmetry properties of positive entire solutions of yamabe-type equations on groups of heisenberg type. Duke Math. J. **3**(106), 411–448 (2001)
7. Mekri, Z., Haken, A.: Left cauchy-riemann operator and dolbeault -grothendieck lemma on the group of heisenberg type. Complex Variables Eliptic Equ. **59**(8), 1185–1199 (2014)
8. Barbas, H.: Riesz Transforms on Groups of Heisenberg Type. J. Geom. Anal. **20**(1), 1–38 (2010)
9. Xianqiang, L., Zhiping, X.: A kind of extension of the famous young inequality. J. Inequalities Appl. **2013**(1), 1–11 (2013)

An Epidemic Propagation Model with Saturated Infection Rate on a Small World Network

Qiao-ling Chen[1], Liang Chen[2(✉)], Zhi-Qiang Sun[3], and Zhi-juan Jia[1]

[1] College of Information Science and Technology,
Zhengzhou Normal University, Zhengzhou, China
[2] National Computer Network Emergency Response Technical
Team/Coordination Center of China, Beijing, China
chenliang@cert.org.cn
[3] The Department of Basic Education, Henan Art Vocational College,
Zhengzhou, China

Abstract. An SIRS model with constant immigration of susceptible and saturated infection rate is established according to the propagation of the infectious viruses in a small world network. By using the mean-field theory and qualitative theory of differential equations, the existence and stability of equilibrium points of the system was analyzed. It also prove that the transmission threshold of this model is not entirely concerned with the topology of networks but also related to other factors such as immunization rate of the susceptible people. By using numerical simulation method to study the different factors which control the viruses, we obtained the conclusion that rewiring probability and the average of the nodes had an evident effect on the computer viruses' propagation in a small world network.

Keywords: Small-world network · SIRS model · Saturated infection rate · Global asymptotical stability

1 Introduction

One of the main struggles between human and disease is about the infectious disease. The spread of smallpox, plague, cholera, H1N1virus has being made people's life suffering a serious threat. And complex network provides a new platform for the study of the spread of the virus in terms of network topology.

The study found that not only social network has features of high concentration, short-distance et al., the computer network also has the characteristics of such small-world network. The regular networks [1, 2] and random networks [3, 4] are not well reflecting the network evolution mechanism. Therefore, small-world network in the study of the propagation of virus get more attention. Liu and Watts [5, 6] used the discrete time to marked the virus state in the cycle of virus propagation status and researched the influences of rewiring probability on the spread of the virus with WS small network. Newman and Watts [7] believe that the propagation of virus is equivalent to physical percolation problems, and find out the scale of the probability

© Springer International Publishing Switzerland 2015
D.-S. Huang and K. Han (Eds.): ICIC 2015, Part III, LNAI 9227, pp. 34–42, 2015.
DOI: 10.1007/978-3-319-22053-6_4

distribution of infectious people. Jin [8] proposed a "De-small world" model which use the average size of a neighborhood to define the node degrees, by limiting the contact between the rate of nodes to achieve control and slow the spread of the virus. Zhien [9] achieve the same purpose by immunizing part of the nodes.

According to the status quo of the propagation of the virus in the small-world network, we can summarize in the following key points:

1. The epidemiological model has been adapted to research the propagation of viruses, and considered the total population of virus, which is not conform to the reality.
2. The former models usually using linear incidence, but in a certain time, an infected contact the other susceptible is limited, thus the infectious ability is limited.
3. The traditional model of virus did not take into account people's subjective behavior affect the spread of the virus.

Based on the above, we introduce the topology of the network to the transmission dynamics equation; assume the susceptible with constant immigration and the number of people is changed during propagation of the virus. Use the saturated infection rate, which depends on the structure of the network and simulate the propagation of the virus in the small-world network. Consider the influence of people's behavior on the propagation of the virus, we not only establish the SIRS model with the saturated infection rate of the form $\frac{\beta <k> I^p S}{1+\alpha I^q}$ but also the stability and dynamic behavior are studied.

2 The SIRS Model in Small-World Network

1. Construction of Small-World Networks

The main mechanism to construct small-world networks is the Watts -Strogatz mechanism. Then NW model [7], JB [10], KA [11] model et al. is proposed. We adapt the small world network model proposed by Newman and Watts [12]:

Start with a nearest-neighbor coupled ring lattice with N nodes, in which each node is connected to its K/2 neighboring nodes, where there is an even integer. Add edges between nodes with probability p at random, of which self-connections and duplicated links are excluded.

2. Model Establishment

$$\begin{cases} \dfrac{dS}{dt} = A - \dfrac{\beta <k> I^p S}{1+\alpha I^q} + \gamma R - \omega S \\ \dfrac{dI}{dt} = \dfrac{\beta <k> I^p S}{1+\alpha I^q} - \delta I \\ \dfrac{dR}{dt} = \delta I - \gamma R + \omega S \\ N = S + I + R \end{cases} \tag{1.1}$$

Where A is the recruitment rate of population, δ is the infections' removal rate, γ is the rate of removed individual who lose immunity and return to susceptible class, ω is the rate of susceptible people who has immunity go into the recovery. The parameters p is positive constant, the incidence rate in this model is $\frac{\beta <k> I^p S}{1+\alpha I^q}$, θ is the rewiring probability, $q \geq 1$, $\alpha = 1/\theta < k>$, $\frac{1}{1+\alpha I^q}$ express that susceptible behavior will hinder the virus' spread while the number of infected people increase. It means that the infected people who contact with the susceptible people will be reduced during the increase of the number of infected people. Therefore, the incidence rate is more reasonable.

Theorem 1. The region $B = \left\{ \begin{array}{l} (S,R,I)|0 \leq S \leq A/_\omega, 0 \leq I, 0 \leq R, \\ I+R \leq N \end{array} \right\}$ is the positively invariant region of system (1.1).

Proof. The region B is positively invariant: any solutions originating within this region remain confined to it. because $S' = A + \gamma R(0) > 0$, if $S(0) = 0$, $R(0) \geq 0$, when $t > 0$, $S(t) > 0$ in (1), no solution can exit the feasible region across the boundary $I + R \leq N$. Similarly, $R'(0) = \delta I(0) > 0$ for $R(0) = 0$ and $I(0) > 0$. Therefore, no solution can exit the feasible region across the boundary $R = 0$, $I > 0$. At the boundary $I = 0$, the situation is somewhat more complicated. Let's discuss it.

Because $0 < P < 1$. in this case replace I by the new variable $U = I^{1-P}$, then the system (1.1) can replaced by

$$\begin{cases} \frac{dU}{dt} = \frac{(1-p)<k>\beta S}{1+\alpha U^{\frac{q}{1-p}}} - (1-p)\delta U \\ \frac{dR}{dt} = \delta U^{\frac{1}{1-p}} - \gamma R + \omega S \\ \frac{dS}{dt} = A - \frac{(1-p)<k>\beta S}{1+\alpha U^{\frac{q}{1-p}}} + \gamma R - \omega S \end{cases} \quad (1.2)$$

By our assumptions of system (1.1), because $q/(1-p) > 1$, if $q > 1$, therefore, when $q \geq 1$, the right hand sides of the first two equation of system (1.2) are continuously partial differential coefficient in the first quadrant of U and R. So system (1.2) satisfy the conditions of Existence and uniqueness theorems in UR plane.(1) $S(0) > 0$ when $U(0) = 0$ and $0 < R(0) < N$, thus $U'(0) = (1-p)<k>\beta S(0) > 0$, no solution of system (1.1) can go beyond the boundary of region B.(2) $U'(0) = (1-p) < k > \beta S(0) > 0$ if $U(0) = R(0) = 0$, When t is sufficiently small $U(t) > 0$ and $I(t) = U(t)1/(1-p) > 0$; $R'(t) > -\gamma R(t)$, $R(t) \geq e-\gamma t$, $R(0) = 0$, so no solution of the point (0,0) of system (1.1) can go beyond the boundary of region B.(3) $S(0) = 0$ when $U(0) = 0$, $R(0) = N$, so if $R'(t) = -\gamma N < 0$ as $t \to 0$, we can get $U''(0) = (1-p) < k>\beta(A + \gamma N) > 0$; similarly, as $t \to 0$, we can get $U'(t) > U'(0) = 0$, $U(t) > U(0) = 0$, $I(t) = U(t)1/(1-p) > 0$, no solution can exit the feasible region across the boundary $I + R = N$ has been proved, so no solution of the point $(I,R) = (0,N)$ of system (1.1) can go beyond the boundary of region B. B is the positively invariant region of system (1.1). All the solution of system (1.1) (S, R, I), ultimately tend to, enter or stay in region B. $\qquad\square$

3 Equilibrium Points and Their Stability

(1) The system (1.1) always exist disease-free equilibrium point $(I_e, R_e) = (0,0)$, and the disease-free equilibrium is unstable in region B.
(2) The system (1.1) has the unique positive equilibrium point (I_e, R_e) in addition to the disease-free equilibrium states. And (I_e, R_e) is globally asymptotically stable in B.

Proof. (1) $(I_e, R_e) = (0, 0)$ is the disease-free equilibrium of system (1.1) is clearly established. Here to prove it in region B instability.

Assuming that $(0,0)$ is stable, note that $0 < \eta_0 < N/2$, make the $\eta_0^{1-p} - \frac{\beta(1-2\eta_0)}{2\delta(1+\alpha\eta_0^q)} < 0$

So there is $\varepsilon_0 > 0$, if $0 < I_0 < \varepsilon_0$, $0 < R_0 < \varepsilon_0$, (1.1)'s solution $(I(t, I_0, R_0), R(t,I_0,R_0))$ with the initial condition $I(0) = I_0$, $R(0) = R_0$ satisfy if $0 \leq I(t, I_0, R_0) < \eta_0$, $0 \leq R(t, I_0, R_0) < \eta_0 (t \geq 0)$, then

$$U(t, I_0, R_0) = (I(t, I_0, R_0))^{1-p} < \eta_0^{1-p}, (t \geq 0) \tag{2.1}$$

On the other hand $S(t, I_0, R_0) > 1 - 2\eta_0$, $(t \geq 0)$. by (1.2) shows that

$$
\begin{aligned}
\bullet \quad \frac{dU(t, I_0, R_0)}{dt} &= (1-p) <k> \beta \frac{S(t, I_0, R_0)}{1 + \alpha(U(t, I_0, R_0))^{\frac{q}{1-p}}} \\
&\quad - (1-p)\delta U(t, I_0, R_0) \\
&> \frac{(1-p) <k> \beta(1-2\eta_0)}{1 + \alpha\eta_0^q} - (1-p)\delta U(t, I_0, R_0)
\end{aligned}
$$

Because $I(0) = 0$, then $U(0) = 0$, solute the differential inequalities

$$U(t, I_0, R_0) > \frac{<k> \beta(1-2\eta_0)}{\delta(1+\alpha\eta_0^q)} (1 - e^{-(1-p)\delta t}), (t \geq 0) \tag{2.2}$$

Therefore, there exists $T > 0$, such that for $t > T$, we deduce that

$$U(t, I_0, R_0) > \frac{<k> \beta(1-2\eta_0)}{2\delta(1+\alpha\eta_0^q)} \tag{2.3}$$

From (2.1), (2.2) we see that $\eta_0^{1-p} > \frac{\beta(1-2\eta_0)}{2\delta(1+\alpha\eta_0^q)}$. This is contradictory with determining η_0, Therefore, the disease-free equilibrium point $(0,0)$ is unstable.

(2) Define (I_e, R_e) is the equilibrium point of (1.1), then (I_e, R_e) satisfy

$$\frac{<k> \beta I_e^p S}{1 + \alpha I_e^q} - \delta I_e = 0 \tag{2.4}$$

$$\delta I_e - \gamma R_e + \omega S_e = 0 \tag{2.5}$$

$$S_e + R_e + I_e = N \tag{2.6}$$

From (2.4), (2.5), (2.6) we see that $\frac{I_e^{p-1}}{1+\alpha I_e^q}\left(1 - \frac{I_e}{H}\right) = \frac{1}{\sigma}$, if $I_e > 0$. Where $\sigma = \frac{<k> \beta\gamma N}{\delta(\omega+\gamma)} > 0$, $H = \frac{\gamma N}{\gamma+\delta}$. It's clear that $I_e < H$. define

$$f(I) = \frac{I^{p-1}(1 - \frac{I}{H})}{1 + \alpha I^p}, I \in (0, H),$$

Then

$$f'(I) = \frac{I^{p-2}g(I)}{(1 + \alpha I^q)^2}, I \in (0, H), \tag{2.7}$$

Where

$$g(I) = p - 1 - \frac{pI}{H} + \alpha(p - q - 1)I^q - \frac{\alpha}{H}(p - q)I^{q+1} \tag{2.8}$$

From (2.8), $g(I) < 0, (I \in (0,H))$ if $0 < p < 1$; From (2.7), $f'(I) < 0$, $(I \in (0,H))$, then $f(H) = 0$, $f(I) \to +\infty$ as $I \to 0+$. Therefore for any $\sigma > 0$, the equation has a unique solution I_e, here $0 < I_e < H$.

Below we prove its stability. Suppose

$$\lambda^2 + b\lambda + c = 0, \tag{2.9}$$

is the Linear approximation equation of equilibrium (I_e, R_e). where

$$b = (\gamma + \omega) + \frac{\delta(1 + \alpha I^q)}{I^p \beta <k>} - (N - I - R)\frac{p + \alpha I^q(p - q)}{1 + \alpha I^q}$$

$$c = (\gamma + \omega)[\frac{\delta(1 + \alpha I^q)}{I^p \beta <k>} - (N - I - R)\frac{p + \alpha I^q(p - q)}{1 + \alpha I^q}]$$

$$+ \frac{<k> \beta I^p(\delta - \omega)}{1 + \alpha I^q}$$

From the above equation, $b > 0$, $c > 0$, if $0 < p \leq 1$. Define the eigenvalues of (2.9) are λ_1, λ_2. Because $\lambda_1 + \lambda_2 = -b < 0$, $\lambda_1\lambda_2 = c > 0$, then $\lambda_1 < 0$, $\lambda_2 < 0$. we see that (I_e, R_e) is locally asymptotically stable according to stability theorem [9]. from (1.1), within the interior of the region B, we obtain

$$\frac{\partial}{\partial I}(\frac{I'(t)}{I^p}) + \frac{\partial}{\partial R}(\frac{R'(t)}{I^p}) =$$

$$- \beta <k> \frac{(1 + \alpha I^q) + \alpha q(N - R - I)I^{q-1}}{(1 + \alpha I^q)^2}$$

$$- \delta(1 - p)I^{-p} - \gamma I^{-p} < 0$$

As $0 < p \leq 1$. From Dulac theorem, no closed trajectory in the interior of B, so for any initial condition $0 < I(0) = I_0 < N$, $0 < R(0) = R_0 < N$, the solution of (1.1) $(I(t, I_0, R_0), R(t, I_0, R_0)) \to (I_e, R_e)$, as $t \to \infty$. Therefore (I_e, R_e) is global attractive in the interior of B. so (I_e, R_e) is globally asymptotically stable. $\square$

Theorem 3. For the case p = 1, then

(1) If $1/\sigma \geq 1$, the disease-free equilibrium $(0, 0)$ is globally asymptotically stable in B.
(2) If $1/\sigma < 1$, then disease-free equilibrium $(0, 0)$ in B is unstable (saddle point), in addition to the disease-free equilibrium exists positive equilibrium point (I_e, R_e), and this positive equilibrium point is globally asymptotically stable in B.

The proof is similar to the Theorem 2's Certification process.

4   Immunization Strategy

1. Effects on the Propagation of the Virus of Parameter K

From Fig. 1, we show the influence of propagation of virus by the change of k, which is the degree of the network, while the nodes of the network and the infectious incidence rate are constant. We can see clearly that the number of infected people change with k. When the K decreases, the infected changes tends to be stable with time, but the threshold increases while k increases. When the K is reduced to a certain extent, the spreading threshold is infinite, the epidemic failure to widely spread. Thus the network structure has a close relationship with the spread of the virus, in other words, the structure of network has a great influence on the spread of the virus. In view of this, we can the change the average degree of node in order to control the virus in network. the Simulation of the network size $N = 4 \times 10^5$, p = 1, $\beta = 0.2$, $\gamma = 0.003$, $\omega = 10^{-3}$, q = 2, A = 0.003, $\delta = 0.01$。

2. The Effects of the Rewiring Probability on Propagation of Virus

We consider different value of θ impact on the propagation of virus under the premise of constant incidence and immigration. Data results show that θ is smaller, the longer time which is the disease break to epidemic state required. And with the increase of θ, the epidemic will soon reach endemic disease state. As shown in Fig. 2. Simulations for network size $N = 6 \times 10^4$, the average degree of $< k > = 4$, $\beta = 0.2$, $\gamma = 0.0003$, $\omega = 10^{-3}$, q = 2, A = 0.03, $\alpha = 1.5$, $\delta = 0.1$, do 50 times and then take the average.

3. The Influence of the Network Clustering Coefficient

The network clustering coefficient is constantly changing when the virus spread in the small-world network. In this subsection, we will investigate the characteristics of virus

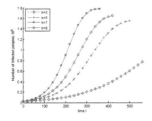

Fig. 1. The change of infected people with time when k changes

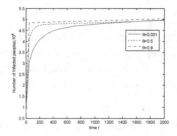

Fig. 2. The change of infected people with time when θ changes

as the clustering coefficient changes. NW small-world network clustering coefficient is defined as [12]:

$$C(\theta) = \frac{3(k-2)}{4(k-1) + 4k\theta(\theta+2)}$$

The literature [1] shows the degree of all nodes approaches the average degree in the small-world network. And C(θ) is the monotone decreasing function of rewiring probability. indicating that the network clustering effect weakened during the enhancement of network randomness. Figure 3 shows that the change of infected people with time under the corresponding clustering coefficient as p = 1, N = 6×104, k = 4, 6, θ = 0.2, and θ = 0.4. It is clear that the spread of the virus is more quickly, the scope is more widely when the clustering coefficient is larger.

In addition, facing the emergent virus, people's behavior can also affect the spread of the virus, Fig. 4 shows that as the premise of the constant incidence and immigration, the rewiring probability of network is reduced as people changes their behavior during the spread of the virus. Just as active isolation, which also reduce the number of infected people obviously? As is shown in Fig. 4. Simulations for network size $N = 6×10^4$, the average degree of < k > = 4, γ = 0.0003, ω = 10^{-3}, A = 0.03, α = 2, δ = 0.1,θ are respectively 0.8, 0.5, 0.2, from top to bottom. It is clear that the number of infected person will be significantly reduced as q increases, when the effect of people's behavior is ignored.

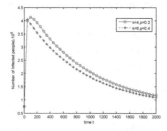

Fig. 3. The change of infected people with time when the Clustering coefficient changes

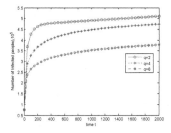

Fig. 4. The change of infected people with time when β remain constant

5   Conclusion

In this work, we have analyzed a SIRS epidemical model with saturated infection rate in a small-world network. By analysis of the model in the small-world network, the propagation of epidemic virus is not totally depends on the topology of the network. From the simulation analysis, we can obtain the propagation of the virus correlation with the average degree of the network, the rewiring probability and people's behaviors and so on. Therefore, to reduce the spread of the virus by changing these parameters values, such as reducing the average degree of nodes, lowing transmission and rewiring probability of network by change people's behavior, just as wearing a mask, reducing travel, we can take active isolation to control or slow down the spread of the virus.

Acknowledgments. This paper is sponsored by the National Natural Science Foundation of China (NSFC, Grant U1204703, U1304614), the Key Scientific and Technological Project of Henan Province (122102310004), the Innovation Scientists and Technicians Troop Construction Projects of Zhengzhou City (10LJRC190, 131PCXTD597), the Key Scientific and Technological Project of The Education Department of Henan Province (13A413355, 13A790352, ITE12001) and 2012 year university subject of Zhengzhou Normal University (2012074).

References

1. Golder, S.A., Huberman, B.A.: Usage patterns of collaborative tagging systems. J. Inf. Sci. **32**(2), 198–208 (2006)
2. Kleczkowski, A., Oleś, K., Gudowska-Nowak, E., et al.: Searching for the most cost-effective strategy for controlling Epidemics spreading on regular and small-world networks. J. Roy. Soc. Interface **9**(66), 158–169 (2012)
3. Sourlas, V., Tassiulas, L.: Effective cache management and performance limits in information-centric Networks. In: 2013 International Conference on Computing, Networking and Communications (ICNC), pp. 955–960. IEEE (2013)
4. Wang, Y., Jin, Z., Yang, Z.: Global analysis of an SIS model with an infective vector on complex networks. Nonlinear Anal.: Real World Appl. **13**(2), 543–557 (2012)
5. Wang, Z.: Instant messaging im virus spread of complex network. In: Zhong, Z. (ed.) Proceedings of the International Conference on Information Engineering and Applications (IEA). Lecture Notes in Electrical Engineering, pp. 185–193. Springer, London (2012)

6. Liu, M., Xiao, Y.: Modeling and analysis of epidemic diffusion within small-world network. J. Appl. Math. **2012**, 14 (2012)
7. Watts, D.J., Strogatz, S.H.: Collective dynamics of 'small-world' networks. Nature **393**, 440–442 (1998)
8. Newman, M.E.J., Watts, D.J.: Scaling and percolation in the small-world network model. Phys. Rev. E **60**, 7332–7342 (1999)
9. Jin, R., Shen, Y., Liu, L.: Limiting the Neighborhood: De-Small-World Network for Outbreak Prevention (2013). arXiv preprint arXiv:1305.0513
10. Zhien, M., Yi Cang, Z.: Ordinary differential equation qualitative and stability method. Science Press, Beijing (2001)
11. Jespersen, S., Blumen, A.: Small-world networks: Links with long-tailed distributions. Phys. Rev. E **62**, 6270–6274 (2000)
12. Kuperman, M., Abramson, G.: Complex structures in generalized small worlds. Phys. Rev. E **64**, 047103 (2001)
13. Xiaofan, W., Xiang, L., Guan Rong, C.: Network Science: An Introduce. Higher education press, Beijing (2012)
14. Abramson, G., Kuperman, M.: Small world effect in an epidemiological model. Phy. Rev. Lett **86**(13), 2909–2912 (2001)
15. Meng, M., Li, S., Ma, H.R.: The transition of epidemic spreading in small world. J. Shanghai Jiao Tong Univ. **40**(5), 869–872 (2006)
16. Hole, K.J.: Toward a practical technique to halt multiple virus outbreaks on computer networks. J. Comput. Netw. Commun. **2012**, 13 (2012)
17. Lusseau, D.: The emergent properties of a dolphin social network. In: Proceedings of the Royal Society B: Biological Sciences, vol. 270, pp. 186–188 (2003)

Granular Twin Support Vector Machines Based on Mixture Kernel Function

Xiuxi Wei[1] and Huajuan Huang[2,3（✉）]

[1] Information Engineering Department,
Guangxi International Business Vocational College, Nanning 530007, China
weixiuxi@163.com
[2] College of Information Science and Engineering,
Guangxi University for Nationalities, Nanning 530006, China
hhj-025@163.com
[3] Guangxi Key Laboratory of Hybrid Computation and Integrated Circuit
Design Analysis, Nanning 530006, China

Abstract. The recently proposed twin support vector machines, denoted by TWSVM, gets perfect classification performance and is suitable for many cases. However, it would reduce its learning performance when it is used to solve the large number of samples. In order to solve this problem, a novel algorithm called Granular Twin Support Vector Machines based on Mixture Kernel Function (GTWSVM-MK) is proposed. Firstly, a grain method including coarse particles and fine particles is propsed and then the judgment and extraction methods of support vector particles are given. On the above basis, we propose a granular twin support vector machine learning model. Secondly, in order to solve the kernel function selection problem, minxture kernel function is introduced. Finally, compared with SVM and TWSVM, the experimental results show that GTWSVM-MK has higher classification performance.

Keywords: Granulation · Mixture kernel function · Twin support vector machines

1 Introduction

As a new machine learning technology, twin support vector machines (TWSVM) is proposed by Jayadeva [1] in 2007. TWSVM would generate two non-parallel planes, such that each plane is closer to one of the two classes and is as far as possible from the other. Theoretically, the speed of TWSVM has approximately 4 times faster than the traditional support vector machine (SVM). Because of its excellent performance, TWSVM has been applied to many areas such as speaker recognition [2], medical detection [3–5], etc. At present, many improved TWSVM algorithms have been proposed. For example, in 2010, M. Arun Kumar et al. [6] brought the prior knowledge into TWSVM and least square TWSVM and then got two improved algorithms. Experimental results showed the proposed algorithms were effective. In 2011, Qiaolin Yu et al. [7] adding the regularization method into the TWSVM model, proposed the TWSVM model based on regularization method. This method ensured that the proposed model

© Springer International Publishing Switzerland 2015
D.-S. Huang and K. Han (Eds.): ICIC 2015, Part III, LNAI 9227, pp. 43–54, 2015.
DOI: 10.1007/978-3-319-22053-6_5

was the strongly convex programming problem. In 2012, Yitian Xu et al. [8] proposed a twin multi-class classification support vector machine. Experimental results demonstrated the proposed algorithm was stable and effective. However, the research on TWSVM is still at the starting stage at present. For example, the theoretical basis of TWSVM has not been enough perfect and the selection parameters methods [9, 10].

In order to further improve the classification performance of TWSVM, a novel algorithm called Granular Twin Support Vector Machines based on Mixture Kernel Function (GTWSVM-MK) is proposed in this paper. Firstly, in order to overcome the blindness of choice kernel function for TWSVM, minxture kernel function which has good generalization ability is introduced as the kernel function of TWSVM. Secondly, the granulation method of samples is designed, the estimation and extraction methods of support vector particle are proposed. Therefore, we can construct a new granular twin support vector machine learning model based on the above idea. Finally, the experimens on the UCI dataset show the effectiveness of the proposed method.

This paper is organized as follows: In Sect. 2, we briefly introduce the basic theory of TWSVM. In Sect. 3, GTWSVM-MK algorithm is detailed introduced and analyzed. Computational comparisons on UCI datasets are done in Sects. 4 and 5 gives concluding remarks.

2 Twin Support Vector Machines

Consider a binary classification problem of classifying m_1 data points belonging to class $+1$ and m_2 data points belonging to class -1. Then let matrix A in $R^{m_1 \times n}$ represent the data points of class $+1$ while matrix B in $R^{m_2 \times n}$ represent the data points of class -1. Two nonparallel hyper-planes of the linear TWSVM can be expressed as follows.

$$x^T w_1 + b_1 = 0 \text{ and } x^T w_2 + b_2 = 0 \tag{1}$$

The target of TWSVM is to generate the above two nonparallel hyper-planes in the n-dimensional real space R^n, such that each plane is closer to one of the two classes and is as far as possible from the other. A new sample point is assigned to class $+1$ or -1 depending upon its proximity to the two nonparallel hyper-planes. The linear classifiers are obtained by solving the following optimization problems.

$$\min_{w^{(1)}, b^{(1)}, \xi^{(2)}} \frac{1}{2} \left\| Aw^{(1)} + e_1 b^{(1)} \right\|^2 + c_1 e_2^T \xi^{(2)}$$
$$s.t. \quad -(Bw^{(1)} + e_2 b^{(1)}) \geq e_2 - \xi^{(2)}, \tag{2}$$
$$\xi^{(2)} \geq 0.$$

$$\min_{w^{(2)}, b^{(2)}, \xi^{(1)}} \frac{1}{2} \left\| Bw^{(2)} + e_2 b^{(2)} \right\|^2 + c_2 e_1^T \xi^{(1)}$$
$$(Aw^{(2)} + e_1 b^{(2)}) \geq e_1 - \xi^{(1)}, \tag{3}$$
$$\xi^{(1)} \geq 0.$$

where c_1 and c_2 are penalty parameters, $\xi^{(1)}$ and $\xi^{(2)}$ are slack vectors, e_1 and e_2 are the vectors of ones of appropriate dimensions. $x_j^{(i)}$ represents the j th sample of the i th class.

For the nonlinear case, the two nonparallel hyperplanes of TWSVM based on kernel can be expressed as follows:

$$K(x^T, C^T)w^{(1)} + b^{(1)} = 0, \ K(x^T, C^T)w^{(2)} + b^{(2)} = 0 \tag{4}$$

where, $C = [A^T, B^T]^T$. So the optimization problem of nonlinear TSVMs can be expressed as follows.

$$\min_{w^{(1)}, b^{(1)}, \xi^{(2)}} \frac{1}{2} \left\| K(A, C^T)w^{(1)} + e_1 b^{(1)} \right\|^2 + c_1 e_2^T \xi^{(2)}$$

$$s.t. \quad -(K(B, C^T)w^{(1)} + e_2 b^{(1)}) \geq e_2 - \xi^{(2)}, \tag{5}$$

$$\xi^{(2)} \geq 0.$$

$$\min_{w^{(1)}, b^{(1)}, \xi^{(2)}} \frac{1}{2} \left\| K(B, C^T)w^{(2)} + e_2 b^{(1)} \right\|^2 + c_2 e_1^T \xi^{(1)}$$

$$s.t. \quad (K(A, C^T)w^{(2)} + e_1 b^{(2)}) \geq e_1 - \xi^{(1)}, \tag{6}$$

$$\xi^{(1)} \geq 0.$$

3   Granular Wavelet Twin Support Vector Machines

3.1   Construction GTWSVM-MK Learning Model

Set a given data: $X = (x_i, y_i), i = 1, 2, \cdots, n$, where n is the number of samples, y_i is the class label of x_i. Use fuzzy clustering method to granulate dataset X, which then are labeled as l coarse particles, and ensure the label of each particle with the sample to be consistent, denoted by:

$$(X_1, Y_1), (X_2, Y_2), \cdots, (X_i, Y_i), \cdots, (X_l, Y_l)$$

Assume that each coarse particle has l_i samples, and Y_i is the label of the ith coarse particle. The original data set can be expressed as:

$$X = \{(X_i, Y_i), i = 1, 2, \cdots, l\}$$

where,

$$X_1 = \begin{Bmatrix} x_1 \\ x_2 \\ \cdots \\ x_{l_1} \end{Bmatrix}, \; X_2 = \begin{Bmatrix} x_{l_1+1} \\ x_{l_1+2} \\ \cdots \\ x_{l_1+l_2} \end{Bmatrix}, \cdots, \; X_l = \begin{Bmatrix} x_{l_1+l_2+\cdots+l_{l-1}+1} \\ x_{l_1+l_2+\cdots+l_{l-1}+2} \\ \cdots \\ x_{l_1+l_2+\cdots+l_{l-1}+l_l} \end{Bmatrix}$$

After finishing the division, the support vector particles consist the new training samples $X' = \{(x_i', y_i'), i = 1, 2, \cdots, m\}$. Where m is the number of samples, y_i' is the class label of x_i'.

Thus we get the optimal separating hyperplane

$$x_i'^T w_1 + b_1 = 0 \text{ and } x_i'^T w_2 + b_2 = 0 \tag{7}$$

For linearly inseparable, the main idea of GWTWSVM is that the input vector is mapped to a high-dimensional feature vector space, and then constructs the optimal surface in the feature space.

Map the input samples x' from the input space R^n into the feature space H:

$$x' \rightarrow \Phi(x') = (\Phi_1(x'), \Phi_2(x'), \cdots, \Phi_l(x'))^T \tag{8}$$

Insteading of input vector x' by feature vector $\Phi(x')$, we can get the optimal separating hyperplane as follows:

$$K(x'^T, C^T)w^{(1)} + b^{(1)} = 0, \quad K(x'^T, C^T)w^{(2)} + b^{(2)} = 0 \tag{9}$$

where, $K(x_i', C^T)$ is the granularity of the kernel function.

3.2   Construction Mixed Kernel Function

At present, the most commonly used kernel functions are as follows:

(1) Linear Function:

$$K(x, x_i) = x \cdot x_i \tag{10}$$

(2) Polynomial Function:

$$K(x, x_i) = (\gamma(x \cdot x_i) + r)^d, \gamma > 0 \tag{11}$$

(3) Gauss Radial Basis Function:

$$K(x, x_i) = \exp(-\frac{\|x - x_i\|^2}{2\sigma^2}) \tag{12}$$

The mercer theorem is the theory of the traditional kernel function construction, which is showed as follows:

Theorem 1. When $g(x) \in L_2(R^N)$ and $k(x, x') \in L_2(R^N \times R^N)$, if

$$\iint k(x, x')g(x)g(x')dxdy \geq 0 \tag{13}$$

There is $k(x, x') = (\Phi(x) \cdot \Phi(x'))$.

The choice of kernel function is a critical problem in the practical application. This is because that the learning ability of kernel function will directly affect the quality of kernel model performance realization.

As we know, the most used kernel function in TWSVM is the Gauss kernel which is a typical local kernel function. For the Gauss kernel function, the sketch map of the testing point 0.1 is shown as Fig. 1. When the values of σ^2 are 0.1, 0.2, 0.3, 0.4, 0.5 respectively.

From Fig. 1 we can see that the Gauss function has good learning ability because of only having a role for the near test point, but its generalization ability is weak.

The polynomial kernel is a typical global kernel function. Compared with the local kernel function, the learning ability of global kernel function is weak, but it has good generalization ability. For polynomial kernel function, the sketch map of the testing point 0.1 is shown as Fig. 2. When the values of q are 1, 2, 3, 4, 5 respectively.

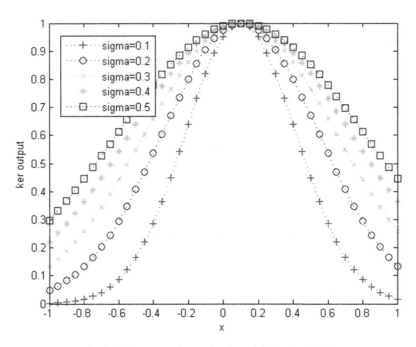

Fig. 1. The curve of gaussian kernel in test point 0.1

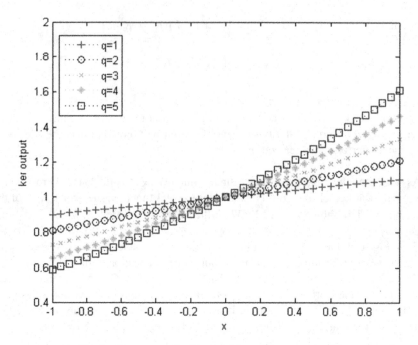

Fig. 2. The curve of polynomial kernel in test point 0.1

From Fig. 2 we can see that the polynomial kernel function has good generalization ability in the appropriate parameters because of having a role for the near test point and far across the data points. But the learning ability in the test point is not obvious, which meas it's learning ability is not only strong.

Based on the above analysis, if the Gauss kernel function and the polynomial kernel function is mixed to generate a new mixed kernel function, which can have better learning ability and better generalization ability.

Theorem 2. $k(x_i, x_j) = a \cdot \exp(-\frac{\|x_i - x_j\|^2}{2\sigma^2}) + (1 - a) \cdot k(x_i, x_j)$, $(0 < a < 1)$ which contains Gauss kernel function $K(x_i, x_j) = \exp(-\frac{\|x_i - x_j\|^2}{2\sigma^2})$ and any kernel function $k(x_i, x_j)$ is a new kernel function.

Proof. According to Mecer theorem, the Gram matrix K of any kernel function $k(x_i, x_j)$ is symmetric and positive semi-definite. So there is orthogonal matrix U, let

$$K = U \begin{bmatrix} \lambda_1 & & & \\ & \ddots & & \\ & & \lambda_k & \\ & & & 0 \end{bmatrix} U^{-1}, \text{ where } \lambda_i > 0. \text{ Therefore, when } \sigma \to 0, \text{ there is}$$

$$k \approx a \cdot I + (1-a) \cdot K = U \begin{bmatrix} a & & & \\ & a & & \\ & & \ddots & \\ & & & a \end{bmatrix} U^{-1} + U \begin{bmatrix} (1-a)\lambda_1 & & & \\ & \ddots & & \\ & & (1-a)\lambda_k & \\ & & & 0 \end{bmatrix} U^{-1}$$

$$= U \begin{bmatrix} a + (1-a) \cdot \lambda_1 & & & & \\ & \ddots & & & \\ & & a + (1-a) \cdot \lambda_k & & \\ & & & \ddots & \\ & & & & a \end{bmatrix} U^{-1}$$

So we can know that k is full rank. According to Theorem 1, $\widetilde{k}$ is a kind of kernel function can linearly separate any training samples.

Based on the above analysis, we can obtain the mixed function as follows:

$$k(x_i, x_j) = a \cdot \exp(-\frac{\|x_i - x_j\|^2}{2\sigma^2}) + (1-a)[(x_i \cdot x_j) + 1]^q \qquad (14)$$

According to Theorem 2, (14) is a kernel function. For the mixed kernel function, the sketch map of the testing point 0.1 when $\sigma^2 = 0.1$, $q = 3$ is shown as Fig. 3.

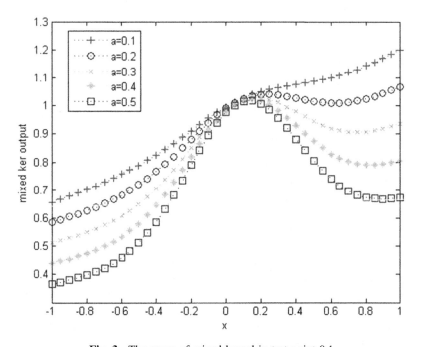

Fig. 3. The curve of mixed kernel in test point 0.1

From the Fig. 3, we can see that the mixed kernel function makes good use of the properties of global and local kernels. It not only has the strong learning capability but also has strong generalization ability.

3.3   The Algorithm Steps of GTWSVM-MK

- **Step1:** Division the samples into coarse particles. For each class, the samples whose distance similar to the class of center distance are composed of a coarse particle.
- **Step2:** Extracting the support vector coarse particles. Using the fuzzy clustering method to cluster the coarse particles according to the degree membership of each coarse particle. Given a threshold, when the degree membership of coarse particle is less than the given threshold, it means that they are support vector coarse particles, and then we extract the support vector coarse particles.
- **Step3**: Division the coarse particle into fine particles. In order to retain the original data information, the principle of division is that each sample in the support vector coarse particle is regarded as each support vector fine particle.
- **Step4**: Extracting the support vector particles. The fine particles of Step3 are the support vector particles which are the input data set of GTWSVM.
- **Step5**: Introducing the mixed kernel function and constructing the model of GTWSVM-MK.
- **Step6**: Training the support vector set. Carrying out the GTWSVM-MK training in the new training samples.
- **Step7**: testing the generalization ability of GTWSVM-MK.

4   Experimental Results and Analysis

In order to verify the efficiency of GTWSVM-MK, two parts of experiments for typical UCI data sets are carried out in Matlab7.11. In the first experiment, we mianly test the influence of the learning ability and generalization ability on the same data set for using different kernel function in TWSVM. At the same time, testing the influence of the accuracy classification using different coefficient weight for the mixed kernel function. In the second experiment, we mainly verify the different performance of SVM, TWSVM and GTWSVM-MK.

4.1   The First Experiment

In order to test the impact on the generalization ability and learning ability of TWSVM using different kernel functions, heart_scale sample set is used in this paper. Heart_-scale sample is the two classification problems, which contains 13 properties and 270 samples. Firstly, 100 samples are taken from samples set as the training samples set and 50 samples are taken as the testing samples set 1. In order to further illustrate the impact on the generalization ability on the TWSVM using different kernel functions,

Table 1. Comparison of classification results using three kernel functions for TWSVM

Kernel Function	Accuracy		
	Test Set 1	Test Set 2	Test Set 3
Gauss kernel function	85 %(150/50)	84.57 %(150/90)	83.14 %(150/120)
polynomial kernel function	77 %(150/50)	83.14 %(150/90)	82.18 %(150/120)
Mixed kernel function	89.53 %(150/50)	88.56 %(150/90)	87.69 %(150/120)

Table 2. The impact to TWSVM using different value a for the mixed kernel function

The value of a, b	Accuracy		
	Test Set 1	Test Set 2	Test Set 3
$a = 0.1, b = 0.9$	92.15 %(150/50)	85.06 %(150/90)	85.26 %(150/120)
$a = 0.2, b = 0.8$	90.08 %(150/50)	87.25 %(150/90)	86.37 %(150/120)
$a = 0.3, b = 0.7$	88.28 %(150/50)	89.39 %(150/90)	88.28 %(150/120)
$a = 0.4, b = 0.6$	88.22 %(150/50)	87.08 %(150/90)	85.03 %(150/120)
$a = 0.5, b = 0.5$	87.77 %(150/50)	89.95 %(150/90)	90.85 %(150/120)
$a = 0.6, b = 0.4$	87.15 %(150/50)	88.57 %(150/90)	91.32 %(150/120)
$a = 0.7, b = 0.3$	86.92 %(150/50)	88.13 %(150/90)	89.21 %(150/120)
$a = 0.8, b = 0.2$	86.37 %(150/50)	87.52 %(150/90)	90.18 %(150/120)
$a = 0.9, b = 0.1$	84.26 %(150/50)	85.77 %(150/90)	86.93 %(150/120)

we continue increasing 40 samples based on the testing samples set 1 as the testing samples set 2. Furthermore, based on the testing samples set 2, 30 samples are added as the testing samples set 3. The parameter values of this experiment are as follows:

$C = 100, p_1 = 100, p_2 = 3, \sigma = 0.5, a = 0.3, b = 0.7$. The experimental results are shown in Table 1.

From the Table 1, we can see that polynomial kernel function as the global kernel function, its generalization ability is strong when its learning ability is weak. On the contrary, Gauss kernel function as the local kernel function, it has strong learning ability but its generalization ability is weak. However, Mixed Kernel Function has better performance than the two kernel functions. We would discuss the impact on the Mixed Kernel Function using different coefficient weight as follows. The experimental results are shown in Table 2. From the Table 2, we can see that the accuracy would higher when a is smaller. At present, the determining vaules of a, b method is usually adopts multiple experimental method to get average value.

4.2 The Second Experiment

In this paper, we use 3 data set in UCI to test the performance of GTWSVM-MK. In the experiment, the parameters of SVM and TWSVM are got by using cross validation method. Meanwhile, the fuzzy particle cluster number is set 20. As we know, affecting the performance of GTWSVM-MK is mainly the value of threshold. Therefore, we analyze the impact on GTWSVM-MK using different thresholds as follows. Table 3 is the data set. When the value of threshold is set as 0.95, 0.9, 0.85 and 0.8, the results are respectively shown as Tables 4, 5, 6 and 7. Can be seen from the experimental results, the classification accuracy of GTWSVM-MK is close to the SVM and is better than TWSVM. Furthermore, its run time is far less than the operating time of the SVM.

Table 3. Datasets used in experiments

The sample set	Training Samples	Test Samples	The Dimension
Abalone	3177	1000	8
Contraceptive Method Choice	1000	473	9
Pen-Based Recognition of Handwritten Digits	6280	3498	16

Table 4. Comparison of classification results using three algorithms in the condition of $k < 0.95$

Data Sets	Algorithm	Accuracy	Time (s)
Abalone	SVM	82.17 %	3.77
	TWSVM	80.96 %	0.26
	GTWSVM-MK	81.45 %	0.38
Contraceptive Method Choice	SVM	86.32 %	1.56
	TWSVM	84.34 %	0.11
	GTWSVM-MK	86.11 %	0.15
Pen-Based Recognition of Handwritten Digits	SVM	84.75 %	15.32
	TWSVM	81.33 %	2.31
	GTWSVM-MK	83.94 %	2.52

Table 5. Comparison of classification results using three algorithms in the condition of $k < 0.9$

Data Sets	Algorithm	Accuracy	Time(s)
Abalone	SVM	82.15 %	3.71
	TWSVM	78.21 %	0.14
	GTWSVM-MK	80.73 %	0.11
Contraceptive Method Choice	SVM	86.92 %	1.53
	TWSVM	81.33 %	0.07
	GTWSVM-MK	84.14 %	0.12
Pen-Based Recognition of Handwritten Digits	SVM	84.72 %	15.21
	TWSVM	79.54 %	1.54
	GTWSVM-MK	82.22 %	1.66

Table 6. Comparison of classification results using three algorithms in the condition of $k < 0.85$

Data Sets	Algorithm	Accuracy	Time(s)
Abalone	SVM	82.14 %	3.71
	TWSVM	75.94 %	0.07
	GTWSVM-MK	80.12 %	0.13
Contraceptive Method Choice	SVM	86.92 %	1.53
	TWSVM	79.65 %	0.06
	GTWSVM-MK	82.43 %	0.07
Pen-Based Recognition of Handwritten Digits	SVM	84.74 %	15.3
	TWSVM	76.63 %	1.21
	GTWSVM-MK	81.32 %	1.43

Table 7. Comparison of classification results using three algorithms in the condition of $k < 0.8$

Data Sets	Algorithm	Accuracy	Time(s)
Abalone	SVM	82.14 %	3.72
	TWSVM	72.85 %	0.03
	GTWSVM-MK	78.54 %	0.05
Contraceptive Method Choice	SVM	86.97 %	1.55
	TWSVM	77.72 %	0.05
	GTWSVM-MK	80.28 %	0.04
Pen-Based Recognition of Handwritten Digits	SVM	84.79 %	15.3
	TWSVM	72.68 %	0.09
	GTWSVM-MK	79.65 %	0.11

5   Conclusion

In order to improve the low efficiency of TWSVM in dealing with large-scale data, a new granular twin support vector machine based on mixed kernel function algorithm is proposed in this paper. Firstly, a new granular support vector machine learning model is constructed through designing granular methods and extracting support vector particles methods. In order to further improve the performance of the proposed algorithm, a new granular TWSVM based on mixed kernel function (GTWSVM-MK) is proposed, which effectively uses the strong generalization ability and strong learning ability of mixed kernel function. Finally, the theoretical analysis and experimental results show the effectiveness of the method. However, the threshold parameter selection still has a certain randomness which would make affect the learning performance of GTWSVM-MK. Therefore, how to adaptively adjust the appropriate threshold will be the next step to study.

Acknowledgment. This work is supported by the Foundation of Guangxi University of science and technology research project (2013YB326), and the Fundation of Guangxi Key Laboratory of Hybrid Computation and Integrated Circuit Design Analysis (No. HCIC201304).

References

1. Jayadeva, R., Reshma, K., Chandra, S.: Twin support vector machines for pattern classification. IEEE Trans. Pattern Anal. Mach. Intell. **29**(5), 905–910 (2007)
2. Cong, H.H., Yang, C.F., Pu, X.R.: Efficient speaker recognition based on multi-class twin support vector machines and GMMs. In: 2008 IEEE Conference on Robotics, Automation and Mechatronics, pp. 348–352 (2008)
3. Zhang, X.S., Gao, X.B., Wang, Y.: Twin support vector machine for mcs detection. J. Electron. **26**(3), 318–325 (2009)
4. Zhang, X.S.: Boosting twin support vector machine approach for MCs detection. Asia-Pacific Conf. Inf. Process. **46**, 149–152 (2009)
5. Zhang, X.S., Gao, X.B.: MCs detection approach using bagging and boosting based twin support vector machine. In: 2009 IEEE International Conference on Systems, Man, and Cybernetics, San Antonio, TX, USA, pp. 5000–5005 (2009)
6. Kumar, M.A., Khemchandani, R., Gopal, M., Chandra, S.: Knowledge based Least Squares Twin support vector machines. Inf. Sci. **180**(23), 4606–4618 (2010)
7. Ye, Q.L., Zhao, C.X., Chen, X.B.: A feature selection method for twsvm via a regularization technique. J. Comput. Res. Dev. **48**(6), 1029–1037 (2011)
8. Xu, Y.T., Guo, R., Wang, L.S.: A twin multi-class classification support vetor machine. Cogn. Comput. (2012). doi:10.1007/s12559-012-9179-7
9. Peng, X.J., Xu, D.: Norm-mixed twin support vector machine classifier and its geometric algorithm. Neurocomputing **99**, 486–495 (2013)
10. Peng, X.J., Xu, D.: Robust minimum class variance twin support vector machine classifier. Neural Comput. Appl. **22**, 999–1011 (2013)

Detecting Multiple Influential Observations in High Dimensional Linear Regression

Junlong Zhao[✉], Ying Zhang, and Lu Niu

LMIB of the Ministry of Education, School of Systems and Mathematics,
Beihang University, Beijing, China
zjlczh@126.com

Abstract. In this paper, we consider the detection of multiple influential observations in high dimensional regression, where the p number of covariates is much larger than sample size n. Detection of influential observations in high dimensional regression is challenging. In the case of single influential observation, Zhao et al. (2013) developed a method called High dimensional Influence Measure (HIM). However, the result of HIM is not applicable to the case of multiple influential observations, where the detection of influential observations is much more complicated than the case of single influential observation. We propose in this paper a new method based on the multiple deletion to detect the multiple influential.

Keywords: High dimensional linear regression · Influential observation · Robust

1 Introduction

Influential observation detection is an important issue in modeling. In linear model, influential observation can distort the estimation of regression coefficient significantly. Let $X = (X_1, \cdots, X_p) \in \mathbf{R}^p$ be the regressor and $Y \in \mathbf{R}$ be the responsor. In the case of p being fixed, Cook's distance [1] has been widely used and proven to be effective. Motivated by the seminal work of Cook, many other methods have been proposed, such as, studentized residuals [2] and DFFITS [3] etc. More works on influential observation detection can be inferred to Chatterjee and Hadi [4], Pena [5, 6], Nurunnabi et al. [7, 8], Imon and Hadi [9, 10], and Zakaria [11].

Compared with the case of p being small or fixed, the influential observation detection in high dimensional linear regression is less developed. Since ordinary least square (OLS) estimate is unstable in the high dimensional setting, Cook's distance and other related methods can't be applied directly. Detecting the influential observations is critical in high dimensional setting where $p \gg n$, since influential observations can greatly decrease the effectiveness of many widely used variable selection methods, such as, least absolute shrinkage and selection operator (LASSO) [12], sure independent screening (SIS) [13] etc. Further discussion is referred to Zhao et al. (2013). Among the literature of high dimensional linear model, some robust methods, using the robust loss function, have been proposed, such as, LAD-Lasso [14]. However, these methods focus on estimating coefficient and prediction that is not designed for the identification of influential observations. [15] proposed a method named Θ–IPOD to

© Springer International Publishing Switzerland 2015
D.-S. Huang and K. Han (Eds.): ICIC 2015, Part III, LNAI 9227, pp. 55–64, 2015.
DOI: 10.1007/978-3-319-22053-6_6

detect the outliers. However, this method, assuming a specific mean shifting structure on outlier, is restrictive and consequently can't deal with the general case. Zhao et al. (2013) proposed a new method called high dimensional influence measure (HIM), based on the marginal correlation. The marginal coefficient method has been widely used in high dimensional variable selection, such as SIS [13]. HIM can be viewed as the generalization of Cook's distance to high dimensional setting. Zhao et al. (2013) showed the effect of dimension blessing and established the asymptotic distribution of the test statistics, while the asymptotic distribution is unavailable for the Cook's distance.

However, the detection of multiple influential observations is more challenging. It is well known that the "masking" or "swamping" effects may exist, when multiple influential observations are presented. Masking means that an influential observation is revealed as influential only after deleting of several other influential observations. Swamping means a non-influential observation is falsely labeled as influential one due to strong effect of influential observations. Due to the effects of "masking" and "swamping", it is generally hard to obtain the asymptotic distribution of testing statistics.

The asymptotic result of HIM is only suitable for the case of single influential observation. Therefore, HIM method, which shares the spirit as Cook's distance by deleting only single observations each time, can't handle effectively the case of multiple influential observations. Therefore, new method is necessary when there are multiple influential observations.

In the case of p fixed, multiple deletion methods are proved to be effective to deal with multiple influential observations. In order to reduce the masking and swamping phenomenon, Hadi and Simonoff (1993) proposed an approach that attempts to separate the data into a set of "clean" points and the set of rest data may contain the potential influential observation. The multiple deletion has been considered in [7–9, 16, 17] etc. for linear model and logistic regression model.

In this paper, combining the strength of multiple deletion and HIM, we propose a new method to deal with multiple influential observations detection in high dimensional setting. Multiple deletion makes it possible to reduce detection of the multiple influential observation to that of the single influential observation. The price we paid is the increase of the computing cost. We propose a new algorithm, which is computational feasible much more effective than HIM when multiple influential observations are presented. The theoretical properties are established.

The main contents of this paper are arranged as follows. In Sect. 2, we briefly review the method of HIM and propose a new method to deal the multiple influential observation detection. The simulation results are presented in Sect. 3.

We introduce some notations. For any sets A, we denote $|A|$ as the cardinality of A. Let I_{inf} and I_{inf}^c be the set of indices of the influential and non-influential observations, respectively.

2 Detection of Multiple Influential Observations in High Dimensional Setting

In this article, we focus on influence diagnosis in the context of the high dimensional linear regression model. Consider the following model

$$Y_i = \beta_0 + X_i^T \beta_1 + \varepsilon_i \tag{1}$$

where the pair (Y_i, X_i), $1 \le i \le n$, denote the observation of the ith subject, $Y_i \in \mathbf{R}$ is the response variable, $X_i = (X_{i1}, \cdots X_{ip})^T \in R^p$ is the associated p-dimensional predictor vector, and $\varepsilon_i \in \mathbf{R}$ is a mean zero normally distributed random noise. Let $\beta = (\beta_0, \beta_1^\top)^\top$ denote the coefficient vector. Under the classical setup of $n > p$ the OLS estimate of β is obtained by minimizing the objective function $\sum_{i=1}^{n} (Y_i - \beta_0 - X_i^T \beta_1)^2$, and the solution is $\widehat{\beta} = (\mathbf{X}^\top \mathbf{X})^{-1} \mathbf{X}^\top \mathbf{Y}$ where $\mathbf{Y} = (Y_1, \cdots, Y_n)^T$ denotes the $n \times 1$ response vector, and $\mathbf{X}$ denotes the $n \times p$ design matrix with the ith row being p + 1 dimensional vector $(1, X_i^T), i = 1, \cdots n$.

2.1   Brief Review of HIM

In the case of $p > n$, the OLS estimator is highly unstable and Cook's distance (CD) does not work well. Zhao et al. (2013) use the leave-one-out principle as in the classical Cook's distance and propose the high dimensional influence measure (HIM). We give a short review of this method. Specifically, we define the marginal correlation as $\rho_j = E\{(X_{ij} - u_{xj})(Y_i - u_y)\}/(\sigma_{xj}\sigma_y)$ where $\mu_{xj} = E(X_j)$, $\mu_y = E(Y_i)$, $\sigma_{xj} = \text{var}(X_j)$, and $\sigma_y = \text{var}(Y_i)$. We then obtain the sample estimate $\hat{\rho}_j = \{\sum_{i=1}^{n}(X_{ij} - \hat{\mu}_{xj})(Y_i - \hat{\mu}_y)\}/ \{n\hat{\sigma}_{xj}\hat{\sigma}_y\}$, for $j = 1, \ldots, p$, where $\hat{u}_{xj}$, $\hat{u}_y$, $\hat{\sigma}_{xj}$ and $\hat{\sigma}_y$ are the sample estimates of u_{xj}, u_y, σ_{xj} and σ_y respectively. Zhao et al. (2013) compute the marginal correlation with the kth observation removed as

$$\mathcal{D}_k = \frac{1}{p}\sum_{j=1}^{p}\left(\hat{\rho}_j - \hat{\rho}_j^{(k)}\right)^2 \tag{2}$$

where $\hat{\rho}^{(k)}$ is the estimate of marginal correlation without kth observation (Y_k, X_k). Under some conditions, we have the following results.

Proposition 1. Under some conditions (conditions (C.1)–(C.3) in Zhao et al. 2013), When there is no influential points and min $\{n_1, p\} \to \infty$, we have

$$n^2 \mathcal{D}_k \to \chi^2(1) \tag{3}$$

where $\chi^2(1)$ is the chi-square distribution with one degrees of freedom.

For the hypothesis that the k-th observation is not influential versus its alternative, p-value $P(\chi^2(1) > n^2\mathcal{D}_k)$ can be used for the testing. As sample size n is large, we should handle the multiple hypotheses testing problem simultaneously, the authors use the B-H procedure [18] of Benjamini and Hochber (2005) to control the false discovery rate (FDR).

2.2 Multiple Deletion for High Dimensional Influence Measure

Although HIM has been shown effective in high dimensional setting, this method uses only leave-one-out approach and consequently can't deal effectively with the case of multiple influential observations. It is well known that multiple influential observation detection is more challenging due the "masking" and "swamping" effects. In this case, multiple deleting is more effective than leave-one-out method. Many works have proposed on multiple deletion such as GFFITS and PP-type outlier identifier. In this paper, we propose a new high dimensional influence measure based on multiple deletion (MHIM) to reduce the effect of "masking" and "swamping". The main idea is to find a clean subset data firstly based on multiple deletion and subset selection.

Let $\mathbb{Z}_n = \{Z_i : Z_i = (X_j, Y_j), i = 1, \cdots, n\}$ be the sample of size n. Let $I_{\inf} = \{i{:}Z_i$ is influential$\}$ and $I_{\inf}^c = \{1, \cdots, n\} \backslash I_{\inf}$. Denote $n_{\inf} = |I_{\inf}|$, the number of influential observations and assume that $n_{\inf} < n/2$. For any $1 \leq i_0 \leq n$, we consider the multiple deletion, choosing at random a subset of sample with indices $A \subseteq \{j{:}j \neq i_0, j = 1, \cdots, n\}$ where $|A| \leq (n - 1)/2$. Then let $A^{(i_0)} = A \cup \{i_0\}$ and compute the HIM for Z_{i_0} in the data set $\{Z_i, i \in A^{(i_0)}\}$

$$\mathcal{D}_{A^{(i_0)}} = \|\hat{\rho}_{A^{(i_0)}} - \hat{\rho}_A\|^2 / p \tag{4}$$

where $\hat{\rho}_A$ and $\hat{\rho}_{A^{(i_0)}}$ are the marginal correlations based on data with indices A and $A^{(i_0)}$ respectively. Recalling the definition of marginal correlation in Sect. 2.1, the mean $\mu_{x,j}$, μ_y and variance $\sigma_{x,j}$, σ_y should be estimated when we compute $\hat{\rho}_A$ and $\hat{\rho}_{A^{(i_0)}}$. Due to the existence of influence observations, it is desirable to use the robust estimate for $\mu_{x,j}$, μ_y and $\sigma_{x,j}$, σ_y. In this paper, we consider sample median to estimate $\mu_{x,j}$, μ_y and median absolute deviation (MAD) to estimate $\sigma_{x,j}$, σ_y based on data $\{Z_i, i \in A^{(i_0)}\}$.

Sine A may contains influential observations, a single subset A can't provide sufficient evidence to determine whether Z_{i_0} is influential or not, due to the randomness of subset A. The multiple subsets will be helpful. Suppose that we have multiple subsets $B_1, \cdots, B_m \subset \{i, 1 \leq i \neq i_0 \leq n\}$ with the cardinality $|B_j|$, $1 \leq j \leq m$ and that Z_{i_0} is influential. Let $B_j^{(i_0)} = B_j \cup \{i_0\}, 1 \leq j \leq m$. For any fixed B_j, Z_{i_0} may be masked by other influential points such that statistic $|B_j^{(i_0)}|^2 \mathcal{D}_{B_j^{(i_0)}}$ is not large enough, if B_j contains other influential observations. However, as m is large, we will have large probability to get a subset B_{j_0} such that it contains no influential observations. Consequently, the statistic $|B_{j_0}^{(i_0)}|^2 \mathcal{D}_{B_{j_0}^{(i_0)}}$ will be large than $1 - \alpha$ quantile of χ_1^2 distribution and consequently Z_i can be detected, if the estimate has large power on the alternative. This motivates us to consider the statistics

$$F_{i_0} = \max_{1 \leq j \leq m} |B_j^{(i_0)}|^2 \mathcal{D}_{B_j^{(i_0)}}, \quad 1 \leq i_0 \leq n. \tag{5}$$

If Z_{i_0} is influential, then F_{i_0} will be larger than $\chi_{1,1-\alpha}^2$ in high probability. However, if Z_{i_0} is not an influential observation, F_{i_0} defined above may also be much larger than $\chi_{1,1-\alpha}^2$ due to the swamping effect and consequently Z_{i_0} is incorrectly identified as influential. Although estimate F_i may lead to some false positive, it gives a good estimate of a subset of clean data $\{Z_i, i \in \hat{S}\}$ as m is sufficiently large where

$$\hat{S} = \{i : F_i \leq \chi^2_{1,1-\alpha}, 1 \leq i \leq n\} \tag{6}$$

And the set $\hat{S}^c = \{i : F_i > \chi^2_{1,1-\alpha}\}$ consists of the estimated influential observations and some non-influential ones due to swamping effect. Based on this estimated clean data, we can remove some false positive observations in the set $\hat{S}^c$.

For any $Z_i, i \in \hat{S}^c$, we apply HIM to data with indices $\hat{S}^{(i)} = \hat{S} \cup \{i\}$ and compute $\mathcal{D}_{\hat{S}^{(i)}}$. If $\hat{S}$ is a good enough such that it consists of only clean data, then $\hat{S}^{(i)}$ contains at most one influential observation, and consequently leave-one-out method HIM will be effective in testing whether Z_i is an influential observation. Note that checking whether $Z_i, i \in \hat{S}^c$ are influential observations is a multiple hypothesis testing problem. Therefore instead simply comparing $|\hat{S}^{(i)}|^2 \mathcal{D}_{\hat{S}^{(i)}}$ and $\chi^2_{1,1-\alpha}$, we can control the false discovery rate (FDR) by the commonly used FDR control methods, such as, the B-H procedure [18]. We summarize our algorithm into following two steps.

(1) For $1 \leq i \leq n$, select m subsets $B_1, \cdots, B_m \subset \{j : 1 \leq j \neq i \leq n\}$ with the same cardinality $|B_1| := n_1 < n/2$. For $i = 1, \cdots, n$, compute $F_i = \max_{1 \leq j \leq m} (n_1 + 1)^2 \mathcal{D}_{B_j^{(i)}}$, where $B_j^{(i)} = B_j \cup \{i\}$. Then $\hat{S} = \{i : F_i \leq \chi^2_{1,1-\alpha}, 1 \leq i \leq n\}$ is the estimate of clean data.

(2) For each observation Z_i with $i \in \hat{S}^c = \{i : F_i > \chi^2_{1,1-\alpha}\}$, consider the testing $H_{i,0}$: Z_i is not influential among data with indices $\hat{S}^{(i)} = \{i\} \cup \hat{S}$. Compute the p-value $P(\chi^2_1 > |\hat{S}^{(i)}|^2 \mathcal{D}_{\hat{S}^{(i)}})$ and control the error using FDR method. The final estimate of influential observations is defined as

$$\hat{I}_{\inf} = \{Z_i, i \in \hat{S}^c \text{ and } H_{i,0} \text{ is rejected by FDR procedure}\} \tag{7}$$

We study the power of MHIM method. We assume that the number the influential observations $n_{\inf} < n/2$. To ensure the success of multiple influential observation detection, HIM must be effective for the case of single influential. We make that following assumption on the power of HIM method for the case of single influential observation.

Assumption 1. (Power for single influential observation). Let $\{Z_i = (X_i, Y_i), i = 1, \cdots, n\}$ be *i.i.d.* non-influential observation and $Z_0 = (X_0, Y_0)$ be the only influential observation. Suppose that Z_0 can be identified as influential observation by HIM method with probability at least $1 - \varepsilon$.

Theorem 1. Suppose that assumptions (C.1)–(C.3) of Zhao et al. (2013) and Assumption 1 holds. For any $\beta \in [0, 1]$ and ε defined in Assumption 1 such that $(1 - n_{\inf}\varepsilon)\beta < \varepsilon_0$, where $\varepsilon_0 > 0$ is sufficiently small. As $m > [\log(1 - \beta)] / \log\left[1 - \left(\frac{n - n_{\inf} - n_1}{n-1}\right)^{n_{\inf}}\right]$ and $\min\{n_1, p\} \to \infty$, it holds that

$$P(\hat{S}^c \supseteq I_{\inf}) > 1 - \varepsilon_0 \tag{8}$$

3   Simulation

In this section, we report the simulation results of the MHIM. In the simulations below, we generate the non-influential observations of size n from the following true model,

$$Y_i = X_i^\top \beta + \varepsilon_i, \quad i = 1, \cdots, n \tag{9}$$

where $\beta = (1, 1, 1, \cdots, 1)$, $\varepsilon_i \sim N(0, 1)$, $X_i = (X_{i1}, \cdots, X_{ip})^T \sim N(0, \Sigma)$ and $\Sigma = (\sigma_{ij})$ with $\sigma_{ij} = 0.5^{|i-j|}$. Denote by $i_0 = \arg\max_{1 \le i \le n} Y_i$ and $\left(X_i^{\inf}, Y_i^{\inf}\right)$, $i = 1, \cdots, n_{\inf}$ (be the influential observations, where $X_i^{\inf} = (X_{i1}^{\inf}, \cdots, X_{ip}^{\inf})^T \in R^p$, $n_{\inf}$ stand for the number of influential observations.

Let TPR stands for the power of influential detection, or equivalently, the proportion of the correctly identified influential observation over $n_{\inf}$. FPR stands for false positive rate, which is the ratio of the number of the non-influential points that are incorrectly identified as influential ones over the $n - n_{\inf}$. F_1 stands for F_1-score, whose computational formula is

$$\mathbf{F}_1 = \frac{2TP}{2TP + FP + FN} \tag{10}$$

where TP stands for true positive number, FP stands for the number of false positive observations, and FN is the number of false negative observations. The larger the F_1-score is, the more efficiently the method does. In the following simulations, we set the number of subsets $m = 300$ and repeat $T = 100$ times to report the mean of FPR, TPR and F_1-score.

3.1   Simulation for Masking Effect

As the deviation of influential observations from the non-influential ones is mall, some influential observations in $\left(X_i^{\inf}, Y_i^{\inf}\right), i = 1, \cdots, n_{\inf}$ tends to be masked. Therefore, we generate the influential observations from the following model 1–3, where ε_i is the same as model (9). We set the sample size $n = 300$, $n_{\inf} = 10$, the number of predictors $p = 1000$.

Example 1. For $i = 1, \cdots, n_{\inf}, Y_i^{\inf} = Y_{i_0} + \varepsilon_i + k, X_i^{\inf} = X_{i_0}$.

Example 2. For $i = 1, \cdots, n_{\inf}, j = 1, \cdots, p$, $X_{ij}^{\inf} = X_{i_0j} + 0.5kX_{i_0j}I_{\{j \in S\}}$, with $S = \{1, 2, \cdots, 10\}$ and $Y_i^{\inf} = (X_i^{\inf})^\top \beta + \varepsilon_i$, where $I_{\{j \in S\}}$ is the indicator function.

Example 3. For $i = 1, \cdots, n_{\inf}, j = 1, \cdots, p$, $X_{ij}^{\inf} = X_{i_0j} + 0.5kX_{i_0j}I_{\{j \in S\}}$, with $S = \{1, 2, \cdots, 10\}$, and $Y_i^{\inf} = (X_i^{\inf})^\top \beta + \varepsilon_i + 0.5k$, where $I_{\{j \in S\}}$ is the indicator function.

From the Table 1, we see that, in all cases, MHIM performs much better than HIM in terms of TPR and F_1-score. HIM has smaller values in terms of FPR, but the price is that the power TPR is very low. As a result, the F_1-score of MHIM is always higher

Table 1. Simulation results for Model 1–3

Example	k	HIM			MHIM		
1		TPR	FPR	F_1	TPR	FPR	F_1
	0.8	0.01	0.0100	0.144	0.56	0.017	0.566
	1.0	0.01	0.0007	0.179	0.60	0.015	0.590
	1.2	0.12	0.0670	0.078	0.82	0.012	0.756
	1.4	0.15	0.0032	0.241	0.88	0.014	0.770
	1.6	0.23	0.0011	0.365	0.97	0.017	0.788
2	0.8	0.07	0.0004	0.130	0.72	0.015	0.668
	1.0	0.12	0.0006	0.211	0.75	0.014	0.696
	1.2	0.13	0.0005	0.227	0.79	0.015	0.710
	1.4	0.14	0.0004	0.243	0.81	0.014	0.731
	1.6	0.23	0.0010	0.365	0.85	0.016	0.735
3	0.8	0.05	0.0003	0.094	0.24	0.010	0.314
	1.0	0.05	0.0130	0.070	0.57	0.015	0.569
	1.2	0.05	0.0003	0.094	0.60	0.013	0.609
	1.4	0.10	0.0008	0.150	0.64	0.014	0.626
	1.6	0.11	0.0006	0.195	0.88	0.015	0.760

than HIM. Therefore, the leave-one-out methods can't detect the influential observations efficiently. With the increase of k, the TPR of MHIM increase much faster than that of HIM. Therefore, MHIM is more effective than HIM.

3.2   Simulation for Swamping Effect

As the deviation of $(X^{\text{inf}}, Y^{\text{inf}})$ is large, there is swamping effect such that some observations in (X_i, Y_i) will be false identified as influential observation. To compare the performance of HIM and MHIM in the presence of swamping, we generate influential observations from the following Examples 4–6. We set the sample size $n = 100$, $p = 1000$ and $n_{\text{inf}} = 10$.

Example 4. For $i = 1, \cdots, n_{\text{inf}}, Y_i^{\text{inf}} = 15kY_{i_0} + \varepsilon_i, X_i^{\text{inf}} = X_{i_0}$.

Example 5. For $i = 1, \cdots, n_{\text{inf}}, X_i^{\text{inf}} = 3kX_{i_0}, Y_i^{\text{inf}} = (X_i^{\text{inf}})^T \beta + \varepsilon_i$.

Example 6. For $i = 1, \cdots, n_{\text{inf}}, X_i^{\text{inf}} = 2kX_{i_0}, Y_i^{\text{inf}} = 3k(X_i^{\text{inf}})^T \beta + \varepsilon_i$.

For the simulation results in Table 2, it can be inferred that MHIM has better performance than HIM. For Example 4–6, we see that both HIM and MHIM have good performance in terms of TPR. However, the FPR of HIM is much larger than that of MHIM, which shows that HIM is too aggressive. On the other hand, MHIM makes a good balance between TPR and FPR. As a result, F_1-score of MHIM is always higher than HIM.

Table 2. Simulation results for Model 4–6

Example	k	HIM			MHIM		
		TPR	FPR	F_1	TPR	FPR	F_1
4							
	0.8	1	0.190	0.539	1	0.030	0.881
	1.0	1	0.740	0.231	1	0.050	0.816
	1.2	1	0.870	0.203	1	0.060	0.787
	1.4	1	1	0.182	1	0.070	0.760
	1.6	1	1	0.182	1	0.070	0.760
5	0.8	1	0.010	0.957	1	0.002	0.917
	1.0	1	0.010	0.975	1	0.004	0.847
	1.2	1	0.110	0.669	1	0.017	0.929
	1.4	1	0.830	0.211	1	0.039	0.851
	1.6	1	1	0.182	1	0.060	0.787
6	0.8	1	0.004	0.982	1	0.003	0.881
	1.0	1	0.162	0.578	1	0.078	0.740
	1.2	1	1	0.182	1	0.039	0.851
	1.4	1	1	0.182	1	0.070	0.760
	1.6	1	1	0.182	1	0.213	0.511

3.3   Simulation on a Real Data

We consider the data set reported in Scheetz et al. (2006). In order to identify the genetic variation relevant to a multisystem human disease known as Bardet-Biedl syndrome (BBS), expression quantitative trait locus mapping was used in laboratory rats to gain a broad perspective of gene regulation. This set contains 120 twelve-week-old male F2 rats that are the offspring of the F1 rats and selected from eyes and microarray analysis for tissue harvesting. The RNA from the eyes of F2 rats was analyzed via microarrays that contain over 31,000 different probes. One of the probes that is for gene TRIM32 was found to cause BBS. The interest of this data set is to find genes whose expressions are correlated with that of TRIM32. Excluding the unexpressed and insufficiently variable one, there are only 18,976 probes as regressors. Then we chose the top 1000 probes that are most correlated with TRIM32 as regressors. As a result, the set contains 120 samples and the number of predictors is 1000.

Because of the existence of influential observations in this data, we identify them by HIM and MIHM simultaneously. When using the MHIM, we set the number of subsets $m = 500$ and repeat $T = 100$. The results of HIM and MIHM is the same. Both of them find 6 influential observations with indices {28, 32, 59, 80, 95, 120}. After removing the influential observations, we denote by Y^{med} and X_j^{med}, $1 \le j \le p$ the median of response and j-th variable in remain sample. We artificially add 10 outliers as followings to the remain sample as follows. Denote by $1_{n_{inf}}$ the vector of size n_{inf} with elements one.

Example 7. For $i = 1, \cdots, n_{\inf}$, $Y_i^{\inf} = Y^{\mathrm{med}} + 0.23k1_{n_{\inf}}$,

Example   8. For $i = 1, \cdots, n_{\inf}, j = 1, \cdots, p$, $X_{ij}^{\inf} = X_j^{\mathrm{med}} + 0.65kI_{\{j \in S\}}$ with $S = \{1, 2, \cdots, p\}$ and $I_{\{j \in S\}}$ is the indicator function.

Example   9. For $i = 1, \cdots, n_{\inf}$, $Y_{ij}^{\inf} = Y^{\mathrm{med}} + 0.23k1_{n_{\inf}}$, $X_{ij}^{\inf} = X_j^{\mathrm{med}} + 0.65kI_{\{j \in S\}}$ with $S = \{1, 2, \cdots, p\}$ and $I_{\{j \in S\}}$ is the indicator function.

Example 10. For $i = 1, \cdots, n_{\inf}$, $Y_i^{\inf} = Y^{\mathrm{med}} + 3.8k1_{n_{\inf}}$.

Example 11. For $i = 1, \cdots, n_{\inf}$ and $j = 1, \cdots, p$, $X_{ij}^{\inf} = X_j^{\mathrm{med}} + 200kI_{\{j \in S\}}$ with $S = \{1, 2, \cdots, p\}$ and $I_{\{j \in S\}}$ is the indicator function.

Example 12. For $i = 1, \cdots, n_{\inf}$, $Y_{ij}^{\inf} = Y^{\mathrm{med}} + 10k1_{n_{\inf}}$, $X_{ij}^{\inf} = X_j^{\mathrm{med}} + 10kI_{\{j \in S\}}$, with $S = \{1, 2, \cdots, p\}$, and $I_{\{j \in S\}}$ is the indicator function.

Simulations results are quite similar to those in Sects. 3.1 and 3.2, which shows the advantages of MHIM over HIM in the presence of multiple influential observations (Table 3).

Table 3. Simulation results for Model 7–12

Example	k	HIM			MHIM		
		TPR	FPR	F_1	TPR	FPR	F_1
7	0.8	0	0	0	0.01	0	0.020
	1.2	0.07	0	0.131	0.13	0	0.230
	1.6	0.10	0	0.182	0.93	0	0.964
8	0.8	0	0	0	0	0	0
	1.2	0	0	0	0.30	0	0.462
	1.6	0.15	0	0.261	0.35	0	0.519
9	0.8	0	0	0	0.01	0	0.020
	1.2	0	0	0	0.15	0	0.261
	1.6	0.13	0	0.230	0.90	0	0.947
10	0.8	1	0	1	1	0	1
	1.2	1	0.058	0.768	1	0.040	0.786
	1.6	1	1	0.161	1	0.090	0.681
11	0.8	1	1	0.161	0.87	0.094	0.613
	1.2	1	1	0.161	0.88	0.160	0.497
	1.6	1	1	0.161	0.91	0.170	0.493
12	0.8	1	1	0.161	1	0.140	0.589
	1.2	1	1	0.161	1	0.170	0.531
	1.6	1	▸1	0.161	1	0.180	0.517

4 Conclusion

In this paper, we consider the multiple influential observations detection in high dimensional linear regression model. We propose a new algorithm based on the idea of multiple deletion. Under some conditions, our method can detect a set containing all of the multiple influential observations with probability tending to 1.

Acknowledgements. The research of Zhao was supposed by National Science Foundation of China (No. 11471030, 11101022) and the Fundamental Research Funds for the Central Universities.

References

1. Cook, R.D.: Detection of influential observation in linear regression. Technometrics **19**, 15–18 (1977)
2. Behnken, D.W., Draper, N.R.: Residuals and their variance patterns. Technometrics **14**, 101–111 (1972)
3. Belsley, D.A., Kuh, E., Welsch, R.E.: The Grid: Regression Diagnostics: Identifying Influential Data and Sources of Collinearity. Wiley, New York (2005)
4. Chatterjee, S., Hadi, A.S.: The Grid: Sensitivity Analysis in Linear Regression. Wiley, New York (1988)
5. Pena, D.: A new statistic for influence in linear regression. Technometrics **47**(1), 1–12 (2005)
6. Pena, D.: Measures of Influence and Sensitivity in Linear Regression, pp. 523–536. Springer, London (2006). Springer Handbook of Engineering Statistics
7. Nurunnabi, A.A.M., Imon, A.H.M.R., Nasser, M.: A diagnostic measure for influential observations in linear regression. Commun. Stat. Theor. Methods **40**, 1169–1183 (2011)
8. Nurunnabi, A.A.M., Hadi, A.S., Imon, A.H.M.R.: Procedures for the identification of multiple influential observations in linear regression. J. Appl. Stat. **41**, 1315–1331 (2014)
9. Imon, A.R., Hadi, A.S.: Identification of multiple outliers in logistic regression. Commun. Stat. Theor. Methods **37**, 1697–1709 (2008)
10. Imon, A.R., Hadi, A.S.: Identification of multiple high leverage points in logistic regression. J. Appl. Stat. **40**, 2601–2616 (2013)
11. Zakaria, A., Howard, N.K., Nkansah, B.K.: On the detection of influential outliers in linear regression analysis. Am. J. Theor. Appl. Stat. **3**, 100–106 (2014)
12. Tibshirani, R.: Regression shrinkage and selection via the Lasso. J. Roy. Stat. Soc. B **58**, 267–288 (1996)
13. Fan, J., Lv, J.: Sure independence screening for ultrahigh dimensional feature space. J. Roy. Stat. Soc. Ser. B (Stat. Methodol.) **70**, 849–911 (2008)
14. Wang, H., Li, G., Jiang, G.: Robust regression shrinkage and consistent variable selection through the LAD-Lasso. J. Bus. Econ. Stat. **25**, 347–355 (2007)
15. She, Y., Owen, A.B.: Outlier detection using nonconvex penalized regression. J. Am. Stat. Assoc. **106**, 626–639 (2011)
16. Rahmatullah Imon, A.H.M.: Identifying multiple influential observations in linear regression. J. Appl. Stat. **32**, 929–946 (2005)
17. Pan, J.X., Fung, W.K., Fang, K.T.: Multiple outlier detection in multivariate data using projection pursuit techniques. J. Stat. Plann. Infer. **83**(1), 153–167 (2000)
18. Benjamini, Y., Hochberg, Y.: Controlling the false discovery rate: a practical and powerful approach to multiple testing. J. Roy. Stat. Soc.: Ser. B (Methodol.) **57**, 289–300 (1995)

Drift Operator for States of Matter Search Algorithm

Yuxiang Zhou, Yongquan Zhou[✉], Qifang Luo, Shilei Qiao, and Rui Wang

College of Information Science and Engineering,
Guangxi University for Nationalities,
Nanning Guangxi 530006, China
yongquanzhou@126.com

Abstract. States of matter search (SMS) algorithm is based on the simulation of the states of matter phenomenon. In SMS, individuals emulate molecules which interact to each other by using evolutionary operations which are based on the physical principle of the thermal-energy motion mechanism. Although the SMS algorithms have been used to solve many optimization problems, there still slow convergence and easy to fall into local optimum in some applications. In this paper, a novel drift operator-based states of matter search algorithm (DSMS) is proposed. The main idea involves using drift operator to keep the concept of location and abandon the concept of velocity for accelerate the convergence speed while simplifying algorithm, meanwhile a new variable differential evolution (DE) strategy is introduced to diversify the individuals in the search space for escape from the local optima. The proposed method is applied to several benchmark problems and is compared to four modern meta-heuristic algorithms. The experimental results show that the proposed algorithm outperforms other peer algorithms.

Keywords: States of matter search algorithm · Drift operator · Thermal-energy motion mechanism · Differential evolution strategy · Meta-heuristic algorithms

1 Introduction

Recently, global optimization problem in real-world is more and more complex and has attracted a lot of researchers to search for efficient problem-solving methods. Evolutionary algorithm is the better solution to solve the global optimization problems and widely applied in various areas of science, engineering, economics and others, where mathematical modeling is used. In general, the goal is to find a goal optimum for an objective function which is defined over a given search space. During the last few decades, several evolutionary algorithms have been suggested that mimics some natural phenomena. Such phenomena include animal-behavior phenomena such as the Particle Swarm Optimization (PSO) algorithm [3], the Bat (BA) algorithm proposed by Yang [4]. Some other methods which are based on physical processes, for example, the Electromagnetism-like Algorithm [5], the Gravitational Search Algorithm (GSA) [6] and the States of Matter Search (SMS) algorithm [1, 2].

The SMS algorithm is based on the simulation of the states of matter phenomenon. In SMS, individuals emulate molecules which interact to each other by using

© Springer International Publishing Switzerland 2015
D.-S. Huang and K. Han (Eds.): ICIC 2015, Part III, LNAI 9227, pp. 65–71, 2015.
DOI: 10.1007/978-3-319-22053-6_7

evolutionary operations based on the physical principles of the thermal-energy motion mechanism. Thus, the evolutionary process is divided into three stages which emulate the three states of matter: gas, liquid and solid. At each state, molecules (individuals) exhibit different behaviors. The differences among such states are based on forces which are exerted among particles composing a material [7]. Even the basic SMS has been shown powerful [1, 2], however it use many certain parameters which complicated the algorithm, thus lead to slow down the convergence rate of the algorithm and lost the randomness and diversity.

In order to solve these problems, we present, a novel, Drift operator-based States of Matter Search algorithm (DSMS) in this paper. The main idea involves using drift operator to keep the concept of location and abandon the concept of velocity for accelerate the convergence speed while simplifying algorithm, meanwhile a new variable Differential Evolution (DE) strategy is introduced to diversify the individuals in the search space for escape from the local optima. The proposed approach has been compared with some other well-known evolutionary algorithms. The obtained results confirm a high performance of the proposed algorithm for solving various benchmark functions.

This paper is organized as follows. In the Sect. 2, the basic SMS algorithm is described. Section 3 gives the DSMS algorithm. The simulation and comparison of this proposed algorithm are presented in Sect. 4. Finally, some remarks and conclusions are provided in Sect. 5.

2 Basic SMS Algorithm

Individuals are considered as molecules whose positions on a multidimensional space are modified as the algorithm evolves. The movement of such molecules is motivated by the analogy to the motion of thermal-energy. The velocity and direction of each molecule's movement are determined by considering the collision, the attraction forces and the random phenomena experimented by the molecule set. Such behaviors have been implemented by defining several operators such as the direction vector, the collision and the random positions operators, all of which emulate the behavior of actual physics laws.

2.1 Direction Vector

The direction vector operator mimics the way in which molecules change their positions as the evolution process develops. For each n-dimensional molecule P_i from the population P, it is assigned an n-dimensional direction vector d_i which stores the vector stores the vector that controls the particle movement. Initially, all the direction vectors $(D = \{d_1, d_2, \ldots d_N\})$ are randomly chosen within the range of [−1, 1].

Therefore, the new direction vector for each molecule is iteratively computed considering the following model:

$$d_i^{k+1} = d_i^k \cdot (1 - \frac{k}{gen}) \cdot 0.5 + a_i \tag{1}$$

Where a_i represents the attraction unitary vector calculated as $a_i = (P^{best} - P_i)/||P^{best} - P||$, being P^{best} the best individual seen so-far, while P_i is the molecule i of population P. k represents the iteration number whereas gen involves the total iteration number that constitutes the complete evolution process.

In order to calculate the new molecule position, it is necessary to compute the velocity V_i of each molecule by using:

$$V_i = d_i \cdot v_{init} \tag{2}$$

Being v_{init} the initial velocity magnitude which is calculated as follows:

$$v_{init} = \frac{\sum_{j=1}^{n}(b_j^{high} - b_j^{low})}{n} \cdot \beta \tag{3}$$

Where b_j^{low} and b_j^{high} are the low j parameter bound and the upper j parameter bound respectively, whereas $\beta \in [0, 1]$. Then, the new position for each molecule is updated by:

$$P_{i,j}^{k+1} = P_{i,j}^{k} + V_{i,j} \cdot rand(0, 1) \cdot \rho \cdot (b_j^{high} - b_j^{low}) \tag{4}$$

$0.5 \leq \rho \leq 1$. (represent the maximum permissible displacement among particles)

2.2 Collision

The collision operator mimics the collision experimented by molecules while they interact to each other. Collisions are calculated if the distance between two molecules is shorter than a determined proximity value. Therefore, if $||P_i - P_q|| < r$, a collision between molecules i and q is assumed; otherwise, there is no collision, considering $i, q \in \{1, ..., N\}$ such that $i \neq q$. If a collision occurs, the direction vector for each particle is modified by interchanging their respective direction vectors as follows:

$$||P_i - P_q|| = \sqrt{\sum_{j=1}^{D}(P_{ij} - P_{qj})^2} \tag{5}$$

$$d_i = d_q \text{ And } d_q = d_i \tag{6}$$

The collision radius is calculated by:

$$r = \frac{\sum_{j=1}^{n}(b_j^{high} - b_j^{low})}{n} \cdot \alpha \tag{7}$$

$\alpha \in [0, 1]$.

2.3 Random Positions

In order to simulate the random behavior of molecules, algorithm generates random positions following a probabilistic criterion within a feasible search space. For this operation, a uniform random number r_m is smaller than a threshold H, a random molecules position is generated; otherwise, and the element remains with no change. Therefore such operation can be modeled as follows:

$$P_{i,j}^{k+1} = \begin{cases} b_j^{low} + rand(0,1) \cdot (b_j^{high} - b_j^{low}) \ with\ probability\ H \\ P_{i,j}^{k+1} \qquad\qquad\qquad\quad with\ probability\ (1-H) \end{cases} \tag{8}$$

Where $i \in \{1, ...N\}$ and $j \in \{1, ...D\}$.

3 Drift Operator-Based States of Matter Search Algorithm (DSMS)

Although in some functions the basic SMS algorithm shows good performance. But as use many certain parameters which complicated the algorithm itself, hence lead to slow down the convergence rate of the algorithm and lost the randomness and diversity. So in this paper, on the basis of the original algorithm, we introduce the drift operator and a new variable differential evolution strategy to solve the above problems. Thus makes the algorithm can quickly converge to a good solution. Simplified algorithm at the same time, strengthen the randomness of the algorithm, and more practical.

3.1 Drift Operator

Drift operator was proposed by Xing Xu in improved Particle Swarm Optimization based on Brownian Motion (BMPSO). The principle is based on ITO process to define the drift represent particles on macroscopic trend, it abandons the concept of velocity but retain the concept of individual's positions. This approach redefines the update formula of molecule position as follows:

$$P_{i,j}^{k+1} = P_{i,j}^k + d_{i,j} \cdot rand \cdot \rho \cdot (b_j^{high} - b_j^{low}) + \rho \cdot (P^{best} - P_{i,j}^k) \tag{9}$$

Through the analysis of (9), we can find that drift operator and the basic SMS have much in common. Similarities: remain the core idea of basic SMS algorithm, which are memory and social cognition. The first part of formula is memory part, represent the past position is good or bad effect on the present. Second part shows that social cognition, each molecule will be attracted by the best molecule and moving towards it. Differentia: the third part of the formula is the molecule's self cognition, expresses the molecule moves a part from his own experience. And there is no concept of velocity in DSMS algorithm, therefore, introduces the concept of drift instead of the velocity update, molecules drift to the best one during the movement.

3.2 Variable Differential Evolution Strategy

As a simple and efficient global optimization strategy, DE algorithm and some improved version DE such as DE with a Modified Neighborhood-Based Mutation Operator (MNDE) shows its powerful search and optimization ability. Whereas that still can not avoid the problem that optimization speed is slow and premature convergence. In the paper, we on the basis of MNDE introduce a new variable differential evolution strategy, the model as follows:

$$P_i^{k+1} = P^{best} + c1 \cdot (P_p^k - P_m^k) + c2 \cdot (P_h^k - P_l^k) \tag{10}$$

The $P_p^k, P_m^k, P_h^k, P_l^k$ are the four molecules which different from P_i^k at k generation. And P^{best} stores the best historical individual found so-far. Different with the MNDE, formula (10) does not use the best neighborhood, but randomly select four different molecules, and use the best individual to mutation. Diversify the individuals in the search space; meanwhile retain the ability of individual memory. Furthermore help the algorithm escape from the local optima and faster convergence to the global optimal.

3.3 The Implementation Steps of DSMS Algorithm

1. Initialize the basic parameters: population size N; population P_i $i \in (1, \dots, N)$; location update impact factor ρ, collision radius impact factor α and threshold H under different states; the maximum number of iterations iterMax.
2. According to the current number of iterations setting the state of matter.
3. Evaluate the fitness value of each particle and find the best element of the population P, compute the new molecules by using (1) and (9).
4. Solve collisions by using (5)–(7).
5. Judge whether random movement of molecules. If moving randomly, select four molecules which different from the molecule itself to mutation by using (10) in Sect. 3.2, otherwise, keep its original position.
6. Update the best population.
7. Determine if termination condition have been met (i.e., reach maximum number of iterations or satisfy the search accuracy), go to step 8; otherwise, go to step 2 and execute the next search.
8. Output the best fitness values and global optimal solution.

4 Simulation Experiments and Results Analysis

4.1 Benchmark Functions

In order to verify the effectiveness of the proposed algorithm, we have applied the DSMS algorithm to four standard benchmark functions in Table 1 whose results have been compared to those produced by the Gravitational Search Algorithm (GSA), The Particle Swarm Optimization (PSO), the bat algorithm (BA) and the basic state of matter search algorithm (SMS).

Table 1. Benchmark functions

Id	Name	Functions	Domain
$f1$	Schwefel 2.21	$f(x) = \max\{|x_i|, 1 \le i \le D\}$	$[-100,100]$
$f2$	Zakharov	$f(x) = \sum\limits_{i=1}^{D} x_i^2 + (\sum\limits_{i=1}^{D} 0.5i \cdot x_i)^2 + (\sum\limits_{i=1}^{D} 0.5i \cdot x_i)^4$	$[-10,10]$
$f3$	Rastrigin	$f(x) = \sum\limits_{i=1}^{D} [x_i^2 - 10\cos 2\pi x_i + 10]$	$[-5.12,5.12]$
$f4$	Ackley	$f(x) = -20\exp(-0.2\sqrt{\frac{1}{D}\sum\limits_{i=1}^{D} x_i^2}) - \exp(\frac{1}{D}\sum\limits_{i=1}^{D} \cos 2\pi x_i) + 20 + e$	$[-32,32]$

4.2 Parameter Setting

In the test, the maximum number of iterations of each algorithm is iterMax = 1000, the population has been set to 50; the dimension of every problem has been set to 30. Other parameters are set in Table 2

Table 2. Parameters of algorithms

Algorithm	Parameters
GSA	$G_0 = 100$, $\alpha = 20$
PSO	$c1 = c2 = 1.4962$, $w = 0.8$
BA	$Q_i \in [0, 2]$, $A^0 = 0.5$, $r^0 = 0.5$, $\alpha = 0.95$, $\gamma = 0.05$
SMS	$\rho \in [0, 1]$, $\alpha \in [0, 0.8]$, $\beta \in [0.1, 0.8]$, $H \in [0, 0.9]$

4.3 Comparison of Experiment Results

The test compares the DSMS to other algorithms such as PSO, GSA, BA and SMS. The mean and standard deviation values are shown in Table 3. The best outcome for each function is boldfaced.

Table 3. Simulation results for test functions $f1 - f4$

Functions		PSO	GSA	BA	SMS	DSMS
$f1$	Mean	7.3709	0.0055	56.7759	0.0221	**0**
	Std	2.4255	0.0301	6.8347	0.0058	**0**
$f2$	Mean	1.24E + 04	1.79E + 03	1.81E + 04	0.0296	**0**
	Std	9.22E + 03	4.68E + 02	6.52E + 03	0.0231	**0**
$f3$	Mean	81.5201	14.3274	1.93E + 02	0.9284	**0**
	Std	23.8657	3.9332	30.3516	0.88	**0**
$f4$	Mean	4.3324	3.59E-09	18.9884	0.004	**8.88E-16**
	Std	1.1704	5.65E-10	0.2343	8.71E-04	**0**

From Table 3, besides $f4$, other functions DSMS can find out the optimal solution and all the standard deviation is 0. Other algorithms can not gain the accurate value in actual and even traps into local optima. Even of $f4$, DSMS has a higher precision of optimization. As a consequence, DSMS is a superb algorithm with outstanding robustness and wonderful accuracy for the functions above.

5   Conclusions

As use many certain parameters which complicated the basic SMS algorithm itself, hence lead to slow down the convergence rate of the algorithm and lost the randomness and diversity. So in this paper, a novel, Drift operator-based States of Matter Search algorithm (DSMS) is proposed. The DSMS algorithm using drift operator to keep the concept of location and abandon the concept of velocity for accelerate the convergence speed while simplifying algorithm, meanwhile a new variable Differential Evolution (DE) strategy is introduced to diversify the individuals in the search space for escape from the local optima. The results of comparison with the PSO, CLSPSO, BA, GSA and the basic SMS show that precision of optimization, convergence speed and robustness of DSMS are all better than other algorithms. The results of simulation test show that the proposed algorithm is effective and feasible.

Acknowledgements.   This work is supported by National Science Foundation of China under Grant No. 61165015; 61463007. Key Project of Guangxi High School Science Foundation under Grant No. 201203YB072, and the Innovation Project of Guangxi University for Nationalities (gxun-chx2014090).

References

1. Cuevas, E., Echavarría, A., Ramírez-Ortegón, M.A.: An optimization algorithm inspired by the states of matter that improves the balance between exploration and exploitation. Appl. Intell. (2014). doi:10.1007/s10489-013-0458-0
2. Cuevas, E., Marte, A.E., Zaldívar, D., Pérez-Cisneros, M.: A novel evolutionary algorithm inspired by the states of matter for template matching. Expert Syst. Appl. **40**, 6359–6373 (2013)
3. Kennedy, J., Eberhart, R.: Particle swarm optimization. 1995 IEEE Proc. Int. Conf. Neural Netw. **4**, 1942–1948 (1995)
4. Yang, X.S.: A new metaheuristic bat-inspired algorithm. In: Gonzalez, J.R., et al. (eds.) Nature inspired cooperative strategies for optimization NICSO, vol. 284, pp. 65–74. Springer, Heidelberg (2010)
5. Ilker, B., Birbil, S., Shu-Cherng, F.: An electromagnetism-like mechanism for global optimization. J Glob Optim. **25**, 263–282 (2003)
6. Rashedia, E., Nezamabadi-pour, H., Saryazdi, S.: Filter modeling using gravitational search algorithm. Eng. Appl. Artif. Intell. **24**, 117–122 (2011)
7. Ceruti, M.G., Rubin, S.H.: Infodynamics: Analogical analysis of states of matter and information. Inf. Sci. **177**, 969–987 (2007)

Classification on Imbalanced Data Sets, Taking Advantage of Errors to Improve Performance

Asdrúbal López-Chau[1]([⊠]), Farid García-Lamont[2], and Jair Cervantes[2]

[1] Centro Universitario UAEM, Universidad Autónoma Del Estado de México,
CP 55600 Zumpango, Estado de Mexico, México
alchau@uaemex.mx
[2] Centro Universitario UAEM, Universidad Autónoma Del Estado de México,
56159 Texcoco, Estado de Mexico, México

Abstract. Classification methods usually exhibit a poor performance when they are applied on imbalanced data sets. In order to overcome this problem, some algorithms have been proposed in the last decade. Most of them generate synthetic instances in order to balance data sets, regardless the classification algorithm. These methods work reasonably well in most cases; however, they tend to cause over-fitting.

In this paper, we propose a method to face the imbalance problem. Our approach, which is very simple to implement, works in two phases; the first one detects instances that are difficult to predict correctly for classification methods. These instances are then categorized into "noisy" and "secure", where the former refers to those instances whose most of their nearest neighbors belong to the opposite class. The second phase of our method, consists in generating a number of synthetic instances for each one of those that are difficult to predict correctly. After applying our method to data sets, the AUC area of classifiers is improved dramatically. We compare our method with others of the state-of-the-art, using more than 10 data sets.

Keywords: Imbalanced · Classification · Synthetic instances

1 Introduction

Achieving a good performance on imbalanced data sets is a challenging task for classification methods [3]. They usually focus on majority class, almost ignoring the opposite class [8]. Currently, there are many real-world applications that generate this type of data sets, for example: software defect detection [6], medical diagnosis [1], fraud detection in telecommunications [4], financial risks [7] and DNA sequencing [9], among others. In this type of applications, there are two objectives in conflict, on the one hand, for the classifier should be more important to predict the minority class instances with the minimal errors, and on the other hand, the classification accuracy for majority class instances should not be severely damaged. The AUC ROC measure is one of the most widely used to capture this requirement.

The problem of classification on imbalanced data sets has attracted the attention of the machine learning and data mining communities in the last past few years [2]. The state-of-the-art methods to deal this problem can be categorized into:

© Springer International Publishing Switzerland 2015
D.-S. Huang and K. Han (Eds.): ICIC 2015, Part III, LNAI 9227, pp. 72–78, 2015.
DOI: 10.1007/978-3-319-22053-6_8

(1) external methods, which pre-processes the data sets to balance them before applying a classification method;
(2) internal methods, which modify the algorithms to make them more suitable to this problem;
(3) ensembles, that use two or more classifiers and then combine their outputs to predict the class;
(4) cost-sensitive methods, which use cost matrices to penalize misclassification, or
(5) other methods, that include combinations of the strategies mentioned, and application of genetic algorithms.

External methods work at the data level, regardless the classifier to be used. These methods are based on two main techniques: *under-sampling* and *over-sampling*, both of them balance the data sets, either by removing objects from the majority class or inserting synthetic minority class objects, respectively. One of the most representative methods is SMOTE. It balances data sets by creating synthetic instances between the line that joins a minority class instance and their nearest neighbors. Variants of SMOTE guide the creation of minority instances towards specific parts of the input space, considering characteristics of the data such as density of minority class instances, the decision boundaries or using ensembles of classifiers.

In this paper, we propose a method to pre-process imbalanced data sets for classification. It works in two phases: the first one identifies instances, which are difficult to predict for a classification method. These instances are important because represent regions in the input space where the classifier is unable to perform adequately, and therefore, it is necessary to clarify the concepts or sub-concepts by generating synthetic instances in such regions. The instances that are difficult to predict, are categorized into "noisy" and "secure" instances, where the former refers to those which most of their nearest neighbors belong to the opposite class. Noisy instances are usually near to decision boundaries, or in overlapped class regions [5]. The second phase of our method, consists in generating a number of synthetic instances considering the noisy ones. Depending on the imbalance ratio, the number of generated instances is adapted. We tested our method on 11 data sets, and compare the performance of C4.5 classifier using other balancing algorithms. According to the results, AUC is improved significantly in most cases.

The rest of this paper is organized as follows. Our proposal is shown in detail in Sect. 2. The experiments, results and a discussion is shown in Sect. 3. The conclusions and references are in the last part of this paper.

2   Method Based on Observations of Errors

The method presented in this paper is effective and very easy to implement. Different from SMOTE and other similar algorithms that generate instances regardless the classification method or class distributions, our approach takes advantage of observations about the correctness of predictions. These are used to identify difficult regions of the input space, and then the generation of synthetic instances focuses on such regions.

Given an imbalanced data set: $X = \{(x_i, y_i)_{i=1}^{N}, y_i \in \{+1, -1\}\}$, where N is the number of instances, $y_i = +1$ is the minority class, and $y_i = -1$ the majority class. In our method, we create some sets, in order to detect the regions of the input space are difficult to predict for the classifier.

Minority $= \{(x_i, y_i), x \in X, y_i = +1\}$, this set contains all the instances of the minority class in X. The following two subsets of X, contain only instances of the majority class:

$$TrM_j = \{(x_i, y_i), x_i \in X \text{ and } x_i \notin TeM_j, y_i = -1\}$$
$$TeM_j = \{(x_i, y_i), x \in X \text{ and } x_i \notin TrM_j, y_i = -1\}$$

such that $TrM_j \cup TeM_j = X -$ Minority, and $TrM_j \cap TeM_k$ is empty.

The elements of TrM_j and TeM_j are chosen randomly. The size of these sets is 60 % of $|X -$ Minority$|$ and 40 % of $|X - Minority|$, respectively.

Algorithm 1. Counter of errors in predictions

Input : X: Training data set, C: Type of classifier, I: Number of iterations
Output: $\mathcal{E}$: Mean of of missclassifications for each instance
begin
    for $j \leftarrow 1$ **to** I **do**
        Create the sets Tr_j and Te_j and Build a classifier C of type C from Tr_j;
        foreach *instance* $x \in Te_j$ **do**
            Use C to predict the class of x;
            if C *incorrectly classifies* x **then**
                Update the counter of errors $\mathcal{E}$ for this instance;
            end
        end
    end
    return $mean(\mathcal{E})$
end

The sub-training set, Tr_j, is composed of all instances of the minority class and the elements of TrM_j: $Tr_j = TrM_k \cup$ Minority. Also, we create the sub-testing set, Te_k, composed of all instances of the minority class, and those instances of the majority class that are not in Tr_j: $Te_j = TeM_j \cup$ Minority. Having these sets created, a classifier is trained and tested several times. The errors in predictions are stored in a vector ε to be analyzed later. Algorithm 2 shows the pseudo code that implements this part of our method

Once ε obtained, those instances which have been classified incorrectly a number of times that exceeds a certain threshold, are categorized into two types:

(a) Noisy instances, difficult to predict instances and most of their k-nearest neighbors have opposite class.
(b) Secure instances, difficult to predict instances and most of their k-nearest neighbors have the same class.

During the experiments, we found that $k = 5$ produces good results for most data sets. Different from other approaches that only take into account a number of nearest neighbors, in our approach, the noisy instances play an important role in the generation of new synthetic ones. The latter are generated in the lines that joins a noisy instance and its nearest L-neighbors.

2.1   Run-Time Complexity

In our method, the separation of majority and minority class instances is realized in linear time, $O(n)$. The creation of sub-training and sub-testing sets is also a linear time task. Training time varies form a type of classifier to other, we represent it with $T(|Tr_j|)$. The prediction of the class for each an instance depends on the classification method, so we represent time with C, therefore, the time to predict all the instances in the sub-testing set is $|Te_j|C$. Updating the vector ε is a constant time task, C_0. In current implementation of the algorithm, the generation of synthetic instances requires a linear search of the L-nearest neighbors for each noisy instance, the worst case is $O(n^2)$. Our method is slow for large data sets. The time-complexity of our method is therefore:

$$O(n) + IO(n) + IT(Tr_j) + I|Te_j|C + IC_0 + IO(n^2) \approx$$
$$IT(0.6n) + I0.4nC + IO(n^2)$$

3   Experiments and Results

In order to observe how the performance of classifiers is improved by pre-processing the data with our method, we select the C4.5 classifier, which is one of the most commonly algorithms chosen to test the performance of balancing methods. The data sets used to test the experiments are publicly available on the Internet,[1] their main features are shown in Table 1.

Table 1. Data sets for experiments

Data set	D	S	IR	Data set	D	S	IR
yeast-2_vs_4	8	514	9.08	glass-0-1-6_vs_2	99	192	10.29
glass2	9	214	11.59	ecoli4	77	336	15.80
page-blocks-1-3_vs_4	10	472	15.86	abalone9-18	88	731	16.4
glass-0-1-6_vs_5	9	184	19.44	glass5	99	214	22.78
car-good	6	1,728	24.04	yeast5	88	1,484	32.73
abalone19	8	4,174	129.44				

In Table 1, D is the number of attributes, S is the number of instances, and IR is the imbalance ratio. In these sets IR varies from 9, up to more than 120. We present the comparative of our method against SMOTE, re-sampling with and without

---

[1] http://sci2s.ugr.es/keel/datasets.php.

replacement. SMOTE algorithm generates synthetic instances using the 5 nearest neighbors, re-sampling makes copies of minority class instances.

All the experiments were conducted on a computer with the following characteristics: 2.6 GHz Intel Core i5 processor, 8 GB RAM, Mavericks Operating System. The size of RAM allocated to the JVM is 256 MB. In the experiments, each data set was partitioned into two subsets, randomly: training and testing. The former contains 60 % of instances of data set; the latter contains the rest. The training set is processed using our method, SMOTE and re-sampling with and without replacement. Then, a classifier C4.5 is trained with the processed data. The testing set is used to test performance of classifier. This process was repeated 30 times and the average is reported in the results.

4   Results

The application of Algorithm 2 provides with the information presented in Table 2, whose column have the following meaning. Data set: Name of data set analyzed; P: Number of minority class instances in the sub-training set; N: Number of majority class instances in the sub-training set; D_p: Number of minority class instances which are difficult to predict for the classifier; D_n: Number of majority class instances which are difficult to predict for the classifier; N_p: Number of noisy minority class instances; N_n: Number of noisy majority class instances. In order to achieve repeatable results for other researchers, the C4.5 (J48 Weka implementation) classifier was used with default parameter values. The threshold used in the experiments was set to one.

Table 2. Identification of difficult instances for the C4.5 classifier

Data set	P	N	D_p	D_n	N_p	N_n	Data set	P	N	D_p	D_n	N_p	N
yeast-2_vs_4	37	323	11	19	9	2	glass-0-1-6_vs_2	14	121	4	18	4	1
glass2	15	135	15	34	15	2	ecoli4	15	221	4	6	3	0
page-blocks-1-3_vs_4	23	308	0	2	0	0	abalone9-18	32	480	19	27	19	1
glass-0-1-6_vs_5	7	122	1	6	1	1	glass5	8	142	1	5	1	0
car-good	53	1,157	53	70	53	20	yeast5	37	1,002	2	22	2	8
abalone19	25	2,897	25	0	25	0							

Based on the average results shown in Table 2, the following can be observed:

(1) Most of the instances that are difficult to predict, belong to majority class. This is probably due to between-class imbalance, because of the large number of majority class instances.

(2) In general, the minority class instances that are difficult to predict, are also noisy instances. We attribute this to within-class imbalance.

(3) Most of majority class instances that are difficult to predict, are secure instances. This result is different from the informed in the literature, further investigation is necessary.

(4) All the minority class instances of data sets glass2, car-good and abalone19 are noisy instances, i.e., these data sets do not contain secure instances of the minority class.

(5) Data sets glass-0-1-6_vs_5, glass5 a yeast 5, contain just a few noisy instances of minority class. This makes difficult for our method to generate many instances.

In our method, we use the noisy instances to generated a number of synthetic instances, such that a balance of approximately 30 % is achieved. The underlying idea is to warn the classifier on regions not considered important, but they are.

Table 3 shows the area under the ROC for classifier C4.5. *None* corresponds to the performance of classifier without a pre-processing step of data. *Proposal* column is the method presented in this paper. *SMOTE* is the classic method with $K = 5$ nearest neighbors. R1 and R2 are re-sampling of minority class instances with and without replacement, respectively. In general, our method outperforms SMOTE, R1 and R2 in the cases where the number of difficult and noisy instances is not too small. In the other cases, our method produces results that are acceptable. Due to space issues, we don't present more results with other classification methods.

Table 3. AUC for classifier C4.5

Data set	None	Proposal	SMOTE	R1	R2
yeast-2_vs_4	0.895	**0.913**	0.880	0.857	0.895
glass-0-1-6_vs_2	0.675	**0.730**	0.712	0.634	0.675
glass2	0.744	**0.778**	0.689	0.657	0.744
ecoli4	0.821	**0.887**	0.875	0.869	0.821
page-blocks-3vs4	0.969	**0.987**	0.978	0.978	0.969
abalone9-18	0.619	0.685	**0.692**	0.622	0.619
glass-0-1-6_vs_5	0.875	0.965	**0.967**	0.903	0.875
glass5	**0.953**	0.862	0.905	0.867	**0.953**
car-good	0.444	0.911	**0.942**	0.904	0.444
yeast5	0.882	**0.921**	0.907	0.873	0.882

5 Conclusions

The performance of classifiers on imbalanced data sets is generally unacceptable. This problem is complex, since there are many factors involved, such as rare instances, between-class imbalanced within-class imbalance and noisy instances.

In this paper, we introduce a method to tackle with the classification task on imbalanced data sets. Different from other state-of-the-art proposals, our method is based on the philosophy that classification algorithms need to be involved in the generation of synthetic instances. We identify those instances that are difficult to predict correctly for a classifier. These instances are considered to detect regions in the input space that need to be reinforced with new synthetic instances. The method proposed in this paper was tested with 11 data sets and compared with other state-of-the-art methods. According to the results, our approach outperforms the current methods in most cases.

References

1. Esfandiari, N., Babavalian, M.R., Moghadam, A.-M.E., Tabar, V.K.: Review: knowledge discovery in medicine: current issue and future trend. Expert Syst. Appl. **41**(9), 4434–4463 (2014)
2. García, V., Sánchez, J.S., Mollineda, R.A.: On the effectiveness of preprocessing methods when dealing with different levels of class imbalance. Knowl. Based Syst. **25**(1), 13–21 (2012). Special Issue on New Trends in Data Mining
3. He, H., Garcia, E.A.: Learning from imbalanced data. IEEE Trans. Knowl. Data Eng. **21**(9), 1263–1284 (2009)
4. Hilas, C.S., Mastorocostas, P.A.: An application of supervised and unsupervised learning approaches to telecommunications fraud detection. Knowl. Based Syst. **21**(7), 721–726 (2008)
5. Lemnaru, C., Potolea, R.: Imbalanced classification problems: systematic study, issues and best practices. In: Zhang, R., Zhang, J., Zhang, Z., Filipe, J., Cordeiro, J. (eds.) ICEIS 2011. LNBIP, vol. 102, pp. 35–50. Springer, Heidelberg (2012)
6. Sheng, V.S., Gu, B., Fang, W., Wu, J.: Cost-sensitive learning for defect escalation. Knowl. Based Syst. **66**, 146–155 (2014)
7. Sun, J., Li, H., Huang, Q.-H., He, K.-Y.: Predicting financial distress and corporate failure: a review from the state-of-the-art definitions, modeling, sampling, and featuring approaches. Knowl. Based Syst. **57**, 41–56 (2014)
8. Tomasev, N., Mladenic, D.: Class imbalance and the curse of minority hubs. Knowl. Based Syst. **53**, 157–172 (2013)

PSO-Based Method for SVM Classification on Skewed Data-Sets

Jair Cervantes[1]([⊠]), Farid García-Lamont[1], Asdrúbal López[3],
Lisbeth Rodriguez[2], José S. Ruiz Castilla[1], and Adrián Trueba[1]

[1] Posgrado e Investigación UAEMEX (Autonomous University of Mexico
State), Av. Jardín Zumpango s/n, Fracc. El Tejocote, 56259 Texcoco, Mexico
jcervantesc@uaemex.mx
[2] Division of Research and Postgraduate Studies, Instituto Tecnológico de
Orizaba, Orizaba, Veracruz, México
[3] Zumpango University Center, University of the State of Mexico,
Texcoco, Mexico

Abstract. Support Vector Machines (SVM) have shown excellent generalization power in classification problems. However, on skewed data-sets, SVM learns a biased model that affects the classifier performance, which is severely damaged when the unbalanced ratio is very large. In this paper, a new external balancing method for applying SVM on skewed data sets is developed. In the first phase of the method, the separating hyperplane is computed. Support vectors are then used to generate the initial population of PSO algorithm, which is used to improve the population of artificial instances and to eliminate noise instances. Experimental results demonstrate the ability of the proposed method to improve the performance of SVM on imbalanced data-sets.

Keywords: Support vector machines · PSO · Imbalanced data sets

1 Introduction

In the past few years, Support Vector Machines (SVM) has shown excellent generalization power in classification problems. However, it has been shown that generalization ability of SVM drops dramatically on skewed data-sets [1, 2], because SVM learns a biased model, which affects the classifier performance. Moreover, the performance of SVM is more affected when the imbalanced ratio is large. Although there are several external techniques to tackle the imbalance in data sets, SMOTE has been one of the most-used approaches among several methods. SMOTE introduces artificial instances in data sets by interpolating features values based on neighbors. In several studies have been shown that SMOTE is better than under-sampling and over-sampling techniques [3–7]. Moreover, SMOTE not cause any information loss and could potentially find hidden minority regions, because SMOTE identify similar but more specific regions in the feature space as the decision region for the minority class. Despite its excellent features, SMOTE is limited to introduce instances with low information because the new instances are obtained using a linear combination between positive examples which increments the density of points. The best artificial examples

© Springer International Publishing Switzerland 2015
D.-S. Huang and K. Han (Eds.): ICIC 2015, Part III, LNAI 9227, pp. 79–86, 2015.
DOI: 10.1007/978-3-319-22053-6_9

(instances with more information of each class) are in the region between positive and negative instances. Introduce instances in this region could increment the discriminative information of positive instances and improve the performance of a classifier on imbalanced data sets. However, this external region is very sensible to artificial instances. Inadequate instances lead to introduce noise and loss of performance in the classifier. Different artificial instances can cause significant differences in performance. Therefore, artificial instances must be generated carefully.

In this paper, we present a novel algorithm which improves the performance of SVM classifiers on imbalanced data sets. The proposed algorithm uses a hybrid system to generate new examples. Hybrid techniques have been widely used in recent years [8, 9]. Some authors have proposed hybrid techniques to address this problem, but this problem is still a challenge today. In contrast to SMOTE, the examples generated with proposed method are derived from the most critical region for SVMs, called the margin. The margin is the distance between the decision boundary and the closest examples. Other techniques generate examples which are located randomly. The proposed algorithm obtains artificial instances from the most important region. However, although generating new instances in the minority class can improve the performance in SVM classification, this process could introduce noise in the data-set. Moreover, it is particularly difficult to introduce good instances in this region because this region is extremely sensitive. Introducing artificial instances in the data-sets must be generated carefully. To find optimal and synthetic instances is an important step in the proposed algorithm. This is the main reason to combine PSO with SVM in this research. In the proposed algorithm, PSO is used to guide the search process of artificial instances that improve the SVM performance. Moreover, the synthetic instances are evolved and improved by following the best particle p_{gi}. Experimental results show that the proposed algorithm can get better performance than traditional models.

The rest of the paper is organized as follows. In Sect. 2, a brief overview to the related work on SVM with imbalanced data sets is presented. Section 3 presents the proposed method. The results of experiments are shown in Sect. 4. Conclusions are in Sect. 5.

2   SVM Classification

Formally, the training of SVM begins with a training set X_{tr} given by

$$X_{tr} = \{(x_i, y_i)\}_{i=1}^{n} \tag{1}$$

with $x_i \in R^d$ and $y_i \in \{-1, +1\}$. The classification function is determined by

$$y_i = sign\left(\sum \alpha_i y_j K\langle x_i \cdot x_j \rangle + b\right) \tag{2}$$

where α_i are the Lagrange multipliers, $K\langle x_i \cdot x_j \rangle$ is the kernel matrix, and b is the bias. The optimal separating hyperplane is computed by solving the following optimization problem:

$$\min \frac{1}{2} w_i^T w_i + C \sum_{t=1}^{l} \eta_i^2 \tag{3}$$

$$s.t. \quad y_i \left(w_i^T K \langle x_i \cdot x_j \rangle + b_i \right) \geq 1 - \eta_i$$

where C is the margin parameter to weight the error penalties η_i. The margin is optimal in the sense of Eq. (3). Formally, given a data-set $\{(x_i, y_i)\}_{i=1}^{n}$ and separating hyperplane $f(x) = w_i^T x + b$ the shortest distance from separating hyperplane to the closest positive example in the non-separable case is

$$\gamma+ = \min \gamma_i, \forall \gamma_i \in class + 1 \tag{4}$$

The shortest distance from separating hyperplane to the closest negative example is

$$\gamma- = \min \gamma_i, \forall \gamma_i \in class - 1 \tag{5}$$

where γ_i is given by

$$\frac{y_i \left(w_i^T K \langle x_i \cdot x_j \rangle + b_i \right)}{\|w\|} \tag{6}$$

3   Proposed Method

In this Section we describe the proposed Imbalanced SVM-PSO system. The proposed Imbalanced SVM-PSO system can be divided into two parts, in the first part is trained a SVM in order to obtain the most important data points from the skewed data set, the second part describe the way of PSO try to optimize the artificial data points generated. The initial population is obtained by generating artificial instances, each instance is defined by $x_{g_i} = (x_1, x_2, \ldots, x_d)$, where d is de dimensionality of each instance. Each PSO particle is defined by $p_i = \left(x_{g1}^i, x_{g2}^i, \ldots, x_{gm}^i \right)$, where m is the number of artificial instances generated which are obtained by

$$v_k = x_{sv+}^i - x_{sv-}^j, \quad k = 1, \ldots, d \tag{7}$$

where x_{sv+}^i is the ith positive support vectors (sv) of X_{tr}^+, and x_{sv-}^j represent the jth sv nearest neighbors of x_{sv+}^i. Initial vector $v_i = 0$, $k = 1$, ..., d and the algorithm picks one or more random entries out of an array. In the experiments only one is selected. The artificial instance is obtained by

$$x_g = x_{sv+}^i + \varepsilon \cdot v_k \tag{8}$$

which modified only the ith dimension of x_{sv+}^i.

Finally, we denote $P = [p_1, p_2, ..., p_{qm}]^T$ as a $(q \times m)$-dimensional vector, where q is the size of initial population. The problem is determining the artificial instances that improve the performance. The $(q \times m)$-dimensional search space Γ is defined by

$$\Gamma = \prod_{i=1}^{p \times m \times d} \left[\Gamma_{i,\min}, \Gamma_{i,\max} \right] \tag{9}$$

The search space of each individual $x = [x_1, x_2, ..., x_d]^T$ is defined by the minimal distance between SV's with different class, i.e.

$$x_{\min,i} = sv_{sv+}^i \cdot 1 \tag{10}$$

$$x_{\max,i} = \min_{1 \leq k \leq n} D\left(sv_i^+, sv_i^-\right) \cdot \varepsilon \tag{11}$$

When applying a PSO to solve the optimization problem, a swarm of the candidate particles $\{P_i^l\}_{i=1}^s$ are moving in the search space Γ in order to find a solution $\hat{x}$ where s is the size of the swarm and $l \in \{0, 1, ..., L\}$ denotes the lth movement of the swarm. Each particle $p(i)$ has a $(q \times m \times d)$-dimensional velocity $v = [v_1, v_2, ..., v_{qmd}]^T$ to direct its search, and $v \in V$ with the velocity space defined by

$$V = \prod_{i=1}^{p \times m \times d} \left[V_{i,\min}, V_{i,\max} \right] \tag{12}$$

where $V_{i,\max} = \frac{1}{2}\left(\Gamma_{i,\max} - \Gamma_{i,\min} \right)$. To start the PSO, the candidate particles $\{X_i^0\}_{i=1}^s$ are initialized randomly within Γ and the velocity of each candidate particle is initialized to zero, $\left\{ v_j^0 = 0 \right\}_{i=1}^s$. The cognitive information $\mathbf{pb}_i^l$ and the social information $\mathbf{gb}^l$ record the best position visited by the particle i and the best position visited by the entire swarm, respectively, during l movements. The cognitive information $\mathbf{pb}_i^l$ and the social information $\mathbf{gb}^l$ are used to update the velocities according to the velocity of particle P_i wich is changed as follows:

$$V_i(t+1) = wV_i(t) + c_1 r_1(t)(\mathbf{pb}(t) - P_i(t)) + c_2 r_2(t)(\mathbf{gb}(t) - P_i(t)) \tag{13}$$

$$P_i(t+1) = P_i(t) + V_i(t) \tag{14}$$

The choice of parameters w, c_1, $r_1(t)$, $r_2(t)$ and the search space are essential to the performance of the PSO. The input space is expressed by the artificial instances generated. Each particle p_i contains m new artificial instances with dimensionality d. The general process of the algorithm proposed is described in Algorithm 1 and Algorithm 2.

Algorithm 1. General process of the proposed algorithm

Input: Skewed dataset

Output: Final hyperplane H_f

1. Divide the input data set in $X^+, x_i \in X : y = +1, i = 1, \ldots, m$ and $X^-, x_j \in X : y = -1, i = 1, \ldots, n$

2. Obtain training and testing data sets from X^+ and X^- obtain $X^+_{tr}, X^-_{tr}, X^+_{tf}, X^-_{tf}, X^+_{te}, X^-_{te}$ with 70%, 15% and 15% respectively.

3. Train the SVM with the training data set, trainSVM $\left(X^+_{tr}, X^-_{tr} \right)$

4. Obtain Support Vectors x^-_{svi} and x^+_{svi} from H_1

5. Obtain H_f from $\left(X^+_{te}, X^-_{te} \right)$ using the PSO algorithm described in Algorithm 2

Algorithm 2.   PSO algorithm

Input: Support vectors x^-_{svi} and x^+_{svi}, number of iterations ρ

Output: Global best particle

1. Generate an initial swarm of size $(q \times m \times d)$ from x^-_{svi} and x^+_{svi} with eqs. (8) and (9).

2. Set initial velocity of vectors $V_i(i = 1,2,\ldots,q \times m \times d)$ associated with the particles

3. For each position p_i of the particle $P_i(i = 1,\ldots,s)$ which contains artificial data points created from support vectors, train an SVM classifier and compute its fitness function φ.

4. Set the best position of each particle with its initial position, i.e., $\mathbf{pb}_i = P_i(i = 1,\ldots,s)$.

5. Obtain the best global particle **gb** in the swarm.

6. Update the speed of each particle using (13).

7. Update the position of each particle using (14).

8. For each candidate particle p_i, train an SVM classifier and compute its fitness function φ.

9. Update the best position **pb** of each particle if its current position has a smaller fitness function.

10. Return to 5 if the pre-specified stopping condition is not yet satisfied

11. Obtain the best global particle.

Once we have obtained the final hyperplane, it gives us a decision function (Eq. (2)). From this decision function we obtain the performance by testing data set X^+_{te} and X^-_{te}. In the proposed algorithm, the population size and number of iterations or stop criterion serve like a mechanism to avoid over-learning in training data. Our empirical studies have determined that size of populations with 10 particles and iterations minor to 100 works appropriately. The fitness function value φ associated with the ith particle P_i is essentially the objective function for the problem. In our case we use the G-mean measure as fitness function which is given by $G - mean = \sqrt{S_n^f \cdot S_n^t}$, where S_n^t and S_n^f represent the Sensitivity and Specificity respectively.

4 Experimental Results

In this section is showed the improvement achieved in SVM by the proposed algorithm. The usefulness of the proposed methodology is checked by means of comparisons using classical implementations to imbalanced data sets. In this study, we have selected a wide benchmark of 18 data-sets selected from the KEEL data-set repository. Keel Data sets are imbalanced ones (Public available at http://sci2s.ugr.es/keel/datasets. php). Table 1 shows the data sets used in the experiments. In the experiments all data sets were normalized and the 10 fold cross validation method was applied for the measurements.

The approach is implemented in Matlab. In all the experiments presented were used k fold cross validation. The results of the experiments on skewed data sets are reported in Table 1. In this table the first column indicates the data set, and the other columns report the corresponding AUC, G-mean measure. In the Table, σ represents standard deviations of the proposed method.

The experimental results show that the performance of the proposed method is better than classical implementations when imbalance ratio is large. In all data-sets, the

Table 1. Detailed results table for the algorithm proposed

Dataset	Under-sampling		Over-sampling		SMOTE		PM		
	AUC	G	AUC	G	AUC	G	AUC	G	σ
Liver_disorders	0.786	0.737	0.754	0.691	0.837	0.792	0.871	0.856	0.005
glass1	0.765	0.624	0.741	0.673	0.746	0.636	0.802	0.779	0.047
Glass0	0.805	0.761	0.801	0.768	0.765	0.725	0.839	0.817	0.023
vehicle2	0.944	0.939	0.945	0.898	0.953	0.945	0.993	0.971	0.054
vehicle3	0.593	0.675	0.635	0.678	0.658	0.706	0.734	0.715	0.002
ecoli1	0.852	0.871	0.806	0.877	0.886	0.877	0.944	0.936	0.021
ecoli3	0.809	0.787	0.798	0.780	0.741	0.817	0.869	0.836	0.071
new-thyroid1	0.989	0.981	0.983	0.964	0.977	0.959	0.995	0.991	0.014
new-thyroid2	0.978	0.963	0.917	0.973	0.972	0.969	0.986	0.977	0.009
yeast4	0.793	0.781	0.786	0.729	0.791	0.761	0.847	0.824	0.062
yeast6	0.845	0.817	0.841	0.816	0.837	0.812	0.848	0.826	0.085
German	0.753	0.728	0.735	0.641	0.785	0.710	0.806	0.74	0.004
Haberman	0.683	0.632	0.652	0.600	0.689	0.634	0.742	0.683	0.089
Abalone	0.835	0.776	0.821	0.781	0.845	0.783	0.872	0.814	0.001
Letter	0.996	0.952	0.954	0.842	0.998	0.993	0.997	0.954	0.017
pima	0.696	0.725	0.647	0.718	0.714	0.735	0.742	0.785	0.042
glass2	0.607	0.639	0.624	0.652	0.674	0.725	0.738	0.742	0.020
shuttle	0.950	0.871	0.921	0.853	0.950	0.877	0.961	0.891	0.082

proposed method achieves better measures performance than classical competent methods. These results allow us to highlight the goodness of the proposed model to evolve synthetic instances. The improvement provided by proposed methodology, proves that a right management of the PSO algorithm associated with SVM has a positive synergy with the tuning of artificial instances, leading to an improvement in the global behavior of the system.

5 Conclusions

In this paper a novel method that enhances the performance of SVM for skewed data sets was presented. The method reduces the effect of imbalance ratio by exciting and evolving SV and moving separating hyper plane toward majority class. The method is different from other state of the art methods for two reasons, the new instances are added close to optimal separating hyperplane, and they are evolved to improve the classifier's performance. According to the experiments, the proposed method produces the most noticeable results when the imbalance ratio is big. The principal advantage of the proposed method is the performance improvement on imbalanced data-sets by adding artificial examples. However, the first disadvantage is the computational cost. The proposed method can be used only on small data sets. The computational complexity of the proposed method on medium and large data-sets is prohibitive. In comparison with under-sampling, over-sampling and SMOTE the proposed algorithm is computationally very expensive.

Acknowledgements. This research is supported in part by the UAEM Grant No. 3771/2014/CIB.

References

1. Koknar-tezel, S., Latecki, L.J.: Improving SVM classification on imbalanced data sets in distance spaces. In: IEEE International Conference on Data Mining, pp. 259–267 (2009)
2. Cai, Q., He, H., Man, H.: Imbalanced evolving self-organizing learning. Neurocomput. **133**, 258–270 (2014)
3. Batuwita, R., Palade, V.: Class imbalance learning methods for support vector machines. In: He, H., Ma, Y. (eds.) Imbalanced Learning: Foundations Algorithms and Applications. Wiley, New York (2013)
4. Batista, G.E.A.P.A., Prati, R.C., Monard, M.C.: A study of the behavior of several methods for balancing machine learning training data. ACM Explor. Newsl. **6**(1), 20–29 (2004)
5. Chawla, N.V., Lazarevic, A., Hall, L.O., Bowyer, K.W.: SMOTEBoost: improving prediction of the minority class in boosting. In: Lavrač, N., Gamberger, D., Todorovski, L., Blockeel, H. (eds.) PKDD 2003. LNCS (LNAI), vol. 2838, pp. 107–119. Springer, Heidelberg (2003)
6. Wu, G., Chang, E.: KBA: Kernel boundary alignment considering imbalanced data distribution. IEEE Trans. Knowl. Data Eng. **17**(6), 786–795 (2005)

7. Nguyen, H.M., Cooper, E.W., Kamei, K.: Borderline over-sampling for imbalanced data classification. Int. J. Knowl. Eng. Soft Data Paradig. **3**(1), 4–21 (2011)
8. Du, J.-X., Huang, D.S., Wang, X.-F., Gu, X.: Shape recognition based on neural networks trained by differential evolution algorithm. Neurocomput. **70**, 896–903 (2007)
9. Huang, D.S., Du, J.-X.: A constructive hybrid structure optimization methodology for radial basis probabilistic neural networks. IEEE Trans. Neural Netw. **19**(12), 2099–2115 (2008)

Set-Based Many-Objective Optimization Guided by Preferred Regions

Dunwei Gong[1], Fenglin Sun[1(✉)], Jing Sun[2(✉)], and Xiaoyan Sun[1(✉)]

[1] School of Information and Electrical Engineering, China University of Mining
and Technology, Xuzhou 221116, Jiangsu, China
dwgong@vip.163.com, fenglinsun0509@163.com,
xysun78@126.Com
[2] College of Science, Huai Hai Institute of Technology, Lianyungang 222005,
Jiangsu, China
Jing8880@sina.com

Abstract. Set-based evolutionary optimization based on the performance indicators is one of the effective methods to solve many-objective optimization problems; however, previous researches didn't make full use of the preference information of a high-dimensional objective space to guide the evolution of a population. In this study, we propose a set-based many-objective evolutionary optimization algorithm guided by preferred regions. In the mode of set-based evolution, the proposed method dynamically determines a preferred region of the high-dimensional objective space, designs a selection strategy on sets by combining Pareto dominance relation on sets with the above preferred region, and develops a crossover operator on sets guided by the above preferred region to produce a Pareto front with superior performance. The proposed method is applied to four benchmark many-objective optimization problems, and the experimental results empirically demonstrate its effectiveness.

Keywords: Many-objective optimization · Set-based evolution · Achievement scalarizing function · Preferred region · Pareto dominance on sets

1 Introduction

In the real world applications, many optimization problems can be formulated as many-objective optimization problems (MaOPs), such as automobile cabin and flight control system design [1]. Without loss of generality, we consider the following minimization problems in this study:

$$\min f(\mathbf{x}) = (f_1(\mathbf{x}), f_2(\mathbf{x}), \ldots, f_m(\mathbf{x})) \tag{1}$$
$$\text{s.t. } \mathbf{x} \in S \subset R^n$$

For MOPs with two or three objectives, a mass of Pareto dominance based multi-objective evolutionary algorithms (MOEAs) have been proposed, such as NSGA-II [2] and MOPSO [3]. The main difficulties faced by these methods in solving MaOPs are as follows: (1) almost all solutions are non-dominated by each other with the increasing of

© Springer International Publishing Switzerland 2015
D.-S. Huang and K. Han (Eds.): ICIC 2015, Part III, LNAI 9227, pp. 87–93, 2015.
DOI: 10.1007/978-3-319-22053-6_10

the objectives; (2) the required number of the Pareto optimal solutions increases exponentially; (3) the optimal solutions are difficult to display in the Cartesian coordinate, which makes it hard to choose from these solutions for decision makers (DMs).

To overcome the first and the second difficulties, four types of evolutionary methods for MaOPs have been proposed at present: (1) modifying the dominance relation to enhance the selection pressure [4]; (2) omitting the redundant objectives according to the correlations among the objectives to reduce the hardness of dominant comparison [5, 6]; (3) transforming MaOPs into single-objective optimization problems by decomposing or weighting the objectives [7]; (4) taking the performance indicators of a Pareto solution set as the optimized objectives [8–10].

In the last few years, there have been a lot of researches about multi-objective evolutionary optimization integrating preferences [11–13]. The preferences can effectively contract the search area and increase the selection pressure, and it is thus helpful to improve the efficiency of many-objective evolutionary optimization. In view of this, it is very necessary to study many-objective evolutionary optimization integrating preferences. Although the mode of set-based evolution based on performance indicators has certain practicability, it didn't fully consider the guidance of the preferences in the objective space for the evolutionary process of a population. Consequently, in this study, we consider many-objective evolutionary optimization algorithm guided by preferred regions in the mode of set-based evolution.

The remainder of this paper is organized as follows. Section 2 provides a brief introduction for the proposed method. The proposed method is applied to four benchmark MaOPs and compared with two classical methods in Sect. 3. Section 4 summarizes the main results of our work and suggests some new research directions.

2   Set-Based Many-Objective Optimization Guided by Preferred Regions

We chooses the first two performance indicators in [8], namely hypervolume and distribution, to convert problem (1) into bi-objective optimization problem. This method is based on NSGA – II, and the population formed by several set-based evolutionary individuals is evolved under the guidance of the preferred region. It mainly contains the following three core techniques: (1) determining the preferred region; (2) comparing set-based evolutionary individuals based on the preferred region; (3) developing a crossover operator on sets based on the preferred region.

2.1   Determining the Preferred Region in the Mode of Set-Based Evolution

Achievement scalarizing function is a special function based on a reference point, and can reflect preferences [14]. López et al. defined preferred region using the achievement function [15]:

$$N(z^{\text{ref}}, \delta) = \{z | s_\infty(z|z^{\text{ref}}) \le s_\infty^{\min} + \delta\} \tag{2}$$

where $s_\infty(z|z^{\text{ref}}) = \max\limits_{i=1,2,\ldots,m} \left\{ \lambda_i(z_i - z_i^{\text{ref}}) \right\} + \rho \sum\limits_{i=1}^{m} \lambda_i(z_i - z_i^{\text{ref}})$, $\delta = \tau \cdot (s_\infty^{\max}(P(t)) -$

$s_\infty^{\min}(P(t)))$ is a threshold that impacts the size of the preferred region. $\tau \in [0, 1]$ is a pre-specified value, reflecting the coverage rate of the preferred region on the current Pareto front.

On account of the big differences between the set-based evolution employed by this study and the traditional evolution employed by López et al., two problems should be addressed to determine the preferred region in this study; one is determining the parameters of achievement function in mode of set-based evolution, and the other is assigning the value of λ.

The values of three parameters involved in achievement function are determined. The first parameter λ, its value of each component is $1/(z_i^{\max} - z_i^{\min})$, where $z_i^{\max}$ and $z_i^{\min}$ are the maximum and minimum values of the ith objective function in the current population, the second parameter ρ equals to 10^{-6}, and the last is z^{ref}, an evolutionary individual is a set containing multiple solutions in the mode of set-based evolution; therefore, each evolutionary individual must have a minimum value for each objective, implying that each objective has multiple minimum values for the whole population. We can select the best one among these minimum values as the optimal value for each objective of the whole population, in this way, the vector formed by the optimal values of the population on all objectives can be taken as a reference point.

The value of τ changes dynamically along with the generation is proposed in this study. We use the mean of the achievement scalarizing function of a population, denoted as $s_\infty^{avg}(P(t))$, to measure the distance of the population to reference point, and adopt the variance of the achievement function of the population, denoted as $s_\infty^{std}(P(t))$, to reflect the distribution of the evolutionary individuals. The value of τ changes along with the generation can be defined as follows.

$$\tau = e^{-s_\infty^{avg}(P(t)) \cdot s_\infty^{std}(P(t))} \tag{3}$$

It can be seen from formula (3) that when the mean and the variance of the achievement function of an evolutionary population are large, the population is far away from the reference point or is dispersed, which leads to the decrease of the value of τ; on the contrary, the value of τ increases. It thus can reasonably adjust the size of the preferred region. Therefore the preferred region can dynamically change according to the evolutionary characteristic of the population, and full use of the preference information is realized.

2.2 Comparing Set-Based Evolutionary Individuals Based on the Preferred Region

As we all know, once the preferred region is determined, for each set-based evolutionary individual, the number of its solutions located in the preferred region can be calculated. It is easily to understand that the larger the number, the better the performance of the individual. A mean of the achievement function of each evolutionary individual can be used to further compare their qualities when the numbers of the different set-based individuals are equal It is obvious that the smaller the mean of the achievement function, the nearer the most of the solutions in the set-based individual trend to the reference point, suggesting that the quality of the set-based individual is better.

Based on the above analysis, this subsection proposes the following strategy for comparing the set-based evolutionary individuals based on the preferred region. For two set-based individuals,X_i and X_j, Pareto dominance on sets proposed by Gong et al. [8] is first adopted to compare them. If $X_i \succ_{spar} X_j$, X_i is better than X_j. If $X_j \succ_{spar} X_i$, X_j is better than X_i. If $X_i\|_{spar}X_j$, the numbers of the solutions in the preferred region for X_i and X_j are compared; the larger the number, the better the performance of the individual. If the two numbers X_i and X_j are equal, the individual with a smaller mean of the achievement function will win.

2.3   Crossover Operator on Sets Guided by Preferred Regions

We still adopt the traditional crossover mechanism with two steps, crossover between sets and crossover within a set. The modification is that the crossover objectives are no longer paired randomly.

The crossover between sets is suggested as follows. First, for each set-based evolutionary individual, the number of its solutions inside the preferred region is calculated; second, all set-based evolutionary individuals, $X_1, X_2, X_3, \cdots, X_N$, in the population $P(t)$ is reordered according to the numbers, and the reordered population is denoted as $P'(t) = \{X_1', X_2', X_3', \cdots, X_N'\}$; finally, the first individual in $P'(t)$ is paired with the last one, and the second one is paired with the penultimate one, and so on; the solutions in the paired individuals are migrated randomly. Exchanging information between two individuals can share the resource of the population, can avoid loss in the diversity of a single set-based evolutionary individual, and also can ensure the population search in a larger scope so as to maintain the diversity of the population. The crossover within a set is developed. First, the solutions in the individual are grouped into two classes, i.e., one is located inside the preferred region and the other is located outside the preferred region. Second, two solutions are randomly chosen from class 1 and class 2, respectively, and are paired to cross over. Finally, the rest solutions in the individual are randomly paired to cross over. In this way, the newly generated set-based evolutionary individuals after crossover approach the preferred region.

3   Experiments

The proposed method was applied to solve benchmark optimization problems, to evaluate its performance. The performance of the proposed method, called P-SEA, was evaluated by comparing it with other two methods. The first one, called SetEA, was employing simulated binary crossover and polynomial mutation operators to solutions belonging to an individual, and adopting Pareto dominance on sets to compare set-based individuals. The second one was proposed in [10], called STD, whose unique objective is hypervolume and in which no preference was incorporated into the optimization. The hypervolume (H indicator, for short), distribution (D indicator, for short) and the mean of the achievement scalarizing function which measures the distance of the final Pareto front to the reference point(A indicator, for short) were used for comparing the above methods.

Table 1 reports the values of H, D and A indicators of different methods, where the boldface data are the best among all the data, and the underlined data indicate that P-SEA is superior to SetEA. Table 2 lists their non-parametric test results on H, D and A indicators, where '+' and '−' represent the proposed method is significantly superior and inferior to the other two methods, respectively, and '0' indicates no significant difference between them.

Table 1. Performance of final Pareto front obtained by different methods

		STD			Set-EA			P-SEA		
		H	D	A	H	D	A	H	D	A
DTLZ1	5	**0.9966**	0.0692	0.7105	0.9689	0.0608	0.6878	0.9883	**0.0553**	**0.6251**
	10	**0.9957**	0.0798	0.7263	0.9553	0.0777	0.7262	0.9899	**0.0652**	**0.6861**
	20	**0.9971**	0.0790	0.7005	0.9652	0.0586	0.7452	0.9855	**0.0580**	**0.6999**
DTLZ2	5	**0.9671**	0.0501	0.8573	0.8853	0.0521	**0.8403**	0.9035	**0.0415**	0.8445
	10	**0.9561**	0.0639	0.8607	0.8648	0.0661	0.8498	0.8808	**0.0482**	**0.8481**
	20	**0.9539**	0.0588	0.8795	0.8698	0.0580	0.8637	0.8862	**0.0536**	**0.8499**
DTLZ3	5	**0.9829**	0.1209	0.6584	0.8875	0.1206	0.7351	0.9478	**0.0949**	**0.6293**
	10	**0.9847**	0.1434	**0.6662**	0.8815	0.1194	0.7498	0.9507	**0.0910**	0.6736
	20	**0.9815**	0.1353	**0.6783**	0.8987	0.1021	0.7324	0.9569	**0.0966**	0.6805
DTLZ7	5	**0.7141**	0.0983	2.9115	0.4897	0.0998	3.1955	0.5597	**0.0788**	2.8747
	10	**0.6099**	0.3840	2.5669	0.4553	0.2901	1.5898	0.5258	**0.1585**	**1.5135**
	20	**0.5847**	0.2884	1.8330	0.4102	0.2746	1.6703	0.5326	**0.2507**	**1.6370**

Table 2. Non-parametric test results on H, D and A indicator

		DTLZ1			DTLZ2			DTLZ3			DTLZ7		
		H	D	A	H	D	A	H	D	A	H	D	A
STD	5	0	+	+	−	+	0	−	+	+	−	+	0
	10	0	+	+	−	+	+	−	+	0	−	+	+
	20	−	+	0	−	0	+	−	+	0	−	0	+
SetEA	5	+	+	+	+	+	0	+	+	+	+	+	+
	10	+	0	+	0	+	0	+	+	+	+	+	0
	20	+	0	+	0	0	+	+	0	+	+	0	0

Tables 1 and 2 show that (1) except for 5- and 10-objective DTLZ1, the value of H indicator has no significant difference between P-SEA and STD; for the other optimization problems, P-SEA is significantly inferior to STD on H indicator. The reason is that P-SEA considers both of the hypervolume and distribution indicators, and STD only consider the hypervolume indicator when comparing individuals based on performance indicators; (2) except for 10- and 20-objective, there is no significant difference between P-SEA and SetEA on H indicator; for the other optimization problems, P-SEA is significantly superior to SetEA on H indicator, which indicates that both P-SEA and SetEA consider the hypervolume and distribution indicators at the same time, nevertheless, P-SEA can obtain optimal solution set with better convergence by using the preference information to guide the evolution of the population; (3) except for 20-objective DTLZ2 and DTLZ7, there is no significant difference between P-SEA and STD on D indicator; for the other optimization problems, P-SEA is significantly superior to STD on D indicator, indicating that the proposed set-based many-objective optimization guided by the preferred region can improve the distribution of the final Pareto front; (4) except for 5-objective DTLZ2 and DTLZ7, 10-objective DTLZ3 and 20-objective DTLZ1and DTLZ3, there is no significant difference between P-SEA and STD on A indicator; for the other optimization problems, P-SEA is significantly superior to STD on A indicator; except for 5- and 10-objective DTLZ2 and 10- and 20-objective DTLZ7, there is no significant difference between P-SEA and SetEA on A indicator; for the other optimization problems, P-SEA is significantly superior to SetEA on A indicator, suggesting that the preferred region guided evolutionary strategy on sets can get the optimal solutions that is closer to the reference point. In conclusion, the proposed method can produce a Pareto front with more superior performance for many optimization problems.

The above experimental results and analyses indicate that the proposed method can efficiently guide the evolution of the population, eventually get a Pareto optimal set with better convergence and distribution by the utilization of the preference information.

4   Conclusions

In this method, the preferred region was first depicted based on the achievement scalarizing function and its size could change dynamically; then the qualities of set-based evolutionary individuals were compared based on Pareto dominance on sets and the preferred region, furthermore, superior individuals were selected based on the above comparison strategy; finally, the individuals used to perform the crossover operators between sets and within a set were selected based on the preferred region so as to balance the diversity and the convergence of the evolutionary population. The experimental results show that the proposed method can improve the convergence and the distribution of the final Pareto solution set. Integrating the preference information into the mutation on sets to drive the evolution of the population and determining the preferred region dynamically using different methods are our future research topics.

Acknowledgements. This work was jointly supported by National Natural Science Foundation of China with grant No. 61375067 and 61403155, National Basic Research Program of China (973 Program) with grant No. 2014CB046306-2, and Natural Science Foundation of Jiangsu Province with grant No. BK2012566

References

1. Deb, K., Jain, H.: An Improved NSGA-II Procedure for Many-objective Optimization, Part I: Solving Problems with Box Constraints. KanGAL Report (2012)
2. Deb, K., Prata, P.A., Agarwal, S.: A fast and elitist multi-objective genetic algorithm: NSGA-II. IEEE Trans. Evol. Comput. 2(6), 182–197 (2002)
3. Reyes-Sierra, M., Coello, C.C.: Multi-objective particle swarm optimizers: a survey of the state-of-the-art. Int. J. Comput. Intell. Res. 2(3), 287–308 (2006)
4. Sato, H., Aguirre, H.E., Tanaka, K.: Controlling dominance area of solutions and its impact on the performance of MOEAs. In: Obayashi, S., Deb, K., Poloni, C., Hiroyasu, T., Murata, T. (eds.) EMO 2007. LNCS, vol. 4403, pp. 5–20. Springer, Heidelberg (2007)
5. Deb, K., Saxena, D.K.: On Finding Pareto-optimal Solutions through Dimensionality Reduction for Certain Large-dimensional Multi-objective Optimization Problems. KanGAL Report (2005)
6. Deb, K., Saxena, D.: Searching for pareto-optimal solutions through dimensionality reduction for certain large-dimensional multi-objective optimization problems. In: Proceedings of the World Congress on Computational Intelligence (2006)
7. Zhang, Q.F., Li, H.: MOEA/D: a multi-objective evolutionary algorithm based on decomposition. IEEE Trans. Evol. Comput. 11(6), 712–731 (2007)
8. Gong, D., Ji, X.: Solving many-objective optimization problems using set-based evolutionary algorithm. Chin. J. Electron. 42(1), 77–83 (2014)
9. Zitzler, E., Thiele, L., Bader, J.: On set-based multi-objective optimization. IEEE Trans. Evol. Comput. 14(1), 58–79 (2010)
10. Bader, J., Brockhoff, D., Welten, S., Zitzler, E.: On using populations of sets in multiobjective optimization. In: Ehrgott, M., Fonseca, C.M., Gandibleux, X., Hao, J.-K., Sevaux, M. (eds.) EMO 2009. LNCS, vol. 5467, pp. 140–154. Springer, Heidelberg (2009)
11. Gong, D., Ji, X., Sun, J.: Interactive evolutionary algorithms with decision-maker's preferences for solving interval multi-objective optimization problems. Neurocomputing 137, 241–251 (2014)
12. Yang, D., Jiao, L., Gong, M., Yu, H.: Clone selection algorithm to solve preference multi-objective optimization. J. Softw. 21(1), 14–33 (2010)
13. Liu, R., Wang, X., Liu, J.: A preference multi-objective optimization based on adaptive rank clone and differential evolution. Nat. Comput. 12(1), 109–132 (2013)
14. Ehrgott, M.: Multicriteria optimization. Springer Science and Business Media (2006)
15. López-Jaimes, A., CoelloCoello, C.A.: Including preferences into a multi-objective evolutionary algorithm to deal with many-objective engineering optimization problems. Inf. Sci. 277, 1–20 (2014)

An Improved Incremental Error Minimized Extreme Learning Machine for Regression Problem Based on Particle Swarm Optimization

Fei Han, Min-Ru Zhao$^{(\boxtimes)}$, and Jian-Ming Zhang

School of Computer Science and Communication Engineering,
Jiangsu University, Zhenjiang, Jiangsu, China
{hanfei,zhjm}@ujs.edu.cn, zhaominru_2007@sina.com

Abstract. Error minimized extreme learning machine (EM-ELM) is a simple and efficient approach to determine the number of hidden nodes. However, EM-ELM lays much emphasis on the convergence accuracy, which may obtain a single-hidden-layer feedforward neural network (SLFN) with good convergence performance but bad condition. In this paper, an effective approach based on error minimized ELM and particle swarm optimization (PSO) is proposed to automatically determine the structure of SLFN for regression problem. In the new method, the hidden node optimized by PSO is added to the SLFN one by one. Experimental results verify that the proposed algorithm achieves better generalization performance with better condition than other constructive ELM.

Keywords: Extreme learning machine · Particle swarm optimization · Generalization performance · Condition value

1 Introduction

Because of randomly choosing the input weights and the hidden biases, the traditional ELM inevitably requires more hidden nodes and the network structure becomes more complex. Thus, a class of constructive ELMs were proposed to design the network architecture [1–4]. In incremental ELM (I-ELM) [1], the hidden nodes were added one by one, while some nodes playing a very minor role in the network output were added. In enhanced I-ELM (EI-ELM) [2] several hidden nodes were randomly generated. An error minimized ELM (EM-ELM) [3], added random hidden nodes with varying size to the existing network, and the output weights updated incrementally during the network growth, which significantly reduced the computational complexity. In dynamic ELM (D-ELM) [4], the hidden nodes can be recruited dynamically according to their significance to network performance.

The above constructive ELMs laid more emphasis on the convergence accuracy of the ELM, while numerical stability is generally ignored [5]. An ill-conditioned system may have its solutions very sensitive to perturbation in data.

© Springer International Publishing Switzerland 2015
D.-S. Huang and K. Han (Eds.): ICIC 2015, Part III, LNAI 9227, pp. 94–100, 2015.
DOI: 10.1007/978-3-319-22053-6_11

Particle swarm optimization (PSO) [6], as an effective evolutionary computation technique, is widely used as a global searching method, because it has good search ability, fast convergence and no complicated evolutionary operators. In this paper, a new method combining incremental EM-ELM and PSO called as IPSO-EM-ELM is proposed. In the proposed algorithm, not only the root mean squared error (RMSE) of training data but also the condition value of the hidden output matrix of the SLFN is considered to select the optimal hidden nodes, which may establish more compact and well-conditioned SLFN.

2   EM-ELM

Assume that an SLFN with d inputs, L hidden nodes, $G(\mathbf{a}_i, b_i, \mathbf{x})$ denotes the output of the ith hidden node with the randomly generated input weights and hidden bias $(\mathbf{a}_i, b_i) \in R^d \times R$. $\beta_i \in R$ is the weight connecting the i-th hidden node and the output node. Given a set of training data $\{(\mathbf{x}_i, \mathbf{t}_i)\}_{i=1}^{N} \subset R^d \times R$, the output of the network equals to the target means

$$\mathbf{H}\beta = \mathbf{T} \tag{1}$$

where

$$\mathbf{H} = \begin{pmatrix} G(\mathbf{a}_1, b_1, \mathbf{x}_1) & \cdots & G(\mathbf{a}_L, b_L, \mathbf{x}_1) \\ G(\mathbf{a}_1, b_1, \mathbf{x}_2) & \cdots & G(\mathbf{a}_L, b_L, \mathbf{x}_2) \\ \vdots & \vdots & \vdots \\ G(\mathbf{a}_1, b_1, \mathbf{x}_N) & \cdots & G(\mathbf{a}_L, b_L, \mathbf{x}_N) \end{pmatrix}_{N \times L}, \beta = \begin{pmatrix} \beta_1 \\ \beta_2 \\ \vdots \\ \beta_L \end{pmatrix}_{L \times 1}, \mathbf{T} = \begin{pmatrix} t_1 \\ t_2 \\ \vdots \\ t_N \end{pmatrix}_{N \times 1}.$$

The output weights vector β, which can be calculated as follows:

$$\hat{\beta} = \mathbf{H}^+ \mathbf{T} \tag{2}$$

where $\mathbf{H}^+$ is the Moore-Penrose generalized inverse of $\mathbf{H}$.

In EM-ELM, given target error $\varepsilon > 0$, an SLFN $f_{n_0}(\mathbf{x}) = \sum_{i=1}^{n_0} \beta_i G(\mathbf{a}_i, b_i, \mathbf{x})$ with n_0 hidden nodes is first initialized. $\mathbf{H}_1$ denotes the hidden-layer output matrix of this network. Then, the output weight matrix can be calculated by $\beta_1 = \mathbf{H}_1^+ \mathbf{T}$.

If the network output error $E(\mathbf{H}_1) = \|\mathbf{H}_1\beta_1 - \mathbf{T}\| > \varepsilon$, then new hidden nodes $\delta n_0 = n_1 - n_0$ are added to the existing SLFN, and the new hidden-layer output matrix become $\mathbf{H}_2 = [\mathbf{H}_1, \delta\mathbf{H}_1]$. EM-ELM proposes a fast output weights updating method to calculate $\mathbf{H}_2^+$ as follows:

$$\mathbf{H}_2^+ = \begin{bmatrix} \mathbf{U}_1 \\ \mathbf{D}_1 \end{bmatrix} \tag{3}$$

where $\mathbf{D}_1 = ((\mathbf{I} - \mathbf{H}_1\mathbf{H}_1^+)\delta\mathbf{H}_1^+)^+$, $\mathbf{U}_1 = \mathbf{H}_1^+(\mathbf{I} - \delta\mathbf{H}_1\mathbf{D}_1)$.

Similarly, the output weights are updated incrementally as follows:

$$\beta_{k+1} = \mathbf{H}_{k+1}{}^{+}\mathbf{T} = \begin{bmatrix} \mathbf{U}_k \\ \mathbf{D}_k \end{bmatrix}\mathbf{T} \tag{4}$$

where $\mathbf{D}_k = ((\mathbf{I} - \mathbf{H}_k\mathbf{H}_k{}^{+})\delta\mathbf{H}_k{}^{+})^{+}$, $\mathbf{U}_k = \mathbf{H}_k{}^{+}(\mathbf{I} - \delta\mathbf{H}_k\mathbf{D}_k)$,

$$\delta H_k = \begin{bmatrix} G(\mathbf{a}_{L_{k-1}+1}, b_{L_{k-1}+1}, \mathbf{X}_1) & \cdots & G(\mathbf{a}_{L_k}, b_{L_k}, \mathbf{X}_1) \\ \vdots & \cdots & \vdots \\ G(\mathbf{a}_{L_{k-1}+1}, b_{L_{k-1}+1}, \mathbf{X}_N) & \cdots & G(\mathbf{a}_{L_k}, b_{L_k}, \mathbf{X}_N) \end{bmatrix}_{N \times \delta L_{k-1}}.$$

3 The Proposed Method

In this study, PSO is used to select the randomly generated nodes based on the EM-ELM. In addition to the training RMSE of the system being chosen as the fitness function to gain a high accuracy, the condition value of the hidden output matrix is also considered in the optimization process to ensure that the constructed SLFN is well-conditioned. The impoved method is named IPSO-EM-ELM, and the detailed steps are listed as follows.

Given a set of training data $\{(\mathbf{x}_i, t_i)\}_{i=1}^{N} \subset R^d \times R$, the maximum number of the hidden nodes $L_{\max}$, the expected training accuracy ε, and $L = 0$.

Step 1: Increase the number of hidden nodes $L = L + 1$.

Step 2: Use PSO to select the new hidden node.

Substep 2.1: Randomly generate the swarm within the range of $[-1, 1]$. Each particle is composed of input weights from the new added hidden node to the input nodes and the hidden bias for the new added hidden node: $P_{Li} = (\mathbf{a}_{Li}, b_{Li})$.

Substep 2.2: According to Eq. (2), calculate the output weights β_{Li}, $(1 \le i \le n, L = 1)$ as $L = 1$. According to Eq. (4), calculate the output weights β_{Li}, $(1 \le i \le n, L > 1)$ as $L > 1$.

Substep 2.3: The fitness of each particle is calculated by Eq. (5). With the fitness of all particles, the p_{ib} of each particle and the p_g of the swarm are computed by Eqs. (6) and (7), respectively.

$$f() = \|\mathbf{H}_{Li}\beta_{Li} - \mathbf{T}\| \tag{5}$$

$$p_{ib} = \begin{cases} p_i & (f(p_{ib}) - f(p_i) > \eta f(p_{ib})) \text{ or } (f(p_{ib}) - f(p_i) < \eta f(p_{ib})) \\ & \text{and } (K_i < K_{ib})) \\ p_{ib} & \text{else} \end{cases} \tag{6}$$

$$p_g = \begin{cases} p_i & (f(p_g) - f(p_i) > \eta f(p_g)) \text{ or } (f(p_g) - f(p_i) < \eta f(p_g)) \\ & \text{and } (K_i < K_g)) \\ p_g & \text{else} \end{cases} \tag{7}$$

where $f(p_i), f(p_{ib})$ and $f(p_g)$ are the corresponding fitness values for the i-th particle, the best position of the i-th particle and global best position of all particle, respectively; $\eta > 0$ is the tolerance rate; K_i, K_{ib} and K_g are the condition values of the hidden output matrix of the SLFNs related to the i-th particle, the best position of the i-th particle and global best particle, respectively. According to the literature [5], the 2-norm condition value of the matrix $\mathbf{H}$ can be computed as (8).

$$K_2(\mathbf{H}) = \sqrt{\frac{\lambda_{\max}(\mathbf{H}^T\mathbf{H})}{\lambda_{\min}(\mathbf{H}^T\mathbf{H})}} \tag{8}$$

where $\lambda_{\max}(\mathbf{H}^T\mathbf{H})$ and $\lambda_{\min}(\mathbf{H}^T\mathbf{H})$ are the largest and the smaller eigenvalues of the $\mathbf{H}^T\mathbf{H}$.

Substep 2.4: Each particle updates its position according to PSO algorithm. All components in the particle limited within the range of $[-1, 1]$.

Substep 2.5: Repeat the above substeps until the goal is met or the maximum epoch is completed. Then the optimal weights and hidden bias $(\mathbf{a}_{L^*}, b_{L^*})$ are obtained from the global best of the swarm.

Step 3: According to Eq. (2), calculate the output weight of the SLFN after adding the new hidden node $(\mathbf{a}_{L^*}, b_{L^*})$ as $L = 1$. According to Eq. (4), calculate the output weight of the network after adding the new hidden node $(\mathbf{a}_{L^*}, b_{L^*})$ as $L > 1$.

Step 4: Calculate the output error $E(\mathbf{H}_L) = \|\mathbf{H}_L\mathbf{H}_L^+\mathbf{T} - \mathbf{T}\|$.

Step 5: Go to Step 1 until $L > L_{\max}$ or $E(\mathbf{H}_L) < \varepsilon$, and the constructed ELM is obtained.

4 Experimental Results

To verify the effectiveness and efficiency of the proposed algorithm, IPSO-EM-ELM is compared with other constructive ELMs including I-ELM [1], EI-ELM [2], EM-ELM [3] and D-ELM [4] on five benchmark regression problems from UCI database [7]. The specifications of the problems are shown in Table 1.

Table 1. Specification of five benchmark datasets

Datasets	Attributes	Case	Training data	Testing data
Abalone	8	4177	2000	2177
Boston housing	13	506	250	256
California housing	8	20640	8000	12640
Census (House8L)	8	22784	10000	12784
Delta ailerons	5	7129	3000	4129

Table 2. Mean performance of the five ELMs on the five benchmark regression data

Algorithm	Data	Abalone	Boston housing	California housing	Census (House8L)	Deltaailerons
	Stop RMSE (ε)	0.09	0.12	0.16	0.09	0.05
I-ELM	Test RMSE ± Dev	0.0988 ± 0.0020	0.1450 ± 0.0055	0.1680 ± 0.0055	0.0914 ± 0.0021	0.0538 ± 0.0056
	Train times(s)	0.0546	0.0125	0.1538	0.716	0.0546
EI-ELM	Test RMSE ± Dev	0.0973 ± 0.0006	0.1451 ± 0.0049	0.1606 ± 0.0004	0.0875 ± 0.0002	0.0513 ± 0.0012
	Train times(s)	0.0499	0.0125	0.3752	0.4664	0.1084
EM-ELM	Test RMSE ± Dev	0.0938 ± 0.0033	0.1502 ± 0.0119	0.1572 ± 0.0037	0.0859 ± 0.0020	0.0465 ± 0.0036
	Train times(s)	0.0218	0.0016	0.0109	0.0086	0.0023
D-ELM	Test RMSE ± Dev	0.0940 ± 0.0031	0.1492 ± 0.0095	0.1556 ± 0.0032	0.0864 ± 0.0018	0.0468 ± 0.0023
	Train times(s)	0.0156	0.0585	0.0507	0.0874	0.0133
IPSO-EM-ELM	Test RMSE ± Dev	**0.0919** ± 0.0037	**0.1444** ± 0.0099	**.1549** ± 0.0025	**.0857** ± 0.0022	**0.0432** ± 0.0019
	Train times(s)	0.6162	0.3549	1.4095	3.1465	0.4672

In the experiments, the inputs and outputs have been normalized in the range of [–1, 1] and [0, 1], respectively. The sigmoid function $G(\mathbf{a}, b, \mathbf{x}) = 1/(1 + \exp(-(\mathbf{a} \cdot \mathbf{x} + b)))$ is selected as the activation function of the hidden layer in all algorithms. The population size and maximum iteration number are set as 30 and 10, respectively; the tolerance rate is selected as 0.02. According to [8], the initial inertial weight w_{ini} and the final inertial weight w_{end}, are selected as 0.9 and 0.4 respectively; the acceleration constants c_1 and c_2 are both selected as 1.6. All the results shown in this paper are the mean values of 20 trials.

Table 2 shows the average test RMSE and train time of I-ELM, EI-ELM, EM-ELM, D-ELM, IPSO-EM-ELM. The IPSO-EM-ELM obtains the least test RMSE of all ELMs, which indicates the proposed method has the better generalization performance than the other four ELMs. However, the proposed method requires most time to train SLFN than other ELMs, since the IPSO-EM-ELM use PSO to select the optimal hidden nodes.

Figure 1 shows the corresponding condition value of the hidden output matrix in the five ELMs on the five datasets. From Fig. 1(a), the condition values in the EM-ELM, D-ELM, IPSO-EM-ELM are much smaller than I-ELM and EI-ELM. From Fig. 1(b), the condition value in IPSO-EM-ELM is smaller than that in EM-ELM and D-ELM. Moreover, the condition curve for IPSO-EM-ELM is more stable than the other four ELMs. Therefore, the SLFN established by the IPSO-EM-ELM is better-conditioned than the ones constructed by the other four ELMs.

Figure 2 shows the convergence curves of three constructive ELMs including EM-ELM, D-ELM, and IPSO-EM-ELM. Obviously, the IPSO-EM-ELM requires much less hidden nodes with less train RMSE than the EM-ELM and D-ELM.

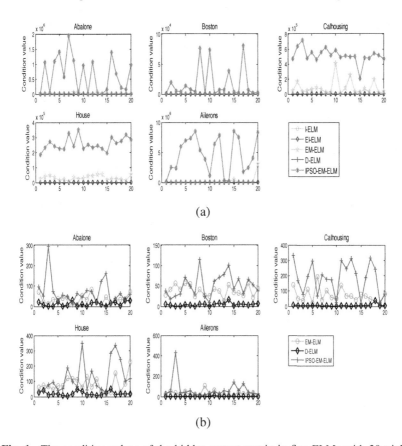

(a)

(b)

Fig. 1. The condition values of the hidden output matrix in five ELMs with 20 trials

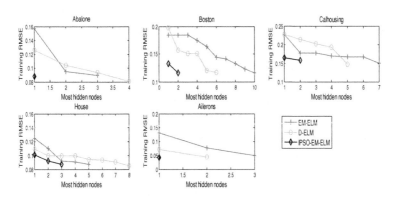

Fig. 2. The convergence curves of three constructive ELMs on the five data

5 Conclusions

In this paper, an incremental error minimization extreme learning machine based on particle swarm optimization (IPSO-EM-ELM) was proposed. Different from traditional incremental error minimization extreme learning machine, the proposed method not only considers the training RMSE but also the condition performance of the SLFN, so the SLFN established by the proposed method is well-conditioned. Because of using PSO to select the weight parameters, IPSO-EM-ELM inevitably spends more time than other ELMs. Future research work will include how to solve this problem and apply the IPSO-EM-ELM to complex classification problems.

Acknowledgements. This work was supported by the National Science Foundation of China (No. 61271385).

References

1. Huang, G.B., Chen, L., Siew, C.K.: Universal approximation using incremental constructive feedforward networks with random hidden nodes. IEEE Trans. Neural Netw. **17**(4), 879–892 (2006)
2. Huang, G.B., Chen, L.: Enhanced random search based incremental extreme learning machine. Neurocomputing **71**(16–18), 3460–3468 (2008)
3. Feng, G., Huang, G.B., Lin, Q., Gay, R.: Error minimized extreme learning machine with growth of hidden nodes and incremental learning. IEEE Trans. Neural Netw. **20**(8), 1352–1357 (2009)
4. Zhang, R., Lan, Y., Huang, G.B., Xu, Z.B., Soh, Y.C.: Dynamic extreme learning machine and its approximation capability. IEEE Trans. Syst. Man and Cybern.-Part B **43**, 2054–2065 (2013)
5. Zhao, G.P., Shen, Z.Q., Miao, C.Y., Man, Z.H.: On improving the conditioning of extreme learning machine: a linear case. In: The 7th International Conference on Information, Communications and Signal Processing, pp. 1−5. Macau, China (2009)
6. Kennedy, J., Eberhart, R.: Particle swarm optimization. IEEE International Conference on Neural Networks 4, 1942–1948 (1995). Perth, Australia
7. Black, C.L., Merz, C.J.: UCI Repository of Machine Learning Databases. Dept. Inf. Comput. Sci., Univ. California, Irvine, http://www.ics.uci.edu/∼mlearn/mlrepository.html
8. Shi, Y., Eberhart, R.C.: A modified particle swarm optimizer. In: IEEE World Conference on Computation Intelligence, pp. 69–73. Anchorage, AK (1998)

A Systematic Comparison and Evaluation of Supervised Machine Learning Classifiers Using Headache Dataset

Ahmed J. Aljaaf[1(✉)], Dhiya Al-Jumeily[1], Abir J. Hussain[1], Paul Fergus[1], Mohammed Al-Jumaily[2], and Naeem Radi[3]

[1] Applied Computing Research Group, Liverpool John Moores University, Byrom Street, Liverpool, L3 3AF, UK
A.J.Kaky@2013.ljmu.ac.uk,
{d.aljumeily,a.hussain,p.fergus}@ljmu.ac.uk
[2] Department of Neurosurgery, Dr. Sulaiman Al Habib Hospital, Dubai Healthcare City, UAE
[3] Al Khawarizmi International College, Abu Dhabi, UAE
n.radi@khawarizmi.com

Abstract. The massive growth of data volume within the healthcare sector pushes the current classical systems that were adapted to the limit. Recent studies have focused on the use of machine learning methods to develop healthcare systems to extract knowledge from data by means of analysing, mining, pattern recognition, classification and prediction. Our research study reviews and examines different supervised machine learning classifiers using headache dataset. Different statistical measures have been used to evaluate the performance of seven well-known classifiers. The experimental study indicated that Decision Tree classifier achieved a better overall performance, followed by Artificial Neural Network, Support Vector Machine and k-Nearest Neighbor. This would determine the most suitable classifier for developing a particular classification system that is capable of identifying primary headache disorders.

Keywords: Machine learning · Performance analysis · Primary headache

1 Introduction

The launch of electronic health records (EHRs) has provided a robust platform for consistent data collection. On the other hand, a massive quantity of data is accumulated nowadays. The healthcare sector has never faced such a huge amount of electronic data before [1]. Therefore, extracting information and knowledge from that huge amount of data is considered to be a challenge. The healthcare sector can benefit from the applications of machine learning and data mining techniques to improve the use of knowledge for supporting decision making, and therefore improving the quality of healthcare service being delivered to the patient [2, 3]. In recent years, there has been a dramatic increase in the use of machine learning techniques to analyse, predict and classify medical data and images.

© Springer International Publishing Switzerland 2015
D.-S. Huang and K. Han (Eds.): ICIC 2015, Part III, LNAI 9227, pp. 101–108, 2015.
DOI: 10.1007/978-3-319-22053-6_12

The classification is a machine learning approach, which aims to separate a given dataset into two or more classes based on attributes measured in each instance of that dataset. However, within the healthcare sector, the classification is used for medical data analysis and detection or diagnosis a particular disease [2, 4]. Recently, several studies [10–13] have employed machine learning classifiers within the healthcare systems with the purpose of enhancing the accuracy and effectiveness of disease diagnosis and consequently preventing or at least minimising the medical errors. Many others [5–9] have focused on evaluating the overall performance of different classifiers within a particular classification domain. In this paper, we aimed to (a) review many related studies that have focused on the comparison and performance evaluation of various classifiers, and (b) examine seven well-known classifiers using headache dataset, with the aim of identifying the most appropriate classifier for developing a particular classification system that is capable of identifying primary headache disorders.

2 Related Works

This section reviews a range of recent studies that aims to analyse, compare and evaluate the overall performance of supervised machine learning classifiers. These studies have used a variety of data sets and proposed several classification enhancement methods. However, they followed approximately the same evaluation methods by measuring the sensitivity, specificity and classification accuracy.

The authors in [7] have evaluated the performance of different classifiers. The aim was to identify the most appropriate classifier to be used in the classification of eye gaze direction. K-Nearest Neighbor (kNN), Artificial Neural Network (ANN), Linear Support Vector Machine (SVM) and Decision Tree (DT) classifiers have been examined. Viola-Jones algorithm was used for face detection. Then Circular Hough Transform for eye detection and features extraction. Finally, the extracted features were employed in eye gaze direction classification by the mentioned classifiers. 92.1 % was the highest classification accuracy and obtained by linear LSVM classifier.

Thepade and Kalbhor [8] examined twelve different classifiers for content based image classification, which is the process of clustering images under a comparatively similar group. To improve the classification accuracy, three common transformation methods were applied for features extraction, which were Discrete Cosine Transform (DCT), Discrete Sine Transform (DST) and Walsh Transform (WT). The experimental outcomes have confirmed that the Simple Logistic classifier with WT can obtain better classification accuracy for image classification [8]. However, the authors in [9], have used a different method to improve the performance of content based image classification and retrieval for biological image. They adopted two step approaches for features selection. The first was partitioning the canonical feature set into four distinct feature sets, while the second was using the Principal Component Analysis (PCA) and Fisher Score for features selection. Finally, Bayesian, SVM and Wndchrm Classifiers were applied for the classification of biological image dataset. Comparison results showed that SVM classifier with Gaussian Kernel yielded better classification accuracy.

3   Experimental Study and Results

3.1   The Used Dataset

A sample dataset was synthetically generated based on the international classification of headache disorders (ICHD-2) [14]. This study is targeting headache dataset due to the following fact; The diagnosis of headache considered to be a challenge, as it can be caused by more than hundred diseases, with a variety of forms, frequency and severity from mild that disappear easily to severe and repeated disabling headache that can be debilitating in some individuals [15, 16]. Therefore, the aim is to evaluate different machine learning classifiers using headache dataset, and find the most suitable classifier that can predict or diagnose primary headache disorders correctly. Where primary headache disorders are the most common in the community, they are not related to any underlying medical condition and the headache itself is the disorder [14].

The data set consists of 900 instances and 8 attributes 66.67 % of them are for patients diagnosed with primary headache disorders (e.g. migraine with and without aura, tension type headache and cluster headache), while the remaining 33.33 % for normal individuals suffering with common headaches. The dataset would be represented as a set $S = \{<number\ of\ attacks_1,\ pain\ location_1,\ pain\ quality_1,\ pain\ intensity_1,\ headache\ exacerbated\ by_1,\ aura\ symptoms_1,\ gender_1>,...,<number\ of\ attacks_n,\ pain\ location_n,\ pain\ quality_n,\ pain\ intensity_n,\ headache\ exacerbated\ by_n,\ aura\ symptoms_n,\ gender_n>\}$. The size of the set S is n, which is the total number of records.

Table 1. Attributes characteristics

No.	Attributes
1	Number of attacks (per day)
2	Pain location (0: unilateral pain and 1: bilateral)
3	Pain quality (0: pulsing headache and 1: non-pulsing headache)
4	Pain intensity (1: mild, 2: moderate, 3: severe and 4: very severe)
5	Headache exacerbated by (1: non, 2: physical activity and 3: laying down)
6	Aura symptoms (1: non, 2: visual, 3: sensory and 4: speech)
7	Gender (0: female and 1: male)
8	Class (yes for patient with primary headache and no for normal individual)

As shown in Table 1, the dataset attributes (features) includes; the number of attacks, pain location, quality and intensity, actions that exacerbate headaches, signs and symptom that accompany the headache, gender and finally the class attribute. Predefined class label is represented as $K = \{yes,\ no\}$, which indicate a patient with primary headache disorder and normal individual suffering from headache, respectively.

3.2 Validation Methods

In order to examine the overall performance of classifiers, two validation methods were used. The first method is the holdout method, in which, the data set was divided into 80 % for training and 20 % for testing. The training set consist of 720 instances with approximately 47 % males and 53 % females, while the testing set comprises of roughly 48 % males and 52 % females with a total of 180 instances. Partitioning data into training and testing sets is to validate the accuracy of different classifiers.

The 10-fold cross-validation technique was the second method used to measure the classifiers' performance. This method is usually utilised to maximise the use of the data set. The data set is randomly partitioned into 10 equal subsets. Each contains approximately the same proportion of healthy individuals and patients suffering with a headache. Of the 10 subsets, a single subset is retained as a testing data, and the remaining 9 subsets as the training data. The cross-validation process is then repeated 10 times, until each one of the 10 subsets was used exactly once as a testing set. The results can then be averaged to estimate the classifier's performance. The advantage of this method is that all subsets are used for both training and testing, and each subset is used for testing exactly once [17, 18].

3.3 Classifiers Assessment

In this section, we investigate the performance of different classifiers, particularly to find the sensitivity, specificity, mean absolute error and accuracy of various classification methods. Seven well-known classifiers were examined with the same data set. The classifiers considered in this section include Naive-Bayes (NB), Artificial Neural Network (ANN - MLP), Decision Tree (DT - J48), ZeroR (ZR), Support Vector Machine Linear (SVML), k-Nearest Neighbor (kNN), and Logistic Regression (LOGR).

The performance of supervised machine learning classifiers with two possible classes (binary classification) can be evaluated using the number of true positives, false positives, true negatives and false negatives [18, 19]. Several different statistical methods (e.g. Sensitivity, Specificity and Accuracy) are calculated to evaluate the performance of classifiers. Sensitivity, also called the true positive rate (TPR), refers to the classifier's ability to identify a disease correctly, in our case (patients with primary headache disorders), while the specificity refers to the classifier's ability to exclude the disease correctly (identifies the negative cases), in our case (normal individuals suffering from headache). The classification accuracy is the overall correctness of the model, it can be calculated as the sum of true results (both true positives and true negatives) divided by the total number of the examined test set. The Sensitivity, specificity and accuracy can be expressed mathematically as follows [17, 18].

$$Sensitivity = TP / (TP + FN)$$

$$Specificity = TN / (TN + FP)$$

$$Accuracy = (TP + TN) / (TP + FP + TN + TP)$$

3.4 Results and Discussion

Nine hundred instances were included in this study. Among the 900 individuals, 600 patients with primary headache disorders and 300 were healthy individuals suffering from the usual headache. The major contribution of this study is to evaluate the performance of seven different machine-learning classifiers. These classifiers have shown a different range of sensitivity, specificity, classification accuracy and mean absolute error (MAE) as illustrated in Table 3.

Table 2. Performance of different classifier with holdout method

Algorithms	Sensitivity	Specificity	MAE	Accuracy
Naive Bayes (NB)	90 %	100 %	0.068	93.33 %
Artificial Neural Network (ANN)	100 %	90 %	0.039	96.67 %
Decision Tree (DT - J48)	100 %	90 %	0.050	96.67 %
Zero R Classifier (ZeroR)	100 %	0.0 %	0.444	66.67 %
Support Vector Machine (SVM)	96.67 %	96.67 %	0.033	96.67 %
k-Nearest Neighbor (kNN)	100 %	90 %	0.035	96.67 %
Logistic Regression (LOGR)	97.5 %	93.33 %	0.039	96.11 %

The ANN, DT, kNN, and SVM achieved the highest classification accuracy with the holdout method (Table 2), which was 96.67 %. Followed by LOGR with 96.11 % and then NB with 93.33 %. The lowest accuracy registered was 66.67 % by ZR. The first three of the highest accuracy classifiers reached 100 % of sensitivity to predict and classify the patients with primary headache. ZR classifier also obtained this rate, despite the lowest level of specificity, used to predict healthy individuals. Although the NB

Table 3. Performance of different classifier with 10 folds cross-validation

Algorithms	Sensitivity	Specificity	MAE	Accuracy
Naive Bayes (NB)	89 %	100 %	0.076	92.67 %
Artificial Neural Network (ANN)	98.17 %	92 %	0.043	96.11 %
Decision Tree (DT - J48)	100 %	91 %	0.048	97 %
Zero R Classifier (ZeroR)	100 %	0.0 %	0.445	66.67 %
Support Vector Machine (SVM)	94.67 %	98.67 %	0.040	96 %
k-Nearest Neighbor (kNN)	98.83 %	91 %	0.038	96.22 %
Logistic Regression (LOGR)	97.17 %	91.67 %	0.041	95.33 %

classifier achieved 100 % of specificity, it came at the end of the list with 90 % sensitivity. In general, DT, ANN and kNN produced the same results with regard to sensitivity, specificity and accuracy. However, the mean absolute errors were slightly different. Finally, the SVM classifier produced 96.67 % sensitivity, specificity and accuracy.

In the 10 folds cross-validation (Table 3), the majority of the classifiers demonstrated different classification accuracy and overall performance. The disparity rate among their accuracies was roughly 2 %. The DT classifier reached the highest accuracy with 97 %, followed by ANN, kNN and SVM with approximately 96 %. Logistic Regression (LOGR) achieved 95.33 % and then NB with 92.67 %. The ZR classifier has also registered the lowest accuracy rate with 66.67 %, which was the same result within the hold out method. In spite of that, ZR showed 100 % for sensitivity, which was only produced by the DT classifier. ANN and kNN showed around 98 % for sensitivity followed by LOGR, SVM and finally NB with 89 %. NB predicted on the largest number of healthy individuals with 100 % for specificity. Followed by the SVM, with 98.67 %, and the ANN with 92 %. The DT, kNN along with LOGR obtained around 91 % of specificity, while the ZR classifier obtaining the lowest level for specificity.

There were no significant changes regarding the accuracy using two validation methods (hold out and 10 folds cross-validation). However, ANN, SVM and kNN classification accuracy were decreased slightly using the 10 folds cross-validation method. In contrast, the DT classifier showed some improvement in the classification accuracy and specificity. It was stable regarding the prediction of the patients with primary headache as it showed 100 % of sensitivity using both methods. ANN, SVM and kNN sensitivities reduced slightly using 10 folds cross-validation at the expense of enhanced specificities. The ZR classifier achieved the worst performance, with zero rates in predicting a healthy individuals and 66.67 classification accuracy. Finally, there was no noticeable change in the ZR classifier performance using both validation methods.

4   Conclusion

Using synthetically produced data set, we have presented a systematic comparison and assessment of seven well-known supervised machine learning classifiers. Hold-out method and 10-folds cross validation method have been applied to a performance evaluation. Statistical measurements (i.e. sensitivity, specificity, accuracy) were used for classifiers assessment. Overall, the experimental study showed promising results. Using holdout method, ANN, DT, kNN, and SVM equally achieved 96.67 % of classification accuracy with 100 % of sensitivity for the first three of them. However, ANN, SVM and kNN classification accuracies were reduced slightly using the 10-folds cross validation method. In contrast, the DT classifier showed some improvement in the classification accuracy and specificity. It was stable regarding the prediction of the patients with primary headache as it showed 100 % of sensitivity using both methods. Therefore, we can claim that DT classifier would be the most appropriate classification method to address such classification problem domain.

References

1. Jeon, H.S., Lee, W.D.: Performance measurement of decision tree excluding insignificant leaf nodes. In: International Conference on Cyber-Enabled Distributed Computing and Knowledge Discovery (CyberC). IEEE (2014)
2. El-Sappagh, S.H., El-Masri, S., Riad, A.M., Elmogy, M.: Data mining and knowledge discovery: applications, techniques, challenges and process models in healthcare. Int. J. Eng. Res. Appl. **3**(3), 900–906 (2013)
3. Aljaaf, A.J., Al-Jumeily, D., Hussain, A.J., Fergus, P., Al-Jumaily, M., Abdel-Aziz, K.: Toward an optimal use of artificial intelligence techniques within a clinical decision support system. In: The Science and Information (SAI), London (2015)
4. Foster, K.R., Koprowski, R., Skufca, J.D.: Machine learning, medical diagnosis, and biomedical engineering research commentary. Biomed. Eng. **13**, 94 (2014)
5. Sewak, M., Singh, S.: In pursuit of the best artificial neural network for predicting the most complex data. In: International Conference on Communication, Information and Computing Technology (ICCICT). IEEE, Mumbai (2015)
6. Dastanpour, A., Ibrahim, S., Mashinchi, R., Selamat, A.: Comparison of genetic algorithm optimization on artificial neural network and support vector machine in intrusion detection system. In: IEEE Conference on Open Systems (ICOS). IEEE, Malaysia (2014)
7. Al-Rahayfeh, A., Faezipour, M.: Classifiers comparison for a new eye gaze direction classification system. In: LISAT Conference. IEEE, Long Island (2014)
8. Thepade, S.D., Kalbhor, M.M.: Novel data mining based image classification with Bayes, tree, rule, lazy and function classifiers using fractional row mean of cosine, sine and walsh column transformed images. In: International Conference on Communication, Information and Computing Technology (ICCICT). IEEE, Mumbai (2015)
9. Siji, K.K., Mathew, B.S., Chandran, R., Shajeemohan, B.S., Shanthini, K.S.: Feature selection, optimization and performance analysis of classifiers for biological images. In: National Conference on Communication, Signal Processing and Networking. IEEE (2014)
10. Milovic, B., Milovic, M.: Prediction and decision making in health care using data mining. Int. J. Public Health Sci. (IJPHS) **1**(2), 67–78 (2012)
11. Sweety, M.E., Jiji, G.W.: Detection of alzheimer disease in brain images using PSO and decision tree approach. In: International Conference on Advanced Communication Control and Computing Technologies (ICACCCT). IEEE, Ramanathapuram (2014)
12. Sundararaj, G.K., Balamurugan, V.: An expert system based on texture features and decision tree classifier for diagnosis of tumor in brain MR images. In: International Conference on Contemporary Computing and Informatics (IC3I). IEEE, Mysore (2014)
13. Kaya, M., Bilge, H.S.: Classification of pancreas tumor dataset using adaptive weighted k-nearest neighbor algorithm. In: International Symposium on Innovations in Intelligent Systems and Applications (INISTA). IEEE, Alberobello (2014)
14. Headache Classification Subcommittee of the International Headache Society: The international classification of headache disorder: 2nd edition. Cephalalgia: Int. J. Headache **24**(Suppl. 1), 9–160 (2004)
15. O'Flynn, N., Ridsdale, L.: Headache in primary care: how important is diagnosis to management. Br. J. Gen. Pract. **52**(480), 569–573 (2002)
16. National Institute for Health and Care Excellence (NICE): Diagnosis and management of headaches in young people and adults: clinical guideline. Br. J. Gen. Pract. **63**(613), 443–445 (2013)

17. Silva, F.R., Vidotti, V.G., Cremasco, F., Dias, M., Gomi, E.S., Costa, V.P.: Sensitivity and specificity of machine learning classifiers for glaucoma diagnosis using spectral domain OCT and standard automated perimetry. Arq. Bras. Oftalmol. **76**(3), 170–174 (2013)

18. Gupta, N., Rawal, A., Narasimhan, V.L., Shiwani, S.: Accuracy, sensitivity and specificity measurement of various classification techniques on healthcare data. IOSR J. Comput. Eng. (IOSR-JCE) **11**(5), 70–73 (2013)

19. Fergus, P., Cheung, P., Hussain, A., Al-Jumeily, D., Dobbins, C., Iram, S.: Prediction of preterm deliveries from EHG signals using machine learning. PLoS ONE **8**(10), e77154 (2013). doi:10.1371/journal.pone.0077154

A Generative Model with Ensemble Manifold Regularization for Multi-view Clustering

Shaokai Wang[1], Yunming Ye[1(✉)], and Raymond Y.K. Lau[2]

[1] Department of Computer Science, Shenzhen Graduate School,
Harbin Institute of Technology, Shenzhen, China
wangshaokai@gmail.com, yeyunming@hit.edu.cn
[2] Department of Information Systems, City University of Hong Kong,
Kowloon Tong, Hong Kong
raylau@cityu.edu.hk

Abstract. Topic modeling is a powerful tool for discovering the underlying or hidden structure in documents and images. Typical algorithms for topic modeling include probabilistic latent semantic analysis (PLSA) and latent Dirichlet allocation (LDA). More recent topic model approach, multi-view learning via probabilistic latent semantic analysis (MVPLSA), is designed for multi-view learning. These approaches are instances of generative model, whereas the manifold structure of the data is ignored, which is generally informative for nonlinear dimensionality reduction mapping. In this paper, we propose a novel generative model with ensemble manifold regularization for multi-view learning which considers both generative and manifold structure of the data. Experimental results on real-world multi-view data sets demonstrate the effectiveness of our approach.

Keywords: Multi-view clustering · Generative model · Manifold learning

1 Introduction

Exploring the rich information among multiple views arouses vast amount of interest in the past decades [1–3, 5, 6, 10]. Co-training introduced by Blum and Mitchell [1] is one of the earliest schemes for multi-view learning. Many variants following the initial idea of co-training have since been developed [2, 3]. However, the idea of co-training requires the conditional independent assumption to work well, but conditional independence assumption is usually too strong to be satisfied in practice, such that these methods may not effectively work [4].

Unlike the most previous works following the idea of co-training, recently, Zhuang et al. [5] proposed MVPLSA algorithm which is a multi-view learning algorithm via Probabilistic Latent Semantic Analysis. However, this algorithm does not take into account the manifold structure of the data, which is generally informative for nonlinear dimensionality reduction mapping.

Cai et al. recently proposed two topic models, Laplacian pLSI (LapPLSI) [9] and Locally-consistent Topic Modeling (LTM) [8], which use manifold structure information

© Springer International Publishing Switzerland 2015
D.-S. Huang and K. Han (Eds.): ICIC 2015, Part III, LNAI 9227, pp. 109–114, 2015.
DOI: 10.1007/978-3-319-22053-6_13

based on PLSA. However, both models are designed for solving single-view data clustering problem.

In this paper, we propose a new multi-view clustering method via generative model with ensemble manifold regularization (GMEMR) which takes into account the manifold structure of the data. We construct a nearest neighbor graph to model the underlying manifold structure for each view. Multiple manifold regularization terms are separately constructed for each view. Then, the regularization terms is incorporated with a generative model based on the Probabilistic Latent Semantic Analysis (PLSA) method, resulting in a unified objective function.

The paper is organized as follows. Section 2 provides background and notation. Section 3 presents our proposed GMEMR method. Experimental results are given in Sect. 4. Finally, we conclude the paper in Sect. 5.

2 Background and Notation

Suppose that we have N data points, $X = \{x_1, \ldots, x_N\}$, let $C = \{c_1, \ldots, c_K\}$ be the set of data classes. There are T views $\{f_1, \ldots, f_T\}$ for the data, f_t is the data view label. Each data point $x_i \in X$ has a feature vector $<w_1^t, \ldots, w_{M_t}^t>$ in the t-th view f_t. Each view f_t has M_t features. Our goal in multi-view clustering is to partition $X = \{x_1, \ldots, x_N\}$ into K clusters by exploiting the information stored in all T different views $\{f_1, \ldots, f_T\}$ of the input data.

Zhuang et al. [5] proposed MVPLSA algorithm which is a multi-view learning algorithm via Probabilistic Latent Semantic Analysis. The MVPLSA algorithm is motivated by the following observations. Different features in the context may be grouped together to indicate a high-level topic (e.g., feature clusters), i.e., the words "price", "performance", "announcement" from an enterprise news may present the concept "produce announcement". Let $\left\{z_1^t, \ldots, z_{Q^t}^t\right\}$ be the latent variables for high-level feature topic (e.g., feature clusters) for each view f_t. In the model there are two latent variables z_q^t for the high-level latent topic and c_k for the data class, and three visible variables x_i for the data point, w_j^t for the feature, and f_t for the view label. In the model, the parameters are $P(c_k \mid x_i)$, $P\left(z_q^t \mid c_k, f_t\right)$ and $P\left(w_j^t \mid z_q^t, f_t\right)$, they can be estimated by maximizing the log-likelihood

$$
\begin{aligned}
L &= \sum_{i=1}^{N} \sum_{t=1}^{T} \sum_{j=1}^{M_t} n\left(x_i, w_j^t, f_t\right) log P\left(x_i, w_j^t, f_t\right) \\
&\propto \sum_{i=1}^{N} \sum_{t=1}^{T} \sum_{j=1}^{M_t} n\left(x_i, w_j^t, f_t\right) log \sum_{k=1}^{K} \sum_{q=1}^{Q^t} P\left(w_j^t \mid z_q^t, f_t\right) P\left(z_q^t \mid c_k, f_t\right) P(c_k \mid x_i)
\end{aligned}
\tag{1}
$$

where M_t is the number of features in the t-th view f_t and $n\left(x_i, w_j^t, f_t\right)$ indicates the frequency of feature w_j^t occurring in instance x_i. However, this model do not take into account the manifold structure of the data, which is generally informative for nonlinear dimensionality reduction mapping.

3 Method

Based on the smoothness assumption that if two instances x_i and x_j are close on the manifold, then $P(C|x_i)$ and $P(C|x_j)$ are "similar" to each other, we set up an ensemble manifold regularization term in the generative model to enforce the smoothness. Let $P_i(C) = P(C|x_i)$ and $P_s(C) = P(C|x_s)$. By using the symmetric KL-Divergence, conditional probability distribution $P(C|x_i)$ un-smoothness on the intrinsic manifold of the multi-view data can be written as

$$R = \sum_{i=1}^{N} \sum_{s=1}^{N} D(P_i(C), P_s(C)) W_{is} \tag{2}$$

where $D(P_i(C), P_s(C))$ is the symmetric KL-Divergence, $W \in R^N \times R^N$ is the data adjacency graph, W is the intrinsic manifold approximation of the multi-view data. We'll introduce how to obtain W in the following. The value of R ranges from 0 to ∞. The smaller R is, the smoother conditional probability distribution $P(C|x_i)$ is. $P(c_k|x_i)$ are estimated by the following maximization problem:

$$\max_{P} \quad O = L - \lambda_1 R \tag{3}$$

where L and R are defined in Eqs. 1 and 2 respectively, λ_1 is the regularization parameter. The value of λ_1 is in the range of 0 to ∞. In the following, we'll introduce how to obtain the intrinsic manifold approximation W.

The local manifold structure can be effectively modeled through a nearest neighbor graph [7]. We construct a nearest neighbor graph to model the underlying manifold structure for each view. Define the edge weight matrix U^t of view f_t as follows: $U_{is}^t = 1$, if $x_i \in N_p(x_s)$ or $x_s \in N_p(x_i)$ in view f_t, otherwise $U_{is}^t = 0$, where $N_p(x_i)$ denotes the set of p nearest neighbors of x_i (with respect to the Euclidean distance). Now we have T graphs $\{U^t\}_{t=1}^{T}$, we want to combine them to approximate the intrinsic manifold. Inspired by [11, 12], the intrinsic manifold is approximated by the linear combination of these candidate manifolds, i.e., $W = \sum_{t=1}^{T} \mu_t U^t$, s.t. $\sum_{t=1}^{T} \mu_t = 1$, $\mu_t \geq 0$, for $t = 1, \ldots, T$. Let $\{L^t\}_{t=1}^{T}$ be a set of the corresponding graph Laplacian matrices [14], where $L^t = D^t - U^t$, D and D^t is a diagonal matrix with the i-th diagonal entry $D_{ii} = \sum_{j=1}^{N} W_{ij}$, $D_{ii}^t = \sum_{j=1}^{N} U_{ij}^t$. Based on the smoothness assumption, when P is fixed, the optimal hyperparameters $\{\mu_t\}_{t=1}^{T}$ is the solution of the following minimization problem:

$$\min_{\mu} \quad \sum_{t=1}^{T} \frac{\mu_t}{Z^t} \left(Tr\left(P^T L^t P\right) + \frac{\lambda_2}{2} \|\mu\|^2 \right)$$

$$s.t. \quad \sum_{t=1}^{T} \mu_t = 1, \mu_t \geq 0, for \quad t = 1, \ldots, T. \tag{4}$$

where Z^t is a normalization constant that is defined by $Z^t = \|L^t\|_F$, $\mu = (\mu_1, \ldots, \mu_T)^T$. The regularization term $\|\mu\|^2$ is introduced to avoid μ overfitting to one single manifold, λ_2 are tradeoff parameters. The minimization problem (4) can be solved by the quadric optimization solver in matlab toolbox. The optimal graph hyperparameter μ is obtained, and then we obtain the intrinsic manifold approximation

$$W = \sum_{t=1}^{T} \mu_t U^t. \tag{5}$$

In the generative model, the re-estimation equations for parameters $\{P(c_k \mid x_i)\}$ can be obtained using EM algorithm:

$$Y_k = (\Omega + \lambda_1 L)^{-1} V_k \tag{6}$$

$Y_k = [P(c_k \mid x_1), \ldots, P(c_k \mid x_N)]^T$, Ω denotes a N-by-N diagonal matrix with $\rho_i = \sum_{t=1}^{T} \sum_{j=1}^{M_t} n\left(x_i, w_j^t, f_t\right)$ as entries. V_k is a N-dimensional vector with $\sum_{t=1}^{T} \sum_{j=1}^{M_t} \sum_{q=1}^{Q^t} n\left(x_i, w_j^t, f_t\right) P\left(z_q^t, c_k \mid x_i, w_j^t, f_t\right)$ as entries. L is a N-by-N graph Laplacian matrix, $L = D - W$, D denotes a N-by-N diagonal matrix whose entries are row sums of W, $D_{ii} = \sum_s W_{is}$.

4 Experiments

4.1 Datasets and Comparison Methods

In this section, we experimentally evaluate the proposed multi-view learning framework on two real-world multi-view data sets.

DBLP Dataset. The DBLP dataset is a bibliographic network. This dataset is an adaptation of the original dataset of Ming Ji.[1] It consists of 4057 authors classified into one of four classes. There are two views for each author: the names of an author's papers and the terms of an author's papers.

Reuters Multilingual Dataset.[2] This collection contains feature characteristics of documents originally written in five different languages, and their translations, over a

---

[1] Ming Ji's dataset is available at http://web.engr.illinois.edu/_mingji1/DBLP_four_area.zip.
[2] http://membres-lig.imag.fr/grimal/data.html.

common set of 6 categories. We use documents originally in English as the first view and their French and German translations as the second and third view.

We compare the performance of the proposed GMEMR algorithm with several baseline methods: Locally-consistent Topic Modeling (LTM) [8], ConcatLTM (Concatenating the features of all the views, and then run LTM directly on this concatenated view representation), MVPLSA [5], Pairwise Co-regularized Spectral Clustering (PRSC) [3] and Centroid Co-regularized Spectral Clustering (CRSC) [3]. As adopted in [6, 13], we evaluate clustering performance using two standard measures: Clustering accuracy (AC) and normalized mutual information (NMI) from 10 random runs. In the experiments, we assign the initial value of $P(c_k \mid x_i)$ as the results of ConcatLTM algorithm for the tasks from DBLP and Reuters datasets.

4.2   Results

Results of different models are presented in Table 1. In the table, GMEMR denotes the accuracy of GMEMR when using composite manifold learned by (4) to construct the regularization term Eq. (2). GMEMR_View t denotes the accuracy of GMEMR when using U^t (5NN graph of view f_t) to construct the regularization term Eq. (2). LTM_View t denotes the accuracy of LTM on t-th view.

From this table, we have following observations: Our method GMEMR outperforms all the compared approaches. The methods with manifold regularization outperform the methods without manifold regularization, i.e., multi-view algorithm GMEMR are superior to MVPLSA in terms of Accuracy values and NMI values in all cases. A possible reason is that the manifold regularization term causes the data space locality information to be preserved on the low dimensional representations. Furthermore, it is demonstrated that the data space geometry information is crucial for clustering performance. The performance of GMEMR is better than the best performance of GMEMR_View t. This suggests that the composite manifold learned by ensemble manifold regularization can

Table 1. Performance of the different model on two data sets. Bold performance correspond to the best model.

Method	Accuracy (%)		NMI (%)	
	DBLP	Reuters	DBLP	Reuters
GMEMR	**81.65**	**56.28**	**52.50**	**32.47**
GMEMR_View 1	77.99	50.82	48.43	31.67
GMEMR_View 2	77.71	49.18	47.41	31.42
GMEMR_View 3		47.56		28.52
MVPLSA	74.43	47.62	45.55	29.65
PRSC	70.79	48.08	37.08	28.06
CRSC	70.87	48.11	37.20	27.75
LTM_View 1	72.15	40.59	44.63	23.10
LTM_View 2	77.85	42.61	48.21	23.81
LTM_View 3		42.46		24.49
ConcatLTM	73.30	45.53	45.22	28.13

select the most effective graphs and combine them to approximate the intrinsic manifold, thus, the data space nonlinear structure is preserved effectively.

5 Conclusion

In this paper, we proposed a novel multi-view clustering method via generative model with ensemble manifold regularization which considers both generative and manifold structure of the data. Experiments on the real world datasets demonstrated the effectiveness of our method. Experimental results show the proposed method (1) can jointly model the co-occurrences from multiple views and obtain additional gains, (2) balances the generative model and manifold regularization term, and (3) effectively discovers the underlying structure of data space.

Acknowledgments. This work was supported in part by National Key Technology R&D Program of MOST China under Grant No. 2014BAL05B06, NSFC under Grant No.61272538, Shenzhen Science and Technology Program under Grant No. JCYJ20140417172417128, and the Shenzhen Strategic Emerging Industries Program under Grant No. JCYJ20130329142551746.

References

1. Blum, A., Mitchell, T.: Combining labeled and unlabeled data with co-training. In: COLT, pp. 92–100 (1998)
2. Yu, S., Krishnapuram, B., Rosales, R., Rao, R.B.: Bayesian co-training. J. Mach. Learn. Res. **12**, 2649–2680 (2011)
3. Kumar, A., Rai, P., Daum, H.: Co-regularized multi-view spectral clustering. In: NIPS, pp. 1413–1421 (2011)
4. Belkin, M., Niyogi, P., Sindhwani, V.: Manifold regularization: A geometric framework for learning from labeled and unlabeled examples. J. Mach. Learn. Res. **1**, 1–48 (2006)
5. Zhuang, F., Karypis, G., Ning, X., et al.: Multi-view learning via probabilistic latent semantic analysis. Inf. Sci. **199**, 20–30 (2012)
6. Liu, J., Jiang, Y., Li, Z., et al.: Partially Shared Latent Factor Learning With Multiview Data. IEEE Trans. Neural Netw. Learn. Syst. (2014). doi:10.1109/TNNLS.2014.2335234
7. Belkin, M., Niyogi, P.: Laplacian eigenmaps and spectral techniques for embedding and clustering. In: NIPS, pp. 586–691 (2001)
8. Cai, D., Wang, X., He, X.: Probabilistic dyadic data analysis with local and global consistency. In: ICML, pp. 105–112 (2009)
9. Cai, D., Mei, Q., Han, J., Zhai, C.: Modeling hidden topics on document manifold. In: CIKM, pp. 911–920 (2008)
10. Liu, J., Wang, C., Gao, J., Han, J.: Multi-view clustering via joint nonnegative matrix factorization. SDM **13**, 252–260 (2013)
11. Geng, B., Tao, D., Xu, C., Yang, L., Hua, X.S.: Ensemble manifold regularization. IEEE Trans. Pattern Anal. Mach. Intell. **34**(6), 1227–1233 (2012)
12. Karasuyama, M., Mamitsuka, H.: Multiple graph label propagation by sparse integration. IEEE Trans. Neural Netw. Learn. Syst. **24**(12), 1999–2012 (2013)
13. Wang, H., Weng, C., Yuan, J.: Multi-feature spectral clustering with minimax optimization. In: CVPR, pp. 4106–4113 (2014)
14. He, X., Niyogi, P.: Locality Preserving Projections. In: NIPS, pp. 145–153 (2004)

On Classifying Diabetic Patients' with Proliferative Retinopathies via a Radial Basis Probabilistic Neural Network

Leonarda Carnimeo$^{(\boxtimes)}$ and Rosamaria Nitti

Department of Electrical and Information Engineering,
Technical University of Bari, Bari, Italy
leonarda.carnimeo@poliba.it

Abstract. Diabetic retinopathies have to be detected early and treated to avoid serious damages to patients' retina. A severe progress of diabetes can deteriorate human vision and the effects of a Proliferative Diabetic Retinopathy (PDR) could appear in fundus images, showing a neovascularization that can rise abruptly. Until now only some network models for classifying presence/absence of PDR have been faced by means of PNNs or SVMs. In this paper a first approach to follow diabetic patients affected by early PDR via a novel neural classifier based on a Fundus Image Preprocessing Subsystem and a Radial Basis Probabilistic Neural Network (RBPNN) is presented. The proposed classifier aims at classifying a certain number of diabetic patients by means of their accurately preprocessed digital fundus images and could support their follow-up paths in alerting if variations in retinal vasculature of classified PDR should occur.

Keywords: Human retina · Pattern classification · Diabetic retinopathy · Radial basis probabilistic neural networks

1 Introduction

In the last decades, improvements have been achieved in image processing for ophthalmology [1, 2]. Advances in automated diagnostic retinal tools actually allow ophthalmologists to perform mass screening for the most common diseases, such as diabetes or glaucoma [3–5]. In particular, vessel extraction is important in the analysis of digital fundus images, since it helps in diagnosing retinal diseases. The medical motivation towards the segmentation of blood vessels of retinal images is to suppress the background and accentuate the small vessels so that features such as abnormal neovascularization become more visually enhanced. These clinical markers help ophthalmologists in diagnosing diabetic retinopathy [3–6]. Moreover, an early diagnosis of Proliferative Diabetic Retinopathy (PDR), that is a severe complication of diabetes that damages human retina, is crucial for protecting diabetic patients' vision. The onset of PDR is indicated by the appearance of a neovascularization, which might be identified using proper retinal vessel extraction techniques [6–8]. Furthermore, several innovations in ophthalmic diagnostic tools for the detection of various anatomical or pathological features have been recently developed on the basis of particular neural network models, such as MultiLayer Perceptron (MLPs) and Probabilistic Neural Networks

© Springer International Publishing Switzerland 2015
D.-S. Huang and K. Han (Eds.): ICIC 2015, Part III, LNAI 9227, pp. 115–126, 2015.
DOI: 10.1007/978-3-319-22053-6_14

(PNNs), which are supervised neural networks widely used for pattern recognition [9–13]. Moreover, in [14] an MLP classifier was used to obtain a segmentation of hard exudates in a fundus image, whereas PNNs and MLPs were also considered for the classification of EEG signals [15]. In so far as the authors are aware, only some neural network models for classifying the presence/absence of PDR have been until now approached by means of SVMs or PNNs [16] with interesting results. It has to be considered that PNNs, as well as Radial Basis Function networks, are widely used in various pattern classification tasks due to their robustness and their quicker learning with respect to other neural network models, beside converging always to the Bayes optimal solution if the training set increases [17]. But the increase of PDR in diabetic patients should be specifically supported with adequate tools, due to the fact that this severe eye complication can rise abruptly. Thus, in this paper, a first attempt to classify more diabetic patients affected by early PDRs is presented, via the development of a novel neural classifier based on a fundus image preprocessing subsystem and a Radial Basis Probabilistic Neural Network (RBPNN) [18–21]. The RBPNN is preferred to strengthen some capabilities of PNNs via ad hoc probability density functions given by radial basis ones. In detail, an accurate preprocessing of diabetic patients' retinas is firstly performed to obtain corresponding thinned images; a dedicated RBPNN is subsequently synthesized as a neural classifier. In detail, after enhancing vascular patterns in input fundus images, adequate thresholding and thinning techniques [4, 8, 9] are considered to derive thinned patterns for those patients' classification. On the basis of preprocessed images, the training set is subsequently derived for synthesizing a dedicated RBPNN. It is herein assumed that a lack of matching for the designed neural classifier means that an early PDR is worsening for the corresponding patient, thus an Alert has to be produced for eye doctors, or rather that the patient's identity is not confirmed. A case study is developed and reported.

2   A Neural Classifier of Diabetic Patients Based on a RBPNN

The behavior of the proposed neural classifier is herein described. Each retinal image is firstly preprocessed to enhance vessel vasculature. Then, every obtained image is segmented and successively thinned [4, 5, 8] by means of morphological operators. Finally, each thinned image is extracted and stored. The thinned images of diabetic patients with proliferative retinopathy can be considered as inputs for synthesizing a proper RBPNN able to classify them. If the proliferative disease progressively grows up during time, also the corresponding thinned images will present changes. In the case of a positive matching, the synthesized RBPNN will correctly classify the patient with a proliferative retinopathy; if the matching reveals negative, the system should eventually provide alert signals.

A block diagram of the proposed classification procedure is reported in Fig. 1.

Retinal images of any patient with proliferative retinopathy are firstly captured by a Retinal Scanning Subsystem. Then a Fundus Image Preprocessing Subsystem performs the image processing tasks which are described in the following.

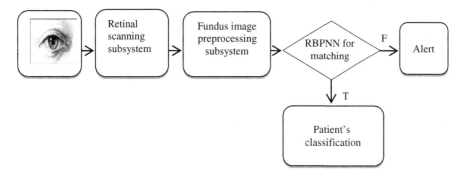

Fig. 1. Block diagram of the proposed neural classifier

Vessel Vasculature Enhancement and Image Segmentation. Detecting the retinal vasculature means, substantially, generating a binary mask in which pixels are labeled as vessel or background. The aim is to capture as much detail as possible, preserving vessel connectivity and avoiding false positives. Thus, the retinal area containing the vessel tree, is properly enhanced. Several methods to segment automatically blood vessels have been until now developed [4, 5, 7, 11, 15]. In particular, the Threshold Probing method in [4, 5] is herein considered, where individual pixel labels are decided using both local and region-based properties. In detail, in [4] an algorithm is proposed to approximate the intensity profile using a Gaussian Curve and gray level profiles are calculated in the perpendicular direction to the length of each vessel. Image enhancement is carried out before segmentation of vessels using a (5 × 5)-windowed mean filter which reduces the effect of spurious noise. Twelve different orientations were chosen between 0° to 180°, each differing by a step of 15°, keeping the length segment L = 9, σ = 2 (σ denotes the spread of intensity profile) and T = 6 (where T = 3σ). This structure is chosen to detect vessels however oriented in direction with varying lengths. A kernel is created corresponding to each orientation. The highest response in one of the directions is then selected for a particular pixel. The algorithm detects vessels despite of the fact that the local contrast is very low but the problem with this algorithm is that other brighter objects, which do not resemble any vessel segment (such as lesions, optic disc, edges etc.), are contemporary obtained.

Image Thinning. This morphological operation progressively erodes away outer pixels of thick stripes of the same gray level until they become one pixel wide. The thinning algorithm performs the required task on each binary image of human retina in two parallel sub-iterations [8] using a (3 × 3)-sized window with a rectangular tessellation in sub-blocks, where x1, .., x8 are the values of close pixels in each window numbered counterclockwise starting from the middle right pixel, being x9 the window central one. In each sub-iteration the generic pixel x9 with its eight nearest neighbors is subsequently examined on the basis of the following four pixel-deletion conditions

Condition G1:

$$x_h = 1 \quad \text{with} \quad x_h = \sum_{i=1}^{4} b_i \quad \text{and} \quad b_i = \begin{cases} 1 & \text{if } x_{2i-1} = 0 \wedge (x_{2i} = 1 \vee x_{2i+1} = 1) \\ 0 & \text{otherwise} \end{cases}$$

Condition G2:

$$2 \le \min(n_1(x_9), n_2(x_9)) \le 3$$

where

$$n_1(x_9) \sum_{i=1}^{4} x_{2i-1} \vee x_{2i} \quad n_2(x_9) = \sum_{i=1}^{4} x_{2i} \vee x_{2i+1}$$

$$\text{Condition } G_3 : \qquad\qquad \text{Condition } G_3' :$$
$$(x_2 \vee x_3 \vee \bar{x}_8) \wedge x_1 = 0 \quad (x_6 \vee x_7 \vee \bar{x}_4) \wedge x_5 = 0$$

The first Sub-iteration 1 can be expressed as follows

$$\text{If } G_1, G_2, G_3 \text{ are satisfied, then } x_9 = 0$$

The latter Sub-iteration 2 can be expressed as follows

$$\text{If } G_1, G_2, G_3' \text{ are satisfied, then } x_9 = 0$$

Sub-iterations continue until no pixel can be deleted any more. The application of this thinning procedure to a binary image of human retina provides a skeletal image, that is a thinned image Ithin to be subsequently considered for training.

The next block in Fig. 1 is a Radial Basis Probabilistic Neural Network (RBPNN), which is a neural network characterized both from features of Radial Basis Function Neural Networks (RBFNNs) and from those of Probabilistic Neural Networks (PNNs)

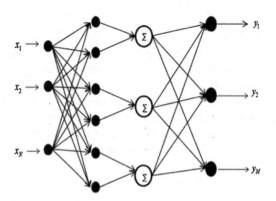

Fig. 2. Network topology of a RBPNN

[18–20] with lowered disadvantages. This network consists of four layers as shown in Fig. 2.

Data are concurrently feed-forwarded from the input layer to the output layer without any feedback connections within the three involved layers. The first hidden layer is a nonlinear processing layer, generally consisting of the centers selected from training samples. The second hidden layer selectively adds the outputs of the first hidden layer, according to the classes to which the hidden centers belong. Connection weights between the first hidden layer and the second hidden layer are equal to 1 or 0. Outputs in the second hidden layer need to be normalized for pattern recognition problems. The last layer for the RBPNN is generally the output layer [19]. If considering a generic input vector x, the actual output value of the i-th output neuron y_i^a of the RBPNN can be expressed as

$$y_i^a = \sum_{k=1}^{M} w_{ik} h_k(x) \tag{1}$$

$$h_k(x) = \sum_{i=1}^{n_k} \phi_i(\|x - c_{ki}\|_2), k = 1, 2, 3, \ldots\ldots M \tag{2}$$

where $h_k(x)$ is the k-th output value of the second hidden layer of the RBPNN; w_{ik} is the synaptic weight between the k-th neuron of the second hidden layer and the i-th neuron of the output layer of the RBPNN; c_{ki} is the i-th hidden center vector for the k-th pattern class of the first hidden layer; n_k represents the number of hidden center vectors for the k-th pattern class of the first hidden layer; the symbol $\|\cdot\|_2$ stays for the Euclidean norm; M denotes the number of the neurons both of the output layer and of the second hidden layer, or the pattern class number for the training set; the Gaussian kernel function $\varphi_i(\cdot)$ is expressed as

$$\phi_i(\|x - c_{ki}\|_2) = \exp\left[-\frac{\|x - c_{ki}\|_2^2}{\sigma_i^2}\right]$$

where σ_i is the corresponding parameter for the Gaussian kernel function.

The Orthogonal Least Square Algorithm (OLSA) is herein preferred for training the RBPNN. For N training samples, corresponding to M pattern classes, considering the form of the matrix, Eq. (1) can be written in vector form as

$$Y^a = H W$$

where Y^a and H are both (N × M)-dimension matrices, the synaptic weight matrix W is an (M × M)-dimension matrix between the output layer and the second layer of the RBPNN and can be determined as:

$$W = R^{-1} \hat{Y}$$

where R is an (M × M) upper triangular matrix with the same rank as H, and $\hat{Y}$ is an (M × M) matrix. The output layer is typically trained using the Least Mean Squares (LMS) algorithm [9, 18–20] and the weights w_{2j} are firstly initialized to small random values for j = 1, 2, ...N_2. Then, the errors $e_j = y_j - d_j$ are computed for all j; finally, the weights $w_j(k + 1) = w_j(k) - \mu e_j u_l$ are updated for all j, until the stop condition is reached.

3 Case Study and Experimental Results

The proposed neural classifier has been synthesized by considering fundus images selected from the publicly available database named DRIVE [3]. In this database there are 40 color eye fundus images taken with a Canon CR5 which is a nonmydriatic 3CCD camera with 45° Field Of View (FOV). Each image has a resolution of (565 × 584) pixels with 8 bits per color plane with a TIFF format. The database consists of two sets, named test set and training set. The first one contains 20 images along with 20 GT images, manually derived by two ophthalmologists. The second set contains 20 images along with the GT images provided by only one eye doctor. Moreover, both sets contain the corresponding FOV masks for the images. In this experiment fundus images of 7 patients characterized by an early proliferative retinopathy have been selected from DRIVE as listed in [6] and have been herein processed by the Fundus Image Preprocessing Subsystem to obtain the patients' thinned images. In Fig. 3 it can be noticed that the vascularization is highlighted and skeletal in thinned images on the right.

At first, the performance of the Fundus Preprocessing Subsystem in determining thinned images, have been investigated by considering the Accuracy defined as

$$\text{Accuracy} = \frac{TP + TN}{TP + TN + FP + FN}$$

with the True Positive rate estimated as

$$TP\,rate \approx \frac{\text{Correctly classified Positives}}{\text{Total Positives}}$$

and the False Positive rate

$$FP\,rate \approx \frac{\text{Incorrectly classified Negatives}}{\text{Total Negatives}}$$

The True Positive TP rate (the True Negative TN rate, respectively) has been evaluated by determining the number of black pixels of vessels, which are correctly classified as black pixels (the number of white pixels of background, which are correctly classified as white pixels, respectively) in thinned images by considering variations with respect to the their ground truth images provided by the database DRIVE. A comparison of percentage Average Accuracy values (AA %) obtained by

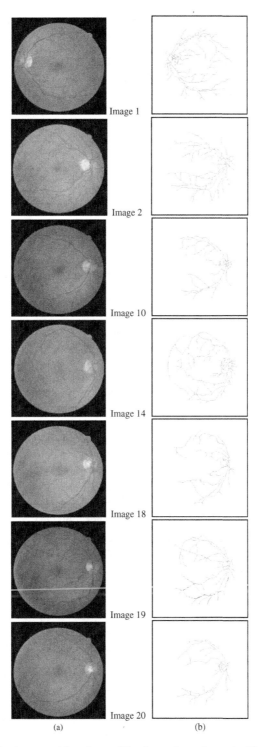

Image 1

Image 2

Image 10

Image 14

Image 18

Image 19

Image 20

(a)   (b)

Fig. 3. Seven Fundus images with early proliferative retinopathy as in [6]: (a) Fundus images; (b) Thinned images

Table 1. Comparison of (A %) on thinned images obtained using the method: (a) reported in [6]; (b) herein proposed

Patient	AA % (a)	AA % (b)
1	93,7904	94,5400
2	93,5918	94,5202
3	93,6670	93,8801
4	93,8912	94,0103
5	93,6106	95,1410
6	94,7460	95,4404
7	93,9676	94,9805
Average value	93,8949	94,6446

preprocessing retinal images with the herein proposed method and following the one proposed in [6] is reported in Table 1.

It can be noticed that an Average value of 94.6446 % better than the one derived in [6] has been obtained on thinned images by the Fundus Image Preprocessing Subsystem. Then, for every (M × M)-size thinned image, a number L of (M/L) × (M/L) submatrices has been considered to form the input set for training a RBPNN. Input neurons are equal to $(M/L)^2$, such as in the first hidden layer. In the second hidden layer eleven neurons have been reported, whereas in the output layer there are seven neurons to codify the seven classes, each providing a value 0 or 1. A radial basis function has been used as the probability density function between the layers [9]. A gradient descent with momentum has been used to train the network. As it is known, RBPNNs resemble to the family of PNNs with probability density functions chosen equal to Radial Basis Functions (RBF). Thus, the training set must be collected through several representatives of the actual population of the intended classification by considering a sufficiently sparse set. Taking into account that it is not easy to have an amount of fundus images for the same patient and that erroneous samples and outliers are acceptable for this kind of network, training has been performed via a training set of 100 binary images for each of the seven classes, collected by considering both the already shown thinned images and distorted ones obtained with a percentage of additional black pixels randomly distributed both in position and in quantity in the range (0−5) % of the total amount. Thus the training set has been composed of a

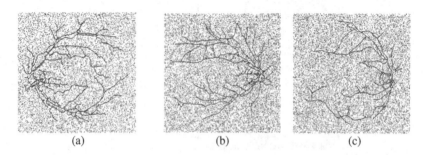

| (a) | (b) | (c) |

Fig. 4. Some distorted images of thinned image: (a) no. 1; (b) no. 2 (c) no. 18

significant number of patterns. For a better comprehension, some distorted training patterns with a 5 % of randomly distributed pixels are reported in Fig. 4.

After the synthesis, a cross validation phase has been performed by considering another set of 700 noisy patterns, obtained by considering a randomly distributed noise on original thinned images with percentage values in the range (0 %−8 %). Results are encouraging as shown in Table 2. In this table the synthesized RBPNN reveals able to correctly classify the seven patients until a percentage value equal to 5 % of random noise affects their thinned images; there is a reduced capability of classification if noise

Table 2. Classification capability of the synthesized RBPNN: Y = Correctly classified; N = Misclassified

Noise %	Patient 1	Patient 2	Patient 3	Patient 4	Patient 5	Patient 6	Patient 7
0	Y	Y	Y	Y	Y	Y	Y
1	Y	Y	Y	Y	Y	Y	Y
2	Y	Y	Y	Y	Y	Y	Y
3	Y	Y	Y	Y	Y	Y	Y
4	Y	Y	Y	Y	Y	Y	Y
5	Y	Y	Y	Y	Y	Y	Y
6	Y	Y	N	N	N	Y	N
7	N	N	N	N	N	Y	N
8	N	N	N	N	N	N	N

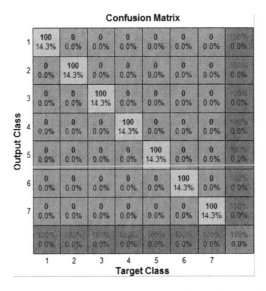

Fig. 5. Confusion matrix for the synthesized network

increases more, as shown in the table. The RBPNN classifies each known patient's vasculature into one of seven classes, whereas unknown vessel trees are misclassified.

Several criteria may be used to evaluate the performance of a classification in supervised learning. A confusion matrix is herein considered, which is a useful tool for analyzing how well a classifier can identify test samples of different classes [22]. Each entry C_{ij}, in the confusion matrix denotes the number of skeletal images of class i predicted to be of class j. In this test, 100 images/person for training and 100 images/person for cross validation have been used. Training has been performed by considering images with a random noise variable both spatially and in intensity in the range (0 %–5 %) of the total amount of pixels. A plot of the corresponding confusion matrix, which highlights if each pattern is correctly classified by the network, is shown in Fig. 5.

Finally, the synthesized neural classifier can provide its support to the diagnostic by performing matching operations between subsequent classifications of patients' thinned images. If the matching is positive, the eye state is unchanged. If the matching is negatively detected by the RBPNN, the retina vascularization is varied much and an Alert is produced by the neural network for eye doctors.

4   Conclusions and Future Work

In this paper a first approach to follow diabetic patients affected by early PDR has been dealt with via the synthesis of a novel neural classifier on the basis of a Fundus Image Preprocessing Subsystem and a RBPNN, aiming at investigating the possibility of supporting ophthalmologists in the follow up of patients' with this severe retinal pathology. In detail, human retina images of patients with early PDRs have been pre-processed to derive the corresponding thinned images, which have been subsequently used to synthesize a RBPNN able to support patients' classification and follow-up. A case study has been developed to classify an amount of diabetic patients, where simulations have been firstly performed to evaluate the performance of Fundus Image Preprocessing Subsystem and later to cross validate the designed RBPNN. Experimental results demonstrate that the presented system performs patients' classification with a satisfying accuracy. Thus, the proposed neural classifier could support ophthalmologists in alerting if variations in retinal vasculature of classified PDR should occur. In future work the capabilities of the neural classifier could be improved in its scale as a diagnostic support for mass screening, taking into account the drawback of a large memory requirement and the need of a significant representative training set.

Acknowledgement. The authors' acknowledgement goes to the financial support to this research given by the F.R.A. 2012 Fund - Technical University of Bari. Sincere thanks must be given to the student F. Digregorio for his useful contribution.

References

1. Osareh, A., Mirmehdi, M., Thomas, B., Markham, R.: Classification and localisation of diabetic-related eye disease. In: Heyden, A., Sparr, G., Nielsen, M., Johansen, P. (eds.) ECCV 2002, Part IV. LNCS, vol. 2353, pp. 502–516. Springer, Heidelberg (2002)
2. Walter, T., Klein, J.-C., Massin, P., Erginay, A.: A contribution of image processing to the diagnosis of diabetic retinopathy – detection of exudates in color fundus images of the human retina. IEEE Trans. Med. Imaging 21(10), 1236–1243 (2002)
3. Staal, J., Abràmoff, M.D., Niemeijer, M., Viergever, M., van Ginneken, B.: Ridge-based vessel segmentation in color images of the retina. IEEE Trans. Med. Imaging 23(4), 501–509 (2004)
4. Chaudhuri, S., Chatterjee, S., Katz, N., Nelson, M., Goldbaum, M.: Detection of blood vessels in retinal images using two-dimensional matched filters. IEEE Trans. Med. Imaging 8(3), 263–269 (1989)
5. Hoover, A.: Locating blood vessels in retinal images by piecewise threshold probing of a matched filter response. IEEE Trans. Med. Imaging 19, 203–210 (2000)
6. Ramlugun, G.S., Nagarajan, V.K., Chakraborty, C.: Small retinal vessels extraction towards proliferative diabetic retinopathy screening. Expert Syst. Appl. 39(1), 1141–1146 (2012). Elsevier
7. Zhang, D., Qin, L., You, J., Zhang, D.: A modified matched filter with double-sided thresholding for screening proliferative diabetic retinopathy. IEEE Trans. Inf. Technol. Biomed. 13(4), 528–534 (2009)
8. Lam, L., Lee, S.-W., Suen, C.Y.: Thinning methodologies-a comprehensive survey. IEEE Trans. Pattern Anal. Mach. Intell. 14(9), 869–885 (1992)
9. Hush, D.R., Horne, B.: Progress in supervised neural networks. IEEE Signal Process. Mag. 5, 8–39 (1993)
10. Bevilacqua, V., Carnimeo, L., Mastronardi, G., Santarcangelo, V., Scaramuzzi, R.: On the comparison of NN-based architectures for diabetic damage detection in retinal images. J. Circ. Syst. Comput. 18(08), 1369–1380 (2008)
11. Carnimeo, L., Bevilacqua, V., Cariello, L., Mastronardi, G.: Retinal vessel extraction by a combined neural network–wavelet enhancement method. In: Huang, D.-S., Jo, K.-H., Lee, H.-H., Kang, H.-J., Bevilacqua, V. (eds.) ICIC 2009. LNCS, vol. 5755, pp. 1106–1116. Springer, Heidelberg (2009)
12. Carnimeo, L., Benedetto, A.C., Mastronardi, G.: A voting procedure supported by a neural validity classifier for optic disk detection. In: Huang, D.-S., Gupta, P., Zhang, X., Premaratne, P. (eds.) ICIC 2012. CCIS, vol. 304, pp. 467–474. Springer, Heidelberg (2012)
13. Specht, D.F.: Probabilistic neural networks for classification, mapping, or associative memory. IEEE Int. Conf. Neural Netw. 1, 525–532 (1998)
14. Garcia, M., Hornero, R., Sanchez, C.I., Lopez, M.I., Diez A.: Feature extraction and selection for the automatic detection of hard exudates in retinal images. In: 29th Annual International Conference of the IEEE EMBS Cite Internationale, Lyon, France (2007)
15. Güler, I., Übeyli, E.D.: Multiclass support vector machines for EEG signals classification. IEEE Trans. Inf. Technol. Biomed. 11(2), 117–126 (2007)
16. Priya, R., Aruna, P.: SVM and neural network based diagnosis of diabetic retinopathy. Int. J. Comput. Appl. 41(1), 6–12 (2012)
17. El Emary, M.M.I., Ramakrishnan, S.: On the application of various probabilistic neural networks in solving different pattern classification problem. World Appl. Sci. J. 4(6), 772–780 (2008)

18. De-Shuang, H.: Radial basis probabilistic neural networks: model and application. Int. J. Pattern Recogn. Artif. Intell. **13**(7), 1083–1101 (1999). World Scientific Publishing Company
19. Shang, L., De-Shuang, H., Dua, J., Zheng, C.: Palmprint recognition using FastICA algorithm and radial basis probabilistic neural network. Neurocomputing **69**, 1782–1786 (2006)
20. De-Shuang, H.: A constructive hybrid structure optimization methodology for radial basis probabilistic neural network. IEEE Trans. Neural Netw. **19**(12), 2099–2115 (2008)
21. Kulkarni, A.H., Rai, H.M., Jahagirdar, K.A., Kadkol, R.J.: A leaf recognition system for classifying plants using RBPNN and pseudo zernike moments. Int. J. Latest Trends Eng. Technol. (IJLTET) **2**(1), 6–11 (2013)
22. Han, J., Kamber, M.: Data Mining concepts and Techniques, 2nd edn. Elsevier publishers, Waltham (2009)

Passivity Analysis of BAM NNs with Mixed Time Delays

Weiyuan Zhang[1]($\boxtimes$) and Yajie Wang[2]

[1] Institute of Nonlinear Science, Xianyang Normal University,
Xianyang 712000, People's Republic of China
ahzwy@163.com
[2] Institute of Mathematics and Applied Mathematics,
Xianyang Normal University, Xianyang 712000, People's Republic of China
195297832@qq.com

Abstract. This paper is concerned with a delay-differential equation modeling a bidirectional associative memory neural networks (BAM NNs) with mixed time delays. By using the inequality techniques, a Lyapunov–Krasovskii functional candidate is introduced to reach the novel sufficient conditions that warrant the passivity of delayed BAM NNs. The novel passivity criterion is proposed in terms of inequalities, which can be checked easily. A numerical example is provided to demonstrate the effectiveness of the proposed results.

Keywords: BAM NNs · Delays · Passivity · Lyapunov functional

1 Introduction

In recent years, the problem of neural networks (NNs) has been widely investigated in the last few decades due to their potential applications in many areas such as pattern recognition, static image processing, associative memory and combinatorial optimization [1, 2]. It is known that stability is one of the most important properties for the designed NNs. On the other hand, both in biological and artificial neural networks, the interactions between neurons are generally asynchronous, which inevitably results in time delays. In electronic implementation of analog neural networks, nevertheless, the delays are usually time-varying due to the finite switching speed of amplifiers. Meanwhile, time delays are inevitable encountered in many practical systems and usually the main reason resulting in instability, therefore, much attention has been focused on the problem of stability analysis for delayed NNs, and many related delay-independent and delay-dependent criteria have been reported in recent literature, see, e.g., [3–8]. Furthermore, as a powerful tool for analyzing the stability of systems, passivity theory has also considerable backgrounds in many control fields, for instance, fuzzy control [9] and signal processing [10].

Since NNs related to BAM have been proposed by Kosko [11], the BAM NNs have been one of the most interesting research topics and extensively studied due to its potential applications in pattern recognition, etc. This class of NNs have been successfully applied to pattern recognition, signal and image processing, artificial intelligence due to its generalization of the single-layer auto-associative Hebbian correlation

© Springer International Publishing Switzerland 2015
D.-S. Huang and K. Han (Eds.): ICIC 2015, Part III, LNAI 9227, pp. 127–135, 2015.
DOI: 10.1007/978-3-319-22053-6_15

to a two-layer pattern matched hetero-associative circuits. These applications are built upon the stability of the equilibrium of neural networks. Thus, the stability analysis is a necessary step for the design and applications of NNs. It is inspiring that the stability analysis of BAM NNs has been widely investigated and various stability conditions for BAM neural networks have been presented in the literature [12–14]. Some of these applications require that the designed network has a unique stable equilibrium point [15]. Recently, this two-layer hetero associative networks called BAM networks with axonal signal transmission delays have been studied in [15], which have been used to obtain important advances in many fields such as pattern recognition and automatic control.

Stability problems are often linked to the theory of dissipative systems, which postulate that the energy dissipated inside a dynamic system is less than that supplied from external source. Passivity is part of a broader and a general theory of dissipativeness [16]. The passivity theory originated from circuit theory plays an important role in the analysis and design of linear and nonlinear systems. It should be pointed out that the essence of the passivity theory is that the passive properties of a system can keep the system internal stability. The passivity theory was firstly proposed in the circuit analysis [17], and since then has found successful applications in diverse areas such as stability [18], complexity [19] and so on. These are the main reasons why passivity theory has become a very hot topic across many fields, and much investigative attention has been focused on this topic. It is noted that research on passivity has attracted so much attention, little of that had been devoted to the passivity properties of delayed NNs until [20] derived the conditions for passivity of delayed NNs. The results on passivity analysis of discrete-time NNs with time-varying delays can be found in [21].

In fact, one remarkable feature of passivity is that the passive system utilizes the product of input and output as the energy provision, and embodies the energy attenuation character. Passive system only burns energy, without energy production, i.e., passivity represents the property of energy consumption of the system. In addition, the passivity analysis for NNs can help us understand the complex brain functionalities. Few authors considered the passivity problem for delayed NNs. It is interesting and important to discuss the passivity of NNs. Based on the above discussion, there are many results about the passivity problem for delayed NNs. To the best of our knowledge, passivity analysis is seldom reported for the class of delayed BAM NNs. This paper will investigate the passivity problem for a class of BAM NNs with mixed delays, which is very important in theories and applications and also is a very challenging problem. Sufficient general condition is derived for the passivity of delayed BAM NNs which are very convenient to verify. Finally, a numerical example is illustrated to show the usefulness of the proposed criteria.

2 Model Description and Preliminaries

In this paper, we consider the BAM NNs model described by the following differential equations:

$$\frac{\partial u_i}{\partial t} = -p_i u_i(t) + \sum_{j=1}^{n} b_{ji} f_j(v_j(t)) + \sum_{j=1}^{n} \tilde{b}_{ji} f_j(v_j(t - \theta_{ji}(t)))$$

$$+ \sum_{j=1}^{n} \bar{b}_{ji} \int_{-\infty}^{t} k_{ji}(t - s) f_j(v_j(s)) ds + \sigma_i(t),$$

$$\frac{\partial v_j}{\partial t} = -q_j v_j(t) + \sum_{i=1}^{m} d_{ij} g_i(u_i(t)) + \sum_{i=1}^{m} \tilde{d}_{ij} g_i(u_i(t - \tau_{ij}(t))) \tag{1}$$

$$+ \sum_{i=1}^{m} \bar{d}_{ij} \int_{-\infty}^{t} \bar{k}_{ij}(t - s) g_i(u_i(s)) ds + \vartheta_j(t),$$

$$y_i(t) = a_i u_i(t) + c_i \sigma_i(t), z_j(t) = \bar{a}_j v_j(t) + \bar{c}_j \vartheta_j(t),$$

where $u = (u_1, u_2, \ldots, u_m)^T \in R^m$, $v = (v_1, v_2, \ldots, v_n)^T \in R^n$, $u_i(t)$ and $v_j(t)$ represent the states of the ith neuron and the jth neuron, respectively. $b_{ji}, \tilde{b}_{ji}, \bar{b}_{ji}, d_{ij}, \tilde{d}_{ij}$ and $\bar{d}_{ij}$ are known constants denoting the synaptic connection strengths between the neurons, respectively; f_j and g_i denote the activation functions; $\sigma_i(t)$ and $\vartheta_j(t)$ denote inputs of the ith neuron and the jth neuron at time t, respectively; p_i and q_j represent the rate with which the ith neuron and the jth neuron will reset its potential to the resting state when disconnected from the networks and external inputs in space, respectively; $\tau_{ij}(t)$ and $\theta_{ji}(t)$ represent continuous time-varying discrete delays, respectively; $y_i(t)$ and $z_j(t)$ denote outputs of the ith neuron and the jth neuron at time t, respectively. $i = 1, 2, \ldots, m$, $k = 1, 2, \ldots, l$ and $j = 1, 2, \ldots, n$.

For system (1), the following assumptions are made for each subsystem in this paper:
(A1) The functions $\tau_{ij}(t), \theta_{ji}(t)$ are piecewise-continuous of class C^1 on the closure of each continuity subinterval and satisfy

$$0 \le \tau_{ij}(t) \le \tau_{ij}, 0 \le \theta_{ji}(t) \le \theta_{ji}, \dot{\tau}_{ij}(t) \le \mu_\tau < 1, \dot{\theta}_{ji}(t) \le \mu_\theta < 1,$$

$$\tau = \max_{1 \le i \le m, 1 \le j \le n} \{\tau_{ij}\}, \theta = \max_{1 \le i \le m, 1 \le j \le n} \{\theta_{ji}\},$$

with some constants $\tau_{ij} \ge 0, \theta_{ji} \ge 0$, $\tau > 0, \theta > 0$, for all $t \ge 0$.
(A2) The activation functions are bounded and Lipschitz continuous, i.e., there exist positive constants L_j^f and L_i^g such that for all $\eta_1, \eta_2 \in R$

$$|f_j(\eta_1) - f_j(\eta_2)| \le L_j^f |\eta_1 - \eta_2|,$$

$$|g_i(\eta_1) - g_i(\eta_2)| \le L_i^g |\eta_1 - \eta_2|.$$

(A3) The delay kernels $K_{ji}(s), \bar{K}_{ij}(s) : [0, \infty) \to [0, \infty), (i = 1, 2, \ldots, m, j = 1, 2, \ldots, n)$ are real-valued non-negative continuous functions that satisfy the following conditions

(i) $\int_0^{+\infty} K_{ji}(s) ds = 1, \int_0^{+\infty} \bar{K}_{ji}(s) ds = 1,$
(ii) $\int_0^{+\infty} s K_{ji}(s) ds < \infty, \int_0^{+\infty} s \bar{K}_{ij}(s) ds < \infty,$

Definition 1. A system with inputs $\sigma(t)$, $\vartheta(t)$ and outputs $y(t), z(t)$ where $\sigma(t) \in R^m$, $\vartheta(t) \in R^n, y(t) \in R^m$, $z(t) \in R^m$ is said to be passive if there are constants $\gamma \geq 0$ and $\beta \in R$ such that

$$2 \int_0^{t_p} \left(y(t)^T \sigma(t) + z(t)^T \vartheta(t) \right) dt$$

$$\geq -\beta^2 - \gamma \int_0^{t_p} \left(\sigma(t)^T \sigma(t) + \vartheta(t)^T \vartheta(t) \right) dt$$

for all $t_p \geq 0$, where *Omega* is a bounded compact set.

3 Passive Analysis

Theorem 1. Under the assumption (A1)-(A3), system (1) is passive if there exist constants $w_i, w_{m+j} > 0$ and $\gamma > 0$ such that

$$\begin{pmatrix} \Xi_i & w_i - a_i \\ w_i - a_i & -\gamma - 2c_i \end{pmatrix} \leq 0, \tag{2}$$

and

$$\begin{pmatrix} \Theta_j & w_{m+j} - \bar{a}_j \\ w_{m+j} - \bar{a}_j & -\gamma - 2\bar{c}_j \end{pmatrix} \leq 0 \tag{3}$$

in which $\Xi_i = w_i \left(-2p_i + \sum_{j=1}^n |b_{ji}| L_j^f + \sum_{j=1}^n |\tilde{b}_{ji}| + \sum_{j=1}^n |\bar{b}_{ji}| \right) + L_i^g \sum_{j=1}^n w_{m+j} |d_{ij}| + \frac{1}{1-\mu_\tau} (L_i^g)^2$
$\sum_{j=1}^n w_{m+j} |\tilde{d}_{ij}| + (L_i^g)^2 \sum_{j=1}^n w_{m+j} |\bar{d}_{ij}|, \Theta_j = \left(L_j^f \right)^2 \sum_{i=1}^m w_i |\bar{b}_{ji}| + L_j^f \sum_{i=1}^m w_i |b_{ji}| + w_{m+j} (-2q_j +$
$\sum_{i=1}^m |d_{ij}| L_i^g + \sum_{i=1}^m |\tilde{d}_{ij}| + 2 \sum_{i=1}^m |\bar{d}_{ij}|) + \frac{1}{1-\mu_\theta} \left(L_j^f \right)^2 \sum_{i=1}^m w_i |\bar{b}_{ji}|$, $i = 1, 2, \ldots, m$, $j = 1, 2, \ldots,$
n, L_j^f and L_i^g are Lipschitz constants.

Proof. Construct the following Lyapunov functional

$$V(t) = \sum_{i=1}^m w_i \left[u_i(t)^2 + \sum_{j=1}^n |\tilde{b}_{ji}| \frac{1}{1-\mu_\theta} \int_{t-\theta_{ji}(t)}^t |f_j(v_j(\xi))|^2 d\xi \right.$$

$$\left. + \sum_{j=1}^n |\bar{b}_{ji}| \int_0^{+\infty} k_{ji}(s) \int_{t-s}^t |f_j(v_j(\xi))|^2 d\xi ds \right] dx + \int_{Omega} \sum_{j=1}^n w_{m+j} \left[v_j(t)^2 \right.$$

$$+ \sum_{i=1}^m |\tilde{d}_{ij}| \frac{1}{1-\mu_\tau} \int_{t-\tau_{ij}(t)}^t |g_i(u_i(\xi))|^2 d\xi + \sum_{i=1}^m |\bar{d}_{ij}| \int_0^{+\infty} \bar{k}_{ij}(s) \int_{t-s}^t |g_i(u_i(\xi))|^2 d\xi ds \right].$$

Its upper Dini-derivative along the solution to system (1) can be calculated as

$$
\begin{aligned}
D^+V(t) \leq \sum_{i=1}^{m} w_i \Bigg[&-2p_i|u_i(t)|^2 + 2\sum_{j=1}^{n} |b_{ji}||u_i(t)|L_j^f|v_j(t)| + 2\sum_{j=1}^{n} |\bar{b}_{ji}||u_i(t)||f_j(v_j(t-\theta_{ji}(t)))| \\
&+ 2\sum_{j=1}^{n} |\tilde{b}_{ji}| \int_{-\infty}^{t} k_{ji}(t-s)|u_i(t)||f_j(v_j(s))|ds + 2u_i(t)\sigma_i(t) + \sum_{j=1}^{n} |\tilde{b}_{ji}|\frac{1}{1-\mu_\theta}|f_j(v_j(t))|^2 \\
&+ \sum_{j=1}^{n} |\tilde{b}_{ji}| \int_{0}^{+\infty} k_{ji}(s)|f_j(v_j(t))|^2 ds - \sum_{j=1}^{n} |\tilde{b}_{ji}|\frac{1}{1-\mu_\theta}\left(1-\dot{\theta}_{ji}(t)\right)|f_j(v_j(t-\theta_{ji}(t)))|^2 \\
&- \sum_{j=1}^{n} |\tilde{b}_{ji}| \int_{0}^{+\infty} k_{ji}(s)|f_j(v_j(t-s))|^2 ds \Bigg] + \sum_{j=1}^{n} w_{m+j} \Bigg[-2q_j|v_j(t)|^2 \\
&+ 2\sum_{i=1}^{m} |d_{ij}|L_i^g|u_i(t)||v_j(t)| + 2\sum_{i=1}^{m} |\bar{d}_{ij}||g_i(u_i(t-\tau_{ij}(t)))||v_j(t)| \\
&+ 2\sum_{i=1}^{m} |\tilde{d}_{ij}| \int_{-\infty}^{t} \bar{k}_{ij}(t-s)|g_i(u_i(s))||v_j(t)|ds + 2v_j(t)\vartheta_j(t) \\
&+ \sum_{i=1}^{m} |\tilde{d}_{ij}|\frac{1}{1-\mu_\tau}|g_i(u_i(t))|^2 - \sum_{i=1}^{m} |\tilde{d}_{ij}|\frac{1}{1-\mu_\tau}\left(1-\dot{\tau}_{ij}(t)\right)|g_i(u_i(t-\tau_{ij}(t)))|^2 \\
&+ \sum_{i=1}^{m} |\tilde{d}_{ij}| \int_{0}^{+\infty} \bar{k}_{ij}(s)|g_i(u_i(t))|^2 ds - \sum_{i=1}^{m} |\tilde{d}_{ij}| \int_{0}^{+\infty} \bar{k}_{ij}(s)|g_i(u_i(t-s))|^2 ds \Bigg] dx
\end{aligned}
$$

(4)

According to (4) and (A1)-(A3), we can drive

$$
\begin{aligned}
D^+V(t) \leq \sum_{i=1}^{m} w_i \Bigg[&-2p_i|u_i(t)|^2 + \sum_{j=1}^{n} |b_{ji}|L_j^f\left(|u_i(t)|^2+|v_j(t)|^2\right) \\
&+ \sum_{j=1}^{n} |\bar{b}_{ji}|\left(|u_i(t)|^2+|f_j(v_j(t-\theta_{ji}(t)))|^2\right) \\
&+ \sum_{j=1}^{n} |\tilde{b}_{ji}| \int_{-\infty}^{t} k_{ji}(t-s)\left(|u_i(t)|^2+|f_j(v_j(s))|^2\right)ds + 2u_i(t)\sigma_i(t) \\
&+ \sum_{j=1}^{n} |\tilde{b}_{ji}|\frac{1}{1-\mu_\theta}|f_j(v_j(t))|^2 + \sum_{j=1}^{n} |\tilde{b}_{ji}| \int_{0}^{+\infty} k_{ji}(s)|f_j(v_j(t))|^2 ds \\
&- \sum_{j=1}^{n} |\tilde{b}_{ji}||f_j(v_j(t-\theta_{ji}(t)))|^2 - \sum_{j=1}^{n} |\tilde{b}_{ji}| \int_{0}^{+\infty} k_{ji}(s)|f_j(v_j(t-s))|^2 ds \Bigg] dx \\
&+ \int_{\Omega} \sum_{j=1}^{n} w_{m+j} \Bigg[-2q_j|v_j(t)|^2 + \sum_{i=1}^{m} |d_{ij}|L_i^g\left(|u_i(t)|^2+|v_j(t)|^2\right) \\
&+ \sum_{i=1}^{m} |\tilde{d}_{ij}|\left(|g_i(u_i(t-\tau_{ij}(t)))|^2+|v_j(t)|^2\right)
\end{aligned}
$$

$$+ 2\sum_{i=1}^{m}|\overline{d}_{ij}|\int_{-\infty}^{t}\overline{k}_{ij}(t-s)(|g_i(u_i(s))|^2+|v_j(t)|^2)ds + 2v_j(t)\vartheta_j(t)$$

$$+ \sum_{i=1}^{m}|\tilde{d}_{ij}|\frac{1}{1-\mu_\tau}|g_i(u_i(t))|^2 - \sum_{i=1}^{m}|\tilde{d}_{ij}||g_i(u_i(t-\tau_{ij}(t)))|^2$$

$$+ \sum_{i=1}^{m}|\overline{d}_{ij}|\int_{0}^{+\infty}\overline{k}_{ij}(s)|g_i(u_i(t))|^2 ds - \sum_{i=1}^{m}|\overline{d}_{ij}|\int_{0}^{+\infty}\overline{k}_{ij}(s)|g_i(u_i(t-s))|^2 ds \Bigg] dx$$

$$\leq \sum_{i=1}^{m}\Bigg\{ w_i\Bigg[-2p_i + \sum_{j=1}^{n}|b_{ji}|L_j^f + \sum_{j=1}^{n}|\tilde{b}_{ji}| + \sum_{j=1}^{n}|\overline{b}_{ji}| \Bigg]$$

$$+ L_i^g\sum_{j=1}^{n}w_{m+j}|d_{ij}| + \frac{1}{1-\mu_\tau}(L_i^g)^2\sum_{j=1}^{n}w_{m+j}|\tilde{d}_{ij}| + (L_i^g)^2\sum_{j=1}^{n}w_{m+j}|\overline{d}_{ij}| \Bigg\}|u_i(t)|^2$$

$$+ \int_{\Omega}\sum_{j=1}^{n}\Bigg\{ w_{m+j}\Bigg[-2q_j + \sum_{i=1}^{m}|d_{ij}|L_i^g + \sum_{i=1}^{m}|\tilde{d}_{ij}| + 2\sum_{i=1}^{m}|\overline{d}_{ij}| \Bigg]$$

$$+ \frac{1}{1-\mu_\theta}(L_j^f)^2\sum_{i=1}^{m}w_i|\tilde{b}_{ji}| + (L_j^f)^2\sum_{i=1}^{m}w_i|\overline{b}_{ji}|\int_{0}^{+\infty}k_{ji}(s)ds + L_j^f\sum_{i=1}^{m}w_i|b_{ji}| \Bigg\}|v_j(t)|^2$$

$$+ 2\int_{\Omega}\Bigg[\sum_{i=1}^{m}w_iu_i(t)\sigma_i(t) + \sum_{j=1}^{n}w_{m+j}v_j(t)\vartheta_j(t) \Bigg]$$

$$\leq \sum_{i=1}^{m}\Bigg\{ \Bigg[w_i\Bigg(-2p_i + \sum_{j=1}^{n}|b_{ji}|L_j^f + \sum_{j=1}^{n}|\tilde{b}_{ji}| + \sum_{j=1}^{n}|\overline{b}_{ji}| \Bigg)$$

$$+ L_i^g\sum_{j=1}^{n}w_{m+j}|d_{ij}| + \frac{1}{1-\mu_\tau}(L_i^g)^2\sum_{j=1}^{n}w_{m+j}|\tilde{d}_{ij}| + (L_i^g)^2\sum_{j=1}^{n}w_{m+j}|\overline{d}_{ij}| \Bigg]|u_i(t)|^2$$

$$+ \Big[2(w_i - a_i)u_i(t)\sigma_i(t) + (-\gamma - 2c_i)\sigma_i(t)^2 \Big] \Bigg\}$$

$$+ \sum_{j=1}^{n}\Bigg\{ \Bigg[w_{m+j}\Bigg(-2q_j + \sum_{i=1}^{m}|d_{ij}|L_i^g + \sum_{i=1}^{m}|\tilde{d}_{ij}| + 2\sum_{i=1}^{m}|\overline{d}_{ij}| \Bigg)$$

$$+ \frac{1}{1-\mu_\theta}(L_j^f)^2\sum_{i=1}^{m}w_i|\tilde{b}_{ji}| + (L_j^f)^2\sum_{i=1}^{m}w_i|\overline{b}_{ji}| + L_j^f\sum_{i=1}^{m}w_i|b_{ji}| \Bigg]|v_j(t)|^2$$

$$+ \Big[2(w_{m+j} - \overline{a}_i)v_j(t)\vartheta_j(t) + (-\gamma - 2\overline{c}_i)\vartheta_j(t)^2 \Big] \Bigg\}$$

$$= \sum_{i=1}^{m}\begin{pmatrix} u_i(t) \\ \sigma_i(t) \end{pmatrix}^T \begin{pmatrix} \Xi_i & w_i - a_i \\ w_i - a_i & -\gamma - 2c_i \end{pmatrix}\begin{pmatrix} u_i(t) \\ \sigma_i(t) \end{pmatrix}$$

$$+ \sum_{j=1}^{n}\begin{pmatrix} v_j(t) \\ \vartheta_j(t) \end{pmatrix}^T \begin{pmatrix} \Theta_j & w_{m+j} - \overline{a}_j \\ w_{m+j} - \overline{a}_j & -\gamma - 2\overline{c}_j \end{pmatrix}\begin{pmatrix} v_j(t) \\ \vartheta_j(t) \end{pmatrix}$$

$$\text{(5)}$$

From (2), (3), (5) $V(t_p) \geq 0$ and $V(0) \geq 0$, we can get

$$2 \int_0^{t_p} \left[\sum_{i=1}^m y_i(t)\sigma_i(t) + \sum_{j=1}^n z_j(t)\vartheta_j(t) \right] dt \geq - \beta^2 - \gamma \int_0^{t_p} \left[\sum_{i=1}^m \sigma_i(t)^2 + \sum_{j=1}^n \vartheta_j(t)^2 \right] dt$$

(6)

for all $t_p \geq 0$, where $\beta = \sqrt{V(0)}$. This completes the proof of Theorem 1.

4 Illustration Example

Example 1. Consider the following delayed BAM NNs:

$$\frac{\partial u_i}{\partial t} = -p_i u_i(t) + \sum_{j=1}^n b_{ji} f_j(v_j(t)) + \sum_{j=1}^n \tilde{b}_{ji} f_j(v_j(t - \theta_{ji}(t)))$$

$$+ \sum_{j=1}^n \bar{b}_{ji} \int_{-\infty}^t k_{ji}(t - s) f_j(v_j(s)) ds + \sigma_i(t),$$

$$\frac{\partial v_j}{\partial t} = -q_j v_j(t) + \sum_{i=1}^m d_{ij} g_i(u_i(t)) + \sum_{i=1}^m \tilde{d}_{ij} g_i(u_i(t - \tau_{ij}(t)))$$

$$+ \sum_{i=1}^m \bar{d}_{ij} \int_{-\infty}^t \bar{k}_{ij}(t - s) g_i(u_i(s)) ds + \vartheta_j(t),$$

$$y_i(t) = a_i u_i(t) + c_i \sigma_i(t), z_j(t) = \bar{a}_j v_j(t) + \bar{c}_j \vartheta_j(t),$$

in which $n = m = r = 2, k_{ji}(t) = \bar{k}_{ij}(t) = te^{-t}, f_j(\eta) = g_i(\eta) = \frac{1}{2}(|\eta + 1| + |\eta - 1|)$, $L_j^f = L_j^{\bar{f}} = L_i^g = L_i^{\bar{g}} = 1, \lambda_1 = 3, \tau = \theta = \ln 2.a_i = \bar{a}_j = 1, c_i = \bar{c}_j = 0.5, p_i = 4, q_j = 3, i, j = 1, 2, \mu_\tau = \mu_\theta = 0.2, d_{11} = 0.5, d_{12} = 1, d_{21} = 0.5, d_{22} = 0.2, \bar{d}_{11} = 0.2, \bar{d}_{12} = 0.6, \bar{d}_{21} = 0.5, \tilde{d}_{11} = 0.3, \tilde{d}_{12} = 0.8, \tilde{d}_{21} = 0.1, \tilde{d}_{22} = 0.2, \bar{d}_{22} = 0.8, b_{11} = 0.5, b_{12} = 0.6, b_{21} = 1, b_{22} = -0.8, \bar{b}_{11} = -1, \bar{b}_{12} = 0.2, \bar{b}_{21} = 0.5, \bar{b}_{22} = 0.4, \tilde{b}_{11} = -0.5, \tilde{b}_{12} = 0.1, \tilde{b}_{21} = 0.3, \tilde{b}_{22} = 0.5$. By simple calculation with $w_1 = w_2 = w_3 = w_4 = 1$ and $\gamma = 1$, we have

$$\begin{pmatrix} \Xi_1 & w_1 - a_1 \\ w_1 - a_1 & -\gamma - 2c_1 \end{pmatrix} = \begin{pmatrix} -4.525 & 0 \\ 0 & -2 \end{pmatrix} \leq 0,$$

$$\begin{pmatrix} \Xi_2 & w_2 - a_2 \\ w_2 - a_2 & -\gamma - 2c_2 \end{pmatrix} = \begin{pmatrix} -7.225 & 0 \\ 0 & -2 \end{pmatrix} \leq 0,$$

and

$$\begin{pmatrix} \Theta_1 & w_3 - \bar{a}_1 \\ w_3 - \bar{a}_1 & -\gamma - 2\bar{c}_1 \end{pmatrix} = \begin{pmatrix} -3.15 & 0 \\ 0 & -2 \end{pmatrix} \leq 0,$$

$$\begin{pmatrix} \Theta_2 & w_4 - \bar{a}_2 \\ w_4 - \bar{a}_2 & -\gamma - 2\bar{c}_2 \end{pmatrix} = \begin{pmatrix} -1.3 & 0 \\ 0 & -2 \end{pmatrix} \leq 0.$$

That is (7) and (8) hold. Therefore, it follows from Theorem 1 that system (35) is passive.

5 Conclusions

In this paper, we have investigated the passivity analysis problem for a class of BAM NNs with mixed time delays. We have developed novel sufficient condition to ensure the passivity of BAM NNs with mixed time delays. In particular, many techniques and approaches such as Lyapunov functional and classical inequalities, have been successfully applied. Hence, the results obtained in this paper are less conservative, and generalize and improve many earlier results. Finally, a numerical example has been presented to show the effectiveness of the derived results.

Acknowledgement. This work is partially supported by the Natural Science Foundation of Shannxi Province under Grant No. 2015JM1015, Doctor Introduced Project of Xianyang Normal University under Grant No. 12XSYK008 and University Innovation and Entrepreneurship Training Program Project of Xianyang Normal University under Grant No.048.

References

1. Xu, S., Lam, J., Ho, D.: A new LMI condition for delayed dependent asymptotic stability of delayed Hopfield neural networks. IEEE Trans. Neural Netw. **52**, 230–234 (2006)
2. He, Y., Liu, G., Rees, D.: New delay-dependent stability criteria for neural networks with time-varying delay. IEEE Trans. Neural Netw. **18**, 310–314 (2007)
3. Kwon, O.M., Lee, S.M., Park, J., Cha, E.J.: New approaches on stability criteria for neural networks with interval time-varying delays. Appl. Math. Comput. **218**, 9953–9964 (2012)
4. Sun, J., Chen, J.: Stability analysis of static recurrent neural networks with interval time-varying delay. Appl. Math. Comput. **221**, 111–120 (2013)
5. Kwon, O., Park, J., Lee, S., Cha, E.: Analysis on delay-dependent stability for neural networks with time-varying delays. Neurocomputing **103**, 114–120 (2013)
6. Li, T., Yang, X., Yang, P., Fei, S.: New delay-variation-dependent stability for neural networks with time-varying delay. Neurocomputing **101**, 361–369 (2013)
7. Yang, R., Zhang, Z., Shi, P.: Exponential stability on stochastic neural networks with discrete interval and distributed delays. IEEE Trans. Neural Netw. **2**, 169–175 (2010)
8. Zeng, H., He, Y., Wu, M.: Complete delay-decomposing approach to asymptotic stability for neural networks with time-varying delays. IEEE Trans. Neural Netw. **22**, 806–812 (2011)

9. Calcev, G.: Passivity approach to fuzzy control systems. Automatica **33**, 339–344 (1998)
10. Xie, L., Fu, M., Li, H.: Passivity and passification for uncertain signal processing systems. IEEE Trans. Signal Process. **46**, 2394–2403 (1998)
11. Kosko, B.: Bi-directional associative memories. IEEE Trans Syst Man Cybern **18**, 49–60 (1988)
12. Li, C., Liao, X., Zhang, R.: Delay-dependent exponential stability analysis of bi-directional associative memory neural networks with time delay: an LMI approach. Chaos Soliton. Fract. **24**, 1119–1134 (2005)
13. Zhao, H.: Global stability of bidirectional associative memory neural networks with distributed delays. Phys. Lett. A **297**, 182–190 (2002)
14. Zhou, Q., Wan, L.: Global robust asymptotic stability analysis of BAM neural networks with time delay and impulse: An LMI approach. Appl. Math. Comput. **216**, 1538–1545 (2010)
15. Gopalsamy, K., He, X.Z.: Delay-independent stability in bidirectional associative memory networks. IEEE Trans. Neural Networks **5**, 998–1002 (1994)
16. Willems, J.C.: Dissipative dynamical systems part I: General theory. Arch. Rational Mech. Anal. **45**, 321–351 (1972)
17. Bevelevich, V.: Classical Network Synthesis. Van Nostrand, New York (1968)
18. Hill, D.J., Moylan, P.J.: Stability results for nonlinear feedback systems. Automatica **13**, 377–382 (1977)
19. Chua, L.O.: Passivity and complexity. IEEE Trans. Circuits Syst. I: Fundam. Theory Appl. **46**, 71–82 (1999)
20. Li, C., Liao, X.: Passivity analysis of neural networks with time delay. IEEE Trans. Circuits Syst. II **52**, 471–475 (2005)
21. Wu, Z., Shi, P., Su, H., Chu, J.: Passivity analysis for discrete-time stochastic Markovian jump neural networks with mixed time delays. IEEE Trans. Neural Networks **22**, 1566–1575 (2011)

Soft Computing

A Fuzzy Logic Controller for Indirect Matrix Converter Under Abnormal Input Voltage Conditions

Quoc-Hoan Tran and Hong-Hee Lee$^{(\boxtimes)}$

School of Electrical Engineering, University of Ulsan, Ulsan, Korea
tranquochoan06@gmail.com, hhlee@mail.ulsan.ac.kr

Abstract. In this paper, an intelligent control method based on a fuzzy logic controller (FLC) has been presented for an indirect matrix converter under abnormal input voltage conditions. The proposed FLC method is a closed-loop control of the output current, so the three-phase sinusoidal and balanced output current are assured regardless of unbalanced or distorted input voltage. The performance of the MC with the proposed FLC method becomes much better than that with the traditional PI controller. Simulation results are given to verify the effectiveness of the proposed FLC method.

Keywords: Fuzzy logic controller · Indirect matrix converter · Unbalanced voltage · Distorted voltage

1 Introduction

A matrix converter (MC) is a single-stage power converter which can generate an ac voltage with variable amplitude and frequency from constant ac voltage source. The MC has been recently received more considerable attention due to its advantages, such as sinusoidal input/output current waveforms, controllable input power factor, no energy storage devices and bidirectional power flow [1]. The MC is usually classified into two types: direct matrix converter (DMC) and indirect matrix converter (IMC) [2]. The DMC and IMC have similar performance in terms of input/output current wave-forms and voltage transfer ratio. However, the IMC in Fig. 1 has recently become more attractive compared to the DMC because of its additional advantages such as simpler and safer commutation, simple clamping circuit, and reduced number of power switches [3, 4].

However, since there is no energy storage devices inside the topology, the MCs are directly connected from an input voltage source to a load. Hence, the input voltage affects the output performance immediately, and the output voltage waveform can be non-sinusoidal under abnormal input voltage, which makes the output current distorted due to the output voltage harmonics. Several solutions have been reported in order to reduce the influence of the abnormal operating conditions for the MCs. The proportional-integral (PI) controllers based on the closed-loop control of input voltage and output current have been proposed to improve the output performance of the MCs [5, 6]. These PI control methods can attenuate the undesired harmonics components.

© Springer International Publishing Switzerland 2015
D.-S. Huang and K. Han (Eds.): ICIC 2015, Part III, LNAI 9227, pp. 139–150, 2015.
DOI: 10.1007/978-3-319-22053-6_16

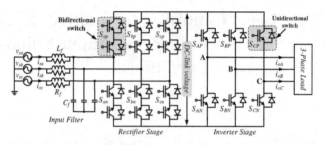

Fig. 1. Indirect matrix converter topology.

However, the control performance significantly depends on PI parameters sensitively. In practice, many non-ideal factors and disturbances may exist together and unavoidably, so the turning of the PI parameters becomes difficult. Moreover, the PI controllers with constant gains cannot guarantee high quality for all load conditions.

Therefore, the intelligent control methods based on fuzzy logic controller (FLC) have recently been investigated for MC control to overcome the drawback of PI control methods. The FLC methods have been adopted to compensate the influence of distorted input voltage [7–9]. In [10], a fuzzy control adaptive method has been used to compensate the undesired harmonics caused by nonlinear loads. On the other work, the adaptive neural fuzzy inference system control method based on MC fed induction motor drive has been employed as an intelligent tool to replace the conventional PI controllers for high performance applications [11]. Furthermore, the comparisons have also proved the superiority of the FLC methods than other methods in complex and nonlinear systems [12, 13]. However, all aforementioned research studies are established for DMC. The application of FLC method in IMC has not been presented in the literature.

This paper presents a feedback compensation method based on the FLC method to improve the output performance of an IMC under abnormal input voltage conditions. The proposed FLC method has a closed-loop to control the output current in order to adapt the voltage transfer ratio to the instantaneous value of input voltages. The proposed FLC method can guarantee three-phase output currents to be sinusoidal and balanced in spite of the abnormal input voltage. The feasibility of the proposed FLC method is verified by simulated results.

2   Control Strategy for IMC Under Normal Condition

As shown in Fig. 1, IMC topology comprises of a rectifier stage and an inverter stage. The rectifier stage is similar to the traditional current source rectifier with bidirectional switches. The purpose of the rectifier stage is to supply a positive dc-link voltage and maintain sinusoidal input currents. While the inverter stage is a conventional two-level voltage source inverter (VSI). The output voltage with variable frequency and amplitude is generated by controlling the inverter stage. The indirect space vector modulation (SVM) method, which was introduced in [2–4], is generally used to control the IMC.

The SVM method is based on the space vector analysis of the input current and output voltage under the constraint of unity input power factor. It can also provide the possibility to obtain the highest voltage transfer ratio and to optimize the switching patterns by combining the switching states between the rectifier stage and the inverter stage.

2.1 Rectifier Stage Control

It is assumed that three-phase voltage source is sinusoidal and balanced as follows:

$$[v_s] = \begin{bmatrix} v_{sa} \\ v_{sb} \\ v_{sc} \end{bmatrix} = V_s \begin{bmatrix} \cos(\omega_s t) \\ \cos(\omega_s t - 2\pi/3) \\ \cos(\omega_s t - 4\pi/3) \end{bmatrix} \tag{1}$$

where V_s is the input phase voltage magnitude and ω_s is the input angular frequency. To explain the SVM method for the rectifier stage, the reference input current space vector is defined as follows:

$$\vec{i}_i = \frac{2}{3}\left(i_{ia} + i_{ib}e^{j2\pi/3} + i_{ic}e^{j4\pi/3}\right) = I_i e^{j\theta_i}. \tag{2}$$

Figure 2(a) shows the space vector diagram of the rectifier stage which is composed of six active vectors and three zero vectors. Each current vector represents the switching state of the bidirectional switches. For example, the current vector I_{ab} represents the input phase "a" and "b" are connected to the positive pole and the negative pole of dc-link bus voltage, respectively. So, the upper switch of phase "a" and the lower switch of phase "b", S_{ap} and S_{bn}, are ON state and all other switches are OFF state when the current vector I_{ab} is applied.

Assume that the reference input current vector is located in sector 1 $(-\pi/6 \leq \theta_i \leq \pi/6)$, the duty cycles of two active vectors I_{ab} and I_{ac} are given by:

$$d_\gamma = m_i \sin(\frac{\pi}{6} - \theta_i); \quad d_\delta = m_i \sin(\frac{\pi}{6} + \theta_i). \tag{3}$$

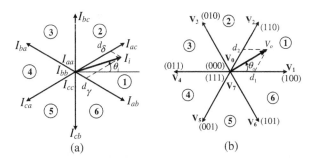

(a)   (b)

Fig. 2. Space vector diagram of the rectifier and inverter stage.

where m_i is the rectifier stage modulation index and θ_i is the angle of the reference input current vector.

In the rectifier stage control, the zero vectors are not considered and the modulation index is unity. Therefore, the duty cycles of two active vectors, I_{ab} and I_{ac}, are recalculated to complete the sampling period:

$$d_{ab} = \frac{d_\gamma}{d_\gamma + d_\delta} = \frac{\cos(\pi/3 + \theta_i)}{\cos\theta_i}; \quad d_{ac} = \frac{d_\delta}{d_\gamma + d_\delta} = \frac{\cos(\pi/3 - \theta_i)}{\cos\theta_i} \qquad (4)$$

The average value of the dc-link voltage in this sector is:

$$\bar{V}_{dc} = d_{ab}(v_{sa} - v_{sb}) + d_{ac}(v_{sa} - v_{sc}) = \frac{3V_s}{2\cos(\theta_i)}. \qquad (5)$$

The maximum and minimum values of the average dc-link voltage are:

$$V_{dc(max)} = \sqrt{3}V_s; \quad V_{dc(min)} = \frac{3}{2}V_s. \qquad (6)$$

Similarly, the duty cycles and the switching states for all sectors can be obtained.

2.2   Inverter Stage Control

In order to control the inverter stage, the conventional SVM for three-phase two-level VSI can be applied. The output voltage vector is synthesized by using six active vectors $V_1 \sim V_6$ and two zero vectors V_0, V_7 as shown in Fig. 2(b). Each voltage vector represents the switching state of the set of three upper switches of three legs, e.g., the voltage vector V_1 (100) represents for switching state of the set of switching functions (S_{AP}, S_{BP}, S_{CP}).

The reference output voltage vector can be expressed as:

$$\vec{v}_o = \frac{2}{3}\left(v_{oA} + v_{oB}e^{j2\pi/3} + v_{oC}e^{j4\pi/3}\right) = V_o e^{j\theta_o}. \qquad (7)$$

Assuming the reference output voltage vector is also located in sector 1, the duty cycles of active vectors V_1, V_2 and zero vectors V_0, V_7 are calculated as follows:

$$d_1 = \sqrt{3}\frac{V_o}{\bar{V}_{dc}}\sin(\pi/3 - \theta_o)$$

$$d_2 = \sqrt{3}\frac{V_o}{\bar{V}_{dc}}\sin(\theta_o) \qquad (8)$$

$$d_0 = d_7 = 0.5(1 - d_1 - d_2)$$

Where θ_o is the angle of reference output voltage vector.

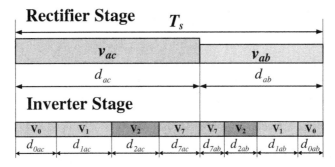

Fig. 3. The switching pattern for input current and output voltage in sector 1 of the IMC.

To obtain balanced output voltages and input currents in the same sampling period, the switching events of the inverter stage should be synchronized with those of the rectifier stage. Figure 3 demonstrates the switching pattern of the IMC when both input current vector and output voltage vector lie in sector 1. So, the duty cycles of the IMC are calculated as follows:

$$d_{1ac} = d_1.d_{ac} = \frac{2q}{\sqrt{3}}\sin(\pi/3 - \theta_o) \times \frac{\cos(\pi/3 - \theta_i)}{\cos(\theta_i)}$$

$$d_{2ac} = d_2.d_{ac} = \frac{2q}{\sqrt{3}}\sin(\theta_o) \times \frac{\cos(\pi/3 - \theta_i)}{\cos(\theta_i)}$$

$$d_{0ac} = d_{7ac} = 0.5[1 - 2q\sin(\pi/6 - \theta_o)] \times \frac{\cos(\pi/3 - \theta_i)}{\cos(\theta_i)}$$

$$d_{1ab} = d_1.d_{ab} = \frac{2q}{\sqrt{3}}\sin(\pi/3 - \theta_o) \times \frac{\cos(\pi/3 + \theta_i)}{\cos(\theta_i)} \quad (9)$$

$$d_{2ab} = d_2.d_{ab} = \frac{2q}{\sqrt{3}}\sin(\theta_o) \times \frac{\cos(\pi/3 + \theta_i)}{\cos(\theta_i)}$$

$$d_{0ab} = d_{7ab} = 0.5[1 - 2q\sin(\pi/6 - \theta_o)] \times \frac{\cos(\pi/3 + \theta_i)}{\cos(\theta_i)}$$

where $q = V_o/V_s$ is the voltage transfer ratio.

Equation (10) shows that the final duty cycles of the IMC depend on the voltage transfer ratio q. In general, this modulation method is usually used under normal condition, i.e., sinusoidal and balanced input voltages. In this case, the voltage transfer ratio q is easily calculated according to the amplitude of the input phase voltage and the desired output phase voltage, which are assumed to be constant values. Thus, the final duty cycles of the IMC are fixed during one sampling period.

3 Proposed Control Method for IMC Under Abnormal Conditions

The control method derived in the previous section is only suitable for the ideal voltage source. When the input voltages are non-sinusoidal and unbalanced, the amplitude of input voltage is not constant so that the voltage transfer ratio q becomes variable. In that case, this control method cannot provide the good output performance. To overcome this problem, a feedback compensation method is developed in this paper.

3.1 Feedback Compensation Method

The feedback control scheme of output currents is shown in Fig. 4. Three phase output currents are used to control the magnitude of the desired output current space vector. In general, when the input voltages are sinusoidal and balanced, the output currents of IMC are also sinusoidal and balanced, i.e., the magnitude of output current space vector are constant. However, if the input voltages are distorted, the magnitude of output current space vector is not constant as shown in Fig. 5.

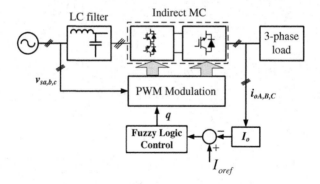

Fig. 4. Feedback control scheme of output currents.

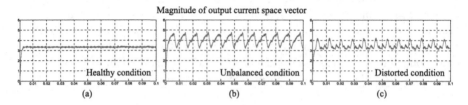

Fig. 5. The magnitude of output current space vector in: (a) healthy condition, (b) unbalanced condition, (c) distorted condition.

The magnitude of output current space vector is calculated as follows:

$$I_o = \sqrt{\frac{2}{3}\left((i_{oA})^2 + (i_{oB})^2 + (i_{oC})^2\right)} \tag{10}$$

The proposed control method tries to make the magnitude of output current space vector constant regardless of any input voltage condition. This magnitude is controlled to track the magnitude of reference output current space vector which is given by:

$$I_{oref} = \frac{q_{ref}.V_s}{\sqrt{R^2 + (\omega_0 L)^2}} \tag{11}$$

Where q_{ref} is the reference voltage transfer ratio, R and L are resistor and inductor of the load, and ω_0 is output angular frequency.

3.2 Fuzzy Logic Controller Design

Block diagram of the proposed FLC method is shown in Fig. 6 [14]. The fuzzy logic controller includes two input variables and one output variable. The two input variables are the error (e) and the change in error (Δe), which are the difference between the measured and reference magnitude of output current space vector. The output variable is the desired voltage transfer ratio (q). As mentioned in previous section, under abnormal conditions, the voltage transfer ratio should be controlled corresponding to the amplitude of input phase voltage.

The error (e) and the change in error (Δe) are calculated as follows:

$$e(k) = I_{oref}(k) - I_o(k) \tag{12}$$

$$\Delta e(k) = e(k) - e(k-1) \tag{13}$$

where k and $(k-1)$ are present and previous values, respectively.

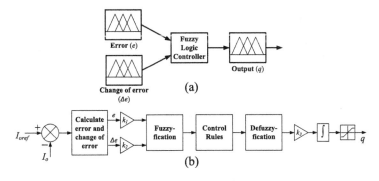

Fig. 6. Structure of the proposed FLC method.

The design of the proposed FLC method is divided into three steps: fuzzyfication, control rules design, and defuzzyfication. The asymmetrical triangular membership functions (MFs) are selected in the design of fuzzyfication and defuzzyfication process which can get more precision, as shown in Fig. 7. It should be noted that the complete range of all the fuzzy variables $e(pu)$, $\Delta e(pu)$, and $q(pu)$ spreads in the region from -1 to +1, and the MFs are symmetrical on both positive and negative sides. The input variables have seven MFs, whereas the output variable has nine MFs. The control rules include 49 rules as shown in Table 1.

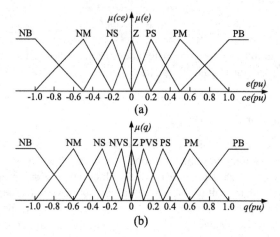

Fig. 7. Membership functions of proposed FLC method.

Table 1. Rule base for proposed FLC method

Δe\\e	NB	NM	NS	Z	PS	PM	PB
NB	NB	NB	NB	NM	NS	NVS	Z
NM	NB	NB	NM	NS	NVS	Z	PVS
NS	NB	NM	NS	NVS	Z	PVS	PS
Z	NM	NS	NVS	Z	PVS	PS	PM
PS	NS	NVS	Z	PVS	PS	PM	PB
PM	NVS	Z	PVS	PS	PM	PB	PB
PB	Z	PVS	PS	PM	PB	PB	PB

4   Simulation Results

In order to verify the effectiveness of the proposed FLC method, some numerical simulations have been carried out by using MATLAB/SIMULINK. The simulation parameters are given as follows:

- Three-phase voltage source: 100 V / 60 Hz (phase-to-neutral)
- Input filter: L_f = 1.4 mH, C_f = 25 μF

- Three-phase R-L load: $R = 20\ \Omega$, $L = 15$ mH
- Output frequency: $f_o = 50$ Hz
- Reference voltage transfer ratio: $q_{ref} = 0.7$
- PWM frequency: 10 kHz ($Ts = 100\ \mu s$)

Figure 8 shows the simulated results of three-phase input current, output line voltage, and three-phase output current in case of the input voltage is sinusoidal and balanced. Under the normal condition, the conventional SVM shows good performance with sinusoidal input and output current waveforms.

In order to investigate the performance under abnormal condition, the following unbalanced input voltages are applied:

$$\begin{cases} v_{sa} = V_s \cos(\omega_s t) \\ v_{sb} = 1.5 V_s \cos(\omega_s t - 2\pi/3) \\ v_{sc} = 0.8 V_s \cos(\omega_s t + 2\pi/3) \end{cases} \tag{14}$$

The waveforms of three-phase input voltage and output line voltage are shown in Figs. 9(a) and (b), respectively. Owing to the absence of energy storage devices, the output voltage is simultaneously affected by the unbalanced input voltages. However, the output currents are kept balanced and sinusoidal by using the proposed FLC method, while they are distorted without the proposed FLC method as shown in Fig. 9 (c) and Fig. 9(d).

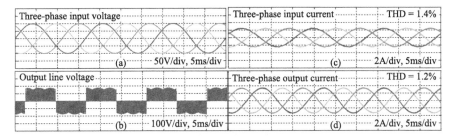

Fig. 8. Simulated results under normal condition: (a) three-phase input voltage, (b) output line voltage, (c) three-phase input current, and (d) three-phase output current.

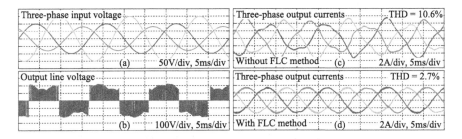

Fig. 9. Simulated results under unbalanced condition: (a) three-phase input voltage, (b) output line voltage, (c) three-phase output current without proposed FLC method, and (d) three-phase output current with proposed FLC method.

$$[v_s] = \begin{bmatrix} v_{sa} \\ v_{sb} \\ v_{sc} \end{bmatrix} = \begin{bmatrix} V_s \cos(\omega_s t) \\ V_s \cos(\omega_s t - 2\pi/3) \\ V_s \cos(\omega_s t + 2\pi/3) \end{bmatrix} + \begin{bmatrix} 0.2V_s \cos(3 * (\omega_s t + \pi/20)) \\ 0.2V_s \cos(3 * (\omega_s t - 2\pi/3 + \pi/20)) \\ 0.2V_s \cos(3 * (\omega_s t + 2\pi/3 + \pi/20)) \end{bmatrix}$$
$$+ \begin{bmatrix} 0.1V_s \cos(5 * (\omega_s t + \pi/10)) \\ 0.1V_s \cos(5 * (\omega_s t - 2\pi/3 + \pi/10)) \\ 0.1V_s \cos(5 * (\omega_s t + 2\pi/3 + \pi/10)) \end{bmatrix}$$

$$(15)$$

Similarly, the distorted input voltages in (15) are applied and the simulated results are shown in Fig. 10. In spite of the distorted input voltage with 22.4 % total harmonic distortion (THD), the output currents are kept balanced and sinusoidal thanks to the proposed FLC method.

In order to compare with the PI control method, the dynamic response of the output current space vector magnitude is examined under variable load conditions. The dynamic response of the output current space vector magnitude and three-phase output currents are shown in Fig. 11 for the PI control method and the proposed FLC method. It is obvious that the performance of the FLC method is better than that of PI control method.

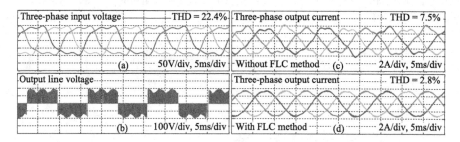

Fig. 10. Simulated results under distorted condition: (a) three-phase input voltage, (b) output line voltage, (c) three-phase output current without proposed FLC method, and (d) three-phase output current with proposed FLC method.

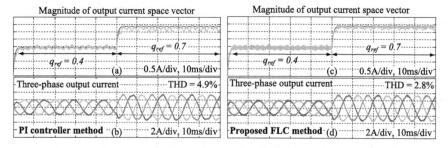

Fig. 11. Dynamic response comparison: (a) & (b) PI controller method, and (c) & (d) Proposed FLC method.

5 Conclusion

This paper has presented an intelligent control method based on the FLC to improve the output current performance of the IMC under the unbalanced and distorted input voltage conditions. The proposed FLC method which uses a closed-loop control of output current can reduce the harmonic components and also control the magnitude of output current. The dynamic response of the proposed FLC method is better than that of the PI control method under the variable load conditions. Simulated results are provided to evaluate the effectiveness of the proposed FLC method.

Acknowledgment. This work was supported by the National Research Foundation of Korea Grant funded by the Korean Government (NRF-2013R1A2A2A01016398).

References

1. Rodriguez, J., Rivera, M., Kolar, J.W., Wheeler, P.W.: A review of control and modulation methods for matrix converters. IEEE Trans. Ind. Electron. **59**(1), 58–70 (2012)
2. Nguyen, T.D., Lee, H.-H.: Modulation Strategies to Reduce Common-Mode Voltage for Indirect Matrix Converters. IEEE Trans. Ind. Electron. **59**(1), 129–140 (2012)
3. Lixiang, W., Lipo T.A.: A novel matrix converter topology with simple commutation. In: Conference Record of the 2001 IEEE Industry Applications Conference, vol.3, pp. 1749–1754 (2001)
4. Tran, Q.-H., Chun, T.-W., Lee, H.-H.: Fault tolerant strategy for inverter stage in indirect matrix converter. In: IECON'2013 – 39th Annual Conference of the IEEE Industrial Electronics Society, pp. 4844–4849 (2013)
5. Jussila, M., Tuusa, H.: Comparison of simple control strategies of space-vector modulated indirect matrix converter under distorted supply voltage. IEEE Trans. Power Electron. **22**(1), 139–148 (2007)
6. Song, W., Zhong, Y., Zhang, H., Sun, X., Zhang, Q., Wang, W.: A study of Z-source dual-bridge matrix converter immune to abnormal input voltage disturbance and with high voltage transfer ratio. IEEE Trans. Ind. Inf. **9**(2), 828–838 (2013)
7. Karaca H., Akkaya R., Dogan H.: A novel compensation method based on fuzzy logic control for matrix converter under distorted input voltage conditions. In: ICEM'2008 – 18th International Conference on Electrical Machines, pp. 1–5 (2008)
8. Park, K., Lee, E.-S., Lee, K.-B.:A Z-source sparse matrix converter with a fuzzy logic controller based compensation method under abnormal input voltage conditions. In: IEEE International Symposium on Industrial Electronics (ISIE), pp. 614–619 (2010)
9. Karaca, H., Akkaya, R.: Modelling and simulation of matrix converter under distorted input voltage conditions. Simulation Modelling Practice and Theory **19**(2), 673–684 (2011). ISSN 1569-190X
10. Boukadoum, A., Bahi, T., Oudina, S., Souf, Y., Lekhchine, S.: Fuzzy Control adaptive of a matrix converter for harmonic compensation caused by nonlinear loads. Energy Procedia **18**, 715–723 (2012). ISSN 1876-6102
11. Venugopal C.: ANFIS based Field Oriented Control for Matrix Converter fed Induction Motor. In: 2010 IEEE International Conference on Power and Energy (PECon), pp. 74–78 (2010)

12. Hamane B., Doumbia M.L., Cheriti A., Belmokhtar K.: Comparative analysis of PI and Fuzzy Logic Controllers for Matrix Converter. In: 2014 Ninth International Conference on Ecological Vehicles and Renewable Energies (EVER), pp. 1–7, 25–27 March 2014
13. Calvillo C.F., Olloqui A., Martell F., Elizondo J.L., Avila A., Macias M.E., Rivera M., Rodriguez J.: Comparison of Model Based Predictive Control and Fuzzy Logic Control of a DFIG with an Indirect Matrix Converter. In: IECON 2012 – 38th Annual Conference on IEEE Industrial Electronics Society, pp. 6063–6068, 25-28 January (2012)
14. Bose, B.: Power Electronics and Motor Drives. Academic, New York (2006)

On Denjoy-McShane-Stieltjes Integral of Fuzzy-Number-Valued Functions

Wenkai Shao$^{(\boxtimes)}$, Jiechang Ruan, and Shu Gong

Yibin Vocational and Technical College, Yibin 644003, Sichuan, China
yb-shao@163.com

Abstract. In this paper, we introduce the concepts of the McShane-Stieltjes integral and the Denjoy-McShane-Stieltjes integral for fuzzy-number-valued functions and give a characterization of the McShane-Stieltjes integrability and investigate some properties of the Denjoy-McShane-Stieltjes integral.

Keywords: Fuzzy number · Fuzzy-number-valued function · Fuzzy McShane-Stieltjes integral · Fuzzy Denjoy-McShane-Stieltjes integral

1 Introduction

The integrals of fuzzy-valued functions have been discussed in recent papers. It is well known that the notion of the Stieltjes integral of fuzzy-number-valued functions was originally proposed by Nanda [3] in 1989. In 1998, Wu [4] proposed the concept of the fuzzy Riemann-Stieltjes integral by means of the representation theorem of fuzzy-number-valued functions, whose membership function could be obtained by solving a nonlinear programming problem, but it is difficult to calculate and extend to the higher dimensional space. In 2006, Ren et al. introduced the concept of two kinds of fuzzy Riemann-Stieltjes integral for fuzzy-number-valued functions [5] and showed that a continuous fuzzy-number-valued function was fuzzy fuzzy Riemann-Stieltjes integrable with respect to a real-valued increasing function. However, we note that if a fuzzy-number-valued function has some kind of discontinuity, the existing methods have been restricted. In real analysis, The Henstock integral is designed to integrate highly oscillatory functions which the Lebesgue integral fails to do. It is known as nonabsolute integration and is a powerful tool. It is well-known that the Henstock integral includes the Riemann, improper Riemann, Lebesgue and Newton integrals [6–8]. Though such an integral was defined by Denjoy in 1912 and also by Perron in 1914, it was difficult to handle using their definitions. But with the Riemann-type definition introduced more recently by Henstock in 1963 and also independently by Kurzweil, the definition is now simple and furthermore the proof involving the integral also turns out to be easy. Wu and Gong [9–11] have combined the fuzzy set theory and nonabsolute theory and discussed the fuzzy Henstock integrals of fuzzy-number-valued functions which is extended Kaleva integration.

In this paper, we introduce the concept of the fuzzy McShane-Stieltjes integral and fuzzy Denjoy-McShane-Stieltjes integral. And we give a characterization of the fuzzy McShane-Stieltjes integrability and investigate some properties of fuzzy Denjoy-McShane-Stieltjes integral.

© Springer International Publishing Switzerland 2015
D.-S. Huang and K. Han (Eds.): ICIC 2015, Part III, LNAI 9227, pp. 151–158, 2015.
DOI: 10.1007/978-3-319-22053-6_17

2 Preliminaries

Let $P_k(R^n)$ denote the family of all nonempty compact convex subset of R^n and define the addition and scalar multiplication in $P_k(R^n)$ as usual. Let A and B be two nonempty bounded subset of R^n. The distance between A and B is defined by the Hausdorff metric [2]:

$$d_H(A, B) = \max\{\sup_{a \in A} \inf_{b \in B} |a - b|, \sup_{b \in B} \inf_{a \in A} |b - a|\}.$$

Denote $E^n = \{u : R^n \to [0, 1], u$ satisfies (1)–(4) below $\}$ is a fuzzy number space. Where

(1) u is normal, i.e. there exists an $x_0 \in R^n$ such that $u(x_0) = 1$;
(2) u is fuzzy convex, i.e. $u(\lambda x + (1 - \lambda)y) \geq \min\{u(x), u(y)\}$ for any $x, y \in R^n$ and $0 \leq \lambda \leq 1$,
(3) u is upper semi-continuous;
(4) $[u]^0 = cl\{x \in R^n | u(x) > 0\}$ is compact.

Define $D : E^n \times E^n \to (0, +\infty)$

$$D(u, v) = \sup\{d_H([u]^\alpha, [v]^\alpha) : \alpha \in [0, 1]\},$$

where d_H is the Hausdorff metric defined in $P_k(R^n)$. Then it is easy see that D is a metric in E^n. Using the results [3], we know that the metric space (E^n, D) has a linear structure, it can imbedded isomorphically as a cone in a Banach space of function $u^* : I \times S^{n-1}$, where S^{n-1} is the unit sphere in E^n, which an imbedding function $u^* = j(u)$ defined by $u^*(r, x) = \sup_{\alpha \in [u]^\alpha} \langle \alpha, x \rangle$.

Since Hausdorff metric is a kind of stronger metric, much problems could not be characterized. It is well known, the supremum (infimum) is a main concept in analysis, and how to characterized the supremum (infimum) of fuzzy number is an important problem in fuzzy analysis. Refer to (see [2, 10]), if $\{u_n\}$ is a bounded fuzzy number sequence, then it has supremum and infimum and if u is supremum (infimum) of $\{u_n\}$, $D(u_n, u) \to 0$ is not correct generally. The integral metric between two fuzzy numbers by using support functions of fuzzy numbers is defined by Gong (see [12]),

$$D^*(u, v) = \sqrt{\sup_{x \in S^{n-1}, \|x\|=1} \int_0^1 (u^*(r, x) - u^*(r, x))^2 dr}$$

We easy see that (E^n, D^*) is a metric space, and for each fuzzy number sequence $\{u_n\} \subset E^n$ and fuzzy number $u \in E^n$, if $D(u_n, u) \to 0$, then $D^*(u_n, u) \to 0$. The converse result does not hold.

Definition 2.1 [14]. A fuzzy number valued function f has H-difference property, i.e., for any $t_1, t_2 \in [a, b]$ satisfying $t_1 < t_2$ there exist a fuzzy number $u \in E^n$ such that $f(t_2) = f(t_1) + u$, u is called H-difference of $f(t_1)$ and $f(t_2)$, we denote $f(t_2) -_H f(t_1) = u$.

Definition 2.2 [12]. A McShane partition of $[a, b]$ is a finite collection $P = \{([c_i, d_i], t_i) : 1 \leq i \leq n\}$ such that $\{[c_i, d_i] : 1 \leq i \leq 2\}$ is a non-overlapping family of subintervals of $[a, b]$ covering $[a, b]$ and $t_i \in [a, b]$ for each $i \leq n$. A gauge on $[a, b]$ is a positive real valued function $\delta : [a, b] \to (0, \infty)$. A McShane partition $P = \{([c_i, d_i], t_i) : 1 \leq i \leq n\}$ is subordinate to a gauge δ if $[c_i, d_i] \subset (t_i - \delta(t_i), t_i + \delta(t_i))$ for every $i \leq n$. If $f : [a, b] \to$ E and if $P = \{([c_i, d_i], t_i) : 1 \leq i \leq n\}$ is a McShane partition of $[a, b]$, we will denote $f(P)$ for $\sum_{i=1}^{n} f(t_i)(d_i - c_i)$.

Definition 2.3 [13]. A fuzzy number valued function $f : [a, b] \to$ E is McShane integrable on $[a, b]$, with a fuzzy number A, if for each $\varepsilon > 0$ there exists a gauge $\delta : [a, b] \to (0, \infty)$ such that $D(f(P), A) < \varepsilon$ whenever $P = \{([c_i, d_i], t_i) : 1 \leq i \leq n\}$ is a McShane partition of $[a, b]$ subordinate to δ.

Definition 2.4. Let $F : [a, b] \to E$ be a fuzzy-number-valued function and let $t \in (a, b)$. A fuzzy number A in E is the approximate derivative of F at t if there exists a measurable set $S \subset [a, b]$ that has t as a point of density such that $\lim\limits_{\substack{s \to t \\ s \in E}} \frac{F(s) - F(t)}{s - t} = z$.

we will write $F'_{ap}(t) = A$

Definition 2.5. A fuzzy number valued function $f : [a, b] \to E$ is fuzzy Denjoy integrable on $[a, b]$ if there exists an ACG function $F : [a, b] \to E$ such that $F'_{ap} = f$ almost everywhere on $[a, b]$.

Definition 2.6. Let a fuzzy number valued function $F : [a, b] \to E$ and let $\alpha : [a, b] \to \Re$ be a strictly increasing function and let $E \subset [a, b]$.

(a) The fuzzy-number-valued function F is BV with respect to α on E if $V(F, \alpha, E) =$

$\sup \left\{ \sum\limits_{i=1}^{n} D(F(d_i), F(c_i)) \frac{\alpha(d_i) - \alpha(c_i)}{d_i - c_i} \right\}$ is finite where the supremum is taken over all finite collections $\{[c_i, d_i] : 1 \leq i \leq n\}$ of non-overlapping intervals that have endpoints in E.

(b) The fuzzy valued function F is AC with respect to α on X if for each $\varepsilon > 0$ there exists $\delta > 0$ such that $\sum\limits_{i=1}^{n} D(F(d_i), F(c_i)) < \varepsilon$ whenever $\{[c_i, d_i] : 1 \leq i \leq n\}$ is a finite collection of non-overlapping intervals that have endpoints in E and satisfy $\sum\limits_{i=1}^{n} [\alpha(d_i) - \alpha(c_i)] < \delta$.

(c) The fuzzy-number-valued function F is BVG with respect to α on E if E can be expressed as a countable union of sets on each of which F is BVG with respect to α.

(d) The fuzzy-number-valued function F is ACG with respect to α on E if F is continuous on E and if E can be expressed as a countable union of sets on each of which F is AC with respect to α.

Definition 2.7. Let $F : [a, b] \to E, t \in (a, b)$ and let $\alpha : [a, b] \to \Re$ be a strictly increasing function such that $\alpha \in C^1([a, b])$. A fuzzy number $A \in E$ is the approximate derivative of F with respect to α at t if there exists a measurable set $E \subset [a, b]$ that has t as a point of density such that $\lim\limits_{\substack{s \to t \\ s \in E}} \frac{F(s) - F(t)}{\alpha(s) - \alpha(t)} = z$. we will write $F'_{\alpha, \alpha \rho}(t) = z$.

Definition 2.8. A fuzzy-number-valued function $f : [a, b] \to E$ is Denjoy-Stieltjes integrable with respect to α on $[a, b]$ if there exists an ACG function $F : [a, b] \to E$ with respect to α such that $F'_{a, ap} = f$ almost everywhere on $[a, b]$.

3 McShane-Stieltjes Integral of Fuzzy-Number-Valued Functions

In this section we introduce the concept of the fuzzy McShane-Stieltjes integral and give a characterization of the fuzzy McShane-Stieltjes integrability.

Let $\alpha : [a, b] \to \Re$ be an increasing function. If a fuzzy number valued function $f : [a, b] \to E$ and if $P = \{([c_i, d_i], t_i) : 1 \leq i \leq n\}$ is a McShane partition of $[a, b]$, we will denote $f_\alpha(P)$ for $\sum\limits_{i=1}^{n} f(t_i)[\alpha(d_i) - \alpha(c_i)]$.

Definition 3.1. Let $\alpha : [a, b] \to \Re$ be an increasing function. A fuzzy-number -valued function $f : [a, b] \to E$ is fuzzy McShane-Stieltjes integrable with respect to α on $[a, b]$, with fuzzy McShane-Stieltjes integral A, if for each $\varepsilon > 0$ there exists a gauge $\delta : [a, b] \to (0, \infty)$ such that $D(f_\alpha(p), A) < \varepsilon$ whenever $P = \{([c_i, d_i], t_i) : 1 \leq i \leq n\}$ is a McShane partition of $[a, b]$ subordinate to δ.

Theorem 3.2. Let $\alpha : [a, b] \to \Re$ be a strictly increasing function such that $\alpha \in C^1([a, b])$ and let $f : [a, b] \to E$ be a bounded function. Then f is fuzzy McShane-Stieltjes integrable with respect to α on $[a, b]$ if and only if $\alpha' f$ is McShane integrable on $[a, b]$.

Proof. Since fuzzy number valued function $f : [a, b] \to E$ is a bounded function, there exists $M > 0$ such that $D(f(x), \tilde{0}) \leq M$ for all $x \in [a, b]$. Continuity of α' on $[a, b]$ implies uniform continuity on $[a, b]$. Hence for each $\varepsilon > 0$ there exists $\eta > 0$ such that for all $x, y \in [a, b]$ and $|x - y| < \eta$, we have $|\alpha'(x) - \alpha'(y)| < \frac{\varepsilon}{3M(b-a)}$.

Choose a gauge δ_1 on $[a, b]$ with $\delta_1(x) < \eta$ for all $x \in [a, b]$ Let $P = \{([c_i, d_i], t_i) : 1 \leq i \leq n\}$ be a McShane partition of $[a, b]$ subordinate to δ_1. Then by the Mean Value Theorem, there exists $x_i \in (c_i, d_i)$ such that $\alpha(d_i) - \alpha(c_i) = \alpha'(x_i)(d_i - c_i)$ for $1 \leq i \leq n$ since $|t_i - x_i| < \delta_i(t_i) < \eta$ for all $1 \leq i \leq n, |\alpha'(t_i) - \alpha'(x_i)| < \frac{\varepsilon}{3M(b-a)}$ for $1 \leq i \leq n$. Hence we have

$$D(f_\alpha(P), (\alpha' f)(P)) = D(\sum_{i=1}^{n} f(t_i)[\alpha(d_i) - \alpha(c_i)], \sum_{i=1}^{n} \alpha'(t_i) f(t_i)(d_i - c_i))$$

$$= D(\sum_{i=1}^{n} f(t_i)[\alpha'(x_i) - \alpha'(t_i)](d_i - c_i), \tilde{0}) = \frac{\varepsilon}{3}$$

Whenever $P = \{([c_i, d_i], t_i) : 1 \leq i \leq n\}$ is a McShane partition of $[a, b]$ subordinate to δ_1.

If f is fuzzy McShane-Stieltjes integrable with respect to α on $[a, b]$, then there exists a gauge δ_2 on $[a, b]$ such that

$$D(f_\alpha(P_1), f_\alpha(P_2)) < \varepsilon/3$$

Whenever P_1 and P_2 are McShane partitions of $[a, b]$ subordinate to δ_2. Define δ on $[a, b]$ by $\delta(x) = min\{\delta_1(x), \delta_2(x)\}$ for $x \in [a, b]$. Then δ is a gauge on $[a, b]$ and

$$D((\alpha'f)(P_1), (\alpha'f)(P_2)) < D((\alpha'f)(P_1), f_\alpha(P_1))$$
$$+ D(f_\alpha(P_1), f_\alpha(P_2))$$
$$+ D(f_\alpha(P_2), (\alpha'f)(P_2)) < \varepsilon$$

Whenever P_1 and P_2 are McShane partitions of $[a, b]$ subordinate to δ. Hence $\alpha'f$ is fuzzy McShane integrable on $[a, b]$.

Conversely, if $\alpha'f$ is fuzzy McShane integrable on $[a, b]$, then for each $\varepsilon > 0$ there exists a gauge δ_3 on $[a, b]$ such that $D((\alpha'f)(P_1), (\alpha'f)(P_2)) < \varepsilon/3$ whenever P_1 and P_2 are McShane partitions of $[a, b]$ subordinate to δ_3. Define δ on $[a, b]$ by $\delta(x) = min\{\delta_1(x), \delta_3(x)\}$ for $x \in [a, b]$. Then δ is a gauge on $[a, b]$ and

$$D(f_\alpha(P_1), f_\alpha(P_2)) \leq D(f_\alpha(P_1), (\alpha'f)(P_1))$$
$$+ D((\alpha'f)(P_1), (\alpha'f)(P_2))$$
$$+ D((\alpha'f)(P_2), f_\alpha(P_2)) < \varepsilon$$

Whenever P_1 and P_2 are McShane partitions of $[a, b]$ subordinate to δ. Hence f is fuzzy McShane-Stieltjes integrable with respect to α on $[a, b]$.

Theorem 3.3. Let $\alpha : [a, b] \to \Re$ be an increasing function. A fuzzy number valued function $f : [a, b] \to E$ is fuzzy McShane-Stieltjes integrable with respect to α on $[a, b]$, if and only if for each $\varepsilon > 0$ there exists a gauge $\delta : [a, b] \to (0, \infty)$ such that $D(f_\alpha(P_1), f_\alpha(P_2)) < \varepsilon$ whenever P_1 and P_2 are McShane partitions of $[a, b]$ subordinate to δ.

4 Denjoy-McShane-Stieltjes Integral of Fuzzy-Number-Valued Functions

In this section we introduce the concept of the fuzzy Denjoy-McShane-Stieltjes integral and investigate some properties of this integral.

Definition 4.1. Let $\alpha : [a, b] \to \Re$ be a strictly increasing function such that $\alpha \in C^1([a, b])$. A fuzzy-number-valued function $f : [a, b] \to E$ is fuzzy Denjoy-

McShane-Stieltjes integrable with respect to α on $[a,b]$ if there exists a continuous function $F : [a,b] \to F$ such that F is ACG with respect to α on $[a,b]$ and is approximately differentiable with respect to α almost everywhere on $[a,b]$ and $(F)'_{\alpha,ap} = f$ almost everywhere on $[a,b]$.

Theorem 4.2. Let $\alpha : [a,b] \to \Re$ be a strictly increasing function such that $\alpha \in C^1$ $([a,b])$. Then a fuzzy-number-valued function $f : [a,b] \to E$ is Denjoy-McShane-Stieltjes integrable with respect to α on $[a,b]$ if and only if $\alpha' f$ is Denjoy- McShane integrable on $[a,b]$.

The following three corollaries are obtained from Theorem 4.2.

Corollary 4.3. Let $\alpha : [a,b] \to \Re$ be a strictly increasing function such that $\alpha \in C^1([a,b])$ and let $f : [a,b] \to E$ be a fuzzy-number-valued function. If $\alpha' f$ is fuzzy McShane integrable on $[a,b]$, then f is fuzzy Denjoy-McShane-Stieltjes integrable with respect to α on $[a,b]$.

Corollary 4.4. Let $\alpha : [a,b] \to \Re$ be a strictly increasing function such that $\alpha \in C^1([a,b])$ and let $f : [a,b] \to E$ be a fuzzy-number-valued function. If $\alpha' f$ is Denjoy-Bochner integrable on $[a,b]$, then f is Denjoy-McShane-Stieltjes integrable with respect to α on $[a,b]$.

Corollary 4.5. Let $\alpha : [a,b] \to \Re$ be a strictly increasing function such that $\alpha \in C^1([a,b])$ and let $f : [a,b] \to E$ If f is fuzzy Denjoy-McShane-Stieltjes integrable with respect to α on $[a,b]$ then $\alpha' f$ is fuzzy Denjoy-Pettis integrable on $[a,b]$.

Theorem 4.6. Let $\alpha : [a,b] \to \Re$ be a strictly increasing function such that $\alpha \in C^1([a,b])$. If $f : [a,b] \to E$ is a bounded fuzzy McShane-Stieltjes integrable with respect to α on $[a,b]$, then f is fuzzy Denjoy-McShane-Stieltjes integrable with respect to α on $[a,b]$.

Proof. If $f : [a,b] \to E$ is a bounded fuzzy McShane-Stieltjes integrable with respect to α on $[a,b]$, then by Theorem 3.2 $\alpha' f$ is fuzzy McShane integrable on $[a,b]$. Then we have $\alpha' f$ is fuzzy Denjoy-McShane integrable on $[a,b]$. By Theorem 4.2, f is fuzzy Denjoy-McShane-Stieltjes integrable with respect to α on $[a,b]$.

Definition 4.7. A fuzzy-number-valued function $f : [a,b] \to E$ is Denjoy-Stieltjes-Bochner integrable with respect to α on $[a,b]$ if there exists an ACG function $F : [a,b] \to E$ with respect to α such that F is approximately differentiable with respect to α almost everywhere on $[a,b]$ and $F'_{a,ap} = f$ almost everywhere on $[a,b]$.

Theorem 4.8. Let $\alpha : [a,b] \to \Re$ be a strictly increasing function such that $\alpha \in C^1([a,b])$. If $f : [a,b] \to E$ is fuzzy Denjoy-Stieltjes-Bochner integrable with respect to α on $[a,b]$, then $f : [a,b] \to E$ is fuzzy Denjoy-McShane-Stieltjes integrable with respect to α on $[a,b]$.

Proof. If $f : [a,b] \to E$ is Denjoy-Stieltes-Bocner integrable with respect to α on $[a,b]$, then there exists an ACG function $F : [a,b] \to E$ with respect to α such that F is approximately differentiable with respect to α almost everywhere on $[a,b]$ and $F'_{a,ap} =$

f almost everywhere on $[a, b]$. It is easy to show that for each $j \circ F$ is ACG with respect to α on $[a, b]$ and $j \circ F$ is approximately differentiable with respect to α almost everywhere on $[a, b]$ and $(j \circ F)'_{ap} = j \circ f$ almost everywhere on $[a, b]$. Hence f is Denjoy-McShane-Stieltjes integrable with respect to α on $[a, b]$.

Theorem 4.9. Let $\alpha : [a, b] \to \Re$ be a strictly increasing function such that $\alpha \in C^1([a, b])$. If $f : [a, b] \to E$ is Denjoy-McShane-Stieltjes integrable with respect to α on $[a, b]$, then $f : [a, b] \to E$ is Denjoy-Stieltjes-Pettis integrable with respect to α on $[a, b]$.

Proof. Suppose that $f : [a, b] \to E$ is fuzzy Denjoy-McShane-Stieltjes integrable with respect to α on $[a, b]$. Let $F(t) = (DMS) \int_a^t f d\alpha$ Since $j \circ F$ is ACG with respect to α on $[a, b]$ and $(j \circ F)'_{ap} = j \circ f$ almost everywhere on [a,b]. $j \circ f$ is Denjoy-Stieltjes integrable with respect to α on $[a, b]$.

For every interval $[c, d]$ in $[a, b]$ we have

$$j \circ (F(d) - F(c)) = j \circ F(d) - j \circ F(c)$$
$$= (DS) \int_a^b j \circ f d\alpha - (DS) \int_c^d j \circ f d\alpha$$
$$= (DS) \int_c^d j \circ f d\alpha$$

Since $F(d) - F(c) \in E, f$ is fuzzy Denjoy-Stieltjes-Pettis integrable with respect to α on $[a, b]$.

Theorem 4.10. Let $\alpha : [a, b] \to \Re$ be a strictly increasing function such that $\alpha \in C^1([a, b])$ If fuzzy number valued function $f : [a, b] \to E$ is fuzzy Denjoy-McShane-Stieltjes integrable with respect to α on $[a, b]$ and $T : E \to E$ is a fuzzy bounded linear operator, then $T \circ f : [a, b] \to E$ is fuzzy Denjoy-McShane-Stieltjes integrable with respect to α on $[a, b]$.

Proof. If fuzzy-number-valued function $f : [a, b] \to E$ is fuzzy Denjoy-McShane-Stieltjes integrable with respect to α on $[a, b]$, then there exists a continuous fuzzy valued function $F : [a, b] \to E$ such that

(i) for each $j \circ F$ is with respect to α on $[a, b]$ and
(ii) for each $j \circ F$ is approximately differentiable with respect to α almost everywhere on $[a, b]$ and $(j \circ F)'_{ap} = j \circ f$ almost everywhere on $[a, b]$.

Let $G = T \circ F$. Then $G : [a, b] \to Y$ is a continuous function such that

(i) for each $j \circ G = j \circ (T \circ F) = (j \circ T)F$ is ACG with respect to α on $[a, b]$ and
(ii) for each $j \circ G = j \circ (T \circ F) = (j \circ T)F$ is approximately differentiable which respect to α almost everywhere on $[a, b]$ and

$$(j \circ G)'_{\alpha,ap} = ((j \circ T)F)'_{\alpha'ap} = (j \circ T)f = j \circ (T \circ f).$$

Hence $T \circ f : [a, b] \to E$ is fuzzy Denjoy-McShane-Stieltjes integrable with respect to α on $[a, b]$.

Acknowledgments. This work is supported by National Natural Science Foundation of China under grant No. 11161041.

References

1. Riesz, F., Sz.-Nagy, B.: Functional Analysis. Ungar, New York (1955)
2. Zadeh, L.A.: Probability measure of fuzzy events. J. Math. Anal. Appl. **23**, 421–427 (1968)
3. Nanda, S.: On Fuzzy Integral. Fuzzy Sets Syst. **32**, 95–101 (1989)
4. Wu, H.C.: The Fuzzy Riemann-Stieltjes Integral. Int. J. Uncertainty Fuzziness Knowl.-Based Syst. **6**, 51–67 (1998)
5. Ren, X.K., Wu, C.X.: A new kind of fuzzy Riemann-Stieltjes integral. In: Proceeding of ICMLC 2002 Conference, Dalian. 1885–1888 (2006)
6. Henstock, R.: Theory of Integration. Butterworth, London (1963)
7. Lee, P.: Lanzhou Lectures on Henstock Integration. World Scientific, Singapore (1989)
8. Kurzweil, J.: Generalized ordinary differential equations and continuous dependence on a parameter. Czechoslovak Math. J. **7**, 418–446 (1957)
9. Wu, C., Gong, Z.: On Henstock intergrals of fuzzy-valued functions (I). Fuzzy Sets Syst. **120**, 523–532 (2001)
10. Gong, Z., Shao, Y.: The controlled convergence theorems for the strong Henstock integrals of fuzzy-number-valued functions. Fuzzy Sets Syst. **160**, 1528–1546 (2009)
11. McShane, E.J.: A Riemann-type Integral that Includes Lebesgue-Stieltjes, Bochner and Stochastic Integrals, vol. 88. Memoirs of the American Mathematical Society, Providence (1969)
12. Gong, Z., Wu, C.: The McShane integral of fuzzy-valued functions. Southeast Asian Bull. Math. **24**, 365–373 (2000)
13. Gong, Z.: The Convergence Theorems of the McShane Integral of Fuzzy-Valued Functions. Southeast Asian Bull. Math. **27**, 55–62 (2003)
14. Diamond, P., Kloeden, P.: Metric Space of Fuzzy Sets. Theory and Applications. World Scientific, Singapore (1994)

A Brief Survey on Fuzzy Cognitive Maps Research

Yajie Wang[(⊠)] and Weiyuan Zhang

School of Mathematics and Information Science,
Xianyang Normal University, Xianyang, China
195297832@qq.com, ahzwy@163.com

Abstract. The Fuzzy Cognitive Map (FCM) has emerged as a convenient and powerful soft modeling tool since its proposal. During the last nearly 30 years, Fuzzy Cognitive Maps have gained considerable research interests and have been applied to many areas. The advantageous modeling characteristics of FCMs encourage us to investigate the FCM structure, attempting to broaden the FCM functionality and applicability in real world. In this paper, the main representation and inference characteristics of conventional Fuzzy Cognitive Maps are investigated, and also the current state of the extensions of FCMs, learning algorithms for FCMs is introduced and summarized briefly.

Keywords: Fuzzy Cognitive Map · Learning algorithm · Intelligent computing · Fuzzy logic · Neural network

1 Introduction

Fuzzy Cognitive Maps (FCMs) are a soft computing methodology introduced by Kosko in 1987 to represent the causal relationships between concepts and analyze inference patterns [1, 2]. The attractiveness of the FCM lies in its natural knowledge representation, relatively simple structure and operations, ease of implementation, and its well adaptability. FCMs have gained considerable research interests and have been applied to many areas [1, 2].

In a graphical form, the FCMs are typically signed fuzzy weighted digraphs, usually involving feedbacks, consisting of nodes and directed edges between them. The nodes represent descriptive behavioral concepts of the system, and the edges represent cause–effect relations between the concepts. In the context of FCM theory, the concept's fuzzy value (state) denotes the degree to which the fixed concept is active in the general system, usually bounded in a normalized range of [0, 1] or [−1, +1]. Furthermore, the weights of the edges show the degree of causal influence between the presynaptic and postsynaptic concepts [3].

The FCM has emerged as a convenient and powerful soft modeling tool since its proposal. However, certain problems restrict its applications, such as: monotonic or symmetric causal relationships; synchronous inference, or lack of temporal concept; unable to simulate AND/OR combinations between the cause nodes; poor stability, etc. To solve these shortcomings and to improve the performance of FCMs, several methodologies were explored. Extensions to the FCMs theory are more than anything

© Springer International Publishing Switzerland 2015
D.-S. Huang and K. Han (Eds.): ICIC 2015, Part III, LNAI 9227, pp. 159–166, 2015.
DOI: 10.1007/978-3-319-22053-6_18

needed because of the feeble mathematical structure of FCMs and, mostly, the desire to assign advanced characteristics that are not met in other computational methodologies.

It is worth noting that another important methodology to improve the performance of FCMs is learning algorithms. Learning methodologies for FCMs have been developed in order to update the initial knowledge of human experts and/or include any knowledge from historical data to produce learned weight matrices. The adaptive Hebbian-based learning algorithms, the evolutionary-based algorithms(such as genetic algorithms), and the hybrid approaches composed of Hebbian-type and genetic algorithms were established to handle the task of FCM training [5]. These algorithms are the most efficient and widely used for training FCMs.

2   Conventional Fuzzy Cognitive Maps

Fuzzy Cognitive Maps are a modeling technique originated from the combination and synergism of fuzzy logic and neural network. Particularly, a FCM is a fuzzy signed oriented graph with feedback that models the worlds as a collection of concepts (usually depicted as circles or nodes) and causal relations (depicted as directed arcs or edges) between concepts.

Nodes represent system variables, whose values change over time, usually in the unit interval $[0, 1]$. Edges represent causality from the cause nodes (or antecedent nodes) to the effect nodes (or consequent nodes). Edges take on values, also called weights, representing degrees of causality, usually in the interval $[-1, 1]$, or also in the unit interval $[0, 1]$.

The value $c_j(t)$ for each concept C_j at every time step is calculated by the following equation [1–3]:

$$c_j(t + 1) = T(\sum_{i=1}^{n} e_{ij} c_i(t)) \tag{1}$$

Where $e_{ij}(i \neq j)$ is the weight value of the edge from concept C_i to concept C_j. $e_{ij} > 0$ indicates positive causality between concepts C_i and C_j; $e_{ij} < 0$ indicates negative or inverse causality between concepts C_i and C_j; $e_{ij} = 0$ indicates no (direct) relationship between concepts C_i and C_j. e_{jj} indicates influence of the previous state of C_j on C_j; that is, the C_j itself is also regarded as a cause if $e_{jj} \neq 0$[5]. T is a threshold function, which forces the unbounded inputs, or the weighted sum into a strict range. An example of FCM is shown in Fig. 1.

FCMs have gained considerable research interest and accepted as useful technique in many diverse scientific fields from knowledge modeling and decision making.

Anyway, conventional FCMs have several drawbacks, such as listed below [6].

(1)   FCM models lack time delay in the interactions between nodes.
(2)   The theoretical basis of threshold function selection is unclear.
(3)   FCMs cannot represent logical operators (AND, OR, NOT, etc.) between ingoing nodes.

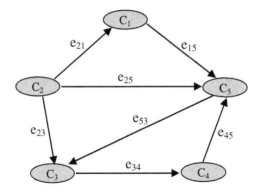

Fig. 1. A simple example of conventional fuzzy cognitive maps

(4) FCMs cannot model multi-meaning (gray) environments.
(5) It does not include a possible multistate (quantum) of the concepts.
(6) It cannot handle more than one relationship between nodes.
(7) FCMs can represent only symmetric or monotonic causal relations.
(8) FCMs do not handle randomness associated with complex domains.

Several extensions have been proposed and more research is needed to overcome the above limitations of conventional FCMs.

According to the FCM construction process, the number and kind of concepts are determined by a group of experts, which comprise the FCM model. Then, each interconnection is described by a domain expert with an IF – THEN rule that infers a fuzzy linguistic variable from a determined set, which associates the relationship between the two concepts and determines the grade of causality between the two concepts. The information about these cause-result relations is generated and enriched over time with either the experience of experts in the related field or the knowledge of historical data. The construction of FCM requires a large amount of concepts and connections that need to be established, which substantially add to the difficulty of manual development process. Therefore, learning mechanisms for FCMs must be investigated.

3   Extensions of Fuzzy Cognitive Maps

The FCM has emerged as a convenient and powerful soft modeling tool since its proposal. However, certain problems restrict its applications, such as: monotonic or symmetric causal relationships; synchronous inference, or lack of temporal concept; unable to simulate AND/OR combinations between the cause nodes; poor stability, etc.

Due to its representation or inference defects, the conventional FCM needs further enhancement and stronger mathematical justification. To improve the performance of FCMs, many extensions have been proposed, such as Extended Fuzzy Cognitive Maps (EFCM) [7], Contextual Fuzzy Cognitive Maps (CFCM) [8], Probabilistic Fuzzy

Cognitive Maps (PFCM) [9], Random Fuzzy Cognitive Maps (RFCM) [10], and so on. Most of the extensions are numeric FCMs.

3.1 Dynamical Cognitive Networks

Dynamic cognitive network (DCN) proposed in [11], enables the definition of dynamic causal relationships between the FCM concepts, and it is able to handle complex dynamic causal systems. Theoretically, DCN can support a full set of time-related features. It is the first paper in the literature studying the two aspects of the causality separately, i.e., causes and effects. However, the DCN relies on the Laplacian framework to describe the causal relationships. The transformation between fuzzy knowledge and Laplacian functions imposes more modeling efforts to system designers. It is not easy for domain experts to model their cognitive knowledge by this way.

Each DCN concept can have its own value set, depending on how accurately it needs to be described in the network. It allows that DCNs describe the strength of causes and the degree of effects that are crucial to conducting relevant inferences. The DCN's edges establish dynamic causal relationships between concepts. Structurally, DCNs are scalable and more flexible than FCMs.

A DCN can be as simple as a cognitive map, i.e., an FCM, or as complex as a nonlinear dynamic system. DCNs take into account the three major causal inference factors: the strength of the cause, the strength of the causal relationship, and the degrees of the effect. It is also able to model dynamic cognitive processes. As a result, DCNs improve FCMs by quantifying the concepts and introducing nonlinear dynamic functions to the arcs. Recently, a simplified DCN extending the modeling capability of an FCM has been proposed in [12], where the equivalence of the DCN and the FCM models has also been addressed.

3.2 Rule-Based Fuzzy Cognitive Maps

Rule-based fuzzy cognitive maps (RBFCMs) as an evolution of FCMs including relations other than monotonic causality are proposed in [13]. RBFCMs are iterative fuzzy rule-based systems with fuzzy mechanisms to deal with feedback, including timing and new methods with uncertainty propagation. RBFCMs provide a representation of the dynamics of complex real-world qualitative systems with feedback and allow the simulation of the occurrence of events and their influence in the system. They are fuzzy directed graphs with feedback, which are composed of fuzzy nodes (Concepts), and fuzzy links (Relations). RBFCMs are a tool to predict the evolution through time caused by those changes. RBFCMs provide two kinds of Concepts (Levels and Variations), several kinds of Relations (Causal, Inference Common, similarity, ill-defined, Crisp, Level-Variation, and other common relations), and mechanisms that support invariant and time-variant probabilities, the possibility of subsystems including decision support systems to simulate the process of decision making by "actors" within the system.

3.3 Fuzzy Cognitive Maps Based on WOWA Aggregation

A common drawback of the existing FCMs is that they can not handle the various AND/OR relationships among the antecedent nodes. Besides, the conventional or numeric Fuzzy Cognitive Maps can only represent monotonic or symmetric causal relationships; the Rule Based or linguistic Fuzzy Cognitive Maps (RBFCMs) usually suffer from the well known combinatorial rule explosion problem as multiple-antecedent fuzzy rules are adopted.

The Weighted Ordered Weighted Averaging (Weighted OWA, WOWA) operator is introduced into the Fuzzy Cognitive Maps in [14] to simulate the explicit, partially explicit, or fuzzy combinations of the antecedent nodes. Under the framework of WOWA aggregation based Fuzzy Cognitive Maps, furthermore, a hybrid fuzzy cognitive model based on single-antecedent fuzzy rules is proposed, which combines the advantages of numeric FCMs and linguistic FCMs.

Without loss of generality, assuming there are $k \, (k = 1, 2, \ldots, n)$ causes for an effect node, the framework for Fuzzy Cognitive Maps is proposed as follows:

(1) For the case of $c_p(t) \in [0, 1]$ and $e_{pj} \in [0, 1]$, $p = 1, 2, \ldots, k$:

$$c_j(t+1) = F_{WOWA}(a_{1j}(t), a_{2j}(t), \ldots, a_{kj}(t))$$

(2) For the case of $c_p(t) \in [0, 1]$ and $e_{pj} \in [-1, 1]$, $p = 1, 2, \ldots, k$:

$$c_j(t+1) = \begin{cases} F_{WOWA}(a_{1j}(t), a_{2j}(t), \ldots, a_{kj}(t)) \\ 0 \end{cases}$$

$$\text{if} \begin{array}{l} F_{WOWA}(a_{1j}(t), a_{2j}(t), \ldots, a_{kj}(t)) \geq 0 \\ F_{WOWA}(a_{1j}(t), a_{2j}(t), \ldots, a_{kj}(t)) < 0 \end{array}$$

$a_{pj}(t) = e_{pj}a_p(t)$ is the individual effect on the consequent node C_j of each antecedent C_p. F_{WOWA} is the Weighted OWA Operator. The Weighted OWA operator is based on the Ordered Weighted Averaging (OWA) operator [15, 16] and combines the advantages of the OWA operator and the ones of the Weighted Mean (WM).

Basically, for the WOWA based FCMs, the threshold functions and the straight sum of the conventional FCMs are replaced with the WOWA operators. By means of WOWA or OWA operators, FCMs can go from conjunction (intersection) to disjunction (union) in a continuous way. Therefore, the use of the WOWA operators allows the representations of sophisticated combinations between the arguments of FCMs.

4   Learning Algorithms for Fuzzy Cognitive Maps

The main incentive that leads to further research and development on FCMs is the wide recognition of FCM as a promising modeling and simulation methodology with remarkable characteristics such as abstraction, flexibility, adaptability, and fuzzy reasoning. FCM simulates its evolution over time when an initial state vector and initial weight matrix are given. This way, it provides straightforward answers to causal "what-if" questions. However, the robustness and accuracy of answers highly depend on the weight matrix of a given FCM. Thus, the determination of the weight matrix, i.e., strength of relationships among concepts in an FCM, is crucial [5].

Learning algorithms, by modifying the FCM weight matrix, have been proven to be essential components for FCMs in order to improve their operation and accuracy in a number of modeling and prediction tasks. The analysis and development of these algorithms were concentrated mainly on weights, which update on the basis of experts' knowledge and/or historical data. So far, there have been attempts to investigate and propose learning techniques that are suitable for FCMs. Mainly, the strength connections among concepts in a given FCM can be modified/updated from three knowledge sources: from historical data, from experts' knowledge, and from a combination of experts' knowledge and historical data. Accordingly, there are generally three types of learning approaches to handle the task of FCM training: Hebbian-based, evolutionary-based, and hybrid that subsequently combine the main aspects of Hebbian-based and evolutionary-type learning algorithms [19].

In the case of unsupervised learning algorithms, Hebbian-based methods use available data and a learning formula, which is based on several modifications of Hebbian law, to iteratively adjust FCM weights. During Hebbian-based learning, the values of weights are iteratively updated until the desired structure is found. The weights of outgoing edges for each concept in the connection matrix are modified only when the corresponding concept value changes. The main drawback of this approach is that the formula updates weights between each pair of concepts, thus taking into account only these two concepts and ignoring the influence that comes from other concepts. In the case of evolution -based algorithms, evolution-based learning uses available input datasets and is oriented toward finding models that mimic the input data. They are optimization techniques and for this reason, they are computationally quite demanding. The evolution-based methods provide better quality of the learned models in the context of similarity of their dynamic behavior, which is defined as the simulation error. The main drawback of the evolution-based learning methods is that they provide solutions that are hard or impossible to interpret and which may lead to incorrect static analysis. Some experimental results that concern both static and dynamic properties of the FCMs, learned with the RCGA-based methods, are promising [20]. However, further research toward a systematic approach to develop FCMs from data could be still carried out.

The hybrid approach could be implemented by using a combination of the first two mentioned learning types for FCMs: the Hebbian-based learning and the evolution-based learning. A few hybrid learning methods were proposed till now for training FCMs, which are mainly composed of a two-stage learning process [19, 21].

Papageorgiou and Groumpos investigated a coupling of DE algorithm and NHL algorithm by using both the global search capabilities of evolutionary strategies and the effectiveness of the NHL rule [19]. This hybrid learning module was applied successfully in real-world problems, and through the experimental analysis, the results shown that this hybrid strategy was capable to train FCM effectively, thus leading the system to desired states and determining an appropriate weight matrix for each specific problem. Later on, Zhu and Zhang presented another hybrid scheme by using RCGA and NHL algorithm, and implemented it in problem of partner selection [21]. Their algorithm inherits the main features of each learning technique, RCGA evolution -basd algorithm and NHL type, thus combining expert and data input. The first results were encouraging for future research.

5 Conclusion

FCMs have some specific advantageous characteristics over traditional mapping methods: they capture more information in the relationships between concepts, are dynamic, combinable, and tunable, and express hidden relationships. The resulting fuzzy model can be used to analyze, simulate, and test the influence of parameters and predict the behavior of the system. Although the research on fuzzy cognitive maps started late, in the nearly 30 years, scholars still spare no effort to expand and improve it. The achievements are remarkable, both in the theoretical research and in the application area. Many models are proposed, but on the other hand, only a few of these models have been further in-depth studied, and much less analyzed mathematically and comprehensively. However, due to the outstanding performance on describing and reasoning of causality, the fuzzy cognitive map still has a lot of research and application value.

Acknowledgment. This work is partially supported by the Scientific Research Project of Education Department of Shanxi Province (No. 2013JK0578).

References

1. Kosko, B.: Fuzzy engineering. Prentice-Hall, New Jersey (1997)
2. Aguilar, J.: A survey about fuzzy cognitive Maps papers. Int. J. Comput. Cognit. 3(2), 27–33 (2005)
3. Glykas, G.: Fuzzy Cognitive Maps: Theory, Methodologies, Tools and Applications. Springer, Heidelberg (2010)
4. Stach, W., Kurgan, L., Pedrycz, W.: Expert-based and computational methods for developing fuzzy cognitive maps. In: Glykas, M. (ed.) Fuzzy Cognitive Maps. STUDFUZZ, vol. 247, pp. 23–41. Springer, Heidelberg (2010)
5. Stach W., Kurgan L. A., Pedrycz W., A survey of fuzzy cognitive map learning methods. In: Grzegorzewski, P., Krawczak, M., Zadrozny, S. (eds.) Issues in Soft Computing: Theory and Applications. Akademicka Oficyna Wydawnicza (2005)

6. Papageorgiou E.I.: A review study of FCMs applications during the last decade. In: Proceedings of IEEE International Conference on Fuzzy System, pp. 828–835, 27–30 June 2011

7. Hagiwara M.: Extended fuzzy cognitive maps. In: Proceedings of the 1st IEEE International Conference on Fuzzy Systems, pp. 795–801. IEEE, New York (1992)

8. Satur, R., Liu, Z.Q.: A contextual fuzzy cognitive map framework for geographic information systems. IEEE Trans. Fuzzy Syst. 7(10), 481–494 (1999)

9. Luo, X.F., Gao, J.: Probabilistic fuzzy cognitive maps [J]. J. Univ. Sci. Technol. China 33(1), 26–33 (2003). (in Chinese)

10. Aguilar, J.: A dynamic fuzzy cognitive map approach based on random neural networks. Int. J. Comput. Cognit. 1(4), 91–107 (2003)

11. Miao, Y., Liu, Z.Q., et al.: Dynamical cognitive network: An extension of fuzzy cognitive map. IEEE Trans. Fuzzy Syst. 9(5), 760–770 (2001)

12. Miao, Y., Miao, C., et al.: Transformation of cognitive maps. IEEE Trans. Fuzzy Syst. 18(1), 114–124 (2010)

13. Carvalho J P, Tomé JAB., Rule based fuzzy cognitive maps and fuzzy cognitive maps - a comparative study. In: Proceedings of the 18th International Conference of the North American Fuzzy Information Processing Society. IEEE, New York, pp. 115–119 (1999)

14. Lv Z.B., Zhou L.H.,: Fuzzy cognitive maps based on WOWA aggregation. Journal of Sichuan University (Natural Science Edition), pp. 43–47 (2008)

15. Torra, V.: The weighted OWA operator [J]. Int. J. Intell. Syst. 12(2), 153–166 (1997)

16. Calvo, T., Mesiar, R., Yager, R.R.: Quantitative weights and aggregation. IEEE Trans. Fuzzy Syst. 12(1), 62–69 (2004)

17. Ruan D., Hardeman F., Mkrtchyan L.: Using belief degreedistributed fuzzy cognitive maps in nuclear safety culture assessment. In: Proceedings of Annual Meeting North American Fuzzy Information Processing Society, pp. 1–6 (2011)

18. Papageorgiou, E.I., Stylios, C.D., Groumpos, P.P.: Unsupervised learning techniques for fine-tuning fuzzy cognitive map causal links. Int. J. Human-Comput. Stud. 64, 727–743 (2006)

19. Papageorgiou, E.I., Groumpos, P.P.: A new hybrid learning algorithm for fuzzy cognitive maps learning. Appl. Soft Comput. 5, 409–431 (2005)

20. Stach, W., Kurgan, L., Pedrycz, W.: A divide and conquer method for learning large fuzzy cognitive maps. Fuzzy Sets Syst. 161(19), 2515–2532 (2010)

21. Zhu Y., Zhang W.: An integrated framework for learning fuzzy cognitive map using RCGA and NHL algorithm. In: The International Conference on Wireless Communication, Network and Mobile Computing. Dalian (2008)

Sensor Data Driven Modeling and Control of Personalized Thermal Comfort Using Interval Type-2 Fuzzy Sets

Chengdong Li[1(✉)], Weina Ren[1], Huidong Wang[2], and Jianqiang Yi[3]

[1] School of Information and Electrical Engineering,
Shandong Jianzhu University, Jinan, China
chengdong.li@foxmail.com
[2] School of Management Science and Engineering,
Shandong University of Finance and Economics, Jinan, China
[3] Institute of Automation, Chinese Academy of Sciences, Beijing, China

Abstract. As people in different rooms usually have different thermal comfort feelings or demands, it is valuable to study the modeling and control of thermal comfort to meet the personalized requirements. This paper tries to solve this issue using the data collected by the temperature and humidity sensors in the working or living time periods in the room being studied. We firstly present a statistic method based sensor data preprocessing strategy to discard noisy data and obtain the reasonable intervals for the temperature and humidity of each day. Then, we construct the Gaussian interval type-2 fuzzy set models to depict the personalized temperature and humidity comfort through measuring the uncertainty degrees of the obtained intervals. At last, we propose a control scheme to realize the personalized thermal comfort regulation. Our results show that the constructed thermal comfort models can recommend a reasonable temperature and humidity range for the demand in a specific room.

Keywords: Type-2 fuzzy set · Uncertainty degree · Sensor data · Thermal comfort · Data driven

1 Introduction

With the growth of economy and the improvement of living standard, our demands on the indoor environmental comfort are also increasing. Comfortable environment can not only bring living pleasures, but also improve working efficiency in some sense. Hence, in the research domain of intelligent buildings and smart homes, the control of the indoor environment is attracting more and more concerns and becomes one of the most popular research directions.

The indoor comfort research mainly refers to the study on human thermal comfort. In 1970s, Fanger established the indices: Predicted Mean Vote (PMV), Predicted Percentage of Dissatisfied (PPD) to evaluate the thermal comfort degree [1]. After that, lots of studies [2–4] on the modeling and control of thermal comfort are based on such indices. The PMV and PPD are proposed based on the experimental results in several specific regions and used to describe the comfort degree in the stable environments. So the models

© Springer International Publishing Switzerland 2015
D.-S. Huang and K. Han (Eds.): ICIC 2015, Part III, LNAI 9227, pp. 167–178, 2015.
DOI: 10.1007/978-3-319-22053-6_19

and control strategies based on such indices have difficulties to adjust the changing environment to dynamically satisfy the personalized demands on thermal comfort.

On the other aspect, with the development of intelligent buildings, different kinds of sensors have been installed in the buildings and a large amount of data about the environmental parameters such as temperature and humidity can be collected. Such data can reflect the personal demands and/or preferences in different rooms. Hence, we can utilize the hidden information from the collected data to realize the modeling and control of the personalized thermal comfort. Recently, data driven modeling and control methods [5–7] provide us powerful tools to accomplish our objective.

In this real-world application, a variety of physical factors, such as the sensor errors and the outdoor environmental changes, will influence the accuracy of the sensor data in the process of data collection. Fuzzy sets [8], including type-1 and type-2 fuzzy sets, provide us useful tools to handle such kinds of uncertainties. On the other hand, different people in the same room have different feelings even on the same temperature and humidity. The crisp membership grades of type-1 fuzzy sets (T1FSs) limit their capabilities to deal with such kinds of physiological uncertainties. In order to tackle different kinds of uncertainties caused by both the physical factors and the physiological factors, type-2 fuzzy sets (T2FSs), including interval type-2 fuzzy sets (IT2FSs), are much more suitable to be adopted to model the thermal comfort words, as the membership grades of T2FSs are fuzzy and have greater freedom than T1FSs [9–12]. Thus, in this study, we adopt T2FSs to model such uncertainties. As general T2FSs are much more complex than IT2FSs, in this paper, IT2FSs are adopted.

In brief, this study presents a sensor data driven type-2 fuzzy modeling and control method to realize the modeling and control of the personalized thermal comfort. Firstly, data will be collected every day by the sensors at particular time intervals in the room being studied. Then, the proposed data preprocessing method will be utilized to tackle the sensor data to discard noisy ones and generate reasonable intervals which can reflect the thermal preferences in the room. Further, the thermal comfort models for the temperature and humidity will be established using IT2FSs and their uncertainty degrees calculated from the sensor data. The constructed thermal comfort model for every specific room can recommend a reasonable temperature and humidity range to satisfy the comfort requirement and preference in this room.

2 Backgrounds

In this section, we will firstly give a brief introduction of IT2FSs and Gaussian IT2FSs, then, we will discuss how to construct an IT2FS model.

2.1 Interval Type-2 Fuzzy Sets

A type-1 fuzzy set A is denoted as [8]:

$$A = \int_X \frac{\mu_A(x)}{x} \tag{1}$$

where $\mu_A(x)$ is a crisp value which indicates the membership function grade of the element x, and $\int$ denotes union to all acceptable x.

A type-2 fuzzy set (T2FS) $\tilde{A}$ on the domain X is described as [8]:

$$\tilde{A} = \int_{x \in X} \mu_{\tilde{A}}(x)/x = \int_{x \in X} \int_{u \in J_x \in [0,1]} f_x(u)/u \Big/ x \qquad (2)$$

where the membership grade $\mu_{\tilde{A}}(x)$ of each element x is a T1FS. J_x is an interval in $[0,1]$ and is called the primary membership function of the primary variable x, while $f_x(u)$ is the secondary membership function of the secondary variable u in J_x.

When the value of the secondary membership function of the T2FS equals to 1, i.e. $f_x(u) = 1$, the T2FS $\tilde{A}$ can be simplified as an interval type-2 fuzzy set (IT2FS) defined as [8]:

$$\tilde{A} = \int_{x \in X} [\int_{u \in J \in [0,1]} 1/u] \Big/ x \qquad (3)$$

An IT2FS can be completely characterized by its upper membership function (UMF) $\bar{\mu}_{\tilde{A}}(x)$ and its lower membership (LMF) $\underline{\mu}_{\tilde{A}}(x)$. The region between the LMF and UMF is named as the footprint of uncertainty (FOU) which contains all the uncertainties of an IT2FS and can also be used to model different kinds of uncertainties in real-world applications.

Figure 1 shows us a pictorial description of the conception of the LMF, UMF and FOU.

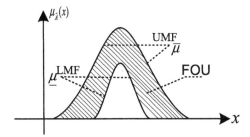

Fig. 1. The LMF, UMF and FOU of an IT2FS

One of the most popular IT2FSs is the Gaussian IT2FS whose LMF and UMF can be described as [8]:

$$\underline{\mu}_{\tilde{A}}(x) = \exp\left[-\frac{1}{2} \left(\frac{x - m}{\sigma_1} \right)^2 \right] \qquad (4)$$

$$\bar{\mu}_{\tilde{A}}(x) = \exp\left[-\frac{1}{2}\left(\frac{x-m}{\sigma_2}\right)^2\right] \tag{5}$$

where m is the center of the Gaussian IT2 FS, and $0 < \sigma_1^2 \leq \sigma_2^2$ are the uncertain width of the Gaussian IT2 FS.

2.2 Uncertainty Degree and Modeling of IT2FS

Obviously, to construct the Gaussian IT2FS model, three parameters need to be determined. In recent years, several data driven methods have been proposed for the IT2FS modeling, for example the interval approach [13, 14], the fuzzistics approach [15–17] and the fuzzy information-theoretic kernels method [18]. In [19], through combing the fuzzistics approach, we proposed an uncertainty degree based method for the IT2FS modeling.

In this study, we will adopt the uncertainty degree based method to construct the Gaussian IT2FS model. Similar to the moments estimation method in Statistics, the uncertainty degree based method first computes the means, right centers and uncertainty degrees of both the Gaussian IT2FS and the collected interval data. Then, it lets the mean, right center and uncertainty degree of the Gaussian IT2FS be equal to the ones of the collected interval data to obtain three equations for the three parameters of the Gaussian IT2FS. At last, through solving the three equations, we can determine the three parameters of the Gaussian IT2FS.

In this subsection, according to the results in [19], we will only review the formulas for compute the mean, right center and uncertainty degree of the Gaussian IT2FS, while computation of the mean, right center and uncertainty degree of the collected interval data will be discussed in the next section.

According to the results in [19], the mean of the Gaussian IT2FS is obviously the parameter m, and the right center and the uncertainty degree of the Gaussian IT2FS can be determined as:

$$c_r(\tilde{A}) \approx \left(\underline{c}_r(\tilde{A}) + \bar{c}_r(\tilde{A})\right)/2 = m + \sigma_1\left[(\sigma_2/\sigma_1)^2 + 2\sigma_2/\sigma_1 - 3\right]/\sqrt{2\pi}, \tag{6}$$

$$\rho(\tilde{A}) = 1 - \frac{\sigma_1}{\sigma_2}. \tag{7}$$

For more details on how to obtain these equations, please refer to [19].

3 IT2FS Modeling Method for Thermal Comfort

In this section, we will show how to establish a Gaussian IT2FS model based on the sensor data. The proposed method is shown in Fig. 2. Following this figure, in this section, we will first discuss the data collection by the sensing devices, then we will show how to preliminarily preprocess the gathered data to generate daily interval data which reflects the daily preference, further we will adopt the method proposed by Liu

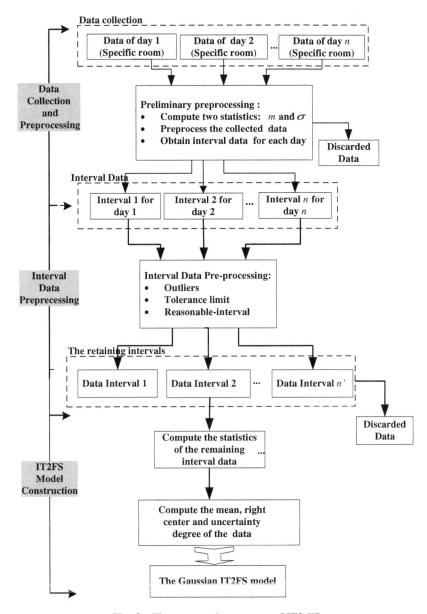

Fig. 2. The construction process of IT2 FS

[13] and Wu [14] to preprocess such interval data, at last we will present how to construct the Gaussian IT2FS models using the uncertainty degree method.

3.1 Data Collection and Preliminary Preprocessing

This section consists of two stages, *data collection and Preliminary Preprocessing*, which are the foundation of the IT2FS modeling.

3.1.1 Data Collection

At present, questionnaire-based surveys are wildly utilized [13, 14, 19–21] to gather the data for IT2FS modeling. However, with the development of intelligent buildings, lots of sensors have been installed in the buildings. The collected sensor data can reflect the personal demands and/or preferences in different rooms. Therefore, in this study, we adopt the sensor data to construct the IT2FS models. There are three major superiorities that support the utilization of the sensor data for modeling: (1) Data can be monitored and collected in real-time; (2) A simple and low-cost solution can be offered to acquire a high data density which enhances the accuracy of the constructed model; (3) the sensor data driven model can reflect personal preference.

3.1.2 Preliminary Data Preprocessing

The proposed modeling approach is based on the statistics of the intervals for each day, so we first need to convert these real-time monitoring data to corresponding intervals. To realize this objective, we use the following five stages to preliminarily preprocess the collected data for each day.

(1) *Compute the statistics of the collected daily data:* First, compute the sample mean m_i and the sample standard deviation σ_i for the i-th day as:

$$m_i = \sum_{j=1}^{n_i} data_{i,j} \Big/ n_i \qquad (8)$$

$$\sigma_i = \sqrt{\sum_{j=1}^{n_i} \left(data_{i,j} - m_i\right)^2 \Big/ n_i} \qquad (9)$$

where $data_{i,j}$ is the j − th data collected in i-th day, n_i is the total number of the collected data in i-th day.

(2) *Daily data preprocessing:* Based on the above statistics, we first compute the following formula for each $data_{i,j}$:

$$\left|data_{i,j} - m_i\right| \leq k * \sigma_i \qquad (10)$$

if the data $data_{i,j}$ satisfies (10), it is accepted; otherwise, it is rejected. Generally, k is chosen according to the situation of the problem. After this step, in the i-th day, n_i' $(n_i' \leq n_i)$ data will be left.

(3) *Compute the statistics of the remaining data in all the n days:* compute the sample mean m and the sample standard deviation σ for the remaining data in all the n days as:

$$m = \sum_{i=1}^{n} \sum_{j=1}^{n_i'} data_{i,j} \Big/ \sum_{i=1}^{n} n_i' \qquad (11)$$

$$\sigma = \sqrt{\sum_{i=1}^{n} \sum_{j=1}^{n_i'} \left(data_{i,j} - m_i\right)^2 \Big/ \sum_{i=1}^{n} n_i'} \qquad (12)$$

(4) **n-days' data preprocessing:** Again, we compute the following formula for each $data_{i,j}$:

$$\left| data_{i,j} - m \right| \leq k * \sigma \tag{13}$$

if the $data_{i,j}$ satisfies (13), it is accepted; otherwise, it is rejected.

(5) **Obtain the daily interval:** The maximum and minimum values of the daily collected data can reflect the bounds of the personalized comfort feeling in that day, so we use the interval between the minimum and maximum data to represent the personalized comfort feeling in the specific day. In other words, the interval representing the i-th day can be computed as:

$$[c_i, d_i] = [\min_{j \in I} data_{i,j}, \max_{j \in I} data_{i,j}] \tag{14}$$

where $i = 1, \ldots, n$. I denotes the daily collected data left after preprocessing.

3.2   Interval Data Preprocessing

Once the intervals are obtained, the interval data preprocessing method proposed by Liu [13] and Wu [14] will be adopted to dispose these intervals. This preprocessing method mainly consists of three major stages: (1) *Outlier Processing*, (2) *Tolerance limit Processing*, and (3) *Reasonable-interval Processing*.

(1) **Outlier processing:** We perform Box and Whisker tests on c_i and d_i, and then $L_i = d_i - c_i$. Some data intervals that can satisfy Eq. (2) in [13] are kept. After this process, $m' \leq n$ data intervals are remained, and we re-compute the sample mean and standard deviation on c_i, d_i and L_i respectively, as (m_c, σ_c), (m_d, σ_d), (m_L, σ_L).

(2) **Tolerance limit processing:** If the endpoint values of the retaining m' data intervals can satisfy the Eq. (3) in [13], then the data intervals are kept, otherwise, the interval will be deleted. After this process, the $m'' \leq n$ data intervals are kept, we re-compute its sample mean and the standard deviation on c_i, d_i and L_i respectively, as (m_c, σ_c), (m_d, σ_d), (m_L, σ_L).

(3) **Reasonable-interval processing:** Only the interval that satisfies the Eq. (1) in [14] will be kept, otherwise, it will be discarded.
    After the reasonable-interval processing, the left n' ($1 \leq n' \leq n$) data intervals are renumbered as $1, 2, \ldots, n'$, and denoted as $\left[t_i^l, t_i^r \right]$, $(i = 1, 2, \ldots, n')$.

For more details on the interval data preprocessing, please refers to [13, 14].

3.3   Construction of the Gaussian IT2FS

In this subsection, computation of the mean, right center and uncertainty degree of the interval data will be firstly given. Then, the equations for the IT2FS modeling will be derived.

First, we compute the four statistics of the remaining interval data as [19]:

$$x_m^l = \sum_{i=1}^{n'} t_i^l / n',$$ (15)

$$s^l = \sqrt{\sum_{i=1}^{n'} \left(t_i^l - x_m^l\right)^2 / n'},$$ (16)

$$x_m^r = \sum_{i=1}^{n'} t_i^r / n',$$ (17)

$$s^r = \sqrt{\sum_{i=1}^{n'} \left(t_i^r - x_m^r\right)^2 / n'}$$ (18)

where t_i^l, t_i^r are the left endpoints and the right endpoints of the remaining i th data interval; x_m^l, x_m^r are the sample mean of all the left endpoints and all right endpoints; s^l, s^r are the standard deviation of the left endpoints and right endpoints.

From the results in [19], the uncertainty degree of the interval data can be computed as:

$$\rho_X = \frac{2\Delta x}{x_m^r - x_m^l + 2\Delta x}$$ (19)

where Δx can be evaluated as $\Delta x = \Delta x(\alpha, \gamma) = ks$. The results on how to choose k, α and γ can be found in [13, 21].

Then, through letting the mean, right center and uncertainty degree of the Gaussian IT2FS be equal to the ones of the collected interval data, we can determine the three parameters of the Gaussian IT2FS as [19]:

$$m = \frac{1}{2}[x_m^l + x_m^r],$$ (20)

$$\sigma_1 = \left(2\sqrt{2\pi}(1 - \rho_X)^2(x_m^r - m)\right) \Big/ \left(4\rho_X - 3\rho_X^2\right),$$ (21)

$$\sigma_2 = \left(2\sqrt{2\pi}(1 - \rho_X)(x_m^r - m)\right) \Big/ \left(4\rho_X - 3\rho_X^2\right).$$ (22)

4 Experimental Results in a Specific Room

In this section, we will demonstrate how to construct the thermal comfort IT2FS models for a specific room.

4.1 Data Collection and Preprocessing

In our experiments, we use the temperature and humidity sensors to collect the temperature and humidity data. In the working time of each day in summer, one air conditioner was opened to adjust the temperature and humidity to satisfy the thermal comfort demand of the people in this room. And, simultaneously, the temperature and humidity sensors worked to collect the environment data in this room. The experiment continued for 16 days and 2438 data pairs were collected. Notice that the numbers of the data in different days are not the same.

After the *Preliminary Data Pre-procession*, the left data intervals for the temperature and humidity are shown in Tables 1 and 2 respectively. One thing that needs to be mentioned is that, in this specific room, the humidity is a little high. The possible reasons for this phenomenon may be that (1) the user in this room like a little high level of humidity, (2) the data collection period was in the rainy season in summer.

Table 1. Intervals for Temperature after the preliminary data pre-procession

Day	Number of Data		c_i	d_i	Day	Number of Data		c_i	d_i
	original	After pre-procession				original	After pre-procession		
1	139	133	24.57	26.56	9	120	67	28.50	30.50
2	157	157	23.97	27.12	10	126	123	25.24	27.31
3	50	50	24.84	25.38	11	146	135	26.30	29.00
4	165	155	25.63	26.77	12	238	189	23.80	28.30
5	82	81	26.54	27.35	13	86	60	23.90	28.20
6	416	390	26.20	30.50	14	22	22	25.10	28.90
7	241	241	27.33	29.06	15	11	11	28.70	29.50
8	336	329	26.59	28.16	16	103	81	28.40	30.40

Table 2. Intervals for Humidity after the preliminary data pre-procession

Day	Number of data		c_i	d_i	Day	Number of data		c_i	d_i
	original	After pre-procession				original	After pre-procession		
1	139	120	59.43	68.02	9	120	105	62.30	74.00
2	157	133	56.54	63.03	10	126	101	60.20	67.85
3	50	45	57.90	62.82	11	146	127	58.20	72.30
4	165	1137	56.80	63.31	12	238	44	62.40	75.50
5	82	75	63.31	71.07	13	86	58	57.20	75.20
6	416	198	48.10	52.40	14	22	20	51.90	69.40
7	241	209	59.97	66.24	15	11	9	59.70	61.80
8	336	285	62.79	67.54	16	103	84	48.10	52.70

Further, the *Interval Data Preprocessing* was taken to further clean the data intervals. After this step, only 9 of the 16 intervals for the temperature are left, and 10 of the 16 intervals for the humidity are remained. The four statistics of the remaining interval data, i.e. the sample mean of left endpoints x_m^l and right endpoints x_m^r, and the sample standard deviation of left endpoints s^l and right endpoints s^r, are computed and shown in Table 3. These statistics are utilized to establish the Gaussian IT2 FS thermal comfort models.

Table 3. Statistics of the remaining data intervals after interval data pre-procession

Statistics	x_m^l	x_m^r	s^l	s^r
Temperature	25.30	28.31	1.16	1.06
Humidity	59.77	70.71	3.44	3.40

4.2   Constructed IT2FS Models for Thermal Comfort

Using the proposed modeling method, the IT2FS models for the temperature and humidity comfort are constructed. The parameters of the two IT2FSs are listed in Table 4. From Table 4, we can observe that the constructed models can reasonably reflect the comfortable temperature and humidity and can handle the uncertainties by the FOUs between the LMFs and UMFs.

Table 4. The parameters of the constructed IT2FS thermal comfort models

Parameters	m	σ_1	σ_2
Temperature	26.81	0.56	1.78
Humidity	65.24	2.68	7.43

4.3   Control Strategy to Realize Personalized Thermal Comfort

The constructed IT2FS comfort models can be utilized to realize the indoor environment control to satisfy peoples' demand or preference in the room being studied. A control strategy as shown in Fig. 3 is proposed in this study to realize personalized thermal comfort.

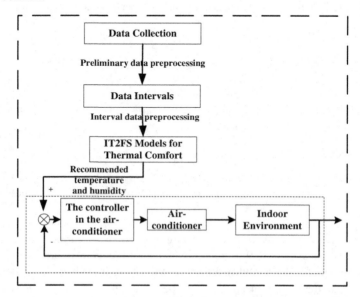

Fig. 3. Control strategy to realize personalized thermal comfort

In the proposed strategy, the constructed IT2FS models are used to recommend reasonable or best temperature and humidity. The recommended values are chosen as the control objectives and input to the environment control system. For example, in the specific room studied in our experiments, the recommended temperature and humidity values are 27 °C and 65 %, which can be the control objective of the air-conditioner system in this specific room. On the other hand, the IT2FS models can also provide a reasonable temperature and humidity range for every room in a building. In order to satisfy the personalized comfort or preference, the temperature and humidity should lie in this range when the air-conditioner system is working. Of course, to obtain such range, the Centroids of the constructed IT2FSs need to be computed first.

5 Conclusions

People in different rooms have different thermal comfort demands or preferences. To satisfy such personalized comfort, in this study, we have proposed a data driven modeling method which adopts IT2FS to model the uncertainties caused by the physical factors and the physiological factors. Through the proposed method, we have constructed the IT2FS models to measure the comfort degree on the temperature and humidity. The models can provide a reasonable control objective and range for the air-conditioner control system. However, in this paper, we have only considered the comfort of the indoor environment. In fact, with the increasing demands for building comfort, energy consumption of electrical equipments in the buildings becomes serious and causes great pressure on the society, energy and environment, so energy-saving control problems should also be considered. In the future, we will focus on such issues to achieve high building comfort and low energy consumption.

Acknowledgments. This work is supported by National Natural Science Foundation of China (61473176, 61402260, and 61273149), the Open Program from the State Key Laboratory of Management and Control for Complex Systems (20140102) and the Excellent Young and Middle-Aged Scientist Award Grant of Shandong Province of China (BS2013 DX043)

References

1. Fanger, P.O., et al.: Thermal Comfort: Analysis and Applications in Environmental Engineering. Danish Technical Press, Copenhagen (1970)
2. Yun, J., Won, K.H.: Building environment analysis based on temperature and humidity for smart energy systems. Sensors **12**(10), 13458–13470 (2012)
3. Djongyang, N., Tchinda, R., Njomo, D.: Thermal comfort: a review paper. Renew. Sustain. Energy Rev. **14**, 2626–2640 (2010)
4. Li, C., Wang, M., Zhang, G.: Prediction of thermal comfort using SIRMs connected type-2 fuzzy reasoning method. ICIC Express Lett. **7**(4), 1401–1406 (2013)
5. Wang, D.: Robust data-driven modeling approach for real-time final product quality prediction in batch process operation. IEEE Trans. Ind. Inform. **7**(2), 371–377 (2011)

6. Wang, Z., Liu, D.: A data-based state feedback control method for a class of nonlinear systems. IEEE Trans. Ind. Inform. **9**(4), 2284–2292 (2013)
7. Hou, Z.S., Wang, Z.: From model-based control to data-driven control: survey. Classif. Perspect. Inf. Sci. **235**, 3–35 (2013)
8. Mendel, J.M.: Uncertain Rule-Based Fuzzy Logic Systems: Introduction and New Directions. Prentice-Hall, Upper Saddle River (2001)
9. Li, C., Yi, J., Zhang, G.: On the monotonicity of interval type-2 fuzzy logic systems. IEEE Trans. Fuzzy Syst. **22**(5), 1197–1212 (2014)
10. Mo, H., Wang, F.: Linguistic dynamic systems based on computing with words and their stabilities. Sci. China-F Ser.: Inf. Sci. **52**(5), 780–796 (2009)
11. Liang, Q., Mendel, J.M.: Interval type-2 fuzzy logic systems: theory and design. IEEE Trans. Fuzzy Syst. **8**(5), 535–549 (2000)
12. Mendel, J.M., John, R.I., Liu, F.: Interval type-2 fuzzy logic systems made simple. IEEE Trans. Fuzzy Syst. **14**(6), 808–821 (2006)
13. Liu, F., Mendel, J.M.: Encoding words into interval type-2 fuzzy sets using an interval approach. IEEE Trans. Fuzzy Syst. **16**(6), 1503–1521 (2008)
14. Wu, D., Mendel, J.M., Coupland, S.: Enhanced interval approach for encoding words into interval type-2 fuzzy sets and its convergence analysis. IEEE Trans. Fuzzy Syst. **20**(3), 499–513 (2012)
15. Mendel, J.M., Wu, H.: Type-2 fuzzistics for symmetric interval type-2 fuzzy sets: part 1 Forward Problems. IEEE Trans. Fuzzy Syst. **14**(6), 781–792 (2006)
16. Mendel, J.M., Wu, H.: Type-2 fuzzistics for symmetric interval type-2 fuzzy sets: part 2 inverse problems. IEEE Trans. Fuzzy Syst. **15**(2), 301–308 (2007)
17. Mendel, J.M., Wu, H.: Type-2 fuzzistics for non-symmetric interval type-2 fuzzy sets: forward problems. IEEE Trans. Fuzzy Syst. **15**(5), 916–930 (2007)
18. Tahayori H., Livi L., Sadeghian A., Rizzi A.: Interval Type-2 Fuzzy Set Reconstruction Based on Fuzzy Information-Theoretic Kernels. IEEE Trans. Fuzzy Syst. (2014) (to be published) doi:10.1109/TFUZZ.2336673
19. Li, C., Zhang, G., Yi, J., Wang, M.: Uncertainty degree and modeling of interval type-2 fuzzy sets: definition method and application. Comput. Math. Appl. **66**(10), 1822–1835 (2013)
20. Li, C., Zhang, G., Wang, M., Yi, J.: Data-driven modeling and optimization of thermal comfort and energy consumption using type-2 fuzzy method. Soft. Comput. **17**(11), 2075–2088 (2013)
21. Mendel, J.M.: Computing with words and its relationships with fuzzistics. Inf. Sci. **177**, 988–1006 (2007)

Knowledge Evaluation with Rough Sets

Sylvia Encheva[1,2(⊠)] and Torleiv Ese[1]

[1] Stord/Haugesund University College,
Bjørnsonsg. 45, 5528 Haugesund, Norway
{sbe, te}@hsh.no
[2] Polytec, Sørhauggata 128, 5527 Haugesund, Norway

Abstract. Students often experience significant difficulties while studying mathematical subjects. In this work we focus on a course in calculus given to third year bachelor engineering students. The course is optional with respect to completing a bachelor degree and compulsory for taking a master degree in engineering. Our intention is to find out whether early identification of students in danger to fail the subject is possible and if affirmative which factors can be used to support the identification process. Methods from rough set theory are applied for selection of important attributes and factors influencing learning.

Keywords: Rough Sets · Logic · Knowledge Evaluation

1 Introduction

A number of students have serious difficulties studying mathematics at all levels. Researches from different fields have been working on this problem for over a century. In [6] we find a nonexhaustive list including cognitive factors, metacognitive factors, habits of learning and previous experiences related to studying mathematics. In [8] it is pointed out that if a student is unable to understand a difficult mathematics class because it is at a level above student's ability to respond to the instruction, the student may not progress to the affective level of valuing the instruction.

In this work we consider problems which third year bachelor engineering students have while completing a course in mathematics that is optional with respect to obtaining a bachelor degree and a prerequisite for taking a master degree in engineering. Our intention is to find out whether early identification of students in danger to fail their final exam is possible and which factors can be used in that identification process.

The volume of the course corresponds to one third of a study load per semester. Students are offered fourteen weeks face to face classroom lectures and tutorials. The former takes four hours a week and the latter is two hours a week. Students have to deliver one course work, take a two hours midterm exam, and a five hours final exam. Both the midterm exam and the final exam are written and taken in a controlled location. The course work is to be completed at home. This means that they can collaborate even though they are supposed to work individually. The course work is marked as 'pass' or 'fail'.

Methods from rough sets theory [10, 11] are employed in the course of our investigations.

© Springer International Publishing Switzerland 2015
D.-S. Huang and K. Han (Eds.): ICIC 2015, Part III, LNAI 9227, pp. 179–186, 2015.
DOI: 10.1007/978-3-319-22053-6_20

The rest of the paper goes as follows. Theoretical supporting the study is presented in Sect. 2, our approach can be found in Sect. 3 while the conclusion is placed in Sect. 4.

2 Preliminaries

Rough Sets were originally introduced in [10]. The presented approach provides exact mathematical formulation of the concept of approximative (rough) equality of sets in a given approximation space.

An *approximation space* is a pair $A = (U, R)$, where U is a set called universe, and $R \subset U \times U$ is an indiscernibility relation [11].

Equivalence classes of R are called *elementary sets* (atoms) in A. The equivalence class of R determined by an element $x \in U$ is denoted by $R(x)$. Equivalence classes of R are called *granules* generated by R.

The following definitions are often used while describing a rough set $X, X \subset U$:
the *R-upper approximation* of X

$$R^*(x) := x \in UR(x) : R(x) \cap X \neq \emptyset\},$$

the *R-lower approximation* of X

$$R_*(x) := x \in UR(x) : R(x) \subseteq X\},$$

the *R-boundary region* of X

$$RN_R(X) := R^*(X) - R_*(X).$$

Elements in the index set $A = a_1, a_2, \ldots, a_m$ are the importance degree of attribute set where each index in the system is determined by:

$$S_A(a_i) = \frac{|POS_A(A)| - |POS_{A-a_i}(A)|}{|U|}$$

where $i = 1, 2, 3, \ldots, m$ and the weight of index a_i is given by

$$w_i = \frac{S_A(a_i)}{\sum_{i=1}^{m}(a_i).}$$

The assessment model is defined by

$$P_j = \sum_{i=1}^{m} f_i$$

where P_j is the comprehensive assessment value of assessed jth object, f_i is the assessment value of i th index a_i according to the comprehensive assessment value,

[12]. Rough set theory software can be downloaded from [7]. Attributes degree of importance is discussed in [15].

Let P be a non-empty ordered set. If $sup\{x,y\}$ and $inf\{x,y\}$ exist for all $x,y \in P$, then P is called a *lattice* [5]. In a lattice illustrating partial ordering of knowledge values, the logical conjunction is identified with the meet operation and the logical disjunction with the join operation.

A *context* is a triple (G,M,I) where G and M are sets and $I \subset G \times M$. The elements of G and M are called *objects* and *attributes* respectively, [16].

For $A \subseteq G$ and $B \subseteq M$, define

$$A' = \{m \in M | (\forall g \in A)gIm\},$$

$$B' = \{g \in G | (\forall m \in B)gIm\}.$$

where A' is the set of attributes common to all the objects in A and B' is the set of objects possessing the attributes in B.

A *concept* of the context (G,M,I) is defined to be a pair (A,B) where

$$A \subseteq G,\ B \subseteq M,\ A' = B \text{ and } B' = A.$$

The *extent* of the concept (A,B) is A while its *intent* is B. A subset A of G is the extent of some concept if and only if $A'' = A$ in which case the unique concept of the which A is an extent is (A,A'). The corresponding statement applies to those subsets $B \in M$ which is the intent of some concepts. The set of all concepts of the context (G,M,I) is denoted by $B(G,M,I)$. $\langle B(G,M,I); \leq \rangle$ is a complete lattice and it is known as the *concept lattice* of the context (G,M,I).

Students' prior mathematical knowledge are of utmost importance when it comes to building of higher order mathematical understanding. Students mathematical competencies are discussed in [14]. Some problems related to building mathematical concepts are listed in [3]. Various problems concerning development of higher level thinking are presented in [1, 2, 9, 14]. The authors state that students' higher level thinking occurs when students are exposed to active learning. The latter can take place when "educators must give up the belief that students will be unable to learn the subject at hand unless the teacher "covers it"", [9]. In [4] higher-order thinking is divided in three categories - higher-order thinking in terms of transfer, in terms of critical thinking, and in terms of problem solving.

3   Data Evaluation

Data used in this study is taken from students results over a period of three years where 115 undergraduate engineering students have been enrolled in a mathematical course. The amount of students taking the course is usually about 40 and varies slightly from year to year.

The course is based on the following chapters in [13]:

12. Vectors and the Geometry of Space,
13. Vector-Valued Functions and Motion in Space,
14. Partial Derivatives,
15. Multiple Integrals,
16. Integrals and Vector Fields.

To illustrate the approach we present data for about halve of the students being subjects of this study, see Tables 1 and 2. Under 'Gender' notations are 'f' for female and 'm' for male. Note that no other descriptions of gender are used in students register.

Table 1. Students results, part 1

	Gender	M 1	M 2	Test	G 1	G 2	G 3
S1	f	a	a	h	h	h	h ·
S2	f	a	a	a	l	h	h
S3	f	a	a	h	h	a	h
S4	m	h	h	h	h	h	h
S5	m	h	h	h	h	a	h
S6	m	l	l	a	l	l	l
S7	m	l	a	a	a	a	l
S8	m	a	a	a	a	a	l
S9	m	a	l	a	a	l	a
S10	f	a	l	h	a	a	h
S11	m	a	l	h	h	a	h
S12	f	a	a	a	l	l	l
S13	f	a	l	a	l	l	l
S14	m	l	a	a	l	l	l
S15	m	a	h	a	a	l	a
S16	f	l	l	h	h	l	a
S17	m	h	a	h	l	a	a
S18	f	h	h	h	a	a	l
S19	f	a	a	a	h	a	a
S20	m	l	l	a	l	l	l
S21	m	a	a	a	l	h	a
S22	f	a	a	h	a	h	a
S23	f	a	a	h	h	a	l
S24	m	l	l	h	h	a	a
S25	m	l	l	a	a	l	l
S26	f	a	a	h	a	a	h
S27	m	a	l	h	h	h	a
S28	m	a	a	a	h	a	a
S29	f	a	l	a	l	l	l

Table 2. Students results, part 2

	Gender	M 1	M 2	Test	G 1	G 2	G 3
S30	f	l	a	a	l	l	l
S31	f	h	h	h	h	h	a
S32	m	h	a	h	h	h	a
S33	f	l	a	h	h	h	a
S34	f	a	l	h	a	h	h
S35	m	a	a	h	h	h	h
S36	f	h	h	h	h	a	h
S37	f	a	a	h	h	a	h
S38	m	a	h	h	h	a	h
S39	m	l	l	a	l	l	l
S40	f	h	h	h	h	h	h
S41	m	h	h	h	h	h	h
S42	m	h	h	h	h	h	a
S43	m	h	h	h	h	h	h
S44	f	h	h	h	h	h	a
S45	m	h	h	h	h	h	h
S46	f	h	h	h	h	h	h
S47	f	h	h	h	h	h	h
S48	m	h	a	h	a	a	h
S49	f	h	a	a	h	a	h
S50	m	h	h	h	h	a	h
S51	m	h	a	h	a	h	h
S52	f	a	a	h	h	a	h
S53	m	a	a	h	a	h	h
S54	f	l	l	h	l	l	l
S55	m	h	h	h	h	h	h

Final grades in two previously taken courses in mathematics at the same university are placed under columns 'M1' and 'M2', where:

- 'M1' stands for Mathematics 1 and
- 'M2' stands for Mathematics 2.
- Group 1 denoted 'G1' refers to application of triple integrals, moments and centers of mass, cylindrical and spherical coordinates;
- Group 2 denoted 'G2' refers to line integrals, surface integrals, Stokes' theorem and Gauss theorem, and
- Group 3 denoted 'sssG3' refers to studying correlations between curves, surfaces and given functions.

Originally the data comes in both text and numerical form. The data in Tables 1 and 2 is converted to text form with scaling grades:

$\{A,B\}$ - high (h),
$\{C,D\}$ - average (a),

{*E,F*} - low (l);
tests and written evaluations
[0 %, 40 %] - low (l),
[41 %, 70 %] - average (a), and
[71 %, 100 %] - high (h).

Suppose students in that course belong to a rough set X. When the data set described as in Tables 1 and 2 is used we obtain the following distribution

$$R_*(X) = \{St1, \ldots, S5, S7, S10, S11, S18, S19, S22, S23, S24, S26, S27, S28, S31, \ldots,$$
$$S38, S40, \ldots, S53, S55\},$$

$$RN_R(X) = \{S8, S9, S12, S15, S16, S17, S21, S25, S39\},$$

$$X - R_*(X) = \{St6, S13, S14, S20, S29, S30, S54\}$$

where $R_*(X)$ is the set of students who have obtained sufficient knowledge and skills, $RN_R(X)$ is the set of students who have obtained somewhat insufficient knowledge and skills, and the set of students who definitely have not obtained sufficient knowledge and skills is $X - R_*(X)$.

Below we present findings from working with data where all 115 students are included. Students who have not obtained sufficient knowledge and skills or have obtained somewhat insufficient knowledge and skills have lower grades from previously taken mathematical courses and their scores from the first test is in the interval [50 %, 60 %] nearly without exceptions. In the future such comparisons can be performed after the results from the first test are available and students who are in danger to fail the subject will be notified. This early warning will encourage them to spent more efforts on studying this subject for the rest of the semester.

The two groups of students,

$$RN_R(X) \text{ and } X - R_*(X),$$

have particular problems while working with line integrals, surface integrals, Stokes' theorem and Gauss theorem, as well as establishing correlations between curves, surfaces and given functions. An indept study of weaker students' performance indicates not only a gap in their mathematical knowledge but even more importantly, lack of higher order skills.

In general it seems that the majority of students have some problems with the above listed topics. However, students who have obtained sufficient knowledge and skills have scores on problem solving in the interval [60 %, 90 %] at average, while the group of students with insufficient or somewhat insufficient knowledge have scores in the interval [0 %, 30 %]. It turns out that very few students with pour grades from previously taken mathematical courses have managed to pass this course. Therefore, students, who for some reasons make slower progress than what previous experience indicates, should be notified as soon as possible and some further actions will be suggested to them.

In future work we will consider the following question: what can be done in the first two courses that will help students from potentially belonging to set $RN_R(X)$ to complete the course belonging to set $R_*(X)$, Fig. 1.

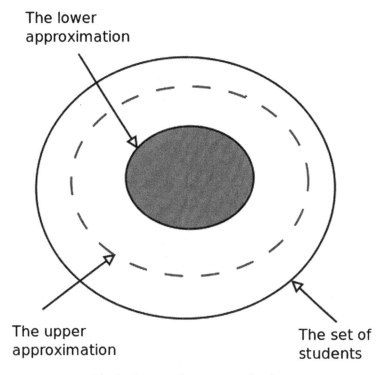

The lower approximation

The upper approximation

The set of students

Fig. 1. Lower and upper approximations

A concept lattice can be depicted from data as presented in Tables 1 and 2 where fifteen students belonging to the three groups 'high', 'average', and 'low' are involved. The idea is to illustrate how formal concept analysis can be used in similar studies. Note that a lattice representing formal concepts for 115 students would need much larger space than the one we can use here. Concepts are presented in gray boxes where the lower row lists concepts objects and the upper row lists concepts attributes. When two concepts are connected with a straight line one can see that the lower concept contains less objects and more attributes while the upper concepts has less objects and more attributes.

Concepts in the lattice show which groups of students have the same results related to particular topics and where they differ in their performance. We can see what 'high and average' groups learn equally good and where exactly are the differences between 'high' and 'average' groups. This can support provision of tailored advise to students in group 'average' that will help them to join the 'high' group.

Similarly, students from the 'low' group can receive personal advise on how to join the 'average' group and possibly the 'high' group. Students with the same results related to particular topics can be found in concepts placed on the third and fourth rows. Their differences can be seen by following superconcepts placed in the fifth and sixth rows.

4 Conclusion

In this work we study correlations between students' knowledge obtained in preliminary courses in mathematics on undergraduate level and their results from a mathematical course on a lower graduate level. Their test results combined with previous history gives an indication on whether they are going to experience difficulties on some particular parts of the curriculum. Such tendencies can also be used to provide support to new students taking the same course.

References

1. Anderson, L.W., Krathwohl, D.R.: A taxonomy for learning, teaching, and assessing: a revision of bloom's taxonomy of educational objectives. Longman, New York (2001)
2. Ball, A.L., Garton, B.L.: Modeling higher order thinking: the alignment between objectives, classroom discourse, and assessments. J. Agric. Educ. **46**(2), 1–9 (2005)
3. Berger, M.: Making mathematical meaning: from preconcepts to pseudoconcepts to concepts. Pythagoras **63**, 14–21 (2006)
4. Brookhart, S.M.: How to Assess Higher-Order Thinking Skills In Your Classroom. Association for Supervision and Curriculum Development, New York (2010)
5. Davey, B.A., Priestley, H.A.: Introduction to lattices and order. Cambridge University Press, Cambridge (2005)
6. Gurganus, S.P.: Characteristics of Students' Mathematics Learning Problems. http://www.education.com/reference/article/students-math-learning-problems/
7. http://www.lcb.uu.se/tools/rosetta/
8. LeMire, S.D., Melby, M.L., Haskins, A.M., Williams, T.: The devalued student: misalignment of current mathematics knowledge and level of instruction. Math. Educ. **22**, 63–83 (2012)
9. Limbach, B., Waugh, W.: Developing higher level thinking. J. Instr. Pedagogies **3**, 1–9 (2010)
10. Pawlak, Z.: Rough sets. Int. J. Comput. Inform. Sci. **11**, 341–356 (1982)
11. Pawlak, Z.: Rough Sets: Theoretical Aspects of Reasoning About Data. Kluwer Academic Publishing, Dordrecht (1991)
12. Ramasubramanian, P., Banu Priya, S., Dhanalakshmi, T.: Study on assessment method for technical institutions using rough sets. Int. J. Res Rev. Inf. Sci. (IJRRIS) **1**(4), 99–103 (2011)
13. Thomas Jr., G.B., Weir, M.D., Hass, J.R.: Thomas' Calculus, 13th edn. Pearson, New York (2014)
14. Turner R.: Exploring mathematical competencies, Research Developments **24**(5), 1–6 (2011). http://research.acer.edu.au/resdev/vol24/iss24/5
15. Yuan, L., Xu, F.: Research on the multiple combination weight based on rough set and clustering analysis-the knowledge transfer risk in it outsourcing taken as an example. Procedia Comput. Sci. **17**, 274–281 (2011)
16. Wille, R.: Concept lattices and conceptual knowledge systems. Comput. Math. Appl. **23**(6–9), 493–515 (1992)

On Solving CCR-DEA Problems Involving Type-2 Fuzzy Uncertainty Using Centroid-Based Optimization

Juan Carlos Figueroa-García[✉] and Carlos Eduardo Castro-Cabrera

Universidad Distrital Francisco José de Caldas, Bogotá, Colombia
jcfigueroag@udistrital.edu.co, jedaca13@hotmail.com

Abstract. In this paper we propose a method for solving Data Envelopment Analysis (DEA) problems involving uncertainty generated by the opinion of multiple experts. Experts opinions define the values of inputs and outputs, and they are handled with interval Type-2 fuzzy sets. The proposed method is an extension of the classic CCR model, solved using a centroid-based strategy to reduce computations.

1 Introduction

DEA models are useful tools for solving productivity related decision-making problems; the first DEA model, namely CCR was proposed by Charnes, Cooper and Rhones [1], the CCR model requires deterministic input information in order to obtain a measurement of the efficiency of specific elements within any organization. When experts are asked to determine the parameters of the model, then uncertainty comes from the opinion of each expert.

To involve all those perceptions, fuzzy sets are used to represent linguistic uncertainty that exists in natural language (such as the language used by multiple experts). This kind of models that include fuzzy uncertainty have been addressed by Sengupta [2], Kahraman and Tolga [3], Triantis and Girod [4], Girod and Triantis [5].

Our proposal solves a DEA problem with fuzzy parameters using Interval Type-2 fuzzy sets to handle expert opinions through computing their centroids and later solving the DEA problem to see its central behavior, which can be seen as the average behavior of the system.

The paper is organized into 6 sections; Sect. 1 introduces the main problem; Sect. 2 presents the CCR classic model; Sect. 3 presents some basics on Interval Type-2 fuzzy sets; Sect. 4 presents the proposed Interval Type-2 fuzzy CCR model; in Sect. 5, an example is presented and solved, and finally Sect. 6 presents some concluding remarks of the study.

Juan Carlos Figueroa-García is Assistant Professor at the Universidad Distrital Francisco José de Caldas, Bogotá - Colombia.
Carlos Eduardo Castro-Cabrera is undergraduate student of the Industrial Engineering Dept. of the Universidad Distrital Francisco José de Caldas, Bogotá - Colombia

© Springer International Publishing Switzerland 2015
D.-S. Huang and K. Han (Eds.): ICIC 2015, Part III, LNAI 9227, pp. 187–195, 2015.
DOI: 10.1007/978-3-319-22053-6_21

2   The Crisp Model CCR

The use of DMU (decision making units) is focused on nonprofit entities rather than industries or physical ones (such as production costs vs. profit margin). They have been designed to relate concepts of economic development such as engineering, social, life standards, etc. where there is a great interest for measuring efficiency (see Charnes and Cooper [1]). The optimal efficiency of any DMU is the maximum ratio between weighted outputs and weighted inputs subject to each DMU should be less or equal to 1, in order to be compared to other DMUs:

$$\max h_o = \frac{\sum_{r=1}^{s} u_r y_{ro}}{\sum_{i=1}^{m} v_i x_{io}}$$

$$s.t. \tag{1}$$

$$\frac{\sum_{r=1}^{s} u_r y_{rj}}{\sum_{i=1}^{m} v_i x_{ij}}$$

$$v_r, u_i \geq 0; i = 1, \dots, m; r = 1, \dots, s$$

where y_{rj}, x_{rj} are known outputs and inputs known of the j_{th} DMU and u_r, v_{io} are the weights of their associated variables which are decision making variables themselves (DMUs used as reference). The efficiency of a member of the reference set $j = 1, \cdots, n$ DMUs is classified relatively to other members of the set, so a DMU is evaluated using the subscript "o" using all original subscripts.

2.1   Reduction to a Linear Programming Model

The model (1) is a non-linear fractional problem, so it is necessary to do a simpler model (computationally speaking) for many observations $j(n)$, smaller amount of entries $i(m)$, and outputs $r(s)$ as well. The Linear Programming (LP) version of this problem has been proposed by Charnes and Cooper [1], as follows:

$$\max \theta = \sum_{r=1}^{s} u_r y_{ro}$$

$$s.t.$$

$$\sum_{i=1}^{m} v_i x_{io} = 1$$

$$\sum_{r=1}^{s} u_r y_{rj} \leq \sum_{i=1}^{m} v_i x_{ij} \quad \forall j \in \mathbb{N}^n$$

$$v_i, \mu_r \geq 0, \quad v, \mu \in \mathbb{R}$$

The Evaluation of a DMU has been widely recognized as a problem of considerable complexity. This evaluation is made more difficult when taking into account multiple inputs and multiple outputs, where a set of weights must be decided to form a relationship and determine the efficiency.

3 Basics on Interval Type-2 Fuzzy Sets

An Interval Type-2 Fuzzy Set (IT2FS) is denoted by emphasized capital letters $\tilde{A}$ with a membership function $\mu_{\tilde{A}}(x)$ defined over $x \in X$. $\mu_{\tilde{A}}(x)$ measures the uncertainty degree of a value $x \in X$ regarding A, so $\tilde{A}$ measures uncertainty around A. This way, an Interval Type-2 fuzzy set is an ordered pair

$$\tilde{A} = \{((x, u), J_x) \mid x \in X; u \in J_x \subseteq [0, 1]\},$$

where $\tilde{A}$ represents uncertainty around the word A, J_x is the *primary membership* of x, u is its *domain of uncertainty*, and $\mathcal{F}_2(\mathbb{X})$ is the class of all Type-2 fuzzy sets (see Mendel [6, 7]). $\mu_{\tilde{A}}(x)$ is composed by an infinite amount of embedded Type-1 fuzzy sets namely A_e. Every element x has associated a set of primary memberships J_x where u is the domain of uncertainty of x, $u \in J_x \subseteq [0, 1]$.

Uncertainty about the word A is conveyed by the union of all of J_x into the *Footprint Of Uncertainty* of $\tilde{A}$, namely *FOU* $(\tilde{A})$, which is bounded by two functions: A *Upper* membership function $UMF(\tilde{A}) = \bar{\mu}_{\tilde{A}}(x) \equiv \bar{A}$ and a *Lower* membership function $LMF(\tilde{A}) = \underline{\mu}_{\tilde{A}}(x) \equiv \underline{A}$. *FOU* $(\tilde{A})$ is shown in Fig. 1.

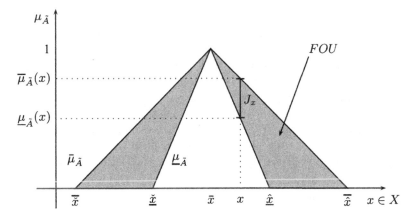

Fig. 1. Interval type-2 fuzzy set $\tilde{A}$

3.1 Type Reduction of An IT2FS: Centroid

Given a set $\tilde{A}$, its centroid $C(\tilde{A})$ is composed by an interval set of centroids bounded by two values $\min\{C(\tilde{A})\} = C_l(\tilde{A})$ and $\max\{C(\tilde{A})\} = C_u(\tilde{A})$. Every point enclosed into

$C(\tilde{A}) \in [C_l(\tilde{A}), C_u(\tilde{A})]$ is also a possible centroid of $\tilde{A}$, so there is an infinite amount of centroids enclosed into $C(\tilde{A})$ (see Wu and Mendel [8, 9], Mendel and Liu [10], Karnik and Mendel [11], and Melgarejo [12]).

The main equations of the Enhanced Karnik-Mendel (EKM) algorithm for computing $C(\tilde{A})$ (see Wu and Mendel [8]) are provided as follows:

$$c_l(\tilde{A}) = \frac{\sum_{i=1}^{N} x_i \underline{f}_i + \sum_{i=N+1}^{k} x_i \left(\bar{f}_i - \underline{f}_i \right)}{\sum_{i=1}^{N} \underline{f}_i + \sum_{i=N+1}^{k} \left(\bar{f}_i - \underline{f}_i \right)} \tag{2}$$

$$c_r(\tilde{A}) = \frac{\sum_{i=1}^{N} x_i \bar{f}_i - \sum_{i=1}^{k} x_i \left(\bar{f}_i - \underline{f}_i \right)}{\sum_{i=1}^{N} \bar{f}_i - \sum_{i=1}^{k} \left(\bar{f}_i - \underline{f}_i \right)} \tag{3}$$

where $\bar{f}$ and $\underline{f}$ are the UMF and LMF of $\tilde{A}$, and the main interest is to find the point N in which $C_l(\tilde{A})$ and $C_u(\tilde{A})$ are located. Improved routines for computing $C(\tilde{A})$ are proposed by Melgarejo and Bernal [13] through the IASCO (Iterative Algorithm With Stop Condition) algorithm.

4   Centroid-Based CCR Model with Type-2 Fuzzy Parameters

Applications of fuzzy sets in DEA problems are classified by Lertworasirikul et al. [14, 15], and Karsak [16] into four groups: 1- tolerance approaches, 2- α-level approach, 3- fuzzy classification, and 4- possibility approach. Girod and Triantis [17] proposed a fuzzy BCC with free disposal hull (FDH) which converges to radial efficiency measures (see Hatami, Emrouznejad and Tavana [18]).

The main idea is to incorporate the opinion of multiple experts to define the inputs and outputs of each DMU using Type-2 fuzzy sets, so we modify the crisp CCR model to include Interval Type-2 fuzzy sets. Based on the crisp CCR model, a model that incorporates Interval Type-2 fuzzy inputs and outputs, is presented as follows:

$$\max_{u,v} \{ u\tilde{y}_o \}$$
$$s.t.$$
$$v\tilde{x}_o = 1 \tag{4}$$
$$u\tilde{Y} \leq v\tilde{X}$$
$$v \geq 0, u \geq 0$$

where $\tilde{y}_o \in \tilde{Y}$, $\tilde{x}_o \in \tilde{X}$, and u, v are weights of the DMUs.

Now, model (4) is still a nonlinear problem that needs special treatment, so we propose to reduce its complexity using the centroids of every input-output pair as a priori information to solve it using a two-step LP method, as shown in next section.

4.1 A Centroid-Based Two-Step LP Proposal

1. Compute $C(\tilde{X})$, $C(\tilde{Y})$ of every Type-2 fuzzy inputs-outputs the problem.
2. Set $[c_l(\tilde{X}), c_u(\tilde{X})]$ and $[c_l(\tilde{Y}), c_u(\tilde{Y})]$ as the expected value of fuzzy inputs-outputs.
3. Solve the following interval-valued linear programming problem using $[c_l(\tilde{X}), c_u(\tilde{X})]$ and $[c_l(\tilde{Y}), c_u(\tilde{Y})]$ of each DMU:

$$\max_{u,v}\{uC(\tilde{y}_o)\}$$

$$s.t.$$

$$vC(\tilde{x}_o) = 1 \tag{5}$$

$$uC(\tilde{Y}) \leq vC(\tilde{X})$$

$$v \geq 0, u \geq 0$$

where $C(\tilde{y}_o) \in C(\tilde{Y})$, $C(\tilde{x}_o) \in C(\tilde{X})$, and u, v are weights of the DMUs.

4. This implies to solve a two-level LP model where $c_l(\tilde{z})$ comes from next LP:

$$c_l(\tilde{z}) = \max_{u,v}\{uc_l(\tilde{y}_o)\}$$

$$s.t.$$

$$vc_l(\tilde{x}_o) = 1 \tag{6}$$

$$uc_l(\tilde{Y}) \leq vc_l(\tilde{X})$$

$$v \geq 0, u \geq 0$$

And $c_r(\tilde{z})$ comes from next LP:

$$c_r(\tilde{z}) = \max_{u,v}\{uc_r(\tilde{y}_o)\}$$

$$s.t.$$

$$vc_r(\tilde{x}_o) = 1 \tag{7}$$

$$uc_r(\tilde{Y}) \leq vc_r(\tilde{X})$$

$$v \geq 0, u \geq 0$$

5. Return $[c_l(\tilde{z}), c_r(\tilde{z})]$ as the set of expected solutions of each $r \in \mathbb{N}^n$ output y_o.

After computing $[c_l(\tilde{z}), c_r(\tilde{z})]$ the most efficient and less efficient DMUs have to be identified and further decisions can be performed over this information. Usually most efficient DMUs are encouraged to be relative operation points while other DMUs are encouraged to be improved in order to go up their performance.

5   An Example

This case study is an example proposed by Cooper et al. [20] as an input-to-output problem. Assuming that there are 8 shops (see the first column of Table 1), this problem can be solved with the classical multiplier CCR-DEA model.

Table 1. Crisp data. Taken from Cooper et al. [19]

	Shops							
	1	2	3	4	5	6	7	8
Staff	2	3	3	4	5	5	6	8
Sales	1	3	2	3	4	2	3	5

To include multiple expert opinions into this fuzzy problem, we used triangular membership functions for their UMFs and LMFs as shown in Table 2.

Table 2. Parameters of the interval type- fuzzy inputs

Shop	1	2	3	4	5	6	7	8
$\bar{\check{X}}$	1.2	2.2	2.2	3.2	4.2	4.2	5.2	7.2
$\bar{\check{Y}}$	0.2	2.2	1.2	2.2	3.2	1.2	2.2	4.2
$\bar{X}$	2	3	3	4	5	5	6	8
$\bar{Y}$	1	3	2	3	4	2	3	5
$\bar{\hat{X}}$	2.8	3.8	3.8	4.8	5.8	5.8	6.8	8.8
$\bar{\hat{Y}}$	1.8	3.8	2.8	3.8	4.8	2.8	3.8	5.8
$\check{X}$	1.6	2.6	2.6	3.6	4.6	4.6	5.6	7.6
$\check{Y}$	0.6	2.6	1.6	2.6	3.6	1.6	2.6	4.6
$\hat{\underline{X}}$	2.4	3.4	3.4	4.4	5.4	5.4	6.4	8.4
$\hat{\underline{Y}}$	1.4	3.4	2.4	3.4	4.4	2.4	3.4	5.4

where all parameters are defined as shown in Fig. 1. The results of computing the centroids of all inputs-outputs are shown in Table 3.

The computation of the centroids of the parameters can be an expensive task, but it simplifies optimization process in the sense that it provides a priori information to see the central behavior of the problem, which finally leads to solve only two LPs per output. Now, we apply Algorithm 4.1 using the centroids shown in Table 3 to solve the two-step LP procedure. Final results are shown in Table 4:

Table 3. Centroids of the inputs X and outputs Y of the example

Shop	1	2	3	4	5	6	7	8
$c_l(\tilde{X})$	2.1	3.1	3.1	4.1	5.1	5.1	6.1	8.1
$c_u(\tilde{X})$	1.9	2.9	2.9	3.9	4.9	4.9	5.9	7.9
$c_l(\tilde{Y})$	1.1	3.1	2.1	3.1	4.1	2.1	3.1	5.1
$c_u(\tilde{Y})$	0.9	2.9	1.9	2.9	3.9	1.9	2.9	4.9

Table 4. Results of centroid-based optimization process (productivity)

Shop	1	2	3	4	5	6	7	8
$c_l(\tilde{z})$	0,4745	1	0,6555	0,7438	0,7960	0,3881	0,4918	0,6203
$c_r(\tilde{z})$	0,5237	1	0,6774	0,7561	0,8039	0,4117	0,5082	0,6296

As seen in Table 4, the efficiency (via DMUs) of the store 2 is the largest among all DMUs for both boundaries of the centroids, which means that store 2 is relatively the most productive store among all stores. In this case, store 2 is (in average) the most productive store when individually compared to all remaining stores.

On the other hand, shop 6 is the less efficient shop among all them (in average). This leads us to think that the central behavior of the whole DEA problem puts shop 2 as the most efficient and shop 6 as the less efficient among all shops. Although we cannot see the whole behavior of the DMUs, we can see its central behavior which is an additional information source for decision making.

Note that lower productivities $c_l(\tilde{z})$ are always smaller than $c_r(\tilde{z})$ except for shop 2 which is always the most productive. This gives us an idea about relative efficiency of other shops regarding shop 2, which becomes a reference point in terms of productivity. Finally, the set $[c_l(\tilde{z}), c_r(\tilde{z})]$ for every shop shows us the set of expected productivities which means that any productivity into $[c_l(\tilde{z}), c_r(\tilde{z})]$ can be considered as normal.

6   Concluding Remarks

The uncertainty generated by the opinion of multiple experts can be managed through Interval Type-2 fuzzy sets in order to represent linguistic information translated to the inputs-outputs of a DEA problem.

The proposed two-step LP methodology (see Algorithm 4.1) provides a simpler way to compute the central behavior of a DEA problem without computing the entire set of possible choices of inputs-outputs. Although our proposal is a Type-reduced method, it provides a good idea of the average value of the DEA problem in advance to help decision making.

It is important to note that the proposed method predefuzzifies all fuzzy parameters before finding an optimal solution in order to simplify computations. Other methods based on α-levels compute optimal solutions based on fuzzy sets and defuzzify its results after optimization process which implies much more computations. Evidently, the obtained solutions are not the same, so analysts have to do a good interpretation of the obtained results.

Finally, our a priori method provides additional information for decision makers, so its applicability in cases where no statistical information exists, is wide. Also we provide a tool for handling information coming from experts which is plenty of linguistic uncertainty, so our proposal is useful in cases where multiple experts try to find a joint solution of the problem.

Future Work

The use of constructive approaches (Figueroa-García [20]) for finding optimal solutions are the next step in fuzzy optimization. This kind of procedures offer complementary information that combined to the presented approach can enrich decision making.

References

1. Charnes, A., Cooper, W., Rhones, E.: Measuring the efficiency of decision making units. Eur. J. Oper. Res. **6**, 429–444 (1978)
2. Sengupta, J.: A fuzzy systems approach in data envelopment analysis. Comput. Math. Appl. **24**(8–9), 259–266 (1992)
3. Kahraman, C., Tolga, E.: Data envelopment analysis using fuzzy concept. In: 28th IEEE International Symposium on Multiple-Valued Logic, pp. 338–342. IEEE (1998)
4. Triantis, K., Girod, O.: A mathematical programming approach for measuring technical efficiency in a fuzzy environment. J. Prod. Anal. **10**(1), 85–102 (1998)
5. Girod, O., Triantis, E.K.: The evaluation of productive efficiency using a fuzzy mathematical programming approach: the case of the newspaper preprint insertion process. IEEE Trans. Eng. Manage. **46**(4), 429–443 (1999)
6. Mendel, J.M.: Uncertain Rule-Based Fuzzy Logic Systems: Introduction and New Directions. Prentice Hall, Englewood cliffs (2001)
7. Mendel, J.M., Wu, D.: Perceptual Computing: Aiding People in Making Subjective Judgments. Wiley, Hoboken (2010)
8. Wu, D., Mendel, J.M.: Enhanced Karnik-Mendel algorithms for Interval Type-2 fuzzy sets and systems. In: Annual Meeting of the North American Fuzzy Information Processing Society (NAFIPS), vol. 26, pp. 184–189. IEEE (2007)
9. Wu, D., Mendel, J.M.: Enhanced Karnik-Mendel Algorithms. IEEE Trans. Fuzzy Syst. **17**(4), 923–934 (2009)
10. Mendel, J.M., Liu, F.: Super-exponential convergence of the Karnik-Mendel algorithms for computing the centroid of an interval type-2 fuzzy set. IEEE Trans. Fuzzy Syst. **15**(2), 309–320 (2007)
11. Karnik, N.N., Mendel, J.M.: Centroid of a type-2 fuzzy set. Inf. Sci. **132**(1), 195–220 (2001)
12. Celemin, C., Melgarejo, M.: A proposal to speed up the computation of the centroid of an interval Type-2 fuzzy set. Adv. Fuzzy Syst **2013**, 17 (2013). (Article ID 158969)
13. Melgarejo, M., Bernal, K.: A fast recursive method to compute the generalized centroid of an interval type-2 fuzzy set. In: Annual Meeting of the North American Fuzzy Information Processing Society (NAFIPS), vol. 26, pp. 190–194. IEEE (2007)

14. Lertworasirikul, S., Fang, S.C., Joines, J.A., Nuttle, H.L.: Fuzzy data envelopment analysis (DEA): a possibility approach. Fuzzy Sets Syst. **139**(2), 379–394 (2003)
15. Lertworasirikul, S., Fang, S.C., Joines, J.A., Nuttle, H.L.: A possibility approach to fuzzy data envelopment analysis. In: 6th Joint Conference on Information Science, pp. 176–179 (2002)
16. Karsak, E.: Using data envelopment analysis for evaluating flexible manufacturing systems in the presence of imprecise data. Int. J. Adv. Manuf. Technol. **35**(9–10), 867–874 (2008)
17. Girod, O., Triantis E,K.: Measuring technical efficiency in a fuzzy environment. Ph.D. dissertation, Department of Industrial and Systems Engineering, Virginia Polytechnic Institute and State University (1996)
18. Hatami-Marbini, A., Emrouznejad, A., Tavana, M.: A taxonomy and review of the fuzzy data envelopment analysis literature: two decades in the making. Eur. J. Oper. Res. **214**(3), 457–472 (2011)
19. Cooper, W., Lawrence, M.S., Kaouru, T.: Introduction to Data Envelopment Analysis and Its Uses. Springer-Verlag, Heidelberg (2006)
20. Figueroa-García J.C. A general model for linear programming with interval type-2 fuzzy technological coefficients. In: 2012 Annual Meeting of the North American Fuzzy Information Processing Society (NAFIPS), pp. 1–6. IEEE (2012)

Image Processing and Computer Vision

Adaptive Quantization of Local Directional Responses for Infrared Face Recognition

Zhihua Xie[✉], Zhengzi Wang, and Guodong Liu

Key Lab of Optic-Electronic and Communication,
Jiangxi Sciences and Technology Normal University, Nanchang Jiangxi, China
xie_zhihua68@aliyun.com

Abstract. Local feature extraction is one key step in infrared face recognition system. In previous local features extraction (local binary pattern and its variants) on infrared face recognition, a fixed threshold is binary encoded, which consider limited structure information. An infrared face recognition method based on adaptive quantization of local directional responses pattern (AQLDRP) is proposed in this paper. Firstly, the normalized infrared face images are directional filtered to generate local directional responses, which represent the local structures distinctively and are robust to the impacts of imaging conditions. Then, each local feature vector is quantized by adaptive quantization thresholds to preserve distinct information. Finally, the partition histograms concatenation representation is used for final recognition. The experimental results show the recognition rates of proposed infrared face recognition method can reach 93.1 % under variable ambient temperatures, outperform the state-of-the-art methods based on LBP variants.

Keywords: Adaptive quantization · Infrared face recognition · Local directional patterns · Feature extraction · Adaptive threshold

1 Introduction

Compared with the visible spectrum imaging, infrared imaging can acquire the interior subcutaneous anatomical information of a face, which is invariance to the changes in illumination changes and disguises [1]. Therefore, infrared face recognition is an important branch of automatic face recognition [2]. The popular Eigenfaces feature extraction (PCA, principal component analysis) is firstly applied to infrared face recognition and get favorable recognition rate. Following these promising results, many appearance based feature extractions methods are reported to get whole statistical information in infrared face: LDA (Linear Discriminant Analysis) and ICA (Independent Component Analysis) [7, 8]. The experiments based on those existing algorithms demonstrated that the performance of infrared face recognition system greatly declines when the database contains time-lapse data, i.e., the test samples are not collected at the same time with the training samples (the challenges from the external environment temperature, low resolution, and other imaging conditions) [1, 2]. Recently, local feature extraction methods, which capture the useful local structure in a face, are appreciated for infrared

© Springer International Publishing Switzerland 2015
D.-S. Huang and K. Han (Eds.): ICIC 2015, Part III, LNAI 9227, pp. 199–207, 2015.
DOI: 10.1007/978-3-319-22053-6_22

face recognition research [2]. In a series of influential researches, Local Binary Pattern (LBP) and its variants is introduced to local–matching of infrared face [3]. Other reported infrared face recognition approaches are based on the use of multi-scale transform: wavelet transform and curvelet transform. However, most of researches focused on near infrared face imaging, which is unsuitable for uncooperative user applications.

This paper focuses on the far infrared face recognition. Local binary pattern was applied to far infrared face feature extraction by Xie et al. in 2011 [9], which got a better performance than appearance based methods such as PCA, LDA and ICA. In last decade years LBP-based texture analysis has gotten more and more attentions for local feature representation. LBP encodes the relative intensity magnitude between central pixel and its neighboring pixels, which can capture the discriminative micro-structures in an image such as flat areas, spots, lines and edges [5, 10]. Since the impact of external environment temperature on infrared face image is almost a monotonic transform, the LBP can extract robust features for infrared face recognition, which is little sensitive to the environment temperatures.

Although LBP performs efficiently in the aspect of feature extraction, they still suffer from the intensity change problem [6]. Abundant LBP-based modification methods were proposed to increase the discriminative capability and remedy the intensity change problem [11–13]. Recently, Jabid et al. [11] presented a Local Directional Pattern (LDP) by taking the edge responding information in different directions into account for noise alleviation. However, LDP operator used the number of maximum responses to generate a decimal pattern, which is different from the LBP operation. Furthermore, the local binary pattern (LBP) operator used a fixed threshold to generate a decimal number to label each local structure, which considering limited infrared face statistical characters.

Motivated by LBP operation using binary coding for quantization and LDP noise alleviation, in this paper we proposed an adaptive quantization method of local directional responses for infrared face recognition. The local directional responses are used to handle infrared imaging noises. The adaptive quantization measure of local directional responses vector is employed to reduce the quantization loss and thus preserve more local structure information in infrared face images. As will be shown in the experiments, our feature representation method is robust to noise and discriminative for infrared face recognition.

2   Related Works

Local binary patterns were introduced by Ojala [10] which has a low computational complexity and a low sensitivity to monotonic photometric changes. It has been widely used in biometrics such as face recognition, gender classification, iris recognition and infrared face recognition [4]. In its simplest form, an LBP description of a pixel is created by threshold the values of the 3×3 neighborhood of the pixel against the central pixel and encoding the result by a binary number. The parameters of the original LBP operator with one pixel radius and 8 neighborhood points are demonstrated in Fig. 1. LBP code for center point g_c can be defined as:

$$LBP_{P,R}(x_c, y_c) = \sum_{i=0}^{P-1} 2^i \cdot S(g_i - g_c) \tag{1}$$

$$S(g_i - g_c) = \begin{cases} 1, g_i - g_c \geq 0 \\ 0, g_i - g_c < 0 \end{cases} \tag{2}$$

Where (x_c, y_c) is the coordinate of the central pixel, g_c is the gray value of the central pixel, g_i is the value of its neighbors, P is the total number of sampling points and R is the radius of the neighborhood. As shown in Fig. 1, the LBP patterns with different radiuses characterize different size local structures. In other words, a radius R stands for a scale. The parameters (P, R) can be (8,1), (8, 2), (16, 2) and (8, 3) etc.

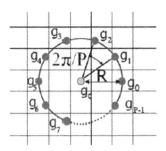

Fig. 1. The parameters P and R of LBP

Local Directional Pattern (LDP) calculates the responses between the masks in 8 different directions and the center block to substitute the pixel values in the corresponding positions of the center block [11]. Kirsch masks with 8 directions (M0–M7) are used for convolution in calculating the responses (m0– m7) (Fig. 2).

$$\begin{bmatrix} -3 & -3 & 5 \\ -3 & 0 & 5 \\ -3 & -3 & 5 \end{bmatrix} \begin{bmatrix} -3 & 5 & 5 \\ -3 & 0 & 5 \\ -3 & -3 & -3 \end{bmatrix} \begin{bmatrix} 5 & 5 & 5 \\ -3 & 0 & -3 \\ -3 & -3 & -3 \end{bmatrix} \begin{bmatrix} 5 & 5 & -3 \\ 5 & 0 & -3 \\ -3 & -3 & -3 \end{bmatrix} \begin{bmatrix} 5 & -3 & -3 \\ 5 & 0 & -3 \\ 5 & -3 & -3 \end{bmatrix} \begin{bmatrix} -3 & -3 & -3 \\ 5 & 0 & -3 \\ 5 & 5 & -3 \end{bmatrix} \begin{bmatrix} -3 & -3 & -3 \\ -3 & 0 & -3 \\ 5 & 5 & 5 \end{bmatrix} \begin{bmatrix} -3 & -3 & -3 \\ -3 & 0 & 5 \\ -3 & 5 & 5 \end{bmatrix}$$
$$\quad M0 \qquad\quad M1 \qquad\quad M2 \qquad\quad M3 \qquad\quad M4 \qquad\quad M5 \qquad\quad M6 \qquad\quad M7$$

Fig. 2. Kirsch edge masks in all eight directions.

Then, the top k bits of 8 directional responses are set to 1. The remaining bits are set to 0. Finally, similar to LBP coding pattern, the LDP code is derived using (3)

$$LDP_k = \sum_{i=0}^{7} S(m_i - m_k) \cdot 2^i \tag{3}$$

where m_k is the k-th biggest directional response.

The LDP feature overcomes the limitations of LBP features since LDP is derived from the edge responses which are less sensitive to illumination changes and noises [6, 13]. In this paper, we make full use of the local directional responses. The vector of local directional responses V can be expressed by

$$V = [m_0^2, m_1^2, m_2^2, \cdots, m_7^2] \tag{4}$$

An important shortcoming in LDP is the empirical binary encoding of directional response which has limited distinguishing ability. It is interesting to find a reasonable way to quantize the directional response vector.

3   Adaptive Quantization of Local Directional Responses

In order to transform the local directional response vector to a code number, the elements of each local directional response vector should be first quantized into N states. We need some quantization thresholds to ensure the quantized local feature vectors are distinct and preserve enough information [12]. The learning stage is to obtain some self-adaptive thresholds. Inspired by the classic histogram specification in image processing [14], the thresholds for N levels are set so that 1/N of the histogram area is within each bin (between two adjacent thresholds). The adaptive quantization for infrared face data has two steps:

3.1   Normalization of Directional Responses Vector

The normalization of local feature is one important step to reduce the effects of noise and imaging conditions. In this paper, we propose a way of normalization and describe as following. Let the local directional responses vector be $[v_0, v_1, v_2, \cdots, v_7]$.

The normalized local feature vector is illustrated as:

$$V = [v_{n0}, v_{n1}, v_{n2}, \cdots, v_{n7}] \tag{5}$$

where v_{ni} is the normalized local directional response defined as:

$$v_{ni} = v_i/(v_{sum} + v_{median}) \tag{6}$$

$$v_{sum} = \sum_{i=1}^{7} v_i \tag{7}$$

where v_{median} is the mean value of v_{sum} for an infrared image, and it is easy to find out that $0 \leq v_{ni} < 1$.

3.2   Quantization

In order to transform the local feature vector to a code number, the elements of each local feature vector should be first quantized into N states. We need some quantization thresholds to ensure the quantized local feature vectors are distinct and preserve enough information. The learning stage based on the distribution of v_{ni} is applied to obtain some self-adaptive thresholds. This process is similar to classic histogram equalization in image processing. Let $p(\bullet)$ denote the probability density function (PDF) of v_{ni} in one infrared image [15]. The monotonous transformation function f (v) can be gotten by:

$$f(v) = \int_0^v p(x)dx \qquad (8)$$

Where v stands for the possible values of elements in the normalized local directional responses vector V. The thresholds for v_{ni} could be determined by $f^{-1}(v)$. Accordingly, the self-adaptive thresholds T are

$$T = \{f^{-1}(1/N), f^{-1}(2/N), \cdots, f^{-1}(N-1/N)\} \qquad (9)$$

As a result, the vector of local directional responses V can be quantized to V_Q by the adaptive thresholds T.

$$V_Q = [v_Q(1), v_Q(2), \cdots, v_Q(7)] \qquad (10)$$

4   The Proposed Infrared Face Recognition

In this section, the detail realization of the proposed infrared face recognition is listed as follow:

Stage one: Infrared face detection and normalization [1]. After normalization, the resolution of infrared face images is the same.

Stage two: The normalized infrared face image is filtered to the local directional responses vector V for each pixel.

Stage three: Adaptive quantization proposed in Sect. 3 is applied to describe the adaptive quantization of local directional responses pattern (AQLDRP) by the following formula

$$AQLDRP = \sum_{i=1}^{7} v_Q(i) \cdot N^i \qquad (11)$$

Stage four: The histogram representation H of $AQLDRP$ is used to get final features by partitioning modes. Suppose the partition block face image is of size $(M \times N)$. After identifying the $AQLDRP$ code of each pixel (x_c, y_c), by computing the $AQLDRP$ patterns histogram, our infrared face recognition method achieves the final features.

$$H(r) = \sum_{x_c=2}^{N-1} \sum_{y_c=2}^{M-1} f(AQLDRP(x_c, y_c), r) \qquad (12)$$

$$f(AQLDRP(x_c, y_c), r) = \begin{cases} 1, & AQLDRP(x_c, y_c) = r \\ 0, & otherwise \end{cases} \tag{13}$$

where the value r ranges from 0 to $N^8 - 1$.

Last stage: The nearest neighborhood classifier (NNC) is employed to perform the recognition task. In our infrared face recognition method, NNC is based on dissimilarity of final features between training datasets and test face. As traditional criterion based on LBP histogram, the metric based on chi-square statistic is used [13, 14]. The dissimilarity of two histograms $(H1, H2)$ can be gotten by:

$$Sim(H1, H2) = \sum_{bin=1}^{n} \frac{(H1(bin) - H2(bin))^2}{H1(bin) + H2(bin)} \tag{14}$$

Where n is the dimension of histogram representation of AQLDRP.

5   Results and Discussion

Our infrared dataset in this paper were captured by an infrared camera Thermo Vision A40 supplied by FLIR Systems Inc [1, 9], which can be divided into same-time data and time-lapse data. The training same-time data comprises 500 thermal images of 50 individuals which were carefully collected under the similar conditions controlled by the air conditioner. The time-lapse data consists of 165 thermal images of one individual which were collected under ambient temperatures from 24.3 to 28.4 °C. The original size for each image is 240 × 320. After preprocess and normalization, it becomes 80 × 60. Our experiments take the nearest neighbor distance for final classifier (Fig. 3).

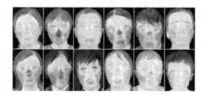

Fig. 3.   Part of infrared face database

In our experiments, AQLDRP based on adaptive vector quantization is occupied for infrared face feature extraction, the parameter N can be 2, 3 and 4. To consider the space location information, the partitioning AQLDRP is applied to get final connect histogram features. In this paper, we use the 4 × 2 partition mode. The dimension number of histogram extracted by our proposed AQLDRP is N^8. The recognition rates with different parameters N are shown in Table 1.

Table 1. The Recognition Rates with different N

Parameter N	Recognition Rates
2	88.6 %
3	92.5 %
4	93.1 %
5	93.1 %

Table 1 shows that the classification accuracies increase with N. A larger N indeed makes the feature more accurate but bring some redundancy as well, so the improvement of performance is limited. However, the larger N would introduce more features. The N can be too larger. According to Table 1, when N equals 3, satisfactory performance without too high feature dimensionality can be achieved. Therefore, in this paper the parameter N is 3.

Liking the local feature descriptors based on histogram, the partitioning modes of AQLDRP is an important factor for discriminative information representation. Five modes of partitioning are used to the effect on infrared face recognition: 1 is non-partitioning, 2 is 2×2, 3 is 4×2, 4 is 2×4, and 5 is 4×4. The recognition results based on AQLDRP and LBP with different partitioning modes are demonstrated in Fig. 4.

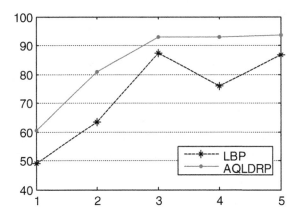

Fig. 4. Recognition results with different partitioning modes

It can be seen from Fig. 4 that feature extraction method based on AQLDRP can get best recognition rate at partitioning mode 4×2. When the number of patches reaches a threshold, the more patches not always contribute to the better recognition performance.

To verify the effectiveness of the proposed features extraction method for infrared face recognition, we conducted experiments to compare different approaches by the AQLDRP and three existing features extraction algorithms: traditional LBP [9], traditional LDP [11] and PCA + LDA [8].

The top recognition results are demonstrated in Table 2. It can be seen from the Table 2 that adaptive quantization of local directional responses can improve the recognition performance of traditional LDP. It is also revealed from Table 2, compared with traditional LBP, our proposed AQLDRP can preserve more local structure information in infrared face representation, which contributes to better recognition performance.

Table 2. The Top Recognition Rates of Different Methods

Methods	Recognition Rates
AQLDRP	**93.1 %**
LDP	88.0 %
LBP	87.4 %
PCA + LDA	33.6 %

6   Conclusions

The conventional LBP-based feature as represented by the empirical binary encoding has limited distinguishing ability. To overcome this limitation, a novel feature extraction based on AQLDRP is proposed for infrared face recognition. The proposed feature is based on adaptive quantization of local directional responses vector could provide much more information than traditional empirical encoding measure. Our experiments illustrate that the proposed AQLDRP can extract accurate discriminative information and decrease the influence of infrared imaging factors, the performance of the proposed infrared face recognition method outperforms state-of-the-art methods based on LBP and LDP.

Acknowledgments. This paper is supported by National Nature Science Foundation of China (No. 61201456), Natural Science Foundation of Jiangxi Province of China (No. 20142BAB207029), the Science & Technology Project of Education Bureau of Jiangxi Province (No.GJJ14581) and Nature Science Project of Jiangxi Science and Technology Normal University (2013QNBJRC005, 2013ZDPYJD04).

References

1. Wu, S.Q., Li, W.S., Xie, S.L.: Skin heat transfer model of facial thermograms and its application in face recognition. Pattern Recogn. **41**(8), 2718–2729 (2008)
2. Ghiass, R.S., Arandjelovic, O., Bendada, A., Maldague, X.: Infrared face recognition: a comprehensive review of methodologies and databases. Pattern Recogn. **47**(9), 2807–2824 (2014)
3. Li, S.Z., Chu, R., Liao, S., Zhang, L.: Illumination invariant face recognition using near-infrared images. IEEE Trans. Pattern Anal. Mach. Intell. **29**(12), 627–639 (2007)

4. Ahonen, T., Hadid, A., Pietikäinen, M.: Face description with local binary patterns: application to face recognition. IEEE Trans. Pattern Anal. Mach. Intell. **18**(12), 2037–2041 (2006)
5. Huang, D., Shan, C., Ardabilian, M., Wang, Y., Chen, L.: Local binary patterns and its application to facial image analysis: a survey. IEEE Trans. Syst. Man Cybern. - part C **41**(6), 765–781 (2011)
6. Ren, J., Jiang, X., Yuan, J.: Noise-resistant local binary pattern with an embedded error correction mechanism. IEEE Trans. Image Process. **22**(10), 4049–4060 (2013)
7. Chen, L., Liao, H., Ko, M.: A new LDA based Face recognition system which can solve the small sample size problem. Pattern Recogn. **33**(10), 1713–1726 (2000)
8. Hua, S.G., Zhou, Y., Liu, T.: PCA + LDA based thermal infrared imaging face recognition. Pattern Recog. Artif. Intell. **21**(2), 160–164 (2008)
9. Xie, Z.H., Zeng, J., Liu, G.D.: A novel infrared face recognition based on local binary pattern. In: Proceedings of 2011 International Conference on Wavelet Analysis and Pattern Recognition (ICWAPR), Qingdao, China, pp. 55–59. IEEE, USA (2011)
10. Ojala, T., Pietikainen, M., Maenpaa, T.: Multi-resolution gray-scale and rotation invariant texture classification with local binary patterns. IEEE Trans. Pattern Anal. Mach. Intell. **14**(7), 971–987 (2002)
11. Jabid, T., Kabir, M.H., Chae, O.: Gender classification using local directional pattern(LDP). In: Proceeding of International Conference on Pattern Recognition (ICPR2010), Istanbul, Turkey, pp. 2162–2164. IEEE, USA (2010)
12. Zhang, J., Liang, J.M., Zhao, H.: Local energy pattern for texture classification using self-adaptive quantization thresholds. IEEE Trans. Image Process. **22**(1), 31–42 (2013)
13. Song, T., Li, H., Meng, F., Wu, Q., Luo, B., Zeng, B., Gabbouj, M.: Noise-robust texture description using local contrast patterns via global measures. IEEE Signal Process. Lett. **21**(1), 93–96 (2014)
14. Gonzalez, R., Woods, R.: Digital Image Processing, 2nd edn. Prentice Hall, Upper Saddle River (2002)
15. Ojala, T., Pietikäinen, M., Harwood, D.: A comparative study of texture measures with classification based on featured distributions. Pattern Recogn. **29**(1), 51–59 (1996)

Preliminary Study of Tongue Image Classification Based on Multi-label Learning

XinFeng Zhang[(⊠)], Jing Zhang, GuangQin Hu, and YaZhen Wang

Signal and Information Processing Lab, College of Electronic Information
and Control Engineering, Beijing University of Technology,
No. 100 Pingleyuan Chaoyang District, Beijing 100124, China
zxf@bjut.edu.cn, {yiyijiuzai,wyzsnowing}@126.com,
hgmh@163.com.cn

Abstract. Tongue diagnosis characterization is a key research issue in the development of Traditional Chinese Medicine (TCM). Many kinds of information, such as tongue body color, coat color and coat thickness, can be reflected from a tongue image. That is, tongue images are multi-label data. However, traditional supervised learning is used to model single-label data. In this paper, multi-label learning is applied to the tongue image classification. Color features and texture features are extracted after separation of tongue coat and body, and multi-label learning algorithms are used for classification. Results showed LEAD (Multi-Label Learning by Exploiting Label Dependency), a multi-label learning algorithm demonstrating to exploit correlations among labels, is superior to the other multi-label algorithms. At last, the iteration algorithm is used to set an optimal threshold for each label to improve the results of LEAD. In this paper, we have provided an effective way for computer aided TCM diagnosis.

Keywords: Tongue diagnosis · Tongue image · Multi-label learning

1 Introduction

In recent years, automatic classification of tongue image in TCM gradually becomes a hot research. Automatic classification of tongue coat based on neural network integration was used in [1]. Supervised FCM clustering algorithm was put forward in [2], to classify the color tongue images. Weighted Support Vector Machine (WSVM) was adopted in the classification of large number of tongue image samples in [3], thus improving the recognition accuracy. An improved BP algorithm was applied to construct neural network model of TCM tongue intelligent diagnosis in [4]. A series of weak classifiers were promoted to strong classifiers, to carry on an in-depth study on the classification of tongue color in [5]. Considerable progress in tongue image classification has been made and the development of tongue diagnosis characterization has been promoted through these studies. But the above classification algorithms works under the single-label scenario, i.e. each example is associated with one single label characterizing its property. However a tongue image is associated with multiple labels of tongue body color, coat color and coat thickness simultaneously. Aiming at this

© Springer International Publishing Switzerland 2015
D.-S. Huang and K. Han (Eds.): ICIC 2015, Part III, LNAI 9227, pp. 208–220, 2015.
DOI: 10.1007/978-3-319-22053-6_23

problem, multi-label learning is applied in this paper. In multi-label learning, each example in the training set is represented by a feature vector and associated with a set of labels. The task is then to predict the label sets of unseen examples through analyzing training examples with known label sets.

The paper is organized as followed. Section 2 introduces tongue image acquisition and labeling. Section 3 shows the extraction of color features and texture features. Section 4 represents the evaluation metrics and algorithms of multi-label learning. Section 5 reports the experimental results of different algorithms. Section 6 concludes.

2 Related Works

2.1 Tongue Image Acquisition

The tongue images were collected by the tongue image instrument in Hospital of Beijing University of Technology. The light environment is unified in the tongue image instrument and the camera is CANON1100D with AF. The subjects were asked to sit straight, open the mouth, and stretch out the tongue as naturally as possible.

2.2 Tongue Image Preprocessing

Segmentation is done on the original tongue images collected by tongue image instrument, to get the tongue area only. The original image and the tongue area are shown in Fig. 1.

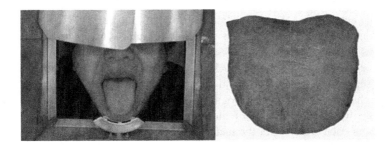

Fig. 1. The original image and the tongue area

2.3 Tongue Image Labeling

According to the suggestion given by TCM experts,we used the most common 7 labels to build the label space, namely pink tongue, red tongue, dark red tongue, thin white coat, thin yellow coat, thick white coat and thick yellow coat. TCM experts chose 702 typical tongue images as samples, and assigned proper labels for each tongue image from the label space as the image's label set. If the tongue body color is between two kinds of color, the two kinds of color are both included in its label set. The schematic diagram of tongue image labeling is shown in Fig. 2.

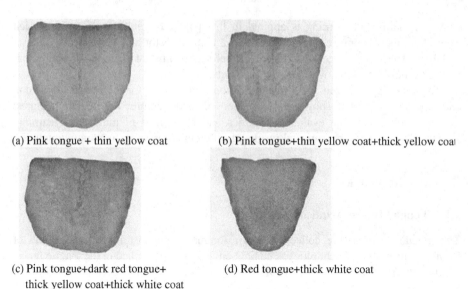

(a) Pink tongue + thin yellow coat   (b) Pink tongue+thin yellow coat+thick yellow coat

(c) Pink tongue+dark red tongue+   (d) Red tongue+thick white coat
thick yellow coat+thick white coat

Fig. 2. The schematic diagram of tongue image labeling

3   Feature Extractions

A series of sub-block images were selected and the RGB values of each pixel in the sub-block image were taken as the characteristics of the tongue image in [3]. The chromaticity coordinates mean of each sub-block image was calculated as the feature vector of the network in [1]. Accumulative histogram based on HSV color space was used for tongue retrieval in [6]. The selection of sub-block images in the above methods has certain subjectivity. Besides, tongue body color and coat color are not discussed separately, and tongue blocking is not taken into consideration either. Therefore, separation of tongue coat and body was done first in this paper, and then the main color characteristic of tongue coat and body was extracted separately. Finally, the tongue image was divided into five regions, and four-dimensional texture feature was extracted from each region with the method of GLCM [7].

3.1   Separation of Tongue Coat and Body

At present, the methods of the separation of tongue coat and body are K-means clustering [8], fuzzy clustering [9], threshold value method [10] and so on. Clustering methods has weak adaptively, and the traditional single threshold value method usually has poor effect. Consequently, the algorithm of dynamic threshold combing multiple channels in [11] was used for separation in this paper.

The results of experiments show that the algorithm can achieve good effects. One of the separation results is as follows: the original image, tongue body and tongue coat are shown in Figs. 3, 4 and 5 respectively.

3.2   Color Feature Extraction

Often a few colors can cover most pixels of an image and the occurrence probability of different colors is different, therefore, several frequent colors can be selected as the main colors through the statistical probability of all colors in the image. For the above reasons, main color histogram was adopted to extract color features in this article. Since the HSV [12] color space is for visual perception and is closer to the way humans perceive color, HSV color space was used in this paper.

We chose pink tongue images with thin white coat, red tongue images with thin yellow coat, dark red tongue images with thick yellow coat and dark red tongue images with thick white coat from the samples, and the number of each kind is 15. After separation of tongue coat and body on those 60 images, tongue coat image and tongue body images were converted from RGB into HSV space separately. Then we extracted the H, S and V on each coat image and body image. Results show that about 88 % of the 60 tongue images meet the corresponding relationship reflected in Tables 1 and 2. From these two tables, we can see that tongue coat color and tongue body color can be distinguished with the methods used in this paper.

Fig. 3. The original image

Fig. 4. Tongue body

Fig. 5. Tongue coat

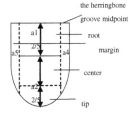

Fig. 6. Tongue blocking

Table 1. Correspondence between the coat color and H, S, V values

V				0-100			
H		18-45		45-325		325-360	
S	0-22	25-50	51-100	none	0-22	25-50	51-100
coat color	thick white coat	thin yellow coat (thick yellow coat)	none	none	none	thin white coat	none

Table 2. Correspondence between the tongue body color and H, S, V values

V				0-100			
H		18-45		45-325		325-360	
S	0-22	25-50	51-100	none	0-22	25-50	51-100
tongue body color	none	none	none	none	dark red tongue	pink tongue	red tongue

On the basis of visual resolving power and different scope of color, we quantified hue H, saturation S and luminance V into 16 parts, 4 parts and 4 parts respectively with unequal interval. Then H, S and V were weighted into a one-dimensional histogram L according to L = 16H + 4S + V. Statistics was made on L and then L was sorted in descending order. We chose the sorted L' first five values and their bins as the color feature both on tongue coat and body images. There are 20 dimensions of color features.

3.3   Texture Feature Extraction

As can be seen from Tables 1 and 2, the H, S value of thin yellow coat and thick yellow coat has the same range. So it is difficult to identify coat thickness just by color features. Through the observation and analysis on thick coat images and thin ones, we can find a big difference on roughness between most of them: thin coat has a rough texture and the thick is on the contrary. For this reason, the texture features were added in this paper. Texture is formed owing to the recurring of gray distribution at spatial position, so there is a certain gray relationship between two pixels at certain intervals, i.e. the spatial correlation characteristics of gray in an image. Gray level co-occurrence matrix (GLCM) is a common way to describe texture by studying the spatial correlation characteristics of gray. GLCM were utilized to extract the texture features of tongue images. In this paper, 4 scalars of energy, entropy, contrast and correlation are used to characterize GLCM. The 4 scalars are expressed as follows [13]:

$$\text{Energy:} \quad ASM = \sum_i \sum_j (\mathbf{P}(i,j))^2 \tag{1}$$

$$\text{Entropy:} \quad ENT = -\sum_i \sum_j \mathbf{P}(i,j) \log \mathbf{P}(i,j) \tag{2}$$

$$\text{Contrast:} \quad CON = \sum_i \sum_j (i-j)^2 \mathbf{P}(i,j) \tag{3}$$

$$\text{Correlation:} \quad COR = \frac{\sum_i \sum_j i \times j \times \mathbf{P}(i,j)^2 - \mu_x \mu_y}{\delta_x \delta_y} \tag{4}$$

Table 3. The texture features of four kinds of tongue images

	thin yellow coat	thick yellow coat	thick white coat	thin white coat
Energy	0.1221	0.1617	0.1744	0.1026
Entropy	2.6863	2.4635	2.2699	2.6445
Contrast	0.3867	0.5685	0.3350	0.2925
Correlation	0.2427	0.3793	0.4246	0.1391

where $\mathbf{P}(i,j)$ is the normalized matrix of GLCM, μ_x and μ_y are the horizontal and vertical mean of the image respectively. δ_x and δ_y are the horizontal and vertical variances of the image respectively.

In order to compare the texture features of thin coat and thick coat, we selected 40 thin coat tongue images (20 thin yellow coat and 20 thin white coat) and 40 thick coat tongue images (20 thick yellow coat and 20 thick white coat) from the samples. Then we extracted energy, entropy, contrast and correlation on the tongue coat part of each image after separation of tongue coat and body. The average results are shown in Table 3. From these figures, we may see that the thin coat has smaller energy and correlation, bigger entropy, while the thick coat is on the contrary. The results are consistent with the actual situation.

In the light of the theory of TCM, tongue has a close relationship with internal organ channels. According to [14], the diseases of internal organs are reflected on tongue surface. There is a certain distribution: cardiopulmonary corresponds to the tip of tongue, spleen and stomach to the center of tongue, kidney to the root, liver and gall to the margin. Tongue surface is divided into five parts in this paper: the root of tongue, the center, the tip, the left and right margin. According to [15], from the tip of tongue to the herringbone groove midpoint are divided into 5 equal parts, first 1/5 is the tip of the tongue, the second 2/5 is the center of the tongue, and the third 2/5 is the root of the tongue. After drawing a midline between the center line and the edge of the tongue image, the outside of the midline is called tongue margin. The division result is shown in Fig. 6.

The 4 values of energy, entropy, contrast and correlation are extracted from the above 5 regions respectively as texture features. There are 20 dimensions of texture features.

The all 40 dimensions of color features and texture features are shown in Table 4.

Table 4. The composition of 40 features

40 features	Color (20)				Texture(20)				
Region	Tongue body		Tongue coat		Root	Left margin	Right margin	Center	Tip
detail features	main color(5)	Bins (5)	main color(5)	Bins (5)	GLMC (4)	GLMC (4)	GLMC (4)	GLMC (4)	GLMC (4)

4 Multi-label Learning

What's more, there is some correlation between the labels of tongue images. For example, the probability of a tongue image be annotated with label "thin white coat" would be high if we know it has label "pink tongue", because the pink tongue with thin white coat is regarded as the normal tongue in TCM [14]. Therefore, effectively exploiting correlations between different labels can improve the classification accuracy. There are already many multi-label learning algorithms aiming to exploiting label correlations. In this paper, many different multi-label algorithms were used to classify tongue images. When the multi-label learning is applied to tongue classification, there are several key issues to be addressed:

4.1 Evaluation Metrics

Since each example is associated with multiple labels at the same time, namely the prediction result of multi-label learning is a vector, the evaluation metrics in traditional single-label learning, such as accuracy, precision, recall, etc., can't be used in multi-label learning directly. Currently there is no universal multi-label evaluation metric that is applicable to all problems, and the choice of metrics depends on specific learning tasks. Based on the particular case of tongue image classification, the following three evaluation metrics are adopted in this paper: hamming loss [16], average precision [16] and $\partial - Evaluation$ [17]. Hamming loss is used to investigate the misclassification on a single label. Average precision is used to investigate the average accuracy of multi-label learning. $\partial - Evaluation$ is used to investigate the average ratio of correct labels on a test sample. The smaller the value of hamming loss is, the better the performance of the classifier, while the average precision and $\partial - Evaluation$ are on the contrary.

4.2 Learning Algorithms

The multi-label learning algorithms are divided into two categories in [16]: problem transformation method and algorithm adaptation method. The former category transforms multi-label learning problems into multiple single-label classification problems by processing the data set. The latter category tackles multi-label learning problem by expanding a certain single-label classification algorithm to deal with multi-label data directly. Both the size of tongue image samples and labels is big, so a lot of time would be spent on the conversion of data sets if problem transformation method was employed. Therefore, algorithm adaptation method is used to classify tongue images in this paper.

Six commonly used algorithms of algorithm adaptation method were chosen, and they are ML-kNN [18], Rank-SVM [19], LEAD [20], BP-MLL [21], ML-RBF and MLNB. The results of these algorithms were compared in this paper.

The parameter neighbor number k is set from 5 to 10 in ML-kNN. Linear kernel, RBF kernel and polynomial kernel are adopted in the SVM kernel function of LEAD respectively. We can get the best g 0.1 with PSO parameter optimization in the case of

RBF kernel function, and then three different g are used. When the kernel is polynomial, parameters gamma and coefficient are set to 1/k (k is the number of label, i.e. 7) and 0 separately, parameter degree is assigned to three values 2, 3 and 4. In ML-RBF, the parameter ∂ and μ are given 5 pairs of different values. Six different values of parameter ratio are set from 0.1 to 0.6 in MLNB. In BP_MLL, parameters γ and epochs are assigned to 5 pairs of different values. In RankSVM, SVM kernel function is discussed in three cases of linear kernel, RBF kernel and polynomial kernel. Take optimal parameter (g = 0.1, C = 0.8) after PSO parameter optimization and two pairs of nearby values to compare in the case of RBF. Parameters gamma, coefficient and c are set to 1/k (k is the number of label, i.e. 7), 0 and 1 separately, and parameter degree is assigned to three values 2, 3 and 4 when the kernel is polynomial.

4.3   Threshold Value Selection

Given a multi-label training set $D = \{(x_i, Y_i) \mid 1 \leq i \leq m\}$, $h(x_i, y)$ returns the confidence of labeling x_i with y. Given a threshold function $t : X \rightarrow$ R, we have $r(x_i) = \{y \mid h(x_i, y) > t(x_i)\}$. Here, $t(x_i)$ produces a bipartition of label space Y into relevant label set and irrelevant label set. In some algorithms, $t(x_i)$ is set to 0.5. If $h(x_i, y)$, the probability of example x_i belonging to a certain label y, is higher than 0.5, then y is added to the label set of x_i, and the label y is set to 1. Whether the threshold value is appropriate or not, has a relationship with the algorithm accuracy. If the threshold value is too small, the tongue image not belonging to the label may be mislabeled 1. If the value is too big, the tongue image belonging to the label may be mislabeled 0, which causing the omission of label set. What's more, the distribution of each label is different, so providing the same threshold value for all labels is not appropriate. In this paper, we proposed to apply the iteration algorithm to determinate the optimal threshold in multi-label learning. We compared the result of the iteration algorithm with another threshold determination algorithm DTML in [22].

DTML threshold determination algorithm The classification model is used to train training data set in DTML firstly. Then, the confidences of every sample belonging to each label are obtained, and these confidences are stored in corresponding sets, as shown in Eqs. 5 and 6. After that, these confidences are learned to determine a threshold value for every label.

$$\Lambda_l^+ = \{f(x_i, y_l) \mid x_i \in D_l^+\}, 1 \leq l \leq |Y| \tag{5}$$

$$\Lambda_l^- = \{f(x_i, y_l) \mid x_i \in D_l^-\}, 1 \leq l \leq |Y| \tag{6}$$

where $D_l^+(D_l^-)$ represents set consisting of all the samples whose label sets contain (not contain) label y_l. All the samples that are predicted to belong to label y_l are chosen, and the confidence sets of these samples are represented by $\Lambda_l^\pm$. Constitute two intervals by the minimum and maximum values in $\Lambda_l^\pm$. In fact, the confidence $h(x_i, y_l)$ in $\Lambda_l^\pm$ is concentrated on a certain section of interval. We suppose that $h(x_i, y_l)$ both in Λ_l^+ and Λ_l^- approximately obeys the normal distribution.

In order to determine the minimum threshold, unbiased estimation is used to estimate parameter μ and δ in normal distribution, as shown in Eqs. 7 and 8.

$$\hat{\mu}_l^{\pm} = \frac{1}{|D_l^{\pm}|} \sum_{i=1}^{|D_l^{\pm}|} f(x_i, y_l), x_i \in D_l^{\pm} \tag{7}$$

$$\hat{\delta}_l^{\pm} = \sqrt{\frac{1}{|D_l^{\pm}| - 1} \sum_{i=1}^{|D_l^{\pm}|} (f(x_i, y_l) - \hat{\mu}_l^{\pm})^2}, x_i \in D_l^{\pm} \tag{8}$$

where $\hat{\mu}_l^+(\hat{\mu}_l^-)$ is the mean value of all confidence in $\Lambda_l^+(\Lambda_l^-)$, and $\hat{\delta}_l^+(\hat{\delta}_l^-)$ is the standard deviation of all fractions in $\Lambda_l^+(\Lambda_l^-)$. According to the 3δ standard of normal distribution, the threshold values are set to the following three kinds:

$$Min_l = \mu_l^+ - i * \delta_l^+, 1 \leq i \leq 3 \tag{9}$$

$$Max_l = \mu_l^- + i * \delta_l^-, 1 \leq i \leq 3 \tag{10}$$

$$Mid_l = (Min_l + Max_l)/2, 1 \leq i \leq 3 \tag{11}$$

The iteration algorithm of the optimal threshold The DTML threshold determination algorithm not only has complicated procedure, but also has 9 kinds of possible threshold values (Eqs. 9, 10, 11) for each label, and it is quite troublesome to find the optimal threshold. In this paper, for the first time we applied the iteration algorithm to the determination of the optimal threshold in multi-label learning.

The iterative algorithm is based on the idea of approximation. It is a process about how to get the optimal threshold value for multi-label learning algorithms with the iterative algorithm. $D = \{(x_i, Y_i) \mid 1 \leq i \leq m\}$ is the multi-label data, m is the number of data. $Y = \{y_1, y_2, \cdots, y_q\}$ is the finite set of q possible labels. The output is $T_l(1 \leq l \leq q)$, representing the optimal threshold of each label.

5   Experimental Results and Discussion

First, the 702 tongue image samples were labeled according to the label space "pink tongue, red tongue, dark red tongue, thin white coat, thin yellow coat, thick white coat and thick yellow coat", labeled 1 if meeting the corresponding category, otherwise labeled 0. Then 20 dimensions color features and 20 dimensions texture features were extracted. After that, hamming loss, average precision and $\partial - Evaluation$ were used to compare the classification result of the six different algorithms. Parameters β and γ in $\partial - Evaluation$ are set to 1, thus $\partial - Evaluation$ becoming the simplest form. Parameter ∂ reflects how much to forgive errors made in predicting labels, ∂ is set to

0.5 in this paper. At last, DTML and the iteration algorithm were used to improve the classification results.

5.1 Classification Results

Ten-fold cross-validation is performed and the mean metric value as well as the standard deviation of each algorithm is recorded, and the detailed results in term of different evaluation metrics with different parameters. By careful experiments, it is shown that ML-kNN works best when k is 9. We can find that LEAD gets the best with linear kernel function. The experiments indicate ML-RBF achieves the best performance with $\partial = 0.01$, $\mu = 1.0$. Results show that the best parameter ratio is 0.6. BP_MLL gets the best result when $\gamma = 20$ % and epochs = 100. Linear kernel outperforms the other two for Rank-SVM. LEAD is significantly superior to the other five algorithms. DTML and the iteration algorithm were used in LEAD to improve the classification results further. Since average precision is not affected by the threshold value, hamming loss and $\partial - Evaluation$ were chosen to assess the classification results under different threshold values. Figure 7 is the result of DTML. Hamming loss is the minimum and $\partial - Evaluation$ is the maximum when the value is Max_l and i is 1, which is superior to the classification result of 0.5. Table 5 is the threshold values of the iterative algorithm compared with the best result of DTML (the threshold value is Max_l and i is 1) and the original threshold 0.5. We can see from this table that the iterative algorithm and DTML can both provide a threshold for each of the labels. Figure 8 is the comparison values of hamming loss and $\partial - Evaluation$ in DTML and iteration algorithm. We can see that iteration algorithm achieves better result both on Hamming loss and $\partial - Evaluation$ than DTML from Fig. 8.

5.2 Results Interpretation

LEAD is superior to the others in tongue image classification can reflect that exploiting label correlation contribute to the improvement of the classification accuracy. The pink tongue with thin white coat is regarded as the normal tongue and the dark red tongue with thick yellow coat is related with blood stasis in TCM [14]. These underlying joint label dependences could be well represented by a DAG with the feature vector as a common parent in LEAD; the correlations between different labels are effectively exploited with the Bayesian network. As a result, learning with LEAD in tongue image classification would give excellent performance.

The iteration algorithm tries to find an optimal threshold for each label by learning from the training data set. An optimal threshold will be set for each label, according to different distributions for each label. One label will be predicted only if the value of real valued function is bigger than the threshold which was set for this label. The iteration algorithm has better result both than 0.5 and DTML.

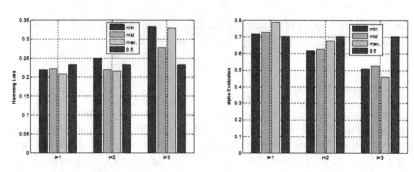

Fig. 7. The values of Hamming loss and $\partial - Evaluation$ in DTML

Table 5. The threshold values under different algorithms

algorithm \ labels	T1	T2	T3	T4	T5	T6	T7
iteration algorithm	0.7821	0.4434	0.4351	0.3984	0.6158	0.3785	0.4730
DTML	0.8203	0.3889	0.3904	0.3702	0.5954	0.3586	0.3799
Original	0.5	0.5	0.5	0.5	0.5	0.5	0.5

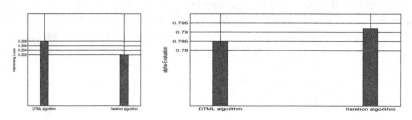

Fig. 8. Comparison of Hamming loss and $\partial - Evaluation$ in DTML and iteration algorithm

6 Conclusions

When multi-label learning is applied in tongue image classification, multiple labels of tongue body color, coat color and coat thickness can be predicted simultaneously. Moreover, multi-label learning takes full advantage of the correlation between the various labels, more consistent with the actual situation. At present, there is almost no literature on the multi-label learning of tongue images at home and abroad. In this paper, we firstly extracted 20 d color features and 20 d texture features on the basis of separation of tongue coat and body as well as tongue blocking. Then six kinds of multi-label learning algorithms were adopted to classify tongue images. At last, the iteration algorithm was applied in LEAD, which has a better performance than 0.5 and DTML. Multiple colors are used to describe the tongue body color in an image can better reflect the diversity of colors. But if the weight values of different colors are taken into account in the subsequent work, the description of tongue images can be more objective.

Acknowledgements. The work is supported by the National Natural Science Foundation of China (No. 61201360), the Importation and Development of High-Caliber Talents Project of Beijing Municipal Institutions (CIT&TCD201504018), and General projects of Beijing Municipal Education Commission (No. JE334001201201). China.

References

1. Liu, G.S., Xu, J.G.: A method for tongue coat classification based on neural network integration. Comput. Eng. **30**(4), 26–27 (2008). (in Chinese)
2. Wang, A.M., Shen, L.S., Zhao, Z.X.: Supervised FCM Clustering Algorithm and its application to automatic classification for Chinese medical tongue images. PR & AI **12**(4), 480–485 (1999). (in Chinese)
3. Jiao, Y., Zhang, X.F., Zhuo, L.: Weighted support vector machine for classification of tongue images. Meas. Control Technol. **29**(5), 1–13 (2010)
4. Xie, Z.G., Qiu, L.Q.: Tongue diagnosis in TCM based on the improved BP neural network. Modern Comput. **30**(4), 26–27 (2008)
5. Hu, S.N., Li, W.S., Shi, G.S., et al.: Color classification of tongue based on PCA-AdaBoost in traditional Chinese medicine. J. Guangxi Normal Univ. Nat. Sci. Edn. **27**(3), 158–161 (2009). (in Chinese)
6. Ni, H.: Application and Research of Traditional Chinese Medicine Tongue Image Retrieval Techniques Based on Color and Texture. Master thesis, Guangdong University of Technology (2011) (in Chinese)
7. Gao, C.C., Hui, X.W.: GLCM-based texture feature extraction. Comput. Syst. Appl. **19**(6), 195–199 (2010). (in Chinese)
8. Guo, Z., Yang X.Z., Si, Y.C., Zhu, Q.W.: Segmentation of tongue body and fur based on common color spaces of k-means clustering. J. Beijing Univ. Tradit. Chin. Med. **32**(12), 819–821 (2009) (in Chinese)
9. Dunn, J.C.: A fuzzy relative of the ISODATA process and its use in detecting compact well-separated clusters. J. Cybern. **3**(3), 32–57 (1973)
10. Bai, L.Y., Shi, Y., Wu, J., Zhang, Y., Weng, W., Wu, Y., Bai, J.: A novel automatic tongue coating extraction method in tongue diagnose of traditional Chinese medicine. In: Magjarevic, R., Nagel, J.H. (eds.) World Congress on Medical Physics and Biomedical Engineering 2006, vol. 14, Track 14, pp. 2624–2628. Spinger, Berlin (2008)
11. Han, F.: Tongue Color Space Analysis and Color Feature Study. Master thesis, Harbin Institute of Technology (2011) (in Chinese)
12. Cui, N.H., Liu, L.P., Li, C.Z.: The spatial dominant color descriptor for feature extraction algorithms of image. J. Shenyang Univ. Technol. **30**(4), 19–22 (2011). (in Chinese)
13. Feng, J.H., Yang, Y.J.: Study of texture images extraction based on gray level co-occurrence matrix. Beijing Surv. Mapp. **3**, 19–22 (2007). (in Chinese)
14. Beijing College of Traditional Chinese Medicine.: Tongue Diagnosis of Traditional Chinese Medicine. People's Medical Publishing House (1978) (in Chinese)
15. Shen, L.S., Cai, Y.H., Zhang, X.F.: Collection and analysis of TCM tongue. Beijing industrial university press (2007) (in Chinese)

16. Tsousmakas, G., Zhang, M.L., Zhou, Z.H.: Learning from multi-label data. In: ECML PKDD (2009)
17. Boutell, M.R., Luo, J., Shen, X., et al.: Learning multi-label scene classification. Pattern Recogn. **37**(9), 1757–1771 (2004)
18. Zhang, M.L., Zhou, Z.H.: ML-kNN: a lazy learning approach to multi-label learning. Pattern Recogn. **40**(7), 2038–2048 (2007)
19. Elisseeff, A., Weston, J.: A kernel method for multi-labeled classification. In: Proceedings of International Conference on Advances in Neural Information Processing Systems 14, Cambridge, MA, pp. 681–687 (2002)
20. Zhang, M.L., Zhang, K.: Multi-label learning by exploiting label dependency. In: Proceedings of International Conference on Knowledge Discovery and Data Mining, Washington, DC, USA, pp. 999–1007 (2010)
21. Zhang, M.L., Zhou, Z.H.: Multi-label neural networks with applications to functional genomics and text categorization. IEEE Trans. Knowl. Data Eng. **18**(10), 1338–1351 (2006)
22. Qin, F., Huang, J., Cheng, Z.K.: Threshold determination algorithm for multi-label learning. Comput. Eng. **36**(21), 214–216 (2010)

Online Kernel-Based Multimodal Similarity Learning with Application to Image Retrieval

Wenping Zhang[1,2] and Hong Zhang[1,2(✉)]

[1] School of Computer Science and Technology, Wuhan University of Science and Technology, Wuhan 430081, China
[2] Hubei Province Key Laboratory of Intelligent Information Processing and Real-Time Industrial System, Wuhan, China
46476522@qq.com

Abstract. A challenging problem of image retrieval is the similarity learning between images. To improve similarity search in Content-Based Image Retrieval (CBIR), many studies on distance metric learning have been published. Despite their success, most existing methods are limited in two aspects: (i) they usually attempt to learn a linear distance metric, which limits their capacity of measuring similarity for complex applications; (ii) they are often designed for learning metrics on unique-modal data, which could be suboptimal for similarity learning on multimedia objects with multiple feature representations. To overcome these limitations, in this paper, we investigate the online kernel-based multimodal similarity learning, which aims to integrate multiple kernels for learning multimodal similarity functions, and conduct experiments to evaluate the performance of the proposed method for CBIR on several different image datasets. The experiment results are encouraging and verify the effectiveness and the superiority of the proposed method.

Keywords: Kernel function · Similarity learning · Online learning · Image retrieval

1 Introduction

Similarity measuring is a basic issue in multimedia retrieval tasks [1, 3], which has been actively studied in the fields of data analysis, signal processing, computer vision, etc. [5, 11, 12]. The performance of similarity search crucially depends on both effective feature representation and proper similarity measurement. On one hand, feature descriptors vary from global features, such as color features [22], edge features [22], texture features [13] and GIST [14, 23], to local feature representations, such as bag-of-words models [24–26, 29] using local descriptors (e.g. SIFT [16] and SURF [17] features). On the other hand, various similarity functions have been proposed for similarity learning, such as Euclidean distance and cosine similarity, etc. In conventional CBIR systems [4, 8], images are represented in a feature vector space, and the typical choices of similarity functions are Euclidean distance and its variants, which may not be always optimal to measure the similarity of multimedia objects with multimodal representations.

© Springer International Publishing Switzerland 2015
D.-S. Huang and K. Han (Eds.): ICIC 2015, Part III, LNAI 9227, pp. 221–232, 2015.
DOI: 10.1007/978-3-319-22053-6_24

Recent years, to improve image similarity search in CBIR, researchers have conducted a number of studies on distance metric learning (DML) [2, 5, 9], which usually learn to optimize the distance metric of similarity. Although a variety of DML algorithms have been proposed, most existing DML methods suffer from two critical limitations. First, they usually try to learn a linear distance metric follows the forms of Mahalanobis distances, which can be viewed as learning a linear function to map feature vectors into another feature space; heir linearity assumption however may limit their capacity of learning similarity for complex image patterns in real applications. Second, even though there are a few existing works for learning nonlinear similarity function with kernel, they usually do not measure the similarity with multiple kernels, and they are often designed for learning metrics on unique-modal data, i.e., either a single type of features or a combined feature space by a simple concatenation of multiple types of features, which could be not optimal for measuring similarity of multimedia objects with multimodal representations.

To overcome the above limitations of existing works, in this paper, we introduce a Online Kernel-Based Multimodal Similarity (OKBMS) learning scheme, which is inspired by [6]. OKBMS method uses multiple kernels to learn a flexible nonlinear similarity function for images that are represented in multiple features via an efficient and scalable online learning scheme. Unlike conventional methods, OKBMS learns nonlinear similarity function with multiple kernels in an online learning fashion. Thus, it is able to learn more optimal similarity measurement to improve image similarity search in CBIR.

The rest of the paper is organized as follow. Section 2 briefly reviews the related techniques. Section 3 gives problem setting. Section 4 describes our OKBMS algorithm in detail. Section 5 shows experimental results. Finally, Sect. 6 gives a conclusion.

2 Related Work

As previously analyzed, OKBMS method is mainly about kernel-based multimodal image similarity learning. Therefore, in this section, we discuss the related works from the following two aspects.

2.1 Kernel-Based Metric Learning

Kernel methods typically consist of two part. The first part maps the input feature space into another space which is often much higher or even infinite dimensionality by applying a nonlinear function; the second part usually applies a linear method in the high dimension space. Kernel-based methods are not new for image retrieval, for example, kernel SVM algorithms have been successfully introduced into the CBIR tasks [20]. Our technique differs from the existing kernel-based methods proposed for image retrieval which address different types of problems.

In kernel-based distance metric learning literature, some algorithms were proposed for similarity learning in CBIR. Also, some metric learning algorithms including (e.g., LMNN [15] and NCA) have proved to be able be kernelized by KPCA [18]. Connections between metric and kernel learning, which can provide kernelization for a set of metric learning methods, have been revealed in recent studies [10]. Our method is different from these approaches in two aspects. First, our method learns with multiple kernels while they are designed to learn with a single kernel. Second, they usually apply a batch learning method, which is not scalable for large scale data, in contrast, we present online learning algorithm for learning a similarity measurement with multiple kernels.

2.2 Multiple Kernel Learning

Multiple kernel learning (MKL) [27, 28] allows the practitioner to optimize over linear combinations of kernels, research in MKL has focused on both learning the MKL formulations as well as their optimization. Different applications need different formulations, the existing MKL methods use different learning functions for determining the kernel combinations. According to different principles, existing MKL studies can be group into different categories [7].

In terms of different combination functions, most MKL studies often work with linear combinations which have two basic categories: unweighted sum and weighted sum. In the unweighted sum case, we use sum or mean of the kernels as the combined kernel; in the weighted case, we can linearly optimize weight for each kernel. There also studies use the nonlinear combination which apply nonlinear functions of kernel (e.g., multiplication, power and exponentiation).

In terms of different target functions, MKL algorithms are typically categorized into three groups: the similarity-based functions; the structural risk functions and the Bayesian functions. All MKL algorithms have the same goal of learning the optimum combination of multiple kernels, but the differences between our methods with others lie in that we aim to learn a kernel-based similarity function for image retrieval while conventional MKL studies often handle classification tasks.

3 Problem Setting

To formulate the learning task, we define the similarity function $S(x_1, x_2)$ for any two images $x_1, x_2 \in \mathbb{R}^n$, and the training data is given sequentially in the forms of triplet $\{(x_t, x_t^+, x_t^-), t = 1, \ldots, T\}$ where each triple contains a similar pair (x_t and x_t^+) and a dissimilar pair (x_t and x_t^-), and T is the number of total triples. The goal of this problem is to learn a similarity function $S(\cdot, \cdot)$ that can always assign higher similarity scores to pairs of more relevant images, i.e.,

$$S(x_t, x_t^+) > S(x_t, x_t^-) \quad (x_t, x_t^+, x_t^-); \forall i; \tag{1}$$

The above discussion assumes similarity is performed on unique-modal data. We aim to extend it to multimodal data, where each image is represented by different types of feature descriptors (e.g., color, edge, or GIST) and the similarity of two images is computed by different kernel functions. Specially, we adopt m different feature descriptors to extract feature for representing images and n kernel functions to build kernels on each kind of feature, which thus result in a total m * n different modalities.

The general idea of extending it to multi-modal is to learn an optimal kernel-based similarity function for each kernel in each modality, and determine an optimal combination of multiple kernels in building the final similarity function with all modalities:

$$S(x_1, x_2) = \sum_{i=1}^{m} \theta_i S_i(x_1, x_2) \tag{2}$$

where m denotes the number of modalities, $\theta_i \in [0, 1] s.t. \sum_{i=1}^{m} \theta_i = 1$ denotes the optional combination of for $i\text{-}th$ modality.

4 Online Kernel-Based Multimodal Similarity Learning

Figure 1 illustrates the flowchart of the proposed OKBMS method. We aim to learn a similarity function in the learning phase which used to rank images in the retrieval phase. During the learning phase, after training a set of OKBS models for each modality, we apply online learning techniques to update the similarity function on each modality, i.e., once receiving a triplet training data, we refine each model, and then find the optimal combination weights of multiple kernels.

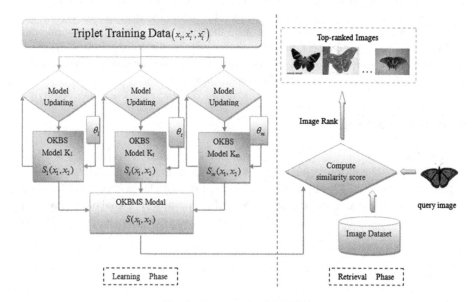

Fig. 1. Framework of OKBMS

In this section, we first introduce online kernel-based similarity learning performed on unique-modal data, and then extend it to multimodal similarity.

4.1 Online Kernel-Based Similarity

Similar to the W in the similarity function of OASIS algorithm in paper [19], $S_w(p_i, p_j) = p_i^T W p_j$, we define a linear operator g that maps a function $f \in H$ to another function $g[f] \in H$, in which the H is the corresponding Hilbert space of a given kernel $k(\cdot, \cdot)$. Based on g, the similarity $S_g(x_1, x_2)$ can be defined as:

$$S_g(x_1, x_2) = \langle k(x_1, \cdot), g[k(x_2, \cdot)] \rangle_H \tag{3}$$

The score computed by Eq. (3) states how strongly x_1 is related to x_2, and the linear operator g acts the same way as matrix W.

The online kernel-based similarity aims to learn S_g that can always satisfy the triplet-wise constrains as follows:

$$S_g(x_t, x_t^+) > S_g(x_t, x_t^-) + 1 \tag{4}$$

which puts a restrict in Eq. (1) by introducing a safety margin of 1. By adopting the framework of OASIS, we can cast the similarity problem into finding the optimum linear operator through minimizing the following loss function:

$$L = \arg\min_g \|g\|_{HS}^2 + C \sum_{t=1}^{T} \ell_g(x_t, x_t^+, x_t^-) \tag{5}$$

where T is the total number of triplets, $\|\cdot\|_{HS}$ is the Hilbert Schmidt norm, and $\ell_g(x_t, x_t^+, x_t^-) = \max(0, S_g(x_t, x_t^-) - S_g(x_t, x_t^+) + 1)$.

Similar to the OASIS, we apply the online Passive-Aggressive (PA) algorithm iteratively over each triplet to optimize g. First, we initialize g_o to be an identity operator, then, at each training iteration, given the training triplet (x_t, x_t^+, x_t^-), we update the operator by solving the follow optimization problem:

$$g_t = \arg\min_g \frac{1}{2} \|g - g_{t-1}\|_{HS}^2 + C\xi \tag{6}$$

s. t. $1 - S_g(x_t, x_t^+) + S_g(x_t, x_t^-) \leq \xi$ and $\xi \geq 0$
To solve the problem in Eq. (6), define the Lagrangian as:

$$L(g, \xi, \tau, \lambda) = \frac{1}{2} \|g - g_{t-1}\|_{HS}^2 + C\xi + \tau(1 - \xi - tr(gV_t^*)) - \lambda\xi$$

where $\tau \geq 0$ and $\lambda \geq 0$ are Lagrangian multipliers, and V_t is a rank one linear operator defined as:

$$V_t[f](\cdot) = k(x_t, \cdot)\langle k(x_t^+, \cdot) - k(x_t^-, \cdot), f\rangle_H$$

and V_t^* is the adjoint of V_t. The optimal solution satisfies that the gradient vanishes $\frac{\partial L(g,\xi,\tau,\lambda)}{\partial g} = 0$, thus

$$\frac{\partial L(g, \xi, \tau, \lambda)}{\partial g} = g - g_{t-1} - \tau V_t = 0$$

therefore, the optimal solution to the Eq. (6) is

$$g_t = g_{t-1} + \tau V_t \tag{7}$$

By setting $\frac{\partial L(g,\xi,\tau,\lambda)}{\partial \xi} = 0$:

$$C - \tau - \lambda = 0$$

which, since $\lambda \geq 0$, means that $\tau \leq C$. Thus, we obtain

$$L(\tau) = \frac{1}{2}\tau^2 \|V_t\|_{HS}^2 + \tau\left(1 - tr\left(gV_t^*\right)\right)$$

$$= -\frac{1}{2}\tau^2 \|V_t\|_{HS}^2 + \tau\left(1 - tr\left(g_{t-1}V_t^*\right)\right)$$

Differentiating the equation with respect to τ and setting it to zero, we have:

$$\frac{\partial L(\tau)}{\partial \tau} = -\tau \|V_t\|_{HS}^2 + \ell_{g_{t-1}}\left(x_t, x_t^+, x_t^-\right) = 0$$

which yields

$$\tau = \frac{\ell_{g_{t-1}}\left(x_t, x_t^+, x_t^-\right)}{\|V_t\|_{HS}^2}$$

since $\tau \leq C$, we obtain

$$\tau = \min\left\{C, \frac{\ell_{g_{t-1}}\left(x_t, x_t^+, x_t^-\right)}{\|V_t\|_{HS}^2}\right\} \tag{8}$$

$$\|V_t\|_{HS}^2 = k(x_t, x_t)(k(x_t^+, x_t^+) + k(x_t^-, x_t^-) - 2k(x_t^+, x_t^-)). \tag{9}$$

Finally, we can obtain the computational formula of

$$S_{g_t}(x_1, x_2) = \langle k(x_1, \cdot), g_t[k(x_2, \cdot)] \rangle_H$$

$$= k(x_1, x_2) + \sum_{j=1}^{t} \tau_j k(x_1, x_j) \left(k\left(x_j^+, x_2\right) - k\left(x_j^-, x_2\right) \right) \qquad (10)$$

Algorithm 1. Online Kernel-Based Similarity Learning

Input: training triplets (x_t, x_t^+, x_t^-); an input kernel $k(\cdot, \cdot)$

Output: the similarity score $S(x_1, x_2) = \langle k(x_1, \cdot), g_T[(k(x_2, \cdot))] \rangle$

1: Initialization $g_0 = I$

2: for $t = 1, 2, ..., T$ do

3:   Receive a training triplet (x_t, x_t^+, x_t^-)

4:   Computer τ_t in (8)

5:   Update g_t as (7)

6: end for

4.2 Extension to Multimodal Similarity

The above discussion learns similarity function based on unique-modal data, now, we extend it to multi-modal data.

Assume $K_i, i = 1, 2, \cdots, m$, are m kernel functions, each of them being associated with Hilbert space H_i. OKBMS aims to learn the set of coefficients $\theta = (\theta_1, \theta_2, \cdots, \theta_m)$, and consequentially learn the final similarity function:

$$S(x_1, x_2) = \sum_{i=1}^{m} \theta_i S_i(x_1, x_2) = \sum_{i=1}^{m} \theta_i \langle k_i(x_1, \cdot), g_i[k_i(x_2, \cdot)] \rangle_{H_i} \qquad (11)$$

where g_i is the linear operator in H_i. Thus, the OKBMS has two sets of variables to learn, i.e., the set of linear operator with respect to each different kernels and the set of combination coefficients. To simultaneously learn both of them, the OBKMS can be casted into a optimization problem:

$$\min_{\theta \in \Delta} \min_{g_i} \frac{1}{2} \sum_{i=1}^{m} \theta_i \|g_i\|_{HS}^2 + C \sum_{t=1}^{T} \ell\left(S(x_t, x_t^+) - S\left(x_t, x_t^-\right)\right) \qquad (12)$$

where $S(x_1, x_2)$ is given in (11), and $\Delta = \{\theta | \sum_{i=1}^{m} \theta_i = 1, \theta_i \in [0, 1]\}$. Specially, for each kernel, i.e., k_i, for each triplet, we apply methods proposed previously in

OKBS to find the optimal similarity function with respect to kernel k_i, and then apply Hedging algorithm to update the combination weights as follows:

$$\theta_i(t) = \theta_i(t-1)\beta^{z_i(t)} \tag{13}$$

$$z_i(t) = \prod \left(S_{g_{t-1,i}}(x_t, x_t^+) - S_{g_{t-1,i}}(x_t, x_t^-) \le 0 \right)$$

where $\beta \in (0,1)$ is a discounting parameter, and $\prod(\cdot)$ is an indicator function that output 1 when statement holds and 0 otherwise. Algorithm 2 summarizes the details of the OKBMS method.

Algorithm 2. Online Kernel-Based Multimodal Similarity

Input: kernels $K_i, i = 1, 2, \cdots, m$; combination coefficients $\theta = (\theta_1, \theta_2, \cdots, \theta_m)$;
     discounting parameter β

Output: the multimodal similarity score $S(x_1, x_2) = \sum_{i=1}^{m} \theta_i \left\langle k_i(x_1, \cdot), g_i[k_i(x_2, \cdot)] \right\rangle_{H_i}$

1: Initialization $g_{0,i} = I$; $\theta_i(0) = 1$
2: for $t = 1, 2, \cdots, T$ do
3:    Receive a training triplet (x_t, x_t^+, x_t^-)
4:    Compute $\tau_{t,i}$ in (8) with $g_{t-1,i}$ instead of g_{t-1}
5:    Update $g_{t,i}$ in (7) with $V_{t,i}[f](\cdot) = k_i(x_t, \cdot)\left\langle k_i(x_t^+, \cdot) - k_i(x_t^-, \cdot), f \right\rangle_{H,i}$
6:    Compute $z_i(t)$
7:    Update $\theta_i(t)$ in   (13)
8: end for

---

5   Experiment

In this section, we conduct a set of experiments to evaluate the performance of the proposed algorithm in CBIR. To measure the retrieval performance, we adopt the Mean Average Precision (MAP) and top-k retrieval accuracy.

5.1   Image Datasets and Sampling Strategy

In the experiment, we adopt four publicly image datasets, which are widely used in some previous works, including Caltech256,[1] Corel5000 [9],[2] Indoor[3] and ImageCLEF.[4] For each dataset, we randomly select a subset from each class to make sure that all

---

[1] http://www.vision.caltech.edu/Image_Datasets/Caltech256/.

[2] http://OMKS.stevenhoi.org/.

[3] http://www.web.mit.edu/torralba/www/indoor.html.

[4] http://imageclef.org/.

classes have the same number of images, and then randomly split it into four disjoint partitions: a training set of 50 % images from each classes, a validation set of 10 % images from each classes, a query set of 10 % images from each classes, and the rest for test. To sample a training triplet $\left(x_t, x_t^+, x_t^-\right)$, we first uniformly select x_t from one class, and then we uniformly select x_t^+ from the same class, lastly, we select x_t^- from another class.

5.2 Features and Kernel Functions

Before represent images, we resize all the images to the scale of 500×500 pixels. Then, we adopt both global and local feature descriptors to extract feature for representing images. For global features, we extract five features which have been widely used in previous CBIR studies, including (1) color histogram and color moments, (2) edge histogram, (3) LBP, (4) Gabor wavelets transformation, (5) GIST. For local features, we extract the bag-of-visual-words representation using two kinds of descriptors: the SIFT feature and Hessian Affine interest region detector with threshold 500; SURF feature and SURF detector with threshold 500. By choosing different vocabulary size and different descriptors, we obtain: SIFT200, SIFT1000, SURF200, SURF 1000. Thus, pictures are represented into 9 different features [6].

For this features, we can build kernels on them, and we adopt 4 kernels on each kind of feature, which thus results in a total 36 different kernels. The 4 kernels are:

(1) linear kernel function: $k(x_1, x_2) = \langle x_1, x_2 \rangle$,
(2) polynomial kernel function: $k(x_1, x_2) = (\gamma \langle x_1, x_2 \rangle + c)^n$,
(3) radial basis function: $k(x_1, x_2) = \exp\left(-\gamma \|x_1 - x_2\|^2\right)$,
(4) sigmoid   function: $k(x_1, x_2) = \tanh(\gamma \langle x_1, x_2 \rangle + c)$ and   we   select $\gamma = 1$, $c = 0, n = 2$.

5.3 Comparison Results

To evaluate the efficacy of the proposed algorithm, we compare OKBMS for image retrieval, against several distance metric learning algorithms, include Euclidean, OASIS [19], LMNN [21], RCA. To adapt these existing distance metric learning algorithms for multimodal image retrieval, we apply each of them for learning similarity for each modality individually and select the best modality of the highest MAP.

Table 1 shows the Mean Average Precision (MAP) of each algorithm on different image datasets. and Fig. 2 shows top-k retrieval accuracy of each algorithm on Indoor dataset and Caltech256 dataset. From both Table 1 and Fig. 2 we can have the similar observation that our proposed algorithm outperforms other algorithms. This is primarily because a single type feature may fail to exploit the potential of all modalities, meanwhile multiple types of feature for retrieval could better explore the potential of all features, which validates the importance of the proposed method.

Table 1. Evaluation of MAP performance

Algorithms	Corel	Caltech256	Indoor	ImageCLEF
Euclidean	0.1901	0.1856	0.0486	0.5164
OASIS	0.2186	0.2437	0.0502	0.5903
LMNN	0.1936	0.1658	0.0428	0.4764
RCA	0.2037	0.1985	0.0447	0.5842
OKBMS	**0.2385**	**0.3726**	**0.0962**	**0.7243**

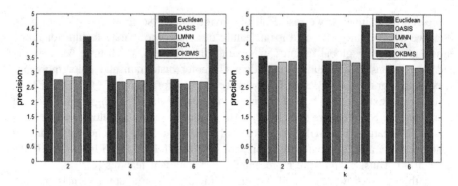

Fig. 2. Top-k precision on Indoor (left) and Caltech256 (right)

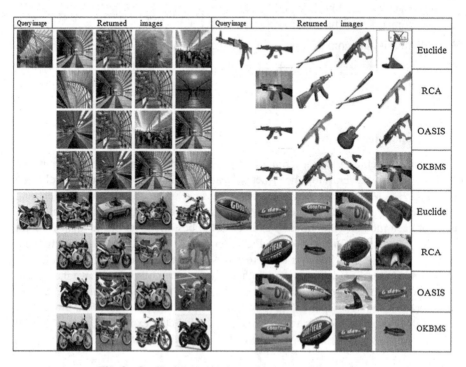

Fig. 3. Qualitative comparison of image ranking results

5.4   Qualitative Comparison

Finally, to examine the qualitative retrieval performance, we sample several query images from the query set and compare the top ranked images retrieved by different methods, including Euclidean, OASIS, LMNN, RCA and OKBMS. Figure 3 shows the results, for each row, the image on the left is the query, others are query results, and each row represents a method. The result shows that OKBMS generally returns more relevant results than others and verify the effectiveness of the proposed method.

6   Conclusions

In this paper, we introduced OKBMS, an algorithm of learning similarity functions for image retrieval. OKBMS learns nonlinear similarity functions by integrating with multiple kernels through a coherent and scalable online learning scheme which can learn both the similarity function with each kernel and the combination of multiple kernels, the nonlinear similarity function can effectively improve image similarity search. And convincing experimental results demonstrate the effectiveness of the proposed method. In the future, we will design more effective methods for image retrieval through other new important breakthrough techniques, such as deep learning, which has been successfully introduced to computer vision and other applications.

Acknowledgment. This work is supported by National Natural Science Foundation of China (No.61373109, No.61003127, No.61273303), State Key Laboratory of Software Engineering (SKLSE2012-09-31) and Program for Outstanding Young Science and Technology Innovation Teams in Higher Education Institutions of Hubei Province, China (No.T201202).

References

1. Jing, Y., Baluja, S.: Visualrank: applying pagerank to large-scale image search. IEEE Trans. Pattern Anal. Mach. Intell. **30**(11), 1877–1890 (2008)
2. Lee, J.-E., Jin, R., Jain, A.K.: Rank-based distance metric learning: an application to image retrieval. In: Proceedings of IEEE Conference on Computer Vision and Pattern Recognition, Anchorage, AK (2008)
3. Grangier, D., Bengio, S.: A discriminative kernel-based approach to rank images from text queries. IEEE Trans. Pattern Anal. Mach. Intell. **30**(8), 1371–1384 (2008)
4. Rahmani, R., Goldman, S.A., Zhang, H., Cholleti, S.R., Fritts, J.E.: Localized content-based image retrieval. TPAMI **30**(11), 1902–1912 (2008)
5. Yang, L., Jin, R.: Distance metric learning: a comprehensive survey. Michigan State Universiy, pp. 1–51 (2006)
6. Xia, H., Hoi, S.C.H., Jin, R., Zhao, P.: Online multiple kernel similarity learning for visual search. IEEE Trans. Pattern Anal. Mach. Intell. **36**(3), 536–549 (2014)
7. Gönen, M., Alpaydın, E.: Multiple kernel learning algorithms. J. Mach. Learn. Res. **12**, 2211–2268 (2011)
8. Smeulders, A.W.M., Worring, M., Santini, S., Gupta, A., Jain, R.: Content-based image retrieval at the end of the early years. TPAMI **22**(12), 1349–1380 (2000)

9. Hoi, S.C., Liu, W., Lyu, M.R., Ma, W.-Y.: Learning distance metrics with contextual constraints for image retrieval. In: CVPR, 17–22 June 2006

10. Jain, P., Kulis, B., Davis, J.V., Dhillon, I.S.: Metric and kernel learning using a linear transformation. J. Mach. Learn. Res. **13**, 519–547 (2012)

11. Jing, Y., Baluja, S.: Visualrank: applying pagerank to large-scale image search. TPAMI **30** (11), 1877–1890 (2008)

12. Lew, M.S., Sebe, N., Djeraba, C., Jain, R.: Content-based multimedia information retrieval: State of the art and challenges. TOMCCAP **2**(1), 1–19 (2006)

13. Manjunath, B.S., Ma, W.Y.: Texture features for browsing and retrieval of image data. IEEE Trans. Pattern Anal. Mach. Intell. **18**(8), 837–842 (1996)

14. Oliva, A., Torralba, A.: Modeling the shape of the scene: a holistic representation of the spatial envelope. Int. J. Comput. Vis. **42**(3), 145–175 (2001)

15. Weinberger, K., Blitzer, J., Saul, L.: Distance metric learning for large margin nearest neighbor classification. In: Advances in Neural Information Processing Systems, pp. 1473–1480 (2006)

16. Lowe, D.G.: Object recognition from local scale-invariant features. In: ICCV, pp. 1150–\1157 (1999)

17. Bay, H., Tuytelaars, T., Van Gool, L.: SURF: speeded up robust features. In: Leonardis, A., Bischof, H., Pinz, A. (eds.) ECCV 2006, Part I. LNCS, vol. 3951, pp. 404–417. Springer, Heidelberg (2006)

18. Chatpatanasiri, R., Korsrilabutr, T., Tangchanachaianan, P., Kijsirikul, B.: A new kernelization framework for mahalanobis distance learning algorithms. Neurocomputing **73**(10-12), 1570–1579 (2010)

19. Chechik, G., Sharma, V., Shalit, U., Bengio, S.: Large scale online learning of image similarity through ranking. JMLR **11**, 1109–1135 (2010)

20. Tong, S., Chang, E.: Support vector machine active learning for image retrieval. In: Proceedings of the Ninth ACM International Conference on Multimedia, Ottawa, Canada, pp. 107–118 (2001)

21. Weinberger, K.Q., Blitzer, J., Saul, L.K.: Distance metric learning for large margin nearest neighbor classification. In: NIPS (2005)

22. Jain, A.K., Vailaya, A.: Image retrieval using color and shape. Pattern Recogn. **29**(8), 1233–1244 (1996)

23. Oliva, A., Torralba, A.: Scene-centered description from spatial envelope properties. In: Bülthoff, H.H., Lee, S.-W., Poggio, T.A., Wallraven, C. (eds.) BMCV 2002. LNCS, vol. 2525, pp. 263–272. Springer, Heidelberg (2002)

24. Sivic, J., Russell, B.C., Efros, A.A., Zisserman, A., Freeman, W.T.: Discovering objects and their localization in images. In ICCV, pp. 370–377 (2005)

25. Wu, L., Hoi, S.C.H.: Enhancing bag-of-words models with semantics-preserving metric learning. IEEE Multimedia **18**(1), 24–37 (2011)

26. Wu, L., Hoi, S.C.H., Yu, N.: Semantics-preserving bag-of-words models and applications. IEEE Trans. Image Process. **19**(7), 1908–1920 (2010)

27. Lanckriet, G.R.G., Cristianini, N., Bartlett, P., Ghaoui, L.E., Jordan, M.I.: Learning the kernel matrix with semidefinite programming. J. Mach. Learn. Res. **5**, 27–72 (2004)

28. Sonnenburg, S., Rätsch, G., Schäfer, C., Schölkopf, B.: Largescale multiple kernel learning. J. Mach. Learn. Res. **7**, 1531–1565 (2006)

29. Yang, J., Jiang, Y.-G., Hauptmann, A.G., Ngo, C.-W.: Evaluating bag-of-visual-words representations in scene classification. In: Multimedia Information Retrieval, pp. 197–206 (2007)

Reverse Training for Leaf Image Set Classification

Yu-Hui Zhang, Ji-Xiang Du$^{(\boxtimes)}$, Jing Wang, and Chuan-Min Zhai

Department of Computer Science and Technology, Huaqiao University,
Xiamen 361021, China
janeyuhui@sina.cn, jxdu@hqu.edu.cn

Abstract. This paper presents a new approach for leaf image set classification, where each training and testing set contains many image instances of a leaf. This approach efficiently extends binary classifiers for the task of multi-class image set classification. First, the training set is divided into two part using clustering algorithms: one will train a classifier with the images of the query set; the rest of the training set will evaluate the trained classifier and then predict the class of the query image set. The PHOG feature and Gist feature of leaf image set are merged into the whole feature of leaf image sets. Extensive experiments and comparisons with existing methods show that the proposed approach achieves state of the art performance for leaf image set recognition.

Keywords: Leaf classification · Reverse training · Multi-features · Image set classification

1 Introduction

There are more than 300,000 plants species on the earth, and each leaf is special. The leaves of rhizoma imperatae have sharp blade on the edge, coconut trees in the sea have big and wide leaves with waves, and plantain leaves are spirally arranged, and so on. Flower, leaf, fruit or peel can be a part of feature information to recognize a plant. To handle such volumes of information, development of a quick and efficient classification method has become an area of active research. Leaf classification is an important component of plant recognition system [1–4, 21]. Leaf features contain significant information that can improve the recognition rate. Leaf color is not recognized as an important aspect to the identification, and the type of the vein is an important morphological characteristic of the leaf, and the shape, size and texture of the leaves also play an important role in leaf classification. Du et al. [21] used a new method of describing the characteristics of plant leaves based on the outline fractal dimension and venation fractal dimension. In this paper, we have carried out multi-feature fusion [2] for plant leaf image set classification.

The training data and test data in traditional classification system have usually single or a little number of objects image. With the emergence of big data, micro SD with big memory and mobile phone with high quality camera, it is very easy to obtain images in our daily life. Thus, the train set and test set become sets of instants image

© Springer International Publishing Switzerland 2015
D.-S. Huang and K. Han (Eds.): ICIC 2015, Part III, LNAI 9227, pp. 233–242, 2015.
DOI: 10.1007/978-3-319-22053-6_25

instead of single instant image, namely object recognition with image set. The images may have various on view, illumination, non-rigid deformation, obscuration, etc. In recent years, there are many methods to improve recognition rate of image set classification, and video classification is special case of image set classification with time sequence [7]. This paper focus on usual image set classification problem without temporal information.

There are numerous approaches [9–19] to solve image set recognition problem which broadly fall into two categories: parametric model based methods and non-parametric model free methods. Parametric modeling methods represent each image set with some parametric distribution function, and then measure the similarity between two distributions in terms of the Kullback-Leibler Divergence (KLD). However, parametric modeling methods need to solve parameter problem and require that train set and test set have strong statistical correlations. In comparison, nonparametric methods try to representing an image set as a linear/affine subspace, mixture of subspaces, nonlinear manifolds, image set covariance, and dictionary. The between-set distance is re-defined based on the representation of image set. Those nonparametric modeling methods may lose some information contained in the images of the image set. In this paper, the proposed approach is inspired by Hayat [17], and has not assumed distribution of image set.

2 Proposed Framework

Problem Description:

Given k training set: $X_1, X_2, \ldots, X_k$, and their corresponding class labels $y_c \in [1, 2, \ldots, k]$. The class c training data $X_c = \{x_t | y_t = c : t = 1, 2, \ldots, N_c\}$ has N_c images. Given a query image set $Y = \{Y_t\}_{t=1}^{N_q}$, and the output is the class label y of Y.

2.1 Image Set Classification Algorithm

The proposed algorithm trains only one binary classifier for the task of multi-class image set classification. First, cluster analysis divides training set X into two set X^1 and X^2, and X^1 consist of some images from all class of training set, and the total number of images in X^1 approximates the number of images of the query image set. The rest images compose of X^2, and X^2 contains image sets. Next, a binary classifier is trained to optimally separate images of the query set from X^1. Note that X^1 has some images which belong to the class of the query set, and the classifier regards them as the outliers. Next, the classifier is evaluated on the images in X^2, and the class that the images of X^2 are classified to belong to the query image set is the class of query image set. An illustration of the image set classification algorithm is presented in Fig. 1.

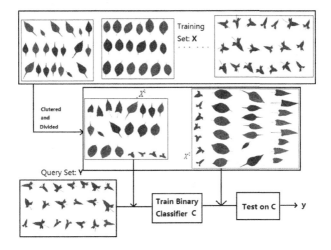

Fig. 1. Structure of the proposed method

Problem Description:
Input: Training image set $X_1, X_2, \ldots, X_k$, training data class labels $y_c \in [1, 2, \ldots, k]$; query image set $Y = \{Y_t\}_{t=1}^{N_q}$;
Output: Label y of Y

A detailed description of the algorithm is presented below.

1. Cluster analysis divides training set X into two set X^1 and X^2. The cluster algorithm splits k images from each class training set, and then those images compose of the set X^1, and the rest is the set X^2. The number of image set X_c^1 is $N_{X_c^1}$, and the number of images in set is $N_Y = \sum_{c=1}^{k} N_{X_c^1}$.

$$X^1 = \bigcup_c X_c^1, c = 1, 2, \ldots, k; X^2 = X \backslash X^1,$$

$$y_{X^2} = \left\{ y^{(t)} \in [1, 2, \ldots, k], t = 1, 2, \ldots, N_{X^2} \right\}$$

2. Train a binary classifier C_1. Training is done on the images of the query image set Y and X^1, and the classifier aims at separating images of Y from the images of other classes. It is worth mentioning that some images of X^1 from the same class as of Y are treated as outliners.

3. The trained classifier C_1 is tested on the images of X^2, and then the images in X^2 are classified as +1, namely $y_{X^2}^+$, of which the label is same as the label of the query image set. And $y_{X^2}^+ \subset y_{X^2}$.

4. Compute a normalized frequency histogram h of class labels in $y_{X^2}^+$. The cth value of the histogram, h_c, is given by the ratio of the number of images of X^2 belonging to

class c and classified as +1 to the total number of images of X^2 belonging to class c. This is given by,

$$h_c = \frac{\sum\limits_{y^{(t)} \in y_{X_2}^+} \delta_c(y^{(t)})}{\sum\limits_{y^{(t)} \in y_{X_2}} \delta_c(y^{(t)})}, where$$

$$\delta_c(y^{(t)}) = \begin{cases} 1, & y^{(t)} = c \\ 0, & otherwise \end{cases}$$

5. Predict the class of Y, and output the label y. This is given by,

$$y = \arg \max_c h_c$$

2.2   The Choice of Cluster Analysis

The proposed framework chooses K-means clustering algorithm [8] which is one of the oldest and most widely used clustering algorithms. The K-means clustering technique is simple, and it partitions the images in each training image set into k clusters. K-means is formally described by follow (Table 1).

Table 1. Basic K-means algorithm

Algorithm: Basic K-means algorithm
1.   Select K points as initial centroids.
2.   Repeat
a)   From K clusters by assigning each point to its closest centroid.
b)   Recomputed the centroid of each cluster.
3.   Until Centroids do not change.

In this work, the only issue is how to choose k. For the total number of images in X^1 equaled to the total number of query images, the value of k is located by: $k = N_q/c$.

2.3   The Choice of the Binary Classifiers

The binary classifier distinguish between images of the query set Y and X^1. Thus the classifier should treat some images in X^1, which have the same class label as of query image set, as outliners. The classifier should generalize well to unseen data. For these reasons, Linear Support Vector Machine (SVM) classifier [9] is a good choice. It is known to can effectively handle outliers and show excellent generalization to unknown data. In this work, C_1 is the linear SVM with L2 regularization and L2 loss function.

3 Leaf Feature Extraction Methods

First step for leaf image set classification is digital images acquisition. The second step is to preprocess images to enhance the important features. In this work, we have chosen an existing leaf image database, and the first two steps should be omitted. The next step is to extract features. The leaf feature extraction includes color feature, texture feature, shape feature, local feature etc. In this work, we apply multi-feature [2] to the leaf image set classification, and the linear combination of multi-feature can be calculated using

$$F = \alpha F_1 + \beta F_2, \text{ St. } 0 < = \alpha < = 1, 0 < = \beta < = 1 \text{ and } \alpha + \beta = 1$$

Where F is equaled to the linear combination of F_1 and F_2, α and β are coefficient.

3.1 Pyramid of Histograms of Orientation Gradients (PHOG)

Bosch introduced pyramid of histograms of orientation gradients (PHOG) [6] that represents an image by its local shape and the spatial layout of the shape with a spatial pyramid kernel. The local shape is captured by distribution over edge orientations within a region, and spatial layout is captured by tiling the image into regions at multiple resolutions. The PHOG descriptor consists of a histogram of orientation gradients over each image subregion at each resolution level. Furthermore, the distance between two descriptors reflects the extent to which the images contain similar shapes and correspond in their spatial layout. In this work, the descriptor first extracts canny edges, and then we set $L = 3$ and $K = 8$ bins, and that's it, the descriptor is a 680-vector. The diagrams shown in Fig. 2 depict the PHOG descriptor of a leaf image.

3.2 Gist Descriptor

The gist feature [5] is a low dimensional representation of an image region and has been shown to achieve good performance for some recognition tack when applied to an entire image. Given an input image, a GIST descriptor is computed by

a) Convolve the image with 32 Gabor filters at 4 scales, 8 orientations, producing 32 feature maps of the same size of the input image.
b) Divide each feature map into 16 regions (by a 4×4 grid), and then average the feature values within each region.
c) Concatenate the 16 averaged values of all 32 feature maps, resulting in a $16 \times 32 = 512$ GIST descriptor.

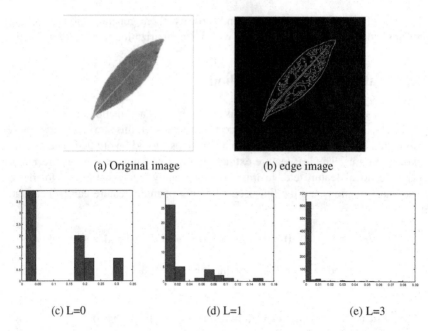

(a) Original image   (b) edge image

(c) L=0   (d) L=1   (e) L=3

Fig. 2. The PHOG descriptor

Intuitively, GIST summarizes the gradient information (scales and orientations) for different parts of an image, which provides a rough description of the image. And Fig. 3 shows the Gist descriptor of a leaf image.

4   Experimental Results

In this section, we show qualitative and quantitative results for our model. We use a subset of the ICL dataset from intelligent computing laboratory of Chinese academy of sciences. The dataset contains 17000 plant leaf images from 200 species, and the subset has 80

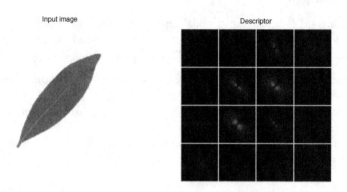

Fig. 3. The Gist descriptor of a leaf image

Fig. 4. Some instances of subset

classes and at least 200 images in each class. In our experiment, the subset is randomly split into training and test sets. Figure 4 shows some instances of leaf. Below, we first present a performance comparison of proposed method with different features. We then compare our method with the existing state of the art image set classification method.

4.1 Comparison with Different Feature Extraction

We present a comparison of the proposed method with different feature extraction, and the compared features include HSV, LBP, GIST, PHOG and multi-feature fusion. In this experiment, K-means cluster method has the same configuration-$K = 10$. Experimental results in terms of identification rates on the leaf dataset are Table 2.

Table 2. Performance comparison with different features

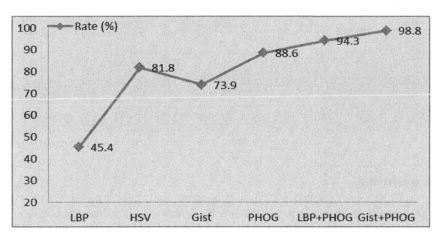

The results show that, amongst the compared features, the Gist and PHOG features fusion performs slightly best than other features. Because the HSV feature dimension is 2700, it is too big to decrease run time, and the multi-feature fusion leaves out the HSV feature. Thus, we choose Gist and PHOG features in the next experiment.

4.2   Comparison with Existing Image Set Classification Methods

In the section, we present a comparison of the proposed method with a number of recently proposed state of art image set classification methods. The compared methods include Mutual Subspace Method (MSM) [7], Manifold-to-Manifold Distance (MMD) [15], the linear version of the Affine Hull-based Image Set Distance (AHISD) [11], the Convex Hull-based Image Set Distance (CHISD) [11], Sparse Approximated Nearest Points (SANP) [12], and Covariance Discriminative Learning (CDL) [13].

The experimental results in terms of the identification rates along with different methods on the leaf dataset are presented in Table 3. The parameters for all methods are optimized for the best performance.

With this table, we can see this proposed method achieves the best identification rate. The AHISD also perform good, but its run-time is a 1000 % increase to our method. Once the total number of images in the set is reduced, the identification rates also decrease. This suggests the robustness of the methods in the relation to the number of image in the set.

Table 3. Performance on leaf ICL datasets

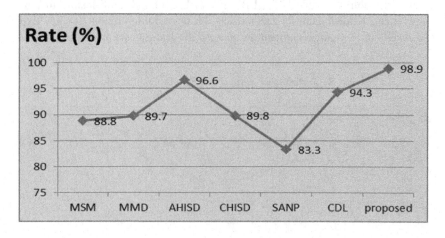

5   Conclusion

The proposed method, in the paper, extends the binary classifiers for multi-class leaf image set classification and is so easy to understand. K-means clustering is popular for cluster analysis, and it performs efficient in our work. The proposed approach uses very few images for training and is very efficient as it trains only one binary classifier. The proposed method has been evaluated for the task of leaf image set classification, and the experimental results shows that the proposed method achieves the state of the art performance compared with the existing methods. The weakness of the proposed method is demand for plenty of query image set and training set.

Acknowledgement. This work was supported by the Grant of the National Science Foundation of China (No. 61175121), the Program for New Century Excellent Talents in University (No. NCET-10-0117), the Grant of the National Science Foundation of Fujian Province (No. 2013J06014), the Promotion Program for Young and Middle-aged Teacher in Science and Technology Research of Huaqiao University (No. ZQN-YX108).

References

1. Beghin, T., Cope, J.S., Remagnino, P., Barman, S.: Shape and texture based plant leaf classification. In: Blanc-Talon, J., Bone, D., Philips, W., Popescu, D., Scheunders, P. (eds.) ACIVS 2010, Part II. LNCS, vol. 6475, pp. 345–353. Springer, Heidelberg (2010)
2. Shabanzade, M., Zahedi, M., Aghvami, S.A.: Combination of local descriptors and global features for leaf recognition. Sig. Image Process. Int. J. **2**(3), 23–31 (2011)
3. Rashad, M.Z., El-Desouky, B.S., Khawasik, M.S.: Plants images classification based on textural features using combined classifier. Int. J. Comput. Sci. Inf. Technol. (IJCSIT) **3**(4), 93–100 (2011)
4. Zheng, X., Wang, X.: Leaf vein extraction based on gray-scale morphology. Int. J. Image Graph. Sig. Process. (IJIGSP) **2**(2), 25 (2010)
5. Oliva, A., Torralba, A.: Modeling the shape of the scene: a holistic representation of the spatial envelope. Int. J. Comput. Vision **42**(3), 145–175 (2001)
6. Bosch, A., Zisserman, A., Munoz, X.: Representing shape with a spatial pyramid kernel. In: Proceedings of the 6th ACM international conference on Image and video retrieval. ACM, pp. 401–408 (2007)
7. Yamaguchi, O., Fukui, K., Maeda, K.: Face recognition using temporal image sequence. In: Proceedings. Third IEEE International Conference on Automatic Face and Gesture Recognition, 1998. IEEE, pp. 318–323 (1998)
8. Wagstaff, K., Cardie, C., Rogers, S., et al.: Constrained k-means clustering with background knowledge. In: ICML. vol. 1, pp. 577–584 (2001)
9. Hsu, C.W., Chang, C.C., Lin, C.J.: A practical guide to support vector classification (2003)
10. Kim, T.K., Kittler, J., Cipolla, R.: Discriminative learning and recognition of image set classes using canonical correlations. IEEE Trans. Pattern Anal. Mach. Intell. **29**(6), 1005–1018 (2007)
11. Cevikalp, H., Triggs, B.: Face recognition based on image sets. In: IEEE Conference on Computer Vision and Pattern Recognition (CVPR), 2010. IEEE, pp. 2567–2573 (2010)
12. Hu, Y., Mian, A.S., Owens, R.: Sparse approximated nearest points for image set classification. In: IEEE Conference on Computer Vision and Pattern Recognition (CVPR), 2011. IEEE, pp. 121–128 (2011)
13. Wang, R., Guo, H., Davis, L.S, et al.: Covariance discriminative learning: A natural and efficient approach to image set classification. In: IEEE Conference on Computer Vision and Pattern Recognition (CVPR), 2012. IEEE, pp. 2496–2503 (2012)

14. Wang, R., Chen, X.: Manifold discriminant analysis. In: IEEE Conference on. Computer Vision and Pattern Recognition, CVPR 2009. IEEE, pp. 429–436 (2009)
15. Wang, R., Shan, S., Chen, X., et al.: Manifold–manifold distance and its application to face recognition with image sets. IEEE Trans. Image Process. 21(10), 4466–4479 (2012)
16. Ortiz, E.G., Wright, A., Shah, M.: Face recognition in movie trailers via mean sequence sparse representation-based classification. In: IEEE Conference on Computer Vision and Pattern Recognition (CVPR), 2013. IEEE, pp. 3531–3538 (2013)
17. Yang, M., Zhu, P., Van Gool, L. et al.: Face recognition based on regularized nearest points between image sets. In: 10th IEEE International Conference and Workshops on Automatic Face and Gesture Recognition (FG), 2013. IEEE, pp.1–7 (2013)
18. Hayat, M., Bennamoun, M., An, S.: Reverse Training: An Efficient Approach for Image Set Classification. In: Fleet, D., Pajdla, T., Schiele, B., Tuytelaars, T. (eds.) ECCV 2014, Part VI. LNCS, vol. 8694, pp. 784–799. Springer, Heidelberg (2014)
19. Lu, J., Wang, G., Deng, W., Moulin, P.: Simultaneous feature and dictionary learning for image set based face recognition. In: Fleet, D., Pajdla, T., Schiele, B., Tuytelaars, T. (eds.) ECCV 2014, Part I. LNCS, vol. 8689, pp. 265–280. Springer, Heidelberg (2014)
20. Hayat, M., Bennamoun, M., An, S.: Learning non-linear reconstruction models for image set classification. In: IEEE Conference on Computer Vision and Pattern Recognition (CVPR), 2014. IEEE, pp. 1915–1922 (2014)
21. Du, J.-X., Zhai, C.-M., Wang, Q.-P.: Recognition of plant leaf image based on fractal dimension features. Neurocomputing 116, 150–156 (2013)

Visual Saliency Detection Based on Color Contrast and Distribution

Yanbang Zhang[1]([⊠]) and Guolong Fan[2]

[1] College of Mathematics and Information Science, Xianyang Normal University,
Xianyang 712000, Shaanxi, China
zhyb@mail.nwpu.edu.cn

[2] China Electronics Technology Group Corporation, No. 39 Research Institute,
Xi'an 712000, Shaanxi, China

Abstract. This paper is concerned with color contrast and distribution for detecting salient regions. First, in order to improve the computational efficiency and reduce the disturbance of noise, the input image is pre-segmented into super-pixels. Next, color contrast features are considered in Lab color space and opponency color space. The color distances between a superpixel and other superpixels are calculated, but we do not choose all superpixels to participate the difference. In the meanwhile, the distribution feature is shown by considering the rarity and position of pixels. Finally, we select 2D entropy to measure the performance of salient maps, and select the proper features to fuse. Experimental results show that the proposed method outperforms the state-of-the-art methods on salient region detection.

Keywords: Visual attention · Saliency region detection · Color contrast · Color distribution

1 Introduction

When we look at the image, in generally, we just pay attention to one or several objects, and aren't interested in other regions. This is because our visual system is an intelligent processing system, which can select the some parts referred to as the important objects while ignoring others. The important object in an image is defined as salient objects which are sufficiently distinct from their neighborhood in terms of color, luminance, texture, and so on. Salient object detection can benefit several computer vision tasks including object detection [1–5], object recognition [6–8], image retrieval [9, 10], image segmentation [11], object tracking [12], and so on.

In 1980, Treisman and Gelade [13] proposed a feature integration theory, in which they stated that visual features are important and how to combined, which is considered to be the original theory about visual attention. Later, Koch and Ullman [14] introduced the concept of a saliency map which is a topographic map that represents conspicuousness of scene locations. In 1998, Itti et al. [16] proposed the first complete model, which extracted multiscale low-level features including intensity, color and orientation and used center-surround operation to calculate image saliency. Motivated

© Springer International Publishing Switzerland 2015
D.-S. Huang and K. Han (Eds.): ICIC 2015, Part III, LNAI 9227, pp. 243–250, 2015.
DOI: 10.1007/978-3-319-22053-6_26

by this work, Harel et al. [17] introduced several new features to characterize image content which included subband pyramids based features, 3D color histogram, horizon line detector, etc.

Studies on the human vision have shown that the human eye is very sensitive to color, which is an important feature for detecting saliency. Therefore, most of the existing algorithm for detecting the salient objects use color features. Itti et al. [16] computed the contrast about the green/red opponency color features and yellow/blue opponency color features. In Lab color space, Achanta et al. [18] proposed a frequency-tuned approach, which defined the color difference between every pixel and the mean value of whole image as saliency value. In 2012, Borji [15] showed a saliency model by considering local and global image patch rarities in both the RGB and Lab color spaces, and found that the results outperformed the results computed in only one space.

Instead of processing an image in the spatial domain, some models derive saliency in the frequency domain. Hou and Zhang [20] developed the spectral residual saliency model based on the idea that similarities imply redundancies. They proposed a saliency detection method by using the spectral residual of the log spectrum of an image. Guo et al. [21] incorporated the phase spectrum of the Fourier transform to replace the amplitude transform and proposed a quaternion representation of an image combining intensity, color, and motion features. In 2013, we analyzed the contrast in the spectral and spatial domain simultaneously [22].

Motivated by aforementioned discussions, in this paper, we study the features of color contrast and the color distribution, and propose a bottom-up model to detect the salient regions. The contrast features contain color difference and space distance, simultaneously. The color distribution features reflect the location and spatial sparsity of pixels.

The rest of this paper is organized as follows. Preprocessing is introduced in Sect. 2. Section 3 states the saliency detection model in detail, which includes contrast model, distribution model and feature combination. Experiment results are given in Sect. 4. Some concluding remarks are drawn in the last section.

2   Preprocessing

As discussed above, color is an important factor to attract our attention. We convert the input image to Lab color space and opponency color space, firstly. Then, in each color channel, we use superpixel algorithm to group pixels into perceptually meaningful atomic regions which have the advantages of replacement of the pixel, redundancy reduction, and decrease in complexity. In 2012, Achanta [19] propose a simple superpixel model called SLIC superpixels, which adopts k-means clustering to generate superpixels, and performs an empirical comparison of the existing superpixel methods. So in this paper, we use this superpixel method to preprocess the original images (Fig. 1).

Fig. 1. Results of SLIC superpixel.

3   The Proposed Method

3.1   Color Contrast Feature

In general, an object can attract our attention, whose color features stand out relative to their neighboring regions or the rest of the image. So, we define color feature of a superpixel as

$$F_C(i, n) = \sum_{\forall j \in I} W_{i,j}^n D(C_{i,n}, C_{j,n}) \tag{1}$$

where $F(i,n)$ denotes the contrast-based salient feature of the superpixel i in color channel n ($n \in \{RG, BY, I, L, a, b\}$). $D(C_{i,n}, C_{j,n})$ means the color distance between the pixels $C_{i,n}$ and $C_{j,n}$. $W_{i,j}^n \psi$ denotes the control parameter which is defined as following.

$$W_{i,j}^n = \frac{1}{k} \exp(-\frac{|D(P_i, P_j)|^2}{\sigma_1^2}) \tag{2}$$

where $D(P_i, P_j)$ denotes the distance between superpixel P_i and P_j, which usually defined as the Euclidean distance of centers in two superpixels. As we known, salient pixels usually get together. So we recalculate the color feature as

$$F_C(i, n) = g(i, n) \sum_{\forall j \in I} W_{i,j}^n D(C_{i,n}, C_{j,n}) \tag{3}$$

where $g(i, n)$ is filter, which is obtained from the feature map of (1). In the existing literatures, most of models emphasize the center of image on the assumption that salient objects located at the center of the image. However, in some cases, salient objects located near the edge of the image. The filter not emphasizes the central area of the image, but emphasizes the regions where salient pixels gather.

3.2   Color Distribution Feature

As discusses above, salient pixels usually locate together, while unimportant pixels scatter over the whole image. So we define the another salient feature as

$$S_{SP}(i, n) = \exp(-V(I(i, n))) \tag{4}$$

where $V(I(i, n))$ denotes the 2 dimensional variance of $I(i, n)$.
With

$$I(i, n) = \left\{ \begin{array}{ll} 1 & I(\cdot) = i \\ 0 & others \end{array} \right. \tag{5}$$

$$V(I(i, n)) = \frac{1}{N} \sum_{j}^{N} (p_j - mean(p))^2 \tag{6}$$

p_j is the position of pixels where $I(i, n) = 1$, whose mean position is $mean(p)$ and the number is N.

3.3   Feature Combination

How to compute salient features often receives attention, however, feature combination also plays an important role in improving the performance of saliency detection. In our existing research results have shown that a salient map contains useful information but also contains noise which greatly affects the effect of detection. Hence, we adopt the idea of [22], and choose two groups of maps with smaller entropy, each of which are combined into the color contrast features $F_C \psi$ and color distribution features S_{SP}, respectively. The final result is obtained by nonlinear fusion algorithm, which is expressed as

$$Smap = F_C \times (1 + S_{SP}) \tag{7}$$

The framework of our proposed method is shown in Fig. 2.

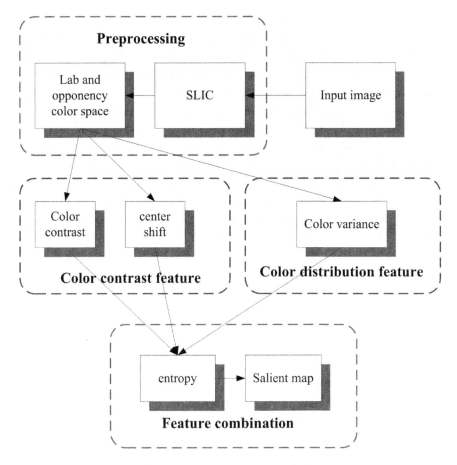

Fig. 2. Architecture of the proposed saliency.

4   Experimental Results

In order to verify the effectiveness of the proposed algorithm, we take a test on Microsoft Research Asia (MSRA) dataset [18], which is publically available dataset and contains 1000 color images with ground truth using binary masks to indicate salient regions. First, we introduce the parameters setting in the experiment. In preprocessing, the given image is segmented into $N = 400$ atomic regions by using SLIC superpixels.

We compare the proposed method with ten state-of-the-art saliency detection methods, including spectral residual (SR) [20], frequency tuned (FT) [18], histogram-based contrast (HC) [23], region contrast (RC) [23] and hypercomplex Fourier transform (HFT) [24]. Some comparison results from our proposed model and the others are shown in Fig. 3, from which we can see that the saliency maps from the proposed model are better than those from other existing ones.

MSRA Dataset

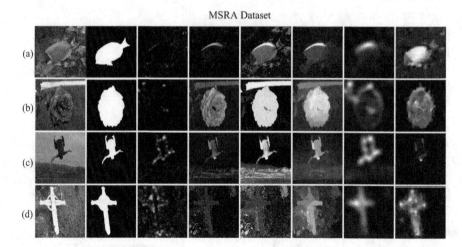

Fig. 3. Examples of saliency maps on the MSRA dataset. (a)–(d) show four examples, including original images, ground truth masks, and saliency maps achieved by the methods FT [17], HC [23], RC [23], HFT [24], and the proposed method, respectively.

To evaluate the performance of the proposed model, we use the receiver operating characteristics (ROC) curve [17] and precision-recall curve as the quantitative evaluation metric. In Figs. 4 and 5, we can see that the saliency maps obtained by the proposed method have the best performance in highlighting salient regions in the MSRA dataset.

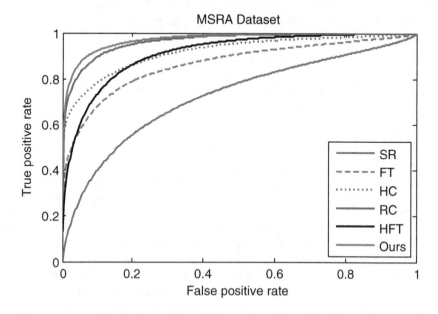

Fig. 4. Comparison of different algorithms by ROC.

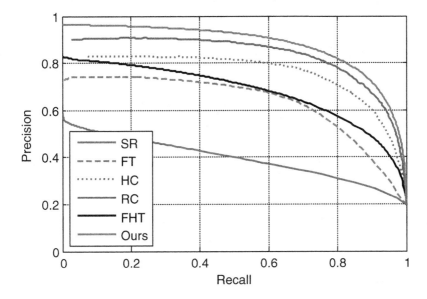

Fig. 5. Comparison of different algorithms by precision-recall curves.

5 Conclusion

In this paper, we present a novel bottom-up saliency detection model. The contrast, distribution, sparsity of color is considered, simultaneously. In fusion, we choose the 2D entropy as an evaluation criterion to select the proper subsets of saliency maps to fuse together. Finally, experimental results are provided to show that the proposed method outperforms the state-of-the-art methods on salient region detection.

Acknowledgments. This research was supported by the National Science Foundation of China (61401357), the Shaanxi Provincial Natural Science Foundation of China (2013JM1014, 2014JM1032), the Shaanxi Educational Committee Foundation of China (14JK1797), the Specialized Research Fund of Xianyang Normal University (13XSYK009, 14XSYK005).

References

1. Frintrop, S.: VOCUS: A Visual Attention System for Object Detection and Goal-Directed Search. Springer, Heidelberg (2006)
2. Navalpakkam, V., Itti, L.: An integrated model of top-down and bottom-up attention for optimizing detection speed. In: 2006 IEEE Computer Society Conference on Computer Vision and Pattern Recognition, vol. 2, pp. 2049–2056. IEEE (2006)
3. Fritz, G., Seifert, C., Paletta, L., Bischof, H.: Attentive object detection using an information theoretic saliency measure. In: Paletta, L., Tsotsos, J.K., Rome, E., Humphreys, G.W. (eds.) WAPCV 2004. LNCS, vol. 3368, pp. 29–41. Springer, Heidelberg (2005)

4. Butko, N.J., Movellan, J.R.: Optimal scanning for faster object detection. In: IEEE Conference on Computer Vision and Pattern Recognition, CVPR 2009, pp. 2751–2758. IEEE (2009)
5. Judd, T., Ehinger, K., Durand, F., et al.: Learning to predict where humans look. In: 2009 IEEE 12th International Conference on Computer Vision, pp. 2106–2113. IEEE (2009)
6. Salah, A.A., Alpaydin, E., Akarun, L.: A selective attention-based method for visual pattern recognition with application to handwritten digit recognition and face recognition. IEEE Trans. Pattern Anal. Mach. Intell. **24**(3), 420–425 (2002)
7. Walther, D., Koch, C.: Modeling attention to salient proto-objects. Neural Netw. **19**(9), 1395–1407 (2006)
8. Han, S., Vasconcelos, N.: Biologically plausible saliency mechanisms improve feedforward object recognition. Vis. Res. **50**(22), 2295–2307 (2010)
9. Tsai, Y.: Hierarchical salient point selection for image retrieval. Pattern Recognit. Lett. **33**(12), 1587–1593 (2012)
10. Lin, L., Shan, J., Zheng, Z.: Partial-duplicate image retrieval via saliency-guided visually matching. IEEE Trans. Multimedia **20**(3), 13–23 (2013)
11. Bai, X., Wang, W.: Saliency-SVM: an automatic approach for image segmentation. Neurocomputing **136**, 243–255 (2014)
12. Guo, M., Zhao, Y., Zhang, C., Chen, Z.: Fast object detection based on selective visual attention. Neurocomputing **144**, 184–197 (2014)
13. Treisman, A., Gelade, G.: A feature-integration theory of attention. Cogn. Psychol. **12**(1), 97–136 (1980)
14. Koch, C., Ullman, S.: Shifts in selective visual attention: towards the underlying neural circuitry. In: Vaina, L.M. (ed.) Matters of Intelligence, pp. 115–141. Springer, Netherlands (1987)
15. Borji, A.: Boosting bottom-up and top-down visual features for saliency estimation. In: IEEE Conference on Computer Vision and Pattern Recognition, pp. 438–445 (2012)
16. Itti, L., Koch, C., Niebur, E.: A model of saliencybased visual attention for rapid scene analysis. IEEE Trans. Pattern Anal. Mach. Intell. **20**, 1254–1259 (1998)
17. Harel, J., Koch, C., Perona, P.: Graph-based visual saliency. In: Advances in Neural Information Processing Systems, vol. 19, pp. 545–552 (2007)
18. Achanta, R., Hemami, S., Estrada, F., Susstrunk, S.: Frequency-tuned salient region detection. In: IEEE Conference on Computer Vision and Pattern Recognition, pp. 1597–1604 (2009)
19. Achanta, R., Shaji, A., Smith, K., Lucchi, A., Fua, P., Susstrunk, S.: SLIC superpixels compared to state-of-the-art superpixel methods. IEEE Trans. Pattern Anal. Mach. Intell. **34**(11), 2274–2281 (2012)
20. Hou, X., Zhang, L.: Saliency detection: a spectral residual approach. In: IEEE Conference on Computer Vision and Pattern Recognition, pp. 1–8 (2007)
21. Guo, C., Ma, Q., Zhang, L.: Spatio-temporal saliency detection using phase spectrum of quaternion Fourier transform. In: IEEE Conference on Computer Vision and Pattern Recognition, pp. 1–8 (2008)
22. Zhang, Y., Han, J., Guo, L.: Saliency detection by combining spatial and spectral information. Opt. Lett. **38**(11), 1987–1989 (2013)
23. Cheng, M., Zhang, G., Mitra, N., Huang, X., Hu, S.: Global contrast based salient region detection. In: IEEE Conference on Computer Vision and Pattern Recognition, pp. 409–416 (2011)
24. Li, J., Levine, M., An, X., Xu, X., He, H.: Visual saliency based on scale-space analysis in the frequency domain. IEEE Trans. Pattern Anal. Mach. Intell. **35**(4), 996–1010 (2013)

Moving Vehicle Detection Based on Visual Processing Mechanism with Multiple Pathways

YanFeng Chen, QingXiang Wu[(⊠)], HaiHui Xie, SanLiang Hong,
and Xue Li

Key Laboratory of OptoElectronic Science and Technology for Medicine of
Ministry of Education, College of Photonic and Electronic Engineering,
Fujian Normal University, Fuzhou 350007, Fujian, China
{474741702,568082212,913452977,1026044914}@qq.com,
qxwu@fjnu.edu.cn

Abstract. In this paper a moving vehicle detection algorithm based on visual processing mechanism with multiple pathways is proposed, in which the multiple pathways visual processing mechanism is inspired by the biological visual system. According to the different moving directions of front vehicles, orientation selectivity of visual cortex cells is used to construct a visual processing model with three pathways. In each pathway, an AdaBoost cascade classifier is trained using a set of special samples for detection of moving vehicles. The AdaBoost cascade classifier is response to multi-block local binary patterns (MB-LBP) of vehicles. The experimental results show that the multiple pathways visual processing mechanism, compared with the single pathway Ada-Boost cascade classifier and the conventional method, not only can reduce the complexity of the classifier and training time, but also can improve the recognition rate of moving vehicle.

Keywords: The multiple pathways visual processing mechanism · Multi-block local binary pattern · Adaboost cascade classifier · Moving vehicle detection

1 Introduction

As increasing number of motor vehicles and influence of natural weather, the traffic environment is becoming more and more complex. In order to improve the road traffic safety, a set of vehicle detection system for complex traffic environment has attracted attention of scientists in the world. In this paper a moving vehicle detection algorithm based on multiple pathways visual processing mechanism is proposed, in which the multiple pathways visual processing mechanism is inspired by the biological visual system.

In early studies of biological visual information processing, Hubel and Wiesel found that almost all of the visual cortex cells are able to distinguish the spatial orientation of a line, i.e. has strong orientation selectivity [1]. Cells in 17 and 18 district of the visual cortex can be divided into two categories: simple cells and complex cells. The visual cortex cells are sensitive to the orientation of the space geometric elements [2]. For each simple cell, there is an optimal orientation and there has the strongest cell responses on

© Springer International Publishing Switzerland 2015
D.-S. Huang and K. Han (Eds.): ICIC 2015, Part III, LNAI 9227, pp. 251–263, 2015.
DOI: 10.1007/978-3-319-22053-6_27

this orientation. Inspiration comes from processing mechanism of spatial information of biological vision. We collected a set of moving vehicle samples in the complex environment, and the sample set of moving vehicles is divided into three data subsets in three different orientations. The vehicles of three different orientations include the moving vehicles on the left and right of the host vehicle and the vehicles in front of the host vehicle. Then the multi-block local binary pattern features are extracted from each sample subset. Finally, the AdaBoost algorithm [3] is used to train cascade classifier for each sample subset. Experiments show that, in the complex traffic environment, the moving vehicle detection system of multi-pathway can effectively detect moving vehicles in different directions. Meanwhile, the system reduces the complexity of the cascade classifier training process and improves the accuracy of vehicle detection.

The remainder of this paper is organized as follows: The algorithm of multi-block local binary pattern is detailed in Sect. 2. The AdaBoost cascade classifier is used as vehicle detector for each pathway. The model of AdaBoost cascade classifier is detailed in Sect. 3. The multi-pathway visual processing mechanism and the multi-scale scanning detection mechanism are proposed in Sect. 4. Experimental results for multi-pathway vehicle detection and comparison with single pathway vehicle detection are shown in Sect. 5. Conclusions about the multi-pathway vehicle detection system are presented in Sect. 6.

2   The Extraction of MB-LBP

In 1996, Ojala et al. [4] proposed the local binary patterns (LBP) operator. This feature has characters of rotational invariance, gray invariance and simple calculation. So LBP features can effectively overcome the impact of vehicle movement, uneven illumination and other complex environments. The basic LBP algorithm as follows:

1. Select a fixed rectangular block with size 3*3. Let g_c is the gray value of center pixel. The value of eight neighborhood pixels is g_i, $i = \{0, 1, \cdots, 7\}$;
2. Take the gray value of center pixel as the threshold, binaryzation of eight neighborhood pixels are as follows: $s_i = \begin{cases} 1 & g_i > g_c \\ 0 & g_i < g_c \end{cases}$, $i = \{0, 1, \cdots, 7\}$;
3. The LBP value of this rectangular block is $LBP = \sum_{i=0}^{7} s_i \times 2^i$;

In [5] the local histogram of oriented gradients (HOG) feature is used to describe the features of vehicles. The calculation of HOG feature is complex so that this feature will reduce the real-time of vehicle detection. In [6] Haar-Like and HOG features are used to detect vehicles. Jin et al. [7] proposed a vehicle detection method based on Haar- like features and Adaboost algorithm. However, these Haar-like rectangle features seem too simple, and the detector often contains thousands of rectangle features for considerable performance. The large number of selected features leads to high computation costs both in training and test phases [8].

In this paper MB-LBP features are used to describe the vehicle. The calculation of traditional LBP texture features pixel is based on a single isolated pixel. With the limitations of the neighborhood, this algorithm can not accurately describe the global characteristics of vehicles. In order to avoid the limitation of LBP features, Zhang [9] proposed Multi-block local binary pattern. The calculation process of MB-LBP features are as follows:

1. An arbitrary size of 3s*3t neighborhood window is selected in the image. And then the window is divided into 9 sub-regions, the size of sub-region is s*t;
2. A 3*3 integer matrix is obtained following the calculation of the average gray value for each sub-region;
3. Then the MB-LBP value of the window is calculated using the traditional LBP algorithm;
4. Finally, the MB-LBP gray value is used to replace the value of central region, then the MB-LBP features of image is gained.

Select different values (s, t), the MB-LBP feature images of different scales can be calculated. When the Adaboost classifier is trained, the Adaboost algorithm will select effective MB-LBP feature series as the feature description of vehicles. The MB-LBP feature images of the vehicle in different directions are shown in the Fig. 1. Set the two parameters s = 4, t = 4.

Fig. 1. The MB-LBP feature images of the vehicle

3   The Adaboost Cascade Classifier

The Adaboost algorithm, which using the ensemble learning method to solve the problem, is a classical algorithm of machine learning. The ensemble learning is a method that using multiple learning devices (usually homogeneous) to solve the same problem. This method can significantly improve the generalization ability of learning systems. The basic idea of ensemble learning algorithm is used in Adaboost algorithm. And the Adaboost algorithm has been successfully applied in many fields, such as the face detection, pedestrian detection and digital recognition and so on. In this paper, the Adaboost cascade classifier is selected as vehicle detector for each pathway.

The Adaboost algorithm is an effective decision tree classification algorithm. The main steps of the algorithm are described as follows. Firstly, set the initial weights of training samples, and then extract the features of training samples, using the features to train weak classifiers. Next, based on the performance of weak classifiers, update the weights of training samples. Then keep on training weak classifier. Finally, a set of weak classifiers is obtained. Then according to the weight of each weak classifier, a set of weak classifiers is integrated into a strong classifier. The function expression of weak classifier is as follows:

$$h(x, f, p, \theta) = \begin{cases} 1 & pf(x) < p\theta \\ 0 & otherwise \end{cases},$$ (1)

where x is a window area in the input image. $f(x)$ denotes the feature of x. θ is the threshold of weak classifier. p is a polarity.

The specific training steps of Adaboost algorithm is listed below:

1. Collect positive and negative training samples $\{(x_1, y_1), \cdots, (x_n, y_n)\}$, where $y_i \in \{0, 1\}$, 0 and 1 denote non-vehicle and vehicle samples respectively;

2. Initialize weights of all training samples $w_{1,i} = \begin{cases} 1/2m & y_i = 0 \\ 1/2l & y_i = 1 \end{cases}$, where m and l are the number of negatives and positives respectively;

3. For $t = 1, \cdots, T$

4. Normalize the sample weights $w_{t,i} = \dfrac{w_{t,i}}{\sum\limits_{j=1}^{n} w_{t,j}}$;

5. Train weak classifier h_k for each MB-LBP feature k. The error of the weak classifier h_k is

$$\varepsilon_k = \sum_i w_i |h_k(x_i, f_k, p_k, \theta_k) - y_i|;$$ (2)

6. Select the weak classifier $h_t(x, f_t, p_t, \theta_t)$ with respect to the minimum weighted error $\varepsilon_t = \min \varepsilon_k$ and determine the value of parameters f_t, p_t, θ_t;

7. Update the weights of training samples: $w_{t+1,i} = w_{t,i}\beta_t^{1-e_i}$, where $e_i = \begin{cases} 0 & example\ x_i\ is\ classified\ correctly \\ 1 & otherwise \end{cases}$, $\beta_t = \dfrac{\varepsilon_t}{1-\varepsilon_t}$;

8. End For;

9. The final strong classifier is:

$$H(x) = \begin{cases} 1 & \sum\limits_{t=1}^{T} \alpha_t h_t(x) \geq \frac{1}{2}\sum\limits_{t=1}^{T} \alpha_t \\ 0 & otherwise \end{cases}, \text{ where } \alpha_t = \log\frac{1}{\beta_t}.$$ (3)

When just using a single Adaboost classifier to detect the targets, it is hard to ensure the accuracy of the vehicle detection. With increase of the number of weak classifiers and features, it will reduce the speed of target detection. Viloa and Jones [10] proposed an algorithm for constructing a cascade of classifiers which achieves increased detection performance while radically reducing computation time. Each stage of Adaboost cascade classifier has a strong classifier. The strong classifiers are constructed as a cascade classifier by cascade algorithm. In the detection process, the initial classifier eliminates a large number of non-vehicle areas with very little processing. Then the amount of computation of target detection is reduced by reducing the number of judgment of target area. Vehicle areas will trigger the evaluation of every classifier in the cascade so that the accuracy of vehicle detection is improved. The model of Adaboost cascade classifier is shown in the Fig. 2.

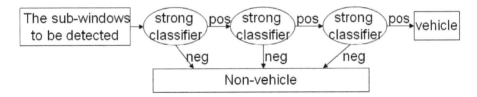

Fig. 2. The model of Adaboost cascade Classifier

4   Multiple Pathway System of Moving Vehicle Detection

Currently, the Adaboost cascade classifier is most suitable for the classification problems of binary-class. Although some multi-classification algorithms of the Adaboost cascade classifier have been proposed, it is hard to be widely used because of the high complexity of training process. In the complex traffic environment, due to the various orientations of moving vehicles is different, this paper proposes a moving vehicle detection system based on multi pathways visual processing mechanism.

4.1   Multi-pathway Visual Processing Mechanism

According to the different moving directions of front vehicles, orientation selectivity [11] of visual cortex cells is used to construct a visual processing model with three pathways. In each pathway, an AdaBoost cascade classifier is trained using a set of special samples for detection of moving vehicles. We collected a sample set of moving vehicles in complex environments, and the sample set of moving vehicles is divided into three data subsets in three different orientations. The vehicles of three different orientations include the moving vehicles on the left and right of the host vehicle and the vehicles in front of the host vehicle. Then extract the effective features of each sample subset. Next, train the different cascade classifiers for each sample subset. The different cascade classifiers take attention to the different orientation vehicles [12]. This algorithm not only reduces the requirements of training samples, but also reduces the complexity training process of the cascade classifiers.

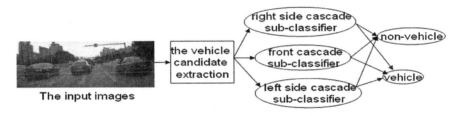

Fig. 3. Multi-pathway moving vehicle detection system

In order to improve the real-time and accuracy of multi-pathway vehicle detection system, the vehicle candidates are extracted using a prior knowledge such as optical flow feature, shadow, symmetry, color, edge and texture. And then the different cascade sub-classifiers are used to detect moving vehicles in different directions. The model of multi-pathway moving vehicle detection system is shown in the Fig. 3.

4.2 Multi-scale Scanning Detection Mechanism

When the cascade classifier detects the moving vehicles for the input images, there are two traditional algorithms of target scanning detection. (1) Keep the size of scanning window invariant while the size of input image is scaled down proportionally. (2) Keep the size of the input image invariant while the size of scanning window is expanded proportionally. The detection speed of these methods is slow for large scale input image. Target scanning detection algorithm based on multi-scale has been improved in this paper. The input image is scaled down proportionally. Meanwhile, the size of scanning window is expanded proportionally. Then detector uses the magnified window to scan the reduced image. So the moving vehicles which have different scales will be detected. And this multi-scale scanning detection mechanism improves the speed of vehicle detection. The specific steps of multi-scale scanning detection mechanism are as follows:

1. Let $IW \times IH$ be the size of the input images;
2. The size of the training window is $winW \times winH$ which is initial size of the scanning window, let the scaling factor $scale = 1.2$;
3. Do
4. Use the window of which the size is $winW \times winH$ to scan the input image.
5. Extract the MB-LBP feature of the window in the image;
6. The MB-LBP feature as input of the cascade classifier, to determine the scanning area whether is the moving vehicle;
7. Use bilinear interpolation to adjust the size of input image $IW = IW/scale$, $IH = IH/scale$;
8. Adjust the scanning window size $winW = winW \times scale$, $winH = winH \times scale$;
9. while $IW \geq winW$ and $IH \geq winH$ goto step 3

5 Experimental Process and Results

PC and VS2010 software are used as the experimental platform in this paper. For training a classifier, a set of training samples must be collected. There are two sample types: negative samples and positive samples. Negative samples correspond to non-object images. Positive samples correspond to object images.

In this paper, 1200 pieces of vehicle sample come from a vehicle traveling data recorder video in AVI format. Among the vehicle samples, there are 400 pieces of front vehicle samples, 400 samples of left moving vehicles and 400 pieces of right moving vehicle samples. Meanwhile, 2000 pieces of negative samples is collected from the vehicle traveling data recorder video. And then the sample pictures are normalized using cubic interpolation and converted to grayscale images. Try to keep the width to height ratio of original image invariant when normalized image. Examples of the training sample images are shown in the Fig. 4.

(1) Positive Samples of Front Vehicle

(2) Positive Samples of Left Vehicle

(3) Positive Samples of Right Vehicle

(4) Negative Samples

Fig. 4. Part of positive and negative samples

5.1   Experiments

In the first set of experiments, a single pathway AdaBoost classifier is used to detect moving vehicles. The training data includes 1200 pieces of moving vehicle samples and 2000 negative samples. Firstly, the vehicle samples were normalized. The vehicle samples are resized as gray scale images of which the size is 33*25. "opencv_createsamples" utility which is provided by opencv is used to create a vec-file for the vehicle sample dataset. And then the description file is created for the negative samples. Finally, "opencv_traincascade" is used to train a cascade classifier. The training parameters of cascade classifier are shown in Table 1. The training results of single cascade classifier are shown in Table 2.

Table 1. The training parameters of single cascade classifier

NumStages	Stage type	Boost type	Min hit rate	Max false alarm rate
16	BOOST	GAB	0.995	0.5

Table 2. The training results of single cascade classifier

NumStages	Features quantity	Training time	Hit rate	False alarm rate
15	89	78 min	0.942	1.30908e-005

In the second set of experiments, a conventional vehicle detection method [7] based on Haar-like features and Adaboost algorithm is used to compare with the algorithm proposed in this paper. The training data is the same as the training data used in the first set of experiments. The training parameters of cascade classifier are shown in Table 3. The training results of single cascade classifier are shown in Table 4.

Table 3. The training parameters of conventional method

NumStages	Stage type	Boost type	Min hit rate	Max false alarm rate
16	BOOST	GAB	0.995	0.5

Table 4. The training results of conventional method

NumStages	Features quantity	Training time	Hit rate	False alarm rate
15	167	636 min	0.939	1.12121e-005

In the third set of experiments, the multiple pathways processing mechanism is used to detect moving vehicles. The training data is the same as the training data used in the first set of experiments. 1200 pieces of moving vehicle samples are divided into three kinds of sample subsets. These sample subsets include 400 pieces of front vehicle samples, 400 samples of left moving vehicles and 400 pieces of right moving vehicle samples. And then the sample subsets are resized to the size 22*20, 30*20 and 30*20 respectively and converted to grayscale image. Finally, three kinds of cascade sub-classifier are created by "opencv_traincascade" utility. The training parameters of the cascade sub-classifiers are shown in Table 5 respectively. The training results of multiple cascade classifiers are shown in Table 6. The experimental results show that the characteristics of vehicles can be more accurately described by feature MB-LBP. The method proposed in this paper not only enhances the hit rate, but also reduces the false alarm rate. And compared with the conventional vehicle detection method in [7], the method proposed in this paper reduces the training time of classifier.

Table 5. The training parameters of multiple cascade classifiers

Parameters / Sub-classifiers	NumStages	Stage type	Boost type	Min hit rate	Max false alarm rate
The front sub classifier	16	BOOST	GAB	0.995	0.5
The left sub classifier	16	BOOST	GAB	0.995	0.5
The right sub classifier	16	BOOST	GAB	0.995	0.5

Table 6. The training results of multiple cascade classifiers

Parameters / Sub-classifiers	NumStages	Features quantity	Training time	Hit rate	False alarm rate
The front sub classifier	13	52	54min	0.961	8.85765e-006
The left sub classifier	12	45	53min	0.971	1.13215e-005
The right sub classifier	13	47	52min	0.980	1.20382e-005

5.2 Experimental Results

In order to test the performance of the multiple pathways detector and the single pathway detector, we select a number of short driving record video in different environments. The different environments include rainy day, cloudy, fine day and the strong light. The test results are shown in Tables 7 and 8. The vehicle detection results of single pathway cascade classifier and multiple pathways cascade classifier under different environments are shown in the Figs. 5, 6, 7 and 8.

Table 7. The test results of multiple pathways cascade classifier

Parameters / Weather	Vehicles quantity	Recognition quantity	Missing quantity	Mistake quantity	Recognition rate	Mistake rate
Cloudy	233	227	6	5	97.42%	2.15%
Rainy	186	178	8	4	95.69%	2.15%
Sunny	173	163	10	8	94.22%	4.62%

Table 8. The test results of single pathway cascade classifiers

Parameters / Weather	Vehicles quantity	Recognition quantity	Missing quantity	Mistake quantity	Recognition rate	Mistake rate
Cloudy	233	223	10	9	95.71%	3.86%
Rainy	186	173	13	11	93.01%	5.91%
Sunny	173	161	12	12	93.06%	6.94%

(a) 3 cars are detected by single pathway conventional method

(b) 3 cars are detected by

(c) 4 cars are detected by multiple pathway

Fig. 5. Cloudy day

(d) 2 car is detected by single pathway (e) 2 cars are detected by conventional method

(f) 4 cars are detected by multiple pathway

Fig. 6. Rainy day

(g) 1 cars is detected by single pathway (h) 2 cars are detected by conventional method

(i) 2 cars are detected by multiple pathway

Fig. 7. Sunny day

(j) 0 car is detected by single pathway   (k) 0 cars is detected by conventional method

(l) 2 cars are detected by multiple pathway

Fig. 8. The strong light

In the figure, the different color circles are detected by different sub-classifiers. Due to the size of training window is 22*20, it can't detect the moving vehicle of which scale is less than 22*20. So this situation does not belong to the missing vehicle. Obviously the algorithm based on three pathways achieves the highest recognition rate and lowest mistake rate. Furthermore, the algorithm proposed in this paper reduces the requirements of the quantity of training samples and the training time.

6   Conclusion

The experimental data shows that the vehicle detection results based on multiple pathways detection mechanism are better than the detection results of single pathway detector and the conventional vehicle detection method. In the experiment, the false detection and miss detection area are mainly related to the sample database. The algorithm in this paper also has some shortcomings: (1) can not detect the overlapping vehicles; (2) multi-pathway algorithm will affect the real-time of vehicle detection. In order to improve the real time of vehicle detection, the vehicle candidates are extracted using a prior knowledge, such as optical flow feature, shadow, symmetry, color, edge and texture. In this paper, a single feature of MB-LBP is used to describe the features of the vehicle. The future research is the vehicle detection based on multi-feature, so as to improve the robustness of the vehicle detection system. Meanwhile, a useful vehicle tracking algorithm will be researched and the algorithm will be implemented on a DSP chip in the future. The algorithm in this paper can be applied in the safety systems of vehicle driving.

Acknowledgments. The authors gratefully acknowledge supports from Fujian Provincial Key Laboratory for Photonics Technology, and the fund from the Natural Science Foundation of China (Grant No. 61179011) and Science and Technology Major Projects for Industry-academic Cooperation of Universities in Fujian Province (Grant No. 2013H6008), and supports from Innovation Team of the Ministry of Education (IRT1115).

References

1. Nicholls, J.G., Martin, A.R., Wallace, B.G.: From Neuron to Brain, 4th edn. Sinauer Associates Inc., Sunderland (2001)
2. Nieuwenhuys, R., Huijzen, C.V., Voogd, J.: The Human Central Nervous System. A Synopsis and Atlas. Springer, Heidelberg (1979)
3. Schapire, R.E., Singer, Y.: Improved boosting algorithms using confidence-rated predictions. Mach. Learn. **37**(3), 297–336 (1999)
4. Ojala, T., Pietikainen, M., Harwood, D.: A comparative study of texture measures with classification based on feature distributions. Pattern Recogn. **29**(1), 51–59 (1996)
5. Yeul-Min, B., Whoi-Yul, K.: Forward vehicle detection using cluster-based AdaBoost. Opt. Eng. **53**(10), 1021103 (2014)
6. Cai, Y.H.: Fusing multiple features to detect on-road vehicles. Comput. Technol. Autom. **32**(1), 98–102 (2013)
7. Jin, L.S., Wang, Y., Liu, J.H., Wang, Y.L., Zheng, Y.: Front vehicle detection based on Adaboost algorithm in daytime. J. Jilin Univ. (Engineering and Technology Edition) **44**(6), 1604–1608 (2014)
8. Mita, T., Kaneko, T., Hori, O.: Joint haar-like features for face detection. In: ICCV (2005)
9. Zhang, L., Chu, R., Xiang, S., Liao, S., Li, S.Z.: Face detection based on multi-block LBP representation. In: Lee, S.-W., Li, S.Z. (eds.) ICB 2007. LNCS, vol. 4642, pp. 11–18. Springer, Heidelberg (2007)
10. Viola, P., Jones, M.J.: Robust real-time face detection. Int. J. Comput. Vis. **57**, 137–154 (2004)
11. Wu, Q.X., McGinnity, T.M., Maguire, L.P., Belatreche, A., Glackin, B.: Processing visual stimuli using hierarchical spiking neural networks. Neurocomputing **71**(10-12), 2055–2068 (2008)
12. Wu, Q.X., McGinnity, T.M., Maguire, L.P., Cai, R.T., Chen, M.G.: A visual attention model based on hierarchical spiking neural networks. Neurocomputing **116**, 3–12 (2013)

A Computer Vision Method for the Italian Finger Spelling Recognition

Vitoantonio Bevilacqua[1](✉), Luigi Biasi[1], Antonio Pepe[1],
Giuseppe Mastronardi[1], and Nicholas Caporusso[2]

[1] Dipartimento di Ingegneria Elettrica e dell'Informazione,
Politecnico di Bari, Bari, Italy
vitoantonio.bevilacqua@poliba.it
[2] INTACT healthcare, Bari, Italy

Abstract. Sign Language Recognition opens to a wide research field with the aim of solving problems for the integration of deaf people in society. The goal of this research is to reduce the communication gap between hearing impaired users and other subjects, building an educational system for hearing impaired children. This project uses computer vision and machine learning algorithms to reach this objective. In this paper we analyze the image processing techniques for detecting hand gestures in video and we compare two approaches based on machine learning to achieve gesture recognition.

Keywords: Image processing · Computer vision · Machine learning · SVM · MLP · Gaussian Mixture Model · Sign language · LIS

1 Introduction

Hearing-impaired people usually communicate using the visual-gestural channel which is notably different from the vocal-acoustic one. To this end, sign language is a complete language having its own grammar, syntax, vocabulary and morphological rules. Furthermore, each community usually develops and employs its specific language. As a result, there are many different languages based on signs, such as the Italian Sign Language (LIS), the American Sign Language (ASL), or the British Sign Language. Additionally, each language has its vernacular variants and, similarly to spoken languages, a constantly evolving lexicon.

The Italian Sign Language - as other sign languages - is based on an alphabet, commonly referred to as finger spelling. Despite being not largely used in conversations, finger spelling is crucial, both for beginners and in communication. Indeed, it is employed to represent names of people or places, and to replace signs which are harder to remember. The LIS finger spelling represents all the 26 letters of the Italian alphabet, as shown in Fig. 1: some letters are associated with a static gesture, others include hand movement.

The goal of this research is to realize a system for detecting LIS gestures and for translating them into written or spoken language (e.g., with the help of text-to-speech systems), as this could be employed in communication technology or educational tools to provide hearing impaired people with enhanced interaction, simplified communication and, in general, with more opportunities of social inclusion.

© Springer International Publishing Switzerland 2015
D.-S. Huang and K. Han (Eds.): ICIC 2015, Part III, LNAI 9227, pp. 264–274, 2015.
DOI: 10.1007/978-3-319-22053-6_28

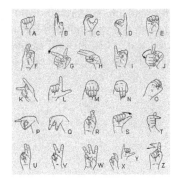

Fig. 1. LIS finger spelling

2 Related Work

In the last years, research showed that hardware/software solutions based on automatic recognition systems can play a crucial role in helping sign language users communicate more easily and efficiently. Also, several projects integrated them in multimedia platforms to address applications accessibility and to increase social inclusion of users having some degree of hearing deficiency.

Recently, promising results have been achieved with RGB-D sensors. However, they require users to be constantly connected to power supply, which is an important drawback especially when mobility is required. Moreover, they are not currently available on commercial mobile devices.

Several experimental projects have been realized in this field; nevertheless, they usually involve very complex equipment and sophisticated settings. For instance, a large number of applications utilize Microsoft Kinect, which is not portable. Others are exploring wearable solutions, such as sensor-equipped gloves [7], which require users to have both their hands busy. Conversely, we adopt an alternative approach based on cameras and low-cost devices with the aim of designing a portable solution especially dedicated to enabling the deaf use sign language in mobility.

In this work, we introduce SignInterpreter, a Sign Language detection system based on RGB sensors, as they are incorporated in the majority of nowadays available mobile devices. Also, we detail an experimental study in which we compare the performance of our solution using different image processing algorithms.

In [6, 9] two methods that provide sign recognition with image processing and machine learning method have been proposed. In [6] a Random Forest Algorithm is trained with Hu moments, in [9] a SVM is trained with Zernike moments and Hu moments too.

3 SignInterpreter

3.1 System Architecture

SignInterpreter is a Sign Language recognition system designed to work with standard RGB cameras, such as webcams, and cameras mounted on mobile and embedded devices: users communicate by realizing gestures in front of the camera. The system acquires the video stream and processes the captured frame in order to extract features that are converted into digitized text. This, in turn, can be utilized to represent messages, or to control applications. The system architecture is shown in Fig. 2.

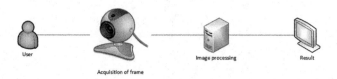

Fig. 2. System architecture

In addition to the recognition hardware, SignInterpreter comprises a client-server architecture (Fig. 3) specifically designed to increase the performances of the system on mobile devices: video streams and image features are acquired on the client; image processing and machine learning tasks are executed on the server, in real-time. In addition to enabling a larger dataset to be collected and used, this improves the overall accuracy of the system, regardless of the computational power of the client.

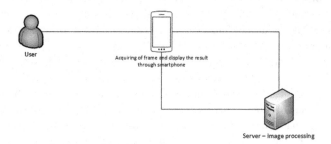

Fig. 3. Distributed architecture

3.2   Software Architecture

The software operates in three main phases: (1) *background acquisition and modeling*; (2) *calibration* to get the user's skin color; (3) *hand detection and gesture recognition*. The first step removes the noise produced by background objects. During calibration, the system collects several color samples of the hand and utilizes them to obtain a precise model of the skin color of user's hand. In the last phase, foreground is extracted from a generic frame: this is realized using information from the background model acquired in the first phase, and the hand model. Then, a segmentation process is executed on the resulting image. Specifically, the algorithm discards all colors considered as different from the color model of user's skin. Subsequently, the Canny edge detection algorithm is employed to extract the contour of the hands. This is the input to a classifier that is trained to recognize gestures. The system workflow is shown in Fig. 4.

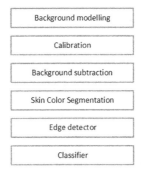

Fig. 4. Workflow

4   Hand Detection

This stage consists in extracting foreground objects from the scene, so that segmentation can be realized [1]. Background subtraction is crucial in order to reduce the number of false negatives in the segmentation phase.

4.1   Background Subtraction

In [4], several background subtraction algorithms are discussed. In our experimental study, we employed the Gaussian Mixture Model (GMM) algorithm [3] to discriminate background. The GMM algorithm models the value of each pixel with a mixture of K Gaussian distributions and it detects the pixel intensity that most likely represents the background, using a heuristic method. If a pixel does not match the intensity value, it is recognized as a foreground pixel.

The likelihood to observe an intensity pixel value x (x_R, x_G, x_B) at the time t is expressed by:

$$p(x_t) = \sum w_{i,t}\, \eta(x_t;\, \mu_{i,t},\, \Sigma_{i,t})$$

Where w is the weight vector, $\eta_{i,t}$ is the Gaussian distribution with an average μ and a covariance matrix Σ.

In order to remove background, we define a threshold discriminates background pixels from foreground pixels.

4.2   Skin Color Detection

In our system, skin color detection plays an essential role. The skin color model proposed in [5] did not show significant results as the authors refer to African-Americans, only. For the purpose of our study, two skin color detection algorithms have been implemented using face color and hand color. The former extracts the color model from the color of user's face. In particular, we define a region of interest, which is located between the eyes and the nose. This region is calculated by extracting the face, by Viola-Jones algorithm, and on it we select a rectangle situated roughly in the middle of face. The choice of this region is due to the fact that we are sure that will contain skin. Extracted color is represented in the HSV color space (Fig. 5).

Fig. 5. Skin color detection: approach 1 (Color figure online)

However, we experienced that this method is not accurate, because face color and hand color can be very different in some individuals.

Therefore, we designed a second algorithm that detects skin color from several points of the hand. In order to do so, we provide users with a window showing the video being captured with overlay markers. By asking the user to place their hands over the markers, the algorithm can evaluate the color model of the skin. This process is shown in Fig. 6.

Fig. 6. Skin color detection: approach 2 (Color figure online)

Once the skin color has been extracted, it is then used as a threshold to segment the foreground image, and a second filter is used to delete the face and all regions with an area bigger than the face area and the regions with a too small area. In this way, the remaining regions are very likely to be the hands. The workflow is shown in Fig. 7.

5   Gesture Recognition

Gesture recognition has been implemented using a supervised learning algorithm. Specifically, we employed *Support Vector Machines*.

With the hypothesis of having a linear binary classifier problem, the SVM algorithm finds the separation level that maximizes the margin between two classes, and maximizes the empty area included between them. The amplitude of this value is

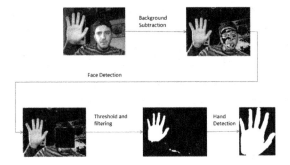

Fig. 7. Hand detection

defined by the distance between the hyperplane that splits the two classes and the samples around that area, i.e., the *support vector* (Fig. 8).

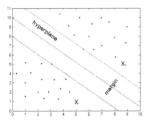

Fig. 8. SVM

In order to train the algorithm, we acquired a training set of 2160 pictures representing all the 26 signs of LIS (80 pictures per sign) (Fig. 9).

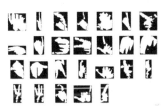

Fig. 9. Dataset

We defined four main classes of signs based on their aspect ratio, in order to improve the recognition performances on letters having similar sign representations. Then, we trained one SVM classifier per group. Particularly, each classifier has been trained using edges extracted from the pictures. All the pictures belonging to the same class have the same dimensions.

- First class: signs {B, R, U, V, Z}, height = 126 px, width = 104 px;
- Second class: signs {A, E, M, N, O, P, Q, T, X}, height = 106 px, width = 141 px;
- Third class: signs {C, D, F, I, L, Y}, height = 179 px, width = 141 px;
- Fourth class: signs {G, H, J, K, S}, height = 159 px, width = 70 px;

The entire image is given as an input to the classifier. As a result, the features vector for each classifier has $h \times w$ components, where w and h represent image width and height, respectively.

A different approach is described in [10]: it consists in training a classifier with a pixel coordinate transformation; specifically, Cartesian coordinates are transformed into polar coordinates, i.e. (ρ, ϑ).

Contour pixels are sampled, and only 50 pixels are taken. The coordinates of each pixel are transformed into polar coordinates, where ρ is calculated from the mass center of the picture.

A multilayer perceptron (MLP) was trained with polar coordinates of 50 pixels, leading to a features vector consisting of 100 components.

6   Experimental Results

In this section the results obtained from both the classifiers are presented. The SVM classifier uses a linear kernel, the maximum number of iterations is set to 100 and the error threshold is set to $\varepsilon = 0,000001$.

The MLP topology has a single hidden layer. The learning rate is set to 0.3 and the number of epochs is set to 500.

The test set consists of 20 pictures per sign. The following confusion matrices show the results for each sign obtained from both classifiers.

Table 1 reports the confusion matrix relative to class 1. The entry "*Neg*" specifies the negative samples, which don't belong to this class (Tables 2, 3 and 4).

Table 1. Confusion matrix for class 1

	classifier	B	R	U	V	W	Z	Neg
B	SVM	80 %	10 %	0 %	5 %	0 %	5 %	0 %
	MLP	85 %	15 %	0 %	0 %	0 %	0 %	0 %
R	SVM	5 %	80 %	5 %	0 %	0 %	10 %	0 %
	MLP	0 %	80 %	15 %	5 %	0 %	0 %	0 %
U	SVM	0 %	0 %	95 %	0 %	0 %	0 %	5 %
	MLP	5 %	5 %	75 %	10 %	0 %	5 %	0 %
V	SVM	0 %	0 %	0 %	95 %	5 %	0 %	0 %
	MLP	0 %	0 %	5 %	85 %	0 %	0 %	10 %
W	SVM	0 %	5 %	0 %	0 %	95 %	0 %	0 %
	MLP	0 %	0 %	0 %	10 %	80 %	0 %	10 %
Z	SVM	0 %	5 %	0 %	0 %	0 %	95 %	0 %
	MLP	0 %	0 %	5 %	0 %	5 %	90 %	0 %
Neg	SVM	0 %	0 %	0 %	0 %	0 %	0 %	100 %
	MLP	0 %	0.8 %	0 %	0 %	1.7 %	0 %	97.5 %

For the first class the accuracy of the SVM classifier is ca. 95 %, while the accuracy of the MLP classifier is ca. 89 %. For the second class the accuracy of the SVM classifier is ca. 95 %, while the accuracy of the MLP classifier is 87 %. For the third class the accuracy of the SVM classifier is ca. 95 %, and the accuracy of the MLP

Table 2. Confusion matrix for class 2

	classifier	A	E	M	N	O	P	Q	T	X	Neg
A	SVM	95 %	0 %	0 %	0 %	0 %	0 %	0 %	0 %	0 %	5 %
	MLP	90 %	0 %	0 %	5 %	0 %	0 %	0 %	0 %	0 %	0 %
E	SVM	0 %	100 %	0 %	0 %	0 %	0 %	0 %	0 %	0 %	0 %
	MLP	0 %	90 %	0 %	0 %	5 %	0 %	0 %	5 %	0 %	0 %
M	SVM	0 %	0 %	100 %	0 %	0 %	0 %	0 %	0 %	0 %	0 %
	MLP	0 %	0 %	70 %	5 %	0 %	15 %	10 %	0 %	0 %	0 %
N	SVM	0 %	0 %	10 %	90 %	0 %	0 %	0 %	0 %	0 %	0 %
	MLP	0 %	5 %	10 %	85 %	0 %	0 %	0 %	0 %	0 %	0 %
O	SVM	0 %	0 %	0 %	0 %	100 %	0 %	0 %	0 %	0 %	0 %
	MLP	0 %	5 %	0 %	0 %	70 %	0 %	0 %	5 %	0 %	20 %
P	SVM	0 %	0 %	0 %	0 %	0 %	100 %	0 %	0 %	0 %	0 %
	MLP	5 %	0 %	0 %	5 %	0 %	70 %	5 %	0 %	0 %	15 %
Q	SVM	0 %	0 %	0 %	0 %	0 %	0 %	100 %	0 %	0 %	0 %
	MLP	0 %	0 %	5 %	0 %	0 %	15 %	75 %	0 %	0 %	5 %
T	SVM	0 %	0 %	0 %	0 %	0 %	0 %	0 %	75 %	0 %	25 %
	MLP	0 %	0 %	0 %	0 %	15 %	0 %	0 %	85 %	0 %	0 %
X	SVM	0 %	0 %	0 %	0 %	0 %	0 %	0 %	0 %	90 %	10 %
	MLP	0 %	0 %	0 %	0 %	0 %	0 %	0 %	0 %	90 %	10 %
Neg	SVM	0 %	0 %	0 %	0 %	0.5 %	0 %	0 %	0 %	0 %	99.5 %
	MLP	0.5 %	0 %	0.5 %	0 %	1.7 %	0 %	0 %	1.7 %	1.1 %	94.5 %

Table 3. Confusion matrix for class 3

	classifier	C	D	F	I	L	Y	Neg
C	SVM	90 %	0 %	5 %	0 %	5 %	0 %	0 %
	MLP	80 %	0 %	0 %	0 %	5 %	0 %	15 %
D	SVM	0 %	100 %	0 %	0 %	0 %	0 %	0 %
	MLP	0 %	95 %	5 %	0 %	0 %	0 %	0 %
F	SVM	0 %	20 %	80 %	0 %	0 %	0 %	0 %
	MLP	0 %	0 %	85 %	0 %	0 %	0 %	15 %
I	SVM	0 %	0 %	0 %	100 %	0 %	0 %	0 %
	MLP	0 %	0 %	0 %	95 %	0 %	0 %	5 %
L	SVM	0 %	5 %	10 %	0 %	85 %	0 %	0 %
	MLP	0 %	0 %	0 %	0 %	100 %	0 %	0 %
Y	SVM	0 %	0 %	0 %	0 %	0 %	100 %	0 %
	MLP	0 %	0 %	0 %	15 %	0 %	85 %	0 %
Neg	SVM	0 %	0 %	0 %	0 %	0 %	0 %	100 %
	MLP	0.7 %	0 %	0.7 %	0.7 %	0.7 %	0 %	97.2 %

Table 4. Confusion matrix for class 4

	classifier	G	H	J	K	S	Neg
G	SVM	95 %	5 %	0 %	0 %	0 %	0 %
	MLP	100 %	0 %	0 %	0 %	0 %	0 %
H	SVM	0 %	100 %	0 %	0 %	0 %	0 %
	MLP	10 %	70 %	0 %	0 %	10 %	10 %
J	SVM	0 %	0 %	100 %	0 %	0 %	0 %
	MLP	0 %	0 %	95 %	5 %	0 %	0 %
K	SVM	0 %	0 %	0 %	100 %	0 %	0 %
	MLP	0 %	5 %	0 %	85 %	0 %	10 %
S	SVM	5 %	0 %	0 %	0 %	95 %	0 %
	MLP	0 %	0 %	0 %	0 %	85 %	15 %
Neg	SVM	0 %	0 %	0 %	0 %	0 %	100 %
	MLP	2.1 %	1 %	1 %	0 %	1 %	94.9 %

classifier is ca. 93 %. For the fourth class, the accuracy of the SVM classifier accuracy is ca. 97 %, while the accuracy of the MLP classifier is ca. 90 %.

The signs *B, C, F, Q, R, S,* and *T* show unreliable results, indeed their accuracy is less than 80 %. Figure 10 shows a screenshot of the software implemented in the experimental study; specifically, the *L* sign is being recognized by our system.

Fig. 10. Test case

7 Conclusion and Future Works

We proposed two methods, the first one trains an SVM classifier with the hand contours, and the second one uses the polar coordinates to train a NN. Even if both the approaches offer good results, which are much higher than 50 % (random classifier), we can conclude that the first approach performs better.

This work is limited to the recognition of the Italian finger spelling only. The segmentation based on the skin color is the hardest problem faced during this study, because it is required that the system is able to work in uncontrolled environments (users within any scene) but can be improved by using in future works the algorithm developed from authors in [12]. SVM and MLP performance could be improved by using several techniques shown in [13] for SVM performance evaluation, or in [14] for MLP pre-processing or in [15] for MLP topology optimization.

A future work could see the system equipped with a most suitable sensor able to recognize hands and fingers, as the Leap Motion, which is a technology equipped with infrared sensor. With the combination of Leap Motion and a RGB sensor, the system could be extended to identify more signs, without having to renounce to portability.

References

1. Alvarez, S., Llorca, D.F.: Spatial Hand Segmentation Using Skin Colour and Background Subtraction, Robesafe Research Group, Department of Automatics, Universidad de Alcala, Madrid, Spain
2. Comparative Study of Statistical Skin Detection Algorithms for Sub-Continental Human Images, Institute of Information Technology, Department of Statistics, Biostatistics and Informatics, University of Dhaka, Bangladesh
3. KaewTraKulPong, P., Bowden, R.: An Improved Adaptive Background Mixture Model for Real-Time Tracking with Shadow Detection, Vision and Virtual Reality group, Department of Systems Engineering, Brunel University
4. Piccardi, M.: Background subtraction techniques: a review, Computer Vision Group, Faculty of Information Technology, University of Technology, Sydney (UTS), Australia
5. Phung, S.L., Bouzerdoum, A., Chai, D.: Skin segmentation using color pixel classification: analysis and comparison. IEEE Trans. Pattern Anal. Mach. Intell. 27, 148–154 (2005)
6. Gebre, B.G., Wittenburg, P., Heskes, T.: Automatic Sign Language Identification, Max Planck Institute for Psycholinguistics, Nijmegen, Radboud University, Nijmegen
7. Sign Language in the Intelligent Sensory, Budapest University of Technology and Economics, Department of Automation and Applied Informatics, Budapest, Hungary, Department of Mechatronics, Optics and Instrumentation Technology
8. Sonka, M., Hlavac, V., Boyle, R.: Image Processing, Analysis, and Machine Vision. Thomson Learning, Toronto (2008)
9. Otiniano-Rodriguez, K.C., Càmara-Chavez, G., Menotti, D.: Hu and Zernike Moments for Sign Language Recognition, Computing Department, Federal University of Ouro Preto, Brazil
10. Yang, M., Kpalma, K., Ronsin, J.: A Survey of Shape Feature Extraction Techniques, IETR-INSA, UMR-CNRS 6164, 35043 Rennes, Shandong University, 250100, Jinan, France, China
11. Stokoe, W.C.: Sign language structure: an outline of the visual communication systems of the American deaf
12. Bevilacqua, V., Filograno, G., Mastronardi, G.: Face detection by means of skin detection. In: Huang, D.-S., Wunsch II, D.C., Levine, D.S., Jo, K.-H. (eds.) ICIC 2008. LNCS (LNAI), vol. 5227, pp. 1210–1220. Springer, Heidelberg (2008)

13. Bevilacqua, V., Pannarale, P., Abbrescia, M., Cava, C., Paradiso, A., Tommasi, S.: Comparison of data-merging methods with SVM attribute selection and classification in breast cancer gene expression. BMC Bioinform. **13**(Suppl. 7), S9 (2012)
14. Bevilacqua, V.: Three-dimensional virtual colonoscopy for automatic polyps detection by artificial neural network approach: new tests on an enlarged cohort of polyps. Neurocomputing (2013). doi:10.1016/j.neucom.2012.03.026. ISSN: 0925-2312 (2012)
15. Bevilacqua, V., Mastronardi, G., Menolascina, F., Pannarale, P., Pedone, A.: A novel multi-objective genetic algorithm approach to artificial neural network topology optimisation: the breast cancer classification problem. In: International Joint Conference on Neural Networks, IJCNN 2006, pp. 1958–1965 (2006)

A Spiking Neural Network for Extraction of Multi-features in Visual Processing Pathways

QiYan Sun[1,2], QingXiang Wu[1(✉)], Xuan Wang[1], and Lei Hou[1]

[1] Key Laboratory of OptoElecronic Science and Technology for Medicine of Ministry of Education, College of Photonic and Electronic Engineering, Fujian Normal University, Fuzhou 350007, Fujian, China
qxwu@fjnu.edu.cn
[2] Fujian Agriculture and Forestry University, Fuzhou 350007, Fujian, China
sunqiyan99168@163.com

Abstract. Based on spiking neural network and colour visual processing mechanism, a hierarchical network is proposed to extract multi-features from a colour image. The network is constructed with a conductance-based integrate-and-fire neuron model and a set of receptive fields. Inspired by visual system, an image can be decomposed into multiple visual image channels and processed in hierarchical structures. The firing rate map of each channel is computed and recorded. Finally, multi-features are obtained from firing rate map. Simulation results show that the proposed method is successfully applied to recognize the target with a higher recognition rate compared with some other methods.

Keywords: Spiking neural network · Multi-features extraction · Visual processing pathways · Integrate-and-fire neuron model

1   Introduction

Colour vision is an important ability of human to perceive and discriminate objects. Quantitative psychophysical and physiological scientists have studied the colour vision mechanism. The hierarchical pathways of colour information processing in visual system are widely reported. They are from retina to LGN (lateral geniculate nucleus), then enter into primary cortex V1 [1]. The landmark studies of De Valois suggested that colour single-opponent neurons existed in the LGN. Many perceptual results indicate that colour perception is dependent on colour contrast at the boundary of the region other than it is on the spectral reflectance of the region [2, 3]. Single-opponent neurons confirm the colour property of sensitivity to context. Furthermore, Livingstone and Hubel reported that double-opponent neurons existed in primary cortex V1 [4]. The double-opponent neurons strongly respond to colour bar, are invariant to illumination and have property of orientation-selective [5]. Based on the above biological principle, many simulation methods are proposed to extract colour features. For example, Yue Zhang established colour histogram feature extraction model based on the hierarchy description of biological vision [6]. Silvio Borer and Sabine Susstrunk introduced an opponent colour space motivated by retinal processing [7]. However, the hierarchical model and colour/spatial

© Springer International Publishing Switzerland 2015
D.-S. Huang and K. Han (Eds.): ICIC 2015, Part III, LNAI 9227, pp. 275–281, 2015.
DOI: 10.1007/978-3-319-22053-6_29

opponency proposed by most of them are conceptual tools rather than computational means, and cannot be applied directly to pattern recognition. How to transform and convert signals among neurons in the visual processing pathways still needs to be discussed. So we combine hierarchical and logical model with spiking neurons.

Spiking neurons are regarded as a computing unit in neural networks. The first biological model of a spiking neuron proposed by Hodgkin and Huxley, is based on experimental recordings from the giant squid axon using a voltage clamp method [8]. The complexity in simulating the model is very high due to the number of differential equations and the large number of parameters. Thus most computer simulations choose to use a simplified neuron model such as integrate-and-fire model [9, 10]. Based on integrate-and-fire model, QingXiang Wu proposed a spiking neural network to extract colour features in different ON/OFF pathways which inspired by the roles of rods and cones in retina [11]. Simei Gomes Wysoski also described and evaluated a spiking neural network based on integrate-and-fire model [12]. But it did not simulate the information processing of rods and cones in retina.

In this paper we simulate the hierarchical visual processing pathways and explore the application of a spiking neuron model and spiking neural network to extract multi-features for pattern recognition. The extracted multi-features contain more information. The simulation results proved that the proposed network can be used complex pattern recognition efficiently.

The remainder of this paper is organized as follows. In Sect. 2, the hierarchical architecture of the neural network is proposed to extract multi-features. The spiking neural network model is based on simplified conductance-based integrate-and-fire neurons. The behaviours of the neural network are governed by a set of equations discussed in Sect. 3. Experimental results and discussions are presented in Sect. 4. Section 5 gives a conclusion and a topic for further study.

2   Hierarchical Architecture of Spiking Neural Network

As mentioned introductions, integrate-and-fire neurons are used as the implementation units in visual processing pathways. Inspired by visual processing mechanism and logical behavior of opponent neurons, a hierarchical architecture of spiking neural network based on integrate-and-fire neuron model is proposed to extract multi-features. For simplicity, only the architecture of the network for red/green opponent pathway is shown in Fig. 1. The blue/yellow opponent pathway has similar architecture.

The spiking neural network as shown in Fig. 1 comprises three layers. Receptors layer shows the mechanism of three type's cones perceiving red, green, blue colours of the input image and transforms a pixel into external current which is fed to neurons at layer1. RF_1 is simulated receptive field for single opponent neurons of Layer1 and can be interpreted as two dimensional difference of Gaussians functions. Each neuron at layer1 has excitatory and inhibitory synapses to transmit spike trains and is connected to a receptive field RF_1 through both of excitatory strength matrix w_{ex} and inhibitory strength matrix w_{ih}. If the membrane potential of neuron reaches the threshold, the neuron will fire. Otherwise, the neuron will be suppressed. Neurons at output layer are used to simulate double opponent neurons of cortex V1 with orientation selectivity. Receptive field RF_G is designed for double opponent neurons of output layer, which can be implemented using Gabor functions. Similarly, each neuron at output layer is connected to a receptive field RF_G through both of excitatory strength matrix w_{Gex} and inhibitory strength matrix w_{Gih}. The details of the neuron model and receptive fields are presented in following sections.

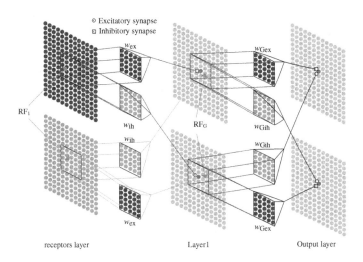

Fig. 1. The architecture of the network for red-green opponent (Color figure online)

3   Spiking Neuron Model and Implementation

A simplified conductance-based integrate-and-fire model is applied in the computation of the network. Let $r_{x,y}(t)$, $g_{x,y}(t)$ represent the normalized strength at a pixel (x,y) of an image. Receptors transfer the image pixel brightness to a synapse current $I_{r(x,y)}$, $I_{g(x,y)}$ by following equations. Where α and β are constants for transformation.

$$I_{r(x,y)} = \alpha r_{x,y}(t), \ I_{g(x,y)} = \beta g_{x,y}(t), \tag{1}$$

In the integrate-and-fire model, the membrane potential $v(t)$ is calculated as:

$$c_m \frac{dv_{r/g}(t)}{dt} = g_l(E_l - v_{r/g}(t)) + \sum_{(x',y')\in RF_1} w_{ex}(x',y')I_{r(x',y')}(t) \\ + w_{ih}(x',y')I_{g(x',y')}(t), \tag{2}$$

g_l is the membrane conductance, E_l is the reverse potential, $v_{r/g}$ is the membrane potential of $r/g(x,y)$ in layer 1, $r/g(x,y)$ represents a red/green single opponent neuron. c_m presents the capacitance of the membrane, w_{ex} and w_{ih} are governed by the following expression.

$$w_{ex}(x',y') = w_0 e^{-\frac{(x_c-x')^2+(y_c-y')^2}{2\delta^2}}, w_{ih}(x',y') = w_1 e^{-\frac{(x_c-x')^2+(y_c-y')^2}{2n\delta^2}}, \tag{3}$$

Where (x_c, y_c) is the center of receptive field RF_1. w_0 and w_1 are used to determine the maximal value of the weight distribution, δ and n are constants.

If the membrane potential reaches a threshold ν_{th}, then the neuron generates a spike. $S_{r/g(x,y)}(t)$ represent the spike train generated by the neuron such as that:

$$S_{r/g(x,y)}(t) = \begin{cases} 1 & \text{if neuron } r/g(x,y) \text{ fires at time } t. \\ 0 & \text{if neuron } r/g(x,y) \text{ does not fire at time } t. \end{cases} \tag{4}$$

The firing rate $F_{r/g(x,y)}(t)$ is calculated by following expression, where T is a time interval for counting the spikes.

$$F_{r/g(x,y)}(t) = \frac{1}{T} \sum_{t}^{t+T} S_{r/g(x,y)}(t), \tag{5}$$

The neuron in output layer is governed by:

$$c_m \frac{dv_{rg}(t)}{dt} = g_l(E_l - v_{rg}(t)) + \sum_{(x',y') \in RF_G} w_{Gex}(x',y')F_{r/g(x',y')}(t)$$
$$+ w_{Gih}(x',y')F_{g/r(x',y')}(t), \tag{6}$$

ν_{rg} is the membrane potential of the $rg(x,y)$ in output layer, $rg(x,y)$ represents a red/green double opponent neuron. The distribution of the weights w_{Gex} and w_{Gih} are simulated by follows, where (x_{cl}, y_{cl}) is the center of receptive field RF_G. λ is wavelength, θ represents orientation selection, γ and σ are constants.

$$w_{Gex}(x',y') = e^{-\frac{((x'-x_{c1})\cos\theta+(y'-y_{c1})\sin\theta)^2 + \gamma^2(-(x'-x_{c1})\sin\theta+(y'-y_{c1})\cos\theta)^2}{2\sigma^2}}$$
$$\times \cos(2\pi \frac{(x'-x_{c1})\cos\theta + (y'-y_{c1})\sin\theta}{\lambda}), \tag{7}$$

$$w_{Gih}(x',y') = -w_{Gex}(x',y') \tag{8}$$

The spike train $S_{rg(x,y)}(t)$ generated by

$$S_{rg(x,y)}(t) = \begin{cases} 1 & \text{if neuron } rg(x,y) \text{ fires at time } t. \\ 0 & \text{if neuron } rg(x,y) \text{ does not fire at time } t. \end{cases} \tag{9}$$

The firing rate $F_{rg(x,y)}(t)$ of neuron in output layer is

$$F_{rg(x,y)}(t) = \frac{1}{T} \sum_{t}^{t+T} S_{rg(x,y)}(t), \tag{10}$$

By analogy, we have the firing rate $F_{gr(x,y)}(t)$, $F_{by(x,y)}(t)$, $F_{yb(x,y)}(t)$. The spiking neural network is simulated in Matlab. Corresponding to biological neurons [13], the following parameters for the network are used in the experiments. The dimension of receptive field RF_1 and RF_G are set to 5×5. $v_{th} = -60$ mv. $E_l = -70$ mv. $g_l = 1.0$ μs/mm^2. $c_m = 8$ nF/mm^2. $T = 1000$ ms. $\alpha = 2$. $\beta = 1$. $\delta = 1$. $n = 25$. $\gamma = 1/3$. $\lambda = 5$. $\theta = $ pi/4, $\sigma = 3$. These parameters can be adjusted to get a good quality output image.

We compute mean and variance colour moments for each fire map in output layer,

$$\mu_i = \frac{1}{N} \sum_{j=1}^{N} p_{i,j} , \ \sigma_i = \left(\frac{1}{N} \sum_{j=1}^{N} (p_{i,j} - \mu_i)^2\right)^{1/2} ,$$ (11)

$$F_{SNN}(t) = [\mu_1, \mu_2, \mu_3, \mu_4, \sigma_1, \sigma_2, \sigma_3, \sigma_4]$$ (12)

4 Experiments and Discussion

In this section, we illustrate the ability of multi-features extracted from the proposed network to test the actual problems in pattern recognition. Flower category dataset comprises 10 flowers categories with similar colour and shape [14]. Each species images also include differences in pose and incomplete flowers owing to shade. Figure 2 shows example of the flower category dataset.

Fig. 2. Typical objects of flower category dataset

Totalizing 804 images are used for experimental purpose. It was divided into two training set and testing set, where 50 % for each group. Experiment is achieved by 5 fold cross-validation of libSVM. We also compare with some other methods, which are from RGB, Retina [6], SNN1 [11], SNN+ colour-opponent respectively, as they both extracted colour features. The RGB method computed colour moments of red, green, blue channel for colour images. The method of Retina is applied in Yue Zhang's paper. It is a logical model based on visual processing mechanism. According to this model, we compute colour moments of output images for each channel. The method SNN1 is introduced by Qing Xiang Wu's paper. A colour image is separated by several

ON/OFF pathways, such as R_ON, R_OFF, B_ON, B_OFF, G_ON, G_OFF. We calculate colour moment of each pathway. The features extracted by the method of SNN+colour-opponent are based on the proposed neural network of this paper, but they only show characteristic of colour contrast sensitive without orientation-selective. SNN+colour-opponent+orientation-selective is the algorithm of multi-features extraction. Obviously, the method of SNN+colour-opponent gets the highest recognition rate based on our networks, SNN+colour-opponent+orientation-selective has a slightly lower recognition rate because some image details are removed after orientation-selective. However, SNN+colour-opponent+orientation-selective will give better performance in terms of illuminant constancy. On the other hand, it shows the powerful function of spiking neural work (Table 1).

Table 1. Comparison of recognition accuracy for flower category dataset

Methods	Recognition rate (%)
Retina	62.19
RGB	78.54
SNN1	89.24
SNN+colour-opponent	93.54
SNN+colour-opponent+orientation-selective	93.28

5 Conclusion

The hierarchical architecture of spiking neural network provides powerful functionalities to perform very complicated computation tasks and intelligent behaviours. This paper has proposed a neural network model that can perform multi-features extraction from a colour image. An integrate-and-fire neuron model and biological visual mechanism are used to construct the network. Simulations show that the proposed network outperforms the alternative methods in recognizing objects. The reason for excellent performance is that multi-feature fusion of colour and spatial information. It also provides a better understanding of how human perceived colours in the visual pathways. In this paper, only the behaviours of retina, LGN and cortex V1 have been discussed. Actually, there are higher levels for colour processing in visual system, such as V2, PIT, IT. This is a topic for further study.

Acknowledgments. The authors gratefully acknowledge the fund from the Natural Science Foundation of China (Grant No. 61179011) and Science and Technology Major Projects for Industry-academic Cooperation of Universities in Fujian Province (Grant No. 2013H6008).

References

1. Chatterjee, S., Callaway, E.M.: Parallel colour-opponent pathways to primary visual cortex. Nature **426**, 668–671 (2003)
2. De Valois, R.L.: Analysis and coding of color vision in the primate visual system. Cold Spring Harb. Symp. Quant. Biol. **38**, 567–580 (1965)
3. Shapley, R., Hawken, M.J.: Color in the Cortex: single-and double-opponent cells. Vis. Res. **51**, 701–717 (2011)
4. Livingstone, M.S., Hubel, D.H.: Anatomy and physiology of a color system in the primate visual cortex. J. Neurosci. **4**, 309–356 (1984)
5. Ratnasingam, S., Robles-Kelly, A.: A spiking neural network for illuminant-invariant colour discrimiantion. In: The 2013 International Joint Conference on Neural Networks (IJCNN), pp. 1–8. IEEE Press, Dallas (2013)
6. Zhang, Y., Huo, H., Fang, T.: Color histogram feature extraction based on Biological visual. Chin. High Technol. Lett. **24**(4), 407–413 (2014)
7. Borer, S., Susstrunk, S.: Opponent color space motivated by retinal processing. In: IS&T First European Conference on Color in Graphics, Imaging and Vision (CGIV), vol. 1, pp. 187–189 (2002)
8. Hodgkin, A.L., Huxley, A.F.: A quantitative description of membrane current and its application to conduction and excitation in nerve. J. Physiol. **117**(4), 500–544 (1952)
9. Wu, Q.X., McGinnity, T.M., Maguire, L.P., Belatreche, A., Glackin, B.: Processing visual stimuli using hierarchical spiking neural networks. Int. J. Neurocomput. **71**(10–12), 2055–2068 (2008)
10. Wu, Q.X., McGinnity, T.M., Maguire, L.P., Cai, R.T., Chen, M.G.: A visual attention model based on hierarchical spiking neural networks. Int. J. Neurocomput. **116**, 3–12 (2013)
11. Wu, Q.X., McGinnity, T.M., Maguire, L., Valderrama-Gonzalez, G.D., Dempster, P.: Colour image segmentation based on a spiking neural network model inspired by the visual system. In: Huang, D.-S., Zhao, Z., Bevilacqua, V., Figueroa, J.C. (eds.) ICIC 2010. LNCS, vol. 6215, pp. 49–57. Springer, Heidelberg (2010)
12. Wysoski, S.G., Benuskova, L., Kasabov, N.: Fast and adaptive network of spiking neurons for multi-view visual pattern recognition. Int. J. Neurocomput. **71**, 2563–2575 (2008)
13. Gerstner, W., Kistler, W.: Spiking Neuron Models: Single Neurons, Populations, Plasticity. Cambridge University Press, Cambridge (2002)
14. 102 Category Flower Dataset. http://www.robots.ox.ac.uk/~vgg/data/flowers/102/index.html

Image Set Classification Based on Synthetic Examples and Reverse Training

Qingjun Liang[1], Lin Zhang[1(✉)], Hongyu Li[1], and Jianwei Lu[1,2]

[1] School of Software Engineering, Tongji University, Shanghai, China
{13_qingjunliang, cslinzhang, hyli,
jwlu33}@tongji.edu.cn
[2] The Advanced Institute of Translational Medicine,
Tongji University, Shanghai, China

Abstract. This paper explores a synthetic method to create the unseen face features in the database, thus achieving better performance of image set based face recognition. Image set based classification highly depend on the consistency and coverage of the poses and view point variations of a subject in gallery and probe sets. By considering the high symmetry of human faces, multiple synthetic instances are virtually generated to make up the missing parts, so as to enrich the variety of the database. With respect to the classification framework, we resort to reverse training due to its high efficiency and accuracy. Experiments are performed on benchmark datasets containing facial image sequences. Comparisons with state-of-the-art methods have corroborated the superiority of our Synthetic Examples based Reverse Training (SERT) approach.

Keywords: Face recognition · Image set classification

1 Introduction

Face recognition conducted on multiple images can be formulated as an image set classification problem. The existing image set classification methods can be divided into two categories, parametric model based methods and nonparametric model based methods [1]. Parametric methods utilize a statistical distribution to represent an image set and measure the similarity between two sets by KL-divergence. The main drawback of such methods is that they need to tune the parameters of a distribution function and rely on strong statistical correlation between training and test image sets [2, 3]. Unlike parametric methods seeking for global characteristics of the sets, non-parametric methods put more emphasis on local samples matching. They attempt to find the overlap views between two sets and measure the similarity upon those parts of data. Nearest Neighbor (NN) matching is used to find the common parts. They model the whole image set as local exemplars [4], an affine hull or a convex hull [5], a regularized affine hull [6], or use the sparsity constraint as a means to find the nearest pair of points between two image sets [7]. Then the similarity of two sets can be reflected by the Euclidean distance between their closest points. Since the NN based methods use only a small part of the

© Springer International Publishing Switzerland 2015
D.-S. Huang and K. Han (Eds.): ICIC 2015, Part III, LNAI 9227, pp. 282–288, 2015.
DOI: 10.1007/978-3-319-22053-6_30

data information, they are more vulnerable to outliers. Later, people find that in some cases the structure of the whole image set might be a nonlinear complex manifold and a linear subspace is not sufficient for representation. Thus, researchers start to model an image set as a point on a certain manifold, e.g., a Grassmannian manifold [8, 9] or a Riemannian manifold [10]. The corresponding distance metrics can be geodesic distance [11], projection kernel metric [12] and Log-Euclidean distance (LED) [13].

In [14], Hayat et al. tried to keep each example independent and to remain the image set in its original form rather than seeking a whole representation. They argued that whatever form you use, once you model a set as a single entity, there must be loss of information. Besides, they adopted the reverse training strategy.

The abovementioned methods mainly focus on devising an efficient classifier. They tacitly make an assumption that the distribution of a person's poses and view points in a probe image set are similar to those in the gallery image set. However, it is sometimes the case that there is pose or view point mismatch between the gallery and probe image sets of the same subject. In such case, the probe image set is more easily classified as the class whose gallery set contains the same head pose as the probe set but is indeed from a different subject.

In this paper, to solve such a problem, we propose a simple yet effective approach by synthesizing more samples for each image set. In this way, the variety of poses and viewpoints within an image set can be apparently enriched. In terms of the classification framework, we resort to Reverse Training [14]. The proposed method is named as *Synthetic Examples based Reverse Training*, SERT for short.

2 Image Set Feature Extraction

We propose a face sample synthesizing method in which the symmetry property of the human face is fully exploited. This approach is inspired by [14]. In [14], Hayat et al. pointed out that based on the manual inspection of the most challenging YouTube Celebrities dataset, a great amount of misclassified query image sets have a common characteristic that their head poses are not covered in the training sets. To address this issue, here we present our solution.

2.1 Horizontal Symmetry Synthetic Examples and LBP

We create synthetic examples to enrich the set variations, by operating directly in "data space". For each example in an image set, we flip the image horizontally and get another symmetry version of the original face. To determine the necessity of this flipping step, we use the Euclidean distance metric to measure the similarity between the original face and the flipped one. A threshold is empirically set. If the distance is less than the threshold, we neglect the flipped face since the original face itself has a good symmetry. Otherwise, we add the new flipped face column to the image set and therefore augment the number of instances in all the sets.

Next, we use Local Binary Patterns (LBP) [15] for face feature extraction. It has three classical mapping table: (1) uniform LBP ('u2'), (2) rotation-invariant LBP ('ri'), and (3) uniform rotation-invariant LBP ('riu2'). Here we adopt the uniform LBP ('u2'), whose binary pattern contains at most two bitwise transitions from 0 to 1 (or 1 to 0). There are totally 2 conditions for 0 transition and 56 conditions for 2 transitions (1 transition is impossible) in the case of (8, R) neighborhood. All the non-uniform LBP that contains more than two transitions are labeled as the 59th bin. The details of feature extraction are listed in Table 1.

Table 1. Feature extraction based on grid division $LBP_{8,1}^{u2}$

Input:	A face image
1.	Divide the face image into $k \times k$ non-overlapping uniformly spaced grid cells
2.	For each pixel in one cell, sample its 8 neighbors with radius 1 and map its pattern into one of the 59 conditions
3.	Build the histogram over each cell, which counts the frequency of each number (1–59)
4.	Normalize the histograms and concatenate them one after another (either column-wise or row-wise)
Output:	A feature vector whose dimension is 59 k^2

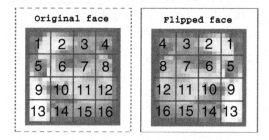

Fig. 1. A synthetic feature and its original feature.

Since we use grid division $LBP_{8,1}^{u2}$ which is not rotation invariant, the flipped image must have a different LBP value from its original one. An intuitive illustration can be seen in Fig. 1. Imagine the case that a training set only composes of left profile faces, while its corresponding upcoming test set only consists of right profile faces. It is obviously that the original gallery and probe set is hard to match to each other. However, after we create synthetic examples, both gallery and probe set contains left and right profile features. It is much easier for the later classification.

2.2 SMOTE and PCA Whitening

The number of instances varies a lot from set to set. Such an uneven distribution will lead to the bias in the classification stage especially for those methods who do not represent the image set as a whole entity. To solve this problem, here we use the

Synthetic Minority Over-sampling Technique (SMOTE) proposed by Chawla et al. [16]. For the image sets whose sizes are smaller than 100, we generate synthetic examples by taking each minority instance and introducing synthetic ones along the line segments between itself and its k nearest neighbors in the same set.

At the training phase, the LBP features are redundant since adjacent pixel intensities are highly correlated. Therefore, we use PCA whitening to make our input features uncorrelated with each other and have unit variance.

3 SERT: Synthetic Examples Based Reverse Training

3.1 Problem Formulation

Denote $X = \{x_1, x_2, ..., x_n\}$ as an image set containing n face examples from a person, where x_i is a feature vector of the ith single image, and is in the form of LBP. A subject can have multiple image sets. Given k training image sets $X_1, X_2, ..., X_k$ that belong to c classes ($k >= c$) and their corresponding labels $y = \{1, 2, ..., k\}$, when there comes in a query image set X_q, our task is to find out which class it belongs to.

3.2 Reverse Training and the Proposed Framework

After the preparation for features, we use the Reverse Training algorithm proposed in [14] to do the classification work.

Suppose a coming query set X_q has 200 images. The 20 training sets that belong to 20 classes (multiple sets per subject are combined as a whole) are marked as $D = \{X_1, X_2, ..., X_{20}\}$. 10 images per set in D are randomly selected to form a set D_1 containing 200 images and the rest of images in D form the set D_2. As the name "Reverse training" suggests, we treat the 200 images in X_q as training data while the images in D_2 as test data. Specifically, 200 features in X_q are labeled as +1 and 200 features in D_1 are labeled as −1. A binary classifier *Liblinear* [17] is trained on these 400 instances and D_2 is tested on the linear decision boundary. Those who are classified as +1 (same side as X_q) are denoted as D_2^+. A normalized histogram h is computed on the $y_{D_2^+}$ (labels of D_2^+) over the 20 class bins. Intuitively, h_i ($i = 1, 2, ... 20$) indicates the percentage of the number of label i in $y_{D_2^+}$ over the number of label i in y_{D_2}.

$$h_i = \sum\nolimits_{y \in y_{D_2^+}} f_i(y) / \sum\nolimits_{y \in y_{D_2}} f_i(y), \quad \text{where } f_i(y) = \begin{cases} 1, y = i \\ 0, y \neq i \end{cases} \tag{1}$$

Finally, the label of the query set X_q is assigned according to h_i that has the largest occurrence,

$$y_q = \arg\max_i h_i \tag{2}$$

The flowchart of the proposed SERT approach is presented in Fig. 2.

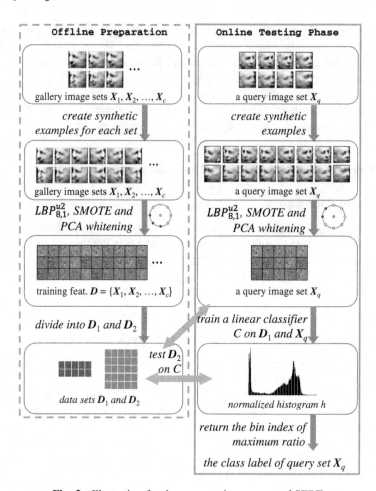

Fig. 2. Illustration for the computation process of SERT.

4 Experimental Results

4.1 Datasets and Settings

Honda/UCSD Dataset. The Honda/UCSD dataset [18] contains 59 video sequences involving 20 different persons. The face in each frame is first automatically extracted using Viola and Jones face detection algorithm [19] and then resized to the size of 20×20. For our experiment, one video is considered as an image set. Specifically, each person has one image set as the gallery and the remaining sets as the probes. We repeat our experiment for 10 times with randomly selected training and testing combinations.

CMU Mobo Dataset. The CMU Mobo dataset [20] consists of 96 video sequences of 24 different subjects. The number of frames for each video is about 300. Similar to the Honda, the faces are detected using [19] and resized to 40×40. As a convention, we consider one video as an image set and select one set per person for training and the rest sets for testing (24 sets for training and 72 for testing). $k = 5$ for LBP grid division.

YouTube Celebrities Dataset. The YouTube Celebrities [21] has 1910 video clips of 47 celebrities. We utilized the method in [22] to track the face region across the entire video, in which the face bounding boxes in initial frame is manually marked and provided along with the dataset. The cropped face region is then resized to 30 × 30. Specifically, we divided the whole dataset into five equal folds with minimal overlapping. From the aspect of fold division, for subjects who have more than 45 videos, we randomly select 45 from them. As for subjects who don't have 45 videos, some videos are selected more than once. Then we divide 45 videos per person into 5 fold.

4.2 Comparisons with Existing Methods

We compare our proposed framework with several recently proposed state-of-the-art methods which include DCC [1], MMD [4], MDA [8], AHISD [5], CHISD [5], SANP [7], CDL [10] and RT [14]. Table 2 tabulates the recognition results for our approach and all the other methods listed above on the three datasets.

Table 2. Average recognition rates (%) with standard deviation of different methods on the three benchmark datasets

Method	Honda/UCSD	CMU Mobo	YouTube
DCC [1]	92.6 ± 2.3	88.9 ± 2.5	64.8 ± 2.1
MMD [4]	92.1 ± 2.3	92.5 ± 2.9	62.9 ± 1.8
MDA [8]	94.4 ± 3.4	90.3 ± 2.6	66.5 ± 1.1
AHISD [5]	91.3 ± 1.8	88.5 ± 3.3	64.4 ± 2.4
CHISD [5]	93.6 ± 1.6	95.7 ± 1.0	63.4 ± 2.9
SANP [7]	95.1 ± 3.1	95.6 ± 0.9	65.6 ± 2.4
CDL [10]	98.9 ± 1.3	88.7 ± 2.2	68.5 ± 3.3
RT [14]	100 ± 0.0	97.3 ± 0.6	76.9 ± 2.0
SERT	**100 ± 0.0**	**98.2 ± 1.1**	**80.5 ± 2.4**

5   Conclusions

In this paper, we examined the value of using synthetic examples combined with reverse training, namely SERT, to increase the recognition rate of set based face recognition. SERT is simple in concept and can be implemented easily. Experimental results indicate that SERT could yield better performance than the other competitors.

References

1. Kim, T.K., Kittler, J., Cipolla, R.: Discriminative learning and recognition of image set classes using canonical correlations. IEEE PAMI **29**, 1005–1018 (2007)

2. Arandjelovic, O., Shakhnarovich, G., Fisher, J., Cipolla, R., Darrell, T.: Face recognition with image sets using manifold density divergence. In: CVPR, pp. 581–588 (2005)
3. Shakhnarovich, G., Fisher III, J.W., Darrell, T.: Face recognition from long-term observations. In: Heyden, A., Sparr, G., Nielsen, M., Johansen, P. (eds.) ECCV 2002, Part III. LNCS, vol. 2352, pp. 851–865. Springer, Heidelberg (2002)
4. Wang, R., Shan, S., Chen, X., Gao, W.: Manifold-manifold distance with application to face recognition based on image set. In: CVPR, pp. 1–8 (2008)
5. Cevikalp, H., Triggs, B.: Face recognition based on image sets. In: CVPR, pp. 2567–2573 (2010)
6. Yang, M., Zhu, P., Gool, L., Zhang, L.: Face recognition based on regularized nearest points between image sets. In: IEEE FG, pp. 1–7 (2013)
7. Hu, Y., Mian, A.S., Owens, R.: Sparse approximated nearest points for image set classification. In: CVPR, pp. 121–128 (2011)
8. Wang, R., Chen, X.: Manifold discriminant analysis. In: CVPR, pp. 429–436 (2009)
9. Harandi, M.T., Sanderson, C., Shirazi, S., Lovell, B.C.: Graph embedding discriminant analysis on grassmannian manifolds for improved image set matching. In: CVPR, pp. 2705–2712 (2011)
10. Wang, R., Guo, H., Davis, Larry S., Dai, Q.: Covariance discriminative learning: a natural and efficient approach to image set classification. In: CVPR, pp. 2496–2503 (2012)
11. Turaga, P., Veeraraghavan, A., Srivastava, A., Chellappa, R.: Statistical computations on grassmann and stiefel manifolds for image and video-based recognition. IEEE PAMI **33**, 2273–2286 (2011)
12. Hamm, J., Lee, D.: Grassmann discriminant analysis: a unifying view on subspace-based learning. In: ICML, pp. 376–383 (2008)
13. Harandi, M.T., Sanderson, C., Wiliem, A., Lovell, B.C.: Kernel analysis over riemannian manifolds for visual recognition of actions, pedestrians and textures. In: WACV, pp. 433–439 (2012)
14. Hayat, M., Bennamoun, M., An, S.: Reverse training: an efficient approach for image set classification. In: Fleet, D., Pajdla, T., Schiele, B., Tuytelaars, T. (eds.) ECCV 2014, Part VI. LNCS, vol. 8694, pp. 784–799. Springer, Heidelberg (2014)
15. Ojala, T., Pietikainen, M., Maenpaa, T.: Multiresolution gray-scale and rotation invariant texture classification with local binary patterns. IEEE PAMI **24**, 971–987 (2002)
16. Chawla, N.V., Bowyer, K.W., Hall, L.O., Kegelmeyer, W.P.: Reverse training: an efficient approach for image set classification. J. Artif. Intell. Res. **16**, 321–357 (2002)
17. Fan, R.E., Chang, K.W., Hsieh, C.J., Wang, X.R., Lin, C.J.: LIBLINEAR: a library for large linear classification. J. Mach. Learn. Res. **9**, 1871–1874 (2008)
18. Lee, K., Ho, J., Yang, M., Kriegman, D.: Video based face recognition using probabilistic appearance manifolds. In: CVPR, pp. 313–320 (2003)
19. Viola, P., Jones, M.J.: Robust real-time face detection. IJCV **57**, 137–154 (2004)
20. Gross, R., Shi, J.: The CMU motion of body (mobo) database. Technical report CMU-RI-TR-01-18 (2001)
21. Kim, M., Kumar, S., Pavlovic, V., Rowley, H.: Face tracking and recognition with visual constraints in real-world videos. In: CVPR, pp. 1–8 (2008)
22. Ross, D.A., Lim, J., Lin, R., Yang, M.: Incremental learning for robust visual tracking. IJCV **77**, 125–141 (2008)

Carried Baggage Detection and Classification Using Part-Based Model

Wahyono and Kang-Hyun Jo$^{(\boxtimes)}$

Intelligent Systems Laboratory, Graduate School of Electrical Engineering,
University of Ulsan, Ulsan 680-749, Korea
wahyono@islab.ulsan.ac.kr, acejo@ulsan.ac.kr

Abstract. This paper introduces a new approach for detecting carried baggage by constructing human-baggage detector. It utilizes the spatial information of baggage in relevance to the human body carrying it. Human-baggage detector is modeled by body part of human, such as head, torso, leg and bag. The SVM is then used for training each part model. The boosting approach is constructing a strong classification by combining a set of weak classifier for each body part. Specify for bag part, the mixture model is built for overcoming strong variation of shape, color, and size. The proposed method has been extensively tested using public dataset. The experimental results suggest that the proposed method can be alternative method for state-of-the art baggage detection and classification algorithm.

Keywords: Carried baggage detection and classification · Part-based model · Video surveillance · HOG · Boosting SVM · Mixture model

1 Introduction

In the last decade, automatic video surveillance (AVS) system has more attention from the computer vision research community. Detecting of carried object is one of important parts of AVS. This task is potentially important objective for security and monitoring in public space. However, the task is inherently difficult due to the wide range of baggage that can be carried by a person and the different ways in which they can be carried. In the literature, there have been several approaches proposed for detecting baggage that abandoned by the owners [1, 2] or still being carried [3–5] by them. Tian et al. [1] proposed a method to detect abandoned and removed object using background subtraction and foreground analysis. In their approach, the background is modeled by three Gaussian mixture that combining with texture information in order to handle lighting change conditions. The static region obtained by background subtraction is then analyze using region growing and is classified as abandoned or removed object by some rules. However, in some cases this method produces many false alarm due to imperfect background subtraction. To overcome this problem, Fan et al. [2] proposed relative attributes schema to prioritize alerts by ranking candidate region. However, in real implementation to know who is the owner of abandoned baggage is very important. Therefore, as prior process, the system should capable to detect the person who carried baggage. The authors from [3, 4] proposed same concept to detect

© Springer International Publishing Switzerland 2015
D.-S. Huang and K. Han (Eds.): ICIC 2015, Part III, LNAI 9227, pp. 289–296, 2015.
DOI: 10.1007/978-3-319-22053-6_31

carried object by people. They utilized the sequence of human moving to make spatial temporal template. It was then aligned against view-specific exemplar generated offline to obtain the best match. Carried object was detected from the temporal protrusion. The author in [4] extend the framework such that the system can also classify the baggage type based on the position in relevance to the human body carrying it. However, the method assumes that parts of the carried objects are protruding from the body silhouettes. Due to its dependence on protrusion, the method cannot detect non protruding carried object. The protruding problem can be solved by method from [5]. This method utilized ratio color histogram. Using assumption of the color of carried object is different with clothes, it will achieve good result in accuracy. However, this method is dependence on event where the bag being transferred or left. The assumption of observing the person before and after the change in carrying status is application specific and cannot be used as a general carried object detector.

This paper proposes a novel approach for detecting people carrying baggage. Our approach utilizes the strong connection between baggage and body parts. Instead of constructing model for entire of object, our method build model for each part [8] of body including the bag part according to possible placement. Overall, this paper offers the following major contributions; (1) Part-based model schema and the relationship among them specific for detecting person carrying baggage and classifying the baggage based on our spatial model. (2) Bag part mixture model for solving strong variation problem of baggage (e.g., location, size, shape, and color).

2   System Model

This section presents the detail of our model for detecting human-baggage object on the image. Our model based on spatial information of bag on the human body.

2.1   Human Body Parameter and Baggage Spatial Model

Using human body proportional model, as shown Fig. 1(a), it can be deducted that in average the height and weight of a person are $H = 8h$, $W = 2h$, respectively, where h is the length of head. Therefore $h = H/8$. Bend line B is defined as the center of the body in vertical axis, vertical line C is denoted as center of body in horizontal axis that traverse the centroid of body. Let define T be the position of the top of the head in the image, and L be the most left location of body in the image, then $B = T + 4h$ and $C = L + h$. These all parameters are used for making our spatial model of human-baggage that is described in detail in the next subsection.

The general idea of spatial model is adopted from [4], by placing the bag in certain location according of the body proportion and the viewing direction of the person. Our spatial model is divided in into three major categories of bag, (1) backpack or hand bag, (2) tote bag or duffle bag and (3) rolling luggage. As shown in Fig. 1(b)–(d), spatial models of bag define the set of conditions for checking whether the bag exists or not in front view direction. For instance, if our part model detect that location depicted in Fig. 1(b) as bag with high probability value, then the bag is classified as a backpack; if

(a)   (b)   (c)   (d)

Fig. 1. Human-baggage spatial model of front view direction. The category of bag is divided into three major categories according to body proportions (a); (b) backpack or hand bag, (c) tote or duffle bag, and (d) rolling luggage.

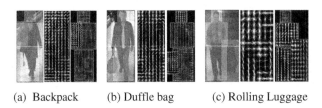

(a) Backpack   (b) Duffle bag   (c) Rolling Luggage

Fig. 2. Visualizations of HOG features from some models on i-Lids data [11]. For each category, first image is input image, the second image is the initial root filter for a human-bag model, and the last image is part filter model from the first image.

not, then it is placed the bag in location of other categories. In addition, if there is no bag identified in all spatial models, it is concluded the human is not carrying any bag.

2.2 Part-Based Model

Many object detection and recognition problems have been successfully implemented using part based model, such as face [6], human [7], and general object detection [8] with incredible result. In our work, the human-baggage is modeled based on observations of the part models and their relative position among them. In this paper, each human-baggage model is divided into four part models; head part, torso part, leg part and bag part, as shown in Fig. 2. Height of head, torso, and leg part are h, $3h$ and $4h$, respectively, while the height of bag part is varying according to our model. The feature vector is then extracted from full body and part model. The framework described here is independent of the specific choice of feature. In our implementation, the histogram of oriented gradient (HOG) [9] feature is used for description the model.

The part detection is represented by $m = \Phi_i(x, y, l, p)$, where specifying an anchor position (x, y) relative to full body in the lth level of the pyramid scale image and part p. The score of part interpolation based on hybrid of boosting Support Vector Machine (SVM) [10] is defined by the formulation as follows:

$$J(m) = \sum_{i=1}^{n} \omega_i \Phi_i(x, y, l, p) \tag{1}$$

where ω_i is the weighting, $\omega_i > 0$ and Φ_i is the output probability of SVM classification of the component p, n is the number of component.

2.3 Mixture Model

Detecting baggage is not easy due to variation of color, size and appearance. Consequently, the bag in certain location may exhibit more intra-class variation for by a single bag part model. Thus, specific for bag part, mixture model is used for handling this problem. More formally, let define a distribution f is a mixture of K component distribution $f_1, f_2, \ldots, f_K$ if

$$f(x) \sum_{k=1}^{K} \lambda_k f_k(x, \theta_k) \tag{2}$$

with the λ_k being the mixing weights, $\lambda_k > 0$, $\sum_k \lambda_k = 1$, x is feature vector of observations, and θ_k specify as parameter vector of the kth component. The overall parameter vector of the mixture model is thus $\theta = (\lambda_1, \ldots, \lambda_K, \theta_1, \ldots, \theta_K)$. Our goal now is to estimate these all parameters using Expectation-Maximization (EM).

2.4 Detection

Let M be the number of possible model (including human object that do not carry any baggage) learned in our framework. The final score of object hypothesis being human-baggage object is the maximum value among the score of each model that formulated as Eq. (3). In addition, if the value of J_{final} is less than a fixed threshold, the object hypothesis is classified as other objects.

$$J_{final} = \max_{m_1, \ldots, m_M} (J(m_1), J(m_2), \ldots, J(m_M)) \tag{3}$$

2.5 Training

A hybrid technique of boosting SVM implemented by [10] is used for training our human-baggage detection system. The boosting technique introduces its ability to extract high discriminative features to construct strong classifier from set of weak classifier. In our model, set of weak classifiers is built according to a set of part models. The standard SVM technique is used to learning of partial part model. The details of the standard SVM can be found in [12]. The SVM is applied to learn part model as a weak

(a) Category 1   (b) Category 2   (c) Category 3

Fig. 3. Some samples used for training. Category 1 includes backpack and handbag, category 2 consists of tote bag and duffle bag, and category 3 contains rolling luggage. First, second and third row for each category are representing viewing direction from front, side and back viewing direction. Thus, in total, nine models of human-baggage are built.

classifier. The boosting technique is then used for interpolating the full body detection. In practical data, the weighting of weak classifiers are automatically decided by our algorithm. The result training indicates that the head part has the largest weighting. Also, the head part is used to distinguish the view direction of object. The output probability of SVM classification is computed by

$$\Phi(x) = P(y = 1|x) = \frac{1}{1 + \exp(-h(x))} \tag{4}$$

where $h(x)$ is the signed distance of feature vector x to the margin of the SVM model.

3   Experiments

Our model was tested on human-baggage dataset. It was collected from small subset of INRIA [9], Caltech [13] and our own images. Since our work was focused on human-baggage detection, only human carrying baggage images were selected. The summary of our dataset is shown in Table 1. Our dataset is divided into two groups, training and testing groups. Each group is classified into 3 categories manually for creating ground truth. The training group contains 338 human-baggage object consisted of 132, 111, and 95 data for category 1, 2, and 3, respectively. The testing group contains 202 human-baggage object distributed as 78, 67, and 57 data for category 1, 2, and 3, respectively. Data from category 1 and 2 is resized to be 128×64 pixels resolution, while for category 3 is more wider becomes 128×72 pixels resolution. Figure 3 shows several samples of our dataset used in our implementation.

Fig. 4. Some typical detection results. The human-baggage object is detected as red bounding box. The baggage is then classified into category 1, 2 and 3 which are represented by blue, yellow and green bounding boxes, respectively (Color figure online).

Our model was evaluated for classified object into human with or without baggage. The human carrying baggage is set to be positive samples. Human without baggage is set to be negative samples. For first evaluation, 338 positives samples and 1,352 negative sample were used. Our method achieves detection rate of 77.02 %. Since, as our knowledge, there is no method researching this specific task, we do not compare our method with others yet. However, we have tried to just use original HOG and SVM on full body [9], but the result is not promising around 45.62 %. Next, our method was evaluated to classify image into three baggage categories. Table 2 summaries the evaluation result on training dataset. Our method obtains classification rate as much as 76.51 %, 77.47 %, 80 % for each category, respectively. In average, it achieves true classification rate of 77.21 %. Table 3 depicts the evaluation result on testing dataset.

Table 1. Dataset specification

Baggage category	#Training	#Testing
1. Backpack/handbag	131	78
2. Tote bag/duffle bag	111	67
3. Rolling luggage	95	57
Total	338	202

Table 2. Evaluation on training data

		Detection		
		C1	C2	C3
Ground truth	C1	101	14	7
	C2	9	86	16
	C3	5	14	76

Table 3. Evaluation on testing data

		Detection		
		C1	C2	C3
Ground truth	C1	46	23	9
	C2	12	39	16
	C3	7	15	35

Our method achieves 59.40 %, which for category 1, 2 and 3 are 58.97 %, 58.2 %, 61.40 %, respectively. Last, our method was evaluated on full image databases. Sliding window mechanism in any position and scale are used to detect human-baggage region. In practical, the same human-baggage region is usually detected with several times with overlapped bounding boxes. Therefore, it is necessary to combine the overlapped regions for unifying detection and rejecting misdetections. Some typical results of our method are shown in Fig. 4.

4 Conclusion

In this paper, part-based model for detecting and classifying baggage carried by human was introduced. First, the human region was modeled into four parts, head part, body part, leg part and baggage part. The model utilized the location information of baggage relative to human body. Histogram of oriented gradient (HOG) features were extracted on each part. The features were then trained using boosting support vector machine (SVM). Gaussian mixture model was also applied for modeling the baggage part for handling variation of baggage size, shape and color. After conducting extensive experiment, our method achieves 77.02 % and 59.40 % for detection and classification rate, respectively. However, our method has some limitations for detecting and classifying baggage carried by human. First, it may fail to detect multiple baggage carried by the same person. The additional model should be considered in our future work for handling this problem. Second, our method fail to detect overlapping human-baggage region. Increasing the number of part body can be one of the solutions for solving this issue.

References

1. Tian, Y.L., Feris, R., Liu, H., Humpapur, A., Sun, M.-T.: Robust detection of abandoned and removed objects in complex surveillance video. IEEE Trans. SMC Part C **41**(5), 565–576 (2011)
2. Fan, Q., Gabbur, P., Pankanti, S.: Relative attributes for large-scale abandoned object detection. In: International Conference on Computer Vision (2013)
3. Damen, D., Hogg, D.: Detecting carried objects from sequences of walking pedestrians. IEEE Trans. PAMI **34**(6), 1056–1067 (2012)
4. Tzanidou, G., Edirishinghe, E.A.: Automatic baggage detection and classification. In: IEEE 11th ICISDA (2011)
5. Chuang, C.-H., Hsieh, J.-W., Chen, S.-Y., Fan, K.-C.: Carried object detection using ratio histogram and its applications to suspicious event analysis. IEEE Trans. Circ. Syst. Video Technol. **19**(6), 911–916 (2009)
6. Tzimiropoulos, G., Pantic, M.: Gauss-Newton deformable part models for face alignment in-the-wild. In: ICIC 2014, Taiyuan, China, 3 Aug 2014
7. Hoang, V.-D., Hernandez, D.C., Jo, K.-H.: Partially obscured human detection based on component detectors using multiple feature descriptors. In: Huang, D.-S., Bevilacqua, V., Premaratne, P. (eds.) ICIC 2014. LNCS, vol. 8588, pp. 338–344. Springer, Heidelberg (2014)

8. Felzenszwalb, P.F., Girshick, R.B., McAllester, D., Ramanan, D.: Object detection with discriminatively trained part based models. IEEE Trans. PAMI **32**(9), 1627–1645 (2010)
9. Dalal, N., Triggs, B.: Histogram of oriented gradients for human detection. In: Conference on Computer Vision and Pattern Recognition, pp. 886–893 (2005)
10. Hoang, V.-D., Le, M.-H., Jo, K.-H.: Hybrid cascade boosting machine using variant scale blocks based HOG features for pedestrian detection. Neurocomputing **135**, 357–366 (2014)
11. Home Office Scientific Development Branch: Imagery library for intelligent detection systems (i-LIDS). In: The Institution of Engineering and Technology Conference on Crime and Security, pp. 445–448 (2006)
12. Chang, C.-C., Lin, C.-J.: LIBSVM: a library for support vector machine. ACM Trans. Intell. Syst. Technol. **2**, 1–27 (2011)
13. Dollar, P., Wojek, C., Schiele, B., Perona, P.: Pedestrian detection: a benchmark. In: International Conference on Computer Vision and Pattern Recognition (2009)

Image Splicing Detection Based on Markov Features in QDCT Domain

Ce Li[1,2(✉)], Qiang Ma[1], Limei Xiao[1], Ming Li[1], and Aihua Zhang[1]

[1] College of Electrical and Information Engineering, Lanzhou University of Technology,
Lanzhou 730050, China
[2] School of Electronic and Information Engineering, Xi'an Jiaotong University,
Xi'an 710049, China
Xjtulice@gmail.com

Abstract. Image splicing is very common and fundamental in image tampering. Therefore, image splicing detection has attracted more and more attention recently in digital forensics. Gray images are used directly, or color images are converted to gray images before processing in previous image splicing detection algorithms. However, most natural images are color images. In order to make use of the color information in images, a classification algorithm is put forward which can use color images directly. In this paper, an algorithm based on Markov in Quaternion discrete cosine transform (QDCT) domain is proposed for image splicing detection. The support vector machine (SVM) is exploited to classify the authentic and spliced images. The experiment results demonstrate that the proposed algorithm not only make use of color information of images, but also can achieve high classification accuracy.

Keywords: Markov model · QDCT · Image-splicing detection · Color image forgery detection

1 Introduction

In recent years, image splicing forgery detection is a hot topic in image processing. Splicing forgery detection need to depend on the concept of hypothesis, although any trace on the vision may not be left in tampering images, the underlying statistics are likely to be changed. It is these inconsistencies that detection techniques are used to detect the tampering. Various passive image splicing detection approaches have been proposed. Shi *et al.* [1] employed wavelet moment characteristics and Markov features to identify splicing images on DVMM dataset [2]. Wang *et al.* used gray level co-occurrence matrix (GLCM) of threshold edge image to classify splicing image [3] and employed Markov chain for tampering detection [4] on CASIA V2.0 Dataset [5]. Sutthiwan *et al.* [6] had pointed out two problems of the data sets [5] and had shown how to rectify the dataset and the experimental results showed the detection rate of the causal Markov model-based features was reduced to 78 % on the rectified dataset. Recently, Zha *et al.* used non-causal Markov model domain on DCT and DWT domains to obtain a high accuracy [7]. He *et al.* [8] proposed a method based on Markov features

© Springer International Publishing Switzerland 2015
D.-S. Huang and K. Han (Eds.): ICIC 2015, Part III, LNAI 9227, pp. 297–303, 2015.
DOI: 10.1007/978-3-319-22053-6_32

in DCT and DWT domain. The detection accuracy was up to 93.42 % on DVMM dataset [2], 89.76 % on CASIA V2.0 dataset [5]. Amerini *et al.* [9] used the DCT coefficients first digit features to distinguish and then localize a single and a double JPEG compression in portions of an image.

These algorithms mainly use the gray information of images or certain color component information to detect whether the images have been spliced, whole color information is not used effectively. In order to make use of the whole color information, QDCT is introduced into image splicing detection algorithm in this paper. The superiority of QDCT in color image processing is fully reflected in these applications [10, 11]. Therefore, an image tamper detection algorithm under QDCT transform domain is proposed in this paper. A new idea for image tamper detection is proposed, which has a certain theoretical and practical significance.

2 The Proposed Approach

In this section, the whole framework of the proposed algorithm is presented with detailed description of each part. The framework of the proposed algorithm based on Markov features is shown in Fig. 1. Compared with literature [8], we extract Markov features in QDCT domain. Expanded Markov features in QDCT domain are obtained in this paper. Besides, reference [1], we introduce main diagonal difference matrices, minor diagonal difference matrices, main diagonal transition probability matrices and minor diagonal transition probability matrices in QDCT. Different the literature [1, 8], We also consider QDCT coefficients correlation between the intra-block correlation and the inter-block in main diagonal and minor diagonal direction. Finally, the feature vector obtained is used to distinguish authentic and spliced images with Primal SVM [12] as the classifier.

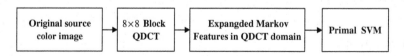

Fig. 1. Illustration of the stages of our algorithm

2.1 Quaternion Discrete Cosine Transform

The research of QDCT is prompted by the precedents of the successful application of real number and complex number domain, the basic principle of QDCT was proposed by Feng and Hu [10], and the actual algorithm was given. $h_q(m, n)$ is a two-dimensional $M \times N$ quaternion matrix, m and n is row and column of the matrix respectively, here, $m \in [0, M - 1]$, $n \in [0, N - 1]$, the definition of L-QDCT as follows:

$$\text{L - QDCT:} \quad J_q^L(p, s) = \alpha(p)\alpha(s) \sum_{m=0}^{M-1} \sum_{n=0}^{N-1} u_q \cdot h_q(m, n) \cdot T(p, s, m, n) \tag{1}$$

In Eq. (1), u_q is a unit pure quaternion, it represents the direction of axis of transformation, and it satisfies $u_q{}^2 = -1$, p and s are row and column of the transform matrix, respectively. It is similar to DCT in real number and complex number domain, the definition of $\alpha(p)$, $\alpha(s)$ and $T(p, s, m, n)$ are shown as follows:

$$\alpha(p) = \begin{cases} \sqrt{1/M} & p = 0 \\ \sqrt{2/M} & p \neq 0 \end{cases} \quad \alpha(s) = \begin{cases} \sqrt{1/N} & s = 0 \\ \sqrt{2/N} & s \neq 0 \end{cases}$$

$$T(p, s, m, n) = \cos\left[\frac{\pi(2m + 1)p}{2M}\right] \cos\left[\frac{\pi(2n + 1)s}{2N}\right]$$

The spectral coefficient of $J(p, s)$ through transformation is still a quaternion matrix of $M \times N$, and its representation is by Eq. (2).

$$J(p, s) = J_0(p, s) + J_1(p, s)i + J_2(p, s)j + J_3(p, s)k \tag{2}$$

2.2 Block QDCT

The Expanded Markov features in DCT domain proposed in [8] are very remarkable in capturing the differences between authentic and spliced images. They can be calculated by seven steps. Unlike the first step, the original color images are blocked into 8×8 non-repeatedly, and each block is still color image. Secondly, three color components R, G and B of blocked images are used to construct quaternion matrix, and the quaternion matrix is processed by QDCT transform to obtain QDCT coefficient matrix of each block, then the square root of the real part (r) and three imaginary parts (i, j, k) are calculated. Then all calculated matrices need to reassemble according to the site of blocking, thus a 8×8 blocked QDCT matrix F of original color image can be acquired. It is shown in Fig. 2.

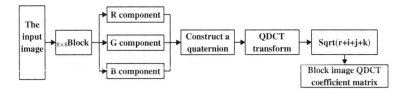

Fig. 2. Illustration of the stages of 8×8 block QDCT.

2.3 Expanded Markov Features in QDCT Domain

Compared with the literature [8], Expanded Markov features in QDCT domains are obtained in this paper. Besides, reference [1], we introduce main diagonal difference matrices, minor diagonal difference matrices, main diagonal transition probability matrices and minor diagonal transition probability matrices in QDCT. Different the literature [1, 8], we also consider QDCT coefficients correlation between intra-block correlation and inter-block in main diagonal and minor diagonal direction. In addition to these different other steps are roughly the same. The following steps are shown.

Firstly, 8×8 block QDCT is applied on the original image pixel array following Part 2.2, and the corresponding QDCT coefficient array G is obtained. Then, the round QDCT coefficients G to integer and take absolute value (denote as arrays F).

Secondly, calculate the horizontal, vertical, main diagonal and minor diagonal intra-block difference 2-D arrays F_h, F_v by applying Eqs. (1) and (2), respectively in [8], F_d and F_{-d} by applying Eqs. (3)–(4) as follow:

$$F_d(u, v) = F(u, v) - F(u + 1, v + 1) \tag{3}$$

$$F_{-d}(u, v) = F(u + 1, v) - F(u, v + 1) \tag{4}$$

and calculate the horizontal, vertical, main diagonal and minor diagonal inter-block difference 2-D arrays G_h, G_v, by applying Eqs. (7) and (8), respectively in [8], G_d and G_{-d} by applying Eqs. (5)–(6) as follow:

$$G_d(u, v) = F(u, v) - F(u + 8, v + 8) \tag{5}$$

$$G_{-d}(u, v) = F(u + 8, v) - F(u, v + 8) \tag{6}$$

Thirdly, introduce a threshold $T(T \in N_+)$, if the value of an element in F_h (or F_v, G_h and G_v) is either greater than T or smaller than $-T$, replace it with T or $-T$.

Fourthly, calculate the horizontal, vertical, main diagonal and minor diagonal transition probability matrices of F_h, F_v, G_h, G_v, F_d, F_{-d}, G_d and G_{-d}. That $P1_h(i,j)$, $P1_v(i,j)$, $P2_h(i,j)$, $P2_v(i,j)$, $P3_h(i,j)$, $P3_v(i,j)$, $P4_h(i,j)$, $P4_v(i,j)$ by applying Eqs. (3)–(6), (9), (10), (11), (12), respectively in [8]. $P1_d(i,j)$, $P1_{-d}(i,j)$, $P3_d(i,j)$, $P3_{-d}(i,j)$ by applying Eqs. (7)–(10) as follow:

$$P1_d(i,j) = \frac{\sum_{u=1}^{S_u-2} \sum_{v=1}^{S_v-2} \delta(F_d(u, v) = i, F_d(u + 1, v + 1) = j)}{\sum_{u=1}^{S_u-2} \sum_{v=1}^{S_v-2} \delta(F_d(u, v) = i)} \tag{7}$$

$$P1_{-d}(i,j) = \frac{\sum_{u=1}^{S_u-2} \sum_{v=1}^{S_v-2} \delta(F_{-d}(u + 1, v) = i, F_{-d}(u, v + 1) = j)}{\sum_{u=1}^{S_u-2} \sum_{v=1}^{S_v-2} \delta(F_{-d}(u, v) = i)} \tag{8}$$

$$P3_d(i,j) = \frac{\sum_{u=1}^{S_u-16} \sum_{v=1}^{S_v-16} \delta(G_d(u, v) = i, G_d(u + 8, v) = j)}{\sum_{u=1}^{S_u-16} \sum_{v=1}^{S_v-16} \delta(G_d(u, v) = i)} \tag{9}$$

$$P3_{-d}(i,j) = \frac{\sum_{u=1}^{S_u-16} \sum_{v=1}^{S_v-16} \delta(G_{-d}(u, v) = i, G_{-d}(u, v + 8) = j)}{\sum_{u=1}^{S_u-16} \sum_{v=1}^{S_v-16} \delta(G_{-d}(u, v) = i)} \tag{10}$$

where $i, j \in \{-T, -T+1, \ldots, 0, \ldots T-1, T\}$, thus $(2T+1) \times (2T+1) \times 12$ dimensionality Markov features in QDCT domain are obtained.

3   Experiments and Results

In this section, we introduce image datasets for experiment, and present a set of experiments to demonstrate high performance and effectiveness of proposed algorithm.

3.1   Image Dataset

Two public image datasets for tampering and splicing detection are provided by DVMM, Columbia University [2]. However, the color information is not provided by their data of gray images. The number of color images is too small. In addition, it doesn't closer to reality tamper image. In order to provide more challenging evaluation database for evaluation, two color image data sets CASIA V1.0 and CASIA V2.0 are chosen, which are shown in Fig. 3.

Fig. 3. Some example images of CASIA dataset (the left from V1.0, and the right from V2.0, authentic images in the top row, their forgery counter parts in the bottom row).

Two problems of the dataset CASAI V1.0 and CASAI V2.0 have been pointed out by Patchara *et al.* [6]. First one is the JPEG compression applied to authentic images is one-time less than that applied to tampered images; the second one is for JPEG images, the size of chrominance components of 7140 authentic images is only one quarter of that of 2061 tampered images. For fairness purpose, we accord to the processing method to modify the database in the article [6]. Because we can't use YCbCr color space, we just only need solve the first problem of CASAI TIDE V2.0. We used Matlab for standard JPEG compression to alleviate the influence of the difference in the number of JPEG compression by the following procedure: (1) Re-compressing 7437 JPEG authentic images with quality factor of 84; (2) Compressing 3059 TIFF tampered images by Matlab with quality factor of 84; (3) Leaving 2064 JPEG tampered images untouched. The data set CASAI V1.0 is also processed by same method.

3.2   Classification

In our experiment, Primal SVM [12] is used for classification. The Primal SVM has more kernel functions, such as RBF, linear, histogram intersection and so on. After the comparison with several experiments, the histogram intersection kernel has the highest classification accuracy. Therefore, histogram intersection is chosen as kernel function. The threshold T is fixed in all of the experiments. For CASIA V1.0, CASIA V2.0 dataset, we choose $T = 4$, which will produce 972-dimensional feature vector. To be fair, all the experiments and comparisons are tested on the dataset mentioned above with the same classifier. The testing platform is Matlab R2012b, and the hardware platform is a PC with a 2G, 32 bit operating system, and Intel i3 processor. In each experiment, the average rate of 30 repeating independent tests is recorded. In each of the 30 runs, 5/6 of authentic images and 5/6 of spliced images in the dataset are randomly selected to train SVM classifier. Then the remaining is used to test.

3.3   The Detection Performance of the Proposed Approach

Some common experiments are conducted first to assess the detection ability of the proposed algorithm. The detailed results are shown in Tables 1 and 2. *TP* (true positive) rate is the ratio of correct classification of authentic images. *TN* (true negative) rate is the ratio of correct classification of spliced images. Accuracy of detection is the weighted average value of *TP* rate and *TN* rate.

3.3.1   Comparison with Other Algorithms

To evaluate the proposed algorithm comprehensively, a comparison between the proposed algorithm and some state-of-the-art image splicing detection methods were performed. The gray image algorithms of NIM [1] and He [8] were chosen to compare. Our color image algorithm obtained a higher accuracy than their algorithms. *Fv, Dn, Ac, FT, FS* and *TT* represent features vector, dimensionality, accuracy, feature extraction time, feature selection time and total time respectively.

Table 1. Experiment results obtained on CASIA V2.0 dataset

Fv	Dn	Ac (%)	FT (s)	FS (s)	TT (s)
NIM [1]	266	84.86	4.479	0	4.479
He [8]	100	89.76	2.218	2.158	4.376
Our method	972	92.38	3.61	0	3.61

Table 2. Results of the proposed algorithm with different threshold T on CASIA dataset

Threshold T	C1-3	C1-4	C1-5	C2-3	C2-4	C2-5
Dimensionality n	588	972	1452	588	972	1452
TP (%)	95.881	95.440	95.735	88.371	89.185	89.426
TN (%)	95.073	96.470	97.131	95.637	95.567	95.914
Accuracy (%)	95.478	95.958	96.435	92.003	92.377	92.668

3.3.2   Choice of Threshold T

Another issue is the choice of the threshold T, which is used to reduce the dimension of Markov features. Based on past experience of Markov threshold selection [1, 8], we choose $T = 3, 4, 5$. In Table 2, we provide the performance of Markov features with three different T. Ci-T (i = 1, 2 represent CASIA V1.0 dataset, CASIA V2.0 dataset respectively, $T = 1, 2, 3$).

4   Conclusion

In the most of current image tamper detection algorithms, color image is converted to gray image before detection. Therefore, the whole color information is not taken into account. Images can be processed by quaternion in a whole manner to improve the accuracy of the image tamper detection algorithm. The experiment results demonstrate that the proposed algorithm not only make use of color information of images, but also can achieve high classification accuracy. Because the tamper images are mostly color in real life, this new idea for image tamper detection research has a certain theoretical and more practical significance.

Acknowledgments. The paper was supported in part by the National Natural Science Foundation (NSFC) of China under Grant Nos. (61365003, 61302116), National High Technology Research and Development Program of China (863 Program) No. 2013AA014601, Natural Science Foundation of China in Gansu Province Grant No. 1308RJZA274.

References

1. Shi, Y.Q., Chen, C., Chen, W.: A natural image model approach to splicing detection. In: ACM Proceedings of the 9th Workshop on Multimedia and Security, pp. 51–62 (2007)
2. Ng, T.T., Chang, S.F., Sun, Q.: A Data Set of Authentic and Spliced Image Blocks. Columbia University, New York (2004)
3. Wang, W., Dong, J., Tan, T.: Effective image splicing detection based on image chroma. In: IEEE International Conference on Image Processing, pp. 1257–1260 (2009)
4. Wang, W., Dong, J., Tan, T.: Image tampering detection based on stationary distribution of Markov chain. In: IEEE international Conference on Image Processing, pp. 2101–2104 (2010)
5. http://forensics.idealtest.org/
6. Sutthiwan, P., Shi, Y.Q., Zhao, H., Ng, T.-T., Su, W.: Markovian rake transform for digital image tampering detection. In: Shi, Y.Q., Emmanuel, S., Kankanhalli, M.S., Chang, S.-F., Radhakrishnan, R., Ma, F., Zhao, L. (eds.) Transactions on Data Hiding and Multimedia Security VI. LNCS, vol. 6730, pp. 1–17. Springer, Heidelberg (2011)
7. Zhao, X., et al.: Passive image-splicing detection by a 2-D noncausal Markov model. IEEE Trans. Circ. Syst. Video Technol. **2**(25), 185–199 (2014)
8. He, Z., Wei, L., Sun, W., et al.: Digital image splicing detection based on Markov features in DCT and DWT domain. Pattern Recogn. **45**(12), 4292–4299 (2012)
9. Amerini, I., Becarelli, R., et al.: Splicing forgeries localization through the use of first digit features. In: IEEE Conference on Parallel Computing Technologies, pp. 143–148 (2015)
10. Feng, W., Hu, B.: Quaternion discrete cosine transform and its application in color template matching. IEEE Congr. Image Sig. Process. **5**(2), 252–256 (2008)
11. Schauerte, B., Stiefelhagen, R.: Quaternion-based spectral saliency detection for eye fixation prediction. In: European Conference on Computer Vision, pp. 116–129 (2012)
12. Chapelle, O.: Training a support vetor machine in the primal. Neural Comput. **19**(5), 1155–1178 (2007)

Facial Expression Recognition
Based on Hybrid Approach

Md. Abdul Mannan[1]([⊠]), Antony Lam[1], Yoshinori Kobayashi[1,2],
and Yoshinori Kuno[1]

[1] Graduate School of Science and Engineering, Saitama University,
Saitama, Japan
{mannan,antonylam,kobayashi,
kuno}@cv.ics.saitama-u.ac.jp
[2] Japan Science and Technology Agency (JST), PRESTO, Kawaguchi, Japan

Abstract. This paper proposes an automatic system for facial expression recognition using a hybrid approach in the feature extraction phase (appearance and geometric). Appearance features are extracted as Local Directional Number (LDN) descriptors while facial landmark points and their displacements are considered as geometric features. Expression recognition is performed using multiple SVMs and decision level fusion. The proposed method was tested on the Extended Cohn-Kanade (CK+) database and obtained an overall 96.36 % recognition rate which outperformed the other state-of-the-art methods for facial expression recognition.

Keywords: Facial expression recognition · Directional number pattern · Feature · Image descriptor · Local pattern

1 Introduction

Automatic facial expression analysis plays a vital role in a wide range of applications such as human computer interaction, data-driven animation, and so on. Due to its wide range of applications, it has drawn much attention and interest in recent years. Though much effort has been made, automatic recognition of facial expression remains difficult.

The facial expressions under examination were defined by psychologists as a set of six basic facial expressions (anger, disgust, fear, happiness, sadness, and surprise) [1]. A survey about recently proposed approaches can be found in [2]. The features employed by most of the existing methods are of two types: geometric features (e.g. [3, 4]) and appearance features (e.g. [5–7]). In addition, there are also hybrid feature based approaches, that use both geometric and appearance features together. As suggested in several studies (e.g. [8]), the best choice for facial expression recognition might be combining the geometric and appearance features. The benefits of combining the two types of features was also demonstrated in [9–11] where the combined features provided superior accuracy over using either feature type alone. We also adopt a hybrid approach. What is unique about our work is that we integrate in the best reported approaches for feature extraction, machine learning, fusion of modalities, into

© Springer International Publishing Switzerland 2015
D.-S. Huang and K. Han (Eds.): ICIC 2015, Part III, LNAI 9227, pp. 304–310, 2015.
DOI: 10.1007/978-3-319-22053-6_33

a coherent system. Through this, we show state-of-the-art performance on a public, challenging dataset.

This paper proposes an automatic system for facial expression recognition. We first address the issue of how to detect the human face and it's salient regions in static images. We then consider how to represent and recognize facial expressions presented in those faces using our hybrid appearance-geometric approach. To extract appearance features, we use the most efficient and robust descriptor in the literature, the Local Directional Number Pattern (LDN) [17], which encodes the structural information and the intensity variation of the face's texture. Specifically, we automatically extract multi-scale LDN features only from the visually salient regions (e.g. eyes, nose, mouth). Thus ensuring we obtain the most informative parts in terms of macro and micro structural details of the human face. In the geometric approach we use automatically detected facial landmark points and their relative distances to construct the feature. Unlike other methods, to extract geometric features we do not use manually annotated information from person specific neutral expressions. Thus our system is fully automatic. After appearance and geometric feature extraction, we classify emotions using independent SVMs for each feature type and perform decision level fusion of the SVM outputs [13] rather than direct feature level fusion approaches such as simple concatenation [17]. This is because the two types of features are of two completely different modalities, so direct concatenation of the two feature vectors makes learning an accurate classifier difficult in practice. In addition, concatenation of two feature vectors leads to high dimensionality, which can cause the "curse of dimensionality" where learning a good classifier is even harder [18]. The benefit of this decision level fusion is that we can jointly consider the effects of appearance features and geometric characteristics, which influence the change of expression greatly without the drawbacks of feature level fusion.

2 Proposed System

Feature selection along with the regions from which the features are to be extracted is one of the most important steps to recognizing expressions. The proposed framework extracts appearance features only from the salient regions while the geometric features are extracted from some feature points on face. A schematic overview of the framework is presented in Fig. 1. The system first takes as input, an image that contains the face. Face detection and segmentation of the salient regions then occurs. Then we extract the feature vectors (geometric and appearance) from the face image and it's salient parts. After feature extraction, independent classification of facial expressions using the geometric and appearance features is performed. The last phase of this system is the decision level fusion method, which will be detailed later.

3 Facial Expression Features

In this paper the appearance and geometric features are obtained from face images independently. The appearance features are extracted from four visually salient regions by applying the LDN descriptor algorithm. These four visually salient regions (eyes,

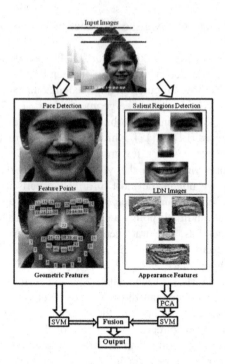

Fig. 1. Block diagram of proposed facial expression recognition system.

nose, mouth) are selected and extracted from the position of face landmarks [12]. At the same time, the facial landmarks are also used to encode a geometry based feature, which is independent of the LDN appearance features.

3.1   Appearance Features

Appearance-based approaches usually use image filters, holistically on the whole face or locally on specific face regions. Local descriptors have gained attention because of their robustness to illumination and pose variations. In particular, the LDN coding method is highly robust and effective for facial expression recognition [17]. Thus we use LDN in this paper but other local descriptors could also be used.

Appearance Feature Extraction. To reduce the effects of illumination in facial expression recognition, energy normalization is used to process the face image after the segmentation of facial salient regions (eyes, nose, mouth). The process of energy normalization is shown in Eq. (1).

$$I'(x, y) = \frac{I(x, y)}{\|I(x, y)\|} \tag{1}$$

where $\|I(x,y)\| = \sqrt{\left(\sum_{x=1}^{M} \sum_{y=1}^{N} I^2(x, y) \right)}$ and the size of image is M × N.

To capture textural micro-structure as well as macro-structure we apply LDN coding on the salient regions at several scales s. Given an original image, a set of cell-based average images are first obtained by calculating the mean values of non-overlapping $s \times s$ cells. At scale s the cell size is $s \times s$. Note that the original image can be considered as the average image at scale 1 because the cell size is 1×1, and the size of the average image at scale s is $1/s$ of the original image. Next, these multi-scale average images are transformed to LDN label images.

We then represent the appearance of a facial expression as a vector of concatenated histograms of the multi-scaled LDN features. Histograms discard spatial coherence so to preserve it, we divide the image into several regions, $\{R_0, \ldots R_N\}$, and compute a histogram, H_i, from each region, R_i, then combine them into single feature vector. Different regions of the mouth and the histogram concatenation process are shown in Fig. 2.

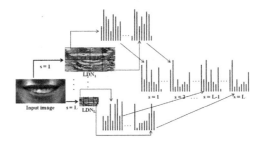

Fig. 2. Fusion procedure of multi-scaled LDN features.

3.2 Geometric Features

It is argued that robust computer vision algorithms for face analysis and recognition are based on con-figural and shape features [14]. These features are defined as distances between facial components (mouth, eye, eyebrow, nose, and jaw line) and we make use of them here. We also note that unlike past work, our extracted geometrical feature does not require prior knowledge of a person's specific neutral expression.

The structure of the proposed facial expression recognition approach is illustrated in Fig. 3. We use a parts mixture and cascaded deformable shape model [12] to automatically locate both the face and 46 facial points. These landmark points are then used to build our geometric feature vector that will be used in later stages of our system.

Feature Extraction. We use 46 facial points. The location of the 46 points relative to the face position and size results in a 92 dimensional feature vector, generated from both the x- and y- coordinates of each point. To encode the relative positions of the facial points to each other. We extracted 45 Euclidean distances D between the 46 points as shown in Fig. 3.

To ensure the features is scale invariant, the distances are normalized to the detected face width and height. Then, the two feature sets (coordinates and distances) are concatenated to produce a vector of length 137.

Fig. 3. Geometric features (feature points and the distance among them).

4   Feature Dimensionality Reduction and Classifier

After feature extraction, the appearance feature's dimensionality was found to be much larger than the geometric feature vector. Thus for the appearance feature, we utilize PCA to reduce dimensionality.

We perform facial expression recognition using Support Vector Machines (SVMs) [15] and decision-level fusion. That is, we use appearance features and geometric features to train separate SVM models and determine class probabilities independently. The probabilities are then combined using the product rule. As shown in [13], the combined score S for a class d using the product rule is given by $S(d) = \dfrac{\prod_m p_m(d)}{z}$ where, $p_m(d)$ is the probability of the test example belonging to class d for mode m and Z is normalizing factor to make $S(d)$ a probability.

5   Experiments and Results

Like in related work, we use the CK+ database [16] to allow for a fair comparison. The CK+ database is composed of 123 subjects with each session, an image sequence that starts from a neutral emotion and gradually ends at a peak prototypical emotion. Unlike most other papers, we do not split multiple images of the same subject with the same emotion label between training and testing sets. Instead, we randomly select 52 of the first neutral frames and the last peak frames were picked from the 309 labeled sequences, resulting in 361 images, including 83 surprise, 28 sadness, 69 happy, 25 fear, 59 disgust, 45 anger and 52 neutral. In this way, the dataset is more challenging and makes the recognition identity independent.

From each emotion label, we randomly select 90 % of the data for training and 10 % for testing. We repeat this protocol for our experiments 1,000 times and calculate the average accuracy. Our framework obtained an average emotion recognition percentage of 96.36 % using an SVM linear kernel. Table 1 shows the confusion matrix of one instance. The performance of state-of-the-art methods are reported in Table 2, including the performance of our method.

Table 1. The confusion matrix obtained by applying our approach on the CK+ database for one train-test split.

	Neutral	Anger	Disgust	Fear	Happy	Sadness	Surprise
Neutral	**100**	0	0	0	0	0	0
Anger	0	**100**	0	0	0	0	0
Disgust	0	0	**100**	0	0	0	0
Fear	20	0	0	**80**	0	0	0
Happy	0	0	0	0	**100**	0	0
Sadness	16.67	0	0	0	0	**83.33**	0
Surprise	0	0	0	0	0	0	**100**

Table 2. Comparison with state-of-the-art methods for the 7-class facial expression recognition on CK/CK+ database.

Method	Experimental environment	Recognition rate (%)
Kabir et al. [17]	Identity dependent	93.1
Khan et al. [19]	6 classes of emotion	96.7
Rivera et al. [20]	Identity dependent	89.3
Saeed et al. [21]	Identity independent	73.63
Zhou et al. [22]	Identity dependent	96.86
Zhou et al. [23]	Identity independent	93.2
Z. Wang et al. [24]	Identity independent	86.3
Ours	Identity independent	**96.36**

6 Conclusion

This paper presented a new method for fully automatic facial expression recognition using a hybrid combination of geometric and appearance features that achieved state-of-the-art performance on the challenging CK+ database. The next step in our research is to generalize facial expression recognition to non-frontal face poses. Real time facial expression recognition from video data will also require a higher level of robustness.

Acknowledgement. This work was supported by JSPS KAKENHI(Grant-in-Aid for Challenging Exploratory Research) Grant Number 26540131 and Saitama Prefecture Leading-edge Industry Design Project.

References

1. Ekman, P., Friesen, W.V.: Emotion in the Human Face. Pergamon Press, Oxford (1975)
2. Zeng, Z., Roisman, G.I., Huang, T.S.: A survey of affect recognition methods: audio, visual, and spontaneous expression. IEEE Trans. Pattern Anal. Mach. Intell. **31**(1), 39–58 (2009)
3. Chang, Y., Hu, C., Feris, R., Turk, M.: Manifold based analysis of facial expression. In: CVPRW, pp. 81–81 (2004)

4. Pantic, M., Rothkrantz, L.J.M.: Facial action recognition for facial expression analysis from static face image. IEEE Trans. Syst. Man Cybern. **34**(3), 1449–1461 (2004)
5. Bartlett, M.S., Littlewort, G., Braathen, P., Sejnowski, T.J., Movellan, J.R.: A prototype for automatic recognition of spontaneous facial action. In: Advances in Neural Information Processing Systems, vol. 15, pp. 1271–1278 (2003)
6. Barlett, M.S., Littlewort, G., Frank, M.G., Lainscsek, C., Fasel, I., Movellan, J.: Fully automatic facial action recognition in spontaneous behavior. In: International Conference on FGR, pp. 223–230 (2006)
7. Shan, C., Gong, S., McOwan, P.W.: Facial expression recognition based on local binary patterns: a comprehensive study. Image Vis. Comput. **27**(6), 803–816 (2009)
8. Pantic, M., Patras, I.: Dynamics of facial expression: recognition of facial actions and their temporal segments form face profile image sequences. IEEE Trans. Syst. Man Cybern. Part B **36**(2), 433–449 (2006)
9. Lucey, P., Cohn, J.F., Kanade, T., Saragih, J., Ambadar, Z., Matthews, I.: The extended Cohn-Kanade dataset (CK+): a complete dataset for action unit and emotion specified expression. In: Proceedings of IEEE Computer Vision and Pattern Recognition Workshops, pp. 94–101 (2010)
10. Mahoor, M.H., Zhou, M., Veon, K.L., Mavadati, S.M., Cohn, J.F.: Facial action unit recognition with sparse representation. In: Proceedings of IEEE International Conference on Automatic Face and Gesture Recognition and Workshops, pp. 336–342 (2011)
11. Kotsia, I., Zafeiriou, S., Nikolaidis, N., Pitas, I.: Texture and shape information fusion for facial action unit recognition. In: Proceedings of First International Conference on Advances in Computer-Human Interaction, pp. 77–82 (2008)
12. Yu, X., Huand, J., Zhang, S., Yan, W., Metaxas, D.N.: Pose-free facial landmark fitting via optimized part mixtures and cascaded deformable shape model. In: ICCV (2013)
13. Meghjani, M., Ferrie, F., Dudek, G.: Bimodal information analysis for emotion recognition. In: IEEE (2009)
14. Martinez, A.M.: Deciphering the face. In: Proceeding of the IEEE Computer Society Conference on Computer Vision and Pattern Recognition Workshops, pp. 7–12 (2011)
15. Cortes, C., Vapnik, V.: Support-vector Networks. Mach. Learn. **20**(3), 273–297 (1995)
16. Kanade, T., Cohn, J.F., Tian, Y.: Comprehensive database for facial expression analysis. In: IEEE Conference on Automatic Face and Gesture Recognition, pp. 46–53. IEEE Computer Society, Grenoble (2000)
17. Kabir, H., Jabid, T., Chae, O.: Local directional pattern variance (LDPv): a robust feature descriptor for facial expression recognition. Int. Arab J. Inf. Technol. **9**(4), 382–391 (2012)
18. Bishop, C.M.: Pattern Recognition and Machine Learning. Springer, New York (2006)
19. Khan, R.A., Meyer, A., Konik, H., Bouakaz, S.: Framework for reliable, real-time facial expression recognition for low resolution images. Pattern Recogn. Lett. **34**, 1159–1168 (2013)
20. Rivera, A.R., Castillo, J.R., Chae, O.: Local directional number pattern for facial analysis: face and expression recognition. IEEE Trans. Image Process. **22**(5), 1740–1752 (2013)
21. Saeed, A., Al-Hamadi, A., Niese, R., Elzobi, M.: Frame-based facial expression recognition using geometrical features. Adv. Hum. Comput. Interact. **2014**, Article ID 408953 (2014)
22. Zhou, J., Xu, T., Gan, J.: Facial expression recognition based on local directional pattern using SVM decision-level fusion. In: Proceeding of the 2nd International Conference on Computer and Applications, vol. 17, pp. 126–132 (2013)
23. Zhou, L., Wang, H.: Person-independent facial expression analysis by fusing multiscale cell features. Opt. Eng. **52**(3), 037201 (2013)
24. Wang, Z., Wang, S., Ji, Q.: Capturing complex spatio-temporal relations among facial muscles for facial expression recognition. In: CVPR 2013, pp. 3422–3429 (2013)

A Local Feature Descriptor Based on Energy Information for Human Activity Recognition

Yubo Shi and Yongxiong Wang$^{(\boxtimes)}$

Shanghai Key Lab of Modern Optical System and Engineering Research Center
of Optical Instrument and System, Ministry of Education, University of Shanghai
for Science and Technology, Shanghai 200093, China
wyxiong@usst.edu.cn

Abstract. A local feature descriptor based on energy information is presented which combines kinetic energy, potential energy and the position information of 3D skeleton joints etc. These features conform to not only kinematics and biology of human action, but also the natural visual saliency for action recognition. The semantic features is obtained by the bag of word (BOW) based on k-means clustering. Finally, SVM based on kernel function is used to carry out human activity recognition. The experimental results show that the accuracy of human activity recognition based on low dimensional features is higher than several state-of-the-art algorithms.

Keywords: Activity recognition · 3D skeleton · Local feature descriptor · Bag of word · Energy information

1 Introduction

Human activity recognition plays an extremely important role in many applications, such as in intelligent furniture, service robot, and intelligent surveillance [1–3]. Because of the affordable depth image devices, image segmentation and object recognition become more simple, and has strong robustness, therefore many researchers have pay more attention to it in recent years. We can conveniently and easily acquire color image and depth information (RGB-D) from a 3D sensor device. And human skeleton can be precisely extracted from this RGB-D data [4].

Feature description or feature selection of human behavior is a key problem in the human behavior recognition. Human actions, from a biological standpoint, could be modeled by the motion of a set of skeleton joints. Most representative skeleton-based methods of human actions describe the positions, velocities, or trajectory of a set of 3D joints [5]. In [5], the motion trajectories of joints are obtained by pair-wise SIFT features. Then they compute a velocity description based on Markov chain model. Gu et al. [6] propose a feature description of 3D joints which includes global movement feature and local configuration feature in the 4D spatial-temporal space. And the recognition rate in ref. [6] is obviously higher than that in ref. [5]. Yang et al. [7] apply 3D position differences of skeleton joints to characterize action information by combining static posture, consecutive motion feature, and overall dynamics feature in each frame. To reduce dimensionality of feature they employ PCA to form the final features called

© Springer International Publishing Switzerland 2015
D.-S. Huang and K. Han (Eds.): ICIC 2015, Part III, LNAI 9227, pp. 311–317, 2015.
DOI: 10.1007/978-3-319-22053-6_34

EigenJoints. And they firstly propose a concept of Accumulated Motion Energy (AME) to select informative frame, which can remove noisy frames and reduce computational cost. It is necessary to choose the more informative joints of skeleton for the action description. In order to represent the most informative joints of actions quantitatively, Ofli et al. [8] employ the entropy of its joint angle time series, i.e. the mean or variance joint angle trajectories, to express as the informativeness of joints.

From a biological standpoint, people always focus on the moving parts of human body and gestures containing information. It is a natural response of human being. In other words, it is the natural features of visual saliency for human action. In view of the idea, we propose a new local feature descriptor for human action recognition. We consider the kinetic energy and potential energy of human body as a new the important features of human activity recognition. The kinetic energy of the human skeleton is introduced as a representation of motion information of human actions. To represent the position information of human skeleton, a new representation method of human pose, the potential energy of human skeleton, are used to measure the spatial information of human pose quantitatively.

2   Methodology of Feature Representation for Activity Recognition

Human activity is often described by 3D skeleton obtained from depth information or RGB image. The human skeleton model consists of 15 joints and the depth image and the corresponding color image combined with the position of joint based on the Cornell Activity Dataset: CAD-60 [9] can be extracted by Kinect.

2.1   Feature Representation

We first extract four types of features as candidate features, and then we use clustering method to construct the local feature descriptors which are a low-dimension feature and are advantageous to achieve activity recognition.

When performing different actions or different phases within an action, the participative parts of each human body part are often various. Considering human action characteristics and the human anatomy, we firstly select the feature of local kinetic energy of all 15 skeleton joints. In this paper we define the feature of local kinetic energy as follows:

$$Ek_{i,t} = kv_{i,t}^2 = \frac{1}{\Delta t^2} k \left| P_{i,t} - P_{i,t-1} \right|^2$$
$$= \frac{k}{\Delta t^2} \left[(x_{i,t} - x_{i,t-1})^2 + (y_{i,t} - y_{i,t-1})^2 + (z_{i,t} - z_{i,t-1})^2 \right] \tag{2.1}$$

where the subscript i represents ith joint ($i = 1, 2, ..., 15$). $Ek_{i,t}$ is the overall kinetic energy of each joint of human skeleton at frame F_t, and $v_{i,t}$ is the velocity of ith joint at frame F_t, $P_{i,t}(x_{i,t}, y_{i,t}, z_{i,t})$ is the position of ith joint at the frame F_t, k is a necessary coefficient of kinetic energy.

Considering the gesture of human is related to the relative position between two different joints of human body, we choose the feature of local potential energy to represent the gesture of human. According to the characteristics of human biology, the head or the joint of trunk can be chosen as a reference point of zero potential energy. Through experimental contrast, we find that choosing the head point as the reference point of zero potential energy shows good performance. The potential energy is defined as follows:

$$E_{i,t} = L\left(\left|P_{i,t} - P_{1,t}\right|\right)$$
$$= L\left(\sqrt{(x_{i,t} - x_{i,t-1})^2 + (y_{i,t} - y_{i,t-1})^2 + (z_{i,t} - z_{i,t-1})^2}\right) \tag{2.2}$$

where $E_{i,t}$ is potential energy of each joint ($i = 1, 2, \ldots, 15$), $P_{1,t}$ is the zero potential energy joint of human skeleton.

The speed and direction are usually considered to describe the moving state of vehicles which are also important parameters for human activity recognition. Thus direction vector is calculated according to the change of the joint position in the before and after frame. The feature of direction vector is defined as follows:

$$\varphi_{i,t} = (x_{i,t} - x_{i,t-1}, y_{i,t} - y_{i,t-1}, z_{i,t} - z_{i,t-1}) \tag{2.3}$$

where $\varphi_{i,t}$ represents the direction vector in F_t frame, $x_{i,t}$, $y_{i,t}$, $z_{i,t}$ indicate the spatial coordinate in F_t frame respectively.

According to the rule of the human body movement we define six most representative body joint angles. θ_1 and θ_2 are the left and right elbow angles respectively, θ_3 and θ_4 are the left and right knee angles respectively, θ_5 and θ_6 are the variant of left and right shoulders angles or armpit angles, respectively. The number of joint angle is only six, so we combine the joint angle feature with BOW feature to obtain the final feature vectors. Joint angle is defined as follows:

$$\theta_{i,t} = \cos^{-1}\left\{\frac{\alpha \cdot \beta}{|\alpha||\beta|}\right\} \tag{2.4}$$

where $\theta_{i,t}$ represents the ith joint in the F_tth frame, the "·" indicates the vector inner product. "||" is the vector norm. α and β represent two vectors of the skeleton.

2.2 Construction of BOW

3D human body skeleton contains 15 joint points, so more than one hundred feature data are extracted from each frame. In order to effectively combine multiple feature data, reduce the dimension of feature vector, the k-means clustering algorithm is used to build BOW of feature. The local feature matrix Y_t comprises above four features, as defined as follows:

$$Y_t = \begin{bmatrix} Ek_{1,t} & \phi_{1,t} & P_{1,t} & E_{1,t} \\ Ek_{2,t} & \phi_{2,t} & P_{2,t} & E_{2,t} \\ \vdots & \vdots & \vdots & \vdots \\ Ek_{15,t} & \phi_{15,t} & P_{15,t} & E_{15,t} \end{bmatrix} \qquad (2.5)$$

The columns of local feature matrix are four types of features, and the rows of matrix are numbers of joints. Then we obtain the k-dimension feature vectors by mapping them to cluster center. To ensure the scale consistency of feature vectors, the local feature matrix should be normalized before clustering based on following equations.

$$X_i^* = \frac{X - M}{S} \qquad (2.6)$$

$$M = \sum_{i=1}^{n} \frac{X_i}{n}, S = \sqrt{\frac{1}{n} \sum_{i=1}^{n} (X_i - M)^2}, \, n = 15 \qquad (2.7)$$

where n represents the number of joints. M represents the average value of the features. X_i is the feature data of the ith joint. S is the standard deviation of the features. X_i^* is the standard feature and the standard feature matrix Y_t^* is:

$$Y_t^* = \begin{bmatrix} Ek_{1,t}^* & \phi_{1,t}^* & P_{1,t}^* & E_{1,t}^* \\ Ek_{2,t}^* & \phi_{2,t}^* & P_{2,t}^* & E_{2,t}^* \\ \vdots & \vdots & \vdots & \vdots \\ Ek_{15,t}^* & \phi_{15,t}^* & P_{15,t}^* & E_{15,t}^* \end{bmatrix} = \begin{bmatrix} v_1 \\ v_2 \\ \vdots \\ v_{15} \end{bmatrix} \qquad (2.8)$$

The number of clustering k is acquired by experience. In this paper we do the experiment many times based on different value of k, finally we find that 5 is the best value of k. Local feature matrix contains 15 lines, each row represents a feature vector v_i. Five clustering centers C_1, C_2, C_3, C_4, C_5 are obtained after clustering, Then all the feature vectors are mapped to the five clustering. The BOW_t vector is defined as Eq. (2.9):

$$BOW_t = \begin{bmatrix} bin_1 & bin_2 & bin_3 & bin_4 & bin_5 \end{bmatrix} \qquad (2.9)$$

The pseudo code of BOW_t vector shows as follows:

1. Initialization: $BOW_t = [0 \ \ 0 \ \ 0 \ \ 0 \ \ 0]$, $bin_k = 0$ ($k=1,2,3,4,5$)

2. Operation process:

    (i) Construct

$$Y_t^* = [vector_1, vector_2, vector_3 \cdots\cdots, vector_{15}]^T ; vector_i = [Ek_{i,t}^* \ \ \phi_{i,t}^* \ \ P_{i,t}^* \ \ E_{i,t}^*] \tag{2.10}$$

    (ii) Using k-means method to cluster the $vector_i$ ($i=1, 2\ldots\ldots, 15$) and then five centers C_1、C_2、C_3、C_4、C_5 namely C_k ($k=1, 2, 3, 4, 5$) will be achieved.

    (iii) for i: 15,

        for k: 5,

      calculate the Euclidean distance between vector and Ck such as type:

$$D[k] = \left\| vector_i - C_k \right\|$$
$$= \sqrt{\left\| vector_i \right\|^2 + \left\| C_k \right\|^2 - 2 * vector_i \cdot C_k} \tag{2.11}$$

```
        end
    if D[index] is the minimum value
        bin_index = bin_index +1
    end
```

3   Experiment and Results

The activity recognition process is divided into three steps: firstly, we extract the features, such as kinetic energy, direction changing vector, joint angle and so on from 3D coordinate information of the human skeleton, and establish the features matrix of the key features; the secondly, a semantic dictionary is established by clustering, then the final feature vector consists of limb joint angles and BOW, and the dimension of the final feature is $k + 6$. The thirdly, training multi-class SVM is modeled to test the result of classification.

3.1   Database

In this paper we adopt CAD-60 as the experimental data. This dataset contains 12 daily actions data of four human. These 12 actions include: still, talking on the phone, writing on whiteboard, and so on. These actions are carried out by four people (one left-handed man, and one left-handed woman). Each participant in five different scenarios (office, kitchen, bedroom, bathroom, living room) takes these 12 behaviors.

3.2   Experimental Results and Analysis

Experiments will adapt 70 % frames of each action in the database data as SVM training set, the remaining 30 % of the data as a test set. Confusion matrix of the classification results of each person are shown in Fig. 1.

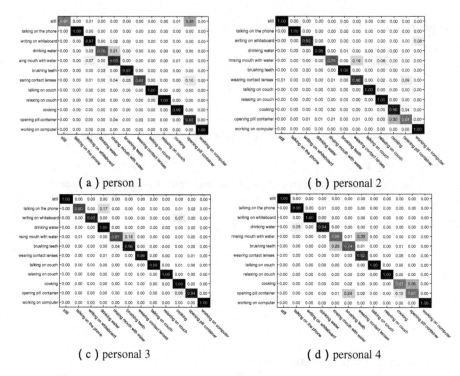

(a) person 1   (b) personal 2

(c) personal 3   (d) personal 4

Fig. 1. Confusion matrix of the classification results based on four dataset.

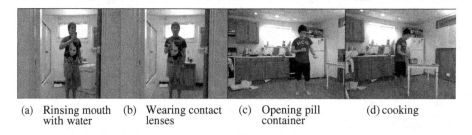

(a) Rinsing mouth   (b) Wearing contact   (c) Opening pill   (d) cooking
     with water   lenses   container

Fig. 2. Comparison of similar postures belong to different actions

It can be found from Fig. 1 that the recognition accuracy of "talking on couch" and "relaxing on couch" is better than other nine actions, and the accuracy of "rinsing mouth with water" and "opening pill container" is the lowest among these 12 actions. By analyzing the experiment, when the people are doing this two actions (regardless of the objects in the hands), in some ways their postures are very similar. As shown in Fig. 2(a) and (b) which are "rinsing mouth with water" and "wear contact lenses" and in Fig. 2(c) and (d) which are "opening pill container" and "cooking", it can be found that the postures at the some point are roughly similar, thus it will lead to error classification. The total recognition rates of four dataset are 93.67 %, 94.49 %, 95.49 %, 85.27 % respectively. The average accuracy of all action is 92.23 %.

4 Conclusion

This paper proposes a local feature descriptor from the human skeleton joint information for human action recognition. Firstly, these features like pose potential energy, kinetic energy, direction variation characteristics and spatial location information characteristic are extracted based on human biology and kinematics. And local characteristic matrix is established. Secondly, k-means method is used to convert the local feature matrix to the BOW feature. The feature vector for final recognition is constructed by combining BOW feature and joint Angle feature. Finally we use the RBF kernel based SVM classifier to recognize human actions. The direction of future work is learning person–object interaction to improve the behavior recognition accuracy and applicability.

Acknowledgments. Supported by the National Natural Science Foundation of China under Grant 61374039, 61403254, the Hujiang Foundation of China (C14002, B1402/D1402, D15009).

References

1. Turaga, P., Chellappa, R., Subrahmanian, V.S., et al.: Machine recognition of human activities: a survey J]. IEEE Trans. Circuits Syst. Video Technol. **18**(11), 1473–1488 (2008)
2. Uddin, M.Z., Thang, N.D., Kim, J.T., et al.: Human activity recognition using body joint-angle features and hidden markov model. ETRI J. **33**(4), 569–579 (2011)
3. Dalal, N., Triggs, B.: Histograms of oriented gradients for human detection. In: Computer Society Conference on Computer Vision and Pattern Recognition (CVPR), vol. 1, pp. 886–893. IEEE (2005)
4. Xia, L., Chen, C.C., Aggarwal, J.K.: Human detection using depth information by kinect. In: Computer Society Conference on Computer Vision and Pattern Recognition Workshops CVPRW 2011, pp. 15–22. IEEE (2011)
5. Sun, J., Wu, X., Yan, S., Cheong, L., Chua, T., Li, J.: Hierarchical spatio-temporal context modeling for action recognition. In: Proceedings of IEEE Conference on Computer Vision and Pattern Recognition (CVPR), pp. 2004–2011 (2009)
6. Junxia, G., Xiaoqing, D., Shengjin, W., et al.: Full body tracking-based human action recognition. In: 19th International Conference on Pattern Recognition (ICPR), pp. 1–4. IEEE (2008)
7. Yang, X., Tian, Y.L.: Effective 3D action recognition using eigenjoints. J. Vis. Commun. Image Represent. **25**(1), 2–11 (2014)
8. Ofli, F., Chaudhry, R., Kurillo, G., et al.: Sequence of the most informative joints: a new representation for human skeletal action recognition. J. Vis. Commun. Image Represent. **25**(1), 24–38 (2014)
9. Sung, J., Ponce, C., Selman, B., et al.: Unstructured human activity detection from RGB-D images. In: International Conference on Robotics and Automation (ICRA), pp. 842–849. IEEE (2012)

An Accurate Online Non-rigid Structure from Motion Algorithm

Ya-Ping Wang, Zhan-Li Sun[(✉)], Yang Qian, Yun Jing, and De-Xiang Zhang

School of Electrical Engineering and Automation, Anhui University, Hefei, China
zhlsun2006@126.com

Abstract. So far, most existing non-rigid structure from motion (NRSFM) problems are solved by the batch algorithm. In this paper, a more accurate online NRSFM is proposed based on the differential evolution (DE) algorithm. Experiment results on several widely used image sequences demonstrate the effectiveness and feasibility of the proposed algorithm.

Keywords: Non-rigid structure from motion · Differential evolution algorithm · 3D reconstruction

1 Introduction

The task of structure from motion (SFM) is to jointly reconstruct 3D shapes, and estimate the corresponding camera external parameters from a set of 2D images. In terms of the assumption on shape deformation, SFM is generally studied under two models, i.e. rigid SFM and non-rigid SFM (NRSFM). Compared to rigid SFM, NRSFM is a more intractable problem due to the deformation [1, 2].

No matter rigid SFM or NRSFM, the solutions are mostly obtained via the off-line model in the past decade [3–5]. In some real situation, e.g. surveillance video, the 3D model should be established in real time as the video is captured. Therefore, it is necessary to develop some effective online SFM algorithms.

As a relative early online SFM algorithm, a sequential factorization method is proposed in [6] by regarding the feature positions as a vector time series. Compared to the existed factorization methods, the sequential factorization method is able to handle infinite sequences because the singular value decomposition is replaced with an updating computation of only three dominant eigenvectors. Instead of global bundle adjustment, a local bundle adjustment is proposed to estimate the motion of a calibrated camera and the three-dimensional geometry of the filmed environment [7].

The work was supported by grants from National Natural Science Foundation of China (Nos. 61370109, 61272025), a grant from Natural Science Foundation of Anhui Province (No. 1308085MF85), and 2013 Zhan-Li Sun's Technology Foundation for Selected Overseas Chinese Scholars from department of human resources and social security of Anhui Province (Project name: Research on structure from motion and its application on 3D face reconstruction).

© Springer International Publishing Switzerland 2015
D.-S. Huang and K. Han (Eds.): ICIC 2015, Part III, LNAI 9227, pp. 318–322, 2015.
DOI: 10.1007/978-3-319-22053-6_35

Except for non-rigid SFM, the online algorithms have been gradually developed for NRSFM in recent years. In [8], a sequential Bayesian estimate is computed via Navier's equations, which model linear elastic solid deformations and are embedded within an extended Kalman filter. In [9], the shape deformation is represented as a 3D-implicit low-rank model. An incremental approach is proposed to estimate the deformable models. The superior performance is verified via the experiments on motion capture sequences with ground truth 3D data. Nevertheless, we found that the rank value selected via the rank-growing strategy does not always correspond to the optimal parameter value.

In this paper, a more accurate online NRSFM is proposed based on the sequential NRSFM (OL-NRSFM) algorithm [9]. Experiment results on several widely used image sequences demonstrate the effectiveness and feasibility of our proposed algorithm.

The remainder of the paper is organized as follows. In Sect. 2, we present the proposed DE-based OL-NRSFM approach. Experimental results are given in Sect. 3, and our concluding remarks are presented in Sect. 4.

2 Methodology

The proposed algorithm (denoted as DE-OL-NRSFM) is composed of two main parts: construct the OL-NRSFM-based weaker estimators; compute the weighting coefficients with the differential evolution algorithm. A detailed description of these two parts is presented in the following subsections.

2.1 The OL-NRSFM Based Weaker Estimator

The first step of our proposed algorithm is to construct the OL-NRSFM-based weaker estimators by varying the rank values. For the fth frame, the observation matrix $\mathbf{w}_f$ can be represented as a $2 \times p$ matrix, i.e.,

$$\mathbf{w}_f = \begin{pmatrix} x_{f,1} & x_{f,2} & \cdots & x_{f,p} \\ y_{f,1} & y_{f,2} & \cdots & y_{f,p} \end{pmatrix}, \tag{1}$$

where p is the number of feature points. In order to compute the mean shape $\bar{\mathbf{S}}$, the observations from the $(f - m)$th frame to $(f - 1)$th frame are put together to form a sub-sequence matrix $\mathbf{W}_s$,

$$\mathbf{W}_s = \begin{bmatrix} \mathbf{w}_{f-m}^T, & \cdots & \mathbf{w}_{f-1}^T \end{bmatrix}^T. \tag{2}$$

Given $\mathbf{W}_s$, the initial motion matrix $\mathbf{R}_{f-1}$ and the mean shape $\bar{\mathbf{S}}$ are computed via an optimal factorization algorithm.

According to the orthographic projection model, $\mathbf{w}_f$ is factorized as follows [9]:

$$\mathbf{w}_f = \mathbf{R}_f \mathbf{S}_f = \mathbf{R}_f(\bar{\mathbf{S}} + \mathbf{U}_f \mathbf{V}), \tag{3}$$

where $\mathbf{U}_f$ and $\mathbf{V}$ denote the mixing coefficient matrix and shape basis matrix. Given $\mathbf{R}_{f-1}$ and $\bar{\mathbf{S}}$, $\mathbf{U}_f$ and $\mathbf{V}$ can be computed via the linear estimation method [9].

In terms of the following projection error,

$$E = \left\| \mathbf{W}_f - \mathbf{R}_f (\bar{\mathbf{S}} + \mathbf{U}_f \mathbf{V}) \right\|_F^2 , \tag{4}$$

the rank is increased gradually until the projection error is lower than a given threshold, or the rank reaches the pre-setting maximum value r_m [9]. When r_m is set as 0, 1, $\cdots$, l in sequence, we can get a set of mixing coefficient matrix $\mathbf{U}_{fi}$ and shape basis matrix $\mathbf{V}_i$.

For a given r_m, the weaker estimator can be constructed by inputting the observations to the OL-NRSFM algorithm. For the ith weaker estimator, the estimated 3D shape of the fth frame can be given as follows:

$$\tilde{\mathbf{S}}_{fi} = \bar{\mathbf{S}} + \mathbf{U}_{fi} + \mathbf{V}_i \tag{5}$$

2.2 Compute Weighting Coefficients with the DE Algorithm

By varying the maximum rank $r_m (\in [0,l])$, l weaker estimators are constructed via the OL-NRSFM algorithm. Further, the final estimated 3D shape $\tilde{\mathbf{S}}_f$ of the fth frame is computed as a linear weighted sum of the outputs ($\tilde{\mathbf{S}}_{fi}$, ($i = 0, \cdots, l$)) of the weaker estimators, i.e.,

$$\tilde{\mathbf{S}}_f = \frac{x_0 \tilde{\mathbf{S}}_{f0} + x_1 \tilde{\mathbf{S}}_{f1} + \ldots + x_l \tilde{\mathbf{S}}_{fl}}{x_0 + x_1 + \ldots + x_l}, \tag{6}$$

where the weighted coefficients $x_0, x_1, \cdots, x_l$ are obtained via the DE algorithm.

The strategy of the evolutionary algorithm is to search for the minimization $x^* \in X$ of a real value objective function $f(\mathbf{x})$, such that

$$f(\mathbf{x}^*) \leq f(\mathbf{x}), \forall \mathbf{x} \in \mathbf{X}, \tag{7}$$

where $\mathbf{X}$ is the feasible region. A common constraint is $\mathbf{X} \neq \emptyset$, and $f(\mathbf{x}^*) \neq -\infty$.

As the task of the DE algorithm is to search for the optimal weighted coefficients, the individual $\mathbf{x}$ is defined as:

$$\mathbf{x} = \left[x_0, x_1, \cdots, x_l, \right]^T . \tag{8}$$

Correspondingly, the object function is defined as the 2D error:

$$E_{2D} = w_f - \frac{x_0 \tilde{w}_{f0} + x_1 \tilde{w}_{f1} + \ldots + x_l \tilde{w}_{fl}}{x_0 + x_1 + \ldots + x_l}, \tag{9}$$

where $\mathbf{w}_f$ is the observation matrix of the test frame f, $\tilde{\mathbf{w}}_{f0}, \tilde{\mathbf{w}}_{f1}, \cdots, \tilde{\mathbf{w}}_{fl}$ are the estimated 2D coordinates when the maximum rank r_m is set as $0, \cdots, l$, respectively.

Assume that the size of the population is N, the individuals of the Gth generation can be represented as:

$$\mathbf{x}_{i,G} = [x_{1,i,G}, x_{2,i,G}, , x_{l,i,G}], i = 1, 2, \cdots N, \tag{10}$$

where each weighted coefficient satisfies the bound constraints:

$$x_j^L \leq x_{j,i,G} \leq x_j^U. \tag{11}$$

The initial parameter values are randomly drawn from the interval $[x_j^L, x_j^U]$. Each individual undergoes the mutation, recombination and selection from generation to generation until some stopping criterion is reached.

After the optimal weighting coefficients $[x_0, x_1, \cdots x_l]^T$ are obtained, the final estimated 3D shape $\tilde{\mathbf{S}}_f$ of the fth frame can be computed via (6). Correspondingly, we can extract the estimated z-coordinate $\tilde{\mathbf{z}}_f$ and the shape $\tilde{\mathbf{S}}_f$.

3   Experimental Results

We evaluate the performance of our proposed method on two widely used motion datasets: matrix and dinosaur. The performance of the reconstruction accuracy is measured by the estimation error ε between the estimated z-coordinate $\tilde{\mathbf{z}}_f$ and the true z-coordinate $\mathbf{z}_f$,

$$\varepsilon = \frac{1}{F - m} \sum_{f=m+1}^{F} \frac{\left\| \mathbf{z}_f - \tilde{\mathbf{z}}_f \right\|}{\left\| \mathbf{z}_f \right\|}. \tag{12}$$

Table 1 shows the mean and standard deviation of the z-coordinate errors of the online NRSFM algorithm [9] (denoted as OL-NRSFM) and our proposed method (denoted as DE-OLNRSFM). It can be seen that the z-coordinate errors of DE-OL-NRSFM are less than those of OL-NRSFM. Therefore, the proposed DE-based OL-NRSFM approach can effectively decrease the estimation errors of the original OL-NRSFM algorithm.

Table 1. The mean and standard deviation of the z-coordinate errors of OL-NRSFM and DE-OL-NRSFM.

	OL-NRSFM	DE-OL-NRSFM
Matrix	0.5161 ± 0.1750	0.3501 ± 0.1149
Dinosaur	0.2585 ± 0.0603	0.2301 ± 0.0959

Moreover, Table 2 shows the computation times of OL-NRSFM and DE-OL-NRSFM. We can see that the training times of DE-OL-NRSFM are less than those of OL-NRSFM. Thus, the proposed method is more efficient than OL-NRSFM.

Table 2. The computation times (seconds) of OL-NRSFM and DE-OL-NRSFM

	OL-NRSFM	DE-OL-NRSFM
Matrix	524.00	19.64
Dinosaur	3690.20	93.70

4 Conclusion

In this paper, a DE-based OL-NRSFM is proposed to estimate the 3D shape of 2D images. Experimental results on several widely used sequences demonstrated that, compared to the existing method, the proposed method not only has higher estimation accuracy, but has a less training time.

References

1. Bregler, C., Hertzmann, A., Biermann, H.: Recovering non-rigid 3D shape from image streams. In: IEEE Conference on Computer Vision and Pattern Recognition, vol. 2, pp. 690–696 (2000)
2. Akhter, I., Sheikh, Y., Khan, S., Kanade, T.: Trajectory space: a dual representation for non-rigid structure from motion. IEEE Trans. Pattern Anal. Mach. Intell. **33**(7), 1442–1456 (2011)
3. Gotardo, P.F.U., Martinez, A.M.: Computing smooth time-trajectories for camera and deformable shape in structure from motion with occlusion. IEEE Trans. Pattern Anal. Mach. Intell. **33**(10), 2051–2065 (2011)
4. Hamsici, O.C., Gotardo, P.F.U., Martinez, A.M.: Learning spatially-smooth mappings in non-rigid structure from motion. In: Fitzgibbon, A., Lazebnik, S., Perona, P., Sato, Y., Schmid, C. (eds.) ECCV 2012, Part IV. LNCS, vol. 7575, pp. 260–273. Springer, Heidelberg (2012)
5. Gotardo P.F.U., Martinez A.M.: Kernel non-rigid structure from motion. In: IEEE International Conference on Computer Vision, pp. 802–809 (2011)
6. Morita, T., Kanade, T.: A Sequential factorization method for recovering shape and motion from image streams. IEEE Trans. Pattern Anal. Mach. Intell. **19**(8), 858–867 (1997)
7. Mouragnon, E., Lhuillier, M., Dhome, M., Dekeyser, F., Sayd, P.: Generic and real-time structure from motion using local bundle adjustment. Image Vis. Comput. **27**(8), 1178–1193 (2009)
8. Agudo A., Montiel J.M.M.: Finite element based sequential Bayesian non-rigid structure from motion. In: Computer Vision and Pattern Recognition, pp. 1418–1425 (2012)
9. Paladini, M., Bartoli, A., Agapito, L.: Sequential non-rigid structure-from-motion with the 3D-implicit low-rank shape model. In: Daniilidis, K., Maragos, P., Paragios, N. (eds.) ECCV 2010, Part II. LNCS, vol. 6312, pp. 15–28. Springer, Heidelberg (2010)

A Multi-feature Fusion Method for Automatic Multi-label Image Annotation with Weighted Histogram Integral and Closure Regions Counting

Sen Xia[1,2,3,6], Peng Chen[4], Jun Zhang[5], Xiao-Ping Li[6], and Bing Wang[1,2,3(✉)]

[1] School of Electronics and Information Engineering, Tongji University, Shanghai China
wangbing@ustc.edu
[2] The Advanced Research Institute of Intelligent Sensing Network, Tongji University,
Shanghai China
[3] The Key Laboratory of Embedded System and Service Computing, Tongji University,
Shanghai China
[4] Institute of Health Sciences, Anhui University, Hefei Anhui, China
pchen.ustc10@gmail.com
[5] College of Electrical Engineering and Automation, Anhui University, Hefei Anhui, China
wwwzhangjun@gmail.com
[6] Faculty of Computer Engineering, Huaiyin Institute of Technology, Huaian 223002, China
laughlee7468@sina.com

Abstract. In order to reduce the semantics gap and improve the consistency between structural difference among images and similarity of semantics from the corner of cognitive, many approaches on automatic image annotation have been developed vigorously. In addition to making use of both the features of n-order color moment and texture information, a multi-feature fusion method for automatic image annotation was proposed by using weighted histogram integral and closure regions counting in this paper. Based on Corel image data set, it showed in our experiments that the proposed approach can achieve better performance than that of traditional one using multi-label learning k-nearest neighbor algorithm, i.e., it can improve average precision measure index from 0.222 to 0.352 in automatic image annotation.

Keywords: Weighted histogram integral · Closure regions counting · Multi-label learning · Automatic image annotation · Multi-feature fusion

1 Introduction

When image is expressed briefly as a refined set of image features, it is practical to retrieve desired image from a big collection [1]. Statistical features of image from information of color, texture, shape have been used as important parameters in image retrieving [2]. However, it showed that single image feature was far from retrieving the satisfied results [3]. Multi-feature fusion methodology therefore based on content itself gets the place of single feature gradually in image retrieving. For the sake of reducing the semantics gap brought by the incompliance between the machine

© Springer International Publishing Switzerland 2015
D.-S. Huang and K. Han (Eds.): ICIC 2015, Part III, LNAI 9227, pp. 323–330, 2015.
DOI: 10.1007/978-3-319-22053-6_36

understanding of structural image difference via machine learning mechanism and that of human being's perception of difference and similarity in semantic cognitions, accurate image annotation marked by semantic labels can work out the semantic gap puzzle somewhat via whether professional labors or with the help of automatic image semantics annotation in which most of the time can be saved. Many valid methods for multi-instance and multi-label automatic image annotation have been studied and developed in recent years [4–9].

M.R.B and J.L etc. presented a framework to handle classes overlapping in features space for learning multi-label scene classification [5]. ML-KNN, a lazy learning method of multi-label learning was proposed by Zhang and Zhou [8], and, it was performed for automatic image annotation in natural image classification. Of course, the class label's number is only 1.24 for each image on average. However, the semantics gap and the relation between image feature structure and semantic content itself remain still a big challenging problem for us.

In order to make a further clear the relation between image structural components difference and semantic similarity cognition, we proposed in this paper a method for multiple features fusion embodying weighted histogram integral and closure regions counting for automatic image annotation method based on multi-label k-nearest neighbor algorithms for achieving better predicting performance.

2   Method

2.1   Multi-label Learning Used for Image Annotation

Each instance in traditional supervised learning was just treated as belonging only to one class. But, the instance is closely related to two or multi-class actually in real life. In the classification of images, each image has different regional objects representing multiple semantics.

Function 'f' was derived from learning process, $f: x \rightarrow 2^Y$. In mapping of s: $\{(x_1, y_1), (x_2, y_2), \ldots, (x_m, y_m)\}$, x_i is one of the elements in X set which was used to represent one image in data set. y_i is the label set of i-th image and at the same time y_i is one sub set of the full label set Y.

The method of multi-label k-nearest neighbor [8] is adopted in this paper in which principle of maximum a posteriori was used to determine the label set for the image instance tested. Hamming loss, one-error, coverage, ranking loss and average precision [9] are used as main algorithm performance measurement index.

2.2   Transformation of Color Space and Histogram Vector

In HSL (Hue, Saturation, Lightness) space, hue, saturation, and lightness, and at the same time correspondingly, hue, saturation, value in HSV (Hue, Saturation, Value) space are essential variants to characterize the image. For the sake of focusing on and grasping the main distribution information, transformation of color image from RGB color space to gray level space can be implemented. And histogram vector was often used to illustrate the distribution of basic color (Fig. 1).

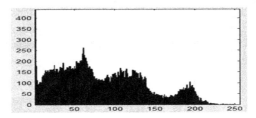

Fig. 1. Histogram of gray level

2.3   Weighted Histogram Integral

In order to reduce the dimensions of features vector, the discrete histogram distributions information of each gray level were replaced by weighted histogram integral area in which gray level is the weight of corresponding frequency.

$$S_{Hist} = \sum_{i=0}^{255} (i + 1)f(i) \tag{1}$$

The original vector composed of each sub gray level from 0 to 255 is now replaced by weighted histogram integral area S_{Hist}.

2.4   Image Semantics and Regions of Interest

Implicit relation between semantics and image regions of interest is assumed to be mined by counting the region numbers and the total area of ROI (Region of Interest) in this paper. One label is clearly correspondent to one specific semantic accordingly. Free version of part of Corel5K image data set was used for the experiments in this paper. The Canny edge detection operator was utilized in the experiments. N_{ROI} and $S_{closure}$ are the representative symbols of numbers of ROI and total area of closed outline in image respectively. And the vector (N_{ROI}, $S_{closure}$) was the comprehensive features of inset targets in one image.

Taking the coarse edge detection as input at low level implemented by differential operator, i.e. Canny edge detection operator, the computing of closure characteristics was conducted among potential target region contour edges on the basis of different and typical relations existed in closure topology [10] in order to judge the relations among various edges.

Counting the number of closure contours or outlines closed and calculating the total closure area, both of them were taken into account being treated as the new features. The number of closure and connected regions in one image represents the discrete distribution that contains the configuration information in the perspective of composition.

Total area of closure and connected regions in one image in this paper is calculated by counting the whole pixel number of these closure and connected regions in one image after the operation of edge detection and binarization. Different labels in Fig. 2 represent different closure and connected regions in one image. The symbol '1' means the first closure and connected region while symbol '2' and '3' represent the second and the third one in the same image.

```
0 0 0 0 0 0 0 0 0 0 0 0 0 0 0 0 0 0 0 0 0 0 0 0
0 0 0 0 0 1 0 0 0 0 0 0 0 0 0 0 0 2 2 2 0 0
0 0 0 0 0 1 1 0 0 0 0 0 0 0 0 0 2 2 2 2 0
0 0 0 0 1 1 1 1 0 0 0 0 0 0 0 2 2 2 2 0 0 0
0 0 0 0 1 1 1 1 0 0 0 0 0 0 2 2 2 2 0 0 0
0 0 0 0 1 1 1 1 1 0 0 0 0 0 0 0 0 0 0 0
0 0 0 0 1 1 1 1 1 0 0 0 0 0 0 0 0 0 0 0
0 0 0 0 1 1 1 1 1 0 0 0 0 0 0 3 0 0 0 0 0
0 0 0 0 1 1 1 1 1 0 0 0 0 0 3 3 3 3 0 0 0
0 0 0 0 0 1 0 0 0 0 0 0 0 0 0 3 3 3 3 0 0
0 0 0 0 0 0 0 0 0 0 0 0 0 0 0 3 3 3 0 0 0
0 0 0 0 0 0 0 0 0 0 0 0 0 0 0 0 0 0 0 0 0
```

Fig. 2. Marking closure and connected regions

2.5   Color Moment and Texture

In one image, n-order color moment of all the sub-color variants can be used as important feature vectors. μ_R, μ_G, μ_B are 1-order color moment which stand for the mean value of all pixel. And $\sigma_R, \sigma_G, \sigma_B, S_R, S_G, S_B$ are accordingly the 2-order and 3-order color moment [3].

$$\mu_R = \frac{1}{N} \sum_{i=1}^{N} P_{iR}, \quad \mu_G = \frac{1}{N} \sum_{i=1}^{N} P_{iG}, \quad \mu_B = \frac{1}{N} \sum_{i=1}^{N} P_{iB} \tag{2}$$

$$\sigma_R = \sqrt{\frac{1}{N} \sum_{i=1}^{N} (P_{iR} - \mu_R)^2}, \quad \sigma_G = \sqrt{\frac{1}{N} \sum_{i=1}^{N} (P_{iG} - \mu_G)^2}, \quad \sigma_B = \sqrt{\frac{1}{N} \sum_{i=1}^{N} (P_{iB} - \mu_B)^2} \tag{3}$$

$$S_R = \sqrt[3]{\frac{1}{N} \sum_{i=1}^{N} (P_{iR} - \mu_R)^3}, \quad S_G = \sqrt[3]{\frac{1}{N} \sum_{i=1}^{N} (P_{iG} - \mu_G)^3}, \quad S_B = \sqrt[3]{\frac{1}{N} \sum_{i=1}^{N} (P_{iB} - \mu_B)^3} \tag{4}$$

Tamura texture features, different kinds of sub parameters including roughness, contrast, direction, resolution lines, regulation and coarseness are used to illustrate the texture of image. Texture parameters derived from gray symbiosis matrix, including energy, inertia moment, entropy and coherence were adopted in this paper to illustrate the image texture distribution characteristics.

2.6   Features Vector

The number of closure and connected regions after edge detection, total area of multiple closure and connected regions, mean value, standard deviation, 1, 2, 3-order color moments, texture information, contrast, correlation, energy, homogeneity, total gray common matrix energy (HSV color space) and gray histogram of image are five main components of features vector.

3   Result

The original images in data set were transformed from color image to gray one and HSV color space separately. The texture information characterized by contrast, correlation, energy, homogeneity and total gray common matrix energy in HSV color space was listed in Table 1.

Table 1. Image texture features

Image no.	Contrast	Correlation	Energy	Homogeneity	Total gray common matrix energy
101005	242.37	2.31	0.02	1.74	10.65

Color moment information characterized by the following n-order variants including mean value, standard deviation and 3-order color moment were given in Tables 2 and 3 separately. Both 3 columns of result data came accordingly from un-normalized image (in Table 2) and normalized one (in Table 3).

Table 2. Un-normalized image's color moment features (Image No.: 101005)

Sub color	Mean	Standard deviation	3-order color moment
R	74.40	58.65	55.07
G	88.79	54.98	40.20
B	75.13	53.94	32.67

Table 3. Normalized image's color moment features (Image No.: 101005)

Sub color	Mean	Standard deviation	3-order color moment
R	0.29	0.23	0.22
G	0.35	0.22	0.16
B	0.36	0.26	0.16

In the process of image multi-label semantics prediction testing, the changing of average precision of prediction on test data set with the adjustment and modification of parameter K which was used to represent the numbers of nearest neighbor were recorded in line chart of Fig. 3.

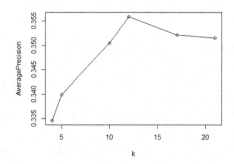

Fig. 3. Average precision with the change of K

Final experimental result of prediction was given in Table 4 with the help of measure index of hamming loss, one-error, coverage, ranking loss and average precision. The total dimensions of features vectors are correspondingly 273, 273 and 17 in three experiments under different conditions. The image pixel data had not been normalized in the first experiment. In both experiments of the second and the third one, all the images pixel data in train set and test set were normalized to 0–1. In the third experiment, the components of image features vector with 17 dimensional features were finally in sequence listed as the following: numbers of ROI, total closure area, weighted histogram area, HSV space's contrast, correlation, energy, homogeneity, total gray common matrix energy and the remaining 9 dimensional features are RGB color space's mean value, standard deviation, 3-order color moment.

Table 4. Comparison of automatic image annotation and performance measure (mean value)

One-error	Coverage	Ranking loss	Average precision
0.789	111.152	0.204	0.159
0.753	110.413	0.192	0.222
0.607	84.645	0.138	0.352

It showed that the prediction performance of image annotation under the condition of image normalized outperforms that of non-normalization in Table 4. And, at the same time, after removing the 256 separate and concrete gray levels histogram distribution information replaced by one single weighted histogram area, hence, the total dimensional degree of features vector was reduced from 273 to 17, the multi-label image annotation performance obtained a notable improving from 0.222 to 0.352 as far as the average precision measure index was concerned. At the same time, one-error decreased from 0.753 to 0.607, coverage form 110.413 to 84.645 and ranking loss from 0.192 to 0.138.

4   Conclusion

To improve the predicting performance, multi-feature fusion embodying weighted histogram integral and closure region counting method for automatic image annotation using multi-label learning k-nearest neighbor method was proposed in this paper. Firstly, the n-order color moment and texture distribution information including contrast, correlation, energy, homogeneity, common gray matrix etc. were calculated. And then, the weighted histogram integral took the place of individual gray level histogram in which the features vector dimensions was cut down in a great deal. Afterwards, counting numbers of closure regions embedded in image worked out in order that the new features vector was appended deliberately. Lastly, the multi-label learning k-nearest neighbor was utilized to test the multi-label predicting performance. It demonstrated that the method with new features vectors fusion embodying weighted histogram integral and closure regions counting for automatic image annotation using ML-KNN has achieved 59 % performance improving from 0.222 to 0.352 in average precision.

Acknowledgement.   This work was funded by the National Science Foundation of China (No. 61472282, No. 61300058 and No. 61271098).

References

1. Amadasun, M., King, R.: Textural features corresponding to texture properties. IEEE Trans. Syst. Man Cybern. **19**, 1264–1274 (1989)
2. Chang, T., Kuo, C.-C.J.: Texture analysis and classification with tree-structured wavelet transform. IEEE Trans. Image Process. **2**(4), 429–441 (1993)
3. Zhang, Y.J.: Image Processing and Analyzing Technology. Higher Education Press, Beijing (2008)
4. Kong, X., Ng, M., Zhou, Z.H.: Transductive multilabel learning via label set propagation. IEEE Trans. Knowl. Data Eng. **25**(3), 704–719 (2013)
5. Boutell, M.R., Luo, J., Shen, X., Brown, C.M.: Learning multi-label scene classification. Pattern Recogn. **37**(9), 1757–1771 (2004)
6. Li, J., Wang, J.Z.: Automatic linguistic indexing of pictures by a statistical modeling approach. IEEE Trans. Pattern Anal. Mach. Intell. **25**(9), 1075–1088 (2003)
7. Carneiro, G., Chan, A.B., Moreno, P.J., Vasconcelos, N.: Supervised learning of semantic classes for image annotation and retrieval. IEEE Trans. Pattern Anal. Mach. Intell. **29**(3), 394–410 (2007)
8. Zhang, M.L., Zhou, Z.H.: ML-kNN: a lazy learning approach to multi-label learning. Pattern Recogn. **40**(7), 2038–2048 (2007)

9. Schapire, R.E., Singer, Y.: BoosTexter: a boosting-based system for text categorization. Mach. Learn. **39**(2/3), 135–168 (2000)
10. Elder, J.H., Zucker, S.W.: Computing contour closure. In: Buxton, B., Cipolla, R. (eds.) ECCV 1996. LNCS, vol. 1064, pp. 399–412. Springer, Berlin Heidelberg (1996)

Diffusion-Based Hybrid Level Set Method for Complex Image Segmentation

Xiao-Feng Wang[(✉)], Le Zou, and Gang Lv

Key Lab of Network and Intelligent Information Processing,
Department of Computer Science and Technology, Hefei University, Hefei 230601, Anhui, China
xfwang@iim.ac.cn,zoule1983@163.com,lvgang119@126.com

Abstract. To capture the weak boundary in complex image, we proposed a new level set method. Different from the exiting methods, our method is performed on diffusion space rather than intensity space. The total energy functional is a linear combination of local part, global part and regularization part. Firstly, the nonlinear diffusion is performed on intensity image to acquire diffused image. Then, the local energy term is formed by implementing a local piecewise constant search on diffused image. To further avoid local minimum, the global energy term is constructed by approximating diffused image in a global piecewise constant way. Besides, the regularization energy term is included to naturally force level set function to be signed distance function. Finally, image segmentation can be performed by minimizing the overall energy functional. The experiments on several complex images with distinct characteristics have shown the powerful boundary approaching ability of our method.

Keywords: Complex image · Global energy · Level set · Local energy · Nonlinear diffusion

1 Introduction

Over the past decades, level set methods (LSMs) have attracted tremendous attentions. Due to the incorporation of different image features into contour evolution, LSMs have been successfully used in various image segmentation applications. Among them, region-based methods [1–3] utilize global image information to drive the level set evolution. These methods are efficient for weak boundaries and are less sensitive to noise. However, they still have some intrinsic limitations, i.e. wrong movement of the evolving contour in complex image and the sensitivity to the placement of initial contour and setting of initial parameters. To improve the performance of region-based methods, local region-based methods were proposed which utilize local image features to guide the contour evolution [4–6]. By statistically examining local region centered in each pixel, they can efficiently segment inhomogeneous objects. Recently, some hybrid methods [7–9] combined global and local information to stabilize and speed up evolution. However, all the above methods are usually performed in intensity space. In complex image, the evolving contours of these methods tend to surround the cluttered background and scattered disturbances, whereas the target object can not be solely separated.

© Springer International Publishing Switzerland 2015
D.-S. Huang and K. Han (Eds.): ICIC 2015, Part III, LNAI 9227, pp. 331–337, 2015.
DOI: 10.1007/978-3-319-22053-6_37

To overcome the aforementioned problems, we proposed performing level set evolution on diffusion space and taking both local and global information into account. Firstly, nonlinear diffusion is performed on intensity image so as to obtain diffused image. Then, the local energy term is constructed by implementing a local piecewise constant search on diffused image. To avoid trapping into local minimum, the global energy term is introduced by approximating diffused image in a global piecewise constant way. The local energy term and global energy term are then combined in a linear form. Further, the regularization energy term is included to naturally force level set function to be a signed distance function.

The remaining of paper is organized as follows: In Sect. 2, we shall discuss the technical details of our method. Then, the experiments of our method on complex images are provided in Sect. 3. Finally, Sect. 4 gives some conclusive remarks.

2   Method

2.1   Local Energy Term

Image diffusion aims to evolve the image as a dynamic system to reduce the random variation and restore the image [10]. Generally, nonlinear diffusion has an impressive effect on image where edges remain well localized and can even be enhanced. Starting with the intensity image $u(t = 0) = I$, the diffused image u is obtained as the solution of the following partial differential equation:

$$\frac{\partial u}{\partial t} = div(f(|\nabla u|)\nabla u), \tag{1}$$

where t and f denote time and diffusivity function, respectively. ∇u is the gradient of u. By decomposing image into piecewise local regions according to their sizes, nonlinear diffusion can efficiently reveal underlying image structures. In other words, cluttered background and scattered disturbances can be smoothed out while object boundary is preserved or even enhanced simultaneously. Hence, the diffused image can be used to strength the description ability of evolving contour.

Firstly, we should select an appropriate diffusivity function. Actually, the choice of the diffusivity function has a rather large influence on the diffusion output. In our application, the TV flow is considered since it can offer a good compromise between the smoothing property and the edge preserving quality.

$$f(|\nabla u|) = \frac{1}{|\nabla u| + \eta}, \tag{2}$$

where η is a small positive constant to maintain the stability of diffusion equation. Here, we use additive operator splitting (AOS) scheme [10] to solve (1) as follows:

$$u_{k+1} = \frac{1}{2}((I - 2\tau A_x(u_k))^{-1} + (I - 2\tau A_y(u_k))^{-1})u_k, \tag{3}$$

where k is iteration number and τ is time-step. A_x and A_y denote the diffusion matrices in horizontal and vertical directions. The most appreciated advantage of AOS scheme is that it is unconditionally stable even a large time-step τ is used.

Next, we shall statistically analyze the local regions in diffusion image. In our implementation, the homomorphic unsharp masking (HUM) method [11] is adopted. At each pixel in the diffused image, the local mean is calculated, and the value of the pixel is multiplied by the ratio of the global mean to this local value.

$$u_{HUM}(x) = u(x)N/M_r(x),\qquad (4)$$

where $M_r(x)$ denotes the mean of $u(x)$ in a local circular window with radius r. N is the global mean of $u(x)$ which acts as a normalized constant.

By performing HUM operation, the remaining low-frequency components in $u(x)$ are reduced while high-frequency boundary can be strengthened. In other words, the sharpness and contrast of $u_{HUM}(x)$ is greatly enhanced. Further, we can construct the local energy term by implementing a local piecewise constant search as follows:

$$E_D^L(l_1, l_2, C) = \int_{inside(C)} |u_{HUM}(x) - l_1|^2\, dx + \int_{outside(C)} |u_{HUM}(x) - l_2|^2\, dx,\qquad (5)$$

where l_1 and l_2 are the averages of $u_{HUM}(x)$ inside and outside C, respectively.

2.2 Global Energy Term

Using only local energy term may trap into local minimum since local statistics is not enough to provide accurate information or even provides incorrect information to misguide the evolution. Besides, the radius r of HUM operation is generally defined with a larger scale (e.g. $r = 32$) to increase the local search ability. Thus, too much detailed information may be introduced and in turn influence contour evolution. To avoid local minimum and increase the segmentation accuracy and speed, we construct the global energy term by approximating $u(x)$ in a global piecewise constant way:

$$E_D^G(g_1, g_2, C) = \int_{inside(C)} |u(x) - g_1|^2\, dx + \int_{outside(C)} |u(x) - g_2|^2\, dx,\qquad (6)$$

where g_1 and g_2 are the averages of $u(x)$ inside and outside C, respectively. It should be noted that local energy term is responsible for accurately detecting object boundaries with low contrast while global energy term aims to segment region with relatively homogeneous feature. To making two energy terms work complementarily, they are combined with a balancing parameter $\alpha \in [0, 1]$ in a linear way as follows:

$$E = \alpha \cdot E_D^L(l_1, l_2, C) + (1 - \alpha) \cdot E_D^G(g_1, g_2, C).\qquad (7)$$

2.3   Regularization Energy Term

During the contour evolution, level set function (LSF) should be firstly formatted to a signed distance function (SDF) and then maintain an SDF to keep the numerical stability and accuracy of contour evolution. Different from the traditional time-consuming re-initialization solution scheme, the following energy term proposed in our recent work [9] is adopted to naturally keep LSF as a SDF.

$$R(\phi) = \int_{\Omega} P(|\nabla \phi(x)|)dx, \quad P(s) = \begin{cases} \dfrac{1}{2}s^2(s-1)^2 + \dfrac{1}{2}s^3(s-1)^3, & \text{if } s \le 1 \\[2mm] \dfrac{1}{2}(s-1)^2, & \text{if } s > 1 \end{cases}, \quad (8)$$

where $P(s)$ is a double-well potential function based on polynomial. By utilizing this energy term, LSF can be preserved as an SDF near the position of zero level set and become a constant far away from the zero level set. Furthermore, to avoid the emergence of small and isolated curves, the frequently used energy term $L(\phi)$ is also included. Thus, the final regularization energy term E^R is described as follows:

$$E^R(\phi) = \mu \cdot L(\phi) + v \cdot R(\phi) = \mu \cdot \int_{\Omega} \delta(\phi(x)) |\nabla \phi(x)| \, dx + v \cdot \int_{\Omega} P(|\nabla \phi(x)|)dx, \quad (9)$$

where μ and v is positive constants to separately control the length penalization effect and SDF penalization effect.

2.4   Level Set Formulation

Finally, the evolving contour C is implicitly represented as the zero level set of LSF ϕ. The total energy functional can be reformulated in terms of ϕ:

$$\begin{aligned} E^T(l_1, l_2, g_1, g_2, \phi) &= \alpha \cdot E_D^L(l_1, l_2, \phi) + (1-\alpha) \cdot E_D^G(g_1, g_2, \phi) + E^R(\phi) \\ &= \alpha \cdot \left(\int_{\Omega} |u_{HUM}(x) - l_1|^2 H_\varepsilon(\phi(x))dx + \int_{\Omega} |u_{HUM}(x) - l_2|^2 (1 - H_\varepsilon(\phi(x)))dx \right) \\ &\quad + (1-\alpha) \cdot \left(\int_{\Omega} |u(x) - g_1|^2 H_\varepsilon(\phi(x))dx + \int_{\Omega} |u(x) - g_2|^2 (1 - H_\varepsilon(\phi(x)))dx \right) \\ &\quad + \mu \cdot \int_{\Omega} \delta_\varepsilon(\phi(x)) |\nabla \phi(x)| \, dx + v \cdot \int_{\Omega} P(|\nabla \phi(x)|)dx, \end{aligned} \quad (10)$$

where $H_\varepsilon(x)$ and $\delta_\varepsilon(x)$ are defined as follows:

$$H_\varepsilon(x) = \frac{1}{2}\left| 1 + \frac{2}{\pi} \arctan\left|\frac{x}{\varepsilon}\right| \right|, \quad \delta_\varepsilon(x) = \frac{1}{\pi} \cdot \frac{\varepsilon}{\varepsilon^2 + x^2}. \quad (11)$$

Fixing ϕ, we can minimize the energy functional in (10) with respect to two pairs of constant functions: $l_1(\phi)$ and $l_2(\phi)$, $g_1(\phi)$ and $g_2(\phi)$. By calculus of variations, the four constant functions are given in (12) and (13).

$$l_1(\phi) = \frac{\int_\Omega u_{HUM}(x) H_\varepsilon(\phi(x)) dx}{\int_\Omega H_\varepsilon(\phi(x)) dx}, \ l_2(\phi) = \frac{\int_\Omega u_{HUM}(x)(1 - H_\varepsilon(\phi(x))) dx}{\int_\Omega (1 - H_\varepsilon(\phi(x))) dx}. \quad (12)$$

$$g_1(\phi) = \frac{\int_\Omega u(x) H_\varepsilon(\phi(x)) dx}{\int_\Omega H_\varepsilon(\phi(x)) dx}, \ g_2(\phi) = \frac{\int_\Omega u(x)(1 - H_\varepsilon(\phi(x))) dx}{\int_\Omega (1 - H_\varepsilon(\phi(x))) dx}. \quad (13)$$

Keeping $l_1(\phi)$, $l_2(\phi)$, $g_1(\phi)$ and $g_2(\phi)$ fixed and minimizing $E^T(l_1, l_2, g_1, g_2, \phi)$ in terms of ϕ, the following gradient flow equation can be obtained:

$$\frac{\partial \phi}{\partial t} = \delta_\varepsilon(\phi)\{\alpha \cdot ((u_{HUM}(x) - l_2)^2 - (u_{HUM}(x) - l_1)^2) + (1 - \alpha) \cdot ((u(x) - g_2)^2$$
$$- (u(x) - g_1)^2) + \mu \cdot div(\nabla\phi / |\nabla\phi|)\} + v \cdot div(d(|\nabla\phi|)\nabla\phi), \quad (14)$$

where $d(s) = P'(s)/s$. In our implementation, the finite difference scheme is used to solve the above equation.

3 Experiments

In this section, we shall validate our method on different types of complex images. The proposed method was coded and tested by Matlab R2010a. Here, the same parameters, i.e. $k = 15$, $\tau = 25$, $\eta = 0.001$, $\alpha = 0.5$, $\Delta t = 0.1$, $\varepsilon = 1$, $r = 32$, $\mu = 0.001 * 255^2$ and $v = 2$ is used for all experiments.

Firstly, we shall test the performance of our method on three distinct images. The first image is a normal image. The second image is a noisy image. The third image displays a piece of leaf where the slant stripes in background are disturbances. For convince, the initial contours of experiments in this paper were all placed in the center of each image. The final evolving contours of our method are presented in the lower row of Fig. 1. All three images were well segmented despite the existence of strong noise or slant stripes. The success of segmentations should owe to the usage of image diffusion which removes the disturbances existed in original intensity image.

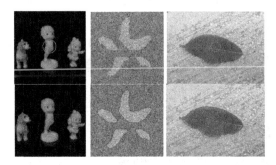

Fig. 1. The segmentation of our method on three distinct images. The upper row: Initial contours. The lower row: Final segmentation results after 8, 10 and 60 iterations.

Next, the efficiency of our method for segmenting images with texture structure was demonstrated in Fig. 2. The first image shows a manhole cover with regular radiated texture structure. The second image contains a starfish with regular natural texture and an inhomogeneous rock background. The third image shows a hawk with irregular natural texture and a cloudy sky background. The final segmentation results are illustrated in the lower row of Fig. 2 where the manhole cover, starfish and hawk were all wholly separated from background. In our method, the powerful boundary approaching ability is due to the usage of local energy term and the robustness for texture structure attributes to the global energy term in diffusion space.

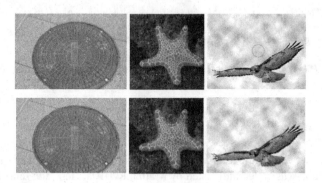

Fig. 2. The segmentation of our method on texture images. The upper row: Initial contours. The lower row: Final segmentation results after 7, 15 and 20 iterations.

Finally, we provide more experiments to demonstrate the segmentation performance of our method on real images with complex background (as shown in Fig. 3). The first image shows a lotus to be separated from the stem and leaf in background. The second image displays a flower surrounded by a mass of disturbed leaves. The third image contains a fire hydrant where the wall, gate and grass are all disturbances. The final segmentation results are presented in the lower row of Fig. 3. Due to the usage of image diffusion, three targets were highlighted in diffused images so that they can be easily separated from respective complex background.

Fig. 3. The segmentation of our method on real images with complex background. The upper row: Initial contours. The lower row: Final segmentation results after 25, 35 and 80 iterations.

4   Conclusion

This paper presented a diffusion-based hybrid level set method to segment complex image. The total energy functional is a linear combination of local, global and regularization energies. Firstly, the nonlinear diffusion is performed to obtain the diffused image. Then, the local energy term is formed by implementing a local piecewise constant search on diffused image. To avoid the local minimum, the global energy term is constructed by approximating diffused image in a global piecewise constant way. The experiments on several complex images can demonstrate that our method has a powerful approaching ability for weak boundary.

Acknowledgements. This work was supported by National Natural Science Foundation of China, No. 61005010, Anhui Provincial Natural Science Foundation, Nos. 1308085MF84, 1408085MF135 and 1508085QF116, Support Project for Excellent Young Talent in College of Anhui Province (X.F. Wang), Key Constructive Discipline Project of Hefei University, No. 2014xk08, Training Object Project for Academic Leader of Hefei University, No. 2014dtr08.

References

1. Chan, T.F., Vese, L.A.: Active contours without edges. IEEE Trans. Image Process. **10**(2), 266–277 (2001)
2. Paragios, N., Deriche, R.: Geodesic active regions and level set methods for supervised texture segmentation. Int. J. Comput. Vis. **46**(4), 223–247 (2002)
3. Shi, Y.G., Karl, W.C.: A real-time algorithm for the approximation of level-set based curve evolution. IEEE Trans. Image Process. **17**(5), 645–656 (2008)
4. Li, C., Kao, C., Gore, J.C., Ding, Z.: Minimization of region-scalable fitting energy for image segmentation. IEEE Trans. Image Process. **17**, 1940–1949 (2008)
5. Wang, X., Huang, D., Xu, H.: An efficient local Chan-Vese model for image segmentation. Pattern Recogn. **43**(3), 603–618 (2010)
6. Zhang, K., Song, H., Zhang, L.: Active contours driven by local image fitting energy. Pattern Recogn. **43**(4), 1199–1206 (2010)
7. Liu, L., Zhang, Q., Wu, M., Li, W., Shang, F.: Adaptive segmentation of magnetic resonance images with intensity inhomogeneity using level set method. Magn. Reson. Imag. **31**, 567–574 (2013)
8. Wang, H., Huang, T., Xu, Z., Wang, Y.: An active contour model and its algorithms with local and global Gaussian distribution fitting energies. Inf. Sci. **263**, 43–59 (2014)
9. Wang, X., Min, H., Zou, L., Zhang, Y.: A novel level set method for image segmentation by incorporating local statistical analysis and global similarity measurement. Pattern Recogn. **48**(1), 189–204 (2015)
10. Weickert, J.: Anisotropic Diffusion in Image Processing. ECMI Series. Teubner-Verlag Stuttgart, Germany (1998)
11. Brinkmann, B.H., Manduca, A., Robb, R.A.: Optimized homomorphic unsharp masking for MR grayscale inhomogeneity correction. IEEE Trans. Med. Imag. **17**(2), 161–171 (1998)

Knowledge Discovery and Data Mining

The Optimization of Resource Allocation Based on Process Mining

Weidong Zhao[1,2(✉)], Liu Yang[1,2], Haitao Liu[1,2], and Ran Wu[1,2]

[1] Software School, Fudan University, Shanghai China
[2] Shanghai Key Laboratory of Data Science, Fudan University, Shanghai China
{wdzhao,12212010021,13212010011,13212010022}@fudan.edu.cn

Abstract. The effectiveness of resource allocation directly affects process performance. In order to optimize resource allocation, this paper proposes a resource allocation model in view of the relationship between resource allocation and process performance, which minimizes process execution time in terms of resource preference, cost constraints and resource availability criteria. Resource coordination is paid less attention in previous resource allocation studies. Therefore, this paper presents the corresponding resource allocation method in consideration of resource coordination, the interval between adjacent activities and distinguishing turnaround time between different resources from event logs. The experiments show that the proposed method can effectively optimize the resource allocation.

Keywords: Resource allocation · Process performance · Process mining · Optimization models · Resource coordination

1 Introduction

Resource allocation, as the key issue of process management, focuses on how to allocate resources efficiently to optimize process performance. Hence, its quality directly affects process performance. Process mining, which can discover useful patterns of resource allocation from process event logs, can optimize the process by eliminating existing bottlenecks in time and is significant in improving both resource execution efficiency and process performance. We only focus on staff (participants) here in view of their special importance [1].

Resource allocation models describe the relationship between process activities and resources. Analysis of resource models can optimize resource allocation. Kumar and Aalst proposed two basic resource allocation modes: push mode and pull mode [2]. The former pushes process tasks to the resources that meet the requirements, while the latter allows resources to request task initiatively from the task pool. The two modes are too simple to deal with complex situations like resource shortage or overload. The purpose of resource allocation models is to allocate the most appropriate resources for process activities [3, 4]. Event logs record the real execution of resource allocation, so mining resource allocation models from event logs is more consistent with actual situations of the process, and more suitable for the optimization of resource allocation. Using the decision

© Springer International Publishing Switzerland 2015
D.-S. Huang and K. Han (Eds.): ICIC 2015, Part III, LNAI 9227, pp. 341–353, 2015.
DOI: 10.1007/978-3-319-22053-6_38

tree algorithm, Ly et al. discovered the task allocation rules from historical data and organizational structure [5]. They treated process actors and the type of activities as input, and whether participants are involved in activities as classification results, to learn the resource allocation model inductively. Task allocation rules reflect information on resources such as their preferences, skills, etc. These rules can be used to adjust resource allocation. Huang et al. used association analysis to mine the dependencies among resources [6]. Utilizing the sequence correlation constraint of process activities, they improved the Apriori algorithm and produced two types of resource allocation rules: the resource dependency rule and activity allocation rule. The resource dependency reflected the relationship between resources, which connected the process activities orderly. For example, the resource r_1 performs the activity a_1, then the follow-up activity a_2 will be performed by r_2; activity allocation rules show that some resources frequently participate in a specific task. Resource dependencies reflect the deep interaction between resources, which helps understand the mode of resource cooperation. On the basis of these models, some researchers tried to use certain resource allocation models to achieve automatic resource allocation, thus improving process efficiency. In addition, by mining event logs to analyze resource allocation rules, we also got preliminary exploration in recommending resources to managers [7]. For example, Yang et al. used the hidden Markov model to build resource allocation models, by mining initialization parameters from event logs, to recommend suitable process resources to activities according to the probability of employees involved in these activities and the transaction between staffs [8].

The researches above seldom concerned about the influence that resource allocation has on process performance. Some studies on performance optimization based on resource allocation have been done. Process execution time and cost are two commonly used metrics for process performance. But it is hard to achieve the metrics best simultaneously. As a result, lowering process cost can increase execution time consumption and speeding up process execution might increase cost. To compromise process cost and time, appropriate balance has to be achieved. Kumar et al. proposed a dynamic resource allocation method that balanced process performance and resource access control [9]. Xu et al. presented a resource allocation method, which minimized process cost under limited time constraint [10].

When allocating resources to activities, most methods only consider the applicability, consistency and availability of resources, etc. [11]. However, in real business processes, employees' productivity usually changes due to variations of some external factors such as work stress, business environments and so on. Resources are the key elements of process performance. Process activities and their logical relations are just external form of resource roles and their coordination [12]. Some scholars highlighted the coordination among resources, ensuring the effective coordination among them. For example, Aalst et al. calculated parameters relevant to resource coordination level according to causality between activities [13]. Huang et al. calculated prior probabilities of activities between resources and made judgments of the correlation between resources on the basis of Aalst's researches [14]. Those studies did not pay enough attention to the effect of resource allocation on process performance. What's more, most of them determined resource correlation subjectively or only considered connections between activities, without analyzing the correlation between staff from process event logs. But rich information on resources is hidden in process event logs, which deserves further researches [15].

In order to solve low process performance and the relationship between resources in existing resource allocation models, this paper proposes a new resource allocation model that can minimize the execution time while meeting the requirements of cost and resource availability constraints for higher process performance. On this basis, it fully describes a real execution situation of the process, considering the effect of correlation between resources on resource allocation.

The remainder of the paper is organized as follows: Sect. 2 generally introduces the goal and constraints of the proposed resource allocation model. Sections 3 and 4 describe the way to judge the resource availability, which is the key constraint in resource allocation. Process cost is discussed in Sect. 5. Section 6 touches on resource coordination, and elaborates the resource allocation method aiming at minimal flow time. Experiments are discussed in Sect. 7, and Sect. 8 concludes this paper.

2　Resource Allocation Model to Minimize Total Process Time

The resource allocation table is used to store the status after a resource is allocated. A record is inserted into the table when certain resource is assigned for an activity. Information contained in this record includes the activity v, the assigned resource s, the corresponding role r, the start time of the activity *start time*, the end time of the activity *end time*. The resource allocation table includes the following subjects:

(1)　For a business process P, we define the activity set SA = { | i = 1,2, ..., k, k is the total number of activities}, resource set SP = { | i = 1,2, ..., n, n is the total number of resources} and role set SR = { | i = 1,2, ..., m, m is the total number of roles}.

(2)　The precursor and sub-sequence relationships between activities and resources: in sequentially connected activities, precursor activity occurs, and then the successor activity occurs after the precursor activity directly. Resources that execute precursor activities are called precursor resources, and while resources that execute subsequence activities are called sub-sequence ones. For example, i is the precursor activity of j, which means j is the subsequence activity of i; If s_i is the precursor resource of s_j, then s_j is the subsequence resource of s_i. Specifically, the process of initiation (termination) activity (resource) does not have precursor (subsequence) activity (resource); there is no precursor or subsequence relationship for parallel activities. In P, RA represents the relationship set among activities, and RP represents the relationship set among resources.

$$P = (SA, SP, SR, RA, RP)$$

The aim of resource allocation model is to minimize the total execution time under three constrains: resource preference, resource availability and total cost.

(1)　**Resource Preference.** With regard to a particular resource, resource preference is a set of activities that have been executed more often and have higher execution efficiency. Although resources have the ability to perform various tasks, they do prefer to do some of them. A resource's willingness and efficiency will be much improved if some activities are allocated to them.

(2) **Resource availability.** The availability of resources for activities is limited by simultaneity and workload. Simultaneity indicates that there is no free time to receive new tasks when a resource is executing certain activity.

(3) **Process Total Cost.** Process cost must be limited within a certain range. Time and cost are two of the most important metrics to measure process performance, yet they cannot be optimized at the same time. Minimizing process time blindly is likely to cause the increase of cost, so we need to develop a method that can control cost within a certain range while minimizing process time.

The resource allocation model is defined as follows:

$$D(P) = min \sum_v Time\,(v) \tag{1}$$

$$\begin{aligned} &preference(s, v) > \beta, t \in SA, \quad s \in SP \\ &availability(tm, s) = true, \quad s \in SP \\ &\sum_v cost(s, v) \le \cos t_{lim}, \quad t \in SA, s \in SP \end{aligned} \tag{2}$$

In Eq. (1), $D(P)$ is the object function which means to minimize the total time of P, and $Time(v)$ denotes time spent by each activity v. Equation (2) includes three constrains. $preference(s,v)$ denotes the preference value of s to v, and β is the threshold for eliminating activities that resources are not interested in; $availability(tm,s)$ denotes whether a resource s is available at tm; $cost(s,v)$ denotes the cost generated when s *is* executing v, and $cost_{max}$ is the cost constraint.

3   Resource Preference Constraints

This section shows the measurement for resource preference constraints. Resource preference *preference(s,t)* indicates the priority of v for s. Higher preference value means that resources are more likely to execute this activity efficiently.

Resource preference support represents the probability that v may be executed by s in a period, as defined in Eq. (3).

$$support(s, v) = \frac{count\,(s, v, tim_1) - count(s, v, tim_2)}{count\,(v, tim_1) - count(v, tim_2)} \tag{3}$$

where $count(s,v,tim_1)$ denotes times that s has executed v until the time tim_1, $count(s,v,tim_2)$ denotes times that s has executed v until tim_2 which is less than tim_1. $count(v,tim_1)$ count (v, tim_1) denotes times that v has been executed up to tim_1, $count(v,tim_2)$ represents times that v has been executed until tim_2.

Resource preference confidence represents the probability that s will execute t in a period of time, as defined in Eq. (4).

$$\text{confidence}\,(s,v) = \frac{\text{count}\,(s,t,tim_1) - \text{count}(s,t,tim_2)}{\text{count}\,(s,tim_1) - \text{count}(s,tim_2)} \tag{4}$$

count(s,tim_1) denotes times of activities that s has executed until tim_1, while *count*(s,tim_2) represents times of activities that s has executed until tim_2.

Resource Preference is related to preference confidence and preference support, as defined in Eq. (5).

$$Preference(s,v) = \alpha\, support(s,v) + (1-\alpha)\, confidence(s,v) \tag{5}$$

In Eq. (5), α is a parameter ($0 < \alpha < 1$), which represents the weight of resource preference support on resource preference. In order to meet the conditions, *preference*(s,v) has to be greater than the threshold β.

4   Resource Availability Constraints

This section discusses the algorithm of the resource availability constraint function *availability(tm,s)*.

(1) ***Resource load***
Resource load represents work pressure of resources. The factors that cause resource overload include the number of activities that has been executed in a certain period (we choose the time window as 1 week) and work time. The *Yerkes-Dodson Law* shows that work efficiency is relevant to work load. With reasonable pressure, resources can work more efficiently, and overload may decrease work efficiency. The curve of the Yerkes-Dodson Law is an inverted U-shape. Inspired by the Yerkes-Dodson Law, resource load can be measured by work efficiency. Since time consumption of v is changing with resources, work efficiency can be calculated by the average time of v divided by time consumption by s. Time consumption start when a resource begins work and end until the start of its subsequence activity. Therefore, through mining event logs, time can be used as the horizontal axis and resource efficiency can be used as the vertical axis to form the curve of resource load. The highest point on the inverted U curve stands for the saturation value of resource load; otherwise, it means that resource efficiency can be improved.

(2) ***Resource potential***
Resource potential signifies the potential of improving work load when the load of resource has already saturated. Higher potential for a resource indicates the larger space to increase the resource capacity, which also means that the resource has the ability to execute more activities. We assume the situation that all the resources are at their saturation value when a new activity arrives. Now we have a young, highly

educated man and an old lady with poor education, whom would you give the task to? It is commonly considered that the young man has more potential, so we allocate the activity to him. The larger the potential is, the higher the probability that a resource can improve its work load.

Clustering analysis, which is a classical data mining method, is used herein to determine the resource potential. Four properties are used to determine the resource potential including age, gender, family status and educational status. Each property has different weights. According to these properties, similarity can be calculated between two resources.

We adopt the K-means algorithm to produce resource groups, which turn out to be three clusters: the high potential set (*hCluster*), the middle potential set (*mCluster*) and the low potential set (*lCluster*) respectively.

(3) **Resource availability**

Resource availability is used to judge whether the resource can execute a new activity. Three steps are needed to estimate resource availability: (1) Examining whether the resource is busy. If the resource is executing the activity, then the resource is unavailable. (2) Check the workload of unoccupied resources. The resource is also unavailable provided that its load has reached the limit value. (3) If all of the unoccupied resources are overloaded, we'll choose resources from the high potential set (*hCluster*).

5   Total Cost Constraints

In this section, we will illustrate cost constraints function *costValue*.

(1) **Transmission probability**

$p(tra_{v,v_s}, s)$ represents the probability that v is transmitted to its subsequence v'. Transmission probability can be calculated by the equation A/B, where A is times that v passes to v' after being executed by s, and B is times that v is executed by s.

(2) **Total cost prediction**

There may be a risk that the total cost will be beyond the constraint if we keep reducing execution time. So we have to predict the total cost of this process after allocating resources. The prediction of total cost contains two parts: the first is the *certain cost*, which is the cost from the activities that have been assigned for resources; the second is *uncertain costs* resulting from activities that have not yet been assigned. We can calculate the certain cost by summing them up, but we have to predict the uncertain one. The prediction of uncertain costs is defined as Eq. (6).

$$forecastc\,(v) = cost\,(s, v) + \sum\nolimits_{v_s} p(tran_{v,v_s}, s) forecastc(v_s) \tag{6}$$

forecastc(v), a recursive equation, denotes the total cost prediction for activities which have not yet been allocated resources. s is the resource that take a minimum time consumption of v. Because *forecastc(v)* is the predicted value, so it allows a deviation compared with the actual value. Meanwhile, resource availability constraints test will spend a lot of time, in order to assure efficiency of the cost constraint test, so we do not take the resource availability constraints test for s. v_s is the subsequence activity selected after v. After predicting the uncertain cost, we can get prediction of the total cost using Eq. (7).

$$overallc\ (s,v) = \sum_{v_p} cost\ (v_p) + forecastc\ (v) \tag{7}$$

where overall $c(s,v)$ represents prediction of the total cost after v has been allocated to resources [16]; $cost(v_p)$ denotes the cost of activities which has been assigned for certain resources; $\sum_{v_p} cost\ (v_p)$ denotes the overall cost of all activities from the initiation activity to the precursor of v.

(3) ***Resource replacement ratio***

If the prediction value of overall cost exceeds the cost constraint, we need to reallocate resources to reduce the cost. To decide which activity should be reallocated to the resource, we need to consider whether the activity is included in the longest path. The longest path is the one that takes the longest execution time [10]. The longest path decides execution time of the whole process. In order to avoid extension of execution time during adjustment, it is necessary to find the most suitable replacement activity from both the longest path and other paths, and then compare their influence on execution time after replacement. The activity that has less influence will be chosen.

Before getting the most suitable replacement activity, we need to find out the most suitable replacement resource for each activity in the process. During adjustment, replacing resources will cause modifications of execution time and processing cost. Let $\Delta time$ represent the time difference when resources change from s_1 to s_2, which has less cost than s_1. $-\Delta time/\Delta cost$ measures the relative change of cost and time due to resource replacement, which is called the resource *replacement ratio*, denoted as *compensation* (v,s_1,s_2). Lower because $\Delta cost < 0$, the resource which has the lowest resource replacement ratio is the best replacement resource. Resource replacement ratio means more reduction of cost with less increase of execution time.

6   Resource Allocation Algorithm

After analyzing resource preference constraints, resource availability constraints and total cost constraints, this section will determine the target of the resource allocation model: the total processing time. Most of researchers only focus on process execution time, ignoring the turnaround time during process execution. Turnaround time is the time from the end of the precursor activity to the start of its subsequence activity. In general, the turnaround time of neighboring activities varies mainly due to the different collaboration relationship between resources. Resources with higher collaboration level usually have shorter turnaround time. To simplify the problem, this paper considers that the collaboration between resources is the main reason for turnaround time difference.

For the reason that we have considered the turnaround time caused by resource collaboration level, we achieve the allocation goal $D(P)$. We need ensure that the execution time *operatetime(v)* for each activity v in the process is as little as possible, while the turnaround time with precursor activities *turntime(v)* also needs to be controlled. Now $D(P)$ has become a multi-objective programming problem. One way to solve the multi-objective programming problem is using linear weighted method to transform the multi-objective programming problem into a single objective programming problem.

In run-time, turnaround time between different resources may vary greatly. From events logs, we find that turnaround time always has larger impact on the total time than process execution time. In some instances, the difference between turn-around time and execution time may reach an order of magnitude, considering the sum of turnaround time in a process may be 30 h while execution time may only be 3 h. A greater order of magnitude that turnaround time is compared to execution time will enhance the impact of turnaround time on process performance. Here is a measurement method using magnitude level k.

$$k = \lg \frac{\overline{tTime}}{\overline{oTime}} \tag{8}$$

where k represents difference order of magnitude between turnaround time and execution time in process instances. $\overline{tTime}$ denotes the average value of turnaround time of completed instances, and $\overline{oTime}$ denotes average value of execution time of process instances. Larger value of k indicates that the turnaround time has greater effect on the total time. For example, turnaround time with magnitude level of 3 has a greater impact on the total time than magnitude level of 1. So in order to embody the impacts of turnaround time on process performance caused by different magnitude level, the weight of execution time is set to $\frac{1}{k+1}$ and the weight of turnaround time is set to $\frac{k}{k+1}$. Meanwhile, in order to achieve the same order of magnitude as that of execution time, we let the turnaround time multiply 10^{-k}. The time spent on each activity v of the resource allocation model's target $D(P) = min \sum_v time(v)$ is:

$$time(v) = \frac{1}{k+1} * operateime(v) + \frac{k}{k+1} * 10^{-k} * turntime(v) \tag{9}$$

In Eq. (9), *operatetime(v)* is time consumption of each activity, and *turntime(v)* is turnaround time between activities and their predecessors.

Collaboration among resources is regarded as mutual adjustments and learning processes between resources. Researches on human learning have shown that the learning curve is a power function curve. The learning curve indicates that more times you study, less time you consume. At first the study rate is relatively fast, but it will gradually diminish to be flat [16]. From this point of view, we can conclude that resources will have relatively low level of correlation and high level of turnaround time at the early stage of cooperation. With the enhancement of correlation, turnaround time decreases and tends towards stability.

On the basis of the learning curve, a regression equation for turnaround time prediction is proposed.

$$turntime = startime * tms^{-r} \tag{10}$$

The *startime* in Eq. (10) is turnaround time at the first collaboration stage between resources; r represents collaboration coefficient between resources; *tms* represents times that two resources collaborate; *turntime* is turnaround time provided that collaboration times of these resources are *tms*. Firstly, evaluating the logarithm of both sides, and then nonlinear regression analysis will be converted into linear regression analysis. We can get a linear regression equation in which $\lg(turntime)$ is the dependent variable and $\lg(tms)$ is the independent variable. By using the sum of squared error and taking the derivative of r, the model will attain the best fitting state that the sum of n deviation is minimum.

$$r = \frac{\lg(starttime) \sum_{i=1}^{n} \lg(tms)_i - \sum_{i=1}^{n} lg(tms)_i * lg(turntime)_i}{\sum_{i=1}^{n} [lg(tms)_i]^2} \tag{11}$$

We can get the collaboration coefficient r by inputting the corresponding data of event logs into Eq. (11), and then predict turnaround time of next collaboration. Not only resources' learning abilities are the determinants of collaboration coefficient r, but also resources' cooperation, operation capacities and even information systems. The collaboration coefficient r will turn out to be larger if the resources have strong cooperation abilities or are familiar with business operations. A good management system and highly efficient staff will lead to larger r as well. Hence, the collaboration coefficient r represents resources' comprehensive abilities.

The resource allocation algorithm of minimum execution cost was put forward by Xu et al., which was to find all the resources that have the lowest execution cost for activities in the process at first, then detect the availability of these resources, and finally check whether the total time of the process exceeds the constraint value after ensuring availability [10]. This method may make the whole process repeatedly. A new method that can detect the availability and cost constraints while allocating resources to each activity is proposed in this section. The steps to allocate resources for each activity are shown as follows:

(1) Find the resource that has the least execution time when performing the target activity v.
(2) Detect whether the resource satisfies the constraints of resource preference. Looking for candidate resources, which have less execution time and satisfy the resource preference constraint.
(3) Detect the availability constraint of the resource given by step (2). If the resource is unavailable, then it will be replaced by other suitable resources, which meet the availability constraint.
(4) Predict the total cost after the target activity has taken this resource, and test whether it exceeds the cost constraint to find the most suitable activity.
(5) If the resource has been replaced in step (4), then repeat step (3) and step (4) until we find out the resource that satisfies both the availability constraint and total cost constraint.

7   Experiments

Some experiments are conducted to show effectiveness of the method proposed above. In the first place, we check whether it is effective to treat resource collaboration as a learning procedure. We have collected some airline compensation process logs and observe variance of average flow time as Table 1.

Table 1. Flow time variance as the number of instances increases

Number of instances	30	40	50	60	70	80	90	100
Average flow time (h)	121	103	90	82	77	75	80	75

It is showed that flow time decreases very fast when the number of process instances increases from 30 to 100, and after that flow time slightly fluctuates in Table 1. The reason is that resource collaboration level improves with increase of cooperation times. Thus, turnover time and flow time will be reduced. However, collaboration capability would remain stable after the number of instances reaches 70. Note that flow time rises a little after the number of instances is 100, which may be explained by the fact that some activities cannot be executed by suitable resources due to the availability issue.

We also check improvement in terms of flow time considering resource collaboration. We choose 8 process log sets of the airline company randomly. The number of process instances of each set is 30. After extracting the original logs, data pre-processing is conducted to filter out resources in which interaction frequencies are less than 3. The remained process instances are used for analysis. A small number of cooperation activities are insufficient to measure resource collaboration. Then, we use the algorithm that only concern *operatetime*(v) in *time*(v) without touching on collaboration. In comparison, we use the resource allocation algorithm proposed in this paper taking account of collaboration as allocation guide. To compare the effect of collaboration, the measure *time*(v) and *operatetime*(v) are used respectively. The average flow time is used as performance metric. Figure 1 depicts the results.

It can be seen that process execution time reduces significantly by taking resource collaboration into account. In addition, we can see more satisfactory results as the number of process instances increases. The underlying reason is that more event logs can generate more accurate collaboration measurement, which in turn provides more effective allocation guideline. It can be noted that improvements can still be made for process data with fewer event records through deleting some process logs. Even a small number of event records can provide sufficient evidence for resource collaboration.

We use the same process log sets as Fig. 1 and compare the resource allocation algorithm proposed in this paper considering resource coordination capacity with the algorithm (named rb) by Kumar [17]. The clearest difference between these two algorithms is that the former predicts collaboration capacity between resources using regression analysis, and considering collaboration to be dynamic as time goes. But the latter calculated the average value as collaboration capacity. It did not pay attention to

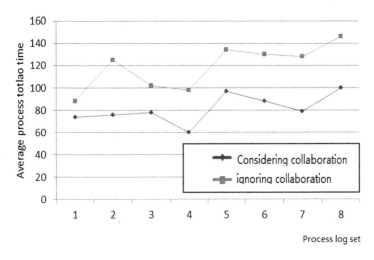

Fig. 1. Considering vs. ignoring collaboration between resources

changing patterns of collaboration between resources, thus lacking of considering collaboration.

Huang also proposed a method (named mc) to compute the degree of collaboration between resources. The main idea of this method was based on the premise of guaranteeing the correlation between resources, and measuring the strength of collaboration between resources by calculating the probabilities [15]. In Fig. 2, process log sets are denoted as the horizontal axis, and the average process total time is denoted as the vertical axis. We can see that dynamic collaboration capacity between resources have more positive effects on performance than the static one.

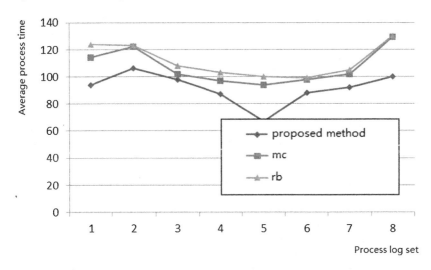

Fig. 2. The influence of different methods in considering collaboration between resources

8 Conclusions

Resource allocation is an important issue in the field of process management. The quality of resource allocation not only determines how the process is executed, but also affects process performance. Process mining has provided new insights and methods for resource allocation. This paper discusses the resource allocation method from the perspective of their effects on process performance, ensuring the process takes less time under the cost constraint. Moreover, this paper also analyzes the effects of collaboration cohesion between resources on process performances. We will do further researches on the coordination among resources, especially on the influence of deep collaboration among resources, in which we will mine further insights from process event logs.

Acknowledgments. Shanghai Pujiang Program no. 14PJC017 and the National Nature and Science Foundation of China under Grants no. 71071038 support this work.

References

1. Van Hee, K., Serebrenik, A., Sidorova, N., et al.: Scheduling-free resource management. Data Knowl. Eng. **61**(1), 59–75 (2007)
2. Wang, J., Kumar, A.: A framework for document-driven workflow systems. In: van der Aalst, W.M., Benatallah, B., Casati, F., Curbera, F. (eds.) BPM 2005. LNCS, vol. 3649, pp. 285–301. Springer, Heidelberg (2005)
3. Russell, N., van der Aalst, W.M., ter Hofstede, A.H., Edmond, D.: Workflow resource patterns: identification, representation and tool support. In: Pastor, Ó., Falcão e Cunha, J. (eds.) CAiSE 2005. LNCS, vol. 3520, pp. 216–232. Springer, Heidelberg (2005)
4. ZurMuehlen, M.: Organizational management in workflow applications–issues and perspectives. Inf. Technol. Manag. **5**(3–4), 271–291 (2004)
5. Ly, L.T., Rinderle, S., Dadam, P., Reichert, M.: Mining staff assignment rules from event-based data. In: Bussler, C.J., Haller, A. (eds.) BPM 2005. LNCS, vol. 3812, pp. 177–190. Springer, Heidelberg (2006)
6. Huang, Z., Lu, X., Duan, H.: Mining association rules to support resource allocation in business process management. Expert Syst. Appl. **38**(8), 9483–9490 (2011)
7. Senkul, P., Toroslu, I.H.: An architecture for workflow scheduling under resource allocation constraints. Inf. Syst. **30**(5), 399–422 (2005)
8. Yang, H., Wang, C., Liu, Y., Wang, J.: An optimal approach for workflow staff assignment based on hidden Markov models. In: Meersman, R., Tari, Z., Herrero, P. (eds.) OTM-WS 2008. LNCS, vol. 5333, pp. 24–26. Springer, Heidelberg (2008)
9. Kumar, A., Van Der Aalst, W.M.P., Verbeek, E.M.W.: Dynamic work distribution in workflow management systems: how to balance quality and performance. J. Manag. Inf. Syst. **18**(3), 157–194 (2002)
10. Xu, J., Liu, C., Zhao, X.: Resource allocation vs. business process improvement: how they impact on each other. In: Dumas, M., Reichert, M., Shan, M.-C. (eds.) BPM 2008. LNCS, vol. 5240, pp. 228–243. Springer, Heidelberg (2008)
11. Reijers, H.A., Jansen-Vullers, M.H., ZurMuehlen, M., Appl, W.: Workflow management systems+swarm intelligence=dynamic task assignment for emergency management applications. In: Alonso, G., Dadam, P., Rosemann, M. (eds.) BPM 2007. LNCS, vol. 4714, pp. 125–140. Springer, Heidelberg (2007)

12. Bozkaya, M., Gabriels, J., Werf, J.: Process diagnostics: a method based on process mining. In: Information, Process, and Knowledge Management, pp. 22–27. IEEE (2009)
13. Van Der Aalst, W.M.P., Reijers, H.A., Song, M.: Discovering social networks from event logs. Comput. Support. Coop. Work (CSCW) **14**(6), 549–593 (2005)
14. Huang, Z., Lu, X., Duan, H.: Resource behavior measure and application in business process management. Expert Syst. Appl. **39**(7), 6458–6468 (2012)
15. Van der Aalst, W.M.P.: Discovery Conformance and Enhancement of Business Processes. Springer, Heidelberg (2011)
16. Bellman, R.E.: Dynamic Programming. Princeton University Press, Princeton (1957)
17. Kumar, A., Dijkman, R., Song, M.: Optimal resource assignment in workflows for maximizing cooperation. Bus. Process Manag. **8094**, 235–250 (2013)

Research on Optimum Weighted Combination GM(1,1) Model with Different Initial Value

Qiumei Chen[✉] and Jin Li

College of Information Science and Technology, Zhejiang Sci-Tech University,
Hangzhou 310018, China
Chenqm909@sohu.com, lj_hsw05@163.com

Abstract. In this paper, a new method of GM(1,1) model based on optimum weighted combination with different initial value is put forward. The new proposed model is comprised of weighted combination models with different initial value of raw data. Weighted coefficients of every model in the combination are derived from a method of minimizing error summation of square. The optimum weighted combination can express the principle of new information priority emphasized on in grey systems theory fully. The result of a numerical example indicates that optimum weighted combination GM (1,1) model presented in this paper can obtain a better prediction performance than that from the original GM(1,1) model.

Keywords: GM(1,1) model · Optimum weighted combination · Initial value · Prediction accuracy

1 Introduction

Since the grey system theory was put forward in 1982, it has been extensively used in many areas of economic management and so on [1, 2]. Extensive studies for grey system including grey correlation analysis, grey decision making and grey prediction have been carried out by scholars [3]. Grey systems theory focuses mainly on such systems as partial information known and partial information unknown. With the rapid development of science and technology more and more systems have the same characteristic as grey systems. Solutions to these uncertain systems have become a great challenge for further development in associated fields. However, one of possible means can be provided by some approaches in grey system.

GM (1,1) is playing an important role in grey prediction. From the procedure of construction of GM(1,1) model we can find that the model is neither a differential equation nor a difference equation. In fact, it is an approximate model which has the characteristic both differential equation and difference equation. It is inevitable for the approximate model to result in some errors in practical applications. To increase prediction accuracy using GM(1,1) model a large number of researchers concentrate upon improvements of GM(1,1) model mainly in three aspects. On the one hand, a number of researchers focus mainly on improvements of grey derivative. Wang proposes a GM (1,1) direct modeling method with a step by step optimizing grey derivative's whitened values to unequal time interval sequence modeling. He has also proves that the new

© Springer International Publishing Switzerland 2015
D.-S. Huang and K. Han (Eds.): ICIC 2015, Part III, LNAI 9227, pp. 354–362, 2015.
DOI: 10.1007/978-3-319-22053-6_39

method still has the same characteristic of linear transformation consistency as the old method [4]. Mu presents a method to optimize the whitened values of grey derivative and constructs an unbiased GM(1,1) model. He also proves that the new model has the characteristic of law of whitened exponent [5, 6].

On the other hand, a number of researchers focus mainly on improvements of background value. Li presented the improvement method of background values to improve the prediction precision of GM(1,1)model with unequal time interval. The author takes advantages of the actual values of generated sequence in point $t_{i+1/2}$ as a background value for GM(1,1) model with unequal time interval [7]. Wang has proved that the generated sequence of applying a first order accumulated generation operator on $X^{(0)}$ is a non-homogeneous exponential function. Then the author utilizes a non-homogeneous exponential function to fit the generated sequence and develops the improvement method of background value [8].

Finally, improvements of the initial condition in the time response function are focused on by some researchers gradually. Xie proposes discretely grey prediction models and corresponding parameter optimization methods. And he also illustrates three classes of grey prediction models such as the starting-point fixed discrete grey model, the middle-point fixed discrete grey model and the ending-point fixed discrete grey model [9]. Liu also presents a method to improve the prediction precision by optimization of the coefficient of exponential function [10]. Wang proposed a novel approach to improve prediction accuracy of GM(1,1) model through optimization of the initial condition. The new initial condition is comprised of the first item and the last item of a sequence generated from applying the first-order accumulative generation operator on the sequence of raw data [11].

Traditional GM (1,1) model and the model in paper [12] come to form their own forecasting formulae by considering $x^{(1)}(1)$ and $x^{(1)}(n)$ as their original condition respectively. But in fact, the fitted curve does not pass any data point according to the least square principles. There is lack of theoretical basis in traditional GM (1,1) model and the model in paper [13].

For view of information theory, optimum weighted combination forecasting method is extracting the useful information of single forecasting method. There is a weighted vector M, the squared forecasting error of combination forecasting method has been the minimum value. Then M is defined as optimum weighted coefficient vectors, and the corresponding combination forecasting method is defined as optimum weighted combination forecasting method [14–17].

As we all know, because of choosing different initial value, we can get different forecasting models and different forecasting precision. In this paper, the new method of GM (1,1) model based on optimum weighted combination with different initial value is put forward. At last, the method is applied into building model of per capita living area of city in China, and the results show the effectiveness and superiority of the method.

2   The Modeling Mechanism of Grey GM (1,1) Model

GM (1,1) is the most frequently used grey forecasting model. It is formed by a first order differential equation with a single variable. Its modeling course is as follows:

1. Let the non-negative original data series be denoted by

$$X^{(0)} = \{x^{(0)}(1), x^{(0)}(2), \ldots, x^{(0)}(n)\} \tag{1}$$

$$x^{(0)}(i) > 0, i = 1, 2, \ldots, n$$

2. The AGO (accumulated generation operation) of original data series is defined as:

$$X^{(1)} = \{x^{(1)}(1), x^{(1)}(2), \ldots, x^{(1)}(n)\} \tag{2}$$

$$x^{(1)}(k) = \sum_{i=1}^{k} x^{(0)}(i), k = 1, 2, \ldots, n$$

3. The grey model can be constructed by establishing a first order differential equation for $X^{(1)}$ as:

$$\frac{dx^{(1)}(t)}{dt} + ax^{(1)}(t) = u \tag{3}$$

The difference equation of GM (1,1) model:

$$x^{(0)}(k) + az^{(1)}(k) = u, k = 2, 3, \ldots, n \tag{4}$$

Unfolding Eq. (4), we can obtain:

$$\begin{bmatrix} x^{(0)}(2) \\ x^{(0)}(3) \\ \vdots \\ x^{(0)}(n) \end{bmatrix} = \begin{bmatrix} -z^{(1)}(2) & 1 \\ -z^{(1)}(3) & 1 \\ \vdots & \vdots \\ -z^{(1)}(n) & 1 \end{bmatrix} \times \begin{bmatrix} a \\ u \end{bmatrix} \tag{5}$$

Let $Y = \left[x^{(0)}(2), x^{(0)}(3), \ldots, x^{(0)}(n)\right]^T$, $\Phi = \left[a \quad u\right]^T$,

$$B = \begin{bmatrix} -z^{(1)}(2) & 1 \\ -z^{(1)}(3) & 1 \\ \vdots & \vdots \\ -z^{(1)}(n) & 1 \end{bmatrix} \tag{6}$$

The background $z^{(1)}(k)$ in Eq. (5) is defined as:

$$z^{(1)}(k+1) = \frac{1}{2}[x^{(1)}(k+1) + x^{(1)}(k)], k = 1, 2, \ldots, n-1$$

4. The least squares solution of identification variable:

$$\hat{\Phi} = (B^T B)^{-1} B^T Y$$

5. We can get the discrete solution of Eq. (3) with the original condition $\hat{x}^{(1)}(1) = x^{(1)}(1) = x^{(0)}(1)$:

$$\hat{x}^{(1)}(k+1) = [x^{(1)}(1) - \frac{u}{a}]e^{-ak} + \frac{u}{a} \tag{7}$$

6. Revert into initial data:

$$\begin{aligned}
\hat{x}^{(0)}(k+1) &= \hat{x}^{(1)}(k+1) - \hat{x}^{(1)}(k) \\
&= (1 - e^a)[x^{(1)}(1) - \frac{u}{a}]e^{-ak}
\end{aligned} \tag{8}$$

3   Defect of Traditional Grey Forecasting Formulae and Its Improvement

Grey GM (1,1) is an extrapolating method virtually by using exponential curve $\hat{X}^{(1)}$ to fit with the data series $X^{(1)}$. Traditional GM (1,1) model and the model of paper [12] come to form their own forecasting formulae by considering $x^{(1)}(1)$ and $x^{(1)}(n)$ as their original condition respectively. But in fact, the fitted curve does not pass any data point according to the least square principles. Therefore, there is lack of theoretical basis in traditional forecasting formulae. Now make a brief analysis as follows.

By solving Eq. (3), we can obtain:

$$x^{(1)}(t) = \frac{u}{a} - \frac{C}{a}e^{-at}$$

We can get by dispersing the equation above:

$$\hat{x}^{(1)}(k+1) = -\frac{C}{a}e^{-ak} + \frac{u}{a}$$

In order to solve out constant C, it is essential to have a definite condition to confirm the solution. Let $\hat{x}^{(1)}(1) = x^{(1)}(1) = x^{(0)}(1)$ be the condition, we can obtain:

$$\hat{x}^{(1)}(1) = -\frac{C}{a} + \frac{u}{a} = x^{(1)}(1)$$

$$-\frac{C}{a} = x^{(1)}(1) - \frac{u}{a}$$

$$\hat{x}^{(1)}(k+1) = (x^{(1)}(1) - \frac{u}{a}e^{-ak}) + \frac{u}{a}$$

This is the expression of the forecasting value of grey GM (1,1) model.

So the fitted curve $\hat{X}^{(1)}$ should pass the data point $(0, x^{(1)}(1))$ on the coordinate plane $(k, \hat{x}^{(1)}(k+1))$. However, according to the least square principles, the fitted curve might not pass the first data point. Therefore, the theoretical basis of considering $\hat{x}^{(1)}(1) = x^{(1)}(1)$ as the known condition does not exist. In addition, $x^{(1)}(1)$ is the oldest datum in terms of time and in far relations with the future. So the traditional method considering $\hat{x}^{(1)}(1) = x^{(1)}(1)$ as the known condition should be generalized.

Let the forecasting value of $x^{(1)}(k+1)$ be denoted by $\hat{x}_m^{(1)}(k+1)$ under the condition of $\hat{x}^{(1)}(1) = x^{(1)}(m)(m = 1, 2, \ldots n)$, it is easy to prove that:

$$\hat{x}_m^{(1)}(k+1) = [x^{(1)}(m) - \frac{u}{a}]e^{-a(k-m+1)} + \frac{u}{a}$$

From above, we can get different forecasting models with different values m. A new model based on optimum weighted combination with n kinds of forecasting results will be put forward, and its fitted precision and forecasting precision will be higher than any model with the single value m.

Let the observed value be denoted by $x^{(1)}(k+1)$ and the weight of $\hat{x}_1^{(1)}(k+1), \hat{x}_2^{(1)}(k+1), \ldots, \hat{x}_m^{(1)}(k+1)$ in combination prediction model be denoted by vector $W = (W_1, W_2, \ldots, W_m)^T$, we can obtain the combination model:

$$\hat{x}^{(1)}(k+1) = \sum_{j=1}^{m} W_j \hat{x}_j^{(1)}(k+1) \tag{9}$$

Let the error be denoted by:

$$e(k+1) = x^{(1)}(k+1) - \hat{x}^{(1)}(k+1) \tag{10}$$

To work out optimum weight of combination prediction model is to solve out the following mathematics programming of sum of square error with the least square principles:

$$\begin{cases} \min Q = \sum_{k=1}^{n} e^2(k) \\ s.t \sum_{j=1}^{m} W_j = 1 \end{cases}$$

The non-negative weight of the optimum weighted combination can be worked out by using the *Qordpro* in *Matlab*.

4 Example

In this section we illustrate the application of optimum weighted combination GM(1,1) model with different initial value. Weighted coefficients of every model in the combination are derived from a method of minimizing error summation of square. The optimum weighted combination can express the principle of new information priority emphasized on in grey systems theory fully. To compare with prediction performances among the original GM(1,1) model, optimum weighted combination GM(1,1) model proposed in this paper and the model in document [2], we utilize the first thirteen data in the sequence of simulation data to construct the three models, respectively.

The city per-capita living area of a country is an important index of the living standard. Of course, it is an important factor to reflect economic strength of this country. Setting up the model of city per capita living area and predicting its development in the future have great realistic significance to country for the arrangement and economic plan in the city. Now build the model of the city per-capita living area of China from 1986 to 1998 (《Statistic almanac of China-2002》) by means of the method proposed by this paper and forecast that of 1999, 2000, and 2001.

Let $x^{(1)}(1)$ be the initial value, the grey GM (1,1) model of city per-capita living area of China is as followings:

$$\hat{x}^{(0)}(k+1) = 8.2553e^{0.0370k}, k > 1$$
$$\hat{x}^{(0)}(1) = 8.8$$

With the different initial value, we can obtain the solution of Eq. (7):

$$\hat{x}_m^{(1)}(k+1) = [x^{(1)}(m) - \frac{u}{a}]e^{-a(k-m+1)} + \frac{u}{a}, \ m = 1, 2, \ldots, 13$$

Let the thirteen kinds of different forecasting results be denoted by

$$\hat{x}_1^{(1)}(k+1), \hat{x}_2^{(1)}(k+1), \ldots, \hat{x}_m^{(1)}(k+1)$$

And let the value of GM (1,1) model based on optimum weighted combination be denoted by

$$\hat{x}^{(1)}(k+1) = \sum_{j=1}^{m} W_j \hat{x}_j^{(1)}(k+1)$$

Then we can obtain the non-negative weight of the optimum weighted combination by using the *Qordpro* in *Matlab*:

$W_5 = 0.4584, W_8 = 0.5416,$
$W_1 = W_2 = W_3 = W_4 = W_6 = W_7 = W_9 = W_{10} = W_{11} = W_{12} = W_{13} = 0$

Therefore, the new forecasting results are worked out:

$$\hat{x}^{(1)}(k+1) = 0.4584\hat{x}_5^{(1)}(k+1) + 0.5416\hat{x}_8^{(1)}(k+1)$$

$$\hat{x}^{(0)}(k+1) = 0.4584\hat{x}_5^{(0)}(k+1) + 0.5416\hat{x}_8^{(0)}(k+1)$$

The new model of city per-capita living area of China is as follows:

$$\hat{x}^{(0)}(k+1) = 0.4584 \times 9.9522e^{0.0370(k-4)} + 0.5416 \times 11.1150e^{0.0370(k-7)}$$

It is to say,

$$\hat{x}^{(0)}(k+1) = 4.4521e^{0.0370(k-4)} + 6.0199e^{0.0370(k-7)}, k > 1$$
$$\hat{x}^{(0)}(1) = 8.8$$

Tables 1 and 2 is the model values and forecasting values of there methods.

Table 1. Comparison of method of this paper and traditional GM (1,1) model (Unit: Square Meter)

Year	No.	City per-capita living area	Method of this paper		Traditional GM (1,1) model	
			Model values	Relative error (%)	Model values	Relative error (%)
1981	1	35.82	35.82	0	8.8	0
1982	2	36.84	38.1463	−3.5458	8.8984	1.1291
1983	3	39.11	40.3691	−3.2195	9.2339	0.7105
1984	4	41.93	42.7216	−1.8878	9.5821	1.2153
1985	5	44.52	45.211	−1.5522	9.9434	−0.4388
1986	6	48.93	47.8456	2.2162	10.3184	−0.1785
1987	7	51.92	50.6337	2.4775	10.7075	−0.0698
1988	8	53.95	53.5842	0.678	11.1112	−1.0112
1989	9	55.5	56.7067	−3.0094	11.5302	−1.1422
1990	10	58.54	60.0111	−2.6709	11.965	−1.3983
1991	11	61.7	63.5081	−2.9305	12.4162	−0.9445
1992	12	69.74	67.2089	3.2548	12.8844	0.8895
1993	13	76	71.1253	6.4141	13.3702	1.6896
1994*	14	77.7	75.27	3.1275	13.8743	2.2937
1995*	15	79.15	79.6561	−0.6394	14.3975	3.3725
1996*	16	83.15	84.2979	−1.3805	14.9404	3.6103

(forecasting value with*)

Table 2. Comparison of method of this paper and method of document [12] (Unit: Square Meter)

Year	No.	City per-capita living area	Method of this paper		Method of document [2]	
			Model values	Relative error (%)	Model values	Relative error (%)
1981	1	35.82	35.82	0	8.8	0
1982	2	36.84	38.1463	−3.5458	8.9244	0.84
1983	3	39.11	40.3691	−3.2195	9.2349	0.7
1984	4	41.93	42.7216	−1.8878	9.5831	1.2048
1985	5	44.52	45.211	−1.5522	9.9445	−0.4494
1986	6	48.93	47.8456	2.2162	10.3195	−0.1891
1987	7	51.92	50.6337	2.4775	10.7086	−0.0805
1988	8	53.95	53.5842	0.678	11.1124	−1.0219
1989	9	55.5	56.7067	−3.0094	11.5314	−1.153
1990	10	58.54	60.0111	−2.6709	11.9663	−1.4091
1991	11	61.7	63.5081	−2.9305	12.4175	−0.9552
1992	12	69.74	67.2089	3.2548	12.8857	0.879
1993	13	76	71.1253	6.4141	13.3716	1.6792
1994*	14	77.7	75.27	3.1275	13.8758	2.2831
1995*	15	79.15	79.6561	−0.6394	14.399	3.3624
1996*	16	83.15	84.2979	−1.3805	14.942	3.6

(forecasting value with*)

It is easy to see from Tables 1 and 2, the relative errors of model proposed by this paper are almost lower than 3 %. The error inspection of post-sample method can be used to inspect a quantified approach. The post-sample error of the model $c_1 = S_1/S_0 = 0.0748$ (where S_1 is variation value of the error and S_0 is variation value of the original series), the post-sample error of traditional GM (1,1) model $c_2 = 0.0771$, and the post-sample error of the model of paper [2] $c_3 = 0.0761$. The probability of the small error $p = \{|e^{(0)}(i) - e^{-(0)}|\} < 0.6745 S_0 = 1$. It is obvious that the method of this paper has the higher fitted precision and forecasting precision than the traditional GM (1,1) model and the model in paper [12].

5 Conclusions

The accuracy of model is key factor of the gray mode. Scholars have made considerable achievements in the aspect of improving the precision of grey model. In this paper, the defect of the traditional grey GM (1, 1) model is analyzed theoretically and the irrationality of choosing initial value is pointed out. The new method of GM (1, 1) model based on optimum weighted combination with different initial value is also put forward. The result of a numerical example indicates that optimum weighted combination GM (1, 1) model presented in this paper can obtain a better prediction performance than that from the original GM(1, 1) model.

References

1. Deng, J.L.: The Basis of Grey Theory, pp. 1–2. Press of Huazhong University of Science & Technology, Wuhan (2002)
2. Liu, S.F., Dang, Y.G., Fang, Z.G.: Grey System Theory and Its Application, pp. 1–3. Science Press, Belling (2004)
3. Chin-Tsai, L., Shih-Yu, Y.: Forecast of the output value of Taiwan's opto-electronics industry using the grey forecasting model. Technol. Forecast. Soc. Change **70**, 177–186 (2003)
4. Wang, Y., Chen, Z., Gao, Z., Chen, M.: A generalization of the GM(1,1) direct modeling method with a step by step optimizing grey derivative's whiten values and its applications. Kybernetes **33**(2), 382–389 (2004)
5. Mu, Y.: An unbiased GM(1,1) model with optimum grey derivative's whitening values. Math. Pract. Theor. **33**(3), 13–16 (2003)
6. Mu, Y.: A direct modeling method of the unbiased GM(1,1). Syst. Eng. Electron. **25**(9), 1094–1095 (2003)
7. Li, C., Dai, W.: Determinator of the background level in the non—equidistant GM(1,1) model. J. Tsinghua Univ. Sci. Technol. **47**(s2), 1729–1732 (2007)
8. Wang, Y.M., Dang, Y.G., Wang, Z.X.: The optimization of background value in non—equidistant GM(1,1)model. Chin. J. Manag. Sci. **16**(4), 159–162 (2008)
9. Xie, N., Liu, S.: Discrete grey forecasting model and its optimization. Appl. Math. Model. (2008). doi:10.1016/j.apm.2008.01.011
10. Liu, S., Lin, Y.: Grey Information Theory, and Practical Applications. Springer, London (2006)
11. Yuhong, W., et al.: An approach to increase prediction precision of GM(1,1) model based on optimization of the initial condition. Expert Syst. Appl. **37**(8), 5640–5644 (2010)
12. Yungui, L., Qingfu, L., Guofan, Z.: The improvement of grey GM (1,1) model. Syst. Eng. **6**, 27–31 (1992)
13. Yang, Z., et al.: Forecast annual electricity consumption of ultra-exponent increase trend by curve fitting. J. Tianjin Univ. **41**(11), 1299–1302 (2008)
14. Geng, K.: The discuss on method of optimum weighted combination. Technol. Econ. Res. Quant. Econ. **11**, 42–44 (1998)
15. Zhang, Y., Zhou, A.: The application of optimum weighted combination prediction. Technol. Econ. Res. Quant. Econ. **10**, 56–60 (1997)
16. Zhang, D., Jiang, S., Shi, K.: The theoretical defect and improvement of grey prediction model. Syst. Eng. Theor. Pract., 140–142 (2002)
17. Ming, H., Jun, L., Xiang, P.: Applications of weighted regression model and weighted GM in slope deformation forecast. J. Yangtze River Sci. Res. Inst. **22**(4), 76–79 (2005)

Spectral Clustering of High-Dimensional Data via k-Nearest Neighbor Based Sparse Representation Coefficients

Fang Chen[1], Shulin Wang[1(✉)], and Jianwen Fang[2]

[1] College of Computer Science and Electronics Engineering,
Hunan University, Changsha 410082, Hunan, China
smartforesting@gmail.com
[2] Division of Cancer Treatment and Diagnosis,
National Cancer Institute, Rockville, MD 20850, USA

Abstract. Recently, subspace clustering has achieved promising clustering quality by performing spectral clustering over an affinity graph. It is a key to construct a robust affinity matrix in graph-oriented subspace clustering. Sparse representation can represent each object as a sparse linear combination of other objects and has been used to cluster high-dimensional data. However, all the coefficients are trusted blindly to construct the affinity matrix which may suffer from noise and decrease the clustering performance. We propose to construct the affinity matrix via k-nearest neighbor (KNN) based sparse representation coefficient vectors for clustering high-dimensional data. For each data object, the sparse representation coefficient vector is computed by sparse representation theory and KNN algorithm is used to find the k nearest neighbors. Instead of using all the coefficients to construct the affinity matrix directly, we update each coefficient vector by remaining the k coefficients of the k neighbors unchanged and set the other coefficients to zero. Experiments on six gene expression profiling (GEP) datasets prove that the proposed algorithm can construct better affinity matrices and result in higher performance for clustering high-dimensional data.

Keywords: Spectral clustering · Affinity matrix · k-nearest neighbor · Sparse representation · Cosine similarity · Gene expression profiling

1 Introduction

Clustering aims at grouping the data objects into several subsets, so that data objects in the same subsets are similar as much as possible and the data objects in the dissimilar subsets are dissimilar as much as possible. The datasets in real-world are getting bigger and bigger with more data objects and attributes, such as multimedia data, document data, and GEP data. It has been a challenging task to cluster high-dimensional data for traditional clustering algorithms such as k-means.

Spectral clustering has attracted increasing attention and been widely used in image segmentation [1], speech separation [2]. It is based on the spectral graph theory and its main tools are graph Laplacian matrices including un-normalized graph Laplacian and

© Springer International Publishing Switzerland 2015
D.-S. Huang and K. Han (Eds.): ICIC 2015, Part III, LNAI 9227, pp. 363–374, 2015.
DOI: 10.1007/978-3-319-22053-6_40

normalized graph Laplacian. Spectral clustering firstly constructs an affinity matrix from data, then derives the graph Laplacian matrix from the affinity matrix and computes the first k eigenvectors corresponding to the k largest eigenvalues of the graph Laplacian matrix, finally uses k-means algorithm to cluster the k dimensional row vectors of the matrix which is formed by the first k eigenvectors of the Laplacian as columns [3]. The performance of spectral clustering heavily depends on the goodness of the affinity graph. Therefore, it is a key to construct an affinity matrix that can faithfully reflects the similarity information between each pair of objects.

Sparse representation (SR) [4] is based on the compressed sensing theory and has been widely applied in various tasks such as classification, subspace learning and spectral clustering. It compresses high-dimensional data by representing each object approximately as a sparse linear combination of other objects. Recently, it has been a useful technique in clustering high-dimensional data. For example, Wright et al. [5] used individual sparse coefficient directly to build the affinity matrix for spectral clustering. However, Wu et al. [6] proved that exploiting complete sparse representation vectors can reflect more truthful similarity among data objects, since more contextual information is taken into consideration. They assumed that the sparse representation vectors corresponding to two similar objects should be similar, since they can be reconstructed in a similar fashion using other data objects. These algorithms trust all the coefficients in coefficient vectors blindly to measure pairwise similarity which may make the affinity matrix sensitive to the noises and outliers, and result in a degraded performance.

To improve the performance of sparse representation based spectral clustering (SRSC) [6], we propose a novel clustering algorithm for high-dimensional data, which constructs affinity matrix via KNN based sparse representation coefficient vectors. We name our proposed algorithm as KNNSRSC. Firstly, each high-dimensional data object is projected onto a lower dimensional vector of sparse coefficients by using sparse representation theory. Then, we find the k nearest neighbors for each object, and remain the k coefficients of the k nearest neighbors unchanged and set others to zero in the coefficient vector. Then these coefficient vectors are used to construct affinity matrix according to the cosine similarity between each pair of coefficient vectors. Finally, spectral clustering is run on the affinity matrix. Extensive experiments on public gene expression profiling (GEP) datasets show that KNNSRSC outperforms SRSC [6], k-means and normalized spectral clustering (NSC) [1] with considerable performance margin.

2 Brief View of Spectral Clustering and Sparse Representation

2.1 Spectral Clustering

Spectral clustering is a widely used graph-based approach for data clustering. For dataset $X = \{x_1, x_2, \ldots, x_i, \ldots, x_n\} \in R^{m \times n}$, where $x_i \in R^m$ represents the i-th object. It can be represented as an undirected graph $G(V, E, W)$, where V, E, W denote the vertex set, the edge set, and the affinity matrix, respectively. Each vertex $v_i \in V$ represents an object x_i, and each $edge(i,j) \in E$ is assigned an affinity weight w_{ij} which represents the similarity between x_i and x_j. And the degree d_i of node i is $d_i = \sum_j w_{ij}$, the degree matrix D is a diagonal matrix $D = diag(d_1, d_2, \ldots, d_n)$.

Spectral clustering algorithms are based on the graph Laplacian matrices. The un-normalized graph Laplacian matrix is defined as $L = D - W$. It is notable that L is symmetric and semi-positive definite, and for any vector $x \in R^n$

$$x^T L x = \frac{1}{2} \sum_{i,j=1}^{n} (x_i - x_j)^2 w_{ij} \tag{1}$$

The symmetric normalized graph Laplacian matrix L_{sym} is defined as

$$L_{sym} = D^{-1/2} L D^{-1/2} = I - D^{-1/2} W D^{-1/2} \tag{2}$$

where I is the identity matrix. Similarly,

$$x^T L_{sym} x = \frac{1}{2} \sum_{i,j}^{n} \left(\frac{x_i}{\sqrt{d_i}} - \frac{x_j}{\sqrt{d_j}} \right)^2 w_{ij} \tag{3}$$

The most popular spectral clustering is NSC algorithm [1] which is described as Algorithm 1.

Algorithm 1. Normalized spectral clustering

Input: Affinity matrix $W \in R^{n \times n}$, the number of clusters C
Output: cluster labels of each sample $L = (l_1, l_2, \cdots, l_n)$

1. Compute the normalized Laplacian L_{sym}

2. Compute the first C eigenvectors $u_1, \cdots u_C$ of L_{sym}

3. Let $U \in R^{n \times C}$ be the matrix contaning the vectors $u_1, \cdots u_C$ as columns

4. Form the matrix $Z \in R^{n \times C}$ from U by normalizing each row by ℓ_2-norm, that is set
   $z_{ij} = u_{ij} / (\sum_C u_{iC}^2)^{1/2}$

5. For $i = 1, \cdots, n$, let $y_i = R^C$ be the vector corresponding to the i-th row of Z

6. Cluster the data objects $(y_i)_{i=1,\cdots,n}$ with the *k-means* algorithm into C clusters

---

We can also understand spectral clustering from the perspective of graph cut. The task of spectral clustering can be transformed to find the best cuts of the graph such that edges between different groups have low weights and edges within each group have high weights. The simplest and most direct way to construct a partition of the graph is to solve the mincut problem. Denote $F(A, B) = \sum_{i \in A, j \in B} w_{ij}$ and $\bar{A}$ for the complement of A. For a given number C of subsets, the mincut approach simply consists in choosing a partition $A_1, A_2, \ldots, A_C$ which minimizes

$$cut(A_1, A_2, \ldots, A_C) = \frac{1}{2} \sum_{i=1}^{C} F(A_i, \bar{A}_i) \tag{4}$$

However, it often does not lead to satisfactory partitions and may simply separate one individual vertex from the rest of the graph. RatioCut (Rcut) [7] and normalized cut (Ncut) [1] are proposed to address this problem by introducing balancing conditions to explicitly request that the sets $A_1, A_2, \ldots, A_C$ should be "reasonably large". In RatioCut, the size of a subset A of a graph is measured by its number of vertices $\|A\|$, while in Ncut the size is measured by the weights of its edges $vol(A)$. They are defined as follows:

$$RatioCut(A_1, A_2, \ldots, A_C) = \sum_{i=1}^{C} \frac{cut(A_i, \bar{A}_i)}{|A_i|} \tag{5}$$

$$NCut(A_1, A_2, \ldots, A_C) = \sum_{i=1}^{C} \frac{cut(A_i, \bar{A}_i)}{vol(A_i)} \tag{6}$$

However, the introduction of balancing conditions makes the previously simple mincut problem an NP hard one. Spectral clustering has been a good way to solve the relaxed versions of these problems, relaxing Ncut leads to NSC [1], while relaxing RatioCut leads to un-normalized spectral clustering [3].

2.2 Sparse Representation

Let $X \in R^{m \times n}$ denote n data objects and each object has m dimensional features, $y \in R^m$ represents a test sample. Sparse representation is an effective algorithm based on ℓ_1-norm minimization, it can represent a sample as a linear combination of others with sparse coefficients. Originally, sparse representation aims to solve the following ℓ_0-norm minimization problem:

$$\alpha_0 = \arg\min_{\alpha} \|\alpha\|_0 \quad subject\ to \quad y = X\alpha \tag{7}$$

where $\|\cdot\|_0$ is the ℓ_0-norm minimization and counts the number of nonzero entries in a vector. As the ℓ_0-norm minimization is a NP-hard problem, it is equal to ℓ_1-norm minimization problem if its solution is highly sparse [8]. Moreover, $Y = Xa$ cannot hold exactly since y may include noise. Thus, the above ℓ_0-norm minimization problem can be relaxed to the following ℓ_1-norm minimization problem:

$$\alpha_1 = \arg\min_{\alpha} \|\alpha\|_1 \quad subject\ to \quad \|y - X\alpha\|_2 \le \varepsilon \tag{8}$$

By exploiting Lagrangian method, (8) can be transformed to the following optimization problem:

$$\alpha^* = \arg\min_{\alpha} \lambda\|\alpha\|_1 + \|y - X\alpha\|_2 \tag{9}$$

where λ is a key parameter which balances the importance between the sparsity and the reconstruction error.

3 The Proposed Method

3.1 KNN Based Sparse Representation Coefficient Vectors

Denote $X = \{x_1, x_2, \ldots, x_i, \ldots, x_n\} \in R^{m \times n}$ a high-dimensional dataset, where $x_i = (x_{i1}, x_{i2}, \ldots, x_{im}) \in R^m$ is the i-th data object. Let $X_i = (x_i, \ldots, x_{i-1}, x_{i+1}, \ldots, x_n)$ represent all the data objects except for object x_i. For each data object x_i, we can use the solution form (9) to represent it as a linear combination of other objects from X_i. The coefficient vector is calculated by solving the following optimization problem:

$$\alpha_i^* = \arg\min_{\alpha_i} \lambda\|\alpha_i\|_1 + \|x_i - X_i\alpha_i\|_2 \tag{10}$$

where $\alpha_i^* = (a_{i,1}, \ldots, a_{i,i-1}, a_{i,i+1}, \ldots, a_{i,n}) \in R^{n-1}$ is the optimal solution, which consists of sparse coefficients corresponding to each data object in $X_i, \forall_i = 1, 2, \ldots, n$. We add a zero coefficient between $\alpha_{i,i-1}$ and $\alpha_{i,i+1}$ in vector α_i^* to obtain a new coefficient vector $\alpha_i = (a_{i,1}, \ldots, a_{i,i-1}, 0, a_{i,i+1}, \ldots, a_{i,n}) \in R^n$, which represents the sparse representation coefficient vector of data object x_i. Then, KNN algorithm is used to determine the k nearest neighbors for x_i from X_i. In coefficient vector a_i, the coefficients corresponding to the k nearest neighbours remain unchanged, but those coefficients corresponding to other objects are set to zero. Thus, for each coefficient vector a_i, there are only k nonzero entries.

According to sparse representation theory, similar objects gives significant coefficients and dissimilar objects gives nearly zero coefficients in a_i. However, some prior works have revealed that sparse representation is not robust enough, some coefficients of dissimilar objects will not be nearly zero when data are noisy or have many outliers. Therefore, it will degrade the discrimination of affinity matrix. KNN based sparse coefficient vector just has k nonzero coefficients, which is more robust to noises and outliers, because the coefficients of the dissimilar objects are set to zero.

3.2 Algorithm Description

Algorithm 2 describes the general procedure for spectral clustering via the KNN based sparse representation coefficient vectors. NSC algorithm [1] is used in line 8.

Algorithm 2. KNNSRSC

Input: A high-dimensional dataset $X = \{x_1, x_2, \cdots, x_n\} \in R^{m \times n}$, where $x_i \in R^m$ represents the i-th data object; the number of clusters C

Param: Sparsity parameter λ; the number of nearest neighbors k

Output: Cluster labels of each data object $P = (p_1, p_2, \cdots, p_n)$

1. **Foreach** data object $x_i \in X$ **do**

2. Set $X_i = X \setminus x_i = x_1, \cdots, x_{i-1}, x_{i+1}, \cdots, x_{in}$

3. $\alpha_i^* = \arg\min_{\alpha_i} \ \lambda \|\alpha_i\|_1 + \|x_i - X_i \alpha_i\|_2$

4. Compute the sparse coefficient vector $\alpha_i = (a_{i,1}, \cdots, a_{i,i-1}, 0, a_{i,i+1}, \cdots, a_{i,n}) \in R^n$ by adding a zero element into α_i^*

5. Find the k nearest neighbors of x_i from X_i, remain the corresponding k coefficients unchanged and set other $(n-k)$ coefficients to zero in α_i

6. **End**

7. Construct affinity matrix $W \in R^{n \times n}$ using cosine similarity between each pair of coefficient vectors as section 3.3

8. $P \leftarrow \text{NSC}(W, C)$

9. Return P

3.3   Constructing Affinity Matrix Using Cosine Similarity

Assuming α_i and a_j are two KNN-based sparse coefficient vectors of data object x_i and x_j, respectively. If x_i and x_j are two similar objects, we assume their coefficient vectors α_i and a_j are also similar. Therefore, the similarity between x_i and x_j can be transformed into the similarity between α_i and a_j. The affinity matrix can be constructed based on the cosine similarity between each pair of coefficient vectors. The similarity between object x_i and x_j is defined as follows:

$$w_{ij} = \max \left\{ 0, \frac{\alpha_i \cdot \alpha_j}{\|\alpha_i\|_2 \times \|\alpha_j\|_2} \right\} \qquad (11)$$

It is notable that $w_{ij} = 1$ when $i = j$. Once the similarity between each pair of data objects are computed, we obtain an affinity matrix $W \in R^{n \times n}$.

4   Experiments

Some experiments are carried out to illustrate the effectiveness and efficiency of KNNSRSC algorithm. Throughout the experiments, all the codes are written in MATLAB and run on a computer with 3.2 GHz CPU and 8 GB RAM. Six public GEP datasets are used in our experiments which are described in Table 1.

Table 1. Datasets description.

Datasets	Size of dataset	Number of dimensions	Number of clusters
DLBCL [9]	77	5469	2
Colon [10]	62	2000	2
Gliomas [11]	50	12625	2
Leukemia [12]	72	5327	3
SRBCT [13]	83	2308	4
Lung [14]	203	12600	4

4.1   Baselines and Evaluation Metrics

We compare KNNSRSC with SRSC [6], NSC [1] and k-means. The last two methods are run on the original data as two baselines. Clustering accuracy (CA), normalized mutual information (NMI) [15], Rand index (RI) [16] and Entropy (E) [17] are used to evaluate the clustering quality. All their value ranges are from 0 to 1. To achieve better clustering, we would like to increase values of CA, NMI and RI, and decrease the value of E. Let $P = (p_1, p_2, \ldots, p_n)$ and $T = (t_1, t_2, \ldots, t_n)$ represent the predicted labels and ground truth labels, respectively. CA can be calculated as:

$$CA(P,T) = \frac{1}{n} \sum_{i=1}^{n} f(t_i, map(p_i)) \tag{12}$$

where $f(x,y)$ equals 1 if $x = y$ and equals 0 otherwise, and $map(p_i)$ is used to map each cluster label p_i to the equivalent label from the ground truth. The KM (Kuhn-Munkres) algorithm [18] is used to compute the best mapping. NMI is defined as:

$$NMI(P,T) = \frac{MI(P,T)}{\max(H(P), H(T))} \tag{13}$$

where $H(P)$ and $H(T)$ represent the entropies of P and T, respectively. $MI(P,T)$ denotes the mutual information between P and T, which is defined as:

$$MI(P,T) = \sum_{p_i \in P} \sum_{t_j \in T} S(p_i, t_j) \log_2 \left(\frac{S(p_i, t_j)}{S(p_i)S(t_j)} \right) \tag{14}$$

where $S(p_i)$ and $S(t_j)$ are the probabilities that an arbitrary sample belongs to the clusters p_i and t_j, respectively. $S(p_i, t_j)$ denotes the joint probability that the arbitrary ample belongs to the clusters p_i and t_j at the same time. RI can be computed with the following formula:

$$RI = \frac{TP + TN}{n(n-1)/2} \tag{15}$$

where TP is the number of pairs of samples with the same true class labels and belonging to the same clusters; TN denotes the number of pairs of samples having different true class labels and belonging to different clusters. The total E for set of clusters is calculated as the sum of entropies for each cluster weighted by the size of each cluster:

$$E = \sum_j \left(\frac{n_j}{n} \times E_j \right) \tag{16}$$

Where n_j is the number of samples of class j, $E_j = -\sum_i p_{ij} \log(p_{ij})$ is the E of each cluster j, p_{ij} is the probability that a member of cluster j belongs to class i.

4.2   Experimental Setup

We repeat NSC [1] and k-means on the original GEP data 100 times and report the average results. For SRSC and KNNSRSC, the process of computing coefficient vectors are repeated 100 times, and for each time of the computation of coefficient vectors, spectral clustering is repeated 100 times on the coefficient vectors. Then their average results are computed.

To examine the influence of parameters combination of λ and k over the clustering result, we carried out some experiments using Colon dataset. Figure 1 illustrates the clustering accuracy of KNNSRSC with varying λ and k. In each case, we varied the value of one parameter and fixed the value of the other.

In the following experiments, the best clustering results of all the methods under different parameter configuration are reported as did in [19]. The values used to find the optimal parameters for KNNSRSC are shown in Fig. 1. For SRSC, the values of λ are the same as those of KNNSRSC. In all of our experiments, the Euclidean distance is used in KNN algorithm.

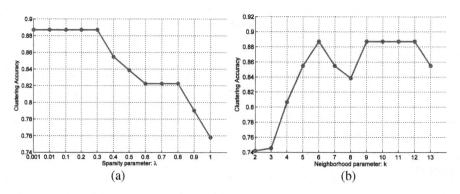

Fig. 1. Clustering accuracy of KNNSRSC on the Colon dataset. (a) The accuracy versus the variation of the sparsity parameter λ, where $k = 10$. (b) The accuracy versus the variation of the neighborhood parameter k, where $\lambda = 0.1$.

4.3   Experimental Results

The affinity matrices of the ground truth, KNNSRSC and SRSC on SRBCT dataset and Leukemia dataset are shown in Figs. 2 and 3, respectively. Each subfigure is a symmetrical matrix with n-by-n elements, where n is the number of data objects of the dataset. Theoretically, the clustering quality will be good if the affinity weights between two objects form different clusters are small, while affinity weights form the same cluster are large. The euclidean distance between Fig. 2(a) and (b) is 39.72, while the distance between Fig. 2(a) and (c) is 40.54. In Fig. 3, the two distance are 41.6 and 42.96, respectively. Therefore, compared with SRSC, KNNSRSC may tend to increase the similarity of data objects belonging to the same cluster and decrease the similarity of objects belonging to different clusters so as to result in more robust affinity matrix.

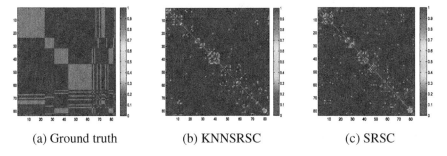

(a) Ground truth   (b) KNNSRSC   (c) SRSC

Fig. 2. The comparison of affinity matrices of KNNSRSC and SRSC on the SRBCT dataset.

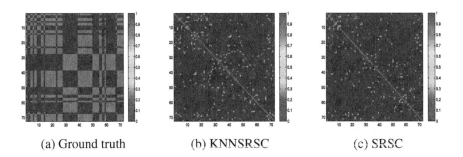

(a) Ground truth   (b) KNNSRSC   (c) SRSC

Fig. 3. The comparison of affinity matrices of KNNSRSC and SRSC on the Leukemia dataset.

Table 2. The comparison of CA.

Dataset	Colon	Gliomas	Leukemia	SRBCT	Lung	DLBCL	Average
k-means	0.7158	0.6100	0.5642	0.4819	0.5894	0.6883	0.6082
NSC	0.5000	0.6002	0.5556	0.3747	0.5030	0.6364	0.5283
SRSC	0.8621	0.7410	0.8490	0.6068	0.7826	0.7273	0.7614
KNNSRSC	**0.8875**	**0.7800**	**0.9084**	**0.7240**	**0.7965**	**0.7403**	**0.8061**

Table 3. The comparison of NMI.

Dataset	Colon	Gliomas	Leukemia	SRBCT	Lung	DLBCL	Average
k-means	0.1923	0.0590	0.3019	0.2262	0.5176	0.1238	0.2368
NSC	0.0012	0.0213	0.1756	0.1111	0.3702	0.0567	0.1226
SRSC	0.3982	0.2164	0.5594	0.3881	**0.6080**	**0.2880**	0.4096
KNNSRSC	**0.4709**	**0.2849**	**0.6862**	**0.5392**	0.5850	0.0188	**0.4308**

Table 4. The comparison of RI.

Dataset	Colon	Gliomas	Leukemia	SRBCT	Lung	DLBCL	Average
k-means	0.6195	0.5275	0.6262	0.6461	0.6525	0.5653	0.6061
NSC	0.4918	0.5102	0.5822	0.6306	0.5679	0.5311	0.5523
SRSC	0.7585	0.6073	0.8036	0.7094	**0.7970**	0.5981	0.7123
KNNSRSC	**0.7964**	**0.6498**	**0.8650**	**0.7900**	0.7798	**0.6104**	**0.7485**

Table 5. The comparison of E.

Dataset	Colon	Gliomas	Leukemia	SRBCT	Lung	DLBCL	Average
k-means	0.6698	0.8682	0.5009	0.6444	**0.0094**	0.5487	0.5402
NSC	0.8726	0.8915	0.6784	0.7893	0.0787	0.6087	0.6532
SRSC	0.5619	0.6821	0.3659	0.5395	0.1418	0.5185	0.4682
KNNSRSC	**0.4926**	**0.6520**	**0.2779**	**0.4035**	0.1822	**0.4710**	**0.4132**

The best clustering results obtained by all algorithms on the 6 GEP datasets with optimal parameter settings are reported in Tables 2, 3, 4 and 5. For each dataset, the best results are marked in bold. The last column in each table presents the average results of each algorithm over all the 6 datasets. Table 2 shows that KNNSRSC achieves the highest CA for all datasets. It is clear form Table 3 that the KNNSRSC obtains the highest NMI value for four datasets, and SRSC achieves the highest NMI value on the DLBCL and Lung dataset, respectively. Moreover, Table 4 shows that KNNSRSC obtains the highest RI for five datasets. The last columns of Tables 2, 3 and 4 show that KNNSRSC achieves the highest average results over all the 6 datasets, which are 4.47 %, 2.12 % and 3.62 % higher than SRSC in terms of AC, NMI and RI, respectively. In Table 5, KNNSRSC has the lowest E on five datasets and the average E for the 6 datasets is 5.50 %, 24 % and 13.70 % better than SRSC, NSC and k-means, respectively. In a word, KNNSRSC outperforms SRSC [6] significantly and has clear advantage over other two compared methods in clustering high-dimensional data.

5   Conclusions

We have presented a novel spectral clustering via KNN based sparse representation coefficient for clustering high-dimensional data. It projects the original high-dimensional data onto a lower and more discriminative coefficient vectors space. The affinity matrix constructed based on the KNN based sparse coefficient vectors is more robust to noise and can boost the clustering quality significantly. The extensive experiments demonstrate the superiority of our proposed algorithm.

Acknowledgments. This work is supported by the National Science Foundation of China (No. 61474267, 60973153 and 61471169) and Collaboration and Innovation Center for Digital Chinese Medicine of 2011 Project of Colleges and Universities in Hunan Province.

References

1. Ng, A.Y., Jordan, M.I., Weiss, Y.: On spectral clustering: analysis and an algorithm. Adv. Neural Inf. Process. Syst. **2**, 849–856 (2002)
2. Bach, F.R., Jordan, M.I.: Learning spectral clustering, with application to speech separation. J. Mach. Learn. Res. **7**, 1963–2001 (2006)
3. von Luxburg, U.: A tutorial on spectral clustering. Stat. Comput. **17**, 395–416 (2007)
4. Hang, X.Y., Wu, F.X.: Sparse representation for classification of tumors using gene expression data. J. Biomed. Biotechnol. (2009)
5. Wright, J., Ma, Y., Mairal, J., Sapiro, G., Huang, T.S., Yan, S.C.: Sparse representation for computer vision and pattern recognition. Proc. IEEE **98**, 1031–1044 (2010)
6. Wu, S., Feng, X.D., Zhou, W.J.: Spectral clustering of high-dimensional data exploiting sparse representation vectors. Neurocomputing **135**, 229–239 (2014)
7. Akram, M.U., Khalid, S., Tariq, A., Khan, S.A., Azam, F.: Detection and classification of retinal lesions for grading of diabetic retinopathy. Comput. Biol. Med. **45**, 161–171 (2014)
8. Donoho, D.L.: For most large underdetermined systems of linear equations the minimal l1-norm solution is also the sparsest solution. Commun. Pure Appl. Math. **59**, 797–829 (2006)
9. Shipp, M.A., Ross, K.N., Tamayo, P., Weng, A.P., Kutok, J.L., Aguiar, R.C.T., Gaasenbeek, M., Angelo, M., Reich, M., Pinkus, G.S., Ray, T.S., Koval, M.A., Last, K.W., Norton, A., Lister, T.A., Mesirov, J., Neuberg, D.S., Lander, E.S., Aster, J.C., Golub, T.R.: Diffuse large B-cell lymphoma outcome prediction by gene-expression profiling and supervised machine learning. Nat. Med. **8**, 68–74 (2002)
10. Alon, U., Barkai, N., Notterman, D.A., Gish, K., Ybarra, S., Mack, D., Levine, A.J.: Broad patterns of gene expression revealed by clustering analysis of tumor and normal colon tissues probed by oligonucleotide arrays. Proc. Natl. Acad. Sci. **96**, 6745–6750 (1999)
11. Nutt, C.L., Mani, D.R., Betensky, R.A., Tamayo, P., Cairncross, J.G., Ladd, C., et al.: Gene expression-based classification of malignant gliomas correlates better with survival than histological classification. Cancer Res. **63**(7), 1602–1607 (2003)
12. Armstrong, S.A., Staunton, J.E., Silverman, L.B., Pieters, R., de Boer, M.L., Minden, M.D., Sallan, S.E., Lander, E.S., Golub, T.R., Korsmeyer, S.J.: MLL translocations specify a distinct gene expression profile that distinguishes a unique leukemia. Nat. Genet. **30**, 41–47 (2002)
13. Khan, J., Wei, J.S., Ringner, M., Saal, L.H., Ladanyi, M., Westermann, F., Berthold, F., Schwab, M., Antonescu, C.R., Peterson, C., Meltzer, P.S.: Classification and diagnostic prediction of cancers using gene expression profiling and artificial neural networks. Nat. Med. **7**, 673–679 (2001)

14. Bhattacharjee, A., Richards, W.G., Staunton, J., Li, C., Monti, S., Vasa, P., Ladd, C., Beheshti, J., Bueno, R., Gillette, M., Loda, M., Weber, G., Mark, E.J., Lander, E.S., Wong, W., Johnson, B.E., Golub, T.R., Sugarbaker, D.J., Meyerson, M.: Classification of human lung carcinomas by mRNA expression profiling reveals distinct adenocarcinoma subclasses. Proc. Natl. Acad. Sci. USA **98**, 13790–13795 (2001)
15. Cai, D., He, X.F., Han, J.W.: Document clustering using locality preserving indexing. IEEE Trans. Knowl. Data Eng. **17**, 1624–1637 (2005)
16. Deng, Z.H., Choi, K.S., Chung, F.L., Wang, S.T.: EEW-SC: enhanced entropy-weighting subspace clustering for high dimensional gene expression data clustering analysis. Appl. Soft Comput. **11**, 4798–4806 (2011)
17. Hammouda, K., Kamel, M.: Collaborative document clustering. In: Proceedings of the Sixth Siam International Conference on Data Mining, pp. 453–463 (2006)
18. Lovász, L., Plummer, M.: Matching Theory. American Mathematical Society, Providence (2009)
19. Peng, X., Zhang, L., Yi, Z., Tan, K.K.: Learning locality-constrained collaborative representation for robust face recognition. Pattern Recogn. **47**, 2794–2806 (2014)

Friend Recommendation by User Similarity Graph Based on Interest in Social Tagging Systems

Bu-Xiao Wu, Jing Xiao[✉], and Jie-Min Chen

School of Computer Science, South China Normal University, Guangzhou 510631, China
xiaojing@scnu.edu.cn

Abstract. Social tagging system has become a hot research topic due to the prevalence of Web2.0 during the past few years. These systems can provide users effective ways to collaboratively annotate and organize items with their own tags. However, the flexibility of annotation brings with large numbers of redundant tags. It is a very difficult task to find users' interest exactly and recommend proper friends to users in social tagging systems. In this paper, we propose a Friend Recommendation algorithm by User similarity Graph (FRUG) to find potential friends with the same interest in social tagging systems. To alleviate the problem of tag redundancy, we utilize Latent Dirichlet Allocation (LDA) to obtain users' interest topics. Moreover, we propose a novel multiview users' similarity measure method to calculate similarity from users' interest topics, co-collected items and co-annotated tags. Then, based on the users' similarities, we build user similarity graph and make interest-based user recommendation by mining the graph. The experimental results on tagging dataset of Delicious validate the good performance of FRUG in terms of precision and recall.

Keywords: Friend recommendation · Social tagging system · Topic modeling · User similarity graph · User interest

1 Introduction

With the development of Web2.0, social tagging system (STS) is growing rapidly. There are many famous social tagging systems, such as Delicious, Last.fm and Flickr. STS allows users to annotate, collect and share items by assigning tags. Moreover, users can use free-form tags to annotate items and need no specific skills. Since tags are associated to both items and users, tagging can represent users' preferences on items and be used for making personalized recommendations.

In social network-based recommendation system, friend recommendation is the essential part. When a user wants to add a new friend, the user would like to choose people whom they may know or have similar interest with. Friend recommendation can be classified into two categories: suggesting friends based on social graphs, like friends of friends; or based on the interest of users which is constructed with the text information in recommendation systems. Following friends of friends on social graphs can mostly recommend people you may already know. Thus it is practical significance to friend recommendation based on interest which can suggest the users with high similarities of

© Springer International Publishing Switzerland 2015
D.-S. Huang and K. Han (Eds.): ICIC 2015, Part III, LNAI 9227, pp. 375–386, 2015.
DOI: 10.1007/978-3-319-22053-6_41

interest as potential friends. On the other hand, users' tagging behavior can represent their hobbies and interest. For example, a user often use "comedy" to annotate the video, it may explain that the user's favorite video genre is comedy. Moreover, if both of two users tag the same items, they could probably have the same interest.

However, social tagging applications may have some disadvantages. Unsupervised annotation brings with a wide variety of tag redundancy, which means that tags may have very similar meanings [1]. Redundant tags can hinder the performance of the friend recommendation algorithms in social tagging systems. In addition, it is very hard to grasp interest of the user who tags a little or gives a little ratings, in other words, it is the problem of cold start in recommender systems.

In order to alleviate the tag redundancy and cold start problems on friend recommendation, we propose a Friend Recommendation algorithm by User similarity Graph (FRUG) in social tagging systems. FRUG recommends potential friends for users with similar interest by the similarity graph based on interest.

Generally speaking, our work is summarized as follows:

- We regard the tags of a user as a document and use LDA to classify tags to users' interest topics. LDA is an efficient way to overcome the problem of tag redundancy.
- To alleviate the cold start problem due to sparse data, we calculate the interest-based user similarity from multiple different views of the interest topics, co-collected items and co-annotated tags. All these information is useful to identify users' interest.
- To recommend potential friends precisely, the user similarity graph is constructed based on users' similarities. And make interest-based user recommendation by mining the graph.

The rest of this paper is organized as follows. Friend recommendation algorithms and social tagging system models are introduced in Sect. 2. The method of building the user similarity graph based on interest is introduced in Sect. 3. The new user similarity by mining the interest-based user similarity graph is introduced in Sect. 4. Experimental results and analysis are described in Sect. 5. We conclude the paper and discuss the future work in Sect. 6.

2   Related Work

With the development of social networks, social networking sites (SNS) provide novel ways for users to communicate and share interests. Users can resort to their friends in SNS to share personal opinions and utilize recommendations of friends before purchasing a product. Thus, friend recommendation has become an integral part of social life and a hot research topic in recent years.

Social graphs-based Friend Recommendation System (FRS) is as same as the predicting new links in social networks [2]. Nowell and Kleinberg [2] made recommendation of friends by considering only the local features of graph on social network sites. Scellato et al. [3] proposed a supervised learning framework which exploits these prediction features to predict new links among friends-of-friends and place-friends. Symeonidis et al. [4] exploited global graph features introducing transitive node similarity that captures adequately the missing local graph characteristics. Interest-based FRS is to find like-minded people. In [5], a friend recommendation algorithm for blogs was proposed and it tried to analyze users' behavior for capturing the users' interest, and recommended the potential friends with the same interest. Xie [6] designed a general friend recommendation framework, which can characterize user interest in two dimensions: context (location, time) and content, as well as combining the domain knowledge to improve recommending quality. The popularity of location acquisition technologies such as GPS enabled people to expediently record the location histories they visited with spatiotemporal data. In [7], a hierarchical graph based similarity measurement was proposed for effectively measuring the similarity among users in geographic information systems.

Tagging is becoming an increasingly important activity to help users to organize various objects, such as bookmarks, music. There are two methods to use tags for recommendation. First, it uses tags to compute user-user or item-item similarities for improving the results. Second, recommendations are completely based on tags and employ the co-occurrence of tags in the user-tag matrix [8]. Zhou et al. [9] attempted to model users' interest by building users' tags graph by community detection. The algorithm utilized Kullback-Leibler divergence (KL-divergence) to measure users' interest similarity and recommended potential friends with low KL-divergence. Social tagging systems contain rich information such as users' tagging behaviors, friends and items. All the heterogeneous information helps to alleviate the cold start problem caused by sparse data. In [10], Feng and Wang modeled a social tagging system as a multi-type graph.

Based on the fact that collaborative tags in social tagging systems contain rich information about personalized preferences and items attributes, we propose a FRUG algorithm to find potential friends with the same interest in social tagging systems. Our method utilizes Latent Dirichlet Allocation (LDA) to obtain users' interest topics. Then we calculate users' similarities by users' interest topics, co-collected items and co-annotated tags from multi different views. Finally, based on the users' similarities, we build user similarity graph and make interest-based friend recommendation by mining the graph.

3 Building User Similarity Graph Based on Interest

3.1 User Similarity Based on Topic Modeling

LDA is proposed by Blei et al. [11], which is an unsupervised algorithm to find latent topic information from document collection. The users' interest can be identified by the topics based on the tags they used in social tagging systems. For this purpose, we consider the tags of a user as a document. Therefore, LDA model can be used to obtain users' interest topics.

Blei presumes LDA is a "bag of words" model, which regards a document as a vector of word counts. Therefore, a document is described as a probability distribution over some topics. In this paper, the topic is described as a probability distribution over a number of tags. The generative process works as shown in Fig. 1. In this figure, given the parameter of Dirichlet prior α and β, the multinomial distribution θ and Φ are drawn from α and β respectively. θ represents the document distribution over topics. Φ represents the topic distribution over words. Topic Z is drawn from θ associated with the document, and the word is drawn from β. D represents the number of documents. Nd is the number of words in document *d*. *w* represents a word in documents.

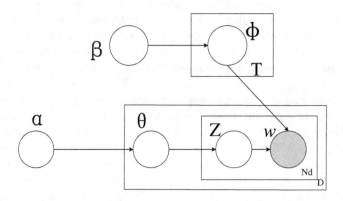

Fig. 1. Plate representation of LDA model

To learn the parameter of LDA model, we adopt Gibbs sampling which is an efficient way to parameter estimation in LDA model. To obtain the topics that users are interested in by LDA model, a document should correspond to the tags that a user has been used. The result is described simply by a D × T matrix, where D is the number of users and T is the number of topics. The number of times a tag used by user has been assigned to a topic, which can represent the topic distribution that users are interested in. The example of topics in Delicious dataset is shown in Table 1.

Table 1. The example of topics in delicious dataset

Topic 1	Thinking virtual_teenage pdf graphics fingertips flat events
Topic 2	Ubuweb documentation webdev tutoriales instant url
Topic 3	Webdevelopment emt brain @css links glitch motivation it
Topic 4	Bridges wiki critical_thinkers factdash wikipedia
Topic 5	Lecture framework learn tutorial tecnologia openid
Topic 6	Reading literacy writing smartboard seo wikipedia

KL-divergence has been widely used to calculate the similarity in LDA model, which can measure the different topic distribution of a document. But it is not an efficient way to measure the similarity of user interest. If both of two users are not interested in a topic at all, it is considered with high similarity in this topic by KL-divergence. Therefore, we propose a new method to calculate the interest-based user similarity based on topics which is defined as following:

$$sim'(u, v) = \frac{\sum\limits_{t \in T} \min(N(u, t), N(v, t))}{\sqrt{\sum\limits_{t \in T} N(u, t)} * \sqrt{\sum\limits_{t \in T} N(v, t)}} \tag{1}$$

T is a set of the topics which are interested by the both of users u and v. $N(u, t)$ is the number of times a tag used by u has been assigned to topic t. $\min(N(u, t), N(v, t))$ means the minimum of user u and v's interest in topic t and is used to measure the interest-based similarity between user u and v in topic t. For example, if $N(u, t)$ is 3 and $N(v, t)$ is 7, the interest-based similarity between user u and v in topic t was 3. We accumulate the similarity in all topics and normalize it as the interest-based similarity between two users.

3.2   User Similarity Based on Co-collected Items and Co-annotated Tags

As the items that users have been collected imply the users' interest, the user similarity can be measured by the co-collected items. The method to get users' similarity based on items is different from the one based on topic modeling. The number of times an item is annotated by tags only represents the attributes of the item (such as color, shape) instead of the likeness of the user who contributes the tags. Accordingly, we use a cosine similarity to calculate the user similarity based on items which only consider the collection of items. It is defined as following:

$$sim''(u, v) = \frac{I(u) \cap I(v)}{\sqrt{I(u) \cup I(v)}} \tag{2}$$

where $I(u)$ is a set of item which have been collected by user u, $I(v)$ is a set of item which have been collected by user v.

Although the tags are redundant, the tagging behavior contains lots of useful information to identify users' interest. Thus, we calculate the users' interest similarity based on tags and the method is similar to the one using topic modeling. The similarity is defined as following:

$$sim'''(u, v) = \frac{\sum\limits_{tg \in TG} \min(TN(u, tg), TN(v, tg))}{\sqrt{\sum\limits_{tg \in TG} TN(u, tg)} * \sqrt{\sum\limits_{tg \in TG} TN(v, tg)}} \tag{3}$$

TG is a set of the co-annotated tags which are used by both of user u and v. $TN(u, tg)$ is the number of times tag tg used by user u.

3.3   Calculating Users' Similarity Based on Interest Topics, Co-collected Items and Co-annotated Tags

To alleviate the cold start problem due to sparse data, we try to use as much information as possible. Therefore, we combine users' similarity based on interest topics, co-collected items and co-annotated tags. The recalculated user similarity is more accurate than the user similarity based on interest topics, co-collected items and co-annotated tags alone. The recalculation is defined as following:

$$sim(u, v) = (1 - P - Y) * sim'(u, v) + Y * sim''(u, v) + P * sim'''(u, v) \qquad (4)$$

where $sim'(u, v)$ is the user similarity based on interest topics (Eq. 1); $sim''(u, v)$ is the user similarity based on co-collected items (Eq. 2), and $sim'''(u, v)$ is the user similarity based on co-annotated tags (Eq. 3) respectively. Y is the parameter which weighs user similarity based on items while P is the weight of user similarity based on tags. $P + Y \in [0, 1]$, $P \in [0, 1]$, $Y \in [0, 1]$.

4   Friend Recommendation by Mining User Similarity Graph

The user similarity graph based on user interest is built by users' interest similarity which is defined in Eq. (4). The nodes in graphs represent users, and the edges in graphs represent the similarity between the users. An example of the interest-based user similarity graph is depicted in Fig. 2. In this figure, five nodes represented five users. Note that the graph is a unidirectional graph, because the user similarity we defined is symmetrical. We can see the interest-based users' similarity between u_1 and u_2 is 0.86 in Fig. 2; the zero values of user similarity are not considerable.

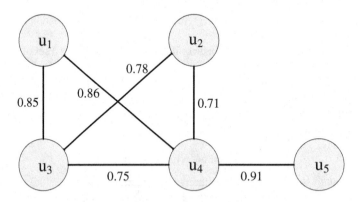

Fig. 2. An example of user similarity graph based on interest

In real world, friends can form a group because of sharing the same interest, and also the members in a group with the same interest are more likely to be friends. In Fig. 2, although the interest-based user similarity between u_4 and u_5 is 0.91 and it is larger than

0.86 which is the user similarity between u_1 and u_4. However, u_1 and u_4 are more likely to become friends than u_4 and u_5 because of u_1 and u_4 have a common neighbor node u_3.

In order to solve the above problem properly and recommend friends more precisely, we propose a new method which takes the common neighbor nodes into consideration to recalculate the user similarity. It is inspired by the similarity propagation [12] which recommends potential friends through the list of friends. In this paper, we propagate the user similarity between common neighbor nodes. The method for calculating the similarity (graph-based, $simgb$) is defined as following:

$$simgb(u, v) = \lambda * sim(u, v) + (1 - \lambda) * \frac{\sum_{i \in I} 0.5 * (sim(u, i) + sim(v, i))}{N(I)} \tag{5}$$

where I is a set of common neighbor nodes which node u and v have. $sim(u, v)$ is the user similarity defined by Eq. (4) and λ is the weight for it. $N(I)$ is the number of common neighbor nodes which node u and v have. $1 - \lambda$ is the weight of user similarity which has been propagated. The common neighbor nodes are considered as the factor of a group with the same interest. Take Fig. 2 for example, u_1 and u_4 have a common neighbor node u_3, assume parameter λ is 0.8, the $simgb(u_1, u_4)$ is 0.88. Since u_4 and u_5 don't have any neighbor common node, the $simgb(u_4, u_5)$ is 0.728. The $simgb(u_1, u_4)$ is larger than $simgb(u_4, u_5)$ according to the Eq. (5), it is a more effective result due to the interest group feature of friendship. The users with high similarity scores by mining the graph are recommended as potential friends.

5   Experimental Evaluation

5.1   Datasets

We evaluate the proposed recommendation algorithm on Delicious dataset which is a famous social tagging resource. hetrec2011-delicious-2k is a Delicious dataset which is published at the workshop of HetRec2011 [13], and it can be downloaded from the website of grouplens [14].

The dataset contains 1867 users, 69226 resources and 53388 tags. There are 437593 times of user annotation; averaged 234.383 tags annotated by a user and averaged 6.321 annotated tags per resource. In addition, it contains 7668 bi-directional user relations, averaged 8.236 friends per user (Table 2).

Table 2. Statistics of the dataset

Users	Resources	Tags	Tag assignments	Bi-directional user relations
1867	69226	53388	437593	7668

Table 3. Density of the dataset

Avg. friends per user	Avg. tags annotated per user	Avg. tags per resource
8.236	234.383	6.321

5.2 Evaluation Methodology

In hetrec2011-delicious-2k dataset, file user_taggedbookmarks contains tag assign-ments of resources provided by users. The data format can be represented by tuples <user, tag, resource>, and we take it as training set in experiments. File user_contacts contains the friend relationship between two users, and we use it as test set in experiments.

To measure the performance of algorithm, we use precision and recall, which are widely used in recommender systems as measurements. Precision is defined as following:

$$Precision = \frac{\sum_{u \in U} |R(u) \cap T(u)|}{\sum_{u \in U} |R(u)|} \tag{6}$$

Recall is defined as following:

$$Recall = \frac{\sum_{u \in U} |R(u) \cap T(u)|}{\sum_{u \in U} |T(u)|} \tag{7}$$

where U is a set of all users. $R(u)$ is the list of potential friends whom recommended to user u by the algorithm. $T(u)$ is the list of friends in test data.

5.3 Parameters

In LDA model, α and β are the parameter of Dirichlet prior; we set α to 1 and β to 0.01. The number of topics represents the granularity of topics. If the number of topics is too large, the correlated tags may not be in a same topic. On the contrary, if the number of topics is too small, the uncorrelated tags may be in a same topic. In our experiments, we set the number of topics to 100 and set the iteration number of LDA model to 400.

The parameter Y in Eq. (4) controls the weight of user similarity based on items; parameter P controls the weight of user similarity based on tags. In our experiments, Y is 0.85 and P is 0.125. λ controls the weight of new defined user similarity. $\lambda \in [0.8, 0.98]$ seems to be a good choice and we set λ to 0.95.

5.4   Comparison of Friend Recommendation Algorithms

To validate the performance of FURG, we compare it with other three methods: the cosine similarity to calculate user similarity based on tags (Tag-cos), the cosine similarity to calculate user similarity based on items (Item-cos), and the method of using user-item-tag tripartite graphs [15] to calculate user similarity (Tri-graphs). We set the parameter of lambda in Tri-graphs as 0.45 in hetrec2011-delicious-2k dataset.

5.5   Experimental Results

Experimental results are shown in Table 4. The performance measured in terms of precision and recall. The number of potential friends recommended to a user is 5 and 10 respectively. We can see Tag-cos and Item-cos have relatively poor performance, worse than Tri-graphs and FRUG. Tri-graph is only worse than FRUG. FRUG is better than the other three methods in Table 4.

Table 4. Precision and Recall on delicious

Algorithm	Precision@5	Precision@10	Recall@5	Recall@10
Tag-cos	0.103696	0.080450	0.063152	0.097991
Item-cos	0.113658	0.080664	0.069220	0.098252
Tri-graphs	0.115529	0.096668	0.070133	0.117367
FRUG	**0.145581**	**0.110927**	**0.079035**	**0.120442**

To further validate the performance of FRUG, the number of potential friends is validated in more details. N is the number of potential friends recommended to one user. Since the average number of friends per user in our dataset is 8.2 (see Table 3), in our experiments, the value of N is set to [5, 12]. The results in terms of precision in Top-N are shown in Fig. 3. In Fig. 3, we can see that FRUG is still better than other three methods in precision when N ∈ [5, 12]. The results in terms of recall in Top-N are also shown in Fig. 4. In Fig. 4, FRUG is observably better than other three methods in recall when N ∈ [5, 10]. It is noted that if N is between [10, 12], although FRUG is better than Tri-graphs in recall, the recall of Tri-graphs is very close to FRUG. Since a user has averagely less than 9 friends, the performance in N ∈ [5, 10] is considered more important than the performance in N ∈ [10, 12].

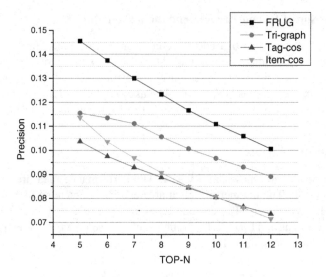

Fig. 3. Comparison of algorithms in precision

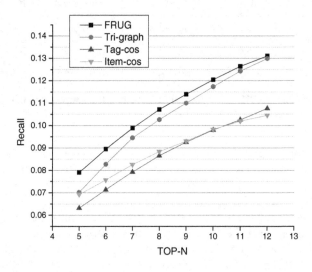

Fig. 4. Comparison of algorithms in recall

6 Conclusion and Future Work

In this paper, we propose a user similarity graph based on user interest for friend recommendation. In order to alleviate the problems of tag redundancy and data sparseness, first, we use LDA to identify users' interest topics. Second, the interest-based user similarity graph was built by using the multi different views of users' interest topics, co-collected items and co-annotated tags. Finally, the algorithm makes interest-based recommendation by mining the graph. There are some possible research topics for future work:

- Reducing the number of parameters. The proposed FRUG has more parameters than other methods, resulting in the complexity of FRUG.
- With the research development of community detection, we will try to find a more accurate method for constructing the similarity graph by community detection for friend recommendation in social tagging systems.

Acknowledgement. This work was partially supported by the National Natural Science Foundation of China (NSFC) projects No. 61202296, the National High-Technology Research and Development Program ("863" program) of China under Grant No. 2013AA01A212, the Natural Science Foundation of Guangdong Province project No. S2012030006242 and the Key Areas of Guangdong-HongKong Breakthrough project No. 2012A090200008.

References

1. Shepitsen, A., Gemmell, J., Mobasher, B., Burke, R.: Personalized recommendation in social tagging systems using hierarchical clustering. In: Proceedings of the 2008 ACM Conference on Recommender Systems, pp. 259–266. ACM, New York (2008)
2. Nowell, L.D., Kleinberg, J.: The link prediction problem for social networks. In: Proceedings of the Twelfth International Conference on Information and Knowledge Management (CIKM), pp. 1019–1031. ACM, New York (2004)
3. Scellato, S., Noulas, A., Mascolo, C.: Exploiting place features in link prediction on location-based social networks. In: Proceedings of the 17th ACM SIGKDD International Conference on Knowledge Discovery and Data Mining, pp. 1046–1054. ACM, New York (2011)
4. Symeonidis, P., Tiakas, E., Manolopoulos, Y.: Transitive node similarity for link prediction in social networks with positive and negative links. In: Proceedings of the Fourth ACM Conference on Recommender Systems (RecSys 2010), pp. 183–190. ACM, New York (2010)
5. Hsu, W.H., King, A.L., Paradesi, M.S., Pydimarri, T., Weninger, T.: Collaborative and structural recommendation of friends using weblog-based social network analysis. In: AAAI Spring Symposium: Computational Approaches to Analyzing Weblogs, pp. 55–60. AAAI, Menlo Park (2006)
6. Xie, X.: Potential friend recommendation in online social network. In: Proceedings of the IEEE/ACM International Conference on Green Computing and Communications and International Conference on Cyber, Physical and Social Computing, pp. 831–835. IEEE, Piscataway (2010)
7. Li, Q., Zheng, Y., Xie, X., Chen, Y., Liu, W., Ma, W.Y.: Mining user similarity based on location history. In: Proceedings of the 16th ACM SIGSPATIAL International Conference on Advances in Geographic Information Systems, pp. 34–42. ACM, New York (2008)
8. Wartena, C., Brussee, R., Wibbels, M.: Using tag co-occurrence for recommendation. In: Intelligent Systems Design and Applications (ISDA 2009), pp. 273–278. IEEE, Piscataway (2009)
9. Zhou, T.C., Ma, H., Lyu, M.R., King, I.: UserRec: a user recommendation framework in social tagging systems. In: Proceedings of the Twenty-Fourth AAAI Conference on Artificial Intelligence, pp. 1486–1491. AAAI, Washington (2010)

10. Feng, W., Wang, J.: Incorporating heterogeneous information for personalized tag recommendation in social tagging systems. In: Proceedings of the 18th ACM SIGKDD International Conference on Knowledge Discovery and Data Mining, pp. 1276–1284. ACM. New York (2012)
11. Blei, D.M., Ng, A.Y., Jordan, M.I.: Latent dirichlet allocation. J. Mach. Learn. Res. **3**, 993–1022 (2003)
12. Zhang, Z.F., Li, Q.D.: Latent friend recommendation in social network services. J. China Soc. Sci. Tech. Inf. **30**, 1319–1325 (2011)
13. Cantador, I., Brusilovsky, P.: 2nd Workshop on Information Heterogeneity and Fusion in Recommender Systems. In: Proceedings of the 5th ACM Conference on Recommender Systems (2011)
14. Delicious dataset form the website Grouplens. http://grouplens.org/datasets/hetrec-2011/
15. Zhang, Z.K., Zhou, T., Zhang, Y.C.: Personalized recommendation via integrated diffusion on user–item–tag tripartite graphs. Phys. A Stat. Mech. Appl. **389**, 179–186 (2010)
16. Weng, J., Lim, E.P., Jiang, J., He, Q.: Twitterrank: finding topic-sensitive influential twitterers. In: Proceedings of the Third ACM International Conference on Web Search and Data Mining, pp. 261–270. ACM, New York (2010)
17. Pennacchiotti, M., Gurumurthy, S.: Investigating topic models for social media user recommendation. In: Proceedings of the 20th International Conference Companion on World Wide Web, pp. 101–102. ACM, New York (2011)

Orchestrating Real-Valued Negative Selection Algorithm with Computational Efficiency for Crude Oil Price

Ayodele Lasisi[1(✉)], Rozaida Ghazali[1], Tutut Herawan[2], and Haruna Chiroma[2]

[1] Faculty of Computer Science and Information Technology,
Universiti Tun Hussein Onn Malaysia,
86400 Parit Raja, Batu Pahat, Johor, Malaysia
lasisiayodele@yahoo.com, rozaida@uthm.edu.my
[2] Faculty of Computer Science and Information Technology,
University of Malaya, 50603 Kuala Lumpur, Malaysia
tutut@um.edu.my, hchiroma@acm.org

Abstract. This paper implements the real-valued negative selection with variable-sized detectors (V-Detectors) for projecting the right decision with respect to crude oil price. The Brent crude oil data is retrieved from US department of energy. Using varying radius values of the V-Detector, comparison in terms of detection rate and false alarm rate, with support vector machine, naïve bayes, multi-layer perceptron, J48, non-nested generalized exemplars, IBk, fuzzy-roughNN, and vaguely quantified nearest neighbor demonstrated that V-Detector is efficient and computationally effective. The experimental outcome can initiate international crude oil market policy making as the V-Detector is able to reach highest detection and lowest false alarm rates.

Keywords: Real-valued negative selection algorithm · V-Detector · Data mining · Crude oil price

1 Introduction

Data mining techniques development paved ways for diffusing several problems encountered in many application areas not limited to science and engineering, management etc. Such data mining techniques include Support Vector Machine (SVM), Genetic Algorithm (GA), and Artificial Neural Network (ANN). In the globalized world of today, crude oil has an enormous effect in the world economic development and growth. With the unstable crude oil prices and for the purpose of making the right decisions, algorithmic methods aforementioned serve as adequate solution. A number of the application of data mining techniques for crude oil price have surfaced in literatures namely; the training of Feed-Forward Neural Network (FFNN) with Levenberg-Marquardt Back-Propagation (LMBP) resulting in the construction of ensemble models for boosting the accuracy of crude oil price forecast [1]. To predict crude oil price, an Orthogonal

© Springer International Publishing Switzerland 2015
D.-S. Huang and K. Han (Eds.): ICIC 2015, Part III, LNAI 9227, pp. 387–396, 2015.
DOI: 10.1007/978-3-319-22053-6_42

Wavelet Support Vector Machine (OSVM) was proposed in [2], and a benchmark with the SVM and Multi-Layer Perceptron (MLP) shows a better performance and robustness of the OSVM. Wang et al. [3] presented a hybrid wavelet decomposition and support vector machine model, and a wavelet neural network model for predicting and fore-casting crude oil price. Also, proposition of a genetic algorithm for training neural network termed GA-NN by [4] was used in predicting crude oil price.

Algorithms that mimic the Biological Immune System (BIS) known as the Artificial Immune System (AIS) inhibit the following properties like diversity, detection, learning, and tolerance which are competitive with the traditional and most common data mining techniques. A comprehensive review of all the AIS algorithms are recorded in Dasgupta et al. [5]. In this study, the Negative Selection Algorithm (NSA) specifically the Vari-able-Sized Detectors (V-Detectors) is adopted for use.

Therefore, the structure of the paper is highlighted as follows: Sect. 2 describes the negative selection algorithm, variable-sized detectors, their characteristics and structure. Experimental simulations, results and analysis are reflected in Sect. 3. The concluding part of Sect. 4 summarizes the contributions of the study.

2   Negative Selection Algorithm

The biological process of *negative selection* inspired the development of Negative Selection Algorithm (NSA). It is solely confined to the adaptive immune system, where the B lymphocytes (B-cells) and T lymphocytes (T-cells) play active roles. Both the B-cells and T-cells act as the second line of defense and are triggered when the first line of defense, controlled by the phagocytes, has been breached. The idea behind negative selection is in the protection of the self cells occupied in the host body and elimination of antigens (non-self cells), and the process emanates from the thymus starting from when the T-cells are immature. For the T-cells to attain maturation, their receptors are generated in a pseudo-random manner and subsequently exposed to the self-peptides of the host body. An elimination process of the T-cells called *apoptosis* emerge when there is a reaction between the T-cells and the self-peptides, and only those T-cells that do not react are granted passage outside of the thymus into the body. These non-reactive T-cells now make up the defense mechanism in fighting against unwanted molecules which could prove harmful to the body.

Thus, resting on the recognition and elimination capabilities of the T-cells, Forrest et al. [6] proposed and developed the Negative Selection Algorithm (NSA) for discrim-inating what is self and non-self in a computer. It employed the use of binary strings. While the binary representation affords some advantages, it failed and underperformed when real-world data is concerned. The data representations for the negative selection algorithm are discussed in [7]. Therefore, to be able to handle real-world data effectively, González et al. [8] proposed the Real-Valued Negative Selection Algorithm (RNSA), that utilizes detectors which are fixed and chosen beforehand. The success of the RNSA in comparison with some artificial immune system algorithms and classification algo-rithm is demonstrated in [9]. In an improvement to the RNSA, Ji and Dasgupta [10] proposed the Variable-Sized Detectors (simply termed V-Detectors) where the detectors are dynamically chosen and terminates after enough detectors to achieve the highest detection rate is achieved. It should be noted that irrespective of the variation of NSA, two important stages known as the generation stage and detection stage constitute the algorithms. Sets of detectors are created and collected at the generation stage, and this pool of detectors is tasked with change detection at the detection stage.

On predicting oil prices, the V-Detectors meet the requirement for use in this study. A description of the V-Detectors algorithm, with its inherent properties is presented in the following section.

2.1 Variable-Sized Detectors (V-Detectors)

The limitation posed by the Real-Valued Negative Selection Algorithm (RNSA) [8] lead to the proposition of the Variable-Sized Detectors by Ji and Dasgupta [10] and termed V-Detectors. Its adopts n-dimensional vectors in real space $[0, 1]^n$ to encode antigens and anti-bodies. The elements to be detected are the antigens represented as $\{e_1, e_2, e_3, e_4, ..., e_n\} \in U$, and detectors for the antibodies. The self elements and detectors have a centre and a radius, denoted as $s = (c_s, r_s)$ and $d = (c_d, r_d)$ respectively. For every detector, there is assigned a radius which is not the same as self radius, and equate to its distance to the self region.

V – Detector – Set (S, T_{max}, r_s, c_0)
S: set of self samples
T_{max}: maximum number of detector
r_s: self radius
c_0: estimated coverage
1: $D \leftarrow \emptyset$
2: Repeat
3:      $t \leftarrow 0$
4:      $T \leftarrow 0$
5:      $r \leftarrow$ infinite
6:      $x \leftarrow$ random sample from $[1, 0]^n$
7:      Repeat for every d_i in $D = \{i = 1, 2, ...\}$
8:            $d_d \leftarrow$ Euclidean distance between d_i and x
9:            if $d_d \leq r(d_i)$ then, where $r(d_i)$ is the radius of d_i
10:                  $t \leftarrow t + 1$
11:                  if $t \geq 1/(1 - c_0)$ then return D
12:                  go to 4 :
13:            Repeat for every s_i in S
14:                  $d \leftarrow$ Euclidean distance between s_i and x
15:                  if $d - r_s \leq r$ then r $\leftarrow d - r_s$:
16:            if $r > r_s$ then $D \leftarrow D \cup \{<x, r>\}$, where $<x, r>$ is a detector
17:            else $T \leftarrow T + 1$
18:            if $T > 1/(1 -$ maximum self coverage) exit
19: Until $|D| = T_{max}$
20: return D

Fig. 1. Detector generation algorithm of V-Detector

Firstly, random detectors are produced; the centre of each detector is determined and must not be in the confines of the hyper-sphere region of the self element. Upon exposure and locating such a centre, the radius of the detector is dynamically resized until distance of the boundary of the region to the self element bridges. Sets of detectors are accumulated upon satisfaction of the above conditions, and radius of the resized detector denoted r becomes greater than self element's radius r_s. The algorithm stops when a predefined number of generated detectors can cover the non-self space.

The pseudo-code for the generation of detectors as regards V-Detector is shown in Fig. 1 proposed in [10].

3   Experimental Investigation and Prediction Results

The aim of this study is to investigate the applicability of negative selection algorithm for predicting crude oil prices. With implementations conducted using MATrix LABoratory (MATLAB), V-Detector algorithm is applied for experimentation. Furthermore, eight benchmarked algorithms namely Naïve Bayes (NB) [11] from the Bayesian family, Multi-Layer Perceptron (MLP) [12] from the neural network, Sequential Minimal Optimization (SMO) [13] from the family of support vector machine, IBk (Instance-based method on k-NN neighbor) [14] from instance-based category, J48 algorithm [15] from decision tree family, Non-Nested Generalized Exemplars (NNGE) [16] from nearest neighbor category, Fuzzy-Rough Nearest Neighbour (FRNN) [17] algorithm, and Vaguely Quantified Nearest Neighbour (VQNN) [18] are all used for comparison. All the benchmarked algorithms rely on Waikato Environment for Knowledge Analysis (WEKA) for their experimentation.

3.1   Dataset

The internationally acclaimed benchmark prices by private and governmental sector are the Brent crude oil prices and West Texas Intermediate crude oil prices. Other core parastatals also benefit from these oil prices. As other crude oil market prices formulate their oil prices, they ultimately refer to both the Brent and West Texas Intermediate crude oil prices as a standard [19]. Based on the fact that two-third of the world make reference to Brent crude oil price [20], the benchmark data employed for use in this study is the Brent crude oil price. The US Department of Energy provide access to the data, in which monthly data from 1987 to 2012 are retrieved. It consist of the following variables/attributes: Organization for Economic Co-operation and Development crude oil Ending Stocks (OECDES), Organization for Economic Co-operation and Development Crude Oil Consumption (OECDCOC), US Crude Oil Production (USCOP), US Crude Oil Stocks at Refineries (USCOSR), US Crude Oil Imports (USCOI), Non OPEC Crude oil Production (NOPECCP), and World Crude Oil Production (WCOP). The Dataset have a class attribute for the oil prices, and there exist five classes ranging from Very Low (VL), Low (L), Medium (M), High (H), to Very High (VH).

3.2 Experimental Procedure

Within the V-Detector algorithm, two important stages are accounted for, and are a training stage and testing stage. For calling the datasets into MATLAB as input for execution, the datasets need to be distinguished based on the class attribute. In the case of a two class dataset, the normal class is employed as the training data and considered as *self* whereas the other class becomes *non-self*. There is a difference when datasets with three or more classes are concerned. With a three class dataset, one of the classes is selected as the *self* and the remaining classes as *non-self*. This procedure is repeated for all the classes, which simply means that each of the class is utilized as self for training, with others as non-self. However, for the testing stage, all the data elements are used in classifying either as *self* or *non-self*. In all the experiments, 100 % of the training data is used and execution of 30 runs each, with the average values recorded. To be able to measure the affinities between the detectors and real-valued elements, the Euclidean distance represented in Eq. 1 is fused into the V-Detector algorithm. The Brent crude oil price is normalized with the *min-max normalization* process in the range [0, 1].

$$D(d, x) = \sqrt{\sum_{i=1}^{n} (d_i - x_i)^2} \qquad (1)$$

where $d = \{d_1, d_2, ..., d_n\}$ are the detectors, $x = \{x_1, x_2, ..., x_n\}$ are the real-valued coordinates, and D is the distance.

In designating parametric values, the same radius value r_s as used in [10] have been utilized. The V-Detector radius values $r_s = \{0.05, 0.1\}$ for the Brent crude oil price. Other parameter settings as in [10] include estimated coverage $c_0 = 99.98$ %, and Maximum Number of Detectors $T_{max} = 1000$.

3.3 Assessment Measures

The detection rate (DR) and false alarm rate (FAR) are the performance metrics to assess the computational effectiveness of the V-Detector algorithm as applied to Brent crude oil price. The definition of both terms follows with respective Equation depicted in (2) and (3).

Definition 1. Detection Rate *DR, which represents the ration of true positive and the total non-self samples identified by detector set, where TP and FN are the tallies of true positive and false negative.*

$$DR = \frac{TP}{TP + FN} \qquad (2)$$

Definition 2. False Alarm Rate *FAR, which represents the ratio of false positive and the total self samples identified by detector set, where FP and TN are the tallies of false positive and true negative.*

$$FAR = \frac{FP}{FP + TN} \tag{3}$$

3.4 Simulation Results and Discussions

The execution of the simulation experiments is performed on 3.40 GHz Intel® Core i7 Processor comprising 4 GB of RAM. Constructed into tables and graphs are the results of experiments on the Brent Crude oil price.

Table 1. Experimental results for Brent Crude oil price

Algorithms	Detection rate (%)	False alarm rate (%)
Naïve Bayes	64.50	11.10
Multi-layer perceptron	81.80	7.90
SMO	66.60	11.20
IBk	68.90	11.00
J48	73.60	9.80
NNGE	74.70	9.10
FRNN	70.30	10.30
VQNN	77.00	8.90
V-Detector$^{r=0.1}$	80.86	0.00
V-Detector$^{r=0.05}$	83.01	0.00

As deducted from the Table 1, it can be seen as a representation of the Brent Crude oil price results which are diagrammatically connoted in Figs. 2 and 3 respectively. From the graph illustrations, the algorithms have been labeled 1 to 10 with Naïve Bayes standing for label 1, Multi-Layer Perceptron for label 2, up until the last algorithm in Table 1 indicating V-Detector$^{r=0.05}$ for label 10 in ascending order. The performance of the compared algorithms in terms of detection rate ranged from 64.50 % for Naïve Bayes to 81.80 % for Multi-Layer Perceptron., and could not surpass that of V-Detector. The detection rate of V-Detector$^{r=0.1}$ at 80.86 % was eclipsed by the Multi-Layer Perceptron at 81.80 %, but not the V-Detector$^{r=0.05}$ at 83.01 %. The numeric figures in percentages signify the superiority of the V-Detector in relation to the other algorithms when detection rate is concerned.

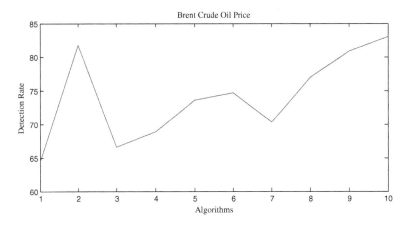

Fig. 2. Graph plots for detection rate

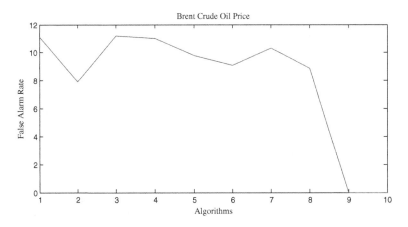

Fig. 3. Graph plots for false alarm rate

Additionally, the motive behind the false alarm rate is to be at its lowest minimum as possible. Resting on this, the V-Detector$^{r=0.1}$ and V-Detector$^{r=0.05}$ generated false alarm rates at 0.00 % respectively. With the other algorithms, SMO produced the highest at 11.20 %, followed by Naïve Bayes and IBk at 11.10 % and 11.00 % accordingly. Still in the descending order, false rates at 10.30 %, 9.80 %, 9.10 %, 8.90 % and 7.90 % are accounted for by FRNN, J48, NNGE, VQNN, and Multi-Layer Perceptron. The graph in Fig. 3 show the slope projection of all the algorithms for false alarm rate.

The number of detectors needed to cover the non-self space for accurate detection becomes paramount for real-valued negative selection algorithm. Thus, the mean number of detectors acquired for Brent Crude oil price is presented in Table 2. The number of detectors plotted against the detection rate is depicted in Fig. 4. For the V-Detector at $r_s = 0.1$, a total of 98 detectors are generated in attaining detection rate of 80.86 %. Also, for V-Detector at $r_s = 0.05$, the number of detectors to give detection rate of 83.01 % amounts to 111 detectors.

Table 2. Generated detectors for V-Detector

Algorithm	Radius r_s	Mean number of detectors
V-Detector	0.1	98
	0.05	111

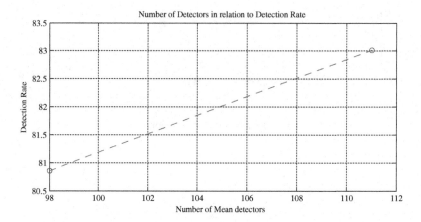

Fig. 4. Graph plots for detection rate with change in number of detectors

4 Conclusion

The formulation of oil market prices ascribed to the data generated and recorded by both the West Texas Intermediate crude oil prices and Brent crude oil prices guided this study in the use of real-valued negative selection algorithm to forecast decision making. The V-Detector is used to train and test Brent crude oil price data by accumulating detectors for effective performance. Eight benchmark algorithms including Naïve Bayes, Multi-Layer Perceptron, Support Vector Machine, IBk, J48, NNGE, FRNN, and VQNN are used as methods of comparison. For evaluation, the detection rates and false alarm rates reveal the superiority of the V-Detector on all other algorithms. Thus, V-Detector has shown to be promising and efficient for future crude oil price analysis. Further insight will dwell on optimizing V-Detector for elevated accuracy.

Acknowledgements.   This work is supported by the Office for Research, Innovation, Commercialization, and Consultancy Management (ORICC), Universiti Tun Hussein Onn Malaysia (UTHM), and Ministry of Higher Education (MOHE) Malaysia under the Fundamental Research Grant Scheme (FRGS) Vote No. 1235.

References

1. He, K., Xie, C., Chen, S., Lai, K.K.: Estimating VaR in crude oil market: a novel multi-scale non-linear ensemble approach incorporating wavelet analysis and neural network. Neurocomputing **72**, 3428–3438 (2009)
2. Chiroma, H., Abdul-Kareem, S., Abubakar, A., Zeki, A.M., Usman, M.J.: Orthogonal wavelet support vector machine for predicting crude oil prices. In: Herawan, T., Deris, M.M., Abawajy, J. (eds.) DaEng-2013. LNEE, vol. 287, pp. 193–201. Springer, Singapore (2014)
3. Wang, J., Xu, W., Zhang, X., Bao, Y., Pang, Y., Wang, S.: Data mining methods for crude oil market analysis and forecast. Data Min. Public Priv. Sect. Organ. Gov. Appl. Organ. Gov. Appl. 184 (2010)
4. Chiroma, H., Abdulkareem, S., Herawan, T.: Evolutionary neural network model for West Texas intermediate crude oil price prediction. Appl. Energy **142**, 266–273 (2015)
5. Dasgupta, D., Yu, S., Nino, F.: Recent advances in artificial immune systems: models and applications. Appl. Soft Comput. **11**, 1574–1587 (2011)
6. Forrest, S., Perelson, A.S., Allen, L., Cherukuri, R.: Self-nonself discrimination in a computer. In: Proceedings of the 1994 IEEE Computer Society Symposium on Research in Security and Privacy, pp. 202–212 (1994)
7. Lasisi, A., Ghazali, A., Herawan, T.: Negative selection algorithm: a survey on the epistemology of generating detectors. In: Herawan, T., Deris, M.M., Abawajy, J. (eds.) DaEng-2013. LNEE, vol. 285, pp. 167–176. Springer, Singapore (2014)
8. Gonzalez, F., Dasgupta, D., Kozma, R.: Combining negative selection and classification techniques for anomaly detection. In: Proceedings of the 2002 Congress on Evolutionary Computation, CEC 2002, pp. 705–710 (2002)
9. Lasisi, A., Ghazali, R., Herawan, T.: Comparative performance analysis of negative selection algorithm with immune and classification algorithms. In: Herawan, T., Ghazali, R., Deris, M.M. (eds.) SCDM 2014. AISC, vol. 287, pp. 441–452. Springer, Heidelberg (2014)
10. Ji, Z., Dasgupta, D.: Real-valued negative selection algorithm with variable-sized detectors. In: Deb, K., Tari, Z. (eds.) GECCO 2004. LNCS, vol. 3102, pp. 287–298. Springer, Heidelberg (2004)
11. Langley, P., Iba, W., Thompson, K.: An analysis of bayesian classifiers. In: AAAI, pp. 223–228 (1992)
12. Fine, T.L.: Feedforward Neural Network Methodology. Springer Science & Business Media, New York (1999)
13. Platt, J., et al.: Sequential minimal optimization: A fast algorithm for training support vector machines. Advances in Kernel Methods, Support Vector Learning, vol. 208, pp. 1–21. (1998)
14. Aha, D.W., Kibler, D., Albert, M.K.: Instance-based learning algorithms. Mach. Learn. **6**, 37–66 (1991)
15. Quinlan, J.R.: Learning with continuous classes. In: 5th Australian Joint Conference on Artificial Intelligence, pp. 343–348 (1992)
16. Martin, B.: Instance-based learning: nearest neighbour with generalisation. Ph.D. thesis. 1–152 (1995)
17. Jensen, R., Cornelis, C.: A new approach to fuzzy-rough nearest neighbour classification. In: Chan, C.-C., Grzymala-Busse, J.W., Ziarko, W.P. (eds.) RSCTC 2008. LNCS (LNAI), vol. 5306, pp. 310–319. Springer, Heidelberg (2008)
18. Jensen, R., Cornelis, C.: Fuzzy-rough nearest neighbour classification and prediction. Theor. Comput. Sci. **412**, 5871–5884 (2011)

19. He, K., Yu, L., Lai, K.K.: Crude oil price analysis and forecasting using wavelet decomposed ensemble model. Energy **46**, 564–574 (2012)

20. Charles, A., Darne, O.: The efficiency of the crude oil markets: evidence from variance ratio tests. Energy Policy **37**, 4267–4272 (2009)

A New Algorithm of Automatic Grading in Computer Paperless Test System

Tian-Lan Liu, Wen-Sheng Tang$^{(\boxtimes)}$, Sheng-Chun Wang, and Jun Qin

The Department of Computer Teaching, HuNan Normal University,
Changsha 410081, China
154976552@qq.com

Abstract. Efficient automatic grading of programs in computer paperless test system is a hotspot in current research. In allusion to the deficiency of scoring computer language in current test system, a kind of machine scoring algorithm that combines static semantic understanding analysis with runtime dynamic analysis is proposed. In order to achieve a reasonable scoring purpose, the scoring function is constructed based on the weighted average method and its function mapping is determined by BP neural network. The experiment result indicates that the effect of this new scoring algorithm is close to the effect of artificial scoring method.

Keywords: Static semantic understanding · Runtime dynamic analysis · Machine scoring

1 Introduction

Paperless computer exam has become a general trend, in recent years, a variety of test systems have been introduced. At present, there are two main methods of machine scoring: static analysis and dynamic testing.

Static analysis uses compiler theory to analyze the source code of the candidates, matches with provided answers and confirms scores according to matched results, which focuses on studying matching algorithm. For example, the literature [1] used LSA (Latent Semantic Analysis) algorithm, and provided a kind of broad program matching method and it is not limited to a certain type of language. Dynamic testing technique is to allow candidates to run and control the input program in a virtual environment, through the way of investigating if the program input and the expected output are consistent. As the literature [2] analyzed some characteristics VFP program, and provided a viable option for VFP automatic marking by black and white box software testing methods in software engineer.

The advantages of static analysis is that it can be applied to a wide range, which can follow the procedural steps to the sub, and artificial scoring is more close to score. The disadvantage is that it cannot completely guarantee score correctly, you need to complete the answer databases, or it is possible to get a correct answer but it does not match the answer in database. The advantage of dynamic testing is quick and easy to implement, the disadvantage is that only a complete program statements matches the subject, it can be applied to determine the input and output, and only scores the results

© Springer International Publishing Switzerland 2015
D.-S. Huang and K. Han (Eds.): ICIC 2015, Part III, LNAI 9227, pp. 397–406, 2015.
DOI: 10.1007/978-3-319-22053-6_43

of the running. It cannot follow the steps to the sub-program. Furthermore, the presence of "infinite loop" and other safety hazards is in the run-out.

Through research and analysis, this article proposed a new algorithm to solve the problem of automatic scoring. The new algorithm combines semantic understanding with analysis of running strategy. Static semantic understanding approaches to the use of compiler theory candidates semantic analysis procedures bypassing unfavorable factors, such as the analysis of visual controls, grammar, algorithms, knowledge of the problem involved only the mastery score, and proposes a labeled tree based on weighted abstract notation key questions and the answers to the key point, making the score points compared to the previous static analysis points more reasonable and accurate; Run-time dynamic analysis is specific to answers which cannot be determined, by analyzing the form source files, by implanting code in the candidates placed in the program code and designing sandbox to run the program saved to a file in the operating results of different stages. Thus it can grade the program result and monitor the program operation. Finally, the grade function constructed by the weighted average method combines two modules to get the end scores. The experiment proves that the proposed algorithm in this thesis can effectively achieve the purpose of automatic scoring with high accuracy rate, and close to manual scoring.

2   Problem Descriptions

The following example draw an object studied in this paper, the topic is produced from a kind of typical problems of simulation test system in database.

Tests: The existing file named formone.scx is in candidates folder, as it is shown in Fig. 1, it contains a list box, a table and a command button.

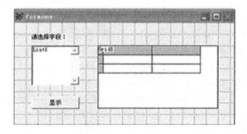

Fig. 1. Exam form

Please complete the appropriate action according to the following requirements:

1. Adding orders table in the form of data environment;
2. The list box List1 sets multiple choice, in addition to its RowSourceType property value will be set to "8- structure", RowSource set to orders;
3. Setting the RecordSourcetype property value of the form Rrid1 to "the 4-SQL statement";

4. Modify the "Display" Click event code of the button. When the button is clicked, displays the specified field in the list box contents selected orders table within a table Grid1.

3 Algorithm of Automatic Grading

3.1 The Overall Framework

In this paper, a new static analysis and semantic understanding runtime dynamic analysis is approached, it can monitor the running of the program and you can also follow the steps to analyze grammar points. The overall frame is shown in Fig. 2.

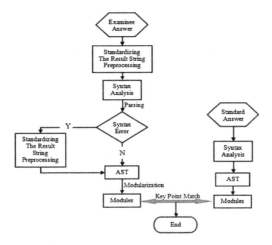

Fig. 2. Auto-grading module

3.2 The Static Semantics Understanding Module

Static semantic analysis to understand the process shown in Fig. 3.

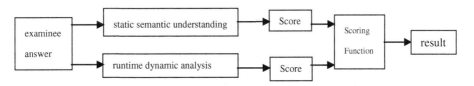

Fig. 3. Static semantic comprehension

400   T.-L. Liu et al.

In real exam, candidates enter the program vary widely, which is not conducive to analyze the source code, so candidates input results need to be standardized before analyzing grammar analysis, (for example: removing extra spaces, comments, and so on). In this thesis, it uses full-text to match the regular expression string on the result, fast filter redundancy information. And it analyzes the standardized results and generates the Token table. Then it analyzes grammar of lexical units according to VFP language description based on BNF grammar (Backus Normal Form). BNF symbol set is used to describe programming language syntax, computer language grammar can be naturally described, each nonterminal corresponds to a function of the process, the following gives a mathematical expression VFP language BNF description example:

Numerical_Exp:: = Term {Add_op Term }
Add_op:: = "+" | "-"
Term:: = Factor {Mul_op Factor }
Mul_op:: = "/" | "*" | "%" | "**" | "^"
Factor:: = ID | NUM | ["("] Numerical_Exp [")"]

Syntax analysis is the process of generating the lexical unit stream syntax tree [3]. In the expression i * (i + i) as an example, after parsing can be drawn as shown in Fig. 4 abstract syntax tree.

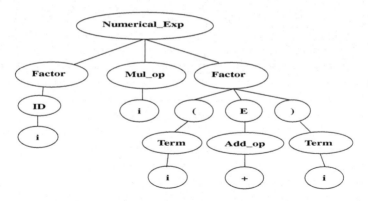

Fig. 4. The expression i* (i + i) corresponding syntax tree

After the source code to generate an abstract syntax tree, it can serve as the basis for the program which is divided into several modules through the tree structure similarity matching algorithm [4], to analyze the similarity of the source and answer process, but this process cannot be done step by step to achieve sub-demand. Divided by points the way to the existing program is the standard answer to mark a piece of code, the code segment is marked as scoring points. Examples to 2.1 example, the solution is as follows.

```
s=""
f=.T.
FOR i=1 TO thisform.List1.ColumnCount              ②
    IF this form.List1.Selected(i)                  ①
        IF f
            s=thisform.List1.value                  ①
            f=.F.
        ELSE
            s=s+thisform.List1.value
        ENDIF
    ENDIF
ENDFOR                                              ①
st="select &s from orders into cursor tmp"         ②
thisform.Grid1.RecordSource=st
```

The correct answer is in marked ①② mark key points scoring commonly used text level, and if this line program that matches the given tag value. However, this approach maybe not produce two paired for-endfor statement matches a key point, or the cycle of in vitro assignment as a loop assignment. That matches only text-based level; semantic point of view cannot be the key point to be judged correctly.

In this regard we propose a weighted syntax tree way of mark key points that is marked on the sentence structure crucial score points, giving the right value for the key syntax tree nodes. When candidates answer program procedures and standards can be matched successfully in the tree nodes with the right values, the candidates can get scores of the node weights, which is similar to artificial marking the steps score. In the above examples, its standard answer with the right syntax tree structure is shown in Fig. 5.

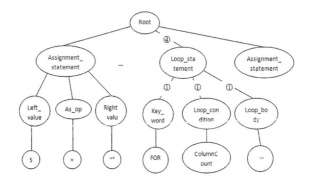

Fig. 5. Example standard answer with the right syntax tree

Process of the algorithm is that after generating candidate program syntax tree, it starts with standard answers tree, makes a breadth-first traversal, each node with the right to call the Levenshtein algorithm [4], if the sub-tree structure is similar to the standard the answer, the node score is obtained, otherwise it selects the sub-tree to the next child node with the right to do the same operation until completing access to all weighted nodes. This dealing makes fuzzy matching on the semantic level and can reduce a wide range of assessment error rate, for more than two matches in plain text may encounter errors are avoided, while for the number of cycles for statement can also be calculated correctly, for cycling conditions, such statements cannot be fully determined, it can be addressed through a flexible set a key weights.

3.3   Runtime Dynamic Analysis Module

Through the analysis of VFP procedure, we can get the form data in the control. Due to the limited controls used in the examination by identifying the name of the form control types, attributes, fields, data can be analyzed to show the way out of control, resulting in a dynamic variable names. By way of further embed code dynamically variable values written to the file the way, run-time data acquisition program. Then through the document analysis to determine the program's operating results are correct, similar to the results in hand scoring points. The specific process shown in Fig. 6.

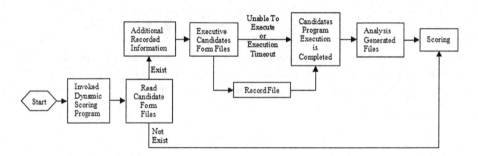

Fig. 6. Run-time dynamic analysis

3.4   Scoring Function Module

10 people answer respectively according to 10 questions drawn from a question bank in this category, then score with the manual marking algorithm and data set M composed of n which is equal to 100. in the vector $X = (x_1, x_2, ..., x_n)^T$ represents a given semantic understanding module static score vector, $Y = (y_1, y_2, ..., y_n)^T$ is represented by the run-time ratings given by the vector analysis module, $Z = (z_1, z_2, ..., z_n)^T$ shows the ratings given by manually marking vector. Because there is a certain gap between the value of the simple average $(X + Y)/2$ and the artificial scoring Z, in order to achieve a reasonable scoring purposes, we construct the following scoring function based on the weighted average method [5]:

$$Z' = \frac{AX + BY}{2} \tag{1}$$

Z' is the final score given by the algorithm, making $A/2 = \alpha$, $B/2 = \beta$, we can get $Z' = \alpha X + \beta Y$. So if we want to make the final score close to hand scoring, as long as the principle of least squares method to determine the coefficient α, β so that the function (2) and artificial marking Z score calculated by scoring the final score function Z' from the value of the minimum sum of squares f [6].

$$f = \sum_{i=1}^{n} |z_i - \alpha x_i - \beta y_i|^2 \tag{2}$$

$$p = \sum_{i=1}^{n} |z_i - \alpha_1 x_i|^2 \tag{3}$$

$$q = \sum_{i=1}^{n} |z_i - \beta_1 y_i|^2 \tag{4}$$

To make the f minimum, the conventional method is through linear equations. First finding the coefficient $\alpha 1$ enable function p value (3) the minimum, and then seeking the coefficient $\beta 1$ to make function q value (4) in the smallest, and then seek α according $\alpha 1$, $\beta 1$, β. Disadvantage of doing so is the number of multi-dimensional computation is too big, but also for the experimental data set it is known and limited, the actual data is unknown and infinite, and therefore it cannot be calculated mathematically precise coefficient, it can only find α, β approximately the optimal solution, we use the BP neural network algorithm.

BP network does not require prior knowledge of mathematical equations and we can learn and store input and output mappings, the main idea of BP algorithm is that it is given training for network input and desired output, layer by layer calculated before the actual output of the network, if there is a deviation between actual output and the desired output, it will be back propagation along the network, which began back layer by layer from the output layer to modify the connection weights and thresholds, until the error meets the requirements [7].

BP network is a learning process right connection between the lower and the upper node weight matrix W_{ij} settings and error correction process [7], there is division into learning and unsupervised learning modes, there are teacher learning needing to set expectations, and no Teachers just entering it. Function (5) is the self-learning model of the network, in where h is the learning factor; Φ_i is a calculation error output node i; O_j is calculated Output node j; a is the momentum factor. Function (6) is a neural network to calculate the size of the desired function of the error between the output and the calculated output. In the function the output value is the desired t_{pi} and the output calculation value is the o_{pi}.

$$W_{ij}(n + 1) = h \times \Phi_i \times O_j + a \times W_{ij}(n) \tag{5}$$

$$E_P = \frac{1}{2} \times \sum (t_{pi} - o_{pi})^2 \tag{6}$$

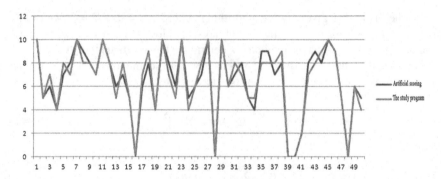

Fig. 7. Z'and Z comparison chart

In this paper, the first 50 rows of the data set is a training set of data, that is, using the foregoing 50 steps points X and Y as a result of sub-input, manual scoring score Z as the expected output to train the network, and then regarding the later 50 rows of data as an experimental set, that is to put back 50 steps into X, results into Y having been already trained network and got a final score Z', finally comparing the Z' and back 50 men who got Z scores by manual scoring, as shown in Fig. 7, you can see the little difference.

4 Experiments

4.1 Experimental Data

We found two types of questions from the question bank and let 30 students answer them. And we give them proper scores by the ways of artificial scoring and study program. The experimental results are shown as Fig. 8. Due to large amount of data, we randomly selected five students, and recorded their scores of each part into Table 1. Every test is 10 points, and 6 points is qualified.

Question one: Question two:

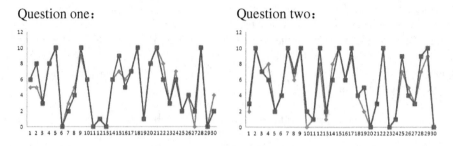

Fig. 8. Experimental results

Table 1. Experiments data comparison

Student number	Method	Question one				
		Grammar	Point	Algorithm	Result	Score
1	The study program	0	1	2	0	2
	Artificial scoring	1	1	2	0	3
2	The study program	3	3	4	10	10
	Artificial scoring	3	3	4	10	10
3	The study program	2	2	3	0	5
	Artificial scoring	2	3	3	0	6
4	The study program	2	3	3	0	6
	Artificial scoring	2	1	3	0	4
5	The study program	1	0	1	0	1
	Artificial scoring	1	0	1	0	1
Student number	Method	Question two				
		Grammar	Point	Algorithm	Result	Score
1	The study program	1	1	2	0	3
	Artificial scoring	1	1	2	0	3
2	The study program	3	3	4	10	10
	Artificial scoring	3	3	4	10	10
3	The study program	2	1	3	0	4
	Artificial scoring	2	2	3	0	5
4	The study program	3	3	4	10	10
	Artificial scoring	3	3	4	10	10
5	The study program	0	0	0	0	0
	Artificial scoring	1	0	1	0	1

Table 2. Analytical formulas table

Sources of variation	Sum of squares	Degrees of freedom	Sample variance	Value F
Factors between groups	SSA	I-1	SSA/(I-1)	SSA/SSE
Intraclass (Errors)	SSE	N-I	SSE/(N-I)	
Summation	SST	N-1		

4.2 Data Inspection

This article uses the method of single factor analysis of variance mentioned in Ref. [8], examining students' achievement in reading method change, namely from manual into automatic marking scheme proposed in this paper and inspect if there exists a significant difference. Formulas are shown in Table 2 and function (7), (8). If $F = SSA/SSE < F$ (I-1, N-1), there was no significant difference.

where

$$\bar{x}_i = \frac{1}{n_i}\sum_{j=1}^{n_i}X_{ij}, \bar{X} = \frac{1}{N}\sum_{i=1}^{I}\sum_{j=1}^{n_i}X_{ij} \qquad (7)$$

$$SST = \sum_{i=1}^{I}\sum_{j=1}^{n_i}(X_{ij} - \bar{X})^2, SSA = \sum_{i=1}^{I}n_i(\overline{X}_i - \bar{X})^2, SSE = \sum_{i=1}^{I}\sum_{j=1}^{n_i}(X_{ij} - \overline{X}_i)^2 \qquad (8)$$

Because the scores are got by comparing 30 students' scores in the two scoring cases, so I = 2, N = 60, from table we can see that F (I-1, N-1) is 4.007. According to the data of test1 and test2, we got that F is 0.155 and 0.002, respectively, two of them are less than F(1, 58), which shows that there is no significant difference between the proposed algorithms and artificial, it can be applied to the real exams.

5 Conclusion

This thesis takes program automatic marking of subjective question as study background. Computer Rank Examination (VFP) program entitled samples takes a class of the form and table tests as the object of study. Since this kind of question types cannot use automatically marking techniques to score, we proposed a new plan, and it proved the feasibility.

Acknowledgment. The authors would like to thank the anonymous reviewers for their helpful and constructive comments. This work is supported by education reform Item of Hunan Normal University (Grant no [2011]75, [2014]85-39),and supported by program for excellent talents in Hunan Normal University (No.ET61008).d

References

1. Zen, K., Iskandar, D.N.F.A, et al.: Using latent semantic analysis for automated grading programming assignment. In: Semantic Technology and Information Retrieval, Putrajaya, Malaysia, pp. 82–88 (2011)
2. Liu-ling: Research and realization of automatic marking technology for visual FoxPro programming papers. Shenyang University of Technology (2005)
3. Yuanyuan, C.: The research and application of automatic scoring system based on abstract syntax tree. Maritime Affairs University of Dalian (2011)
4. Lee, K.-Y., Seo, I.-H., Kim, J.-J., Kang, E.-Y., Park, J.-J.: A study on improved similarity measure algorithm for text-based document. In: Kim, T.-h., Lee, Y.-h., Fang, W.-c. (eds.) FGIT 2012. LNCS, vol. 7709, pp. 181–187. Springer, Heidelberg (2012)
5. Lei, C., Yin, C.: A realization method of fuzzy evaluation. J. Taizhou Univ. **06**, 6–11 (2009)
6. Jigang, X., Xinlu, F., et al.: Polymorphic partial least squares model. Appl. Comput. Syst. **06**, 178–181 (2012)
7. Hui, C., Huicheng, L.: An improved back - propagation NN algorithm and its application. Comput. Simul. **111**(04), 75–77 (2007)
8. Shiquan, S.: Research on automatic marking technology of programming. Central South University (2007)

MLRF: Multi-label Classification Through Random Forest with Label-Set Partition

Feng Liu[1], Xiaofeng Zhang[1], Yunming Ye[1(✉)], Yahong Zhao[1], and Yan Li[2]

[1] Shenzhen Key Laboratory of Internet Information Collaboration, Harbin Institute of Technology Shenzhen Graduate School, Shenzhen, China
{524196009,804676883}@qq.com,
zhangxiaofeng@hitsz.edu.cn, yeyunming@hit.edu.cn
[2] Shenzhen Polytechnic, Shenzhen, China
liyan@szpt.edu.cn

Abstract. Although random forest is one of the best ensemble learning algorithms for single-label classification, exploiting it for multi-label classification problems is still challenging and few method has been investigated in the literature. This paper proposes MLRF, a multi-label classification method based on a variation of random forest. In this algorithm, a new label set partition method is proposed to transform multi-label data sets into multiple single-label data sets, which can effectively discover correlated labels to optimize the label subset partition. For each generated single-label subset, a random forest classifier is learned by an improved random forest algorithm that employs a kNN-like on-line instance sampling method. Experimental results on ten benchmark data sets have demonstrated that MLRF outperforms other state-of-the-art multi-label classification algorithms in terms of classification performance as well as various evaluation criteria widely used for multi-label classification.

Keywords: Multi-label classification · Random forest · Ensemble learning

1 Introduction

Multi-label classification has received significant attentions in machine learning community in the past decade from many researchers all over the world due to the emerging application needs. Many different approaches have been proposed to solve the multi-label classification problems. Tsoumakas and Katakis [1] summarize them into two main categories, i.e., problem transformation methods, and algorithm adaptation methods. Problem transformation methods transform the multi-label problem into several single-label problems and use single label algorithms to solve it, then integrate the optimal results, such as Label Power-set(LP) [2], Pruned Sets(PS) [3], Random k-label-sets(RAkEL) [4], Binary Relevance(BR) [5], Classifier chain(CC) [6], etc. On the other hand, algorithm adaptation methods extend the single-label algorithms to cope with multi-label learning tasks, like ML-C4.5 [2], PCTs [7–9], ML-kNN [10],

D.-S. Huang and K. Han (Eds.): ICIC 2015, Part III, LNAI 9227, pp. 407–418, 2015.
DOI: 10.1007/978-3-319-22053-6_44

BRkNN [11], HOMER [12], AdaBoost.MH and AdaBoost.MR [13], etc. Although there are many multi-label classification algorithms in the literature, designing efficient and accurate algorithms for universal data sets is still a challenge for machine learning researchers.

It's well-known that random forest is one of the best ensemble learning algorithms for single-label classification [14, 15]. However, few attentions have been paid to the research on adapting random forest for multi-label classification tasks. It would be interesting to verify whether random forest can be transformed smoothly for multi-label classification tasks. In this paper, we propose MLRF, a novel multi-label classification algorithm which is a variation of random forest for multi-label learning.

Intrinsically, MLRF is also a problem transformation based method, which adopts a new label set partition method to construct multiple single-label training sets from original multi-label data. The new label set partition method is based on a random permutation approach that can exploit the correlation information of labels to generate coherent label subsets. For each generated label subset, a random forest classifier is learned by an improved random forest algorithm that employs a kNN-like online instance sampling method. Experimental results on ten benchmark data sets have proved that MLRF outperforms other state-of-the-art multi-label classification algorithms with respect to classification performance and various evaluation criteria for multi-label classification.

The rest of this paper is organized as follows. In Sect. 2, we present the proposed MLRF and its detailed implementation. We evaluate the proposed method and compare it with other existing methods in Sects. 3 and 4 concludes the paper.

2 The Proposed Multi-Label Random Forest Algorithm

In this section, we proposed MLRF algorithm which consists of five components, i.e., label set partition, multi-label data transformation, sampling, feature selection, ensemble learning. Details of MLRF are given in Algorithm 1.

```
Algorithm 1 MLRF
Input: Train, Test
Output: Y={Y₁,...,Yₘ}
Begin
  For all Lᵢ∈L do
    Compute the number of unique labels in subset Li
    Use LP algorithm to transform data-set if the number
    of unique labels isn't equal to 1
    Aggregate random forest as C = {C₁,...,Cₘ}
  For all Testᵢ∈Test do
    For all Cᵢ∈C do
      Predict Testᵢ, get the label subset Yᵢ by voting
        strategy
    Integrate the final label set Y={Y₁,...,Yₘ}
End.
```

For most of multi-label classification algorithms, they generally assumed that labels are independent, then, various approaches are proposed based on this assumption. However, these labels are naturally correlated, such as inclusion and overlapping, which make them no longer independent, which consequently deteriorates the classification performance. To incorporate the effect of correlated labels, we proposed MLRF. In MLRF, we first partition existing labels into several label subsets. Labels in the same subset are assumed to have strong correlations, whereas labels from different subsets are assumed to be weakly correlated. For each generated label subset, a random forest is built as the base classifier. If this label subset only contains one unique class label, then this multi-classification task is already degenerated to single-label classification problem. Accordingly, traditional random forest algorithms could be adopted for this subset. If the number of unique labels in a subset is greater than 1, then LP algorithm is adopted to transform the subset into multiple single label subsets, then random forest classifier could be built on this transformed subsets. At last, appropriate voting strategy is adopted to acquire the ensemble results of all base classifiers.

2.1   Label Set Partition

There exist several approaches for discovering label dependency, e.g., Chi-square test [16], Fisher exact test [17], proportions difference test [18], likelihood ratio test [19]. In this paper, we adopt Chi-square test and its steps are given as follows. Given two labels λ_i and λ_j, the contingency table is given as Table 1, and the Chi-square χ^2 is calculated using Eq. 1.

$$\chi^2 = \frac{(ad - bc)^2(a + b + c + d)}{(a + b)(c + d)(b + d)(a + c)} \tag{1}$$

We detail several key steps as follows. In this algorithm, we first calculate the Chi-square value of each pair of labels using Eq. 1, and standardize these values to reveal the data correlation. If the Chi-square value is greater than 6.635, then testing data are assumed to be correlated. Then, we randomly generate partitions N_1 and compute the dependence score of each partition in the next. Then the corresponding partitions are sorted in descending order according to dependence score and we remove the duplicated ones. Now, we can acquire total N_2 label partitions. Generally N_2 could be very large, a parameter N_3 is then defined to reduce the computational complexity by only choosing top N_3 partitions. Then we name N_4 as the smaller value of N_2 and N_3. Finally we choose Parameter N_6 that is determined by the smaller value of N_4 and N_5.

Table 1. Contingency table of λ_i and λ_j

	λ_i	$\neg\lambda_i$	Total
λ_j	a	b	$a + b$
$\neg\lambda_j$	c	d	$c + d$
Total	$a + c$	$b + d$	$a + b + c + d$

Let m denotes the size of label set, we set n = 2 * m − 1 to generate permutations using numbers starting from 0 to n − 1, then choose number whose value is bigger than m − 1 as the split point and split this permutation to generate the label subsets. For example, suppose m = 6, then the permutation is generated using [0, 2 m − 1] as [2, 4, 7, 9, 0, 1, 8, 3, 5, 6, 10, 11]. And the label subsets are then acquired as {[2, 4] [0, 1] [3, 5]}.

To compute the dependence score of the subsets, suppose the partition is given as $\{[l_i,...,l_j],...,[l_m,...l_n]\}$, then the calculation is given as follows:

$$weight = \sum_{i}^{m} \sum_{i}^{j} weight\ matrix[i][j] - \sum_{i}^{m} \sum_{i}^{n} weight\ matrix[i][m]$$

which sums up the weight of label-pairs within the same subset, and minus the weight of label-pairs that are not in the same subset.

Once dependence score is acquired, we partition label subsets to have higher weight. The process is given as follows. First, we set the partition with the highest weight to be selected, then choose the partition that is the farthest to the partition selected. For example, suppose partition 1 is {[0,1] [2, 4] [3, 5]} and partition 2 is {[0,1,3] [2, 4, 5]}, and its co-occurrence matrix is given as Fig. 1.

This co-occurrence matrix gives straightforward representation whether a pair of labels occurs in the same label set or not. If a pair of labels is in the same subset, then the corresponding cell in the matrix is 1, and 0 otherwise. For example, cell(3,5) of both matrix is 1 which means that the 2nd label and the 4th label are in the same subset, i.e., [2, 4] or [2, 4, 5]. While cell(3,6) of both matrix is 0 and 1, respectively. This implies that the 2nd label and the 5th label are not in the same subset in partition 1, and they are in the same set in partition 2. The distance between two partitions can be calculated by counting the number of different cells in the two matrix which is equivalent to perform exclusive OR operation.

2.2 Sampling

A kNN-like online sampling strategy is widely used in multi-label classification tasks. Given a test sample, it generally discovers the K nearest neighbors of each test sample, and these neighbors are collected into set bagNeigh and this set well preserves the characteristics of nearest neighbors. To keep data diversity for learning good random forest classifier, we also randomly sample K data from the nearest neighbor set

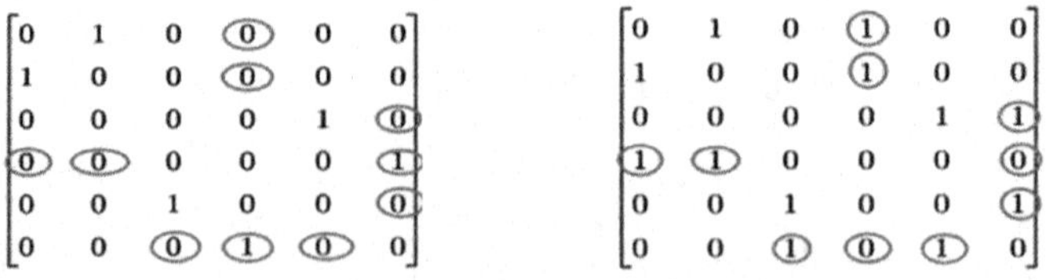

Fig. 1. Matrix representation of partition 1 and 2

bagNeigh to avoid of using too much test data and then sample training set Bag according to proportion ρ, and parameter ρ is carefully tuned from 3.3 % to 6.6 % to achieve the best performance.

2.3 Feature Selection and Ensemble Learning

For feature selection, we also adopt random sampling method as most of random forest do. Suppose the training set contains total F features, we choose $f = \lfloor \log_2 F + 1 \rfloor$ as the candidate split feature set. To determine the optimal split feature, one commonly adopted criterion is Gini index [20]. Gini index tries to assign balanced instances to the same class within each partition. Let D denote the training set having m different labels $C_i (i = 1,2,...,m)$, |D| is the size of data set, and $|C_i,D|$ is the number of label C_i in D. The Gini index can then be calculated using Eq. 2.

$$Gini(D) = 1 - \sum_{i=1}^{m} p_i^2 \qquad (2)$$

Where $p_i = \frac{|C_{i,D}|}{|D|}$ is the probability that instances in D belong to class C_i. If a binary partition split by feature A divides D into D_1 and D_2, then the corresponding Gini index is computed using Eq. 3 as

$$Gini_A(D) = \frac{|D_1|}{|D|} Gini(D_1) + \frac{|D_2|}{|D|} Gini(D_2) \qquad (3)$$

Therefore, to determine the optimal partition feature is to choose the maximal ΔGini(Attr). As for ensemble learning of random forest, one simple strategy is adopted by us which is the majority win strategy. For all base random forest classifier, the vote of all classifiers will be added and the highest vote means the highest probability that the test data sample belongs to this class.

3 Experimental Results and Analysis

For the experiments, we first introduce the benchmark data sets, the evaluation criteria as well as the experimental settings. Then proposed MLRF algorithm and the rest state-of-the-art multi-label classifiers are evaluated on these data sets. Results on different category of evaluation criteria are separately reported.

3.1 Data Sets and Evaluation Criteria

In the literatures, there are several benchmark data sets for multi-label classification task. We choose 10 data sets of 5 categories such as audio, video, image, biology and text. The statistical details of each data set are reported in Table 2.

Table 2. Statistical details of multi-label data sets

Data set	Instance	Attribute	Label	Cardinality	Density	Distinct	Domain
emotions	593	72	6	1.869	0.311	27	Music
birds	645	260	19	1.014	0.053	133	Audio
enron	1702	1001	53	3.378	0.064	753	Text
scene	2407	294	6	1.074	0.179	15	Image
yeast	2417	103	14	4.237	0.303	198	Biology
tmc2007	28596	500	22	2.220	0.101	1172	Text
top90Collection	10789	500	90	1.235	0.014	468	Text
rcv1subset1	6000	500	101	2.880	0.029	1028	Text
rcv1subset2	6000	500	101	2.634	0.026	954	Text
rcv1subset3	6000	500	101	2.614	0.026	939	Text

As for the evaluation criteria, we adopted 13 widely used criteria for the evaluation of multi-label classification task. Note that F_Measure is the harmonic mean of Precision and Recall, and we choose F_Measure instead of Precision and Recall, and the rest evaluation criteria are grouped into Bipartition-based evaluation measures, Rank-based evaluation measures and Confidence-based evaluation measures. For Bipartition-based evaluation measure, it includes Hamming Loss(HL), SubSetAccuracy(SSA), Example Base F_Measure(EB_FM), Example Based Accuracy(EB_A), Micro F-Measure(Mi_FM) and Macro F_Measure(Ma_FM). The Rank-based evaluation measures include Average Precision(AP), Coverage(Cov), One_Error(OE), Is_Error(IE), Error Set Size(ESS), RankingLoss(RL) and Confidence-based evaluation measure is Mean Average Precision(MAP).

3.2   Experimental Settings

To evaluate the effectiveness of the proposed MLRF, we first compare it with the conventional random forest with label subset partition, called MLRF_Trad, then we compare MLRF with standard problem transform method. We also compared our MLRF with several state-of-the-art algorithms such as MLkNN, IBIL and HOMER. MLkNN extends kNN algorithm for multi-label classification and k is set to 10 and we call it MLkNN_10. IBIL was proposed by Cheng et al. in 2009, HOMER is a hierarchical multi-label classification algorithm and its parameter is set as numCluster = 3, and its sub-classifiers are chosen as BR and J48. The parameters of MLRF are set as $N_1 = 50000$, $N_3 = 100$, $N_5 = 1$, $K = 5$, $\rho = 3.3$ % and the experiments are performed using 10-fold cross validation. The distance metric adopted in this paper is Euclidean distance and the standard Chi-square value is 6.635.

3.3   Experimental Results

Results on Bipartition-Based Measures. In this experiment, various multi-label classification algorithms are implemented including HOMER, HMC, IBIL, MLkNN_10, BRkNN_N and etc. Evaluation criteria are Bipartition-based measures. For all index but HL index, the larger the index value, the better the classification performance. Results on different data sets are separately reported in Tables 3, 4, 5 and 6. In these tables, the first column is different index of Bipartition-based measures, and the rest columns are the results of different classifiers.

From these tables, it is obvious that MLRF is much better than the compared algorithms. For data set "emotions", MLRF can achieve the best performance for all evaluation measures reported in Tables 3 and 4. While for the rest data sets, MLRF can also achieve better performance on most of measure index. From the comparison results between MLRF and MLRF_Trad, we can conclude that by integrating label correlation into conventional label partition, the performance of random forest could be further improved. It is also noticed that MLRF_Trad is superior to LP_RF which

Table 3. Bipartition-based measure results on emotions

	HL	SSA	EB_FM	EB_A	Mi_FM	Ma_FM
MLRF	**0.1190**	**0.5714**	**0.7905**	**0.75**	**0.8214**	**0.8348**
MLRF_Trad	0.1508	0.4762	0.7714	0.6984	0.7765	0.7588
LP_RF	0.1472	0.4833	0.7483	0.6792	0.758	0.7232
MLkNN_10	0.1778	0.3	0.675	0.5806	0.68	0.6184
MLkNN_50	0.1917	0.3333	0.6306	0.5556	0.6497	0.5764
BRkNN_N	0.1917	0.3	0.625	0.5444	0.6462	0.5954
BRkNN_A	0.1861	0.3	0.6556	0.5667	0.6633	0.6186
BRkNN_B	0.1889	0.3333	0.6861	0.5958	0.6937	0.6734
IBLR	0.1833	0.3167	0.6589	0.5722	0.6796	0.6328
HOMER	0.2472	0.1833	0.5056	0.425	0.5742	0.5375
HMC	0.2028	0.2667	0.5889	0.5033	0.6368	0.615

Table 4. Bipartition-based measure results on Emotions (cont.)

	HL	SSA	EB_FM	EB_A	Mi_FM	Ma_FM
BR	0.1778	0.3333	0.6339	0.5597	0.6735	0.6119
CLR	0.2056	0.2333	0.5672	0.4861	0.6373	0.6167
CC	0.2389	0.3	0.5944	0.5153	0.6055	0.5847
ILC	0.2028	0.2667	0.5889	0.5033	0.6368	0.615
LP	0.2556	0.2833	0.5333	0.4722	0.5577	0.5386
RAkEL	0.2028	0.2667	0.5889	0.5033	0.6368	0.615
ECC	0.1639	0.2833	0.6217	0.5389	0.7005	0.667
EPS	0.175	0.4	0.6772	0.6083	0.6986	0.6619

Table 5. Bipartition-based measure results on rcv1subset2

	HL	SSA	EB_FM	EB_A	Mi_FM	Ma_FM
MLRF	**0.0144**	**0.4367**	0.6364	**0.5807**	**0.6056**	**0.5734**
MLRF_Trad	0.0144	0.4333	**0.6367**	0.5788	0.6017	0.5249
LP_RF	0.3026	0.2200	0.3470	0.3049	0.3204	0.2461
MLkNN_10	0.026	0.02	0.1024	0.0790	0.1397	0.1275
MLkNN_50	0.026	0.0317	0.1131	0.0893	0.1386	0.1171
BRkNN_N	0.0255	0.0917	0.1553	0.1375	0.1972	0.1545
BRkNN_A	0.0273	0.1317	0.2821	0.2382	0.275	0.1573
BRkNN_B	0.0317	0.105	0.3812	0.3013	0.3737	0.2937
IBLR	0.0286	0.0433	0.2076	0.1598	0.2681	0.1629
HOMER	0.0304	0.0967	0.2747	0.2196	0.2998	0.2075
HMC	0.0271	0.0567	0.1834	0.1472	0.2267	0.191

Table 6. Bipartition-based measure results on rcv1subset2 (cont.)

	HL	SSA	EB_FM	EB_A	Mi_FM	Ma_FM
BR	0.027	0.0817	0.2004	0.1661	0.2339	0.1881
CLR	0.0259	0.0467	0.1754	0.1388	0.2254	0.1637
CC	0.0343	0.1617	0.2495	0.2264	0.2235	0.1883
ILC	0.0272	0.0483	0.1726	0.1374	0.2169	0.1854
LP	0.035	0.1967	0.3244	0.288	0.309	0.2297
RAkEL	0.0271	0.0567	0.1834	0.1472	0.2267	0.191
ECC	0.0266	0.1567	0.2579	0.2298	0.2679	0.183
EPS	0.0285	0.155	0.2953	0.256	0.3124	0.2138

simply transform the original multi-label data sets into multiple single-label data sets. This kind of transformation approach inevitably ignore the inherent relations among labels, and thus its performance is naturally worse than that of those approaches with label partition.

To give more intuitive illustration, we sort the average index value of various algorithms on these 10 data sets, and plot their rank in Fig. 2. Figure 2(a) and (b) show comparison results of multi-label classification algorithms and it is well noticed that MLRF ranks the first on most of data sets. Another observation is that conventional label partition based approach, MLRF_Trad, is not robust as it's performance is getting worse when evaluated on data set "Enron" and "Top90Collection". This observation demonstrates that our approach is more robust which is a key characteristics to be applied on real world applications. From Fig. 2(b), it can be seen that the rest comparison algorithms fluctuate a lot and no robust one could be found except for MLRF. Figure 2(c) and (d) show the ranking results of various problem transformation based algorithms. Similarly, it is easy to see that MLRF can achieve best performance when compared with these algorithms.

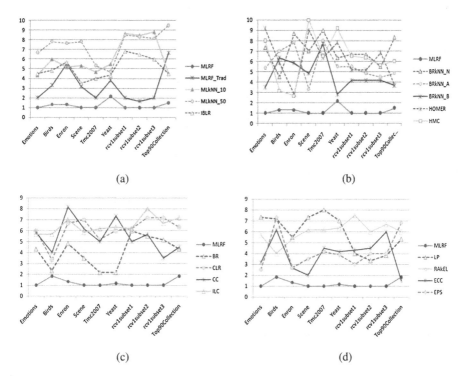

Fig. 2. The rank of various algorithms' performance on bipartition-based measures evaluated on different data sets.

Results on Ranking-Based Measures. In this experiment, similar algorithms are performed on the same data sets but evaluated using ranking-based measures and for all indexes except AP, the smaller the index value, the better the model performance.

Results are reported in the following tables.

Based on the results in Tables 7, 8, 9 and 10, similar observations could be found. It is seen that MLRF is superior to MLRF_Trad as well as the rest state-of-the-art

Table 7. Ranking-based measure results on rcv1subset1

	AP	Cov	OE	IE	ESS	RL	MAP
MLRF	**0.6041**	18.24	**0.233**	**0.607**	36.73	0.0855	0.271
MLRF_Trad	0.5728	38.43	0.256	0.638	43.15	0.2345	0.2
LP_RF	0.1946	61.4033	0.8533	0.8933	154.3667	0.4969	**0.0501**
MLkNN_10	0.5174	17.79	0.525	0.87	28.14	0.0829	0.249
MLkNN_50	0.5113	**15.8**	0.553	0.9	**24.36**	**0.071**	0.266
BRkNN	0.5065	25.12	0.546	0.853	45.18	0.1261	0.256
IBLR	0.501	18.58	0.591	0.885	29.96	0.0846	0.24
HOMER	0.3662	48.58	0.631	0.891	83.08	0.2722	0.138
HMC	0.275	47.74	0.693	0.93	115.4	0.344	0.112

Table 8. Ranking-based measure results on rcv1subset1 (cont.)

	AP	Cov	OE	IE	ESS	RL	MAP
BR	0.5166	19.5183	0.5117	0.8533	29.405	0.0883	0.2678
CLR	0.4415	32.27	0.54	0.8867	47.4433	0.1537	0.1656
CC	0.3991	38.9683	0.6633	0.7967	67.6183	0.2092	0.1425
ILC	0.4415	32.27	0.54	0.8867	47.4433	0.1537	0.1656
LP	0.2571	53.0283	0.7883	0.865	130.568	0.4086	0.0753
RAkEL	0.2704	47.7483	0.7217	0.93	115.55	0.3441	0.117
ECC	0.5349	18.1667	0.4983	0.815	29.7983	0.0838	0.2665
EPS	0.5032	30.645	0.5367	0.7783	59.5467	0.1664	0.2445

Table 9. Ranking-based measure results on yeast

	AP	Cov	OE	IE	ESS	RL	MAP
MLRF	**0.7931**	**5.706**	0.213	**0.663**	5.26	0.1605	0.531
MLRF_Trad	0.69	8.21	0.271	0.712	11.01	0.298	0.407
LP_RF	0.3189	11.0992	0.9256	0.9380	27.6074	0.7239	0.2984
MLkNN_10	0.7655	6.11	**0.21**	0.719	5.6	0.1607	0.484
MLkNN_50	0.7643	6.049	0.227	0.7107	**5.12**	**0.1592**	0.487
BRkNN	0.7586	6.425	0.227	0.6942	6	0.1716	0.49
IBLR	0.7654	6.053	0.223	0.7025	5.4	0.1602	0.501
HOMER	0.6354	8.855	0.281	0.905	11.2	0.3035	0.351
HMC	0.6014	9.578	0.376	0.9215	12.8	0.3442	0.355

Table 10. Ranking-based measure results on yeast (cont.)

	AP	Cov	OE	IE	ESS	RL	MAP
BR	0.7409	6.6818	0.2397	0.7851	6.595	0.1845	0.474
CLR	0.7459	6.6157	0.2231	0.7893	6.3182	0.177	0.439
CC	0.618	8.7355	0.3678	0.8554	12.1901	0.3475	**0.2657**
ILC	0.6216	8.8967	0.376	0.9091	11.4008	0.3022	0.3648
LP	0.5639	9.0455	0.5455	0.8554	14.7645	0.3899	0.3714
RAkEL	0.635	9.562	0.2603	0.9008	12.3182	0.3332	0.3562
ECC	0.7518	6.3471	0.2438	0.7314	6.5413	0.1802	0.4556
EPS	0.7277	6.7231	0.2727	0.7521	7.0041	0.1949	0.4297

multi-label algorithms on most of measures. As MLRF_Trad is also superior to LP_RF which is random forest based classifier with problem transformation, we can conclude that for all evaluation measures we adopted in the experiments, conventional label partition random forest based classifier is better than problem transformation based random forest classifier, and our proposed approach is the best one as it further make use of label correlation.

To summarize again, MLRF is more robust on all evaluation criteria on the average, and is superior to the rest compared algorithms. This implies that by grouping dependent labels into the same label partitions, the performance of random forest classifier can be improved.

4 Conclusion

Most of existing multi-label classifiers simply assume that labels are independent, however, it is not the case as demonstrated in many real world applications. In this paper, we first identify the correlated labels and group these dependent labels into the same label partitions, then we integrate the new label partitions with random forest and propose MLRF algorithm for multi-label classification task. Experiments results have demonstrated that the proposed MLRF is more robust and is superior to several state-of-the-art algorithms when evaluated on 10 data sets based on 13 evaluation criteria. This implies that our MLRF is able to be adopted in many real world applications where training data are more diverse as well as containing noisy data.

Acknowledgement. Yunming Ye's work was supported in part by National Key Technology R&D Program of MOST China under Grant No. 2014BAL05B06, Shenzhen Science and Technology Program under Grant No. JCYJ20140417172417128, and the Shenzhen Strategic Emerging Industries Program under Grant No. JCYJ20130329142551746. Yan Li's work was supported in part by NSFC under Grant No. 61303103, and the Shenzhen Science and Technology Program under Grant No. JCY20130331150354073.

References

1. Tsoumakas G, Katakis I. Multi-label classification: an overview. Department of Informatics, Aristotle University of Thessaloniki, Greece (2006)
2. Madjarov, G., Kocev, D., Gjorgjevikj, D., et al.: An extensive experimental comparison of methods for multi-label learning. Pattern Recogn. **45**(9), 3084–3104 (2012)
3. Read, J., Pfahringer, B., Holmes, G.: Multi-label classification using ensembles of pruned sets. In: Eighth IEEE International Conference on Data Mining, 2008, ICDM 2008, pp. 995–1000. IEEE (2008)
4. Tsoumakas, G., Vlahavas, I.P.: Random k-labelsets: an ensemble method for multilabel classification. In: Kok, J.N., Koronacki, J., Lopez de Mantaras, R., Matwin, S., Mladenič, D., Skowron, A. (eds.) ECML 2007. LNCS (LNAI), vol. 4701, pp. 406–417. Springer, Heidelberg (2007)
5. Montañes, E., Senge, R., Barranquero, J., et al.: Dependent binary relevance models for multi-label classification. Pattern Recogn. **47**(3), 1494–1508 (2014)
6. Read, J., Pfahringer, B., Holmes, G., et al.: Classifier chains for multi-label classification. Mach. Learn. **85**(3), 333–359 (2011)
7. Kocev, D.: Ensembles for predicting structured outputs. Informatica Int. J. Comput. Inf. **36** (1), 113–114 (2012)
8. Blockeel, H., De Raedt, L., Ramon, J.: Top-down induction of clustering trees. arXiv preprint cs/0011032 (2000)

9. Kocev, D., Vens, C., Struyf, J., Džeroski, S.: Ensembles of multi-objective decision trees. In: Kok, J.N., Koronacki, J., Lopez de Mantaras, R., Matwin, S., Mladenič, D., Skowron, A. (eds.) ECML 2007. LNCS (LNAI), vol. 4701, pp. 624–631. Springer, Heidelberg (2007)
10. Zhang, M.L., Zhou, Z.H.: A k-nearest neighbor based algorithm for multi-label classification. In: IEEE International Conference on Granular Computing, 2005, vol. 2, pp. 718–721. IEEE (2005)
11. Spyromitros, E., Tsoumakas, G., Vlahavas, I.P.: An empirical study of lazy multilabel classification algorithms. In: Darzentas, J., Vouros, G.A., Vosinakis, S., Arnellos, A. (eds.) SETN 2008. LNCS (LNAI), vol. 5138, pp. 401–406. Springer, Heidelberg (2008)
12. Tsoumakas, G., Katakis, I., Vlahavas, I.: Effective and efficient multi-label classification in domains with large number of labels. In: Proceedings of ECML/PKDD 2008 Workshop on Mining Multidimensional Data (MMD 2008), pp. 30–44 (2008)
13. Schapire, R.E., Singer, Y.: BoosTexter: a boosting-based system for text categorization. Mach. Learn. 39(2–3), 135–168 (2000)
14. Breiman, L.: Random forests. Mach. Learn. 45(1), 5–32 (2001)
15. Ye, Y., Wu, Q., Huang, J.Z., et al.: Stratified sampling for feature subspace selection in random forests for high dimensional data. Pattern Recogn. 46(3), 769–787 (2013)
16. Wuensch, K.L.: Chi-square tests. In: Lovric, M. (ed.) International Encyclopedia of Statistical Science, pp. 252–253. Springer, Berlin, Heidelberg (2011)
17. Sprent, P.: Fisher exact test. In: Lovric, M. (ed.) International Encyclopedia of Statistical Science, pp. 524–525. Springer, Berlin, Heidelberg (2011)
18. Fleiss, J.L., Levin, B., Paik, M.C.: Statistical Methods for Rates and Proportions. John Wiley & Sons (2013)
19. Anisimova, M., Gascuel, O.: Approximate likelihood-ratio test for branches: a fast, accurate, and powerful alternative. Syst. Biol. 55(4), 539–552 (2006)
20. Raileanu, L.E., Stoffel, K.: Theoretical comparison between the gini index and information gain criteria. Ann. Math. Artif. Intell. 41(1), 77–93 (2004)

Prediction of Oil and Water Layer
by Kernel Local Fisher Discriminant Analysis

Zehao Chen[✉]

l'Ecole centrale de Pékin, Beihang University, Beijing, China
409275072@qq.com

Abstract. The distribution of the oil and water layers in Xinjiang Oilfield is very complex because of the influence by many factors. It is difficult to predict the oil and water layer. In this paper, the oil and water layer of Xinjiang Oilfield was discriminated by kernel local fisher discriminant analysis(KLFDA). The local scatter matrix is defined by a affinity matrix. The original data are projected into the subspace constructed by KLFDA, and the local feature vectors are extracted. Then the prediction (classification) is done in feature subspace by Mahalanobis distance. The results indicate that the performance of KLFDA combining Mahalanobis distance is better than that of LFDA, FDA and ICA; meanwhile, the prediction accuracy of this method is better than that of SVM and ANN.

Keywords: Oil layer · Water layer · Kernel local fisher discriminant analysis · Prediction

1 Introduction

Xinjiang Oilfield Corporation, lying at Junggar Basin, is the biggest oil enterprise of the west China. Since influenced by many factors such as structure, deposition, breaking and diagenesis, the distribution of the oil and water layers is very complex. The reservoir is mainly comprised with fragmentary rock and lava. There are different kinds of minerals in the interval of interest. The rock character is complex with the features of low porosity and hypotonicity. These factors make the characteristic of the oil and water reservoir not obvious, so it is difficult to identify the oil and water layers.

During the process of the oil exploration, some parameters(formation testing data) are obtained by means of well drilling and well logging, then the formation testing data are analyzed to predict the oil and water layers. The economic cost of the oil exploration will be affected by the prediction accuracy. So it is very important to found an effective prediction model.

In recent years, the methods for predicting the reservoir water mainly include the geophysical method, the geochemical method, and the comprehensive method.

Log interpretation is the most commonly used geophysical method for the reservoir water prediction. The reservoir water is explained according the differences among the physical properties of the reservoir fluid. Since affected by many factors such as reservoir rock composition, formation water salinity, mud invasion, and so on, some deviation often appear when interpreting some lower resistivity reservoir, and it is

D.-S. Huang and K. Han (Eds.): ICIC 2015, Part III, LNAI 9227, pp. 419–428, 2015.
DOI: 10.1007/978-3-319-22053-6_45

difficult to recognize the interface of oil and water. Zhenqiang Wu et al. proposed a new method of quantitative evaluation, analyzed the factors that affect the logging environment, obtained the correction coefficients that can eliminate the influence, and built two-factor interpretation map of oiliness-physical property and the soft ware for evaluation. His work is very helpful for the comprehensive evaluation of oil, gas, and water [1–6].

There are different geochemical characters in oil layer and water layer. For example, the output of pyrolytic oil and gas and the content of light aromatic hydrocarbon are high in the oil layer. Geochemical method recognizes the oil, gas and water layer mainly according the chemical characters of the reservoir residual hydrocarbon, it is only related to the chemical characters of reservoir fluid, and doesn't get any influence from the reservoir rock component and the physical characters of the reservoir fluid. This method can cover the shortage of the log interpretation. For example, founding the chemical criterion and recognizing the oil, gas and water according the extract contents, the fluorescence intensity, the distribution of fluorescence and the carbon number of the reservoir hydrocarbon [7].

When using the comprehensive method, some effective feature parameters are first obtained by the geophysical or the geochemical method, then the data are analyzed using multiple regress analysis, support vector machine(SVM), and artificial neural net (ANN), and so on, at last, the discriminant function for reservoir water are founded. These methods can realize recognizing the reservoir water by using several evaluation parameters, and show the results quickly and intuitively, thus the problem of low accuracy are solved when predicting by using a single parameter [8–12].

Since affected by many factors, the actual geologic features are complex, and the well logging curves are indistinct, this makes that the traditional physical geography method and the geochemical method depend more on the genetic origin mechanism of the oilfield, and difficult to generalize. Though the comprehensive method can overcome these faults to some extent, it can't extract the features of the original data enough, so the prediction accuracy is still not satisfied.

To improve the prediction accuracy of oil-water layer, we must extract enough features of the original data, not only the linear features, but also the nonlinear features. In this paper, we predicted the reservoir water of Xinjiang oilfield by using kernel local fisher discriminant analysis(KLFDA) [13]. The original data are projected into the subspace constructed by KLFDA, and the local feature vectors are extracted. Then the prediction (classification) is done in feature subspace by Mahalanobis distance. Since KLFDA extracts the local and nonlinear features of the original data, the prediction performance is improved.

The rest of the paper is organized as follows. In Sect. 2, FDA and LFDA are briefly reviewed. In Sect. 3, we introduce KLFDA and the prediction algorithm. In Sect. 4, we describe the logging data of Xinjiang oilfield and the prediction results. Finally, we give conclusions in Sect. 5.

2   Local Fisher Discriminant Analysis

Consider the problem of classification. Let $x_i \in \mathbf{R}^d (i = 1, 2, \cdots, n)$ denote the d-dimensional samples with corresponding class labels $y_i \in \{1, 2, \cdots, c\}$. For FDA, the within-class scatter matrix and the between-class scatter matrix are defined as

$$S^{(w)} \equiv \frac{1}{2} \sum_{i,j=1}^{n} W_{i,j}^{(w)} (x_i - x_j)(x_i - x_j)^T, \tag{1}$$

$$S^{(b)} \equiv \frac{1}{2} \sum_{i,j=1}^{n} W_{i,j}^{(b)} (x_i - x_j)(x_i - x_j)^T \tag{2}$$

where

$$W_{i,j}^{(w)} \equiv \begin{cases} 1/n_l & if y_i = y_j = l \\ 0 & if y_i \neq y_j \end{cases}, \tag{3}$$

$$W_{i,j}^{(b)} \equiv \begin{cases} 1/n - 1/n_l & if y_i = y_j = l \\ 1/n & if y_i \neq y_j \end{cases}. \tag{4}$$

where n_l is the number of samples in class l. The objective function of FDA is

$$T_{FDA} \equiv \arg \max_{T \in R^{d \times r}} [tr((T^T S^{(w)} T)^{-1} T^T S^{(b)} T)]. \tag{5}$$

The transformation matrix of FDA is

$$T_{FDA} = (\varphi_1 | \varphi_2 | \cdots | \varphi_r) \qquad (r < d)$$

where $\{\varphi_k\}_{k=1}^d$ are the generalized eigenvectors of $(S^{(w)})^{-1} S^{(b)}$ and $\lambda_1 \geq \lambda_2 \geq \cdots \lambda_d$ are the corresponding generalized eigenvalues.

We can see from (3) and (4), the within-class scatter and the between-class scatter are estimated globally, therefore, FDA may give undesirable results for multimodal data.

Local Fisher discriminant analysis (LFDA) [13] overcomes this fault. It defines the between-class scatter and the within-class scatter in a local manner by an affinity matrix A. The element $A_{i,j}$ denotes the affinity between x_i and x_j. The value of $A_{i,j}$ is inversely proportional to the distance between x_i and x_j. In this paper, $A_{i,j}$ is defined as the follow

$$A_{i,j} = \exp(-\frac{\|x_i - x_j\|^2}{s^2}), \tag{6}$$

where $s(> 0)$ is a tuning parameter.

The local within-class scatter matrix $\tilde{S}^{(w)}$ and the local between-class scatter matrix $\tilde{S}^{(b)}$ are defined as follows,

$$\tilde{S}^{(w)} \equiv \frac{1}{2}\sum_{i,j=1}^{n} \tilde{W}_{i,j}^{(w)}(x_i - x_j)(x_i - x_j)^T, \tag{7}$$

$$\tilde{S}^{(b)} \equiv \frac{1}{2}\sum_{i,j=1}^{n} \tilde{W}_{i,j}^{(b)}(x_i - x_j)(x_i - x_j)^T, \tag{8}$$

where

$$\tilde{W}_{i,j}^{(w)} \equiv \begin{cases} A_{i,j}/n_l & if y_i = y_j = l \\ 0 & if y_i \neq y_j \end{cases}, \tag{9}$$

$$\tilde{W}_{i,j}^{(b)} \equiv \begin{cases} A_{i,j}(^1/n - ^1/n_l) & if y_i = y_j = l \\ ^1/n & if y_i \neq y_j \end{cases}. \tag{10}$$

We can see that the values of the sample pairs in the same class are weighted. It has less influence on $\tilde{S}^{(w)}$ and $\tilde{S}^{(b)}$ if the sample pairs are far apart, and it has more influence on $\tilde{S}^{(w)}$ and $\tilde{S}^{(b)}$ if the sample pairs are close. For the sample pairs in the different classes, their values are not weighted because we need to classify them without considering the distance between them in the primary space. Therefore LFDA preserves effectively the local multimodality in each class and obtains the between-class separation simultaneously.

The objective function of LFDA is

$$T_{LFDA} \equiv \underset{T \in R^{d \times r}}{\arg\max}[tr((T^T\tilde{S}^{(w)}T)^{-1}T^T\tilde{S}^{(b)}T)], \tag{11}$$

and the LFDA transformation matrix T_{LFDA} can be got by the following generalized eigenvalue problem

$$\tilde{S}^{(b)}\tilde{\varphi} = \tilde{\lambda}\tilde{S}^{(w)}\tilde{\lambda}\tilde{\varphi}. \tag{12}$$

3   Kernel Local Fisher Discriminant Analysis

LFDA improves the capability of dealing with the multimodal data, but it cannot extract complex non-linear features. In order to compensate for the drawbacks of LFDA, we use KLFDA that extends LFDA to non-linear circumstance by projecting the samples into a kernel feature space.

First, we decompose $\tilde{S}^{(w)}$ as

$$\tilde{S}^{(w)} = X\tilde{L}^{(w)}X^T, \tag{13}$$

where

$$\tilde{L}^{(w)} \equiv \tilde{D}^{(w)} - \tilde{W}^{(w)},$$

and $\tilde{D}^{(w)}$ is the n-dimensional diagonal matrix with the i-th diagonal element being

$$\tilde{D}_{i,j}^{(w)} \equiv \sum_{j=1}^{n} \tilde{W}_{i,j}^{(w)}.$$

Similarly, $\tilde{S}^{(b)}$ can be decomposed as

$$\tilde{S}^{(b)} = X\tilde{L}^{(b)}X^T. \tag{14}$$

where

$$\tilde{L}^{(b)} \equiv \tilde{D}^{(b)} - \tilde{W}^{(b)},$$

and $\tilde{D}^{(b)}$ is the n-dimensional diagonal matrix with the i-th diagonal element being

$$\tilde{D}_{i,j}^{(b)} \equiv \sum_{j=1}^{n} \tilde{W}_{i,j}^{(b)}.$$

Thus, Eq. (12) can be reformulated as

$$X\tilde{L}^{(b)}X^T\tilde{\varphi} = \tilde{\lambda}X\tilde{L}^{(w)}X^T\tilde{\varphi}. \tag{15}$$

Defining n-dimensional vector $\tilde{\varphi} = X\tilde{\alpha}(\tilde{\alpha} \in R^n)$ and n-dimensional matrix K with the elements $K_{i,j} \equiv x_i^T x_j$, we have

$$K\tilde{L}^{(b)}K\tilde{\alpha} = \tilde{\lambda}K\tilde{L}^{(w)}K\tilde{\alpha}. \tag{16}$$

This indicates that $\{x_i\}_{i=1}^{n}$ arise only in forms of their inner products. Therefore, it is convenient to use the kernel trick for extending LFDA to non-linear field.

Let $\phi(x)$ be a non-linear projection from the original space $\mathbf{R}^d$ to a reproducing kernel Hilbert space H, then $\tilde{S}^{(w)}$ and $\tilde{S}^{(b)}$ can be reformulated as

$$\tilde{S}^{(w)} \equiv \frac{1}{2} \sum_{i,j=1}^{n} \tilde{W}_{i,j}^{(w)} (\phi(x_i) - \phi(x_j))(\phi(x_i) - \phi(x_j))^T,$$

$$\tilde{S}^{(b)} \equiv \frac{1}{2} \sum_{i,j=1}^{n} \tilde{W}_{i,j}^{(b)} (\phi(x_i) - \phi(x_j))(\phi(x_i) - \phi(x_j))^T.$$

Defining $\tilde{K}_{i,j} = \langle \phi(x_i), \phi(x_j) \rangle = K(x_i, x_j)$, the following equation is obtained

$$\tilde{K}\tilde{L}^{(b)}\tilde{K}\tilde{\alpha} = \tilde{\lambda}\tilde{K}\tilde{L}^{(w)}\tilde{K}\tilde{\alpha}. \tag{17}$$

In this paper, we use the Gaussian kernel

$$K(x, x') = \exp(-\frac{\|x - x'\|^2}{2\sigma^2}) \, (\sigma > 0).$$

Let $\{\tilde{\alpha}_k\}_{k=1}^{n}$ denote the generalized eigenvectors and the corresponding generalized eigenvalues $\tilde{\lambda}_1 \geq \tilde{\lambda}_2 \geq \ldots \geq \tilde{\lambda}_n$ of Eq. (17). Then the projection of $\phi(x')$ in H is

$$(\sqrt{\tilde{\lambda}_1}\tilde{\alpha}_1 | \sqrt{\tilde{\lambda}_2}\tilde{\alpha}_2 | \cdots | \sqrt{\tilde{\lambda}_r}\tilde{\alpha}_r)^T \begin{pmatrix} K(x_1, x') \\ K(x_2, x') \\ \vdots \\ K(x_n, x') \end{pmatrix}. \tag{18}$$

4 Prediction on the Logging Data

4.1 The Logging Data

We experiment using the logging data from Xinjiang Oil Field Corporation, including the original data of 31 wells. There are 5 features in each sample, such as, formation resistivity(fr), effective porosity(ep), oil saturation(os), interval transit time(itt), and shale content(sc). We use 17 samples as training samples, including 9 "oil" and 8 "water", and the remain 14 samples as the test samples.

4.2 Results and Analysis of Experiment

The prediction can be view as a two-class classification problem (water and oil). We extract features of training and test samples with KLFDA, then classify the test samples by mahalanobis distance. We experiment by using the methods of KLFDA, compared with LFDA, FDA, ICA, ANN, and SVM. The algorithms are implemented in matlab R2013a. The kernel function in KLFDA is Gaussian kernel, and the parameter is $\sigma = 20$. To reduce the dimension, we take r = 1, and take the parameter s = 5 in the affinity matrix when using KLFDA and LFDA.

Mean Vectors and Multimodality. In fact, the water includes the pure water(w), the oily water(w(o)), and the gas water(w(g)), the oil includes the pure oil(o) and the water cut reservoir(o(w)). We calculated the mean vectors of the five clusters separately. We can see from Figs. 1 and 2 that though the pure water, the oily water, and the gas water are all water, they appear different means, meanwhile, the pure oil and the water cut reservoir also appear different means. This indicates that the data are multimodality.

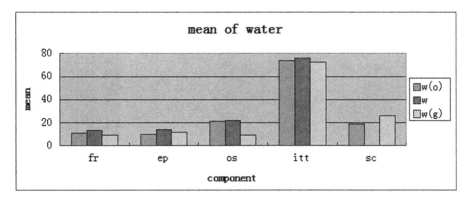

Fig. 1. Mean of water

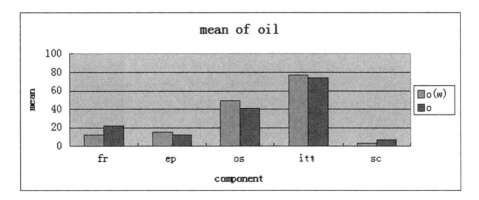

Fig. 2. Mean of oil

Projection. Figure 3 shows the projections of the 14 test samples in the 2-dimensional space when using the method of KLFDA, LFDA, FDA, and ICA. The horizontal axis and the vertical axis represent respectively the first feature and the second feature. We can see that both KLFDA and LFDA can well separate the oil samples from the water samples, and preserve the within-class multimodality simultaneously, but KLFDA preserves the within-class multimodality more clearly due to its nonlinear property. FDA separates the samples in different classes well, but losses the within-class

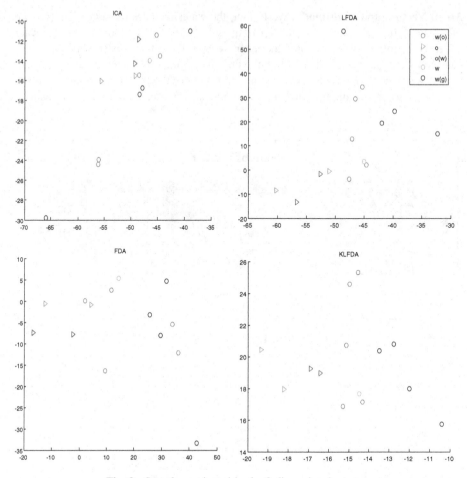

Fig. 3. Samples projected in the 2-dimensional space

multimodality, namely, the red 'o' and the black 'o' are mixed. The situation of ICA is similar to that of FDA.

Result of Prediction. Table 1 shows the prediction results by different methods. We see that KLFDA works the best, owing to its capability to capture the nonlinearity and multimodality. LFDA also works well, since it can deal with multimodality, but it does not compare favorably with KLFDA. Since ICA is a method of high-order statistics, so its performance is better than that of FDA. The performance of SVM matches that of FDA, and the performance of ANN is the weakest.

Table 1. Comparison of recognition performance for different methods

Well No.	Conclusion	FDA	ICA	LFDA	KLFDA	ANN	SVM
1	water(w(o))	oil	water	water	water	oil	water
2	water(w(o))	oil	water	water	water	oil	water
3	water(w(o))	water	water	water	water	oil	water
4	oil(o)	water	water	water	water	water	water
5	oil(o(w))	oil	water	water	oil	oil	oil
6	oil(o(w))	oil	oil	oil	oil	oil	oil
7	water(w)	water	water	water	water	water	water
8	water(w(g))	water	water	water	water	oil	oil
9	water(w(g))	water	water	water	water	water	water
10	water(w(g))	water	water	water	water	water	oil
11	water(w(g))	oil	oil	water	water	water	oil
12	oil(o)	oil	oil	oil	oil	oil	oil
13	water(w(o))	water	water	water	water	water	water
14	water(w(o))	water	water	water	water	water	water
Accuracy(%)		71.4	78.6	85.7	**92.9**	64.3	71.4

5 Conclusion

We predict reservoir water of Xinjiang oilfield corporation by using KLFDA. KLFDA projects the samples onto a nonlinear space H, and LFDA is used in H for dimensionality reduction and feature extraction, by maximizing between-class separability and preserving the within-class local structure simultaneously. The experimental results show that KLFDA is superior to other methods for the classification of multimodal data.

References

1. Lixin, T., Donghong, Z., Jun, M.: A prediction of heavy-oil beds in North Q32-6 area. Bohai sea, China Offshore Oil Gas **22**(1), 12–16 (2010)
2. Hongliang, L., Hongxia, Y., Hongzhu, C.: Research on recognition of low porosity and low permeability reservoir logging fluid in Hongtai area. J. Oil Gas Technol. **29**(03), 34–39 (2007)
3. Guoxin, L., Jian, O., Cancan, Z.: Advancement of petrophysics research and well-logging recognition and evaluation for low-resistivity oil-layer by China petro. China Pet. Explor. **11**(2), 43–50 (2006)
4. Zhenqiang, W., Xinjun, M., Guorong, H.: Research on recognition and evaluation technology of oil and gas in Junggar Basin. Xinjiang Oil Technol. **16**(02), 26–31 (2006)
5. Yanqing, B.: Recognition of oil and water layer from Putaohua oil layer in Taidong Area, Daqing, The full text database of the good master's degree thesis of China, Chengdu University of Technology (2011)

6. Shangming, S., Shuai, G., Jianbin, H.: The research of oil-water layer identification and evaluation of Fuyang formation in Gaotaizi oilfield. Sci. Technol. Eng. **11**(13), 2897–2901 (2011)

7. Youchuan, L., Jipin, J., Xiaoying, X.: Predicting oil and water intervals using geochemistry methods. China Offshore Oil Gas **10**(3), 193–199 (1998)

8. Yinde, Z., Kaijun, T., Jun, Z.: Application of support vector machine meth od for identifying fluid in low-resistivity oil layers. Geophys. Prospect. Petrol. **47**(3), 306–310 (2008)

9. Yu, Z.: Research on recognition of complex lithology reservoir oil-water layer in Surennuoer oilfield. The full text database of the good master's degree thesis of China, Northeast Petroleum University (2012)

10. Liping, Z., Xiongyan, L., Hongqi, L.: Recognition of low resistivity oil layer based on model-driven data mining. J. Daqing Petrol. Inst. **34**(4), 29–34 (2010)

11. Longsheng, J., Xianggong, W., Bowen, X.: Qualitative evaluation of oil-water layer with well logging data. J. Shandong Univ. Technol. **23**(03), 88–90 (2009)

12. Fujun, M., Xiuzhi, C.: Application of fuzzy mathematics in identification of oil bearing layers and water layers. Petrol. Geol. Exploit. Daqing **16**(4), 11–15 (1997)

13. Masashi, S.: Dimensionality reduction of multimodal labeled data by local fisher discriminant analysis. J. Mach. Learn. Res. **8**, 1027–1061 (2007)

Assessment of the Pillar 3 Financial and Risk Information Disclosures Usefulness to the Commercial Banks Users

Anna Pilkova[1], Michal Munk[2(✉)], Peter Svec[2], and Michal Medo[1]

[1] Commenius University, Odbojarov 10, 820 05 Bratislava, Slovakia
anna.pilkova@fm.uniba.sk, miso.medo@gmail.com
[2] Constantine the Philosopher University in Nitra, Tr. A. Hlinku 1,
949 74 Nitra, Slovakia
{mmunk,psvec}@ukf.sk

Abstract. The paper analyses usefulness of the Pillar 3 financial and risk information disclosures to the commercial banks users. The Pillar 3 are specific regulatory disclosures requirements set out in the Basel 2 framework and incorporated into EU law and subsequently laws of the member states. According to Pillar 3 intention market participants should be able to understand and subsequently judge the relevance of the bank risk position and risk management and try to discipline "risky" banks. Due to that the European authorities are focused on control and improvements of the banks' disclosures. However, less is done as far as usefulness of the Pillar 3 risk information to the commercial banks users is. The authors try to assess at which extent is information useful for users of the banks that operate in countries where banking sectors are dominated by foreign-owned entities and depositors (sophisticated and non sophisticated; insured and uninsured; primarily non-financial ones) is a key source of market discipline. The authors focus on modelling of visitor behaviours at website where financial and risk information according to Pillar 3 requirements is available. The results show that there is in general small interest in Pillar 3 information and even financial and risk related information belongs to those where interests is the lowest one.

Keywords: Pillar 3 · Market discipline · Risk management · Web log mining · Data pre-processing · Hadoop

1 Introduction

The last financial crisis has caused, among others, loss of collective respect to financial system supervision and regulation. Reaction on that are numerous initiatives and actions in this field in Europe and around the world after 2009. Core of this focus is on Basel regulatory documents and related EU legal capital requirements directives. Basel III documents have been significantly changed and extended in all three pillars. Pillar 3 and market discipline has deserved special attention by different groups of stakeholders, too.

Market discipline in its broadest terms can be understood as a mechanism via which market participants monitor, assess and discipline a risk taking by financial institutions. In

© Springer International Publishing Switzerland 2015
D.-S. Huang and K. Han (Eds.): ICIC 2015, Part III, LNAI 9227, pp. 429–440, 2015.
DOI: 10.1007/978-3-319-22053-6_46

studies of Bliss et al. [1] is market discipline defined by its distinguishing into two aspects: Market monitoring – market participants' assessment of banks' conditions which are to be reflected in the banks' security prices and deposit rates; Market influence – banks' reaction brought on by market monitoring, or counteract adverse changes in banks' conditions.

Key intention of Pillar 3 is that based on disclosed information market participants should be able to understand and subsequently judge the relevance of the bank risk management and try to "discipline" risky banks by asking higher spreads for deposits or even refusing new funding these banks. So far, no standardised format of information and frequency has been agreed on Basel Committee or European Union. Some national regulators prefer comprehensive annual report and some of them are in favour of more frequent presentations (e.g. on quarterly basis – see National Bank of Slovakia). However, there are still important and so far not addressed research questions related to usage of the disclosed information by key market participants in countries where banking sectors are dominated by foreign-owned entities and depositors discipline is a key source of market discipline. In our paper we addressed two of them: Are disclosed information relevant and useful to market participants? At which extent are key market participants interested in Pillar 3 financial and risk information?

The main objective of this paper is to analyse and study questions of adequacy and usage of financial and risk information by market participants based on analysis of website dedicated to Pillar 3 of commercial bank that operates in CEE country. The analysis of the bank website is based on the analysis of the webserver log file. As we analyse longer period the amount of data in the log file are enormous so new problems in the data analysis arose. We propose approaches to effectively analyse those data in this paper, too.

2 Related Work

Since 2004 is market discipline codified to Basel regulatory requirements under Pillar 3. From that time there are numerous literature and research studies that deal with the topic of market discipline in the banking industry but mostly in the context of mature economies.

Stephanou, who studied Basel's Pillar 3 reports during the financial crisis, revealed that the key issue of market discipline framework is to figure out, how to operationalize and institutionalize it within different financial system structures [2]. Based on the financial crisis events, Freixas et al. [3] emphasize that, what concerns market discipline information publishing, it is important to distinguish between disclosure and transparency. Their research further indicates that as of information transparency, current setup of Pillar 3 is having significant deficiencies, mostly showed during the financial crisis. Due to that, the Basel Committee initiated a review and identified guiding principles to achieve transparent, high quality Pillar 3 risk disclosures [4]. The research of Parwada et al. [5] examined the stock market reaction, in a form of abnormal stock return, on the Pillar 3 reports releases of a subsample of large international banks. EBA in its Pillar 3 guideline on materiality, proprietary and confidentiality and on disclosure frequency [6] is setting that banks should assess their need to publish information more frequently than annually. On the contrary to research of [5] and EBA's indicators of the need of more frequent reporting, is EBF in its comments, questioning whether there is a proven demand among stakeholders for a greater frequency of disclosure [7]. Hasan et al. [8] studied new aspects of market discipline exercised by non-financial depositors on banks operating in CEE countries. In the research is proved that the depositors are reacting in positive correlation much more significantly to rumours concerning the banks' parent companies (especially to negative ones) than to banks own fundamentals (financial reports). Moreover the research results suggest that CEE depositors had ability to differentiate between founded and unfounded rumours.

So far there has been no further study carried out on assessment of the extent at which clients of the commercial banks operating in CEE countries and being part of the European banking group would be interested in key information of market discipline disclosed in Pillar 3: financial and risk information.

We have gathered information related to Pillar 3 from the bank webserver log file. The webserver log files keep information about visitors, which can be used for the analysis of visitor behaviour. Experiments that analyse user's behaviour usually take smaller set of data, e.g. two weeks or one month [9, 10] or just select few reference weeks [11]. As we analysed a quite extensive period, we changed common used methods [12–14] and created the batch application [15] for log pre-processing. This approach is still time consuming as it is run on single machine. There are many solutions to analyse lot of data using high performance computing, e.g. Google's MapReduce [16, 17], Yahoo's PNUTS [18], Microsoft's SCOPE [19], Twitter's Storm [20], LinkedIn's Kafka [21], or WalmartLabs'Muppet [22]. There is also the open-source implementation of Google's MapReduce available, called Apache Hadoop. MapReduce has become the most popular framework for large-scale processing and analysis of vast data sets. The overview of MapReduce focusing on its open-source implementation in Hadoop, weaknesses and limitations of MapReduce, existing approaches that improve the performance of query processing are well described in [23] and its distributed file system used for Big Data storage is well described in [24]. Su et al. [25] focuses on mining web server log files using relaxed biclique enumeration algorithm in MapReduce. Premchaiswadi and Romsaiyud [26] introduced model for efficient web log mining for web users clustering. They compute the similarity measure of any path in a web page, define the k-mean clustering for group and generate the report based on the Hadoop MapReduce Framework. Sakr et al. [27] provides a comprehensive survey for a family of approaches and mechanisms of large-scale data processing mechanisms that have been implemented based on the original idea of the MapReduce framework.

3   Methods

The analysed bank is the third largest by size and belongs to systematically important financial institutions on national level. It is owned almost 80 % by the European foreign financial group. The bank provides the commercial bank services to retail, SME, corporate and private banking customers. It is well recognized within its peer group as innovator and technology leader. It has a lead position in almost all segments of customers.

Based on the methodology described in our previous research [13] the bank webserver log file pre-processing consists of following steps: data cleaning, identification of sessions and reconstruction of activities of a web visitor. The first step in the log file pre-processing is the removal of unnecessary data. These data represents access to graphic files or style sheets [28] so the accesses from web robots. Web robots may be defined as autonomous systems that send requests to web servers across the Internet. A canonical example of a web robot is a search engine indexer while a less common example is an RSS feed crawler or a robot designed to collect web sites for an Internet archive [29]. The difference between the robot and the human can be determined based various metrics [30]. There are many robots or crawlers that cannot be identified based on general crawler attributes. In this case, we can use the method navigational patters analysis [31].

After the cleaning of the log file and removing web crawlers' accesses and unnecessary data, the log file with just the 10 % of the original file length. The main tool for removal of unnecessary data is the Unix grep call. The speed of the grep tools depends on the amount of lines in the log file as the grep process the log file line by line. We can speedup this process using the GNU Parallel or using the Hadoop cluster. Our Hadoop cluster consists of 20 HP Z820 workstations, each with two quad core Xeon processors and 16 GB RAM and 500 GB storage.

Webserver log file from the bank web server do not fit the common log file as there were many version of website and many versions of log with different. We have to search for information usually found in the common log files. This can be achieved using regular expressions.

We created python scripts for the Hadoop (mapper and reducer). We employ the *compile* and *match* functions to filter lines from the original log files. Shortened code of the mapper that search for IP address is in the Table 1.

Table 1. Mapper and Reducer script

Mapper.py	reducer.py
#!/usr/bin/env python import sys, re regex_IP = re.compile('([0-9]{1,3})\.([0-9]{1,3}) \.([0-9]{1,3})\.([0-9]{1,3})') log_line = line.strip() IP_result = regex_IP.match(log_line) print(IP_result.group(0))	#!/usr/bin/env python from operator import itemgetter import sys for line in sys.stdin: line = line.strip() print line

Finding the date, url, referrer and user-agent is made using another regular expression in the *re.compile* function. The reducer is simple as it just prints lines from the mapper.

Using this approach we can filter just those lines, which are essential for our experiment. We already mentioned, that we do not need the accesses to images, style etc. We can also filter some of used IP addresses. In cooperation with the bank, we marked IP addresses that are used in bank local networks. Staff is more familiar with the organisation and has better knowledge of the website structure [32]. We removed these accesses because the majority of accesses are from content creators, web administrators and managers responsible for information disclosure users looking for specific bank information – stakeholders.

We also have to mark what content categories are represented in the sitemap. The sitemap can be generated using the content management system of the bank website or we can use many free tools to create one. Also a bank expert is needed to mark sets of similar content (category) from every page in the sitemap. Our analysed part of the website contained 68 parts of the content from which expert determined 23 categories, which have to be affiliated to Pillar 3. The group of *Pillar 3 disclosures requirements* contained two categories and the group *Pillar 3 related* contained seven categories. Unmarked categories represent the *Other* affiliation. Each category has also *the financial/non-financial* attribute. We track information based on the quarter of the year. The mapping of URL to category can be seen in the Table 2.

Table 2. Mapping URL to categories

URL (from log file)	Category	Basel category	Information
/about/branch.html	17	Other (0)	Financial
/about/the-economic-results	22	Pillar 3 related (1)	Financial
/about/bank-results/quarterly	18	Pillar 3 disclosure requirements (2)	Financial
/about/contacts/	16	Other (0)	Nonfinancial

This pre-processed log file can be directly imported into the database for the session-identification as in our previous experiments [12–15].

4 Results

4.1 The Visit Rate Analysis of *About the Bank* Content

The interaction plot (Fig. 1(a)) shows observed interactive frequencies (*Category* × *YearQuartal*), as well as relative frequencies expressed as a percentage in rows count. The degree of dependence between variables *Category* and *YearQuartal* is represented by contingency coefficients and the significance is tested by Pearson chi-square test. The only requirement (validity assumption) of the use of chi-square test is the high enough expected frequencies. The condition is violated if the expected frequencies are lower than 5. In these tests the validity assumption of chi-square test is not violated as expected frequencies are high enough ($e_{ij} > 6986$).

Contingency coefficients represent the degree of dependency between two nominal variables. The value of coefficient (*Category* × *YearQuartal*) is approximately 0.19 while 1 means a perfect relationship and 0 no relationship. There is a little dependency between the access to *About the bank* content and time (*YearQuartal*), the contingency coefficient is statistically significant (*Chi-square* = 187292.9; $df = 30$; $p = 0.0000$). The zero hypotheses is rejected at the 1 % significance level, i.e. the number of accesses to particular portal category depends on the time of access. Results of the interaction frequencies – *Category* × *YearQuartal* show a low interest for information that is related

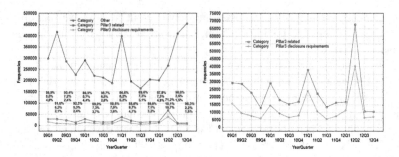

Fig. 1. Interaction plot of frequencies: (a) *Category* × *YearQuartal*; (b) *Pillar3 Category* × *YearQuartal*

to the Pillar 3. Stakeholders are more interested in general bank information; however, they have a higher interest in information to the *Pillar 3 related* than to the *Pillar 3 disclosure requirements* information.

The graph (Fig. 1(b)) depicts interaction frequencies only for the Pillar 3 categories. In this case the curves are copied, they have alike course – which only proves the same interest in both Pillar 3 categories at the time [33, 34].

4.2   Visit Rate Analysis of Web Parts Containing the Pillar 3 Information

We deal with the results of association rules analysis in this session in more detail. This analysis represents the nonsequential approach to the analysed data. We will not analyse sequences but transactions, i.e. we will not include the time variable into the analysis. The transaction thereafter represents the set of visited web parts of *About the bank* content containing the Pillar 3 information. Concerning data, we will consider web parts of the commercial bank portal which contains Pillar 3 information (Category: *Pillar 3 related*, *Pillar 3 disclosure requirements*) which was visited by the stakeholder during one session as one transaction.

Web graph (Fig. 2(a)) depicts the association rules found. The size of node represents *support* of this category item, line width represents *support* of the rule and line brightness stands for the *lift* of the rule. In other words, the *lift* represents the measure of interestingness and offers the most interesting results because it can be interpreted as how many times the categories of content occur together than in the case if they were statistically insignificant.

Considering the web graph we can see that the most visited web parts of the portal lie within *Group, Pillar3 Q-terly Info, Rating, Annual Reports, Information for Banks* and *Pillar3 Semiannualy Info* (*support* > 15 %).

The combinations of these web parts (*Annual Reports ==> Pillar3 Q-terly Info*), (*Group ==> Pillar3 Q-terly Info*), (*Rating ==> Group*) and (*Pillar3 Q-terly Info ==> Pillar3 Semiannualy Info*) in the identified sessions exist with a probability of more than 15 % (*support* > 15 %).

We can also see that the web parts (*Pillar3 Semiannualy Info ==> Emitent Prospects*) are more frequent together as apart in identified sessions (*lift* = 5.02).

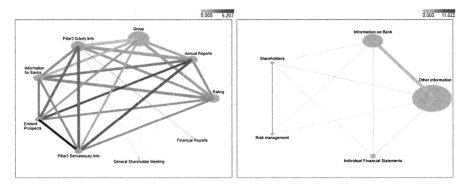

Fig. 2. Web graph: (a) visualisation of found rules for web parts of *About the bank* content related to the Pillar 3; (b) visualisation of found rules for documents of web part *Pillar3 Q-terly Info*

We can claim the same for web parts (*lift*>1.1): (*Information for Banks ==> Emitent Prospects*), (*Emitent Prospects ==> Annual Reports*), (*Pillar3 Semiannualy Info ==> Information for Banks*), (*Pillar3 Semiannualy Info ==> Annual Reports*), (*Emitent Prospects ==> Rating*), (*Financial Reports ==> Pillar3 Q-terly Info*), (*Emitent Prospects ==> Pillar3 Q-terly Info*), (*Pillar3 Semiannualy Info ==> Rating*), (*Pillar3 Semiannualy Info ==> Pillar3 Q-terly Info*), (*Information for Banks ==> Annual Reports*), (*Information for Banks ==> Rating*), (*Annual Reports ==> Rating*), (*Pillar3 Q-terly Info ==> Annual Reports*), (*Information for Banks ==> Pillar3 Q-terly Info*), (*Emitent Prospects ==> Group*), (*Pillar3 Q-terly Info ==> Rating*), (*Group ==> Pillar3 Semiannualy Info*) and (*Information for Banks ==> Group*).

We can observe the greatest measure of interestingness (*lift*) in these cases, because the lift greater than one means that the selected pairs of web parts are more frequent together than apart in the set of web parts visited by the particular stakeholders. But, we have to be aware that the lift does not depend on the rule orientation.

The measure of interestingness (*lift*) of the remaining found rules was less than one.

4.3   Visit Rate Analysis of Documents Containing the *Pillar3 Q-Terly Info*

The transaction represents the set of visited documents of web part *Pillar3 Q-terly Info*. Concerning data, we will consider documents of the commercial bank portal containing the *Pillar3 Q-terly Info* which was visited by the stakeholder during one session as a one transaction.

Considering the web graph (Fig. 2(b)) we see that documents *Other information*, *Information on Bank* and *Individual Financial Statements* (*support* > 5 %) belong to the most visited documents of web part *Pillar3 Q-terly Info*.

We found out that the combinations of documents (*Other information ==> Information on Bank*) and (*Other information ==> Individual Financial Statements*) have *support* > 5 % at the same time.

Documents *Financial Indicators* and *Consolidated Statements* did not meet the minimum *support* (*min support* = 0.5 %).

The other interesting finding is that (*Risk management ==> Shareholders*) exists together more frequently than apart (*lift* = 11.8) in the identified sessions.

Another group of rules, (*Individual Financial Statements ==> Shareholders*) and (*Risk management ==> Information on Bank*) attain also the *lift* > 1.1.

The measure of interestingness (*lift*) of the remaining found rules was less than one.

5 Discussion

We have studied our research questions on data acquired from a typical representative of the universal commercial bank in Slovakia. Its key depositors and creditors come from insured and uninsured (by deposit insurance fund) groups of customers, mortgage bond holders and subordinated debt holders. There is a numerous group (related to size of the country) minority shareholders, too. According to assumptions of regulators and creators of Pillar 3 these key stakeholder groups should be interested in financial results and risk profile of the bank and their dynamics presented on quarterly basis. To confirm this assumption we have analysed and studied website content *About the bank* which consists of three information categories: the first one is *Pillar 3 disclosure requirements* according to valid NBS (National bank of Slovakia) Decrees that are published on quarterly basis. In this part there are obligatory information like on the bank (establishment, structure, licensed activities etc.), structure of shareholders, individual financial statements, risk management information and other information (e.g. on consolidated entities). The second category is *Pillar 3 related information* which contains documents like annual reports, general assembly meetings notes, financial reports, prospects of emitent, information about group, and information for banks. The third category *Other* contains documents like the bank history, awards, mission-vision-values, anti money laundering, social activities, ethical codex, pricing lists, contacts, branches and ATM. According to Fig. 1 is clear that key stakeholders groups have the lowest interest in *Pillar 3* obligatory *disclosure requirements* which varied on quarterly basis in a range from 1.5 % to 10.7 %. They were more interested in *Pillar 3 related information* (from 2.5 % to 8.7 %) but majority of them visited this part of the website due to *Other information* (around 90 %).

As far as frequency of visits is in quarters of the year results confirmed that stakeholders are slightly more interested in both Pillar 3 disclosed and Pillar 3 related data in the first quarter in all analysed years but 2012. Rational explanation of this pattern is that at the end of the first quarter banks are obliged to publish their annual results which might have higher value for stakeholders to follow then quarterly ones. This finding suggests to re-assess frequency of *Pillar 3 information* publishing: rather on annual basis than quarterly one which is practice of the regulators in many other European countries.

The other interesting findings come from Fig. 2(a) web graph containing Pillar 3 information. Visitors have the highest interest in information about *Group*. Then they were interested in *Pillar 3* quarterly, *Ratings and Annual reports*. The least visited were categories *Financial reports and Annual meetings notes*. What is worth to note, too is if they visited category *Pillar 3 Quarterly information* they also most frequently visited in pair information about *Group or Rating or Pillar 3 semiannual* information. The highest interest in information about *Group* (either individual or in pair) in our research is in line with research of Hasan et al. [8]. According to their findings the depositors are reacting in positive correlation much more significantly to rumours concerning the banks' parent companies (especially to negative ones) than to banks own fundamentals (financial reports). This Group focus of stakeholders or one of them – depositors has also rational background: it is related to rules of rating agencies. If they change (mainly decrease) group rating it is usually reflected in decreasing of rating of daughter banks even their financial, risk or deposit positions might be much higher than their group figures.

As we already mentioned Pillar 3 key intention, described in Basel documents, is that market participants should be able to understand and subsequently judge the relevance of the bank risk management and try to discipline "risky" banks. In this context principal agent problem can arise according to Lančarič et al. [35]. However, when we analysed in depth stakeholders behaviour as far as their interest in risk information is we found out very low interest in *Risk information* part as part of *Pillar 3 obligatory* category (see Fig. 2(b) graph). That means that stakeholders of the bank have low interest to understand and judge relevance of the bank risk management. Their highest interest was in *Other information* among which consolidated financial statements are crucial. This finding indirectly supports both our previous finding on Group focus within *Pillar 3 related information* and referred Hasan et al. [8] research. The second highest interest of stakeholders within this category was about *Information on Bank* (establishment, structure, licensed activities etc.), then the *Individual financial statements* and the less interesting for them were information contained in *Risk management and Shareholders parts*. In this respect we have to stress that if stakeholders are interested in *Risk management* information than it is more frequently together with *Shareholders* than apart when they visit website. There are a few reasons why information on risk management has the lowest interest of the bank stakeholders. One of them is that key stakeholders in this type of banks are depositors. Retail depositors´ behaviour is influenced by deposit insurance fund limits which are a key criterion which influence their decisions. On the other hand structure and contain of Risk management information are such complicated that even corporate depositors expertise is not sufficient to understand it. Due to that rational behaviour is not to be interested in this information on the bank website.

6   Conclusion

Pillar 3 plays one of the crucial roles in the new European banking regulation. There have been run numerous initiatives by European Stability Board, European Banking Authority, The European Banking Federation etc., particularly after 2010 with goal to improve Pillar 3 or market discipline mechanism. However, there are still banking systems in countries with foreign ownership domination which would deserve special attention as far as improvement of usage of the disclosed information by key market participants is. In our paper we studied these issues through addressing two research questions: Are disclosed information relevant and useful to market participants? At which extent are key market participants interested in Pillar 3 financial and risk information?

We found out that Pillar 3 in these banking systems needs further improvement and changes, too. According to our findings key stakeholders (primarily depositors) of commercial banks operating in Slovakia and similar CEE regions (it was also confirmed by the other research of BOFIT on CEE market discipline as of August 2013) are very little interested in fundamental financial and risk information disclosed through Pillar 3 documents. Following that we can derive that information presented in Pillar 3 documents are not useful and relevant to this type of stakeholders as valuable sources of market disciplines. It looks like both content and time frequency is not relevant for these regions. We should be aware of that fact and we can expect that if nothing is changed in Pillar 3 current information disclosure in CEE countries then key stakeholders of commercial banks will use the other sources of information and banks will have useless transaction costs. Stakeholders will mainly follow the other sources of information e.g. negative rumours (also negative press rumours) on mother companies or local banks as valuable forms of market discipline. This is also option but we have to be prepared to cover risks coming from these forms of market discipline, measure their impact and find ways how to mitigate risk related to them.

We found also that using the parallel processing of the log file using the MapReduce in Hadoop is beneficial for this kind of problem. We plan to rewrite also the session identification algorithm in the future. We also changed our approach for log file analysis as the bank log files changed its structure several times and the well known structure of common log files could not be used.

Acknowledgments. This paper is supported by the project VEGA 1/0392/13 Modelling of Stakeholders' Behaviour in Commercial Bank during the Recent Financial Crisis and Expectations of Basel Regulations under Pillar 3- Market Discipline.

References

1. Bliss, R.R., Flannery, M.-J.: Market discipline in the governance of US bank holding companies. In: Liu, J., Lean, D., Elroy, E. (eds) European Finance Review, vol. 6, no. 3, pp. 361–395 (2002)
2. Stephanou, C.: Rethinking Market Discipline in Banking. The World Bank (2010)
3. Freixas, X., Laux, C.H.: Disclosure, transparency and market discipline. Vox CEPR's Policy Portal (2012). http://www.voxeu.org/article/market-discipline-disclosure-and-transparency
4. Basel Committee on Banking Supervision: Review of the Pillar 3 requirements (2014). http://www.bis.org/publ/bcbs286.pdf
5. Parwada, J.-T., Ruenzi, S., Saghal, S.: Market discipline and basel Pillar 3 reporting. In: Warren, G.-J. (ed.) 2013 Center for International Finance and Regulation Research Paper Series, pp. 9–11 (2013)
6. EBA: Guidelines on materiality, proprietary and confidentiality and on disclosure frequency under Articles 433(1), 433(2) and 433 of EU Regulation 575/2013 (2014)
7. EBF: Comments on the EBA consultation on draft guidelines on materiality, proprietary and confidentiality and on disclosure frequency under Articles 433(1), 433(2) and 433 of EU Regulation 575/2013 (2014)

8. Hasan, I., Jackowicz, K., Kowalewski, O., Kozlowski, L.: Market discipline during crisis: evidence from bank depositors in transition countries. In: Solanko, L. (ed.) Bank of Finland Discussion Papers (2013)
9. Liu, H., Kešelj, V.: Combined mining of Web server logs and web contents for classifying user navigation patterns and predicting users' future requests. Data Knowl. Eng. **61**(2), 304–330 (2007)
10. Stevanovic, D., An, A., Vlajic, V.: Feature evaluation for web crawler detection with data mining techniques. Expert Syst. Appl. **39**(10), 8707–8717 (2012)
11. Xing, D., Shen, J.: Efficient data mining for web navigation patterns. Inf. Softw. Technol. **46**(1), 55–63 (2004)
12. Munk, M., Drlik, M.: Impact of different pre-processing tasks on effective identification of users' behavioral patterns in web-based educational system. Procedia Comput. Sci. **4**, 1640–1649 (2011)
13. Munk, M., Kapusta, J., Svec, P.: Data preprocessing evaluation for web log mining: reconstruction of activities of a web visitor. In: ICCS 2010 - International Conference on Computational Science, pp. 2267–2274. Elsevier Science Bv, Amsterdam (2010)
14. Munk, M., Kapusta, J., Svec, P.: Data preprocessing dependency for web usage mining based on sequence rule analysis. In: IADIS Multi Conference on Computer Science and Information Systems, MCCSIS, Algarve, Portugal (2009)
15. Kapusta, J., Pilková, A., Munk, M., Švec, P.: Data pre-processing for web log mining: case study of commercial bank website usage analysis. Acta Universitatis Agriculturae et Silviculturae Mendelianae Brunensis **61**(4), 973–979 (2013)
16. Dean, J., Ghemawat, S.: MapReduce: simplified data processing on large clusters. In: Proceedings of USENIX Symposium on Operating Systems Design and Implementation (OSDI) (2004)
17. Dean, J., Ghemawat, S.: MapReduce: simplified data processing on large clusters. Commun. ACM **51**(1), 107–113 (2008)
18. Cooper, B.F., Ramakrishnan, R., Srivastava, U., Silberstein, A., Bohannon, P., Jacobsen, H.-A., Puz, N., Weaver, D., Yerneni, R.: PNUTS: Yahoo!'s hosted data serving platform. Proc. VLDB Endow. (PVLDB) **1**(2), 1277–1288 (2008)
19. Zhou, J., Bruno, N., Wu, M.-C., Larson, P.-Å., Chaiken, R., Shakib, D.: SCOPE: parallel databases meet MapReduce. VLDB J. **21**(5), 611–636 (2012)
20. Leibiusky, J., Eisbruch, G., Simonassi, D.: Getting Started with Storm. O'Reilly, Ireland (2012)
21. Goodhope, K., Koshy, J., Kreps, J., Narkhede, N., Park, R., Rao, J., Ye, V.Y.: Building LinkedIn's real-time activity data pipeline. IEEE Data Eng. Bull. **35**(2), 33–45 (2012)
22. Lam, W., Liu, L., Prasad, S., Rajaraman, A., Vacheri, Z., Doan, A.: Muppet: MapReduce-style processing of fast data. Proc. VLDB Endow. (PVLDB) **5**(12), 1814–1825 (2012)
23. Doulkeridis, C., Nørvåg, K.: A survey of large-scale analytical query processing in MapReduce. VLDB J. **23**(3), 355–380 (2014)
24. Sivaraman, E., Manickachezian, R.: High performance and fault tolerant distributed file system for big data storage and processing using hadoop. In: Proceedings - 2014 International Conference on Intelligent Computing Applications, ICICA 2014, Article no. 6965006, pp. 32–36 (2014)
25. Su, C.-T., Tsao, W.-K., Chu, W.-R., Liao, M.-R.: Mining web browsing log by using relaxed biclique enumeration algorithm in MapReduce. In: Proceedings of the 2012 IEEE/WIC/ACM International Conference on Web Intelligence and Intelligent Agent Technology Workshops, WI-IAT 2012, Article no. 6511648, pp. 54–58 (2012)

26. Premchaiswadi, W., Romsaiyud, W.: Extracting weblog of Siam University for learning user behavior on MapReduce. In: ICIAS 2012 - 2012 4th International Conference on Intelligent and Advanced Systems: A Conference of World Engineering, Science and Technology Congress (ESTCON) - Conference Proceedings, 1, Article no. 6306177, pp. 149–154 (2012)
27. Sakr, S., Liu, A., Fayoumi, A.G.: The family of mapreduce and large-scale data processing systems. ACM Comput. Surv. **46**(1), 11 (2013)
28. Ortega, J.L., Aguillo, I.: Differences between web sessions according to the origin of their visits. J. Informetrics **4**(3), 331–337 (2010)
29. Doran, D., Gokhale, S.S.: Web robot detection techniques: overview and limitations. Data Min. Knowl. Disc. **22**(1–2), 183–210 (2011)
30. Stevanovic, D., An, A., Vlajic, N.: Detecting web crawlers from web server access logs with data mining classifiers. In: Kryszkiewicz, M., Rybinski, H., Skowron, A., Raś, Z.W. (eds.) ISMIS 2011. LNCS, vol. 6804, pp. 483–489. Springer, Heidelberg (2011)
31. Tan, P.-N., Kumar, V.: Discovery of web robot sessions based on their navigational patterns. Data Min. Knowl. Disc. **6**(1), 9–35 (2002)
32. Thomas, P., O'Neill, A., Paris, C.: Interaction differences in web search and browse logs. In: Proceedings of the Fifteenth Australasian Document Computing Symposium, (ADCS) 2010, pp. 52–59 (2010)
33. Cápay, M., Balogh, Z., Boledovičová, M., Mesárošová, M.: Interpretation of questionnaire survey results in comparison with usage analysis in e-learning system for healthcare. Commun. Comput. Inf. Sci. **167**(2), 504–516 (2011)
34. Klocokova, D.: Integration of heuristics elements in the web-based environment: Experimental evaluation and usage analysis. Procedia Soc. Behav. Sci. **15**, 1010–1014 (2011)
35. Lančarič, D., Tóth, M., Savov, R.: Which legal form of agricultural firm based on return on equity should be preferred? A panel data analysis of Slovak agricultural firms. Stud. Agric. Econ. **115**(3), 172–173 (2013)

CKNNI: An Improved KNN-Based Missing Value Handling Technique

Chao Jiang and Zijiang Yang$^{(\boxtimes)}$

School of Information Technology, York University, Toronto, Canada
zyang@york.ca

Abstract. In data mining field, experimental data sets are often incomplete due to the imperfect nature of real world situations. However, the incompleteness of data sets generally leads to biased outcomes. Thus, data completeness is one of the most essential challenges among data mining tasks. In order to achieve better outcome many researchers have explored various techniques to reduce data incompleteness, and some existing methods have been widely used in real world applications. This paper first discusses some existing representative missing data handling techniques with their advantages and drawbacks. Then a new improved KNN based algorithm, *Class-Based K-clusters Nearest Neighbor Imputation* (CKNNI) is proposed, which integrates K-means cluster algorithm and conventional KNN algorithm to impute missing values in data sets. By clustering instances in the same class with K-means algorithm, CKNNI method then applies KNN algorithm to select a closest neighbor from the set of centroids in resulted clusters, and missing values are imputed with the ones from corresponding variables in a selected neighbor. Finally, the comparison based on multiple data sets indicates that CKNNI has improved the performance of KNN imputation significantly on large data sets yet comparative to other superior missing value handling algorithms.

Keywords: Missing value imputation · Classification · Data preprocessing

1 Introduction

In many real world applications, it is shared that the data sets contain errors, invalid values, and missing data, such as unfilled fields in a survey data set, which is considered missing values during later analysis. It is believed that missing data cause biased result when performing data mining tasks because missing data are still considered as valuable representative attributes with respect to the hidden information in data sets. Generally, missing data can be divided into three categories: Missing Completely At Random (MCAR), Missing At Random (MAR) and Not Missing At Random (NMAR). Conrad and Himmelspach [1] concluded that the missing values are considered MCAR, if the absence of data does not rely on data values in the data matrix that are experimented. MAR happens when missingness is related to particular variable, but not related to the value of variable that has missing data. In addition, NMAR happens if the data are missing due to specific reasons. Conrad and Himmelspach [1] also stated that MAR and NMAR are in most cases seen in practice

© Springer International Publishing Switzerland 2015
D.-S. Huang and K. Han (Eds.): ICIC 2015, Part III, LNAI 9227, pp. 441–452, 2015.
DOI: 10.1007/978-3-319-22053-6_47

although MCAR is more general. Moreover, missing data decreases the representation of a data set. Thus, whenever it is possible, missing values should be minimized from the beginning of data collection. In addition, missing values reduce data quality dramatically. Data quality is a concerned aspect in classification process; it may refer to errors, invalidity or missing data among which missing data issue is most challengeable to deal with. Commonly, all of algorithms performed in later steps aim to discover and learn knowledge from the preprocessed data sets. The incomplete data set with low data quality certainly hinders the achievement of better outcome. To properly handle missing values, many techniques have been developed. In practice, there are three major approaches to handle missing data: deleting instances that have missing values, ignoring missing values and replacing missing values based on some rationales, which is called missing value imputation [2]. Some of data mining algorithms are natively compatible with missing values, such as C4.5 and Multi-layerPerceptron which ignore missing values while running classification tasks; some replace missing values with certain techniques simultaneously such as sequential minimal optimization (SMO) algorithm. All approaches aim to overcome data incompleteness, and have their advantages/limitations, but the missing data imputation approach, in most cases, gives advantages over the others.

This paper first introduces some representative missing data handling methods and then proposes a new improved technique, Class-Based K-clusters Nearest Neighbor Imputation (CKNNI), with implementation. The rest of this paper is organized as follows: Sect. 2 discusses the literature on missing data handling along with classification algorithms used in the experiment. Section 3 presents the adopted methodology and the details of the proposed CKNNI method. Finally, Sect. 4 provides evaluation results and Sect. 5 concludes the findings.

2   Literature Review

In reality, less than one percent missing data are generally considered trivial, and one to five percent are manageable. However, five to fifteen percent missing data require some sophisticated technique to handle, and more than fifteen percent missing data may greatly impact any kind of representative characteristics of data set [3].

This section discusses some popular missing data handling methods in real world applications. All missing value imputation techniques are in fact replacing missing values with some estimated ones such as mean or mode values calculated from rows or specific classes. However, whether or not the estimated values are close to the real ones heavily relies on different algorithms to which the final outcome will reflect. In general, missing values imputation often yields better results compared to other approaches mentioned before because the estimated values complete data sets and still represent data characteristics in some degree. Nevertheless, the computational cost is usually a substantial concern and great downside when it comes to certain algorithms such as KNN imputation method, especially when performing data mining tasks on a large data matrix such as [4]. In addition to accuracy concern, the computational cost of some techniques could be much more than others. Therefore, one has to balance the performance and computational cost when analyzing data sets. That is the reason why

CKNNI method is introduced to overcome KNN performance drawback. Its implementation leads to faster running speed with similar performance of other methods.

2.1   K-Nearest Neighbors (KNN) Imputation

KNN imputation is one of the most popular imputation techniques because it is simple, easy to implement and often outputs fair result compared to other algorithms. The rationale is that it fills in missing values of an instance based on given k instances that are closest to the one with missing values. By utilizing distance functions such as Euclidean distance function, the closeness or similarity between target instance and every instance in the data set can be easily calculated. Then the top k closest instances are chosen as candidate to further derive an estimated and weighted value as replacement.

This approach has appealing advantages. It can utilize both qualitative and quantitative data as estimated values for replacement. Moreover, it does not build a predictive model for imputation; therefore, it avoids time consumption and computational cost on modeling. In contrast, because it uses Euclidean distance function, missing attribute values have no impact on similarity calculations. However, it is also true that Euclidean function calculates the sum of every value of all attributes in data set, that is, if data set contains numerous attributes, the time cost will increase exponentially.

In addition to the above mentioned characteristics, KNN imputation loops through all instances in the examined data set to find closest one. This is obviously a time intensive operation when data set has fairly large number of samples.

2.2   Mean Imputation

This approach is one of the most popular missing value imputation methods with straightforward algorithm to implement. It is also called arithmetic mean substitution. The concept is to replace all missing values with mean of some examples, attributes or classes in the observed data set. Generally, mean imputation works relatively well on data sets which have independent variables because this method ignores relationships between attributes. In addition, mean imputation will attenuate the standard deviation and the variance [5] because it fills in missing values with a consistent value across the entire data set. It suggests that if the value of one variable depends on another variable, such correlation is not noted by mean imputation method. Thus, this approach is often used for univariate analysis but not multivariate analysis. Column mean imputation substitutes missing values with the mean of corresponding column or attribute.

Similarly, row mean imputation replaces missing values with the mean of that instance. Both column mean imputation and row mean imputation do not predict missing values. Their straightforward substitution with a constant mean value distorts the hidden pattern in data set and misrepresents it in result. Moreover, this strategy is limited to numeric values due to the mean calculation, and therefore, data sets that contain categorical values or textual values are often required to be converted to numeric values in order to take advantage of this approach. However, conversion of data types eventually raises computational cost especially with large data sets, which minimizes advantage of fast performance of this approach. In that case, not only has it no obvious gain in performance, but also the result is most likely biased. Thus, it is necessary to examine data sets and data type before implementing this approach.

2.3   Singular Value Decomposition (SVD) Imputation

All the above methods perform imputation on local scope, which means that the data set is not treated as a whole during computation. Thus, results are undesirable for some circumstances. On the other hand, SVD imputation method takes the information of the entire data set into account. Originally proposed by Troyanskaya et al. [6], it has been known as a fast and stable algorithm for imputation now. It is clear that the number of predictors required by SVD imputation is much smaller than the number of non-missing observations.

In contrast, SVD imputation is less biased due to the use of global information. More importantly, it takes the distribution of dataset as a whole and outputs better result when global structure exists in expression data [7]. It is known that this approach works better on time-series dataset with low noises.

2.4   Local Least Square (LLSI) Imputation

Similar to the above local-aware algorithms, LLSI is another method that estimates missing values by discovering local similarity. It was proposed to resolve the complexity of other algorithms that utilize complete matrix in bioinformatics field such as microarray data analysis. In fact, LLSI is the combination of both KNNI and SVD algorithms. Simply, there are only two steps involved: finding k closest instances and regression estimation. In LLSI method, KNN algorithm is used as first step to select k instances; the difference is that LLSI calculates similarity by L_2-norm or Pearson correlation coefficients [8].

2.5   Expectation Maximization (EM) Imputation

EM imputation is another popular missing value handling method. It estimates parameters using presented data. The data are described based on models and their parameters, which are estimated by maximum likelihood or maximum a posteriori procedures (MAP) that use variants of EM algorithm [9]. The notion of EM algorithm was proposed long time ago, but it was not widely used until Dempster et al. [10] in 1977 proposed it for computing maximum likelihood estimates of incomplete data set. Since then it became a frequently used algorithm for maximum likelihood estimation in a variety of circumstances with missing value imputation and others. In real world applications, EM algorithm is well generalized and implemented in many types of software such as Weka and SPSS. Not only is EM a well-known missing value handling algorithm in statistics and data mining fields, but also it is popular for being incorporated with other algorithms to accomplish information retrieval tasks.

2.6 Bayesian Principle Component Analysis Imputation (BPCA) Imputation

Similarly, BPCA is also one of those missing value estimation algorithms. Its distinctive feature lies in its assumption that all instances, observable and unobservable, randomly have a joint probability density function that describes their behavior. This characteristic distinguishes BPCA from other non-Bayesian algorithms [11]. In fact, BPCA utilizes principle component analysis (PCA) and EM in the process. This approach improves accuracy by including prior knowledge about the data set and the model. Moreover, it integrates modeling and feature extraction by simultaneously solving parameter estimation and data reconciliation optimization problems [11]. In practice, it requires domain expertise to apply this method efficiently and effectively by fine tuning parameters such as the rank number, the number of principle components, and the number of iterations.

3 Methodology

This section discusses the methodology applied in the experiment and provides details on the proposed CKNNI method. The proposed CKNNI method can improve the performance of KNN imputation significantly on large data sets. The entire experiment consists of two major portions: preprocessing and data analysis (classification). In preprocessing phase, all data sets are initially examined in terms of format, noise, data type, class space and dimensionality to ensure that none contains data types that are unacceptable due to the variation of datasets. Data type conversion is applied beforehand if needed. In addition, file type conversion is also performed to facilitate process with Weka. In each data set missing values are imputed with estimates by the state-of-the-art algorithms and the proposed CKNNI method. The newly generated data sets are stored for analysis phrase. After all data sets are preprocessed, in data analysis step each of four classification algorithms including logistic regression, NaiveBayes, IBk and backpropagation are performed on each of new data sets, and the results of all classification processes are recorded for performance comparisons. The measurement standards are set to be accuracy of prediction model from 10-fold cross-validation and time consumption of each imputation algorithm to fill every data set.

3.1 Preprocessing

First of all, some data sets contain mixed attribute types (both categorical and numeric). Therefore, data type conversions, such as categorical textual data type to numerical, are necessary and accomplished through Weka's "NorminalToBinary" and "NumericToNorminal" filters on some datasets. Then the imputation was run upon the completion of attribute type conversion. This process is implemented with Java and runs randomly so that human handling errors can be minimized since the process involves a large amount of data processing endeavors. After initial examination of the dataset, necessary procedures are applied and next section will discuss the rationale of pre-processing using CKNNI method.

Missing Values Imputation with CKNNI. Given the performance and accuracy consideration, the proposed CKNNI method is the integration of KNN method and K-means cluster technique. In real application, K-means algorithm can converge in a short period while result is still acceptable. Although K-means algorithm is popular for its advantages, it is unfortunately sensitive to outliers. Therefore, any algorithms implemented with K-means need to resolve outliers if there are any. As a result, the proposed CKNNI method which utilizes K-means algorithm is also sensitive to outliers in the process of clustering with K-means. Given such a drawback, CKNNI has been designed to use filters from Weka to remove outliers.

Furthermore, CKNNI is an instance and clustering based supervised imputation method. The difference between k in CKNNI and k in K-means is that k is defined as seeding points in K-means algorithm, but k stands for k clusters in CKNNI method. Moreover, CKNNI only clusters instances within each class other than the entire observation. In fact, when finding the most similar centroid, CKNNI compares with the closest one of target instance's class because centroids are generated by clustering instances of the same class. Although CKNNI also has seeding number, s, as a parameter which is the equivalence to k in K-means, it is preset to 10 if number of observations is larger than 20. In practice, it is set to 2 otherwise for experiment purpose only. The process of CKNNI can be presented as below:

a. Divide observations into their own classes based on each of instance's class;
b. Define k clusters as initial parameter for clustering;
c. Cluster all instances within each class;
d. Run missing value check when there is a missing value;
e. Use KNN algorithm where $k = 1$ to find the closest centroid from the same class of the instance that has missing value;
f. Replace missing value with the one under same attribute of closest centroid
g. Repeat the replacing process until no more missing values are found.

In CKNNI method, KNN algorithm is applied after clustering to find the closest centroid instance. In addition, the number of closest neighbor k for estimation to 1 in CKNNI method. Since there are often a small number of centroid instances to which the distance need to calculate, KNN distance calculation used by CKNNI does not affect computation cost significantly. As a result, the performance of CKNNI in fact rely on first step K-means clustering, which in turn depends on the size of the data set. In the experiment, CKNNI has demonstrated the significant improvement on time efficiency compared to the most examined imputation methods especially KNN imputation.

Outlier Removal. In statistics, an outlier is an observation that is numerically distant from the rest of the data [12]. In the experiment, outliers are investigated and removed using *Interquartile* filter in Weka. After imputation process, outlier instances are removed before classification process. This step ensures that data are the true representation and reduce data size which can be helpful when a data set contains a large number of attributes. Also, it attenuates biases for CKNNI algorithms since CKNNI is sensitive to outliers.

3.2 Classification

In order to be effective and efficient, classification process needs to run on the processed data sets which contain only interesting features. Thus, feature selection is performed immediately after imputation. Then, the result is used for classification subsequently. It is understandable that different feature selection algorithms may yield various outputs, but for the purpose of this experiment, only one state-of-the-art feature selection algorithm from Weka is implemented: *Correlation*-based Feature Selection. Afterwards, all four classifiers are preconfigured so that the settings are ensured to be the same while classification is performed on all the processed data sets. The classification process has no predefined objective on accuracy of predicted model nor time cost concerns but the uniformity of classifiers for all data sets. Thus, classifiers are configured to be as fast as possible, ignoring the accuracy performance for all data sets.

4 Results and Discussions

In order to compare the proposed CKNNI method with other approaches, two major factors (time consumption and accuracy of classification result) are used to evaluate the results from the imputed data sets using various imputation methods. The experiment is implemented in Eclipse Java EE IDE with Weka library support and other software tools including Rstudio, R, and Matlab.

4.1 Data Sets

Data sets are downloaded from UCI machine learning repository. In order to compare various imputation methods comprehensively, 10 data sets with missing values from UCI are used for the experiment. All data sets contain missing values of a certain degree. In fact, the proportion of missing values in all data sets varies significantly from about 0.05 % to 34.6 %. In some case, missing attribute values are univariate, which means missing values in the studied data set only exists in one attribute. Moreover, the chosen data sets also vary in terms of the number of features and sample size significantly. All of those characteristics represent the great diversity in the nature of real world applications, thus, they all contribute to the generalization of the experiment. In addition, the variety of file formats hinders performance and increases cost while conducting the experiment. Therefore, arff file type is used as the standard format through entire experiment because it is typical file type used by Weka and the major parts of experiment are implemented using Weka. This certainly avoids compatibility issues. In addition, using arff type will ease the programming efforts in later implementation phase.

4.2 Results

The results are generated from four classification algorithms. Their accuracy readings are used as one major factor in the experiment; the other factor is the time consumed when imputing each data set. Therefore, this experiment compares different imputation methods based on these two aspects. Accuracy is defined as the number of correctly classified examples divided by the total number of samples in a studied data set. This factor determines the model's ability to correctly predict the unlabeled data. Time consumption is another important factor because time cost can be crucial to real world applications especially when the given time and budget are limited.

Table 1 provides the comparison of accuracy on all missing value handling methods using backpropagation classifier. It can be seen from Table 1 that CKNNI often yields one of the best results among all imputation methods.

Table 2 compares the accuracy of CKNNI with other state-of-the-art imputation methods using NaiveBayes classifier and it shows that the performance of CKNNI

Table 1. CKNNI vs all based on backpropagation classifer

	Column mean	Row mean	CKNNI	KNNI	EM	LLS	BPCA	SVD
agaricus.lepiota.arff	1.000	1.000	1.000	1.000	1.000	1.000	1.000	1.000
anneal.arff	0.688	0.699	0.796	0.695	0.737	0.742	0.794	0.701
Arrhythmia.arff	0.571	0.596	0.581	0.581	0.590	0.571	0.569	0.570
Audiology-Standardized. arff	0.001	0.000	0.000	0.000	0.000	0.000	0.000	0.000
bands.arff	0.377	0.375	0.700	0.460	0.420	0.470	0.414	0.394
bridges.data.version2. arff	0.197	0.231	0.225	0.225	0.223	0.211	0.243	0.206
crx.arff	0.548	0.536	0.561	0.531	0.535	0.480	0.518	0.519
hepatitis.arff	0.429	0.430	0.427	0.382	0.433	0.389	0.365	0.378
horse-colic.arff	0.320	0.469	0.451	0.417	0.469	0.465	0.425	0.456
imports85N2B.arff	0.763	0.773	0.768	0.769	0.772	0.761	0.763	0.763

Table 2. CKNNI vs all based on NaiveBayes classifer

	Column mean	Row mean	CKNNI	KNNI	EM	LLS	BPCA	SVD
agaricus.lepiota.arff	0.947	1.000	0.976	1.000	1.000	1.000	1.000	1.000
anneal.arff	0.908	0.799	1.000	0.912	0.915	0.999	0.892	0.911
Arrhythmia.arff	0.610	0.646	0.756	0.630	0.607	0.611	0.584	0.599
Audiology-Standardized. arff	0.350	0.271	0.229	0.283	0.293	0.394	0.176	0.231
bands.arff	0.414	0.276	0.317	0.303	0.246	0.282	0.192	0.234
bridges.data.version2. arff	0.000	0.000	0.000	0.007	0.006	0.000	0.000	0.000
crx.arff	0.706	0.702	0.623	0.707	0.656	0.692	0.672	0.681
hepatitis.arff	0.229	0.080	0.512	0.000	0.000	0.169	0.157	0.272
horse-colic.arff	0.628	0.585	0.506	0.624	0.514	0.586	0.621	0.548
imports85N2B.arff	0.926	0.915	0.875	0.917	0.901	0.913	0.899	0.904

remains above average. In fact, the data sets anneal.arff, Arrhythmia.arff and hepatitis. arff imputed by CKNNI produce the best result among all. Table 2 also shows there is no one single method that always dominates the others.

The comparison between CKNNI and other methods using logistic regression classifier also yield comparable results as Table 3 suggests. Among all methods the data sets anneal.arff and bands.arff imputed by CKNNI yields the best score. In most cases, the performance of CKNNI remains noticeably higher than others. The overall comparisons suggest that CKNNI imputation produce good and stable outcome.

Table 4 shows that nine out of ten data sets imputed by CKNNI produces the best results compared to other superior methods.

Table 3. CKNNI vs all based on Logistic Regressoin classifer

	Column mean	Row mean	CKNNI	KNNI	EM	LLS	BPCA	SVD
agaricus.lepiota.arff	0.778	1.000	0.810	0.868	0.828	0.962	0.854	0.849
anneal.arff	0.478	0.522	0.936	0.575	0.645	0.594	0.770	0.577
Arrhythmia.arff	0.359	0.498	0.492	0.492	0.527	0.394	0.459	0.466
Audiology-Standardized. arff	0.041	0.000	0.000	0.000	0.000	0.000	0.000	0.000
bands.arff	0.156	0.301	0.487	0.368	0.467	0.124	0.134	0.103
bridges.data.version2. arff	0.000	0.383	0.402	0.347	0.397	0.303	0.209	0.427
crx.arff	0.715	0.756	0.732	0.797	0.785	0.731	0.782	0.784
hepatitis.arff	0.014	0.101	0.287	0.315	0.262	0.146	0.239	0.282
horse-colic.arff	0.654	0.651	0.632	0.714	0.691	0.620	0.716	0.681
imports85N2B.arff	0.560	0.652	0.588	0.689	0.699	0.612	0.668	0.672

Table 4. CKNNI vs all based on IBk classifer

	Column mean	Row mean	CKNNI	KNNI	EM	LLS	BPCA	SVD
agaricus.lepiota.arff	1.000	1.000	1.000	0.921	1.000	1.000	1.000	1.000
anneal.arff	0.378	0.503	1.000	0.550	0.620	0.621	0.670	0.470
Arrhythmia.arff	0.314	0.289	0.520	0.405	0.240	0.325	0.278	0.289
Audiology-Standardized. arff	0.351	0.136	0.434	0.258	0.179	0.358	0.133	0.122
bands.arff	0.178	0.210	0.323	0.325	0.027	0.163	0.127	0.062
bridges.data.version2. arff	0.000	0.000	0.312	0.133	0.271	0.000	0.000	0.000
crx.arff	0.558	0.536	0.671	0.517	0.531	0.548	0.510	0.513
hepatitis.arff	0.058	0.066	0.388	0.000	0.000	0.132	0.054	0.110
horse-colic.arff	0.509	0.430	0.574	0.357	0.330	0.533	0.341	0.512
imports85N2B.arff	0.721	0.701	0.880	0.616	0.854	0.680	0.655	0.653

Table 5. Time performance in milliseconds

	Column mean	Row mean	CKNNI	KNNI	EM	LLS	BPCA	SVD
agaricus.lepiota.arff	682.31	1.256455	2475.49	35186.22	13000	2100	75500	7650
anneal.arff	59.09	0.237155	63.91945	403.2513	3500	2500	64500	1500
Arrhythmia.arff	3.40	0.178304	88.56466	88.32546	2900	900	12500	2400
Audiology-Standardized.arff	3.49	1.664323	80.70222	168.1293	890	1200	9500	1350
bands.arff	34.98	0.931499	934.0842	4139.789	4300	700	12900	600
bridges.data.version2.arff	0.44	0.130676	18.41131	27.76532	700	900	1900	700
crx.arff	2.82	0.108883	74.60306	27.37019	800	700	1700	650
hepatitis.arff	0.27	0.004834	7.027962	2.337078	550	550	800	1200
horse-colic.arff	4.85	0.14474	0.335907	41.79739	82500	680	8600	800
imports85N2B.arff	0.30	0.026818	28.84914	4.220546	420	780	1900	2100

Table 6. Relationship between number of instances and imputation time

	Num. of instances	CKNNI (ms)	KNNI (ms)
agaricus.lepiota.arff	8124	2475.490342	35186.21778
anneal.arff	789	63.919445	403.2513318
Arrhythmia.arff	452	88.564662	88.3254636
Audiology-Standardized.arff	226	80.702222	168.1292722
bands.arff	512	934.0842306	4139.788512
bridges.data.version2.arff	108	18.4113108	27.7653186
crx.arff	690	74.6030566	27.3701912
hepatitis.arff	155	7.0279622	2.3370782
horse-colic.arff	300	0.3359068	41.797388
imports85N2B.arff	205	28.8491368	4.220546

From the above four tables it is evident that the proposed CKNNI method consistently produces better results than KNNI in most cases. In some cases the accuracy is significantly higher than KNNI. This suggests that CKNNI has improved KNNI in term of accuracy.

4.3 Time Cost Comparison

This section concentrates on time efficiency. Since time cost can be impacted by computer hardware, system resource usage during process and etc., time cost is evaluated with an average of five runs in our experiments. Table 5 provides the comparison on time efficiency for all imputation algorithms.

As Table 5 suggests, in most cases CKNNI imputed data sets yield better scores than KNNI, and comparable to other superior imputation methods. Table 6 clearly demonstrates the relationship between the number of instances in a data set and the time cost using both CKNNI and KNNI. Data set agaricus.lepiota.arff has over 8000 instances which results a spike on the time cost when using both KNNI and CKNNI. It takes KNN imputation about 14 times longer than CKNNI. Similarly, it costs more than six times longer to impute anneal.arff using KNNI than CKNNI. Evidently, CKNNI clearly outperforms KNNI in term of time consumption in general.

5 Conclusion

The missing value handling method is one of the most important approaches to preprocess data sets, and it gradually becomes a critical step in entire data mining route because of the proven impact on time efficiency and accuracy improvement. In this paper, a new imputation method, CKNNI is proposed. The suggested method is an integration of K-means clustering algorithm and KNN algorithm. As the experiment indicates, CKNNI is a cost effective, computation efficient and accurate method compared to KNNI method.

References

1. Himmelspach, L., Conrad, S.: Clustering approaches for data with missing values: comparison and evaluation. In: Fifth International Conference on Digital Information Management (ICDIM), pp. 19–28 (2010)
2. Zhu, X., Zhang, S., Jin, Z., Zhang, Z., Xu, Z.: Missing value estimation for mixed-attribute data sets. IEEE Trans. Knowl. Data Eng. **23**(1), 110–121 (2011)
3. Acurna, E., Rodriguez, C: The treatment of missing values and its effect in the classifier accuracy, classification, clustering, and data mining applications. In: Proceedings of the Meeting of the International Federation of Classification Societies (IFCS), pp. 639–647. Illinois Institute of Technology (2004)
4. Kushmerick, N: Learning to remove internet advertisements. In: 3rd International Conference on Autonomous Agents, pp. 175–181 (1999)
5. Enders, C.K.: Applied Missing Data Analysis. Guilford Press, New York (2010)
6. Troyanskaya, O., Cantor, M., et al.: Missing value estimation methods for DNA microarrays. Bioinformatics **17**(6), 520–525 (2001)
7. Gan, X., Liew, W.C., Yan, H.: Microarray missing data imputation based on a set theoretic framework and biological knowledge. Nucleic Acids Res. **34**(5), 1608–1619 (2006)
8. Kim, H., Golub, G.H., Park, H.: Missing value estimation for DNA microarray gene expression data: local least squares imputation. Bioinformatics **21**(2), 187–198 (2005)
9. Farhangfar, A., Kurgan, L., Pedrycz, W.: Experimental analysis of methods for imputation of missing values in databases. In: Intelligent Computing: Theory and Applications II, pp. 172–182 (2004)
10. Dempster, A.P., Laird, N.M., Rubin, D.B.: Maximum likelihood from incomplete data via the EM algorithm. J. Roy. Stat. Soc. Ser. B (Methodol.) **39**(1), 1–38 (1977)
11. Nounou, M.N., Bakshi, B.R., Goel, P.K., Shen, X.: Bayesian principal component analysis. J. Chemom. **16**(11), 576–595 (2002)
12. O'Reilly, C., Gluhak, A., Imran, M., Rajasegarar, S.: Online anomaly rate parameter tracking for anomaly detection in wireless sensor networks. In: 9th Annual IEEE Communications Society Conference on Sensor, Mesh and Ad Hoc Communications and Networks, pp. 191–199 (2012)

An Item Based Collaborative Filtering System Combined with Genetic Algorithms Using Rating Behavior

Jing Xiao, Ming Luo, Jie-Min Chen, and Jing-Jing Li[✉]

School of Computer Science, South China Normal University, Guangzhou
510631, China
jingjingli.1124@gmail.com

Abstract. With the sharp increment of information on the Internet, many technologies have been proposed to solve the problem of information explosion in people's life. Collaborative Filtering (CF) recommendation system is one of the most popular and efficient ways of solutions, especially item based CF systems. While traditional item based CF recommendation algorithms either ignore the diversity of different users' rating behavior or do not deal with it efficiently. In this paper, we present a novel similarity function using the average rating for each user instead of the overall average rating for all users. In order to find the optimal similarity function, we use genetic algorithm (GA) to optimize the weight vectors associated to the similarity function. A series of comparison experiments are conducted to demonstrate the effectiveness in terms of the quality of prediction of the proposed method.

Keywords: Item-based · Rating behavior · Collaborative filtering · Genetic algorithms · Recommendation systems

1 Introduction

With the rapid development of computer hardware and web science, Internet has changed our human's life greatly. Meanwhile, we also step into the era of information overloading. To help people get rid of this, recommender systems (RSs) [1, 2] are emerged to recommend items or services to people automatically based on their personal information or previous behavior.

To the best of our knowledge, traditional CF i.e. user-based [3] and item-base [4], either ignore the diversity of different users' rating behavior or deal with it inefficiently besides suffering some inherent limitations, such as poor scalability, data sparsity and cold-start. To deal with these limitations, genetic algorithms (GAs) [5] have been applied to RSs successfully in many research works. However, there are some problems as well, e.g. some systems may need rich information (manufacture of the item etc.) which is not always available. In this paper, we propose an Item-based Collaborative Filtering recommender system combined with GA (ItemCF_GA) only using rating information. A novel item-based similarity metric using each user's individual average rating to obtain the likeness of users over items to alleviate the inaccuracy caused by the

© Springer International Publishing Switzerland 2015
D.-S. Huang and K. Han (Eds.): ICIC 2015, Part III, LNAI 9227, pp. 453–460, 2015.
DOI: 10.1007/978-3-319-22053-6_48

divergence of users' rating habits is proposed. GA, as an optimization tool, is employed to find an acceptable weight vector associated with the similarity metric.

The remainder of this paper is organized as follows: In Sect. 2, we outline the related work. Then, we depict the new similarity function and the process of the ItemCF$_{GA}$ algorithm. Next, a series of experiments are conducted in Sect. 4 to give a comparison to demonstrate the effectiveness of our method. Conclusion and future work are summarized in Sect. 5.

2   Related Work

In order to improve the performance of the traditional RSs, many researchers incorporated GAs in RSs. Most of GAs in RSs have been used in two aspects: clustering and hybrid models [1]. RSs employing clustering technology are promising strategies to provide more accuracy results and alleviate the poor scalability of massive users. In these systems, GAs are commonly used to improve the efficiency of the clustering algorithms, for example finding the best suitable initial centers of K-means [6, 7]. The hybrid models usually employed the attributes of items or users, such as material of items or demographic information of users [8–11]. A few papers proposed recommend systems for music using GAs [12–14]. A common characteristic of them is that all the algorithms were combined content-based filtering with interactive GAs to design hybrid RSs. Bobadilla et al. [15], breaking up the stereotypes, trained a user-based model by optimizing mean absolute error (MAE) through supervised learning. Good performance was obtained compared with other traditional user-based collaborative filtering RSs. However, this method ignored the divergence between users' different rating habits. Recently, Gong et al. [16] proposed a method considering multiple needs of users using evolutionary algorithm.

In this paper, we propose ItemCF$_{GA}$ to alleviate all the issues mentioned above. This method can be used in any recommender system because it does not need much information except the user-item rating matrices.

3   ItemCF$_{GA}$

In this section, we will present the details of our item based collaborative filter RS combined with GA. It has two stages: an offline stage and an online stage. The offline stage uses the existing database and GA to train a model and the online stage makes recommendation for an active user according to the trained model. In this section, a new kind of similarity measure will be depicted in details. Then we will present each stage of the proposed approach.

3.1   Similarity Measurement

Most similarity measurements operate on users' ratings over the items. While, there is a big difference between two users' rating behavior. In this paper, in order to capture the precise likeness of different users over items, we use each user's individual average rating rather than an overall average rating for all users.

- **Preprocessing.** Firstly, the likeness of each user is obtained through subtracting every rating of user u over item i by the user u's average rating. If the substraction result is greater than or equal to zero, we consider the user likes the item, and vice versa. Given two items, supposing that both of them are rated, a user's likeness i.e. unlike (-1) or like (1) over these two items might have four combinations, Combination$_0$ $(-1, -1)$, Combination$_1$ $(-1, 1)$, Combination$_2$ $(1, -1)$ and Combination$_3$ $(1, 1)$. Lastly, a new vector $\mathbf{V}$, whose dimension is the number of possible combinations, is imported to count up the number of users in each of four combinations. For example, two items $\mathbf{I}_1$ and $\mathbf{I}_2$ are rated by five users, and the likeness of users over them are: $\mathbf{I}_1$ $(-1, -1, 1, 1, 1)$ and $\mathbf{I}_2$ $(-1, 1, -1, 1, -1)$. Thus, $\mathbf{V}$ is $(1, 1, 2, 1)$, since only the first user's likeness of $\mathbf{I}_1$ and $\mathbf{I}_2$ is Combination$_0$ $(-1, -1)$. The remaining components of V can be obtained in the same way.
- **Similarity formula.** Given a weighting vector $\mathbf{W} = \left(w_1, w_2, \ldots, w_{\|V\|}\right)$, $w_i \in \mathbf{W}$ and $w_i \in [0, 1]$, then we can combine each component V_i of the vector $\mathbf{V}$ together, and we regard the combination value as the similarity of two items as following:

$$Similarity(x, y) = \sum_{i=0}^{\|V\|} \frac{w_i \times V_i}{\sqrt{\sum_{j=0}^{\|V\|} w_j^2 \times \sum_{k=0}^{\|V\|} V_k}} \qquad (1)$$

3.2 Description of ItemCF$_{\_GA}$

To find an optimal similarity function of Eq. (1), we employ a GA whose fitness function is MAE of the RS.

Genetic Representation. In our system, we use real number coding scheme in which each weighting vector $\mathbf{W} = \left(w_1, w_2, \ldots, w_{\|V\|}\right)$ is expressed as a real vector of $\|V\|$ genes, each gene corresponds to the importance of each combination in vector V.

Initial Population. In our proposed method, in order to increase the diversity of solutions, the initial population is generated randomly.

Fitness Function. We use the mean absolute error of the database (just the training set) as the fitness function. The following steps are performed:

step1: Obtain the similarity of every pair of items using Eq. (1).
step2: Find the nearest k neighbors of each item i, N(i).
step3: For each user u in the training set, predict the rates of user u over item i $prediction_u^i$ using the following formula:

$$prediction_u^i = \overline{r_i} + \frac{\sum\limits_{n \in N(i)}^{k} Similarity(i,n) \times (r_u^n - \overline{r_n})}{\sum\limits_{n \in N(i)}^{k} |Similarity(i,n)|} \tag{2}$$

where r_u^n is the rate of user u over item n, $\overline{r_n}$ is the average rate of item n.

step4: The fitness function (MAE) of the individual on train set is as following:

$$fitness(individual) = \frac{1}{\#U} \sum_{u=1}^{\#U} \frac{1}{\#I} \sum_{i=1}^{\#I} |prediciton_u^i - r_u^i| \tag{3}$$

Genetic Operators and Termination Condition. Selection: Roulette-wheel selection is chosen in the proposed method. The proportion of selecting an individual depends on its fitness value.

Crossover: We use the one-point crossover method. The crossover probability is 0.8.

Mutation: We randomly replace one element of the chromosome with a random value if the mutation condition is satisfied. Mutation probability is 0.01.

Besides, the elitist strategy is used in ItemCF$_{GA}$ for keeping the best individual in parent population and replacing the worst individual in the offspring generation.

The GA operations repeat until a terminal condition is reached. ItemCF$_{GA}$ stops when there is an individual in the population whose fitness value satisfies a manually set threshold or a predefined maximum number of generation reaches.

4 Experiments

In this section, we will give a description of our experimental design in details to validate the efficiency of the proposed method. We compare the performance of the proposed method with two baseline item based collaborative filter algorithms using popular similarity metrics. Our experiments were implemented in C++ using a computer with an Inter(R) Core 3.6 GHz processor and 8 GB RAM.

4.1 Dataset Description

We employ MovieLens datasets[1] (1M) which contains 1 million anonymous ratings from 6000 users on 4000 movies with a discrete rate scale of 1–5. Each user has rated at least 20 movies. We split the dataset into different ratios of training and test set randomly (7:3, 8:2, 9:1). We repeat each ratio experiment five times and average the results.

---

[1] http://grouplens.org/datasets/movielens/.

4.2   Experimental Design

We compare the proposed algorithm with other traditional similarity measures: adjusted cosine and Pearson which are the most commonly used measurements in the field of CFRS. In the procedure of genetic algorithm, we set the initial population size to 100; the rate of the crossover is 0.8; the mutation rate is 0.01 and the maximum number of generation is 100. We set the number K of neighbors of each item ranging from 10 to 250 and the step is 10. The size of the recommendation list N ranges from 2 to 48, the step is 2. The genetic algorithm is run 10 times to obtain the optimal similarity function and then get the average result for each scenario.

4.3   Results and Discussion

We employ three widely used evaluation metrics in the experiments [1] to validate the efficiency of our proposed method i.e. mean absolute error (MAE), recall and precision.

We first try to evaluate the quality of prediction of the proposed method, compared with the baseline traditional similarity measurements. From Fig. 1, results can be concluded that with the increasing number of neighbors, both ItemCF$_{GA}$'s and baselines' MAE decrease. It is obvious that adjusted cosine and Pearson have a litter higher MAE when the parameter K is small but ItemCF$_{GA}$ has a smaller MAE near 0.78. With the increasing of the number of neighbors, MAE of adjusted cosine and Pearson decrease fast and then become more stable around 100. On the other hand, the MAE of ItemCF$_{GA}$ is a little higher than adjusted Cosine after K is 100 but still lower than Pearson because we fix the threshold at 0.78, which has been studied in [17]. In the stage of optimization, ItemCF$_{GA}$ is exited once there is an individual whose MAE is lower than 0.78 or when the maximum generation number reaches. And that's why ItemCF$_{GA}$ keeps with a more stable MAE value.

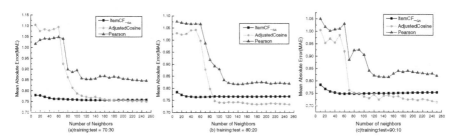

Fig. 1. Mean absolute error of MovieLens 1 M

Then, we validate the recommendation quality of ItemCF$_{GA}$. We set the recommendation list size from 2 to 48 to obtain the quality of recommendation. As seen from Fig. 2, the recall increases with the increasing size of the recommendation list for all methods. At first, when the recommendation list size is below 10, the recall of Item-

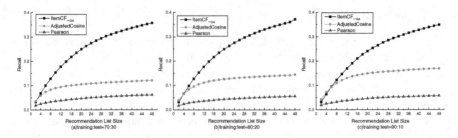

Fig. 2. Recall of MovieLens 1 M

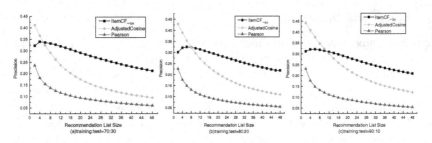

Fig. 3. Precision of MovieLens 1 M

CF$_{GA}$ is almost the same as the baselines methods. But, when the size of the recommendation list reaches to 10, the recall of adjusted Cosine and Pearson becomes stable but the recall of ItemCF$_{GA}$ still increases. From Fig. 3, the precision of all methods decreases when the size of the recommendation list increases. At the beginning of the curves, ItemCF$_{GA}$'s precision is a little lower than adjusted Cosine while it is still much higher than Pearson. In the dataset of MovieLens, each user has rated at least 20 movies. We consider the recommendation list higher than 20 is more important. It is obvious that the precision of ItemCF$_{GA}$ is better than the precision of the other two traditional methods which indicates that ItemCF$_{GA}$ can offer more reliable and interesting items to users than traditional popular methods. Generally speaking, ItemCF$_{GA}$ outperforms the compared approaches in terms of the quality of recommendation for its improving the accuracy of finding the similar items taking users' different rating behavior into account and using GAs as optimization tools.

5   Conclusion and Future Work

In this paper, we propose ItemCF$_{GA}$ with a novel similarity function, taking the fact that different person has his own rating standard into consideration. We also conducted a set of comparison experiments to demonstrate the good performance of our proposed method. In fact, the empirical results do verify that the proposed ItemCF$_{GA}$ outperform other algorithms in terms of MAE, recall and precision on the dataset of

MovieLens 1 M. For future work, more data sets will be used to testify the effectiveness of our proposed ItemCF_$_{GA}$. We will also improve the formula of similarity metric and design modified GA to fit different datasets.

Acknowledgement. This work was partially supported by the National Natural Science Foundation of China (NSFC) projects No. 61202296 and No. 61300044, the National High-Technology Research and Development Program ("863" program) of China under Grant No. 2013AA01A212, the Natural Science Foundation of Guangdong Province project No. S2012030006242 and the Key Areas of Guangdong-HongKong Breakthrough project No. 2012A090200008.

References

1. Bobadilla, J., Ortega, F., Hernando, A., Gutiérrez, A.: Recommender systems survey. Knowl.-Based Syst. **46**, 109–132 (2013)
2. Lü, L., Medo, M., Yeung, C.H., Zhang, Y.C., Zhang, Z.K., Zhou, T.: Recommender systems. Phys. Rep. **519**(1), 1–49 (2012)
3. Goldberg, D., Nichols, D., Oki, B.M., Terry, D.: Using collaborative filtering to weave an information tapestry. Commun. ACM **35**(12), 61–70 (1992)
4. Sarwar, B., Karypis, G., Konstan, J., Riedl, J.: Item-based collaborative filtering recommendation algorithms. In: Proceedings of the 10th International Conference on World Wide Web, pp. 285–295. ACM, April 2001
5. Davis, L. (ed.): Handbook of Genetic Algorithms, vol. 115. Van Nostrand Reinhold, New York (1991)
6. Kim, K.J., Ahn, H.: A recommender system using GA K-means clustering in an online shopping market. Expert Syst. Appl. **34**(2), 1200–1209 (2008)
7. Wang, Z., Yu, X., Feng, N., Wang, Z.: An improved collaborative movie recommendation system using computational intelligence. J. Vis. Lang. Comput. **25**(6), 667–675 (2014)
8. Hwang, C.-S., Su, Y.-C., Tseng, K.-C.: Using genetic algorithms for personalized recommendation. In: P, J.-S., Chen, S.-M., Nguyen, N.T. (eds.) ICCCI 2010, Part II. LNCS, vol. 6422, pp. 104–112. Springer, Heidelberg (2010)
9. Fong, S., Ho, Y., Hang, Y.: Using genetic algorithm for hybrid modes of collaborative filtering in online recommenders. In: Eighth International Conference on Hybrid Intelligent Systems, 2008, HIS 2008, pp. 174–179. IEEE, Sep 2008
10. Salehi, M., Pourzaferani, M., Razavi, S.A.: Hybrid attribute-based recommender system for learning material using genetic algorithm and a multidimensional information model. Egypt. Inf. J. **14**(1), 67–78 (2013)
11. Salehi, M.: Latent feature based recommender system for learning materials using genetic algorithm. Inf. Syst. Telecommun. **2**(3), 137 (2014)
12. Kim, H.T., Kim, E., Lee, J.H., Ahn, C.W.: A recommender system based on genetic algorithm for music data. In: International Conference on 2nd Computer Engineering and Technology (ICCET), 2010, vol. 6, pp. V6–414. IEEE, April 2010
13. Badhe, N., Mishra, D., Joshi, C., Shukla, N.: Recommender system for music data using genetic algorithm (2014)
14. Manjula, A., Neelam, P., Asif, U.K.: Dynamic music recommender system using genetic algorithm. Int. J. Eng. Adv. Technol. (IJEAT) 3(4), April 2014. ISSN 2249 – 8958

15. Bobadilla, J., Ortega, F., Hernando, A., Alcalá, J.: Improving collaborative filtering recommender system results and performance using genetic algorithms. Knowl.-Based Syst. **24**(8), 1310–1316 (2011)
16. Zuo, Y., Gong, M., Zeng, J., Ma, L., Jiao, L.: Personalized recommendation based on evolutionary multi-objective optimization [research frontier]. IEEE Comput. Intell. Magazine **10**(1), 52–62 (2015)
17. Bobadilla, J., Serradilla, F., Bernal, J.: A new collaborative filtering metric that improves the behavior of recommender systems. Knowl.-Based Syst. **23**(6), 520–528 (2010)

Natural Language Processing
and Computational Linguistics

Short Text Classification Based on Semantics

Chenglong Ma[1(✉)], Xin Wan[2], Zhen Zhang[2], Taisong Li[1], and Yan Zhang[1]

[1] The Key Laboratory of Speech Acoustics and Content Understanding,
Institute of Acoustics, Chinese Academy of Sciences, Beijing, China
{machenglong,litaisong,zhangyan}@hccl.ioa.ac.cn
[2] National Computer Network Emergency Response Technical Team/Coordination
Center of China (CNCERT/CC), Beijing, China
{wanxin,zhangzhen}@cert.org.cn

Abstract. Data sparseness and unseen words are two major problems in short text classification. In such a case, it is unsuitable to directly use the vector space model (VSM) which focuses on the statistical occurrence of the terms to represent the text. To solve these problems, we present a novel short text classification method based on semantics. The method of K-Means is used to perform it. In the experiments, we exploit the continuous word embeddings which were trained on very large unrelated corpora to represent the semantic relationships. The experimental results on an open dataset show that the application of semantics greatly improves the performance in short text classification, comparing with a state-of-the-art baseline in VSM; and that the proposed method can reduce the costs of collecting the training data.

Keywords: Short text classification · K-Means · Continuous word embedding · Semantics

1 Introduction

Nowadays, there are more and more short texts produced all over the Internet, including tweets, chat messages, Web search snippets and so on. In order to provide the users with interested information, it is crucial to make out the domain information of the short text documents. However, short texts differ from normal documents because of their shortness (i.e., 140 characters or less in length for Twitter), which brings about some disadvantages in classification. For example, it does not provide enough word occurrence for a good similarity measure between documents [1]. And in general, since the training data are usually not expansive enough to cover adequately all word occurrences, there would be more rare and unseen words in feature extraction during test phase. So, it is unsuitable to directly use the conventional machine learning methods to classify short text documents.

There have been two major directions that attempt to solve the problem of data sparseness in vector space model (VSM). One is to select more useful contextual information to expand the short text, e.g., directly from Wikipedia [2], web search engines [3]. Another is to integrate the short text with a set of hidden topics discovered

© Springer International Publishing Switzerland 2015
D.-S. Huang and K. Han (Eds.): ICIC 2015, Part III, LNAI 9227, pp. 463–470, 2015.
DOI: 10.1007/978-3-319-22053-6_49

from related corpora. Phan et al. [1] manually built a large and rich universal dataset, and derived a set of hidden topics through topic model LDA [4] from the dataset. These researches have made excellent improvement by enriching the original short text, but they rely too much on external repositories which are difficult to collect in some applications, and do not make best use of the semantic meaning that the document expresses.

In the existing works of the literature, a text document is usually represented as a vector of weights in VSM, where each dimension corresponds to a feature and each value encodes the feature weight [5]. The criterion of representation is built on the hypothesis that each feature is independent [6]. Obviously, it disregards the linguistic relationship between words.

In this paper, we raise two questions: if we do not have the related external corpora, how could we solve the data sparseness? If we have got an embedding of words, how could we utilize them in short text classification? In fact, the words "football" and "soccer" are very close in semantics, which come from different documents with the same domain information of "sport", but the VSM can not capture these relationships. Therefore, we try to solve the data sparseness from another direction by utilizing the semantic information that the short text expresses. Given an open dataset, we investigate a novel short text classification method with the K-Means algorithm based on semantics. To do this, we utilize continuous word embeddings to represent the semantic relationships between words, which were trained on very large external corpora (e.g., Google News). Experimental results show that our approach makes a 26.4 % relative accuracy improvement, comparing with a state-of-the-art baseline in VSM. Our main contributions of this paper include the following points:

- K-Means with word embeddings: We first utilize the continuous word embeddings in the K-Means based short text classification.
- Reducing data sparseness in semantics: Instead of counting the features, we represent the short text by their semantics with continuous word embeddings. It is very important that this method can use all the words in the text, especially in the test phase.
- Easy to implement: To classify the short text, what we need is to collect a small set of training data and annotate them. Experimental results show that one tenth of the training data can also achieve great accuracy.

The rest sections are organized as follows: the methodology that was used for the experiments will be presented in Sect. 2. In Sect. 3, experimental results on short text classification are displayed. Finally, we provided some conclusion and directions for future work in Sect. 4.

2   Our Approach

In this section, we introduce the general framework of short text classification based on semantics. The main motivation is to mimic the process thinking of human. More specifically, given two domains "sport" and "computer", one may imagine two distributions of different sets of words. "sport" consists of football, basketball, …, and "computer" has software, IBM, ..., the two sets of words are far away in semantic. Each distribution has its specific set of semantics, and a common set of semantics shared with other domain, as shown in Fig. 1.

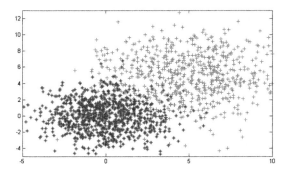

Fig. 1. Semantic distributions in two domains.

Especially, different words with the same semantics are very close and can substitute each other. We can use the average of semantics to represent the property of the distribution.

2.1 Continuous Space Word Embeddings

To represent the semantic relationship between words, we make use of continuous word embeddings. Recently, applications of neural network language model have further made the word embeddings capture syntactic and semantic information [7, 8]. The word is represented by a real-value vector in continuous space. The cosine similarity measures the relationship between a pair of words, the same offset means a particular relation, such as the offset between vec("King") and vec("Man") is very close to that between vec("Woman") and vec("Queen") [8].

2.2 The Semantic K-Means Method

The proposed approach disregards the concept of the VSM, which codifies a document with a vector of terms [9]. This paper presents a novel model of semantic distribution, where different words with the same semantics are very close, and the domain is composed of a set of semantics. The mean of word semantics can describe the distributed situation.

The algorithm of K-Means is one of the most unsupervised techniques for text clustering in data mining. Given a specified number of clusters, it partitions the samples into K clusters in which each sample belongs to the cluster with the nearest mean [10]. In our task of classification, K is specified to be the number of categories.

3 Experiments

To evaluate the effectiveness of our approach, we choose MaxEnt [11] classifier to be the baseline of VSM, because it performs better than SVMs (Support Vector Machine) [12] and other conventional methods in classifying sparse data [1]. In MaxEnt classifier, we use TFIDF to measure the importance of features. In what follows, we evaluate the performance of short text classification on an open web snippet dataset. Firstly, we examine the performance by comparing our method with the baseline in detail. Secondly, we assess the influence of training data size and the ability of coping with unseen words.

3.1 Dataset

We conduct experiments on a web snippet dataset, which has been generally used in [1, 13, 14] and consists of 10,060 training and 2,280 test snippets, as shown in Table 1.

Table 1. Statistics of web snippets data

No.	Domain	Training	Test
1	business	1,200	300
2	computer	1,200	300
3	cul.-arts-ent.	1,880	330
4	engineering	220	150
5	health	880	300
6	politics-soc.	1,200	300
7	sports	1,120	300
8	edu.-sci.	2,360	300
Total		10,060	2,280

The snippets of search results are collected by various phrases belonging to different domains to query the web search engine (Google) and select top 20 and 30 snippets from the search results [1]. The different phrases for training and test data lead to many unseen words in the test data, as shown in Table 2. Column 2 in Table 2 shows that the test data contain about 4,378 words (about 43.62 %) which do not appear in training data, and the column 3 shows the statistics of unseen words after Poter stemming [15]. This table shows that there will be more than 40 % unseen words in test phase and it will be hard to get evidence from those unseen words in VSM and to classify the documents through the rest of words.

To better represent the semantic relationship, we utilize the pre-trained word vectors trained on part of Google News (about 100 billion words) to represent the semantic concepts. The resulting vectors have dimensionality 300, vocabulary size 3 million [7], which have been used in many natural language processing applications.

Table 2. The size of unseen words

	Original	Stem
Training vocabulary	26,265	21,596
Test vocabulary	10,037	8,200
Unseen words	4,378	3,677
Difference (%)	43.62	44.84

3.2 Measurment

The following measurements are adopted to evaluate the performance of the classifier: precision (P), recall (R), F1-measure (F1) and accuracy (Acc).

$$precision = \frac{TP}{TP + FP} \tag{1}$$

$$recall = \frac{TP}{TP + FN} \tag{2}$$

$$F1 - measure = \frac{2 * precision * recall}{precision + recall} \tag{3}$$

$$accuracy = \frac{\#right\_utterances}{\#utterances} \tag{4}$$

where TP is the number of relevant utterances classified as relevant, FN is the number of relevant utterances classified as irrelevant, and FP is the number of irrelevant utterances classified as relevant.

3.3 Experimental Results

In order to assess the quality of our approach, we run experiments to measure the P, R, F1 in each domain test data and Acc in total, as shown in Table 3. These results of the MaxEnt classifier are all lower than our proposed method in F1-measure, and we also can get high performance for difficult domain "engineering" (0.83 versus 0.51). In the table, we can increase the accuracy from 66.80 % to 84.43 % (increasing by 26.4 %). The main reason is that the baseline based on VSM will alleviate or ignore the unseen words which the model can not get any evidence from, but our approach makes the best use of the semantics which all words express in these samples.

The next experiments validate how our framework relies on the size of training data. We vary the size of training data from 1,000 to 10,000 and measure the accuracy on the same test data. For another, the less training samples would bring about the more unseen words occurring in test phase. We show in Fig. 2 that our model is robust to the size of training data, where the experimental results change from 81.32 % to 84.43 %. It shows

Table 3. Evaluations of short text classification

Domain	MaxEnt			Our approach		
	P (%)	R (%)	F1	P (%)	R (%)	F1
1	68.05	60.33	0.64	83.85	81.33	0.83
2	72.26	66.00	0.69	80.23	92.00	0.86
3	62.86	78.48	0.70	74.25	90.00	0.81
4	72.84	39.33	0.51	75.27	93.33	0.83
5	86.94	64.33	0.74	95.24	80.00	0.87
6	70.75	50.00	0.59	95.24	73.33	0.83
7	86.08	78.33	0.82	92.39	89.00	0.91
8	45.93	82.67	0.59	83.97	80.33	0.82
Acc (%)	66.80			84.43		

that our approach can achieve excellent accuracy using fewer data (81.32 % by only one-tenth of training data, increasing 34.26 % absolutely versus MaxEnt), which also proves that the proposed method has a good ability to cope with unseen words. The overall results show that the specific domain can be represented by a few semantics instead of counting the words, and we can reduce the cost of collecting and annotating training samples.

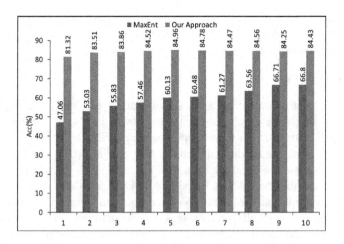

Fig. 2. Evaluation with different sizes of training data.

4   Conclusions

In this work, we proposed to utilize the semantics of document to solve the data sparsity problem in short text classification. This is a significant departure from the conventional methods which classify the document as a regular vector of weights based on statistical machine learning algorithms. It disregards the criterion of VSM which treat words as independent features and utilizes the semantics the most words express including the unseen words. We evaluate our approach on an open dataset and show that the classifier based on semantics outperforms the state-of-the-art baseline based on VSM.

We focus on the word embedding which represents the semantic relationship by mathematical representation of words in continuous space. It is a crucial technique to transform a word into a symbolic vector. In future work, we will try to extend this work to deal with normal documents which contain repetitive sequence.

References

1. Phan, X.H., Nguyen, L.M., Horiguchi, S.: Learning to classify short and sparse text & web with hidden topics from large-scale data collections. In: Proceedings of the 17th International Conference on World Wide Web, pp. 91–100. ACM (2008)
2. Banerjee, S., Ramanathan, K., Gupta, A.: Clustering short texts using wikipedia. In: Proceedings of the 30th Annual International ACM SIGIR Conference on Research and Development in Information Retrieval, pp. 787–788. ACM (2007)
3. Bollegala, D., Matsuo, Y., Ishizuka, M.: Measuring semantic similarity between words using web search engines. WWW **7**, 757–766 (2007)
4. Blei, D.M., Ng, A.Y., Jordan, M.I.: Latent Dirichlet allocation. J. Mach. Learn. Res. **3**, 993–1022 (2003)
5. Clinchant, S., Perronnin, F.: Aggregating continuous word embeddings for information retrieval, p. 100. ACL (2013)
6. Chengqing, Z.: Statistical Natural Language Processing. Tsinghua University Press, Beijing (2008)
7. Mikolov, T., Sutskever, I., Chen, K., Corrado, G.S., Dean, J.: Distributed representations of words and phrases and their compositionality. In: Advances in Neural Information Processing Systems, pp. 3111–3119 (2013)
8. Mikolov, T., Yih, W.T., Zweig, G.: Linguistic regularities in continuous space word representations. In: Proceedings of NAACL-HLT, pp. 746–751 (2013)
9. Rosso, P., Ferretti, E., Jiménez, D., Vidal, V.: Text categorization and information retrieval using wordnet senses. In: Proceedings of the Second International Conference of the Global WordNet Association, pp. 299–304 (2004)
10. Wu, X.D., Kumar, V., Quinlan, J.R., Ghosh, J., Yang, Q., Motoda, H., McLachlan, G.J., Ng, A., Liu, B., Philip, S.Y.: Top 10 algorithms in data mining. Knowl. Inf. Syst. **14**, 1–37 (2008)
11. Berger, A.L., Pietra, V.J.D., Pietra, S.A.D.: A maximum entropy approach to natural language processing. Comput. Linguist. **22**, 39–71 (1996)
12. Joachims, T.: Text Categorization with Support Vector Machines: Learning with Many Relevant Features. Joachims, Thorsten (1998)

13. Sun, A.: Short text classification using very few words. In: Proceedings of the 35th International ACM SIGIR Conference on Research and Development in Information Retrieval, pp. 1145–1146. ACM (2012)
14. Chen, M., Jin, X.M., Shen, D.: Short text classification improved by learning multigranularity topics. In: Proceedings of the Twenty-Second International Joint Conference on Artificial Intelligence, vol. 3, pp. 1776–1781. AAAI Press (2011)
15. Jones, K.S.: Readings in Information Retrieval. Morgan Kaufmann, San Francisco (1997)

Name Disambiguation Using Semi-supervised Topic Model

JinLan Fu, Jie Qiu, Jing Wang, and Li Li[(✉)]

School of Computer and Information Science, Southwest University,
Chongqing 400715, China
lily@swu.edu.cn

Abstract. Name ambiguity is increasingly attracting more attention. With the development of information available on the Web, name disambiguation is becoming one of the most challenging tasks. For example, some persons may share the same personal name. In order to address this problem, topic coherence principle is used to eliminate ambiguity of the name entity. A semi-supervised topic model (STM) is proposed. When we search online, many irrelevant documents always return to users. Wikipedia hierarchical structure information enrich the semantics of the name entity. Information extracted from Wikipedia is sorted out and put in the knowledge base. It is used to match the query entity. By utilizing the context of the given query entity, we attempt to disambiguate various meanings with the proposed model. Experiments on two real-life datasets, show that STM is more superior than baselines (ETM and WPAM) with accuracy 84.75 %. The result shows that our method is promising in name disambiguation as well. Our work can provide invaluable insights into entity disambiguation.

Keywords: Semi-supervised topic models · Name disambiguation

1 Introduction

Name disambiguation [1, 2] refers to disambiguating name entities in natural language text linking them to their corresponding entries in knowledge base. The knowledge base can be Wikipedia, BaiDu encyclopedia, and so on. Name disambiguation has always been a difficult problem that available information is increasing. In many applications, name ambiguity [3, 4] has long been viewed as a challenging problem, such as information retrieval. When searching a people's name in network, many documents such as Wikipedia articles and news containing person's name may be returned to us. Generally speaking, we need judge which object do we want. Although much researches have been conducted and many solutions have been proposed, the problem remains largely unsolved.

Recently, name disambiguation task is widely explored by researchers, many corresponding methods have been proposed. Those methods have three kinds of features. The first feature is an expandable information feature called clustering [5, 6]. However, clustering needs to consider a lot of noisy words, meanwhile wasting time and memory. The second feature is text feature called context similarity [7, 8]. But this method only calculates the number of common words in the two text. The third feature

D.-S. Huang and K. Han (Eds.): ICIC 2015, Part III, LNAI 9227, pp. 471–480, 2015.
DOI: 10.1007/978-3-319-22053-6_50

is topic coherence feature called topic model [9, 10]. It focuses on the topic of the document, and has very strong semantic analysis. In this paper, we focus on the topic model, and develop a semi-supervised topic model.

The rest of the paper is organized as follows: Sect. 2 briefly introduces the related work. The main disambiguation strategies and the development of our semi-supervised topic model are presented in Sect. 3. Section 4 includes experiments and discussion. Finally, Sect. 5 concludes the paper.

2 Related Work

In news articles, the vector space model (VSM) [11] is used to measure similarities between articles [12]. But this method is only suit for a object belongs to one event. Pederson et al. applied the word co-occurrence to calculate word's context vectors [13], and second order context vector [14] is used to regard each document as a context vector. Because word co-occurrence information need collect a large number of documents, this approach is appropriate for people whose name appears in a large number of documents.

Name disambiguation is based on word sense [15]. Many researchers develop name disambiguation based on the context similarity. But this method needs a set of documents, and to-be-tested entities have been annotated manually [16–18]. There are some approaches that never use any knowledge base or ontology [12, 19]. Instead, they just only determine whether these two documents have the same name entity referring to the same instance or not.

The context information of to-be-tested entities and reference entities are available as an input for the disambiguation. Some approaches employ context common words which occur near the to-be-tested entity to disambiguate [16, 18]. Mostly approaches calculate the co-occur words between to-be-tested and candidate entity [17, 20]. The context information includes Wikipedia categories [16, 17], hyperlinks [18] and extra information about the to-be-tested entity.

Recently, topic model is applied to name disambiguation. Because the large annotated corpus unnecessary, name disambiguation apply unsupervised topic model [12, 21]. However, the distribution of "words-topic" and "topic-document" are completely based word co-occurrence of documents in unsupervised topic model. Supervised topic model needs annotate all of the documents, and it is complicated. It is hard for applying supervised model in the name disambiguation.

In this paper, we develop a semi-supervised topic model (STM). It doesn't need annotate all of the documents and have a higher accuracy. Our method partly annotate the Wikipedia articles and the whole news articles. The details of our method are given in the next section.

3 Semi-supervised Topic Models

The standard LDA topic model is unsupervised, and topics and words are completely based on the documents' words co-occurrence. It is likely to be assigned to the same topic, if two documents have many common words. However, the center of the news article is seldom the to-be-tested entity, in other words, unsupervised topic model has trouble in assigning accurate topics for documents. Semi-supervised LDA topic model can solve this problem to some extent.

3.1   Latent Dirichlet Allocation Model

Unlike pLSA [22], the LDA seldom assumes that each document has a certain ratio topic and the rate of each topic word is determined. The LDA is a three-layer Bayesian Model (words, topics, documents), and it is focused on the topic of the potential information in document. It adds the layer "topic" to the "word-document" matrix, then gets the "words-topic" and "topic-document" matrices.

All of documents' words denoted to $Q = \{1, 2, \ldots, q\}$. We assume that a set of documents from the vocabulary Q is denoted to $D = \{d_1, d_2, \ldots, d_i\}$ where $d_i = \{w_1, w_2, \ldots, w_n\}$ denotes the tth documents and contains n words. Let $\overrightarrow{z} = \{1, 2, \ldots, t\}$ to denote the corresponding topics for words, t is t^{th} topic. Therefore, the documentation of the tth word w_i can be describe as follow:

$$P(w_i) = \sum_{t=1}^{T} P(w_i/z_i = t)P(z_i = t) \tag{1}$$

And the probability of word w which is assigned to document d can be described as:

$$P(w/d) = \sum_{t=1}^{T} P(w/z = t)P(z = t) \tag{2}$$

If given the parameter of Dirichlet α and β, the simultaneous distribution of the random variables β, z and w in document d can be described as:

$$P(\theta, z, w/\alpha, \beta) = P(\theta/\alpha) \prod_{i=1}^{W} P(z_i/\theta)P(w_i/z_i, \beta) \tag{3}$$

Because of more than one connotative variables, to some extent, it is difficult in calculating the values directly, and some other solutions are demanded. Gibbs sampling [23] is an inference algorithm. In this paper, we combine LDA with Gibbs sampling to acquire estimated parameter values.

3.2   Using Semi-supervised Topic Models for Entity Disambiguation

The part Wikipedia articles are annotated by its corresponding categories and central entity, and A_1 is to denote it. The whole Wikipedia articles which never contain A_1 is denoted by A, and the whole Wikipedia articles which contain A_1 is denoted by A'. The purpose of annotation is to bias the topic-word distributions $\overrightarrow{\phi}_t$ in favor of words that are frequently annotated with the corresponding entity's topic t in A, and the document-topic distributions $\overrightarrow{\theta}_{dk}$ is in favor of corresponding entities topics that frequently occur in document's annotations in A_1. The news which include to-be-tested entities is denoted by N.

Given the hyper-parameters of Dirichlet $\overrightarrow{\alpha}$ and $\overrightarrow{\beta}$, posterior distribution of $\overrightarrow{\phi}_t$ on A_1 is $P(\overrightarrow{\phi}_t / \overrightarrow{\beta}, A_1)$, and this expression is equivalent to $Dir(\overrightarrow{\phi}_t / \overrightarrow{\beta} + \overrightarrow{\delta}^t)$. δ_q^t denotes to the number of times term q is assigned topic t in A_1. Similarity, δ_t^d is the number of words annotated in document d that are assigned topic t in A_1. The probability that $z_i = t$ on the given observed A_1 regards as:

$$P(z_i = t/z_{-i}, \omega, \alpha, \beta, A_1) \propto \frac{n_{q,-i}^{(t)} + \beta_q + \delta_q^{(t)}}{\sum\limits_{q'=1}^{P} (n_{q',-i}^t + \beta_{q'}) + \delta_{q'}^{(t)}} \cdot \frac{n_{t,-i}^{(d)} + \alpha_t + \delta_t^{(d)}}{\sum\limits_{t'=1}^{T} (n_{t',-i}^{(d)} + \alpha_{t'} + \delta_{t'}^{(m)})} \quad (4)$$

In this paper, part Wikipedia articles A_1 and news N are the training data, and the whole Wikipedia A' and news N are the testing data. In the training phase, we run Gibbs sampling on the collection $A_1 \cup N$, and save the results of the last iteration. Next, in the testing phase, it disambiguates each to-be-tested entity in news by running incremental Gibbs sampling on $A' \cup N$. The results of the last iteration is the final result. In knowledge base, more than one entity has the same name with to-be-tested entity, and we choose the entity which has the same topic with to-be-tested entity to be the result and return KB value; if more than one entity which has the same name and topic with to-be-tested entity, then we judge that result is mistake; if there is without entity having the same name with to-be-tested entity, NIL will be return.

In the process of finding 'KB' value in the knowledge base, this paper isn't mature enough to deal with the case that the knowledge base has more than one entity having same name with the to-be-tested entity, and further investigation will be conducted.

4   Experiments and Discussion

In this section, we discuss that our semi-supervised topic model performs better than baselines on real-life datasets. More specially, we compare STM with entity topic model [9] and wikipedia-based pachinko allocation model [24] on the disambiguation accuracy. We show that STM has a higher precision.

4.1   Data and Preprocessing

In order to construct a semi-supervised model handily, we collect entities and its corresponding categories at the same time. The entity sharing the same category or sub-category is close. The more closer, the more similar entities will be. For example, we get the following character information from Wikipedia. For athlete named Jordan, knowledge base exist the basketball player Jordan and football player Jordan, and we

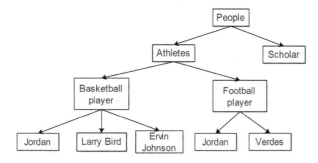

Fig. 1. Category structure

can combine text characters with sub-category to judge which Jordan the text points to. If the text also mentions someone whose name is Larry Bird, it is reliable to infer that the Jordan is a basketball player Jordan. Detail information can be obtained from the hierarchical structure shown in Fig. 1.

The knowledge base has its hierarchy, but the hierarchy contains some spurious categories. These categories can cause incorrectly infer topic correlations, and they are useless for the result of disambiguation. For example, Jay Chou is listed under the categories "1979 births", "Living people". We will manually prune such irrelevant categories.

We have two datasets, the first one is Wikipedia articles (denoted by A), and the second is Sougou News. Wikipedia articles construct the knowledge base, and Sougou news provide disambiguation entities and language environment. The first dataset A is an extract of Wikipedia containing people belonging to four categories - scholars, athletes, singers and actors. Removing the entity which characterization is not obvious, this dataset contains 12720 people, each with an entity page. Because foreigners are infrequent in Chinese news, the Wikipedia dataset is mostly Chinese. The second dataset is a subset of the Sougou news containing a total of 230,000 Sougou news and ranging from the 2012 June to August. In order to reduce the complex of statistics and analysis the experimental result, we select Sougou news 1084 (denoted by N) from 230,000 randomly as an entity disambiguation, and make sure that these 1084 pieces of news occur in A. So we match the entity names that occur together in datasets Wikipedia and Sougou news.

Nouns play an important role in entity disambiguation. We only extract nouns from the web pages and news articles. The tool ICTCLAS[1] is applied for word processing. The input documents of STM are noun terms.

Random sample has its own advantages, such as the Olympic Games in London in August 2012. The news which is during the Olympic Games is mainly related to the athletes, if get the news which is near the Olympic time to be the experimental data, the scope of the entity under test with limitations.

---

[1] http://download.csdn.net/detail/bzbcxwp/311639/#comment.

STM is a disambiguation method proposed in this paper. Wikipedia entity is annotated by central entity and the corresponding categories of central entity, and the annotated Wikipedia is a small part of A. This is the premise of Semi-Supervised learning method. As for the news dataset N, we annotate the news by the to-be-tested entity, and denoted by N'.

4.2 Parameter Setting

The purpose of the experiment is to study the impact of the number of topics for Gibbs sampling algorithm. Therefore, we determine the value of α and β, and select the appropriate values for the T. We select the experience value $\alpha = 50/T$ and $\beta = 0.1$, and select different values of T to run Gibbs sampling algorithm respectively, then detect the change of accuracy.

A proper and rational T helps to allocate entity to the appropriate topic reasonably. Figure 2 is the relationship between topic number T and accuracy of STM.

It is clear from the Fig. 2, when the topic number T is set to 8, STM's greatest precision is $919/1084 = 84.75\,\%$. Therefore, In this paper, the number of topic $T = 8$, and $\alpha = 50/8 = 6.25$.

4.3 Baselines

1. **Entity Topic Model (ETM):** Entity topic model is an unsupervised model. It contains pure text Wikipedia and unmarked news. Entity topic model can uniformly model the text compatibility and the topic coherence can be seen as the statistical dependencies between the mentions, the words, the underlying entities and the underlying topics of a document.
2. **Wikipedia-Based Pachinko Allocation Model (WPAM):** WPAM establishes topic hierarchy based on topic model. It disambiguates entity through the topic of each document on the adjacent topic hierarchy, therefore, it need to create a topic for each document. The more document input, the more topic will be set, and the

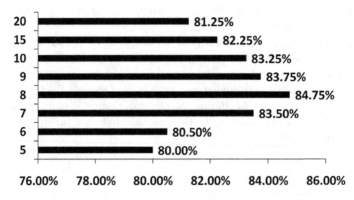

Fig. 2. The number of STM topic

larger CPU memory consume. This experiment needs 32 GB main memory and 8 64-bit 1.87 GHz 4-core Intel Xeon processors each with 4 MB cache.

4.4   Experimental Results and Discussion

(1) Experimental Evaluation

In this paper, we apply the same performance metrics used in the Joel et al. [25], which includes Precision, Recall and F1. We use "In-KB" to denote that the to-be-tested entities is in news N which has a corresponding linking entity in Wikipedia A, and "NIL" to denote that there isn't a corresponding linking entity between A and N. The three evaluation criteria were used to measure In-KB and NIL and we get result values in the Table 1.

Table 1. Evaluation results

Method	Overall correct	In-KB			NIL		
		Precision	Recall	F1	Precision	Recall	F1
STM	**84.75 %**	**88.40 %**	**80.00 %**	**83.99 %**	**81.74 %**	**89.5 %**	**85.44 %**
ETM	82.75 %	90.86 %	73.00 %	80.89 %	77.41 %	92.50 %	84.29 %
WPAM	67.25 %	82.86 %	43.5 %	57.01 %	61.69 %	91.00 %	73.53 %

It is clear from Table 1, the precision of STM is 84.75 %, while ETM and WPAM achieve 82.75 % and 67.25 %, respectively. STM is superior to other methods. The STM's In-KB of F1 is 83.99 %, which is the highest value in the three method. Similarly, STM's NIL of F1 is the highest value, F1 is 85.44 %. Therefore, the method of this paper is optimal.

(2) Experiments on Different Datasets

In addition, we do experiment based on different annotated datasets. A' contains annotated Wikipedia, and A isn't including annotated Wikipedia. N' is Sougou news which is annotated, and N is Sougou news which isn't annotated.

CDA-TE: The datasets both A' and N' are used;
CDA-TD: The datasets both A' and N are used;
CDA-TW: The datasets both A and N' are used;
CDA-TS: The datasets both A and N are used.

The parameter settings and configuration are the same during the experiment on the four datasets. The result is shown in Fig. 3. It is clear from the Fig. 3, The precision of A is 84.75 %, and B is 82.25 %, C is 78.00 %, respectively. Therefore, STM has a highest precision on Wikipedia annotated and news annotated.

(3) Time Cost

In the process of experiment, STM achieves the greatest precision with the least time. The time of STM is 15 min, while ETM and WPAM achieved 60 min and 62 h(3720 minus), respectively. Figure 4 shows the three methods corresponding

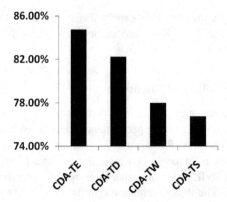

Fig. 3. STM precision based on different annotation dataset

time consumed, and the ordinate represents the time, the unit is "minute", the abscissa represents the type of the method. From Fig. 4 WPAM's unit is "10", the time consumed is $372 * 10 = 3720$ min.

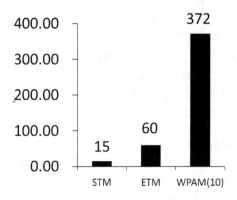

Fig. 4. Time consuming of different models

5   Conclusions

This paper proposes a disambiguation method called Semi-Supervised topic model (STM). It combines topic coherence with text annotated. An optimal number of topics T is set up to analyze the relationship between precision and the number of topics. Comparing with ETM and WPAM, the proposed STM is advantageous. We also evaluate the influences of annotated and non-annotated datasets, respectively.

In this paper, our STM can only link to the previously given entities in a given knowledge base. In the future, we will incorporate data mining techniques into our method to learn the corresponding entities pages dynamically.

Acknowledgments. This work was supported by NSFC (No.61170192) and National College Students' Innovative and Entrepreneurial Training Program (No.201410635029). L. Li is the corresponding author for the paper.

References

1. Li, Y., Wang, C., Han, F., Han, J., Roth D., Yan, X.: Mining evidences for named entity disambiguation. In: Proceedings of the 19th ACM SIGKDD International Conference on Knowledge Discovery and Data Mining, pp. 1070–1078. ACM (2013)
2. Hachey, B., Radford, W., Nothman, J., Honnibal, M., Curran, J.R.: Evaluating entity linking with wikipedia. Artif. Intell. **194**, 130–150 (2013)
3. Wang, F., Tang, J., Li, J., Wang, K.: A constraint-based topic modeling approach for name disambiguation. Front. Comput. Sci. China **4**(1), 100–111 (2010)
4. Peng, H.T., Lu, C.Y., Hsu, W., Ho, J.M.: Disambiguating authors in citations on the web and authorship correlations. Expert Syst. Appl. **39**(12), 10521–10532 (2012)
5. Jun, S., Park, S.S., Jang, D.S.: Document clustering method using dimension reduction and support vector clustering to overcome sparseness. Syst. Appl. **41**(7), 3204–3212 (2014)
6. Kang, I.S., Na, S.H., Lee, S., Jung, H., Kim, P., Sung, W.K., Lee, J.H.: On co-authorship for author disambiguation. Inf. Process. Manag. **45**(1), 84–97 (2009)
7. Hoffart, J., Yosef, M.A., Bordino, I., Furstenau, H., Pinkal, M., Spaniol, M., Taneva, B., Thater, S., Weikum, G.: Robust disambiguation of named entities in text. In: Proceedings of the Conference on Empirical Methods in Natural Language Processing, pp. 782–792. Association for Computational Linguistics (2011)
8. Niu, L., Wu, J., Shi, Y.: Entity disambiguation with textual and connection information. Procedia Comput. Sci. **9**, 1249–1255 (2012)
9. Han, X., Sun, L.: An entity-topic model for entity linking. In: Proceedings of the 2012 Joint Conference on Empirical Methods in Natural Language Processing and Computational Natural Language Learning, pp. 105–115. Association for Computational Lingustics (2012)
10. Sen, P.: Collective context-aware topic models for entity disambiguation. In: Proceedings of the 21st International Conference on World Wide Web, pp. 729–738. ACM (2012)
11. Baeza-Yates, R., Ribeiro-Neto, B., et al.: Modern Information Retrieval, vol. 463. ACM Press, New York (1999)
12. Bagga, A., Baldwin, B.: Entity-based cross-document coreferencing using the vector space model. In: Proceedings of the 36th Annual Meeting of the Association for Computational Linguistics and 17th International Conference on Computational Linguistics, vol. 1. pp. 79–85. Association for Computational Linguistics (1998)
13. Pedersen, T., Purandare, A., Kulkarni, A.: Name discrimination by clustering similar contexts. In: Gelbukh, A. (ed.) CICLing 2005. LNCS, vol. 3406, pp. 226–237. Springer, Heidelberg (2005)
14. Schutze, H.: Automatic word sense discrimination. Comput. Linguist. **24**(1), 97–123 (1998)
15. Fernandez-Amoros, D., Heradio, R.: Understanding the role of conceptual relations in word sense disambiguation. Expert Syst. Appl. **38**(8), 9506–9516 (2011)
16. Bunescu, R.C., Pasca, M.: Using encyclopedic knowledge for named entity disambiguation. EACL **6**, 9–16 (2006)
17. Cucerzan, S.: Large-scale named entity disambiguation based on wikipedia data. EMNLP-CoNLL **7**, 708–716 (2007)

18. Dredze, M., McNamee, P., Rao, D., Gerber, A., Finin, T.: Entity disambiguation for knowledge base population. In: Proceedings of the 23rd International Conference on Computational Linguistics, pp. 277–285. Association for Computational Linguistics (2010)
19. Chen, Y., Martin, J.: Towards robust unsupervised personal name disambiguation. In: EMNLP-CoNLL, pp. 190–198. Citeseer (2007)
20. Nguyen, H.T., Cao, T.H.: A knowledge-based approach to named entity disambiguation in news articles. In: Orgun, M.A., Thornton, J. (eds.) AI 2007. LNCS (LNAI), vol. 4830, pp. 619–624. Springer, Heidelberg (2007)
21. Bhattacharya, I., Getoor, L.: A latent dirichlet model for unsupervised entity resolution. In: SDM, vol. 5, p. 59. SIAM (2006)
22. Lu, Y., Mei, Q., Zhai, C.: Investigating task performance of probabilistic topic models: an empirical study of PLSA and LDA. Inf. Retrieval **14**(2), 178–203 (2011)
23. Heinrich, G.: Parameter estimation for text analysis. Technical report (2005)
24. Kataria, S.S., Kumar, K.S., Rastogi, R.R., Sen, P., Sengamedu, S.H.: Entity disambiguation with hierarchical topic models. In: Proceedings of the 17th ACM SIGKDD International Conference on Knowledge Discovery and Data Mining, pp. 1037–1045. ACM (2011)
25. Nothman, J., Ringland, N., Radford, W., Murphy, T., Curran, J.R.: Learning multilingual named entity recognition from wikipedia. Artif. Intell. **194**, 151–175 (2013)

Automatic Evaluation of Machine Translation Through the Residual Analysis

Daša Munková$^{(\boxtimes)}$ and Michal Munk

Constantine the Philosopher University in Nitra,
Tr. A. Hlinku 1, 949 74 Nitra, Slovakia
{dmunkova, mmunk}@ukf.sk

Abstract. In this study we aim at the automatic evaluation of machine translation through the residual analysis at the sentence level. We created a dataset, which covered one translation direction- a translation from an inflective language (Slovak) into an analytical language (English). BLEU (Bilingual Evaluation Understudy) as a state-of-the-art automatic metric for machine translation evaluation was used. The main contribution consists of rigorous technique (statistical method), novel to research of MT evaluation given by the residual analysis to identify differences between MT output and post-edited machine translation output.

Keywords: Machine translation · MT evaluation · BLEU metric · Regression model · Residual analysis

1 Introduction

Quality of machine translation (MT) system can be evaluated with regard to different aspects, such as feasibility, internal, declarative, usability, operational or comparison [1]. The most common way how to evaluate MT system is through the translation quality. Translation quality assessment plays a crucial task in the field of machine translation. There are two major approaches to MT evaluation: human and automate evaluation (both with a wide range of proposed methodologies). However the manual evaluation is regarded as the most reliable, there are issues which it cannot deal with (e.g. highly time consuming; labor consuming; expensive; not re-usable; trained bilingual evaluators are required) [2]. Papineni et al. [3] stated that methods and metrics of manual evaluation are too slow and time consuming for the development of machine translation systems, for which fast feedback on translation quality is extremely important. Vilar et al. [4] pointed out that subjectivity, which is characterized for manual evaluation causes a problem in terms of biased judgments towards machine translation as well as having no clear definition of a numerical scale by TQA (no clear guidelines on how to assign values to translations). Several methodologies and metrics were proposed on how to assess translation quality in order to decrease "time and labor" spent during evaluation. These are mostly based on the similarity (matches) between translation that is assessing (hypothesis) and human translation(s) (reference(s)). They compare the output of MT system against reference(s) where the causes and the nature of the differences between them are not identified. The advantages of automatic metrics comprise speech, cost and usability,

D.-S. Huang and K. Han (Eds.): ICIC 2015, Part III, LNAI 9227, pp. 481–490, 2015.
DOI: 10.1007/978-3-319-22053-6_51

i.e. they are fast, cheap and re-usable. Metrics of automatic evaluation correlate well with human judgements [3, 5–8]; whereby human judgements are in the form of "adequacy and fluency" quantitative scores [2].

This study demonstrates how residual analysis of the metrics of machine translation can serve as a starting point for discovering significant differences between MT output and post-edited MT output [9], which can help to identify major MT errors type. The residual analysis presented here is a pilot study conducted on source text (informative text from the field of machine translation) translated by free web machine translation service and its comparison with post-edited MT output by translators. It covers one translation direction, from inflective language to analytical language (from Slovak into English). One automatic translation (360 sentences) and 14 post-edited machine translation (5040 sentences), were collected.

Our pilot study proposes an exploratory data technique representing an ideal instrument to evaluate and improve MT systems. The main contribution consists of rigorous technique (statistical method), novel to research of MT evaluation given by the residual analysis to identify differences between MT output and post-edited machine translation output. In addition, this identification can help us to detect the major types of MT errors.

2   Related Work

Statistical machine translation system has been the most widely used for many recent surveys. Specifically, Google Translate (GT), being a free web translation service, is included in almost every research on MT evaluation [10], because it offers translation from/into less widely spoken languages such as Slovak language. GT is also only one online translation service for Slovak language. Dis Brandt [11] in his work presents evaluation of web MT system from Icelandic into English.

Babych et al. [12] investigated the performance of translation from an under-resourced language into English via a closely-related, or cognate, pivot language with well-developed translation resource. Adly and Al Ansary [13] evaluated a machine translation (MT) system based on the interlingua approach, the Universal Network Language (UNL) system, designed for Multilanguage translation.

We propose a statistical analysis framework based on Generalized Linear/Nonlinear Models, which have been applied to several natural language problems (sentiment analysis, automatic speech recognition or spoken language translation).

3   Exploratory Data Technique and Residual Analysis

By assessing individual sentences using metrics for automatic evaluation, we were inspired by exploratory data technique – analysis of residual values [14, 15].

Residual analysis issues from the basic imagination:

$$data = model\ prediction\ function + residual\ value. \tag{1}$$

If we subtract the values (expected values) obtaining from the model from observed data, we capture errors (residual values) and through the data analysis we are able to assess a constructed model.

This type of analysis serves as a means to verify the model validity and its improvement [16], because it helps to detect the aspects of relationships not considered by model.

We used this technique to compare the scores of automatic evaluation of post-edited MT outputs with MT outputs at the sentence level.

In our case the analysis composes of residual analysis defined as follow:

$$(residual\ value)_i = (score\ of\ PE\ sentence)_i - (score\ of\ MT\ sentence)_i, \qquad (2)$$

$i = 1, 2, \ldots, I$ while I is a number of examined sentences in the dataset and PE means post-edited.

We used a rule $\pm 2\sigma$ to identify the extreme values:

$$mean\ of\ residuals\ (PE - MT) \pm 2 \cdot st.dev.of\ residuals\ (PE - MT), \qquad (3)$$

i.e. the residual values outside the interval are considered as extreme values.

By aggregating the scores of automatic metrics by weighted average we created one variable, which represents the automatic metric BLEU in terms of correctness.

4 Method

4.1 Dataset

For our study we used unstructured textual data, particularly a popular-scientific text consisting of 360 sentences from the field of machine translation. We create a dataset, which covers one translation direction- a translation from an inflective language (Slovak) into an analytical language (English). We chose this direction only for one reason, which resulted from our pre-research results. When we translated from foreign language into our mother tongue, we obtain lower scores of automatic metric BLEU. If we wanted to achieve a higher score for automatic metric BLEU we had to examine vice versa translation, i.e. from mother tongue (inflective language) to foreign language-English (BLEU metric is not suitable for inflective languages, e.g. one English noun has 6 forms in Slovak differing only in suffixes and also I comparison to English, Slovak language has a loose word order). These 360 sentences written in Slovak were translated by a statistical machine translation system and consequently post-edited by two professional translators and 12 students of Translation Studies in the master degree. After machine translation and post-editing our dataset composes of 360 machine translated sentences and 5040 post-edited sentences.

4.2   Metrics

In our study two state-of-the-art automatic metrics for machine translation evaluation were used, namely BLEU (Bilingual Evaluation Understudy) and TER (Translation Edit Rate). In this paper we will focus only on an automatic metric BLUE, which is from the first generation of metrics for MT evaluation. It is the most common and most widely used metric. It is used as a standard metric for the evaluation of machine translation during the WMT workshops and others. The standard BLEU score [3] is only meaningful at the corpus level and we investigated a text at the sentence level, we applied smoothed BLEU [14]. BLEU is the geometric mean of n-gram precisions that is scaled by a brevity penalty to prevent very short sentences as a compensation for inappropriately translation. This metric gives a new dimension of MT evaluation – statistical, which arises from the idea that translation quality (quality of hypothesis) can be estimated based on its similarity with reference translation. In other words, translation quality is assessed based on the occurrence of same n-grams various lengths in the hypothesis as well in reference.

Formula for BLEU computation (Corpus-based n-gram precision):

$$BLEU(n) = BP * exp \sum_{n=1}^{N} w_n \cdot \log p_n, \text{where} \tag{4}$$

$$BP = brevity\ penalty = \begin{cases} 1, & if\ hypothesis > reference \\ e^{1 - \frac{reference}{hypothesis}}, & if\ hypothesis \leq reference \end{cases} \text{and} \tag{5}$$

$$p_n = precision_n = \frac{\sum_{S \in C} \sum_{n-gram \in S} count_{matched}(n - gram)}{\sum_{S \in C} \sum_{n-gram \in S} count(n - gram)}. \tag{6}$$

Remark 1. S means hypothesis sentence in the complete corpus C.

The BLEU metric reflects two aspects of human evaluation- adequacy (semantically correct words) and fluency (word order, i.e. well-formed sentence construction). Lin and Och [17] or Papineni et al. [3] proved that shorter n-grams correlates better with adequacy with 1-gram being the best predictor, while longer n-grams has better fluency correlation.

BLUE+1 metric is a smoothed version of BLUE metric. Number 1 is added to each of the n-gram counts before the n-gram precisions will be calculated. It is strictly positive (always positive number). It redefines precision, while the brevity penalty is unchanged.

$$p_n = precision_n = \frac{\sum_{S \in C} \sum_{n-gram \in S} count_{matched}(n - gram) + 1}{\sum_{S \in C} \sum_{n-gram \in S} count(n - gram) + 1}. \tag{7}$$

We compute automatic scores by relaying on one reference. Reference was created by two translators and one native speaker, whose mother tongue is English but also speaks Slovak. We used our own means to align and compute the automatic scores, in this case the metric BLEU. Further we use automatic metrics as variables for residual analysis at the sentence level.

4.3   Residual Analysis Results

The aim of this analysis was to identify sentences, in which the significant differences in the scores of automatic metrics between MT output and post-edited MT output from Slovak into English were found.

As we mentioned before, we will describe only one automatic metric smoothed BLEU-n, i.e. BLEU-1, BLEU-2, BLEU-3 and BLEU-4.

The graphs (Figs. 1, 2, 3, 4 and 5) depict the scores of automatic metrics BLEU-1 to BLEU-4 for post-edited MT output and MT output from inflective language into analytical language and their residual values.

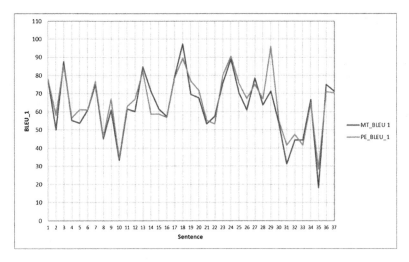

Fig. 1. Scores of automatic metric BLEU-1 for MT outputs and for post-edited MT outputs (only first 37 sentences); Score of automatic metric BLEU-1(y-axis), ID Sentence (x-axis) (Color figure online).

The residual values above the residual mean predicate of extraordinary assessment of post-edited MT output using the automatic metric in comparison to MT output. On the contrary, residual values below the residual mean predicate of substandard assessment of post-edited MT output using the automatic metric in comparison to MT output.

Identification of extreme values helps us to detect sentences, in which the significant differences between machine translation and post-edited machine translation were found.

Concerning the number of examined sentences (360 sentences for evaluation), we introduce only first 37 sentences. As we can see from the Fig. 1, some sentences translated by MT system are more close to reference translation as post-edited ones (e.g. sentences 18 or 14, green color). On the other hand, post-edited sentences like sentences 9, 26 or 29 achieved higher scores of automatic metric than sentences translated by MT system.

Regarding the graphs presenting in this paper (Figs. 2, 3, 4, and 5), post-edited sentence 29 was over average assessed by automatic metrics BLEU-1, BLEU-2,

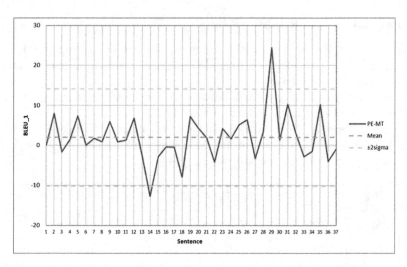

Fig. 2. Results of Residual data analysis of the BLEU-1 scores for both translation (MT and Post-edited MT); Mean differences, ±2sigma - lower and upper bound of extreme differences, ID Sentence

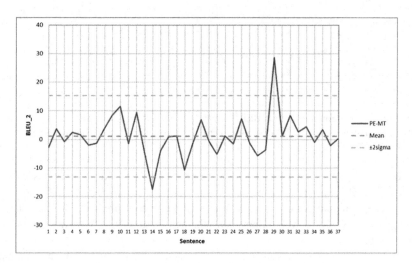

Fig. 3. Results of Residual data analysis of the BLEU-2 scores for both translation (MT and Post-edited MT); Mean differences, ±2sigma - lower and upper bound of extreme differences, ID Sentence.

BLEU-3 and BLEU-4. Besides, based on the score of automatic metrics BLEU-1 post-edited sentences 31 and 35 were over average assessed in the terms of correctness (Fig. 2).

Further based on the automatic metric BLEU-2, post-edited sentence 10 was over average assessed in the terms of correctness (Fig. 3). Similarly, post-edited sentence 25 was over average assessed by the automatic metrics BLEU-2 and BLEU-3.

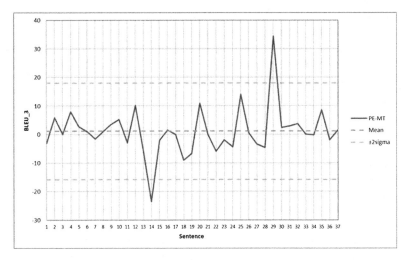

Fig. 4. Results of Residual data analysis of the BLEU-3 scores for both translation (MT and Post-edited MT); Mean differences, ±2sigma - lower and upper bound of extreme differences, ID Sentence.

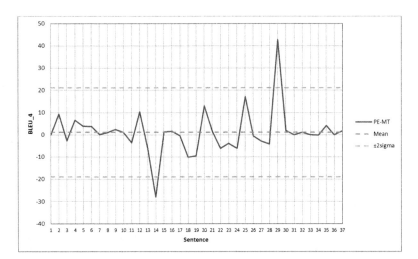

Fig. 5. Results of Residual data analysis of the BLEU-4 scores for both translation (MT and Post-edited MT); Mean differences, ±2sigma - lower and upper bound of extreme differences, ID Sentence.

On the contrary, based on the scores of automatic metrics BLEU-1, BLEU-2, BLEU-3 and BLEU-4, post-edited sentences 14 and 18 were below average assessed in the terms of correctness, i.e. these sentences were more similar to references when they were translate by machine than post-edited. Likewise, the post-edited sentence 19 in respect of the score of automatic metric BLEU-3.

The significant difference was reported in sentence 29 in the scores of automatic metrics BLEU-1, BLEU-2, BLEU-3 and BLEU-4 between post-editing and machine translation in favor of post-edited machine translation (Figs. 2, 3, 4, and 5). On the other hand, significant difference was found in sentence 14 (Figs. 2, 3, 4, and 5) in the scores of automatic metrics BLEU-1, BLEU-2, BLEU-3 and BLEU-4 between MT output and post-edited MT output in favor of MT output. In other words, MT output was more similar to reference than post-edited MT output.

Since the focus of our work is the residual analysis technique rather than the identification of errors occurred in MT output. A closer looked at the reported extreme sentences in the whole dataset we could identify several relevant MT errors. The most frequent MT errors, according to the general error classification and partially overlapping with Vilar's error classification, compose of syntactical errors (agreement and clauses), lexical errors (omitted, mistranslated, untranslated and substituted words), morphological errors (word order, anaphora, articles, passive, number and tense) and punctuation. Errors from several different sub-categories often appear in the same extreme sentence such as morphological errors- word order and articles together with lexical errors like mistranslation or omission. For example: *With the erosion of Use multiple languages is also related to some loss of culture, identity of man and the total change in his thinking and communicating with the surroundings (Ref.: A decrease in the use of various languages is also related to some loss of culture, individual identity and to the total change in individual thinking and communication with the surrounding environment.) or The main reason for use being a quick translation of important information, especially "merits", respectively communication are for ordinary users (Ref.: The main reason to use machine translation is a fast translation of the important information, to get "the gist" of information or communication for ordinary users.).*

Most of these errors come from the differences of both languages. English as an analytical language has a strict word order, on the contrary Slovak as an inflective language has a loose word order, which resulting in omitting the agent or in reordering words within the clause or exchange subject with subject in the sentence structure, which is not possible in English sentence structure.

5 Conclusions

We described another approach how to analyze the performance of machine translation system through the exploratory data technique. In our research we focused on automatic evaluation of machine translation system (online available statistical machine translation service, i.e. Google Translate, for Slovak- English language pair). Based on the reference and automatic metric BLEU we compared machine translation output with post-edited MT output at the sentence level. We investigated the similarity between MT output and post-edited MT output based on the exploratory data technique, namely the residual analysis. In other words, we investigated relations among three variables: reference, machine translation output and post-edited MT output using the metrics of automatic MT evaluation and residual analysis. Based on the scores of automatic metric BLEU and the residual analysis we found the significant differences between MT output and post-edited MT output at the sentence level (Figs. 2, 3, 4 and 5). In some cases the significant difference was in favor of MT output (e.g. sentence 14) and in other cases in favor of post-edited MT output (e.g. sentence 29). This method is very useful if you do not want to make error analysis of the whole MT output which is not only time-consuming but also expensive (it requires a human interaction). Based on the residual analysis we can only analyze the sentence with extremes and then identify the major MT errors like we did. Our method represents an original approach how to evaluate MT system. If improving the system involves the crucial step as the error analysis of MT output, our empirical findings represent a significant contribution in the field of system evaluation and its improvement.

Acknowledgments. This work was supported by the Slovak Research and Development Agency under the contract No. APVV-0451-10 and Scientific Grant Agency of the Ministry of Education of the Slovak Republic (ME SR) and of Slovak Academy of Sciences (SAS) under the contract No. VEGA-1/0559/14.

References

1. White, J.S.: How to evaluate machine translation. In: Somers, H. (ed.) Computers and Translation: a Translator's Guide, pp. 211–244. J. Benjamins, Amsterdam (2003)
2. Munková, D., Munk, M.: An automatic evaluation of machine translation and Slavic languages. In: Proceedings of the 8th International Conference on Application of Information and Communication Technologies (AICT-2014), Astana, pp. 447–451 (2014)
3. Papineni, K., Roukos, S., Ward, T., Zhu, W.: BLEU: a method for automatic evaluation of machine translation. In: Proceedings of the 40th Annual meeting of the Association for Computational Linguistics (ACL-02), Philadelphia, pp. 311–318 (2002)
4. Vilar, D., Xu, J., D'Haro, L.F., Ney, H.: Error analysis of statistical machine translation output. In: Proceedings of the 5th International Conference on Language Resources and Evaluation (LREC-2006), Genoa, pp. 697–702 (2006)
5. Banerjee, S., Lavie, A.: METEOR: an automatic metric for MT evaluation with improved correlation with human judgments. In: Proceedings of the ACL Workshop on Intrinsic and Extrinsic Evaluation Measures for MT and/or Summarization (ACL-05), Michigan, pp. 65–72 (2005)
6. Doddington, G.: Automatic evaluation of machine translation quality using n-gram co-occurrence statistics. In: Proceedings of the 2nd International Conference on Human Language Technology Research (HLT-2002), pp. 138–145 (2002)
7. Koehn, P.: Statistical Machine Translation. Cambridge University Press, Cambridge (2010)
8. Snover, M., Dorr, B., Schwartz, R., Micciulla, L., Makhoul, J.: A study of translation edit rate with targeted human annotation. In: Proceedings of Association for Machine Translation in the Americas (ASMTA-06), pp. 223–231 (2006)
9. Kapusta, J., Munk, M., Drlík, M.: Analysis of differences between expected and observed probability of accesses to web pages. In: Hwang, D., Jung, J.J., Nguyen, N.-T. (eds.) ICCCI 2014. LNCS, vol. 8733, pp. 673–683. Springer, Heidelberg (2014)
10. Seljan, S., Brkić, M., Kučiš, V.: Evaluation of free online machine translations for Croatian-English and English-Croatian language pairs. In: Proceedings of the 3rd International Conference on the Future of Information Sciences: INFuture 2011-Information Sciences and e-Society, Zagreb, pp. 331–345 (2011)

11. Dis Brandt, M.: Developing Icelandic to English shallow transfer machine translation system. Ms. thesis. Reykjavik University (2011)
12. Babych, B., Hartley, A., Sharoff, S.: Translating from under-resourced languages: comparing direct transfer against pivot translation. In: Proceedings of the MT Summit XI, Copenhagen (2007)
13. Adly, N., Al Ansary, S.: Evaluation of Arabic machine translation system based on the universal networking language. In: Horacek, H., Métais, E., Muñoz, R., Wolska, M. (eds.) NLDB 2009. LNCS, vol. 5723, pp. 243–257. Springer, Heidelberg (2010)
14. Munk, M., Drlík, M.: Analysis of stakeholders' behaviour depending on time in virtual learning environment. Appl. Math. Inf. Sci. 8(2), 773–785 (2014)
15. Munk, M., Drlik, M., Vrábelová, M.: Probability modelling of accesses to the course activities in the web-based educational system. In: Murgante, B., Gervasi, O., Iglesias, A., Taniar, D., Apduhan, B.O. (eds.) ICCSA 2011, Part V. LNCS, vol. 6786, pp. 485–499. Springer, Heidelberg (2011)
16. Pilkova, A., Volna, J., Papula, J., Holienka, M.: The influence of intellectual capital on firm performance among slovak SMEs. In: Proceedings of the 10th International Conference on Intellectual Capital, Knowledge Management and Organisational Learning (ICICKM-2013), Washington, DC, pp. 329–338 (2013)
17. Lin, Ch.Y., Och, F.J.: Automatic evaluation of machine translation quality using longest common subsequence and skip-bigram statistics. In: Proceedings of the 42nd Meeting of the Association for Computational Linguistics (ACL-2004), Barcelona, pp. 605–612 (2004)

Is the Most Frequent Sense of a Word Better Connected in a Semantic Network?

Hiram Calvo[(✉)] and Alexander Gelbukh

Centro de Investigación en Computación, Instituto Politécnico Nacional,
Av. Juan de Dios Bátiz s/n, esq. Av. Mendizábal, 07738 Mexico, D.F., Mexico
hcalvo@cic.ipn.mx, gelbukh@gelbukh.com
http://www.gelbukh.com

Abstract. In this paper we show several experiments motivated by the hypothesis that counting the number of relationships each synset has in WordNet 2.0 is related to the senses that are the most frequent (MFS), because MFS usually has a longer gloss, more examples of usage, more relationships with other words (synonyms, hyponyms), etc. We present a comparison of finding the MFS through the relationships in a semantic network (WordNet) versus measuring only the number of characters, words and other features in the gloss of each sense. We found that counting only inbound relationships is different to counting both inbound and outbound relationships, and that second order relationships are not so helpful, despite restricting them to be of the same kind. We analyze the contribution of each different kind of relationship in a synset; and finally, we present an analysis of the different cases where our algorithm is able to find the correct sense in SemCor, being different from the MFS listed in WordNet.

1 Introduction

Word sense disambiguation is required in several natural language applications such as text mining, information retrieval or question answering. For solving this task, several approaches have been proposed. One of the most widely used algorithms is the Lesk algorithm (Lesk 1986), which states that the sense of a word can be selected by calculating the intersection—the number of overlapping words—of its gloss with all the other glosses from other words' senses. For example, if we would like to disambiguate the senses of *pine* and *cone* appearing together in a sentence, we would need to look at each word's glosses:

PINE

1. kinds of evergreen tree with needle-shaped leaves
2. waste away through sorrow or illness

We thank the support of Instituto Politécnico Nacional, and the Mexican Government (CONACyT-SNI, SIP-IPN, COFAA-IPN, and BEIFI-IPN).

© Springer International Publishing Switzerland 2015
D.-S. Huang and K. Han (Eds.): ICIC 2015, Part III, LNAI 9227, pp. 491–499, 2015.
DOI: 10.1007/978-3-319-22053-6_52

CONE

1. solid body which narrows to a point
2. something of this shape whether solid or hollow
3. fruit of certain evergreen trees

Then we select the correct sense of each other by selecting the greatest overlap, which is the first sense of *pine*, and the third sense of *cone*.

Pine#1 ∩ Cone#1 = 0
Pine#2 ∩ Cone#1 = 0
Pine#1 ∩ Cone#2 = 1
Pine#2 ∩ Cone#2 = 0
Pine#1 ∩ Cone#3 = 2
Pine#2 ∩ Cone#3 = 0

Because this algorithm depends on the number of words that intersect amongst each gloss, we can see that the longer the gloss, the higher is the probability for this gloss to have intersections. In fact, we could estimate that a frequent sense is prone to be selected by the Lesk algorithm, given that it will yield a higher number of intersections. Previous works (Calvo and Gelbukh 2014) have shown that indeed, counting the number of characters in a gloss approximates the Most Frequent Sense (MFS) better than a random baseline, and getting close to complex methods requiring a specific corpus for estimating the MFS based on a distributional thesaurus (McCarthy *et al.* 2007). However, to our knowledge, the number and kind of relationships of a synset has ben scantly explored for approximating the MFS given a semantic network such as WordNet.

The aim of this paper is to explore the influence of the amount of information available for each synset in WordNet for finding the Most Frequent Sense (MFS).

2 The Most Frequent Sense

The Most Frequent Sense of a word is calculated in different ways. WordNet itself includes a frequency count for each one of the senses of a word. As reported by Hawker and Honnibal (2006), the sense ranks in WordNet are derived from semantic concordance texts used in the construction of the database. Most senses have explicit counts listed in the database, although sometimes the counts will be reported as 0. In these cases, the senses are presumably ranked by the lexicographer's intuition.

Another important resource used for counting the frequency of senses are sense-tagged texts such as **SemCor** (SEMantic COncoRdance). The SemCor corpus (Miller *et al.* 1994), contains approximately 700,000 English words, each one tagged with a WordNet concept (and thus, a particular sense for a word). In SemCor, all words are grammatically tagged, and more than 200,000 are lemmatized and sense-tagged with the WordNet 1.6 senses inventory. With this resource, it is also possible to calculate the MFS for several words. Usually the frequency counts of senses in WordNet are higher than the frequency of the sense in the SemCor sense-tagged corpus (Miller *et al.* 1993), although not always.

An interesting way of calculating the MFS is proposed by Dianna McCarthy *et al.* (2007). They propose a method for obtaining the most frequent sense using raw text as source of information. Their method consists on two stages: (a) a Lin Thesaurus (Lin 1998) is queried to obtain a list of weighted terms related with the ambiguous word. This list is static; this means that it is always the same for each instance of the ambiguous word, no matter its context; then (b) a maximization algorithm allows each one of the elements of the list to vote for a sense of the ambiguous word, so that the sense with the greatest number of votes is chosen as the predominant sense.

The algorithm of McCarthy *et al.* (2007), consisting on automatically finding the MFSs given a fixed corpus from which a distributional thesaurus is built, yields 48 % for the polysemous words in SemCor.

3   Determining the MFS by Gloss Length

For determining the MFS by the number of relationships it has in WordNet, we initially tried two methods. The first one consisted in counting the number of relationships a word has, and the second one consisted in counting the number of relationships plus the number of relationships of each word related to the original word; that is, *second order relationship count*.

We compare our results with previous work (Calvo and Gelbukh 2014), consisting in considering the longest definition by counting the number of characters and by counting the number of words. Then we expanded the first variant, *measuring by longest char definition* adding the strings of related glosses. The *different chars* feature involved counting chars as types, that is, counting only the variety of characters present in the word, and its related chars itself. Finally, in addition to this, we experimented with removing stop-words from related glosses, and counting the different characters present in them. Another option we tried with the first variant was to remove everything that was not an alphabetic character. We explored independently suppressing stop-words (without adding related glosses or counting chars as types).

For the second variant, *counting by words*, we tried no-preprocessing, considering only different words, *i.e.*, different word types only; removing non-alphabetic characters and word types again, suppressing stop-words and then when adding words from related glosses. We tried different combinations, such as removing all non-alphabetic characters + suppressing stop-words, the same + counting number of types, and finally, this latter + adding words from related glosses.

3.1   Preliminary Results with SemCor

The SemCor test corpus has 88,143 instances (tokens), from which 70,920 are polysemous nouns, and 17, 223 are not polysemous. We will only work with the subset of 70,920 instances of polysemous nouns; that is, 5 577 types of polysemous nouns. Results are shown in Table 1.

First, we calculated the most frequent sense from SemCor itself. This will be the upper bound, as we are trying to disambiguate the entire corpus by selecting always the same sense for each word. This yielded 74.14 %.

Table 1. Results of the longest definition algorithm on SemCor. Methods are additive, *i.e.* adding related glosses + different chars + suppressing stop-words yields 40.59 %

	Method	Prec. (%)
	MFS (SemCor)	74.14
	WordNet first sense	60.45
	Random sense	22.13
	McCarthy *et. al.* (2007)	49.20
	Longest definition measured by number of characters is MFS	
	no-preprocessing	25.72
	adding related glosses	**40.60**
	different chars	**40.57**
	suppressing stop-words	**40.59**
	removing non-alphabetic	25.78
	suppressing stop-words	26.14
	multiplied by the number of relations	**39.68**
	removing non-alphabetic	26.01
	Longest definition counting by words is MFS	
	no-preprocessing	26.22
	different words (number of types)	25.98
	removing non-alphabetic	26.13
	different words (number of types)	26.64
	suppressing stop-words	27.48
	different words (number of types)	28.14
	adding related glosses	**40.58**
	removing non-alphabetic and suppressing stop-words	27.60
	different words (number of types)	27.89
	adding related glosses	**40.45**
Number of relationships	First order	**42.48**
	Second order	30.27

If we always select the MFS provided by WordNet, we obtain a precision of 60.45 %. In contrast, our lower baseline will be given by choosing a random sense (22.13 %). The results of McCarthy *et al.* (2007), are shown afterwards.

Our longest definition algorithm was split in three main variants, one considering the longest definition by counting the number of characters; another by counting the number of words; and lastly, counting the number of relationships.

The first variant, **measuring by longest char definition**, yielded 25.72 % precision with no pre-processing. When adding the strings of related glosses, the results increased above 40 %. The *different chars* feature involved counting chars as types, that is, counting only the variety of characters present in the word, and its related chars itself. Finally, in addition to this, we experimented with removing stop-words from related glosses, and counting the different characters present in them.

Another option we tried with the first variant was to remove everything that was not an alphabetic character. We can see very little difference with previous results. We tried independently suppressing stop-words (without adding related glosses or counting chars as types), observing again that when we involved counting each word's relationships, results improved.

For the second variant, **counting by words**, we tried no-preprocessing, considering only different words, that is, different word types only; removing non-alphabetic characters and word types again, suppressing stop-words, with low results (26–28 % approximately), except when adding words from related glosses (40.58 %). The same happened when we tried different combinations, such as removing all non-alphabetic characters + suppressing stop-words (27.60 %), the same + counting number of types, and finally, this latter + adding words from related glosses. Again, adding related glosses resulted in an increase of more than 12 %.

Finally, for **number of relationships**, we tried two methods. The first one consisted in counting the number of relationships a synset has, and the second one consisted in counting the number of relationships plus the number of relationships of each synset related to the original synset; that is, *second order* relationship count. We obtained 42.48 % and 30.97 % for each one, respectively; apparently second order relationships are not being helpful for finding the MFS.

It can be seen from Table 1 that all the best results (highlighted in bold) involve somehow counting the number of relationships a word has. For example, *adding related glosses* adds the glosses from all synsets related to the original synset, no matter if we count the number of different chars, or if we supress stop-words, results are more or less the same. The same for *supressing stop-words in measuring the longest definition by number of characters*, results change significantly if we multiply this value by the number of relations. The same happens if we select the *longest definition counting by words* as the MFS. Whenever we add related glosses, results get improved, no matter if we *supress stop-words,* or count unique words (*number of types*). Lastly, the best results were obtained when the considering the number of relationships as the MFS. However, up to this point, we have only counted the number of relationships but we have not explored considering the kind of relationships, and if they are *inbound* or *outbound*.

4   Determining the MFS by Its Relationships

In order to determine the MFS by considering its relationship, we first explored in more detail restricting second order relationships to be of the same kind that those who generated them, *i.e.*, we count only hypernyms of the hypernym of a synset, meronyms of the meronym of a synset, etc. Results are shown in Table 2.

Table 2. Results of comparing the MFS vs. SemCor counting the number of relationships, and restricting them to be of the same kind.

Number of relationships	First order	**42.48**
	Second order	30.97
	Second order of the same kind	40.00

Usually, we expect a word to have exactly one hypernym, so that removing them should help to improve results. Table 3 shows results removing hypernyms along with the restrictions of using second order relationships of the same kind. We can see that removing hypernyms for second order relationships, with or without the restriction of counting relationships of the same kind, has a slight positive improvement versus the original results which do not remove hypernyms.

Table 3. Counting both inbound and outbound relationships

Number of relationships				
	First order			42.48
		Bidirectional		**42.81**
			Removing hypernyms	42.61
	Second order			30.97
		Of the same kind		40.00
			Removing hypernyms	40.16
		Removing hypernyms		35.15

So far we have considered so far the number of relationships a synset has *outbound*. This is, we are not counting the number of times a synset is referenced as related by other synset. We call these latter relationships *inbound*. If we count the number of relationships a synset has, both *inbound* and *outbound*, we say we are considering *bidirectional* relationships. See results in Table 3. This represents a slight improvement to the previous best achievement of 42.48, suggesting that we should consider *inbound* relationships as well. Interestingly, in this case removing hypernyms will not improve results. The reason for this will be explained subsequently.

Until now, we have been counting the number of relationships without considering each one separately. In Table 4 we present results for considering separately the number of synonyms, hypernyms, and hyponyms. We can see from this table, that counting the number of different words a synset has (synonyms) has little effect on finding the MFS, while considering the number of hypernyms alone has better results than considering all relationships together (see Table 3). But the best result so far, can be obtained by counting only the number of hyponyms a synset has. The values reported in Table 4 are for outbound relations only.[1] We expected all synsets to have one hypernym, but there are several cases where they have more than one. For example, see Fig. 1, where a word has more than one hypernym in one of its synsets. Moreover, this hypernym is not one of the first ones, meaning it is not close to the MFS, if we consider the first synset of a word as its MFS.

Let us remember that in SemCor, 60.45 % of words are correctly disambiguated if we select always the first sense listed in WordNet, *i.e.*, considering the first synset for a word as the MFS. By counting hypernyms, we found 3 278 cases where the selected

---

[1] Bidirectional count of hyponyms yields 43.20 %, being lower as the unidirectional count shown in Table 4.

Table 4. Results of separately counting relationships by kind

Longest definition counting by words is MFS	
No-preprocessing	26.22
Number of synonyms	28.91
Number of hypernyms	43.36
Number of hyponyms	**44.38**

---

1) day.n:14297391 14266428 14305860 14299333 14279405 14387827
13661216 14347283 14347076 10213836

   14297391n: 1 hypernym   14387827n: 1 hypernym
   14266428n: 1 hypernym   13661216n: 1 hypernym
   14305860n: 1 hypernym   14347283n: 1 hypernym
   14299333n: 1 hypernym   **14347076n: 2 hypernyms**
   14279405n: 1 hypernym   10213836n: 1 hypernym

Fig. 1. A word with 2 hypernyms in one of its synsets.

Table 5. Cases where our strategy of counting the number of hyponyms match the first sense listed in WordNet, and whether it was correct or not in SemCor

		Found first sense	Correct?
4.65 %	3,278	No	Yes
32.52 %	22,915	No	No
39.37 %	27,744	Yes	Yes
23.46 %	16,533	Yes	No
	70,470		

---

1) justice.n:04617818 01072147 09565904 07643831

   04617818n: 2 hyponyms
   01072147n: 1 hyponym
   09565904n: 13 hyponyms
   07643831n: 1 hyponym
justice.n (justice%1:18:00::) selected: 09565904n: (13)
justice.n,1:18:00::, correct!

2) conviction.n:05597125 01122850

   05597125n: 1 hyponym
   01122850n: 3 hyponyms
conviction.n (conviction%1:04:00::) selected: 01122850n: (3)
conviction.n,1:04:00::, correct!

Fig. 2. In some cases, counting the number of hyponyms in WordNet selects the correct sense in SemCor (2 cases out of 3 278 shown here)

synset was not the first one listed in WordNet, but it was the correct sense for SemCor (4.65 %), but 22 915 where the synset was not the first one, but it was incorrect (35.52 %); in 27 744 cases, the count of hypernyms coincided with the first sense listed in WordNet, and it was correct (39.37 %), and finally, 16 533 cases where the selected synset was the first one, and it was not correct (23.46 %). These results are summarized in Table 5.

Information in Table 5 shows that our method is able to find a better solution for WSD in SemCor, than using the first sense listed in WordNet. A particular example of this can be seen in 2 cases (out of 3 278) shown in Fig. 2. Additionally, it is important to note that, despite our method is finding the first sense in WordNet in 23.46 % of the cases, this is not the correct sense tagged in SemCor.

5 Conclusions and Future Work

We have explored finding the Most Frequent Sense (MFS) of a word by counting the number of its relationships in WordNet. We evaluated our results by performing Word Sense Disambiguation (WSD) on SemCor, by always selecting our MFS found. We considered as upper-bound the MFS computed within SemCor itself, *i.e.*, 74.14 %. Since in a real world situation, the MFS is not known *a priori*, we must find a better way to approximate the MFS. An interesting approach was made by McCarthy *et al.* (2007), but they require an external corpus to build their thesaurus. Our aim was to determine to which extent this information is already encoded in a Semantic Network. Of course, information about the MFS is already included in WordNet, by listing the synsets from the most frequent to the less frequent one, but the question we aimed to answer in this paper was, is it possible to approximate the MFS by considering the semantic network structure itself? We were motivated by the hypothesis that, the more frequent a sense is, the greater the number of relationships it will have, since lexicographers would be able to find more hyponyms, more synonyms or even more than one hyponym for most frequent senses.

We tested our method with SemCor, and compared with previous results that tried to approximate the MFS by measuring the length of each synset's gloss. We found better results whenever we included the count of the different relations each synset had, with little direct effect from the length of the gloss definition. The best results were obtained by counting only the number of relationships without considering any word's gloss. Then we proceeded to further examine the kind of relationships that have a greater impact on the WSD of SemCor. We found that when we were considering all relationships, it was important to count bidirectional relationships, that is, outbound and inbound relationships as well. However, we found that when we considered particular relationships, counting outbound relationships was better than counting relationships bidirectionally, probably due to the nature of those relationships—synonyms are a symmetric relationships while hypernymy and hyponymy are not. We unexpectedly found that some words have synsets with more than hyponym.

Finally, we examined the number of cases where our algorithm was able to find the MFS, which was a different one than the one listed by WordNet. This shows that our method could be used to improve WSD performance without considering the MFS that is included in a dictionary, by means of other elements in the structure itself.

We have found that relationship count of each synset provides indeed a good clue for selecting the MFS, giving another measure from WordNet apart from the listed MFS within the same resource. This could yield to other ways of calculating the MFS when a ranked sense inventory is not available. As future work, we plan to experiment with other corpus to assess the extendibility of our proposed method with different corpora.

References

Calvo, H., Gelbukh, A.: Finding the most frequent sense of a word by the length of its definition. In: Gelbukh, A., Espinoza, F.C., Galicia-Haro, S.N. (eds.) MICAI 2014, Part I. LNCS, vol. 8856, pp. 1–8. Springer, Heidelberg (2014)

Hawker, T., Honnibal, M.: Improved default sense selection for word sense disambiguation. In: Proceedings of the 2006 Australasian Language Technology Workshop (ALTW2006), pp 11–17 (2006)

Lesk, M.: Automatic sense disambiguation using machine readable dictionaries: how to tell a pine cone from an ice cream cone. In: Proceedings of the 5th annual International Conference on Systems Documentation, pp. 24–26. ACM (1986)

Lin, D.: An information-theoretic definition of similarity. Int. Conf. Mach. Learn. **98**, 296–304 (1998)

Marcus, M.P., Marcinkiewicz, M.A., Santorini, B.: Building a large annotated corpus of English: the Penn treebank. Comput. Linguist. **19**(2), 313–330 (1993)

Màrquez, L., Taulé, M., Martí, M.A., García, M., Artigas, N., Real, F.J., Ferrés, D.: Senseval-3: the Spanish lexical sample task. In: Senseval-3: Third International Workshop on the Evaluation of Systems for the Semantic Analysis of Text, Barcelona, Spain, Association for Computational Linguistics (2004)

McCarthy, D., Koeling, R., Weeds, J.: Carroll unsupervised acquisition of predominant word senses. Comput. Linguist. **33**(4), 553–590 (2007)

Mihalcea, R., Chklovski, T., Kilgarriff, A.: The Senseval-3 English lexical sample task. In: Senseval-3: Third International Workshop on the Evaluation of Systems for the Semantic Analysis of Text, pp. 25–28 (2004)

Miller, G., Leacock, C., Tengi, R., Bunker, R.T.: A semantic concordance. In: Proceedings of ARPA Workshop on Human Language Technology, pp. 303–308 (1993)

Miller, G.A., Chodorow, M., Landes, S., Leacock, C., Thomas, R.G.: Using a semantic concordance for sense identification. In: Proceedings of the ARPA Human Language Technology Workshop, pp. 240–243 (1994)

Snyder, B., Palmer, M.: The English all-words task. In: ACL 2004 Senseval-3 Workshop, Barcelona, Spain (2004)

Language Processing and Human Cognition

Daša Munková$^{(\boxtimes)}$, Eva Stranovská, and Michal Munk

Constantine the Philosopher University in Nitra,
Tr. A. Hlinku 1, 949 74 Nitra, Slovakia
{dmunkova, estranovska, mmunk}@ukf.sk

Abstract. The order, association and variability of the language of request are different in every language and culture, because it is based not only on different rules in the given culture but also on the dichotomy of language processing-global and detail language processing. The study is focused on the investigation of modelling language processing in terms of human cognition. The transaction/sequence model for text representation was used and an association rules analysis was applied as the research method. Based on the human cognition (category width cognitive style) of examined texts, different models of language processing were being created.

Keywords: Natural language processing · Representation and language model · Human cognition · Social situation

1 Introduction

An examination of language in relation to human cognition has a various interdisciplinary and multidisciplinary character. Even though the research and theory of language cognition stretches deeply back [1–4] is still actual in present [5–8].

Is there a need to examine the relationship between human cognition and language processing? Harris [9] states, that "a hallmark of modern cognitive science is the goal of developing a theory of cognition powerful enough to encompass all human mental abilities, including language abilities". There are two approaches to conceptualize the human cognition: "general purpose" and "modularity of cognition". The first approach refers to universal processes and mechanisms providing the basis for human intelligence (e.g. category induction or pattern completion). Mechanisms using computer algorithms from artificial intelligence (AI) were proposed to prove that the same principles explaining general problem-solving can also explain aspects of language processing [9]. Considering the previous, Rübsamer [10] perceives the role of computer linguistics in language processing, specifically, in stimulation of cognitive processes which is related to the applications of practical models into the artificial intelligence. The second approach derived from the fact that the distinct parts of the brain serve to distinct function (e.g. language processing or memory). To investigate how the human brain processes language can help us to better understand how the language comprehension works and also to make natural language processing more efficiency in terms of accuracy. Furthermore, deeper understanding of human cognition allows building more accurate language models which can explain how difficult it may be for a human

© Springer International Publishing Switzerland 2015
D.-S. Huang and K. Han (Eds.): ICIC 2015, Part III, LNAI 9227, pp. 500–509, 2015.
DOI: 10.1007/978-3-319-22053-6_53

individual to comprehend thoughts. Three different theoretical views on the relationship *language-cognition* were designed: (1) language unique, unlike cognition; (2) subject to same principles; and (3) complex similarities and differences. In other words, the first represents "innate vs learned language competence" paradigm (AI), the second represents "symbolic vs subsymbolic" examination of language cognition (top down vs bottom up model) and the third one refers to dynamical interaction (dynamic language interaction). Additionally, it represents the idea that cognition and language relationship and how they are mutually related, sharing complex similarities and differences refer to genetic factors (nature and abilities) and a certain influence of environment on language formation (cultural learning). Current studies in the field of language processing tend to research interactive as well as autonomous models of language processing [11, 12].

In the present study, we investigate the influence of human cognition on language processing (models formation) through the dimension of social effect- social proximity and social power. We focus on language processing and language modelling of the mother tongue and its dependence on "category width" cognitive style, and on social distance with social power (e.g. *You are preparing a presentation for a key subject and you've just learned there is a new professor at the department specializing in your topic. You don't know the new professor but you decide to pay him a visit and ask him to read the summary of your work and recommend you some literature.*). For this purpose, a transaction/sequence model was formulated, the data – requests in given social situation – was collected and association rules analysis was applied.

1.1 Language Processing

Learning from text and natural language belongs to the big challenges of computational linguistics. Any substantial progress has a strong influence on many applications not only in computational linguistics but also in natural language processing, speech recognition or in machine translation. The fundamental problem, or one of them, is a distinction between the lexical level of what actually has been said or written and the semantical level of what was intended or what was referred to in a text or an utterance, i.e. polysemy (a word may have multiple senses and multiple types of usage in different context), and synonyms or semantically related words (different words may have a similar meaning, they may at least in certain contexts denote the same concept or - in a weaker sense – refer to the same topic [13].

Data pre-processing represents the most time consuming phase in the whole process of knowledge discovery. It converts a document transformation from an original textual data source into a form which is suitable for applying various methods of extraction, in order to transform unstructured form into structured representation, i.e. to create a new collection of documents- texts fully represented by concepts [14]. According to [14], two steps of textual data pre-processing are inevitable. Firstly, it is an identification of features (keywords) in a way that is computationally most efficient and practical for pattern discovery. Secondly, it is an accurate capture of the meaning of an individual text (on the semantic level) [14].

1.2   Human Cognition – Cognitive Style in Social Situations

Language information and communication models are processed in various ways depending on a variety of factors, which co-create choices of linguistic and paralinguistic means of communication as well as strategies of politeness depending on the individuals cognitions [15]. The language processing is discussed widely in the field of academics but it has not been discussed and researched widely in dependence to human cognition – category width – and to the principles of politeness in the speech acts theory, specifically, in the requests formulations [16]. The study also drew from the current studies e.g. [17, 18] which focus on examining and defining modelling of mother tongue language processing. The human cognition in language processing was investigated through the means of category width: global and detail language processing, using the C-W scale in selected social situation, specifically, with social distance and social power.

The dimension 'category width' cognitive style relates to individual differences in the categories width, namely narrow and broad categorizers.

The category "broad categorizer", models global strategies when processing language information as follows [19]:

- Gathering information is a global and holistic speech process, complex comprehension of text for orientation and gist, specifically, organizing and combining information,
- Interpretation of text is a comprehension of relationships and specification of the gist, integration of selected parts of speech and following recognition of the gist,
- Evaluating text for the integration of selected parts of speech and the evaluation of speech, comparison of the relationship between the text and broader knowledge or an explanation of selected parts of the text using personal experience and personal attitudes.

The category 'narrow categorizer' models strategies of detailed analytical processing of language as follows:

- Gathering information happens as detailed processing of speech, as its analyses, showing punctual comprehension of long complex texts in relation to known everyday knowledge, concentration on less general knowledge,
- Interpretation of text is contemplation about individual parts of speech, about its meanings – it is an explanation of the individual word meanings, its phrases, its mutual comparisons and oppositions, explanation of subtle differences and their meanings,
- Evaluating text is a process of critical evaluation or stating hypothesis, it is drawing direct attention to concepts which are in contradiction with one's expectations, connecting or comparing, explaining or evaluating of one specific feature of the text [20].

The requester has many features to formulate a request, which are usually classified according to a specific structure (culturally given). Blum-Kulka et al. [21] defined three elements of a request sequence in addition to the Head Act: alerters, supportive moves (external modifiers) and internal modifications.

The function of alerters is to alert requestee's attention to the upcoming speech act [21]. External modifiers involve: preparators, disarmers, sweeteners, supportive reasons, and cost minimizing [21, 22]. The function of internal modifications is to soften or increase the impact of a request.

The emphasis which the requester makes in carrying out a request can be realized in several perspectives. Blum-Kulka et al. [21] distinguish the following perspectives of a request: (a) Requester (Speaker) – oriented, (b) Requestee (Listener) – oriented, (c) Speaker and Listener – oriented and (d) Impersonal.

Díaz-Pérez [23] defines a request as a having set internal and external elements, whereby the internal and the basic part of a request is at its core, i.e. the minimum unit, which can serve as a particular speech act device.

2   Method

2.1   Dataset

Research was carried out from 2011 to 2014. It was attended by 146 students from different major study programs. We created a dataset consists of 146 utterances (transformed from speech to text) in situation of social distance with social power (*You are preparing a presentation for a key subject and you've just learned there is a new professor at the department specializing in your topic. You don't know the new professor but you decide to pay him a visit and ask him to read the summary of your work and recommend you some literature.*)

2.2   Measure and Procedure

The estimation scale C-W (Category Width) was used. The C-W Scale measures Cognitive Style ′Category Width′ and the real estimation [24] (Slovak translation by Sarmány-Schuller and Jurčová [25]). It contains 20 statements that suggest certain statements in the form of an average value; the respondent has to guess which of the four fixed numerical alternatives corresponds with the highest and lowest number of occurrences of a given phenomenon. The responses were assessed on three scores: C-W1 (in our case a) expresses the average value obtained from the estimates of highest values, C–W2 (in our case b) from the estimates of lowest values and C-W3 is the total questionnaire score (sum of C-W1 + C-W2). As follows from the definition of broad categorization, estimates are far away from the average in both high as well as low values. At the same time, C-W1 and C-W2 values should be in all cases very close or identical.

We used also the classification of request factors (elements) in line with [21–23] and we defined the following 30 factors:

(F1) Title or social role (e.g. *Mr., Mrs., Doctor, Professor XY*)
(F2) Surname or friendly appellation (e.g. *Mr. Smith, Mate*)
(F3) Name (e.g. *Sarah*)
(F4) Attention getter (e.g. *Excuse me, please*)

(F5) Combination of previous

(F6) Indirect perspective – allusion (e.g. *Hello, I was wondering if you've got a minute. I'm studying X and I'm writing a paper on X. I've heard you're an expert. ….*)

(F7) Listener's perspective (e.g. *Could you*)

(F8) Speaker's perspective (e.g. *Could I*)

(F9) Mixed perspective

(F10) Negative formulation (e.g. *Excuse me, you couldn't reach that book for me, could you?*)

(F11) Present tense continuous

(F12) Modal verb question

(F13) Conditional

(F14) Imperative

(F15) Past tense

(F16) Other tenses or ways

(F17) Combination of previous elements

(F18) Correctness of an utterance, in terms of the grammatical structure

(F19) Appropriateness of an utterance, in terms of culture specifics

(F20) Politeness marker (e.g. *thank you, please*)

(F21) Pre-sequences/preparators, elements before the core of a request (e.g. *Hi, could you do me a favor? Could you ...*)

(F22) Post-sequences/supportive reasons, elements after the expressed request (e.g. *Is here any change I could use your phone for a minute? I really need to make a phonecall.*)

(F23) Mitigating devices/disarmers, elements expressing an apology for disturbing (e.g. *I'm sorry to bother you, but could I possibly borrow a book from you to photocopy some pages?*)

(F24) Minimizers, elements minimizing the impact of a request (e.g. *Please, can I borrow your notes? I'll photocopy them and give them back to you in a minute.*)

(F25) Consultative mechanism (e.g. *Do you think I can have a shot of your notes?" "Please, would you mind if I use your telephone?*)

(F26) Compliments/sweeteners, elements intensifying the likelihood of request fulfilment (e.g. *Could you help me prepare for my essay as I know you are very knowledgeable in the subject.*)

(F27) Intensificators (e.g. *important, as soon as possible, quick*)

(F28) Promises, reciprocity (e.g. *Excuse me. Would it be o.k., if I borrowed the book for half an hour to photocopy a couple of chapters? I'll bring it straight back.*)

(F29) Combination of previous

(F30) Others

The first six represent alerters, the following three represent perspectives (in our study they are social factors), factors F10–F19 represent the internal modifications (syntactic and lexical/phrasal downgraders, in our case – language factors) and the rest are external modifiers (supportive moves or in our case – expressive factors).

This classification helps us in the computational modelling and understanding of the culture-specific elements of politeness in speech acts of requesting in particular languages such as a mother tongue.

3   Association Rule Analysis- Results

In our study we used transaction/sequence model for text representation, similar to bag-of-words model. Each text-request is a bag of words, i.e. the order of words is not taken into account (e.g. good boy has the same probability as boy good). It allows us to examine the relationships between the examined attributes and search for associations among the identified words in dataset. The association rule analysis represents a non-sequential approach to the data being analyzed. The sequences were not included, however, transactions were included. Thus, the order of the occurred factors was used in the following analysis. The web graph (Fig. 1) depicts the discovered association rules for the requests, specifically, the size of the node represents the support of incidence of the factor, the thickness of the line represents the support of rule – pairs of the factors (probability of occurrence in pair) and the darkness of the line color presents a lift of the rule – probability of a pair occurrence in transaction separately.

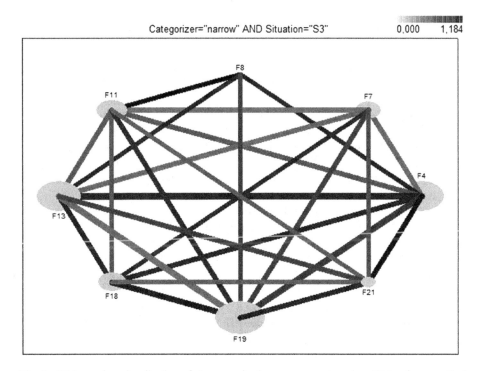

Fig. 1. Web graph – visualization of discovered rules – narrow categorizer (Color figure online)

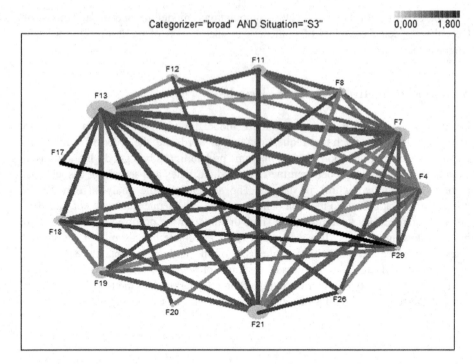

Categorizer="broad" AND Situation="S3"   0,000   1,800

Fig. 2. Web graph – visualization of discovered rules – broad categorizer

The graph (Fig. 1) displays the narrow category and its relation to the factors of request F19 (89.06 %, F4 (85.93 %), F13 (85.93 %), F11 (78.12 %), F18 (76.56 %), F7 (75 %), F21 (68.75 %), F8 (62.50 %), i.e. (support > 61 %), belong to the most frequently used factors. Similarly, combination of the factors` pairs (F4, F19), (F13, F19), (F4, F13), (F18, F19), (F4, F21), (F4, F18), (F7, F19), (F11, F19), (F13, F18), (F4, F7), (F19, F21), (F4, F11), (F13, F21), (F4, F8) and (F7, F21) (support > 52 %), the factors F21 ==>F4, F13, F19, F7, F18,F11, F8 ==>F11, F4,F19, F13 and F4 ==>F21, F8, F19, F13, F18, F7, F11, occur in sets of factors of request more often together than as separate units (lift > 1).

The results displayed different association rules for the utterances of requests in broad category. The web graph (Fig. 2) illustrates discovered association rules. The most frequently used factors were F13 (96.29 %), F4 (88.88 %), F7 (85.18 %), F21 (81.48 %), F19 (74.07 %), F11 (70.37 %), F18 (70.37 %), F8 (62.96 %), F12 (62.96 %), F26 (55.55 %), F29 (55.55 %) and F20 (51.85 %), i.e. (support > 50 %), as well as their pairs (F4,F13), (F7,F13), (F13, F21), (F4,F21), (F4,F7), (F7,F21), (F13, F19), (F13,F18), (F11,F21), (F4,F26), (F4,F8), (F8,F21), (F13,F29), (F21,F26), (F26, F21) (support > 50 %). The factors F29 ==>F18, F19, F13, F7, F8 ==>F11, F21, F19, F4, F7 ==>F20, F11, F13, F29, F21 ==>18 and F4 ==>F26, F21, F12 occurred more often together in transactions of used factors than separately (lift > 1).

4   Conclusions

Language processing in cognitive style category width depends on social situations, for example, situation of social proximity without social power, situation of social proximity with social power, situation of social distance without social power, social distance with social power [18]. For the purpose of the research study, social situation of social distance with social power in request formulations was selected for further examination.

Broad categorizer chose the following factors *Conditional, Attention getter, Listener's perspective, Pre-sequences/preparators, Appropriateness of an utterance, Present tense continuous, Correctness of an utterance, Speaker's perspective, Modal verb question, Compliments, Combination of previous* (combination of expressive factors), *Politeness marker, Combination of previous elements* (combination of language factors) in social situation characterized as social distance with social power. Broad categorizer showed variability in individual factors choices in request formulations, mainly language and expressive factors and their combinations. Furthermore, broad category used expressive and language factors and their combinations in the request modelling, broad categorizer modelled *listener's perspective*, attracting attention of the listener directly without being afraid of negative response. Social connections were directed to recipient by *getting attention* (interpersonal attractiveness) or by using *compliments*. Interpersonal attractiveness was strengthened by double use of language factors in request formulations, specifically, using *questions with modals* or *conditionals*, and correct choice of language means. Broad categorizers strengthened politeness in anticipation of request fulfilment.

Detailed language processing - narrow categorizer - used the following factors in request formulations in social situation of social distance with social power : *Title* or *Social role, Listener's perspective, Speaker's perspective, Mixed perspectives, Correctness of utterance, Appropriateness of utterance, Politeness marker, Pre-sequences/preparators, Post-sequences/supportive reasons, Mitigating devices* and *Intensificators*. Narrow categorizer is usually characterized as rigid, careful, and needy of certainty in cognitive decision-making process, focusing on *correctness* and *appropriateness* of speech, on *pre-sequences* and *post-sequences* explaining reasons and further explanations of requests, strengthening meaning of requests. *Correctness* and *appropriateness* appeared in both categories - narrow and broad. However, in broad category the perspective was different – the intention of broad categorizer was to strengthen social influence and to create impression of confidentiality. In opposition to broad categorizer, narrow categorizer used more difficult language factors in request formulations (questions with modals and use of conditionals). Narrow category did not use the above mentioned factors. The process of requests modelling was more rigid in narrow category compared to broad categorizer, even though the narrow categorizer wished to have the request fulfilled.

Furthermore, the following examination was directed to find out pairs of factors combinations of requests modelling and their mutual dependence or support. The results indicated dominance of pairs of the following factors *Attention getter and Conditional, Listener's perspective and Conditional, Conditional and Pre-sequences, Attention getter and Listener's perspective, Listener's perspective and Pre-sequences,*

Conditional and Appropriateness of utterance, Conditional and Correctness of utterance, Present tense continuous and Pre-sequences, Attention getter and Compliments, Attention getter and Speaker's perspective, Speaker's perspective and Pre-sequences, Conditional and Combination of expressive factors, Pre-sequences and Compliments in broad category language modelling. Broad category used the combination of the above mentioned factors in order to get attention of recipient.

Narrow category modelled the following pairs of factors *Attention getter and Appropriateness of utterance, Correctness of utterance and Appropriateness of utterance, Attention getter and Pre-sequences/preparators, Attention getter and Correctness of utterance, Listener's perspective and Appropriateness of utterance, Present tense continuous and Appropriateness of utterance, Attention getter and Listener's perspective, Appropriateness of utterance and Pre-sequences/preparators, Attention getter and Present tense continuous, Conditional and Pre-sequences/preparators, Attention getter and Speaker's perspective, Listener's perspective and Pre-sequences.*

Broad categorizer showed high level of factors combinations – social factors, language factors, together with expressive factors, language factors and expressive factors, social, expressive and language factors together and expressive factors in mutual combinations. Narrow categorizer showed combinations of social and language factors, language factors mutually as well as social factors mutually. The results indicated that broad categorizers modelled utterances using the factors strengthening meaning and language structure in the formulations. Opposing, the narrow categorizer focused more on language structure and its influence on recipient.

The study intended to find out frequency and combinations of factors in category width. Furthermore, it was the intention to find out sequences of social, language, and expressive factors in category width. The results showed the following factors: *Combination of expressive factors, Correctness of utterance, Appropriateness of utterance, Conditional and Listener's perspective, Speaker's perspective, Present tense continuous, Pre-sequences, Appropriateness of utterance, Attention getter, Listener's perspective, Politeness marker, Present tense continuous, Conditional, Pre-sequences, Attention getter, Compliments, Pre-sequences* and *Modal verb question* in broad category.

When narrow categorizer used *Pre-sequences* then the following factors occurred: *Attention getter, Conditional, Appropriateness of utterance, Listener's perspective, Correctness of utterance, Present tense continuous, Speaker's perspective, Present tense continuous, Attention getter, Appropriateness of utterance, Conditional, Attention getter, Pre-sequences, Speaker's perspective, Appropriateness of utterance, Conditional, Correctness of utterance, Listener's perspective, Present tense continuous.*

We see the further direction of our research in analyzing sequences of individual factors of requests. We consider the analysis of the factors of request modelling as meaningful, actual and necessary because it gives space for the further discourse of reasons and functions of cultural and individual stereotypes in the process of language processing.

Acknowledgments. This work was supported by the Slovak Research and Development Agency under the contract No. APVV-0451-10, and Scientific Grant Agency of the Ministry of Education of the Slovak Republic (ME SR) and of Slovak Academy of Sciences (SAS) under the contract No. VEGA-1/0559/14.

References

1. Chomsky, N.: Three models for the description of language. IRE Trans. Inf. Theory **2**(3), 113–124 (1956)
2. Chomsky, N.: Syntactic Structures. Mouton, The Hague (1957)
3. Neisser, U.: Cognitive Psychology. Prentice-Hall, Englewood Cliffs (1967)
4. Bruner, J.S.: Beyond the Information Given: Studies in the Psychology of Knowing. W. W. Norton & Company, New Yok (1973)
5. Eysenck, M.W., Keane, M.: Kognitivní psychologie. Academia, Praha (2008)
6. Schwarz, M.: Einführung in die Kognitive Linguistik, 3rd edn. Francke, Tübingen (2008)
7. Rickheit, G., Weiss, S., Eikmeyer, H.J.: Kognitive Linguistik: Theorien, Modelle, Methoden. UTB, Tübingen (2010)
8. Traxler, M.J.: Introduction to Psycholinguistics: Understanding Language Science. Wiley, Malden (2012)
9. Harris, C.L.: Language and cognition. In: Encyclopedia of Cognitive Science. MacMillan, London (2003). http://www.bu.edu/psych/charris/papers/Encyclopedia.pdf
10. Rübsamer, R.: Gehirn, Kognition und Sprache. (2015). http://www.zv.uni-leipzig.de/fileadmin/user_upload/Forschung/PDF/gehirn_kognition_sprache.pdf
11. Müller, H.M.: Psycholinguistik – Neurolinguistik. UTB, Tübingen (2013)
12. Höhle, B.: Psycholinguistik. Akademie Verlag, Berlin (2012)
13. Hofmann, T.: Probabilistic latent semantic analysis. In: Uncertainty in Artificial Intelligence (UAI-99), Stockholm (1999)
14. Feldman, R., Sanger, J.: The Text Mining Handbook. Cambridge University Press, Cambridge (2007)
15. Munková, D., Stranovská, E., Munk, M.: How "category width" cognitive style affects language processing. Proc. Soc. Behav. Sci. **171**, 1373–1380 (2015)
16. Munková, D., Munk, M., Ďuračková, B., Fráterová, Z.: Analysis of social and expressive factors of requests by methods of text mining. In: 26th Pacific Asia Conference on Language Information and Computation (PACLIC-26), Bali, pp. 515–524 (2012)
17. Roche, J.: Fremdsprachenerwerb Fremdsprachendidaktik. A. Francke Verlag, Tübingen (2013)
18. Köppel, R.: Deutsch als Fremdsprache – Spracherwerblich reflektierte Unterrichtspraxis. Schneider Verlag, Hohengehren (2013)
19. Munková, D., Stranovská, E., Munk, M.: Language processing dependence on cognitive style "Category Width". In: Procedia Social and Behavioral Sciences: The 9th International Conference on Cognitive Science (ICCS 2013), pp. 122–130, Kuching, Sarawak (2013)
20. Koršáková, P., Tomengová, A.: PISA SK 2003: Národná správa. ŠPU, Bratislava (2004)
21. Blum-Kulka, S., House, J., Kasper, G.: Cross-cultural Pragmatics: Requests and Apologies. Ablex, Norwood (1989)
22. Trosborg, A.: Interlanguage Pragmatics: Requests, Complaints, and Apologies. Mouton de Gruyter, Berlin (1995)
23. Díaz-Pérez, F.J.: La Cortesía Verbal en inglés y en Español. Actos de Habla y Pragmática Intercultural. Universidad de Jaén, Jaén (2003)
24. Pettigrew, T.: The measurement of category width as a cognitive variable. J. Pers. **26**, 532–544 (1958)
25. Jurčová, M., Sarmány-Schuller, I.: Kognitívny štýl "šírka kategorizácie". Československá psychologie **37**(1), 1–13 (1993)

An Ontology-Based Approach for Measuring Semantic Similarity Between Words

Ruiling Zhang[1,2(✉)], Shengwu Xiong[1], and Zhong Chen[1]

[1] School of Computer Science and Technology, Wuhan University of Technology,
Wuhan, Hubei, China
{ruilingzhang,chenzhong}@163.com,xiongsw@whut.edu.cn
[2] Luoyang Normal University, Luoyang 471022, Henan, China

Abstract. The estimation of semantic similarity between words play an important role in many language related applications. In this paper, we survey most of the ontology-based approaches in order to evaluate their advantages and limitations. We also present an approach for measuring semantic similarity. As a kind of feature-based method, proposed method extracts taxonomic features from ontology, aiming to provide a high-efficient, simple and reliable semantic similarity assessment method. We evaluate and compare our approach's results against those reported by related works under a common framework. Result demonstrated that the proposed method has higher correlation with human subjective judgment than most of existing methods.

Keywords: Semantic similarity · Feature-based measure · Wordnet

1 Introduction

In cognitive psychology, similarity refers to psychological proximity between two or more metal representations. In practical application, quantitative index of similarity of terms, vocabularies and concepts is called as semantic similarity. Although human beings are able to easily judge the relative semantic similarity of pairs of concepts, whereas it is a great challenge for computer to estimate semantic similarity quickly and accurately. On one hand, similarity is flexible and depends on contexts, which complicate the understanding of concept content [1]. On the other hand, natural language is for the use of human beings. Computer couldn't understand the natural language and analyze word meanings. This makes semantic similarity assessment methods even more difficult. Semantic Web technology provides a good solution to this problem. Ontology, the core of semantic Web, is a clear specification to conceptual model [2]. Recently, ontology based semantic similarity measurement has been applied in many tasks such as natural language processing [3], information extraction [4], data mining [5], automatic document annotation [6], ontology learning [7].

In This paper, we review and analyze existing measures for semantic similarity computation as well as their advantages and disadvantages. A new method feature-based semantic similarity measures is proposed. Finally, the proposed method using

© Springer International Publishing Switzerland 2015
D.-S. Huang and K. Han (Eds.): ICIC 2015, Part III, LNAI 9227, pp. 510–516, 2015.
DOI: 10.1007/978-3-319-22053-6_54

WordNet [8] as ontology was compared to existing ones on the benchmark dataset, and finding higher accuracy of the proposed method.

2   Methods of Computing Semantic Similarity

In this section, 3 kind of semantic similarity assessment based on ontology will be analyzed. They are edge counting method, information content method and feature-based method.

Edge counting method believes that shorter distance of two concepts on the graph of concepts represents higher similarity. Hence, various methods were proposed. In 1989, Rada et al. [9] suggested to measure the semantic distance between concepts by calculate their minimum path length in ontology. The semantic distance between concepts is defined as follow.

$$Dis(a, b) = \min path_i(a, b) \tag{1}$$

Several variations and improvements of this edge counting approach have been proposed [10].The edge counting methods are based on hypothesis of geometrical distance, which are characteristic of easy-to-implement and low computation complexity. However, they only received limited practical applications due to the low computation accuracy.

The information content (IC), witch firstly proposed by Resnik [14], is a measure quantifying the amount of information a concept expresses. IC has been applied to assess semantic similarity according to the amount of common information of a concept pair. Resnik assumed that common information between concepts a and b is represented by the IC of their Least Common Subsumer (LCS).

$$Sim(a, b) = IC(LCS(a, b)) \tag{2}$$

However, text ambiguity, noise and data sparseness in corpus can influence the IC computation accuracy significantly, even making great mistakes. To address these problems, some scholars suggested computing IC with ontology knowledge [15, 16].

Corpus-independent IC measure or Intrinsic IC models extracts information from ontology directly, thus lowering computation complexity and avoiding many limitations of methods based on corpus. However, such IC metric depends on complete ontology structure. IC calculated from different ontologies of same filed constructed by different experts will differ significantly.

The basic idea of semantic similarity measure based on features is that things are reflected their features, the symbol for distinguishing it from others. Similarity between things is determined by their common and different features Tversky [17] proposed a similarity model based on features. In this model, common features of objects were represented by the intersection of feature sets, while different features were represented by difference sets. Similarity was defined as the function on common and different features.

$$Sim(a, b) = \alpha(\psi(a) \cap \psi(b)) - \beta(\psi(a) - \psi(b)) - \gamma(\psi(b) - \psi(a)) \tag{3}$$

where α, β and γ are three adjustable coefficients; $\Psi(a)$ is the feature set of concept a.Based on Tversky's similarity model, many scholars put forward this mode [18, 19].

Semantic similarity assessment methods based on features involve more semantic information and conform to cognitive characteristics of human beings. However, their accuracy is determined by feature set. Although some scholars have proposed combined methods [21], they still rely on additional information.

3   A New Feature-Based Measure

According to the analysis of existing semantic similarity assessment methods, those based on features involve more semantic information and conform to cognitive characteristics of human beings. However, they depend on complete concept feature set. As a result, the following two aspects shall be considered. Firstly, some general ontology (e.g. WordNet) has complicated structure. They may contain various taxonomy architecture and each taxonomy architecture was constructed by domain experts from different viewpoints and knowledge backgrounds. Secondly, in such one feature set, different elements have different contributions to the feature set.

On this basis, a new semantic similarity assessment method based on the Tversky's model was proposed. Similar with David's method [20], it also extracts taxonomy knowledge of ontology. Relative definitions are introduced as follow.

Definition 1: Let C represent all concept in the ontology O. The binary relation is defined as $<: C \times C$. If c_1 is the parent node of c_2, $c_1 < c_2$. The root concepts set of O is defined as: $R_O(c) = \{c | \neg \exists c' \in C, c < c'\}$.

Definition 2: Let $p \in Path(c)$ represent one path from c to the root node. $Traval(c, p)$ is the set of c and all concepts on p. Therefore, the derivation set of c is: $\theta(c) = \{Traval(c, p) | p \in Path(c)\}$.

For example, in WordNet, there's a concept "lover": $\theta(c)(lover) = \{\{entity, causal agent, person, lover\}, \{entity, object, living thing, organism, person, lover\}\}$

Definition 3: Let $P \subseteq C$, IC of P is defined as follow:

$$\psi(P) = \sum_{c \in P} IC(c) \tag{4}$$

Definition 4: Semantic similarity between concepts defined as follows:

$$s(a, b) = \frac{\alpha\psi(A \cap B)}{\alpha\psi(A \cap B) + \beta\psi(A - B) + (1 - \beta)\psi(B - A)} \tag{5}$$

where $A \in \theta(a)$ and $B \in \theta(b)$. Parameter α is an adjustable coefficient, which is used to adjust weights of common and different features, β is used to balance proportions of two feature sets. Generally, the maximum similarity is used to represent the similarity between two concepts:

$$Sim(a, b) = Max_{A \in \theta(a), B \in \theta(b)} \frac{\alpha\psi(A \cap B)}{\alpha\psi(A \cap B) + \beta\psi(A - B) + (1 - \beta(A, B))\psi(B - A)} \quad (6)$$

Semantic distance between concepts can be expressed as $Dis(a, b) = 1 - Sim(a, b)$.

Property 1: Formula (17) is a semi-metric. It meets:

(1) $Dis(a, b) = 0$ if and only if $a = b$ (identity of indiscernibles);
(2) $Dis(a, b) \geq 0$ (non-negative)
(3) $Dis(a, b) = Dis(b, a)$ (symmetric)

According to Property 1, the proposed similarity computation method can be applied to methods based on similarity metrics (e.g. clustering algorithm).

4   Experimental Results and Discussions

Currently, computation results will be compared with subjective judgment of human beings [22, 23] to evaluate the semantic similarity assessment methods. Pearson correlation coefficient will be computed. Higher Pearson correlation coefficient indicates the computed similarity is in higher accordance with subjective judgment of human beings. Therefore, the computed result of the proposed method was compared to those of recent methods.

Table 1. Correlation values for each measure using WordNet2.0 as ontology

Measure	Type	M&C	R&G
Wu and Palmer [10]	Edge counting	0.803 [24]	/
Li et al. [12]	Edge counting	0.82 [19]	/
Leacock and Chodorow [11]	Edge counting	0.74 [25]	0.77 [25]
Al-Mubaid and Nguyen [13]	Edge counting	/	0.815 [20]
Lin [24]	IC (Corpus)	0.834 [24]	0.72 [25]
Resnik [14]	IC (Corpus)	0.795 [24]	0.72 [25]
Bollegala et al. [26]	IC (Corpus)	0.813 [27]	/
Resnik (IC computed as David et al.) [16]	IC (Intrinsic)	0.84 [16]	/
Lin (IC computed as David et al.) [16]	IC (Intrinsic)	0.85 [16]	/
Jiang and Conrath (IC computed as David et al.) [16]	IC (Intrinsic)	0.87 [16]	/
Resnik (IC computed as Zhou et al.) [15]	IC (Intrinsic)	/	0.842 [15]
Lin (IC computed as Zhou et al.) [15]	IC (Intrinsic)	/	0.866 [15]
Jiang and Conrath (IC computed as Zhou et al.) [15]	IC (Intrinsic)	/	0.858 [15]
Tversky [17]	Feature-based	0.73 [19]	/
Petrakis et al. [19]	Feature-based	0.74 [19]	/
Banerjee and Pedersen [29]	Feature-based	0.80 [25]	0.83 [25]
David et al. [20]	Feature-based	0.83 [20]	0.857 [20]
Our approach ($\alpha = 0.8$)	Feature-based	0.845	0.864

WordNet2.0 was used as ontology for fair comparison. It is important to note that one word may have several semantics and one semantic may have several contexts. Therefore, there may have several similarity values between two words. In this paper, the maximum similarity between two words was taken as the final result. Furthermore, to ensure objectivity and accuracy of the experiment, results of other methods were cited from published articles directly.

Except for the proposed method, results of recent semantic similarity assessment methods and their correlation coefficients with two standard datasets are listed in Table 1. The proposed method gets a group of good results: 0.845 and 0.864, higher than most methods based on IC and features. This demonstrates that classification knowledge in ontology can provide more semantic information and content of easy-to-access classification knowledge is relative reliable. Different Reference [20], the proposed method views all concepts on the path from concept to the root node rather than all upper nodes as the feature set. The experiment found higher accuracy of the proposed method than others. Property 1 reveals that the proposed method is a semi-metric and can be applied to other methods based on semantic similarity metrics.

Moreover, the proposed method is applicable when ontology contains various classification systems. If there's only one classification system (tree classification structure) in the ontology, it will achieve similar result with edge counting methods. Meanwhile, the ontology couldn't cover all knowledge. For example, the similarity between food and fruit is computed 0.03 (range: 0–1) by the proposed method. This differs from expert judgments (3.08, range: 0–4; and 2.69, range: 0–4) greatly. This is caused by the far distance between food and fruit in taxonomy knowledge of WordNet. This can be solved by ontology with rich classification information (e.g. WordNet and Wikipedia) and multiple ontologies.

5   Conclusions

Semantic similarity assessment is the basic issue of many associated fields and one of hotspot recently. In this paper, we introduced a measure of semantic similarity based on features. It extracts taxonomic features from ontology and weights the features through IC of concept based on the Tversky's model. It is simple method because it makes full use of most reliable and easy-to-access taxonomy knowledge in ontology. The comparison experiment finds that the correlation coefficient of the proposed method is higher than most of existing methods. Future research will focus on how to popularize semantic similarity computation method to more fields (e.g. knowledge fusion).

Acknowledgment. This paper is supported by the National Natural Science Funds of China (61272015, 61050004), and also is supported by Henan Province basic and frontier technology research project (142300410303).

References

1. Goldstone, R.L.: The role of similarity in categorization: providing a groundwork. Cognition **52**(2), 125–157 (1994)

2. Gruber, T.R.: Toward principle for the design ontologies used for knowledge sharing. Int. J. Hum. Comput. Stud. **43**(5), 907–928 (1995)
3. Fragos, K.: Modeling WordNet glosses to perform word sense disambiguation. Int. J. Artif. Intell. Tools **22**(2), 1350003 (2013)
4. Chen, P., Lin, S.J., Chu, C.Y.: Using Google latent semantic distance to extract the most relevant information. Expert Syst. Appl. **38**, 7349–7358 (2011)
5. Luo, Q., Chen, E., Xiong, H.: A: semantic term weighting scheme for text categorization. Expert Syst. Appl. **38**, 12708–12716 (2011)
6. Sánchez, D., Isern, D., Millán, M.: Content annotation for the semantic web: An automatic web-based approach. Knowl. Inf. Syst. **27**, 393–418 (2011)
7. Sánchez, D., Moreno, A., Vasto, L.D.: Learning relation axioms from text: an automatic web-based approach. Expert Syst. Appl. **39**, 5792–5805 (2012)
8. Miller, G.A.: WordNet: a lexical database for English. Commun. ACM **38**(11), 39–41 (1995)
9. Rada, R., Mili, H., Bicknell, E., Bletner, M.: Development and application of a metric on semantic nets. IEEE Trans. Syst. Man Cybern. **19**(1), 17–30 (1989)
10. Wu, Z., Palmer, M.: Verb semantics and lexical selection. In: ACL, pp. 133–138 (1994)
11. Leacock, C., Chodorow, M.: Combining local context and WordNet sense similarity for word sense identification. In: WordNet, An Electronic Lexical Database, pp. 265–283. May 1998
12. McLean, D., Li, Y., Bandar, Z.: An approach for measuring semantic similarity between words using multiple information sources. IEEE Trans. Knowl. Data Eng. **15**(4), 871–882 (2003)
13. Al-Mubaid, H., Nguyen, H.A.: A cluster-based approach for semantic similarity in the biomedical domain. In: 28th Annual International Conference of the IEEE Engineering in Medicine and Biology Society, pp. 296–304. New York, USA (2006)
14. Resnik, P.: Using information content to evaluate semantic similarity in ataxonomy. In: 14th International Joint Conference on Artificial Intelligence, pp. 448–453. Montreal, Quebec, Canada (1995)
15. Zhou, Z., Wang, Y., Gu, J.: A new model of information content for semantic similarity in WordNet. In: Second International Conference on Future Generation Communication and Networking Symposia, FGCNS 2008, pp. 85–89. Hainan Island, China (2008)
16. Batet, M., Isern, D., Sánchez, D.: Ontology-based information content computation. Knowl. Based Syst. **24**(2), 297–303 (2011)
17. Tversky, A.: Features of similarity. Psychol. Rev. **84**(4), 327–352 (1997)
18. Rodriguez, M.A., Egenhofer, M.J.: Determine semantic similarity among entity classes from different ontologies. IEEE Trans. Knowl. Data Eng. **15**(2), 442–456 (2003)
19. Petrakis, E.G.M., Varelas, G., Hliaoutakis, A., Raftopoulou, P.: X-similarity: computing semantic similarity between concepts from different ontologies. J. Digit. Inf. Manag. **4**, 233–237 (2006)
20. Batet, M., Isern, D., Valls, A., Sánchez, D.: Ontology-based semantic similarity: a new feature-based approach. Expert Syst. Appl. **39**(9), 7718–7728 (2012)
21. Pirró, G.: A semantic similarity metric combining features and intrinsic information content. Data Knowl. Eng. **68**, 1289–1308 (2009)
22. Rubenstein, H., Goodenough, J.: Contextual correlates of synonymy. Commun. ACM **8**, 627–633 (1965)
23. Miller, G.A., Charles, W.G.: Contextual correlates of semantic similarity. Lang. Cogn. Process. **6**, 1–28 (1991)
24. Lin, D.: An information-theoretic definition of similarity. In: Fifteenth International Conference on Machine Learning, pp. 296–304. Morgan Kaufmann, Madison, Wisconsin, USA (1998)

25. Wan, S., Angryk, R.A.: Measuring semantic similarity using WordNet-based context vectors. In: IEEE International Conference on Systems, Man and Cybernetics, pp. 908–913. Montreal, Quebec, Canada (2007)
26. Bollegala, D., Matsuo, Y., Ishizuka, M.: Measuring semantic similarity between words using web search engines. In: 16th International Conference on World Wide Web, pp. 757–766. Banff, Alberta, Canada (2007)
27. Jiang, J.J., Conrath, D.W.: Semantic similarity based on corpus statistics and lexical taxonomy. In: Proceedings of the International Conference on Research in Computational Linguistics (1998)

Intelligent Control and Automation

Knowledge Bases' Control for Intelligent Professional Activity Automatization

Alexander Kleschev[(⊠)] and Elena Shalfeeva[(⊠)]

Laboratory of Intellectual Systems in the Institute for Automation and Control Processes of the FEB RAS, Vladivostok, Russia
{kleschev,shalf}@iacp.dvo.ru

Abstract. This paper is devoted to organizing of the quality control process of knowledge bases, used for support of daily intelligent activity. The knowledge control is process of expansion, refinement, improvement of used knowledge. The main functions of knowledge bases control systems providing continuous improvement of knowledge are identified. The conceptual scheme for automated support of knowledge bases quality control process is submitted.

Keywords: Automation of intelligent activity · Decision-making and support · Knowledge base · Knowledge quality control · Quality of knowledge · Knowledge base assessments

1 Introduction

One of the ways of improvement of the quality of the professional intelligent activity is its automation by means of computers. Many tools and technologies which are used for automation of different types of activity are created: database management systems, tools for corporate systems, local networks, cloud computing. Since the 1970, with the advent of the knowledge based systems, great expectations for their practical application for automation of intellectual activity appeared.

Despite of considerable achievements in its theory and technology, the knowledge based systems didn't begin to be used everywhere in establishments' or institutions' professional activity. They weren't built into their organizational structure. The most probable explanation of the current situation is the insufficient attention to support of knowledge quality control for an activity, being automated.

The article's purpose is to show the significance of program systems for knowledge base quality control and to define the main tasks of these systems.

2 Knowledge Bases and Support of Intellectual Activity

Intellectual activity in this paper is a process of making of interconnected decisions (on the basis of knowledge) in relation to a real object. Specificity of decision-making on the basis of knowledge is as following: algorithms of such decision-making are unknown; only algorithms of "application of knowledge" for decision-making are known. Thus "quality" of decisions (correctness and accuracy of every decision) depends on the quality of knowledge [1].

© Springer International Publishing Switzerland 2015
D.-S. Huang and K. Han (Eds.): ICIC 2015, Part III, LNAI 9227, pp. 519–525, 2015.
DOI: 10.1007/978-3-319-22053-6_55

Scientific or hi-tech organizations or their certain experts can improve knowledge (when any inaccuracies are detected) and extend it (by introducing the new approved scientific achievements). We will call the activity, connected with expansion, refinement, improvement of knowledge, «*knowledge control*».

Expert systems (ESs) are developed for the automation of intelligent activity (intellectual work). These systems are designed to solve the intelligent problems of certain classes (planning, interpretation, prediction, diagnostics etc.) based on knowledge bases.

Supporting of daily intellectual activity means decision-making by the expert taking into account an explanation of the expert system.

The explanation contains information on the conformity of hypotheses (about potential solutions) to information about an object (input data) and to the knowledge base. For example, in medicine, it is important to explain which hypotheses–diagnoses can be rejected and which hypotheses can be used for some case history considering the specific knowledge of medical diagnosis.

A specialist (i.e. a doctor), who is making a decision (i.e. diagnosing) on the available data about the situation (i.e. clinical records of a patient) relies on his\her own mind (knowledge) and has a chance to read and inspect the explanation formed by expert system (i.e. analysis of possible hypotheses of the diagnosis for this patient).

If the knowledge base is of high quality, expert system (ES) allows to reduce quantity of specialist's mistakes connected with a wrong application of his/her knowledge. An expert system uses the correct algorithm for task decision and can carry out the full analysis of data for any quantity of hypotheses. However the analysis results depend substantially on the quality of knowledge applied during this analysis. The quality and usefulness of the knowledge base are determined by the completeness, accuracy, and correctness of the knowledge that is contained in it.

The knowledge bases generally created by experts [1–4]. The knowledge base (KB) created by the expert may contain defects: may be incomplete or inexact.

Generally the primary ways of evaluation of knowledge bases are an application of means of its formal properties check (entities types and their properties and the extent of the corpus: "two distinct ground truth entities are conflated", "a ground truth entity is split into several entities" etc.) and an involvement of experts for an assessment of "solutions proposed by a system" [2, 5–7].

It is obvious that a more objective evaluation (assessment) of the quality of knowledge bases is necessary.

The assessment of *KB correctness* can be defined in terms of a set of tasks which ES correctly solves on the base of this KB (ES produces some hypotheses about a solution, including the correct solution). The assessment of *KB accuracy* can be defined in terms of a subset of this set of tasks for which ES produces only one solution.

3 Support of Knowledge Quality Control Activity

Each decision made by the specialist during intellectual activity usually undergoes a confirmation procedure. As a result of this procedure the decision has to be validated as right or wrong. The experts from KB management group can compare a task solution made by the specialist, an explanation created by system and a final solution conclusion. In case of discrepancy of these decisions it is necessary to estimate, whether the explanation is wrong.

During professional practice a number of tasks with validated solutions will increase over time. Each increased set of tasks-with-known-solutions is used for checking knowledge base: whether KB satisfies to the accumulated precedents.

Some examples of software tools for handling of the set of precedents (precedents repository) are described in available sources [5, 8].

Fixed KB become outdated over time in the sense that the assessment of its correctness and accuracy won't change while number of precedents is enlarged and the assessment of specialist improves in the course of further practice. Therefore the knowledge base has to be improved over time.

For continuous improvement of the quality of decisions a knowledge bases quality control has to be exercised together with automation of daily intellectual activity. Therefore the quality control of knowledge bases demands integration with electronic "document flow". Every correct and every wrong solution is very useful precedent. All these precedents are the input information for KB improvement process.

KB control system' purpose – to provide usefulness of ES for the specialist during the whole time of its exploitation. For achievement of this purpose the KB control system has to carry out the following functions:

- to accumulate precedents repository permanently;
- to classify all precedents in the repository;
- to find all possible ways of modification of KB for inclusion of new precedents into the assessment of KB correctness and the assessment of KB accuracy;
- to modify KB by one of possible ways.

4 The Main Tasks of Knowledge Base Control System

In order that ES remained useful to the specialist, it is necessary that information on all solved tasks and the validation results was available to ES. Obtaining this information is the function «to accumulate precedents repository».

A natural realization of this function is integration of ES with subsystem of electronic document flow of intellectual activity. Information on the decisions made by specialists, and about results of confirmation of these decisions has to be included in documents in the form allowing its processing (by KB control system). Such integration of ES with subsystem of electronic document flow allows to evaluate the quality of task decisions made by certain specialists or their groups.

Each precedent has to be referred (by KB control system) to one of the following classes (Fig. 1).

(1) ES suggested the correct and exact solution.
(2) ES suggested the correct, but non-exact decision (it gave out some possible alternatives among which there was also the correct solution), while input data of this task provide its exact solution.

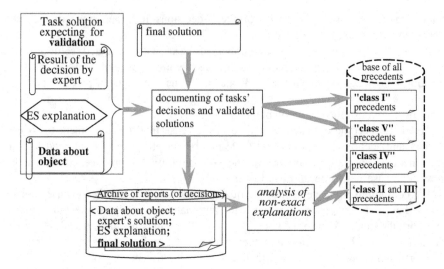

Fig. 1. The scheme of relations of document flow with quality control process

(3) ES suggested the correct, but non-exact solution, however input data of this task provide for more exact solution (reduction of number of alternatives).

(4) ES suggested the correct, but non-exact solution, however input data of this task don't provide for more exact solution.

(5) ES proposed the wrong solution (a set of alternatives without correct solution or the empty set).

The function "to classify all precedents" consists in assignment of each precedent to one of the specified classes. It is obvious that the decision on assignment of precedent to classes 2-4 can't be made automatically therefore it has to be made by the experts from KB management group.

It is shown in Fig. 1 that not only the task' solution and all known data about object are documented, but also the explanation received from ES. In case of discrepancy of the explanation created by system and the final solution conclusion it is necessary to estimate, whether the couple <a known data about object, a final decision> may be used as precedent for knowledge base correction.

The precedents of classes 1 and 4 form *KB correctness assessment* and *accuracy assessment* (the assessment is precedents set). The new precedents assigned to these classes can be included in this assessment without KB modification, unlike the new precedents assigned to classes 2, 3 and 5. Classes 1 and 4 won't change, or some precedents will pass from a class 4 into a class 1.

New precedents from classes 2 and 3 demand improvement (correction) of KB, i.e. its such admissible modification when class-2 precedents pass into a class 1, and class-3 precedents pass into classes 1 or 4. Such improvement of KB aims to add these precedents into correctness and accuracy assessments of modified KB.

New precedents from a class 5 demand correction or expansion of the KB, i.e. its such admissible modification when class-5 precedents pass into classes 1 or 4. KB correction or expansion aims to add these precedents into correctness and accuracy assessments of modified or expanded KB also.

Search of all possible options of such admissible modifications of KB for all new precedents is an essence of the function of search of opportunities for inclusion of new precedents in the assessment of KB (the function "to find all possible ways of modification of KB"). These options of admissible modifications of KB have to be «calculated» automatically by KB control system.

As a result of performance of the previous function some versions of admissible modifications of KB for new group of precedents can be received, or there will not be such versions. Therefore choice of one such version of admissible modification (if there are) and its replacement is an essence of function "to modify KB". If there are not such options a revision of some previous decisions at modification of KB is needed. This function has to be implemented by the experts from group of management of KB with assistance of a control system. For supporting in receiving new formalized knowledge (Fig. 2) the following means are required: convenient software instrument for formation of the learning sample from the precedents of classes 2 and 3, tools for automatic formation of the next version of KB modification and for its estimation. If KB version with an assessment not worse than the specialist's assessment is received, this version becomes new KB for ES (instead of the "current KB" used for this moment).

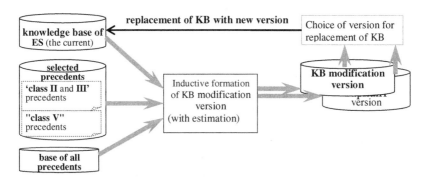

Fig. 2. The scheme of automatic control of knowledge bases quality

Additional source of improvement of the knowledge used during solving of intellectual activity problems are the new scientific results relating to this intellectual activity. It is reasonable that the KB control system has to allow addition of new scientific results in KB (without making worse of its assessment). Such modification of KB can be carried out only by the experts from KB management group (Fig. 3). The automated support for estimation of KB version included scientific results is required. It assist to be convinced that the assessment of the knowledge base has improved. If the assessment of the KB included new scientific knowledge doesn't become worse such version of modification becomes the new KB of ES.

Administrative measures for control of specialists' decisions and of used knowledge are naturally connected with the measures related to matching specialists' knowledge to modern updated knowledge. Software and computer simulators are

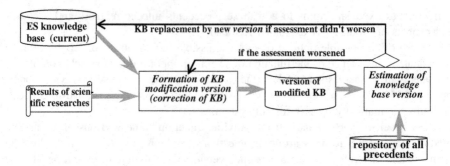

Fig. 3. An inclusion in the knowledge base of results of scientific researches

necessary for support of training process and raising the level of specialists' skill. The knowledge in such computer simulators must completely correspond to the most modern formed knowledge bases.

5   The Advantages of Automation of Knowledge Control

The development of software for KB control support gives benefit both for the specialists carrying out daily intellectual activity and for managers of this activity. The advantages are as following:

- obtaining the knowledge base of ES, whose quality assessment isn't worse, than specialists' knowledge quality assessment,
- achievement of monotonous growth of ES knowledge base quality assessment;
- availability of the knowledge base with the best assessment for use and training;
- possibility of timely bringing the most modern knowledge to specialists;
- objectivity of estimation of specialists' daily activity.

6   Conclusion

For continuous improvement of a decisions' quality, a knowledge bases quality control has to be exercised along with automation of daily intellectual activity. Quality control of knowledge bases demands automation and integration with subsystem of electronic document flow and with expert systems.

The main objectives of software systems for KB quality control are documentation of task decisions data; exercising KB and search for new precedents for it; accumulation and classification of new precedents; support of formation of admissible KB modifications. The documentation of task decisions data includes the task initial data, ES explanation of decision process and final validated solution of the task. The exercising KB by every new validated solution (analysis of ES explanation' correctness in comparison with validated solution) allows KB to be improved over time and remain actual.

Acknowledgements. This work is performed with partial financial support of the RFBR (the projects № 15-07-03193 and № 13-07-00024-a).

References

1. Kleschev, A.S., Chernyakhovskaya, M.Y., Shalfeeva, E.A.: The paradigm of an intellectual professional activity automation. part 1. the features of an intellectual professional activity. Ontology Designing (Russ.) **3**(9), 53–69 (2005)
2. Gavrilova, T., Horoshevskiy, V.: Knowledge bases of intellectual systems. Piter, St. Petersburg (2001)
3. Giarratano, J.C., Riley, G.D.: Expert Systems: Principles and Programming. Course Technology, Boston (2004)
4. Kobrinskiy, B.A.: Retrospective analysis of medical expert systems. Artif. Intell. News (Russ.) **2**, 6–17 (2005)
5. Solovyov, S., Solovyova, G.: Methods of debugging of knowledge bases in system the FIACRE. In: Seminar "Automation and robotization of production with application of microprocessor means", pp. 36–37. Kishinev (1986)
6. Mayfield J., Finin T.: Evaluating the quality of a knowledge base populated from text. In: Proceedings of the NAACL Joint Workshop on Automatic Knowledge Base Construction and Web-scale Knowledge Extraction, pp. 68–73. Montreal (2012)
7. Telnov, Y.F.: Intelligent Information Systems. MESI, Moscow (2004)
8. Popov, E.V., Fominykh, I.B., Kissel, E.B., Shapot, M.D.: Static and Dynamic Expert Systems. Finance and statistics, Moscow (1996)

Maximum Class Separability-Based Discriminant Feature Selection Using a GA for Reliable Fault Diagnosis of Induction Motors

Md. Rashedul Islam, Sheraz A. Khan, and Jong-Myon Kim[✉]

Department of Electrical, Electronics, and Computer Engineering,
University of Ulsan, Ulsan, South Korea
{rashed.cse,sherazalik,jongmyon.kim}@gmail.com

Abstract. Reliable fault diagnosis in bearing elements of induction motors, with high classification performance, is of paramount importance for ensuring steady manufacturing. The performance of any fault diagnosis system largely depends on the selection of a feature vector that represents the most distinctive fault attributes. This paper proposes a maximum class separability (MCS) feature distribution analysis-based feature selection method using a genetic algorithm (GA). The MCS distribution analysis model analyzes and selects an optimal feature vector, which consists of the most distinguishing features from a high dimensional feature space, for reliable multi-fault diagnosis in bearings. The high dimensional feature space is an ensemble of hybrid statistical features calculated from time domain analysis, frequency domain analysis, and envelope spectrum analysis of the acoustic emission (AE) signal. The proposed maximum class separability-based objective function using the GA is used to select the optimal feature set. Finally, k-nearest neighbor (k-NN) algorithm is used to validate our proposed approach in terms of the classification performance. The experimental results validate the superior performance of our proposed model for different datasets under different motor rotational speeds as compared to conventional models that utilize (1) the original feature vector and (2) a state-of-the-art average distance-based feature selection method.

Keywords: Fault diagnosis · Feature distribution analysis · Feature selection · Genetic algorithm · Intra class density · Inter class distance · K-NN

1 Introduction

Modern manufacturing systems rely heavily on the use of induction motors and they have a significant role in the productivity of any industry. Thus, faults in induction motors can lead to their unexpected and sudden failure, thereby adversely affecting production and leading to complete shutdown [1, 2]. Therefore, it is imperative for the purpose of industrial productivity that we develop reliable methods for fault diagnosis of bearings in induction motors in order to avoid potential failure.

Several methods and techniques have been developed over the years to reliably diagnose faults in the bearings of induction motors using vibration, current, and voltage

© Springer International Publishing Switzerland 2015
D.-S. Huang and K. Han (Eds.): ICIC 2015, Part III, LNAI 9227, pp. 526–537, 2015.
DOI: 10.1007/978-3-319-22053-6_56

signals. Recently, acoustic emission (AE) signals processing-based methods have been proposed. All of these methods share the same fundamental steps: (1) extraction of fault features from the original signal, (2) selection of a set of features, and (3) classification of faults. However, the selection of a set of features that can correctly characterize all of the underlying fault conditions is a challenging issue.

Different feature extraction methods are used in different signal analysis domains i.e. time domain [3, 4], frequency domain [4, 5], and the time-frequency domain [6–9], for bearing fault diagnosis. To identify different fault conditions hybrid high dimensional features extracted from different signal domains is becoming increasingly popular [10]. However, a large number of features does not always guarantee high classification performance, because not all of the extracted features carry the same significance in characterizing a given fault condition; some can even degrade the classification performance. Furthermore, a high dimensional feature vector has the added disadvantage of being computationally inefficient. Therefore, in order to select the optimal features, we require an efficient feature selection method.

A feature selection method selects an optimal subset of the original feature set that best represents the underlying fault conditions and accurately performs fault classification. The most representative subset is determined using some suitable criteria that minimize the number of features required for improved performance [11]. There are many approaches that can be used for feature selection. These can be broadly categorized into three groups: (1) the wrapper approach, (2) the filter approach, and (3) the hybrid approach [10, 11]. The wrapper approach can be applied using complete search, sequential search, or the heuristic search. Heuristic search approaches like GA can ensure a compromise between the quality of the feature subset and the computational complexity [11]. In this study, we use a GA for feature selection, which ensures the selection of the best features while maintaining a reasonable computational complexity.

With GA, in addition to other parameters, the fitness or objective function has a direct effect on the selection of the optimal feature set. Depending upon the application environment and signal properties, designing an effective objective function is a challenging task. In [12], the classification accuracy is used as a fitness value in a GA-based feature selection process, which can guarantee higher classification accuracy for a particular classifier. However, embedding a classifier inside the GA process is computationally inefficient. In [13], a GA is used to reduce the number of features with the help of an objective function that incorporates a smaller intra-class distance and a larger inter-class separation. In that study, average distances between samples are used as a measure for both the intra-class density and the inter-class separation. However, this average distance-based approach has several drawbacks that can adversely affect the optimal feature selection process. The average distance-based compactness (density) value of a class ignores samples that are located on the outskirts of a class in less dense areas. These samples may overlap with another class. In the case of average distance-based inter-class separation, a high distance value between two classes can dominate both the distance measures of other nearby classes and the overall class separability value.

To address these critical issues, this paper presents a novel maximum class separability distribution analysis-based objective function (fitness function), which works in full consort with the genetic algorithm to select the best feature subset for fault diagnosis of induction motors. In the proposed model, a hybrid feature vector is extracted from an AE fault signal. Then, the GA is applied to the original feature vector to select the optimal feature set. Finally, in order to validate the optimization of the feature space, a k-nearest neighbors (k-NN) classifier is utilized for classifying the different faults. To evaluate the proposed model, its classification accuracy is compared with two other approaches: one that does not involve any feature selection approach and one that utilizes an average distance-based state-of-the-art objective function. The results show the superior performance of our approach.

The rest of this paper is organized as follows. Section 2 describes the experimental setup for AE fault signal acquisition. Section 3 describes the proposed fault diagnosis model with GA-based optimal feature selection. Section 4 presents the experimental results and analysis. Finally, Sect. 5 concludes this paper.

2 AE Fault Signal Acquisition

Generally, researchers use different types of sensors for collecting data from induction motors, including displacement sensors, current sensors, vibration sensors, thermal sensors, and acoustic emission (AE) sensors [2, 3]. Acoustic emission (AE) is used for low-speed bearing defect diagnosis because it is effective in capturing low-energy signals and detecting incipient bearing defects [14]. Therefore, this paper employs an AE-based technique for incipient bearing fault diagnosis.

To obtain the AE signals of bearing defects, a self-designed experimental environment is setup for the purpose of this study, as depicted in Fig. 1(a). An AE sensor is placed on top of the bearing house of the shaft at the non-drive end, which is connected with an induction motor via a gearbox (1:1.52). A diamond cutter bit is used to generate physical cracks at different physical locations on the surface of a bearing as depicted in Fig. 1(b) along with crack's specifications. In this study, different signals for single and combined fault conditions are collected from a motor operating at different rotational speeds. Moreover, normal signals are acquired for healthy bearings without any faults. A displacement transducer is used to measure the motor speed.

Fig. 1. (a) Signal acquisition environment and (b) Different bearing cracks with specifications.

In this study, three different AE signals for single faults and four different AE signals for combined faults are acquired from different bearings. The single fault signals include signals for an *inner raceway* fault, *outer raceway* fault, and *roller* fault. The combined fault signals include signals for *inner raceway & roller* faults, *outer & inner raceway* faults, *outer raceway & roller* faults, and *inner & outer raceway & roller* faults. AE signals for each fault condition are collected at speeds of 300 RPM, 350 RPM, 400 RPM, 450 RPM, and 500 RPM. In the experimental dataset, there are 90 different AE signals for each motor speed and for each fault type. The length of each signal is 10 s and the sampling frequency is 250 kHz.

3 Proposed Model

The performance of any diagnostic system is strongly dependent on the quality of the feature vectors. The proposed model investigates the optimal feature selection and an efficient fault diagnosis system. Figure 2 shows the block diagram of our proposed model; all of the modules are described in the following sections.

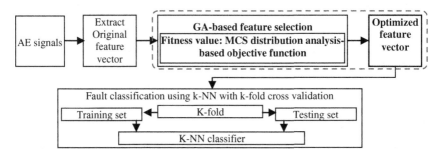

Fig. 2. The proposed fault diagnosis model with the GA-based features selection scheme.

3.1 Hybrid Feature Extraction

To obtain the most distinctive fault information from the AE signals, we start with a large set of extracted features. Our final feature vector has a total of 22 features including 10 time domain statistical parameters of the AE signal, three frequency domain statistical features, and nine RMS features that are calculated from the defect frequency region of the fault using the envelope power spectrum.

The statistical feature elements of the hybrid feature vector for the detection of faults, i.e. different time and frequency domain statistical features of the AE signals, are listed in Tables 1 and 2 along with the mathematical relations to calculate them. The time domain statistical features are root mean square (RMS), square root of the amplitude (SRA), kurtosis value (KV), skewness value (SV), peak-to-peak value (PPV), crest factor (CF), impulse factor (IF), margin factor (MF), shape factor (SF) and kurtosis factor (KF), whereas, the frequency domain statistical features are frequency center (FC), rms frequency (RMSF) and root variance frequency (RVF).

Table 1. Statistical features from the time domain of the AE signal

$$RMS = \left(\frac{1}{N}\sum_{i=1}^{N}x_i^2\right)^{1/2}$$

$$SRA = \left(\frac{1}{N}\sum_{i=1}^{N}\sqrt{|x_i|}\right)^2$$

$$KV = \frac{1}{N}\sum_{i=1}^{N}\left(\frac{x_i-\bar{x}}{\sigma}\right)^4$$

$$SV = \frac{1}{N}\sum_{i=1}^{N}\left(\frac{x_i-\bar{x}}{\sigma}\right)^3$$

$$PPV = \max(x_i)-\min(x_i)$$

$$CF = \frac{\max(|x_i|)}{\left(\frac{1}{N}\sum_{i=1}^{N}x_i^2\right)^{1/2}}$$

$$IF = \frac{\max(|x_i|)}{\frac{1}{N}\sum_{i=1}^{N}|x_i|}$$

$$MF = \frac{\max(|x_i|)}{\left(\frac{1}{N}\sum_{i=1}^{N}\sqrt{|x_i|}\right)^2}$$

$$SF = \frac{\left(\frac{1}{N}\sum_{i=1}^{N}x_i^2\right)^{1/2}}{\frac{1}{N}\sum_{i=1}^{N}|x_i|}$$

$$KF = \frac{\frac{1}{N}\sum_{i=1}^{N}\left(\frac{x_i-\bar{x}}{\sigma}\right)^4}{\left(\frac{1}{N}\sum_{i=1}^{N}x_i^2\right)^2}$$

Table 2. Statistical features from the frequency domain of the AE signal

$$FC = \frac{1}{N}\sum_{i=1}^{N}f_i \qquad RMSF = \left(\frac{1}{N}\sum_{i=1}^{N}f_i^2\right)^{1/2} \qquad RVF = \left(\frac{1}{N}\sum_{i=1}^{N}(f_i-FC)^2\right)^{1/2}$$

To find more specific fault information, envelope power spectrum of the AE fault signal is used to extract RMS features of each fault's defect frequency region. The steps involved in extracting these RMS features are illustrated in Fig. 3.

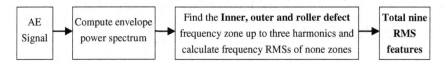

Fig. 3. Process for extracting the envelope spectrum RMS features.

The RMS features extracted from the inner, outer and roller defect frequency ranges of the AE fault signal's envelope power spectrum are shown in Fig. 4 where the green, black, and red rectangular windows represent the inner, outer, and roller defect frequency ranges, respectively. The frequency ranges for each defect are calculated using Eqs. (1), (2), and (3), respectively, where fs is the operating frequency, fc is the case frequency, fi is the *inner* defect frequency, fo is the *outer* defect frequency, and fr is the *roller* defect frequency.

$$range\_inner = 2 \times (number\_of\_sideband \times (fs + fs \times error\_rate) + error\_rate \times fi) \quad (1)$$

$$range\_outer = 2 \times error\_rate \times fo \quad (2)$$

$$range\_roller = 2 \times (number\_of\_sideband \times (fc + fc \times error\_rate) + error\_rate \times fr) \quad (3)$$

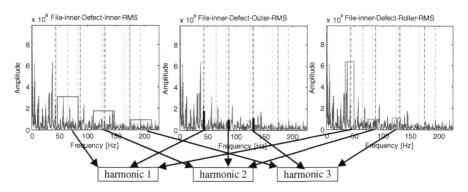

Fig. 4. Defect frequency range of the envelope spectrum RMS feature for inner, outer, and roller defects.

3.2   Maximum Class Separability Distribution Analysis-Based Feature Selection Using the GA

The GA, which is based on natural evolutionary theory, is a robust and effective optimization technique with heuristic search [14–18]. An application of GA to any problem involves five essential steps: 1) *encoding*, 2) *parent selection*, 3) *crossover & mutation*, 4) *fitness evaluation of offspring* and 5) *replacement*. The final solution is a chromosome, which is a combination of genes where each gene represents an element of the optimal feature vector.

In the encoding process, the initial population of chromosomes is created using binary encoding. Strings of 1's and 0's, each representing a chromosome, are constructed as follows: (1) the length of each string, i.e. chromosome, is equal to the number of dimensions of the original feature vector space. (2) if the ith dimension of the original feature space is retained, then the ith element of the chromosome is set to '1'; otherwise, it is set to '0'. In this study, the initial population consists of 500 chromosomes and the number of required generations is set to 500. In parent selection process, fitness values of all chromosomes are calculated using the proposed maximum class separability distribution analysis-based objective function and the best chromosomes are selected for further breeding using the stochastic roulette-wheel selection method.

This study uses generational GA and in each generation, k number of offspring are produced using uniform crossover and one point mutation. Finally using worst-case replacement policy, these new offspring are used to replace the worst chromosomes in the population if the offspring's fitness (calculated using objective function) is higher than theirs; otherwise, the offspring are discarded.

- Maximum class separability analysis-based objective function.

In this study, we propose a novel objective function that calculates the maximum class separability (MCS) value by analyzing the discriminative feature distribution. The proposed objective or fitness function, as given in Eq. (5), is used to evaluate all of the chromosomes in the population. The larger the value of this objective function for a chromosome, the better that chromosome represents the most distinctive features with good feature distribution. In our proposed method, we use the Euclidean distance to calculate the MCS between two samples, as given in Eq. (4) where, x and y are two feature vectors (samples) and n is the total number of feature variables in a vector.

$$d_{x,y} = \sqrt{\sum_{i=1}^{n} (x_i - y_i)} \tag{4}$$

To find the maximum objective value (MCS value), the proposed method finds the maximum inter class distance (i.e., separation) and the minimum intra class density. The MCS value is calculated by Eq. (5).

$$MCS(objective\_value) = \frac{Inter\_class\_distance}{Intra\_class\_density} \tag{5}$$

In general, the fault features are distributed in the multi-dimensional feature space. Depending on the Euclidean distance between samples, the proposed method calculates the inter class distance and intra class density values in an efficient way, enabling the whole process to find the most distinctive features for reliable fault diagnosis. Algorithms 1 and 2 illustrate the process of calculating the inter class distance and intra class density values.

Algorithm 1. Intra class density

INPUT: number of classes N_c, number of samples of each class n, all samples of different fault classes with selected feature dimension
PROCESS:
Repeat for class $n_c \in N_c$
   $center_c \leftarrow$ Calculate the class median of a class ()
        {
           $d\_vector \leftarrow$ Calculate distance vectors between all samples of a class
           $mean\_distance \leftarrow$ Calculate average distance of all samples to others
           $class\_median \leftarrow$ Find a sample that has the minimum average to other samples
        }
   $within\_class\_distance \leftarrow$ Calculate all distance from $center_c$ to all samples in class
   $class\_density_c \leftarrow$ find the maximum distance value
End repeat
$intra\text{-}class\text{-}density_c \leftarrow mean\ (class\_density_c)$
End Process

Algorithm 2. Inter class distance

INPUT: number of classes N_c, number of samples of each class n, all samples of different fault classes with selected feature dimension
PROCESS:
Repeat for class $n_c \in N_c$
   Repeat for class m_c without n_c
        $distance \leftarrow$ Calculate distances between all samples from class n_c to m_c
        $between\_class\_distance_{nm} \leftarrow$ Find the minimum distance between n_c to m_c
   End Repeat
   $Local\_inter\_class\_distance_n \leftarrow$ calculate average of all $between\_class\_distance_{nm}$
End Repeat
$inter\_class\_distance \leftarrow$ calculate average of all $Local\_inter\_class\_distance_n$
End Process

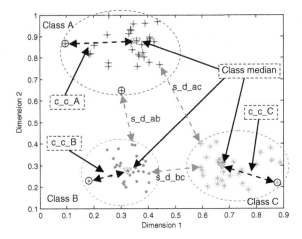

Fig. 5. Calculation of intra class density and inter class distance values

Figure 5 shows the intra class density and inter-class distance values, which are calculated using Algorithms 1 and 2, respectively, where c_c_A, c_c_B, and c_c_C are intra class density values for different classes and s_d_ac, s_d_ab, and s_d_bc are inter class distances among all of the classes.

3.3   Fault Classification Using K-NN

The proposed feature optimization model selects an optimal feature set with the most effective feature elements for fault diagnosis. Once we select the most distinguishable features, we use the k-NN classifier to validate our optimization model in terms of the classification performance. The k-NN classifier is one of the most popular classification methods [19], and it is widely used because of its simplicity and computational efficiency. The basic idea of k-NN is to classify a sample depending on the votes of its k-nearest neighbors. The nearest neighbors are determined by calculating the distance parameter [19].

4   Experimental Results and Analysis

This section presents the experimental results and analysis of the proposed fault diagnosis approach. We utilize five different datasets collected from the motor at rotational speeds of 300 RPM, 350 RPM, 400 RPM, 450 RPM and 500 RPM. Each dataset contains eight types of AE signals including three single fault signals, four combined fault signals and one normal baseline signal.

In the feature extraction step, we extract ten time domain statistical features, three frequency domain statistical features, and nine envelope spectrum RMS features from each AE signal; this yields a feature vector with 22 elements. There is a huge variation among the values of different features, which hampers classification performance as the high valued features tend to dominate the ones with lower values. To overcome this problem, all features are normalized using min-max normalization. In a dataset of N samples and F feature variables per sample, the normalized value of each feature variable $x_{n,f}$ can be calculated by Eq. (6).

$$Normalized(x_{n,f}) = \frac{x_{n,f} - f_{\min}}{f_{\max} - f_{\min}} \qquad (6)$$

Here, $x_{n,f}$ is the value of feature variable f for the nth sample of the dataset and f_{min} and f_{max} are the minimum and maximum values, respectively, of feature variable f for all samples in the dataset.

A high dimensional feature vector does not always guarantee higher classification performance. We can observe that some dimensions overlap with each other for different fault classes. However, some dimensions are well separated, which can help in achieving high classification performance.

In order to select the best set of features that can achieve the highest classification performance, we apply the generational GA to the original feature vector. In the proposed GA-based optimization, the proposed maximum class separability (MCS) distribution analysis-based objective function is used to evaluate the fitness of all chromosomes of the population in each generation. This objective function considers the distribution of different features and tries to find the optimal feature dimensions that yield minimum intra class density values (maximal compact classes) and maximum inter class distances, thereby maximizing the objective value. Figure 6 shows the distribution of some features and the corresponding objective values for the different feature dimensions of the sample dataset.

By applying the GA and the proposed objective function, the redundant and less significant dimensions are discarded, and an optimal feature vector is selected from the original feature space. Table 3 shows the number of selected optimal features for the different datasets.

Finally, we apply the k-NN classifier to validate our feature optimization. With the k-NN classifier, choosing the appropriate value of k is important. However, there is no proper method for the selection of an appropriate value for k. In the experiments, we tried different values of k and finally set k = 3. To attain reliable classification performance, we repeat the classification process 10 times, and each time we use k-fold

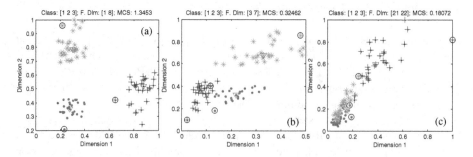

Fig. 6. Sample feature distributions with corresponding objective (MCS) values.

Table 3. Number of selected optimal features for the different datasets

	Dataset-1 300 RPM	Dateset-2 350 RPM	Dataset-3 400 RPM	Dataset-4 450 RPM	Dataset-5 500 RPM
Optimal feature dimensions	2nd, 3rd, 11th, 13th, 14th	2nd, 3rd, 4th, 5th, 9th 11th	2nd, 11th, 13th	2nd, 9th, 11th	2nd, 9th, 11th

Table 4. Classification performance of the three different approaches for individual fault types of the different datasets. Here M-1, M-2 and M-p represents without feature selection, average distance-based feature selection, proposed feature selection.

Dataset	Model	Inner	Outer	Roller	Inner +roller	Outer +inner	Outer +roller	Inner +outer +roller	Normal
Dataset-1 300 RPM	M-1	99.83	81.50	98.78	100.00	96.17	97.56	94.22	99.83
	M-p	99.56	95.67	100.00	100.00	98.89	100.00	99.67	99.61
	M-2	96.94	67.22	96.00	96.67	86.22	92.61	93.94	94.83
Dataset-2 350 RPM	M-1	99.56	89.56	95.44	98.44	95.22	96.78	97.33	100.00
	M-p	99.78	96.22	99.94	99.89	97.89	98.11	96.17	100.00
	M-2	99.17	89.72	98.00	67.67	98.44	88.22	97.39	100.00
Dataset-3 400 RPM	M-1	98.11	88.28	99.39	100.00	94.17	96.39	91.56	100.00
	M-p	100.00	96.50	99.94	100.00	100.00	99.00	92.22	100.00
	M-2	100.00	74.78	100.00	97.11	85.50	56.89	73.61	100.00
Dataset-4 450 RPM	M-1	99.11	91.86	96.22	100.00	94.17	99.00	98.33	100.00
	M-p	100.00	93.17	97.83	100.00	98.56	100.00	99.72	100.00
	M-2	100.00	57.61	84.22	98.67	86.06	89.00	94.89	82.11
Dataset-5 500 RPM	M-1	100.00	100.00	99.94	100.00	100.00	98.06	97.17	100.00
	M-p	100.00	100.00	99.72	100.00	100.00	98.89	100.00	100.00
	M-2	100.00	99.50	99.61	100.00	100.00	98.28	98.33	100.00

(where k = 3) cross validation. The k-fold cross validation splits the dataset randomly into k mutual folds. More specifically, out of the three mutual folds, one fold (30 feature vectors of each fault condition) is used as the training set and the other two folds (60 feature vectors of each fault condition) are used as the testing set.

The experimental results show that our proposed feature selection model, using the GA with a novel objective function, achieves high classification performance for the different datasets. To validate the effectiveness of the proposed approach, we compare its performance with those obtained using the original feature vector (without any feature selection) as well as using an existing average distanced-based state-of-the-art feature selection method (i.e., Algorithm 1 [13]). Table 4 shows the classification precision values for the different fault types and the different datasets using the proposed method and the other two methods for comparison. Additionally, Table 5 shows the averaged classification performance of the three different approaches for the different datasets.

Table 5. Classification performance of the three different approaches

Models	Dataset-1	Dataset-2	Dataset-3	Dataset-4	Dataset-5
Without feature selection	95.99	96.54	95.99	97.34	99.40
Proposed feature selection	99.17	98.50	98.46	98.66	99.83
Ex. Algorithm-1	90.56	92.33	85.99	86.57	99.47

The experimental results indicate that our proposed model shows significant performance improvement for different datasets. In a few cases, the proposed method shows somewhat similar performance to the approach that does not involve feature selection. However, our proposed model drastically reduces the dimensionality of the original feature vector by selecting optimal features, which significantly reduces the computational complexity of the problem. The existing state of the art approach fails to perform as par with the proposed approach, because it does not consider the distribution of features during the feature selection process.

5   Conclusions

This paper presents a maximum class separability-based distinctive feature selection method using a genetic algorithm. In this study, a novel maximum class separability value-based objective function was proposed and evaluated. The proposed objective function analyzes the distribution of features in a high dimensional feature space and helps the GA to select the optimal set of features. The proposed GA-based optimization eliminated unnecessary feature elements and drastically reduced the number of dimensions of the original feature set. Finally, a k-NN classifier with k-fold cross validation was used to validate the feature set optimization. In terms of its classification performance, the proposed feature set optimization outperformed both the original feature vector and also an optimization method using a state-of-the-art average distance-based selection model.

Acknowledgements. This work was supported by the National Research Foundation of Korea (NRF) grant funded by the Korea government (MSIP) (No. NRF-2013R1A2A2A05004566)

References

1. Zhao, M., Jin, X., Zhang, Z., Li, B.: Fault diagnosis of rolling element bearings via discriminative subspace learning: visualization and classification. J. Expert Syst. Appl. **41** (7), 3391–3401 (2014)
2. Uddin, J., Islam, R., Kim, J.: Texture feature extraction techniques for fault diagnosis of induction motors. J. Convergence **5**(2), 15–20 (2014)
3. Prieto, M.D., Cirrincione, G.A., Espinosa, G., Ortega, J.A., Henao, H.: Bearing fault detection by a novel condition-monitoring scheme based on statistical-time features and neural networks. IEEE Trans. Ind. Electron. **30**(8), 3398–3407 (2013)

4. Yu, J.: Local and nonlocal preserving projection for bearing defect classification and performance assessment. IEEE Trans. Ind. Electron. **59**(5), 2363–2376 (2012)
5. Bediaga, I., Mendizabal, X., Arnaiz, A., Munoa, J.: Ball bearing damage detection using traditional signal processing algorithms. IEEE Instrum. Meas. Magz. **16**(2), 20–25 (2013)
6. Lau, E.C.C., Ngan, H.W.: Detection of motor bearing outer raceway defect by wavelet packet transformed motor current signature analysis. IEEE Trans. Instrum. Meas. **59**(10), 2683–2690 (2010)
7. Kankar, P.K., Sharma, S.C., Harsha, S.P.: Fault diagnosis of rolling element bearing using cyclic autocorrelation and wavelet transform. Neurocomput. **110**, 9–17 (2013)
8. Konar, P., Chattopadhyay, P.: Bearing fault detection of induction motor using wavelet and support vector machines (SVMs). Appl. Soft Comput. **11**(6), 4203–4211 (2011)
9. Rafiee, J., Rafiee, M.A., Tse, P.W.: Application of mother wavelet functions for automatic gear and bearing fault diagnosis. Expert Syst. Appl. **37**(6), 4568–4579 (2010)
10. Rauber, T.W., de Assis Boldt, F., Flavio, M.V.: Heterogeneous feature models and feature selection applied to bearing fault diagnosis. IEEE Trans. Industr. Electron. **62**(1), 637–646 (2015)
11. Mahrooghy, M., Nicolas, H.Y.: On the use of the genetic algorithm filter-based feature selection technique for satellite precipitation estimation. IEEE Geosci. Remote Sens. Lett. **9**(5), 963–967 (2012)
12. Kanan, H.R., Faez, K.: GA-based optimal selection of PZMI features for face recognition. Appl. Math. Comput. **205**(2), 706–715 (2008)
13. Nguyen, N.T., Lee, H.H., Kwon, J.: Optimal feature selection using genetic algorithm for mechanical fault detection of induction motor. J. Mech. Sci. Technol. **22**(3), 490–496 (2008)
14. Beasley, D., Bull, D.R., Martin, R.R.: An Overview of genetic algorithms: Part 1, fundamentals. Univ. Comput. **15**(2), 58–69 (1993)
15. Yu, X., Shao, J., Dong, H.: On evolutionary strategy based on hybrid crossover operators. In: International Conference on Electronic and Mechanical Engineering and Information Technology (EMEIT), vol. 5, pp. 2355–2358 (2011)
16. Mudaliar, D.N., Modi, N.K.: Unraveling travelling salesman problem by genetic algorithm using m-crossover operator. In: 2013 International Conference on Signal Processing Image Processing & Pattern Recognition (ICSIPR), pp. 127–130 (2013)
17. Qi-yi, Z., Shu-chun, C.: An improved crossover operator of genetic algorithm. In: International Symposium on Computational Intelligence and Design, ISCID 2009, vol. 2, pp. 82–86 (2009)
18. Ouerfelli, H., Dammak, A.: The genetic algorithm with two point crossover to solve the resource-constrained project scheduling problems. In: 5th International Conference on Modeling, Simulation and Applied Optimization (ICMSAO 2013), pp. 1–4 (2013)
19. Zhang, N., Yang, J., Qian, J.: Component-based global k-NN classifier for small sample size problems. Pattern Recogn. Lett. **33**(13), 1689–1694 (2012)

Multi-fault Diagnosis of Roller Bearings Using Support Vector Machines with an Improved Decision Strategy

M.M. Manjurul Islam, Sheraz A. Khan, and Jong-Myon Kim[✉]

School of Electrical, Electronics, and Computer Engineering,
University of Ulsan, Ulsan, South Korea
{m.m.manjurul, sherazalik, jongmyon.kim}@gmail.com

Abstract. This paper proposes an efficient fault diagnosis methodology based on an improved one-against-all multiclass support vector machine (OAA-MCSVM) for diagnosing faults inherent in rotating machinery. The methodology employs time and frequency domain techniques to extract features of diverse bearing defects. In addition, the proposed method introduces a new reliability measure (SVMReM) for individual SVMs in the multiclass framework. The SVMReM achieves optimum results irrespective of the test sample location by using a new decision strategy for the proposed OAA-MCSVM based method. Finally, each SVM is trained with optimized kernel parameters using a grid search technique to enhance the classification accuracy of the proposed method. Experimental results show that the proposed method is superior to conventional approaches, yielding an average classification accuracy of 97 % for five different rotational speed conditions, eight different fault types and two different crack sizes.

Keywords: Multi-fault diagnosis · Support vector machines · Dempster-Shafer (D-S) theory · Reliability measure · Decision rule

1 Introduction

Reliable fault diagnosis of rolling element bearings (REBs) is a challenging task, especially in industrial machinery. Bearing defects, if not detected in time, can eventually lead to machine failure and cause costly downtime. This has prompted research into vibration analysis [1, 2], motor current signature analysis, and oil debris analysis for fault diagnosis. Vibration analysis has been widely used because it provides essential information about diverse bearing failures [2, 3]. Recently though, the use of acoustic emissions [4] has found prominence in the detection and identification of bearing defects because of its ability to capture low-energy signals that are characteristic of low rotational speeds [4–6]. Specifically, Yoshioka et al. [6] showed that AE could be used to identify bearing defects even before they appear in the range of vibration acceleration, whereas vibration analysis can only to detect defects after they appear on the bearing's surface [3–7].

AE signal based fault diagnosis usually involves fault feature extraction from AE signals, feature selection and fault classification. Fault feature extraction maps the

© Springer International Publishing Switzerland 2015
D.-S. Huang and K. Han (Eds.): ICIC 2015, Part III, LNAI 9227, pp. 538–550, 2015.
DOI: 10.1007/978-3-319-22053-6_57

original signal to statistical parameters that reflect the diverse symptoms of bearing defects. The feature pool in this study includes 10 time-domain statistical parameters, 3 frequency-domain, and 9 complex envelope components. Once we extract features from the AE signal, our problem reduces to achieving better classification accuracy.

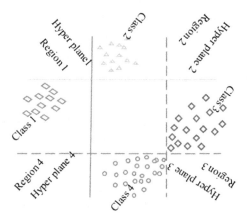

Fig. 1. OAA SVM. Solid lines show true boundary lines and dashed lines linear boundaries

Several classification methods including support vector machines (SVM) [8, 9, 12], artificial neural networks (ANN) [4, 14], fuzzy sets theory, and expert systems have been used in bearing fault diagnosis [3, 5]. SVM is a robust classification tool [13–15] that classifies unknown data into multiple classes [9, 12] by decomposing an L-class into a number of two class problems and constructing a binary classifier for each. In this paper, an improved decision fusion based one-against-all (OAA) multi-class support vector machine is designed for improved fault diagnosis in roller bearings. OAA-MCSVM classifies an unknown feature vector based upon the binary classifier with the largest value for the classifier decision rule. This approach ignores the competence of individual classifiers and relies only on the value of the decision rule, therefore it does not always yield the best results [8–10, 14]. In Fig. 1, which shows the application of the one against all approach to a four class problem, the solid lines separate class 1 and 2 by a definite margin from the remaining classes whereas the dashed lines fail to achieve the same in case of class 3 and 4. Some regions (such as region 3) of the feature space are undecided, where a sample is accepted by more than one class or rejected by all. Individual SVMs are treated equally by the standard OAA-MCSVM method despite differences in their quality, thus compromising the overall classification accuracy. Therefore, it is advantageous to give due weight to each SVM classifier on the basis of classification competence and formulate a new decision strategy to achieve that. The contributions made in this paper are as follows:

- The OAA-MCSVM is modified to improve its classification performance, by introducing a reliability measure, SVMReM, for each SVM and formulating a new decision aggregation rule.

- The effectiveness of the proposed OAA-MCSVM is validated by accurately classify multiple bearing defects under different load conditions and at different rotational speeds.

The remaining parts of the paper is organized as follows. Section 2 explains the experimental setup. Section 3 provides analytical grounds to support our proposition under the framework of Dempster-Shafer (D-S) theory [12] and introduces a reliability measure (SVMReM) for individual SVMs that achieves optimal value irrespective of the test sample location in the feature space. Section 4 describes the proposed method with feature extraction and detailed derivation of reliability measures. Section 5 presents the experimental results, and Sect. 6 concludes this paper.

2 Experimental Setup and Data Acquisition Details

The experimental test rig used for this study is shown in Fig. 2. The AE fault signal is acquired using a general purpose, wideband AE sensor (WS α from Physical Acoustics Corporation) [3–6, 8], placed on top of the bearing housing as shown in Fig. 2. AE sensors are designed to capture high frequency acoustic emissions emanating from periodic impact events involving bearing faults.

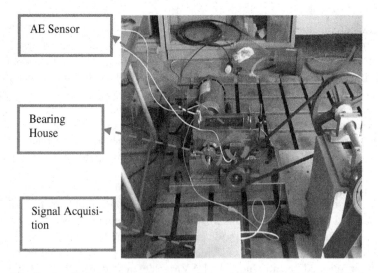

Fig. 2. A self-designed test rig for obtaining AE signals of bearing defects

In this study, one normal and seven different AE fault signals are obtained; (i) (DFN) the normal condition, (ii) (BCO) bearing with outer race crack, (iii) (BCR) bearing with roller crack; (iv) (BCI) bearing with inner race crack, (v) (BCIO) bearing with cracks on inner and outer raceways, (vi) bearing with inner and roller cracks (BCIR), (vii) (BCOR) bearing with roller and outer cracks, and (viii) (BCIOR) bearing with inner, outer, and roller cracks.

3 Reliable One-Against-All (OAA) Multiclass Support Vector Machine (SVM)

The OAA-MCSVM method for an L-class problem creates L binary SVM classifiers to separate each class from the remaining L-1 classes. If we consider a classification problem with P training samples $\{x_1, y_1\}, \ldots, \{x_P, y_P\}$, where $x_j \in R^{Dim}$ is a Dim-dimensional feature vector of the jth training sample and $y_j \in \{1, 2, 3, \ldots, L\}$ is the class to which it belongs. The jth SVM solves the optimization problem given in Eq. (1) [11–14], which provides the jth decision value function.

$$\text{Min } L\left(w, \delta_j^k\right) = \frac{1}{2}\|w\|^2 + c\sum_{j=1}^{p} \delta_j^k, \text{subjected}, y_j'\left(w_j \cdot \phi(x_j) + b_j\right) \geq 1 - \delta_j^k, \delta_j^k \geq 0$$

(1)

Where $y_j' = 1$ if $y_j = L$, otherwise $y_j = -1$. In the classification stage, a test sample x is classified to be in class k such that the decision function Z_k has the highest value as given in Eq. (2).

$$Z_k(x) = (w_k \cdot \varphi(x_k) + b_k), where, k = \frac{\text{arg max}}{k=1,\ldots,L} Z_k(x)$$

(2)

3.1 Improving the Standard OAA-MCSVM Using Dempster-Shafer Belief Theory

The proposed approach improves the standard OAA MCSVM using Dempster-Shafer (D-S) evidence theory [12, 16]. Consider a test sample x, a set of L OAA multiclass classifiers SVMk each with a decision function Zk and a set of hypotheses $\Omega = \{L_K\}$ for $1 \leq k \geq L$, where the kth hypothesis asserts that the sample belongs to class k. When each SVMk is applied to x, the result is a part of evidence supporting a certain proposition [8, 10, 15]. We define a basic belief assignment function (BBA) referred to as belief mass mk on hypothesis Ω [10]. It is actually based on the result of the kth classification. When sign (Zk) = 1, it is logical to enhance belief in SVMk proportional to the value of Zk, because "x fits into class k". This piece of evidence alone does not guarantee the truth of hypothesis Lk, only part of it is committed to the belief in hypothesis Lk. The remaining part of the evidence provided by the value of Zk is assigned to the belief in hypothesis Ω as a whole which asserts that "x does not belong to class k". The basic belief assignment (BBA) function mk is defined for an SVM with a positive response.

$$m_k(\{x\}) = \begin{cases} 1 - \exp(-|Z_k(x)|) = \lambda_k, if, x = \{L_k\} \\ \exp(-|Z_k(x)|) = 1 - \lambda_k, if, x = \Omega \\ 0, else \end{cases}$$

(3)

When $sign(Z_k) = -1$, SVM_k classifies x as not belonging to class k. Then, the belief mass function m_k, for such SVMs with negative responses, is defined by (4).

$$m_k(\{x\}) = \begin{cases} 1 - \exp(-|Z_k(x)| = \lambda_k, if, x = \{\overline{L_k}\} \\ \exp(-|Z_k(x)| = 1 - \lambda_k, if, x = \Omega \\ 0, else \end{cases} \tag{4}$$

After finding the BBA values for all the samples and using D-S rule of combination to obtain the combined BBA, the belief function can be computed using Eq. (5) and finally a test sample x is labeled as class $k*$ with the highest belief.

$$k^* = \begin{array}{c} \text{arg max} \\ k = 1, \ldots, L \end{array} Bel(\{L_k\}) = \begin{array}{c} \text{arg max} \\ k = 1, \ldots, L \end{array} \lambda_k = \begin{array}{c} \text{arg min} \\ k = 1, \ldots, L \end{array} z_k(x)$$

$$= \begin{array}{c} \text{arg max} \\ k = 1, \ldots, L \end{array} |z_k(x)|, \tag{5}$$

Similar to Eq. (5), if at least one or more SVM_k generates a positive response, the belief function can be defined by Eq. (6)

$$k^* = \begin{array}{c} \text{arg max} \\ k = 1, \ldots, L \end{array} Bel(\{L_k\}) = \begin{array}{c} \text{arg max} \\ z_k(x) \geq 0 \end{array} \lambda_k = \begin{array}{c} \text{arg min} \\ k = z_k(x) \geq 0 \end{array} z_k(x)$$

$$= \begin{array}{c} \text{arg max} \\ k = 1, \ldots, L \end{array} |z_k(x)| \tag{6}$$

However, it is not prudent to place the entire belief in decisions made by the SVMk's. When the independent sources of evidence cannot be fully believed, the BBA should be weakened by a certain factor [16]. Thus an appropriate degree to quantify the amount of belief in every SVMk is proposed and defined in the following section.

4 Methodology

The proposed diagnosis method as shown in Fig. 3, includes feature calculation, reliability calculation of each SVMk, and a new decision strategy for the proposed OAA MCSVMs which are then used for fault classification.

4.1 Feature Calculation

The feature pool is produced based on 10 time domain statistical features, 3 frequency domain features, and 9 envelope spectrum RMS features. Such a rich feature pool minimizes the risk of missing important aspects of the data, and it can achieve greater discrimination among various fault conditions. The time and frequency domain parameters are enumerated in the Table 1, where 'x' is the original time-domain signal with N data samples and "f" is a spectral component of the signal 'x'. We divide 10 s AE signals into different samples of one time period length each. In every revolution, bearing faults give rise to impact signals depending upon their defect frequencies.

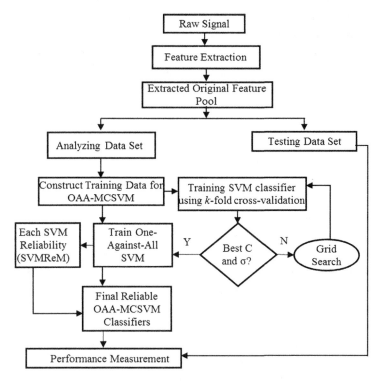

Fig. 3. The proposed fault diagnosis model

Table 1. Multi-domain feature extraction from AE signals

Time-domain feature parameters		
$X_{rms} = \left(\frac{1}{N}\sum_{i=1}^{N} x_i^2\right)^{\frac{1}{2}}$	$X_{sf} = \dfrac{\max(|x_i|)}{\left(\frac{1}{N}\sum_{i=1}^{N} x_i^2\right)^{1/2}}$	$X_{mf} = \dfrac{\max(|x_i|)}{\left(\frac{1}{N}\sum_{i=1}^{N}\sqrt{|x_i|}\right)^2}$
$X_{kv} = \frac{1}{N}\sum_{i=1}^{N}\left(\frac{x_i - \bar{x}}{\sigma}\right)^4$	$X_{sv} = \frac{1}{N}\sum_{i=1}^{N}\left(\frac{x_i - \bar{x}}{\sigma}\right)^3$	$X_{kf} = \dfrac{\frac{1}{N}\sum_{i=1}^{N}\left(\frac{x_i - \bar{x}}{\sigma}\right)^4}{\left(\frac{1}{N}\sum_{i=1}^{N} x_i^2\right)^2}$
$X_{ppv} = \max(x_i) - \min(x_i)$	$X_{cf} = \dfrac{\max(|x_i|)}{\left(\frac{1}{N}\sum_{i=1}^{N} x_i^2\right)^{\frac{1}{2}}}$	$X_{rms} = \left(\frac{1}{N}\sum_{i=1}^{N}\sqrt{|x_i|}\right)^2$
$X_{if} = \dfrac{\max(|x_i|)}{\frac{1}{N}\sum_{i=1}^{N}|x_i|}$		
Frequency-domain feature parameters		
$X_{fc} = \frac{1}{N}\sum_{i=1}^{N} f_i$	$X_{rmsFr} = \left(\frac{1}{N}\sum_{i=1}^{N} f_i^2\right)^{1/2}$	$X_{rvf} = \left(\frac{1}{N}\sum_{i=1}^{N}\left(f_i - X_{fc}\right)^2\right)^{\frac{1}{2}}$

Three types of defects are detected on a bearing depending on their location on the bearing surface. Bearing defect frequency could be measured on the basis of bearing parameters and rotational speed and frequency, for respective defect.

Figure 4 shows the typical signals produced by confined faults in rolling bearing, corresponding envelope signals are calculated using amplitude demodulation [2, 8].

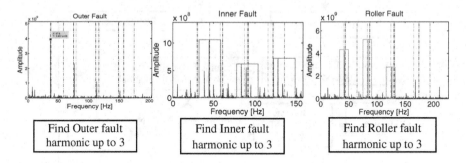

Fig. 4. Complex envelope analysis based on a unique feature extraction method

The defect region for the extraction of features can be determined using Gaussian window method [10, 14], however, it completely ignores the effects of sidebands during defect region calculation [10]. An enveloping spectrum is sensitive to very low impact related events such as sidebands. Thus, we construct a rectangular window around the defect frequency, to overcome problems with Gaussian windows and extract root mean square (RMS) features from the envelope spectrum. A distribution of the extracted features is given in Fig. 5.

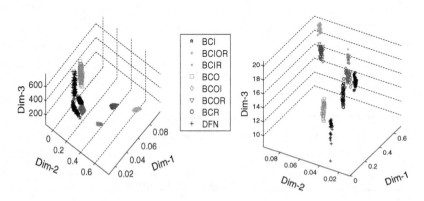

Fig. 5. 3D visualization results of the discriminative fault features: all feature samples of data set #1 (3 MM_300 RPM) for 8 fault conditions (left) and training sample (right)

4.2   Reliability Calculation for Individual SVMs

The effectiveness of a classifier can be calculated using the generalization error R rate, restated as R = E[y = sign (Z(x))], where $y \in \{-1, 1\}$ is the label of true class x and Z is the decision function. A classifier can be measured as more reliable if it provides a smaller value for R [10, 15, 16]. As investigated in [14], when the amount of training sets is relatively small comparing with the dimension of the feature space x, then the small value of Remp will not always ensure a small value regularization error R [11]. In this case, the upper bound of R is defined in [10, 11], and one benefit of SVM is to minimize this objective function value, also minimizes this upper bound of function. In other words, the smaller values of the objective function means smaller regularization errors, and thus it ensures a reliable classifier. So, we restate the objective function for the reliability calculation based on the SVM upper bound optimization equation in (7).

$$ObjFunc = {}^1\!/_2 \left\| w \right\|^2 + C \sum_{k=1}^{n} (1 - y_k \, z(x_k))_+ \tag{7}$$

To calculate the ObjFunc, we introduce the unconstrained optimization [9, 10] problem over 'w' in the training stage. Our problem is quadratic; to get a globally optimum solution for 'w', we utilize QP [9]. The objective function would ensure a minimum value of R with the widest margin for all samples in the training stage.

$$\begin{bmatrix} \delta \\ \\ SVMReM = \exp\left(-ObjcFunc/\theta\right) = \exp\left(-\dfrac{{}^1\!/2\|w\|^2 + c\sum_{k=1}^{P}(1-y_k z(x_k))_+}{2CP}\right) \end{bmatrix} \tag{8}$$

In order to get the reliability measure for each class, and kernel function in Eq. (8), which yields an optimum value for all samples in the feature space. Furthermore, the margin of the classifier is $\frac{2}{\|w\|}$, a smaller value of $\|w\|$ ensures a larger margin, and thus more accurate generalization. The value $\theta = CP$ is used as a regulation factor to compensate the consequence of C, penalty parameter, P, number of training samples.

4.3   Derivation of Proposed Decision Strategy for Reliable OAA-MCSVM

With the reliability measure δ_{SVMReM} in place, we design a new decision function for our proposed one-against-all multiclass SVM (OAA-MCSVM. First, Zk values of the binary SVMs are calculated for all the given test samples using Eq. (2). Then, a decision fusion is done for all samples. After that, the L class BBA mk is generated based on Eqs. (5) and (6) for all the decisions, where all decisions are considered as information from an independent source. The evidence value that is produced by each classifier SVMk is granted more belief if SVMk is also more reliable, and our calculated competence measures are directly used as a belief factor in Eqs. (9) and (10). The final decision rules as follows:

1. If all the SVM_k produce negative responses,

$$k^* = \frac{\text{arg min}}{k = 1, \ldots L} \, \delta_{SVMReM(k)} \left(1 - \exp^{-|Z_k(x)|} \right) \tag{9}$$

2. If one or more SVM_k produce positive responses,

$$k^* = \frac{\text{arg max}}{Z_k(x) \geq 0} \, \delta_{SVMReM(k)} \left(1 - \exp^{-|Z_k(x)|} \right) \tag{10}$$

$\delta_{SVM \, Re \, M \, (k)}$ is the reliability measure for SVM_k, and Eqs. (9) and (10) can be merged into one based on Eq. (11).

$$k^* = \frac{\text{arg max}}{1, \ldots L} \left[\delta_{SVMReM(k)} \left\{ sign(Z_k(x)) \left(1 - \exp^{-|Z_k(x)|} \right) \right\} \right] \tag{11}$$

The value of sign codes the hard decision rule whether "x belongs to class k or not" and the magnitude defines the strength of the decision function value.

4.4   Fault Classification

A SVM discriminates test samples into one of two categories, and subsequently we need multi-class SVMs (MCSVMs) to identify multiple bearing defects. In the OAA multiclass support vector machine approach, each SVM separates one category from the rests, and the final decision is by choosing an SVM that produces the maximum decision value for a given sample. We compare our proposed OAA-MCSVM using a new decision rule with the standard OAA-MCSVM.

5   Experimental Validation

The proposed method is tested on multi-class fault data sets obtained from the experimental setup. The proposed method delivers superior classification performance over its standard counterpart.

5.1   Configuration of the Training and Test Data

The proposed method is evaluated over ten data sets with 90 feature vectors each. Data set for each fault condition, is randomly divided into two subsets for training and testing. The training subset includes 40 randomly-selected feature vectors whereas the remaining 50 feature vectors makeup the testing subset. In the training phase of every SVM, the precision is assessed by several kernel parameters [16] ($C = 2^{-5}, 2^{-3}, 2^{-1} \ldots,$ 2^{15}), and ($\gamma = 2^{-15}, 2^{-13} \ldots, 2^{3}$) and the best one is chosen using a grid search method

[16] for performance comparison. Consequently, classification accuracy is computed for the standard OAA-MCSVMs and the proposed (reliable) OAA-MCSVMs on the testing datasets; the final classification performance is the average value of the accuracies achieved for each feature vector in the testing dataset.

5.2 Performance Evaluation

To ensure the effectiveness of this proposed OAA-MCSVM method, this experimental analysis compares its classification performance with the standard variant. We use confusion matrix [16] and AUC-ROC based performance analysis for the comparison.

Experiment # 1. This analysis was carried out on ten datasets at 5 rotational speeds (300, 350, 400, 450, 500 RPM), two crack sizes (3 mm and 6 mm) and eight different fault conditions. The support vector machines use the Gaussian radial basis kernel. The results show that our proposed OAA-MCSVM is superior to the standard OAA-MCSVM. The classification accuracy for different data sets is given in Table 2. In addition, the results also validate our unique feature extraction process, which yields classification accuracy of almost 100 % at higher RPMs and larger crack sizes, because

Table 2. Classification performance comparison between standard OAA MCSVMs and the proposed OAA MCSVM for eight fault conditions using radial basis kernel machine (Unit: %)

Data set	Method	BCI	BCIOR	BCIR	BCO	BCOI	BCOR	BCR	DFN	Model ACC. (%)
3 MM 300 RPM	Standard	97.57	96.57	94.57	98.51	98.91	99.31	99.44	99.17	**98.01**
	Proposed	100	100	96.00	98.93	99.33	99.73	99.87	99.47	**99.17**
3 MM 350 RPM	Standard	97.57	99.44	96.17	98.91	99.57	99.17	97.69	99.57	**98.51**
	Proposed	100	99.43	99.71	99.14	99.86	99.00	99.86	100	**99.63**
3 MM 400 RMP	Standard	99.44	98.77	94.57	99.17	98.91	99.44	99.44	99.57	**98.67**
	Proposed	99.87	99.20	100	99.60	99.33	99.73	99.87	100	**99.70**
3 MM 450 RPM	Standard	99.31	98.64	99.46	99.44	96.57	99.58	99.27	99.59	**98.98**
	Proposed	99.87	99.20	100	99.73	100	99.73	99.73	100	**99.78**
3 MM 500 RPM	Standard	98.57	99.31	99.39	99.64	95.67	100	100	99.57	**99.02**
	Proposed	100	99.73	100	99.87	100	100	100	100	**99.95**
12 MM 300 RPM	Standard	99.57	98.37	97.57	99.57	99.57	99.57	99.57	99.57	**99.17**
	Proposed	100	100	100	100	100	100	100	100	**100**
12 MM 350 RPM	Standard	98.57	99.07	95.67	99.87	96.87	100	100	100	**98.76**
	Proposed	100	100	100	100	100	100	100	100	**100**
12 MM 400 RPM	Standard	99.57	99.57	99.57	99.41	99.67	97.37	99.86	99.52	**99.32**
	Proposed	100	100	100	100	100	100	100	100	**100**
12 MM 450 RPM	Standard	97.97	97.22	99.69	98.59	100	100	100	100	**99.19**
	Proposed	100	100	100	100	100	100	100	100	**100**
12 MM 500 RPM	Standard	98.57	99.97	98.87	99.57	100	98.22	100	100	**99.40**
	Proposed	100	100	100	100	100	100	100	100	**100**

of good separation between fault features. In Fig. 6, we compare the overall performance of Standard OAA MCSVMs and Proposed OAA MCSVMs over ten data sets (in %) using three different kennel functions.

Experiment # 2. We perform the AUC-ROC based analysis of the proposed and standard approaches for all the datasets to verify the robustness of our proposed algorithm. AUC-ROC graph is actually a way of visualization, organization and selection classifiers based on their performance. An area under the receiver operating characteristics (AUC-ROC) analysis has been extended for use behavior analyzing of fault diagnostic systems [14]. In addition, AUC-ROC curves for the proposed method show that it delivers a more uniform performance in the diagnosis of all fault conditions as compared to the standard approach, as shown in Fig. 7.

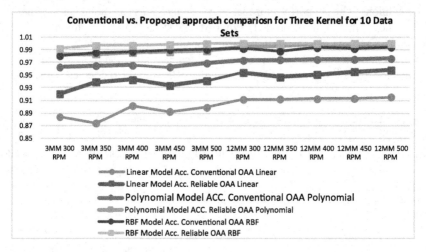

Fig. 6. Relative comparison of the accuracy of proposed and standard methods over three different kennel machines.

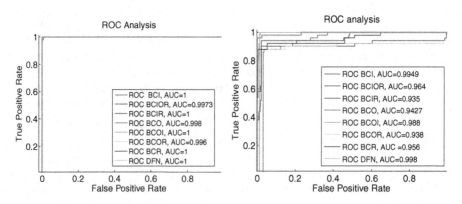

Fig. 7. AUC-ROC characteristic curve of performance comparison between the standard and proposed OAA MCSVM approach.

6   Conclusion

A reliable fault diagnosis methodology for rotating machinery was proposed and evaluated, based upon a modified form of one-against-all multi-class support vector machines, which utilizes individual reliability measures, SVMReM, of the binary SVMs and an improved decision strategy based on the Dempster-Shafer (D-S) theory of evidence. In addition to time and frequency domain analysis techniques, the acquired data from the experiments was preprocessed using envelope analysis methods to extract meaningful features from the fault signals. The experimental analysis demonstrated that the proposed approach yields better classification performance for different fault conditions, at different rotational speeds and different kernel functions for the SVMs. The proposed method yielded an average accuracy exceeding 99 %, 98 %, and 95 % for ten data sets with SVM trained using RBF kernel, polynomial kernel, and linear kernel, respectively.

Acknowledgements.   This work was supported by the National Research Foundation of Korea (NRF) grant funded by the Korean government (MSIP) (No. NRF-2013R1A2A2A05004566).

References

1. Van, M., Kang, H.-J., Shin, K.-S.: Rolling element bearing fault diagnosis based on non-local means de-noising and empirical mode decomposition. Sci. Measur. Technol. IET **8**, 571–578 (2014)
2. Jin, X., Zhao, M., Chow, T.W.S., Pecht, M.: Motor bearing fault diagnosis using trace ratio linear discriminant analysis. IEEE Trans. Ind. Electron. **61**(5), 2441–2451 (2014)
3. Saidi, L., Ben Ali, J., Fnaiech, F., Morello, B.: Bi-spectrum based-EMD applied to the non-stationary vibration signals for bearing faults diagnosis. In: 6th International Conference on Soft Computing and Pattern Recognition, pp. 25–30. IEEE Press (2014)
4. Kang, M., Kim, J., Kim, J.-M.: Reliable fault diagnosis for incipient low-speed bearings using fault feature analysis based on a binary bat algorithm. Inf. Sci. **294**, 423–438 (2015)
5. Ferrando, J.L., Kappatos, V., Balachandran, W., Gan, T.-H.: A novel approach for incipient defect detection in rolling bearings using acoustic emission technique. Appl. Acoust. **89**, 88–100 (2015)
6. Yoshioka, T., Fujiwara, T.: Application of acoustic emission technique to detection of rolling bearing failure. Am. Soc. Mech. Eng. **14**, 55–76 (1984)
7. Harmouche, J., Delpha, C., Diallo, D.: Improved fault diagnosis of ball bearings based on the global spectrum of vibration signals. IEEE Trans. Energy Conserv. **30**, 376–383 (2015)
8. Zhang, X., Chen, W., Wang, B., Chen, X.: Intelligent fault diagnosis of rotating machinery using support vector machine with ant colony algorithm for synchronous feature selection and parameter optimization. Neurocomputing, **167**, 260–279 (2015)
9. Fan, Y., Wang, Z., Zhang, Y., Wang, H.: Fault isolation of non-gaussian processes based on reconstruction. Chemometr. Intel. Lab. Syst. **142**, 9–17 (2014)
10. Soualhi, A., Medjaher, K., Zerhouni, N.: Bearing health monitoring based on Hilbert-Huang transform, support vector machine, and regression. IEEE Trans. Instrum. Measur. Reliab. **64**, 52–62 (2015)

11. Shafer, G.: A Mathematical Theory of Evidence. Princeton University Press, New Jersy (1976)
12. Liu, Y., Zheng, Y.F.: One-against-all multi-class SVM classification using reliability measures. In: IEEE International Joint Conference on Neural Networks, vol. 2, pp. 849–854 (2005)
13. Rauber, T.W., Assis, F., Varejao, F.M.: Heterogeneous feature models and feature selection applied to bearing fault diagnosis. IEEE Trans. Ind. Electron. **62**, 637–646 (2015)
14. Vladimir, N.V.: An overview of statistical learning theory. IEEE Trans. Neural Netw. **10**, 988–999 (1999)
15. Clifton, L., Clifton, D.A., Yang, Z., Watkinson, P., Tarassenko, L., Yin, H.: Probabilistic novelty detection with support vector machines. IEEE Trans. Reliab. **63**, 455–467 (2014)
16. Moreno-Torres, J.G., Saez, J.A., Herrera, F.: Study on the impact of partition-induced dataset shift on -fold cross-validation. IEEE Trans. Neural Netw. Learn. Sys. **23**, 1304–1312 (2012)

A Local Neural Networks Approximation Control of Uncertain Robot Manipulators

Minh-Duc Tran[1] and Hee-Jun Kang[2(✉)]

[1] Graduate School of Electrical Engineering,
University of Ulsan, Ulsan 680-749, South Korea
ductm.ctme@gmail.com
[2] School of Electrical Engineering,
University of Ulsan, Ulsan 680-749, South Korea
hjkang@ulsan.ac.kr

Abstract. In this paper, an adaptive finite-time tracking control scheme is proposed for uncertain robotic manipulators. The controller is developed based on combination of terminal sliding mode control technique and radian basis function neural networks (RBFNNs). The RBFNNs are used to directly approximate individual element of the inertial matrix, the Coriolis matrix and gravity torques vector. The adaptation laws are derived to adjust on-line the parameters of RBFNNs. Finally, the simulation results of a two-link robot manipulator are presented to illustrate the effectiveness of the proposed control method.

Keywords: Nonsingular terminal sliding mode control · Radial basis function neural network · Adaptive control · Finite-time convergence · Robot manipulator

1 Introduction

In recent decades, robot manipulators were applied widely in various fields, in which many tasks require high-speed and high-precision trajectory tracking. However, robotic manipulators are generally nonlinear and involve many uncertainties and external disturbances in their dynamics, such as payload variations, friction, external disturbances, and sensor noise etc. Therefore, designs of a controller that can attenuate the effects of robotic uncertainties have become the subject of many researches. In order to deal with this problem, many control approaches have been proposed such as proportional-integral-derivative (PID) control [1], robust control [2, 3], adaptive control [4, 5], sliding mode control [6–11], and neural network control [12–16].

In this paper, an adaptive terminal sliding mode control based on local approximation method is proposed for trajectory tracking of uncertain robotic manipulators. At first, the controller is developed based on terminal sliding mode technique. Then, in order to improve the system performance and attenuate the effects of uncertainties, the RBFNNs are used to directly approximate individual element of the inertial matrix, the Coriolis matrix and gravity torques vector. Finally, a simulation study is performed on a two-link robot manipulator to prove the effectiveness of the proposed control method.

© Springer International Publishing Switzerland 2015
D.-S. Huang and K. Han (Eds.): ICIC 2015, Part III, LNAI 9227, pp. 551–557, 2015.
DOI: 10.1007/978-3-319-22053-6_58

The rest of this paper is arranged as follows. The system dynamics and problem formulation are described in Sect. 2. The structure of terminal sliding mode neural networks controller is presented in Sect. 3. In Sect. 4, simulation results for a two-link robot manipulator are provided to demonstrate the performance of the proposed controller. Finally, some important remarks are concluded in Sect. 5.

2 System Dynamics and Preliminaries

The dynamics of a serial n-links robot manipulator can be written as [17]

$$M(q)\ddot{q} + C(q,\dot{q})\dot{q} + G(q) + \tau_d = \tau \tag{1}$$

where $q(t), \dot{q}(t), \ddot{q}(t) \in \mathbf{R}^n$ are the vector of joint accelerations, velocities and positions, respectively. $M(q) \in \mathbf{R}^{n \times n}$ is the inertial matrix, $C(q,\dot{q}) \in \mathbf{R}^{n \times n}$ expresses the centripetal and Coriolis matrix, $G(q) \in \mathbf{R}^n$ represents the gravity torques vector, $\tau \in \mathbf{R}^n$ is the control torque, and $\tau_d \in \mathbf{R}^n$ is the bounded external disturbance vector.

For convenience, the above dynamic equation has the following useful structural properties;

Property 1: $M(q)$ is a symmetric positive definite matrix.

$$m_1 \|x\|^2 \le x^T M(q)x \le m_2 \|x\|^2; \forall x \in R^n \tag{2}$$

where m_1, m_2 are known positive scalar constants, $x \in R^n$ is a vector, $\|\|$ denotes the Euclidean vector norm.

Property 2: $M(q) - 2C(q,\dot{q})$ is a skew symmetric matrix.

$$x^T[\dot{M}(q) - 2C(q,\dot{q})]x = 0 \tag{3}$$

for any vector $x \in \mathbf{R}^n$.

The control objective of this paper is to design a stable control law to ensure that the tracking error between joint position vector q and desired joint position vector q_d converge to zero in finite time.

3 Controller Design

In order to apply the terminal sliding mode control, it is necessary to define the terminal sliding surface $s(t)$ for n-link robot manipulator as

$$s = \dot{e} + \beta sig(e)^\varphi \tag{4}$$

where $\beta = diag(\beta_1, \beta_2, \ldots, \beta_n)$, $\beta_1, \beta_2, \ldots \beta_n$ are positive constants, $0 < \varphi < 1$, $sig(\dot{e})^\varphi = (|\dot{e}_1|^\varphi sign(\dot{e}_1), |\dot{e}_2|^\varphi sign(\dot{e}_2), \ldots, |\dot{e}_n|^\varphi sign(\dot{e}_n))$, $s = [s_1, s_2, \ldots, s_n]$, $e(t) = q(t) - q_d(t)$, $\dot{e}(t) = \dot{q}(t) - \dot{q}_d(t)$.

According to the sliding mode design procedure, the control input u consists of the components

$$\tau = u_{eq} - K_{SW} sign(s) \tag{5}$$

where $K_{SW} = diag(k_{SW1}, k_{SW2}, \ldots, k_{SWn})$, $k_{SW1}, k_{SW2}, \ldots, k_{SWn}$ are positive constants. The equivalent control can be interpreted as the continuous control law that is obtained by equation $\dot{s} = 0$ for nominal system in the absence of the uncertainties and external disturbances.

$$\dot{s} = \ddot{e} + \varphi\beta|e|^{\varphi-1}\dot{e} \tag{6}$$

From (1), the $\ddot{e}$ is given by

$$\ddot{e} = \ddot{q}_d - \ddot{q} = \ddot{q}_d - \frac{\tau - \tau_d - C(q,\dot{q})\dot{q} - G(q)}{M(q)} \tag{7}$$

Multiplying both sides of Eq. (6) by $M(q)$ and substituting (7) into it yields

$$M(q)\dot{s} = M(q)\ddot{q}_d + C(q,\dot{q})\dot{q} + G(q) + \varphi\beta M|e|^{\varphi-1}\dot{e} - \tau + \tau_d \tag{8}$$

Define $\dot{q}_s = s + \dot{q}$, then $\ddot{q}_s = \dot{s} + \ddot{q}$, $\dot{q}_s = \dot{q}_d + \beta sig(e)^{\varphi}$, $\ddot{q}_s = \ddot{q}_d + \varphi\beta|e|^{\varphi-1}\dot{e}$. From (8), we have

$$M(q)\dot{s} = -C(q,\dot{q})s - \tau + \tau_d + M(q)\ddot{q}_s + C(q,\dot{q})\dot{q}_s + G(q) \tag{9}$$

If the nonlinear robot dynamic functions $M(q)$, $C(q,\dot{q})$, $G(q)$ are clearly known, then the equivalent control can be defined as

$$u_{eq} = M(q)\ddot{q}_s + C(q,\dot{q})\dot{q}_s + G(q) + Ks \tag{10}$$

where $K = diag(k_1, k_2, \ldots, k_n)$, $k_1, k_2, \ldots, k_n$ are positive constants.

The stability of the close loop system (10) can be easily proved by Lyapunov theory if the gains of the switching controller are bigger than the upper bounds of uncertainties. However, robot manipulators are complex nonlinear systems which involve many uncertainties and external disturbances. Therefore, the RBFNNs are used to directly approximate individual element of the inertial matrix, the Coriolis matrix and gravity torques vector of the robot. Therefore Eq. (9) becomes.

$$M(q)\dot{s} = -C(q,\dot{q})s - \tau + \tau_d + \hat{M}(q)\ddot{q}_s + \hat{C}(q,\dot{q})\dot{q}_s + \hat{G}(q) \tag{11}$$

where $\hat{M}(q)$, $\hat{C}(q,\dot{q})$, and $\hat{G}(q)$ be the estimates of $M(q)$, $C(q,\dot{q})$, and $G(q)$, respectively. For the system (11), the proposed controller is expressed by the following equation

$$u_{eq} = \hat{M}(q)\,\ddot{q}_s + \hat{C}(q,\dot{q})\dot{q}_s + \hat{G}(q) + Ks \tag{12}$$

where $\hat{M}(q) = \left[\hat{W}_M^T \bullet \sigma_M(q)\right]$, $\hat{C}(q,\dot{q}) = \left[\hat{W}_C^T \bullet \sigma_C(q,\dot{q})\right]$, $\hat{G}(q) = \left[\hat{W}_G^T \bullet \sigma_G(q)\right]$. The neural network weight vectors are designed as follows.

$$\dot{\hat{W}}_{Mk} = \Gamma_{Mk}.\{\sigma_{Mk}(q)\}\,\ddot{q}_s\,s_k \tag{13}$$

$$\dot{\hat{W}}_{Ck} = \Gamma_{Ck}.\{\sigma_{Ck}(q,\dot{q})\}\dot{q}_s s_k \tag{14}$$

$$\dot{\hat{W}}_{Gk} = \Gamma_{Gk}.\{\sigma_{Gk}(q)\}s_k \tag{15}$$

where $k = 1, 2, \ldots, n$. Γ_{Mk}, Γ_{Ck}, and Γ_{Gk} are constant symmetric positive definite matrices. $\hat{W}_{Mk}, \hat{W}_{Ck}$, and $\hat{W}_{Gk}$ are column vectors with their elements being W_{Mkj}, W_{Ckj}, W_{Gkj}, respectively.

4   Simulation Results

In this section, to verify the validity and effectiveness of the proposed method, the performance of the proposed controller is tested via simulation on a two-link planar robotic manipulator. The simulations are performed in the MATLAB-Simulink environment using ODE 4 solver with a fixed-step size of 10^{-4} s.

The dynamic equation of the two-link robot is described as follows

$$\begin{bmatrix} M_{11}(q) & M_{12}(q) \\ M_{12}(q) & M_{22}(q) \end{bmatrix}\begin{bmatrix} \ddot{q}_1 \\ \ddot{q}_2 \end{bmatrix} + \begin{bmatrix} C_{11}(q,\dot{q}) & C_{12}(q,\dot{q}) \\ C_{12}(q,\dot{q}) & C_{22}(q,\dot{q}) \end{bmatrix}\begin{bmatrix} \dot{q}_1 \\ \dot{q}_2 \end{bmatrix} + \begin{bmatrix} G_1(q) \\ G_2(q) \end{bmatrix} + \begin{bmatrix} \tau_{d_1} \\ \tau_{d_2} \end{bmatrix}$$
$$= \begin{bmatrix} \tau_1 \\ \tau_2 \end{bmatrix} \tag{16}$$

where the inertia matrix $M_{ij}(q)$ is given by

$$M_{11}(q) = (m_1 + m_2)l_1^2 + m_2 l_2^2 + 2m_2 l_1 l_2 \cos(q_2),$$
$$M_{12}(q) = M_{21}(q) = m_2 l_2^2 + m_2 l_1 l_2 \cos(q_2),$$
$$M_{22}(q) = m_2 l_2^2,$$

The Coriolis and centrifugal matrix $C_{ij}(q,\dot{q})$ is given by

$$C_{11}(q,\dot{q}) = -m_2 l_1 l_2 \sin(q_2)\dot{q}_2 \quad C_{12}(q,\dot{q}) = -m_2 l_1 l_2 \sin(q_2)(\dot{q}_1 + \dot{q}_2)$$
$$C_{21}(q,\dot{q}) = m_2 l_1 l_2 \sin(q_2)\dot{q}_1 \qquad C_{22}(q,\dot{q}) = 0$$

The gravity torques vector $G_i(q)$ is given by

$$G_1(q) = (m_1 + m_2)l_1g\cos(q_2) + m_2l_2g\cos(q_1 + q_2),$$
$$G_2(q) = m_2l_2g\cos(q_1 + q_2),$$

The parameters values employed to simulate the robot are given as of $l_1 = 1\,\text{m}$, $l_2 = 0.8\,\text{m}$, $m_1 = 1\,\text{kg}$, $m_2 = 1\,\text{kg}$, $\beta = diag(12, 12)$, $\varphi = 0.8$, $K = diag(150, 50)$, $K_{SW} = diag(5, 5)$. The external disturbances are selected as

$$\tau_d = \begin{bmatrix} \tau_{d_1} \\ \tau_{d_2} \end{bmatrix} = \begin{bmatrix} 1.7\sin(2t) \\ 1.3\cos(2t) \end{bmatrix} \tag{17}$$

The reference trajectories are given by

$$\begin{aligned} q_{1d} &= 0.3\sin(\pi t) \\ q_{2d} &= 0.5\sin(\pi t) \end{aligned} \tag{18}$$

The initial states are chosen as

$$q_1(0) = 0.4, \ q_2(0) = -0.5, \ \dot{q}_1(0) = 0, \ \dot{q}_2(0) = 0 \tag{19}$$

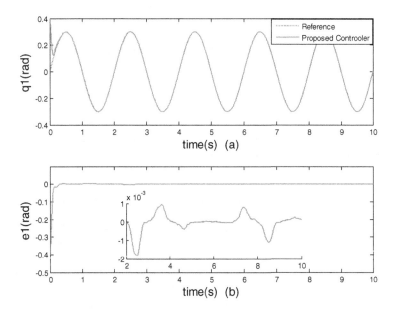

Fig. 1. Responses at joint 1: (a) Output tracking, (b) Tracking error.

The simulation results are shown in Figs. 1 and 2. It can be observed that the controller can track the desired trajectory very well, the controller brings very small tracking errors, and the state trajectories reach to the origin in a finite amount of time.

Thus, the simulation results demonstrate that the proposed controller can effectively control the unknown nonlinear dynamic system.

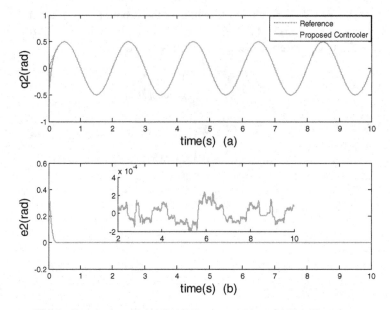

Fig. 2. Responses at joint 2: (a) Output tracking, (b) Tracking error.

5   Conclusion

In this paper, an adaptive terminal sliding mode control based on local approximation method is proposed for trajectory tracking of uncertain robotic manipulators. The controller is developed based on the combination of terminal sliding mode control technique and radian basis function neural networks (RBFNNs). Adaptive learning laws have been derived to adjust on-line the output weights of the RBFNNs during the trajectory tracking control of robot manipulators without any offline training phases. In the simulation example, the proposed control is applied to a two-link robotic manipulator. The simulation results are given to demonstrate the effectiveness of the proposed method.

Acknowledgments. This work was supported by the Research Fund of University of Ulsan.

References

1. Arimoto, S., Miyazaki, F.: Stability and robustness of PID feedback control for robot manipulators of sensory capability. In: Brady, M., Paul, R.P. (eds.) Robotic Research. MIT Press, Cambridge (1984)

2. Spong, M.W.: On the robust control of robot manipulators. IEEE Trans. Autom. Control **37**(11), 1782–1786 (1992)
3. Abdallah, C., Dorato, D.M., Jamishidi, M.: Survey of the robust control robots. Control Syst. Mag. **12**(2) (1991)
4. Slotine, J.J.E., Li, W.: On the adaptive control of robot manipulators. Int. J. Robot. Res. **6**(3), 49–59 (1987)
5. Ortega, R., Spong, M.W.: Adaptive motion control of rigid robot: a tutorial. Automatica **25**(6), 877–888 (1989)
6. Utkin, V.I.: Variable structure systems with sliding modes. IEEE Trans. Autom. Control **22**, 212–222 (1997)
7. Man, Z., Paplinski, A.P., Wu, H.R.: A robust MIMO terminal sliding mode control scheme for rigid robotic manipulators. IEEE Trans. Autom. Control **39**(12), 2465–2469 (1994)
8. Wu, Y., et al.: Terminal sliding mode control design for uncertain dynamic systems. Syst. Control Lett. **34**(5), 281–287 (1998)
9. Yu, X., Man, Z.: Fast terminal sliding-mode control for nonlinear dynamic systems. IEEE Trans. Circ. **49**(2), 261–264 (2002)
10. Yu, S., et al.: Continuous finite-time control for robotic manipulators with terminal sliding mode. Automatica **41**(11), 1957–1964 (2005)
11. Slotine, J.J.E., Li, W.: Applied Nonlinear Control, pp. 41–190. Prentice-Hall, EngleWood Cliffs (1991)
12. Wang, L., Chai, T., Zhai, L.: Neural network-based terminal sliding mode control of robotic manipulators including actuator dynamics. IEEE Trans. Ind. Electron. **56**(9), 3296–3304 (2009)
13. Ge, S.S., Hang, C.C.: Direct adaptive neural network control of robots. Int. J. Syst. Sci. **27**, 533–542 (1996)
14. Ge, S.S., Lee, T.H., Harris, C.J.: Adaptive neural network control of robot manipulators. World Scientific, London (1998)
15. Ge, S.S., Hang, C.C., Woon, L.C.: Adaptive neural network control of robot manipulators in task space. IEEE Trans. Ind. Electron. **44**, 746–752 (1997)
16. Woon, L.C., Ge, S.S., Chen, X.Q., Zhang, C.: Adaptive neural network control of coordinated manipulators. J. Robot. Syst. **16**(4), 195–211 (1999)
17. Craig, J.J.: Introduction to Robotics Mechanics and Control, 3rd edn. Prentice Hall, EngleWood Cliffs (2005)
18. Khalil, H.K.: Nonlinear Systems, 3rd edn. Prentice Hall, EngleWood Cliffs (2002)
19. Girosi, F., Poggio, T.: Networks and the best approximation property. In: Artificial Intelligence Lab. MIT, Cambridge (1989)

Intelligent Communication Networks and Web Applications

A Quantum-Inspired Immune Clonal Algorithm Based Handover Decision Mechanism with ABC Supported

Tingting Liu[✉], Xingwei Wang, Fuliang Li, and Min Huang

College of Information Science and Engineering,
Northeastern University, Shenyang, China
liuting_ting@163.com,
{wangxw,isemhuang}@mail.neu.edu.cn,
lifuliang@ise.neu.edu.cn

Abstract. In this paper, we propose a QIICA (Quantum-Inspired Immune Clonal Algorithm) based handover decision mechanism with ABC (Always Best Connected) supported. We first utilize fuzzy mathematics and microeconomics to describe application types, QoS (Quality of Service) requirements, access networks and terminals. Then, we present the optimal handover solution of assigning N terminals to M access networks based on the QIICA. At last, we evaluate the handover decision mechanism and make a comparison with the existing mechanisms. Evaluation results show that our mechanism is feasible and effective.

Keywords: QIICA · ABC · Qos · Handover decision

1 Introduction

The NGI (Next Generation Internet) is a multi-level heterogeneous network. It needs to ensure QoS, meet ABC (Always Best Connected) and provide high quality services. ABC [1] means at any time and place, once appearing a better access method, transparent handover can be executed. Handover includes horizontal handover in the homogeneous networks and vertical handover in the heterogeneous networks.

With the increase of users, the solution space of multiple users' handover decision mechanism expands, so the traditional optimal handover scheme is not applicable. We need to take the interests of both users and the network providers into account [2], and achieve a win-win situation. At the same time, we also need to avoid the ping-pong effect. Therefore, there is still a greater need for handover decision problem.

In recent years, many scholars have researched the handover decision mechanism [3–9], but they failed to consider the handover decision with ABC. We propose a handover decision mechanism which is based on the QIICA (Quantum-Inspired Immune Clonal Algorithm) with ABC supported, and it makes the utility achieve the Nash equilibrium in the Pareto optimal state [10] in this paper.

The remainder of the paper is organized as follows. The related work is presented in Sect. 2. We describe the method to measure performance in Sect. 3. Section 4 presents

© Springer International Publishing Switzerland 2015
D.-S. Huang and K. Han (Eds.): ICIC 2015, Part III, LNAI 9227, pp. 561–573, 2015.
DOI: 10.1007/978-3-319-22053-6_59

the algorithm design. The simulation and performance evaluation are presented in Sect. 5. Conclusions and future remarks are presented in Sect. 6.

2   Related Work

At present, many scholars have put forward various handover decision algorithms from different angles. Literature [3] proposed a method based on RSS (Received Signal Strength) vertical handover decision scheme. With bandwidth introduced, the algorithm used the average value of RSS. Literature [4] proposed a vertical handover decision method based on fuzzy theory and the method try to reduce the "Ping-Pong Effect" in the handover process. Literature [5] proposed a handover decision method based on QoS requirements, considering the minimum bit rate, delay, delay jitter, error rate, user preferences and other factors. Literature [6] proposed a vertical handoff decision algorithm based on fuzzy rules for QoS sensing, with the use of non birth-death Markov chain as the evaluation model for multiple attribute decision making. Literature [7] proposed a user-network association algorithm based on population games for problems of load balancing for heterogeneous wireless networks. Literature [8] proposed a vertical handoff mechanism based on harmonic oscillator immune optimization algorithm, which treated the load balancing and battery lifetime as two interrelated optimization target. Literature [9] proposed a handoff decision scheme based on fuzzy logic of adaptive vertical, according to the transmission rate, the service fee, the mobile terminal velocity and the received signal strength.

These algorithms studied the handover decision mechanism from the different angles, but failed to consider the ABC situation. In this paper, we attempt to use an intelligent optimization algorithm, QIICA, to get the feasible solution in acceptable time.

3   Model Design

3.1   Application Type and QoS Parameter Weight

Assuming the NGI has I kinds of application types, the set of types can be expressed as $ATS = \{AT_1, AT_2, \ldots, AT_I\}$. Applications of different types have different QoS parameters requirements [11]. For the application type AT_i $(1 \leq i \leq I)$, four QoS parameters are considered in this paper: bandwidth, delay, delay jitter and error rate. In addition, the weight of these parameters is expressed as $W = [\omega_B, \omega_D, \omega_J, \omega_E]^T$.

3.2   Access Network Model

The number of access networks is M. For access network $j(1 \leq j \leq M)$, the provider identifier is $PI_j \in PIS$; type identifier is $TI_j \in TIS$; the set of supported coding schemes is $CS_j \subseteq CIS$; coverage identifier and the maximum movement speed of mobile terminal supported are CA_j and MV_j; the set of supported application type is NAS_j, $NAS_j \subseteq ATS$; the total bandwidth is TB_j; the remaining bandwidth is AB_j; AB_j^{th} represents the remaining bandwidth threshold; spectral range and the lowest intensity signal are FR_j and TP_j respectively.

The QoS parameter range is defined as (1), k is service level.

$$QS_{ji}^k = <[bw_{ji}^{kl}, bw_{ji}^{kh}], [de_{ji}^{kl}, de_{ji}^{kh}], [jt_{ji}^{kl}, jt_{ji}^{kh}], [er_{ji}^{kl}, er_{ji}^{kh}] > \qquad (1)$$

pr_{ji}^k is the price per unit of time and bandwidth, which is defined as (2). λ_k is the adjustment coefficient. The basic price per unit of bandwidth pr_{ji}^0 is defined as (3). ct_{ji}^k is the cost per unit of time and bandwidth.

$$pr_{ji}^k = \lambda_k \cdot pr_{ji} \qquad (2)$$

$$pr_{ji}^0 = ct_{ji}^k \cdot (1 + r) \qquad (3)$$

There may be cooperative collusion in the pursuit of more interests between mobile terminals and access networks. In order to avoid this, we defined as (4) ∼ (6).

$$pr_{ji}^k = pr_{ji}^k - tg_{ji}^{kc} \qquad (4)$$

$$pr_{ji}^k = pr_{ji}^k + ng_{ji}^{kc} \qquad (5)$$

$$pr_{ji}^k = pr_{ji}^k + ng_{ji}^{kc} - tg_{ji}^{kc} \qquad (6)$$

3.3 Terminal Model

The number of terminal models is N. For terminal model t $(1 \leq t \leq N)$, the set of network application types and coding schemes are TAS_t and MCS_t; CV_t is the movement rate of terminal t; CV_h is high speed threshold; RC_t is the remaining battery capacity; RC_{th} is the remaining power threshold; The received signal strength limit and working frequency are RS_t and WF_t respectively; The highest price per unit of time and bandwidth is HP_{ti}.

3.4 QoS Satisfaction

$F_{ji}^k = \{EB_{ji}^k, ED_{ji}^k, EJ_{ji}^k, EE_{ji}^k\}$ represents the evaluation sequence of QoS parameters.

For bandwidth, selecting reference range $[Bw_l, Bw_h]$, the bandwidth range offered by access network j $[bw_{ji}^{kl}, bw_{ji}^{kh}]$ compared with the reference range, we obtain a weighted evaluation function defined as (7).

$$EB_{ji}^k = \omega_B \cdot [\frac{1}{2} \cdot \exp(-\frac{bw_{ji}^{kh} - bw_{ji}^{kl}}{Bw_h - Bw_l}) + \frac{1}{2} \cdot \exp(\frac{bw_{ji}^{kl} + bw_{ji}^{kh}}{2} - Bw_h)] \tag{7}$$

For the delay, like bandwidth, there are $[De_l, De_h]$ and $[de_{ji}^{kl}, de_{ji}^{kh}]$, which is defined as (8).

$$ED_{ji}^k = \omega_D \cdot [\frac{1}{2} \cdot \exp(-\frac{de_{ji}^{kh} - de_{ji}^{kl}}{De_h - De_l}) + \frac{1}{2} \cdot \exp(De_l - \frac{de_{ji}^{kl} + de_{ji}^{kh}}{2})] \tag{8}$$

For the delay jitter and error rate, as same as delay processing, there are $[Jt_l, Jt_h]$ and $[Er_l, Er_h]$. We obtain EJ_{ji}^k and EE_{ji}^k.

The ideal solution sequence is expressed as $F_{IS} = \{EB_{IS}, ED_{IS}, EJ_{IS}, EE_{IS}\}$ and the negative ideal solution sequence is expressed as $F_{NIS} = \{EB_{NIS}, ED_{NIS}, EJ_{NIS}, EE_{NIS}\}$. According to (9) and (10), calculate the distance to ideal solution sequence $d\_is_{ji}^k$ and the distance to the negative ideal solution $d\_nis_{ji}^k$.

$$d\_is_{ji}^k = \sqrt{(EB_{ji}^k - EB_{IS})^2 + (ED_{ji}^k - ED_{IS})^2 + (EJ_{ji}^k - EJ_{IS})^2 + (EE_{ji}^k - EE_{IS})^2} \tag{9}$$

$$d\_nis_{ji}^k = \sqrt{(EB_{ji}^k - EB_{NIS})^2 + (ED_{ji}^k - ED_{NIS})^2 + (EJ_{ji}^k - EJ_{NIS})^2 + (EE_{ji}^k - EE_{NIS})^2} \tag{10}$$

The evaluation coefficient, R_{ji}^k ($R_{ji}^k \in (0, 1]$), is defined as (11).

$$R_{ji}^k = \exp(-(d\_is_{ji}^k + \frac{1}{d\_nis_{ji}^k})) \tag{11}$$

The bandwidth of the mobile terminal demand range is $[BW_i^l, BW_i^h]$, the range corresponding to the bandwidth is $[bw_{ji}^{kl}, bw_{ji}^{kh}]$, service strategy suits the needs of the user level as (12). The evaluation functions are shown as (13) and (14).

$$CB_{ji}^k = \frac{1}{2} EI_{Bw}(bw_{ji}^{kl}, bw_{ji}^{kh}) + \frac{1}{2} Fit_{Bw}(\frac{bw_{ji}^{kl} + bw_{ji}^{kh}}{2}) \tag{12}$$

$$EI_{Bw}(bw_{ji}^{kl}, bw_{ji}^{kh}) = 1 - (\frac{bw_{ji}^{kh} - bw_{ji}^{kl}}{BW_i^h - BW_i^l})^2 \tag{13}$$

$$Fit_{Bw}(bw) = \begin{cases} \frac{2(bw - BW_i^l)^2}{(BW_i^h - BW_i^l)^2} & BW_i^l < bw \leq \frac{BW_i^h + BW_i^l}{2} \\ 1 - \frac{2(BW_i^h - bw)^2}{(BW_i^h - BW_i^l)^2} & \frac{BW_i^h + BW_i^l}{2} < bw \leq BW_i^h \end{cases} \tag{14}$$

For delay of the mobile terminal demand range is $[DE_i^l, DE_i^h]$, the range corresponding to the bandwidth is $[de_{ji}^{kl}, de_{ji}^{kh}]$, and service strategy suits the needs of the user level as (15). The evaluation functions are shown as (16) and (17).

$$CD_{ji}^k = \frac{1}{2}EI_{De}(de_{ji}^{kl}, de_{ji}^{kh}) + \frac{1}{2}Fit_{De}(\frac{de_{ji}^{kl} + de_{ji}^{kh}}{2}) \tag{15}$$

$$EI_{De}(de_{ji}^{kl}, de_{ji}^{kh}) = 1 - (\frac{de_{ji}^{kh} - de_{ji}^{kl}}{DE_i^h - DE_i^l})^2 \tag{16}$$

$$Fit_{De}(de) = \begin{cases} 1 - \frac{2(de-DE_i^l)^2}{(DE_i^h-DE_i^l)^2} & DE_i^l < de \leq \frac{DE_i^h+DE_i^l}{2} \\ \frac{2(DE_i^h-de)^2}{(DE_i^h-DE_i^l)^2} & \frac{DE_i^h+DE_i^l}{2} < de \leq DE_i^h \end{cases} \tag{17}$$

For delay jitter, there are $[JT_i^l, JT_i^h]$, $[jt_{ji}^{kl}, jt_{ji}^{kh}]$ and service strategy as (18). For error rate there are $[ER_i^l, ER_i^h]$, $[er_{ji}^{kl}, er_{ji}^{kh}]$ and service strategy as (19) and (20). Definitions of the two evaluation functions are the same with the delay of mobile terminal.

$$CJ_{ji}^k = \frac{1}{2}EI_{Jt}(jt_{ji}^{kl}, jt_{ji}^{kh}) + \frac{1}{2}Fit_{Jt}(\frac{jt_{ji}^{kl} + jt_{ji}^{kh}}{2}) \tag{18}$$

When $ER_i^l = ER_i^h = 0$,

$$CE_{ji}^k = \begin{cases} 1.0, & er_{ji}^{kl} = er_{ji}^{kh} = 0 \\ 0.0, & er_{ji}^{kl} \neq er_{ji}^{kh} \neq 0 \end{cases} \tag{19}$$

Otherwise,

$$CE_{ji}^k = \frac{1}{2}EI_{Er}(er_{ji}^{kl}, er_{ji}^{kh}) + \frac{1}{2}Fit_{Er}(\frac{er_{ji}^{kl} + er_{ji}^{kh}}{2}) \tag{20}$$

The overall QoS fitness degree is defined as (21).

$$CQ_{ji}^k = \omega_B \cdot CB_{ji}^k + \omega_D \cdot CD_{ji}^k + \omega_J \cdot CJ_{ji}^k + \omega_E \cdot CE_{ji}^k \tag{21}$$

According to (11) and (21), the user overall satisfaction degree is defined as (22).

$$SQ_{ji}^k = R_{ji}^k \cdot CQ_{ji}^k \tag{22}$$

3.5 Other Satisfactions

The user preference satisfaction degree on the Internet supplier is defined as (23). The user preference satisfaction degree on network coding system is defined as (24). The price satisfaction is defined as (25). Fitness of movement velocity is defined as (26). Battery fitness is defined as (27).

$$SR_{tj} = \begin{cases} \left(\frac{q+1-x}{q}\right)^2 & \text{in the sequence} \\ 0 & \text{others} \end{cases} \tag{23}$$

$$SC_{tj} = \begin{cases} \left(\frac{q+1-x}{q}\right)^2 & \text{in the sequence} \\ 0 & \text{others} \end{cases} \tag{24}$$

$$SP_{tj} = \begin{cases} 0 & pr_{ji}^k > HP_{ti} \\ 1 - \frac{1}{2} \times \frac{pr_{ji}^k}{HP_{ti}} & 0 < pr_{ji}^k \leq HP_{ti} \end{cases} \tag{25}$$

$$SV_{tj} = \begin{cases} 1 & CV_t < CV_h \ \& \ CV_t < MV_j \\ \left(\frac{q+1-x}{q}\right) & CV_h \leq CV_t \leq MV_j \\ 0 & MV_j < CV_t \end{cases} \tag{26}$$

$$SY_{tj} = \begin{cases} \left(\frac{q+1-x}{q}\right) & RC_t \leq RC_{th} \\ 1 & \text{others} \end{cases} \tag{27}$$

The definition of load evaluation function is as (28).

$$SL_{tj} = \begin{cases} \exp(-(\eta_i - \eta_0)/2\sigma^2) & \eta_j > \eta_0, AB_j^{\min} < AB_j < AB_j^{th} \\ 1 & \text{others} \end{cases} \tag{28}$$

3.6   Gaming Analysis

In the game process, the two sides of the game for a service strategy are mobile terminal t and access network j. The mobile terminal t has two game strategies, which are accepted strategy (a_1) and refused strategy (a_2). Access network j has two game strategies, which are strategy (b_1) and refused strategy (b_2).

The payoff matrixes are TG and NG as (29) and (30).

$$TG = \begin{bmatrix} HP_{ti} - pr_{ji}^k & 0 \\ -v \cdot (HP_{ti} - pr_{ji}^k) & 0 \end{bmatrix} \tag{29}$$

$$NG = \begin{bmatrix} pr_{ji}^k - ct_{ji}^k & -v \cdot (pr_{ji}^k - ct_{ji}^k) \\ 0 & 0 \end{bmatrix} \tag{30}$$

If the strategy combination (a_{i^*}, b_{j^*}) meets $\begin{cases} TG_{i^*j^*} \geq TG_{ij^*} \\ NG_{i^*j^*} \geq NG_{i^*j} \end{cases}$, (a_{i^*}, b_{j^*}) meets Nash equilibrium. If (a_1, b_1) meets Nash equilibrium, it can be considered that the service strategy is fair for the user side and network side.

3.7 Utility Calculations

Calculate the mobile terminal user utility and the network utility as (31) and (32).

$$uu_{tj} = \Phi \cdot \Omega \cdot [w_{SQ} \cdot SQ_{ji}^k + w_{SR} \cdot SR_{tj} + w_{SC} \cdot SC_{tj} + w_{SP} \cdot SP_{tj}$$
$$+ w_{SV} \cdot SV_{tj} + w_{SY} \cdot SY_{tj} + w_{SL} \cdot SL_{tj}] \cdot \frac{HP_{ti} - pr_{ji}^k}{HP_{ti}} \tag{31}$$

$$nu_{tj} = \Phi \cdot \Omega \cdot [w_{SQ} \cdot SQ_{ji}^k + w_{SR} \cdot SR_{tj} + w_{SC} \cdot SC_{tj} + w_{SP} \cdot SP_{tj}$$
$$+ w_{SV} \cdot SV_{tj} + w_{SY} \cdot SY_{tj} + w_{SL} \cdot SL_{tj}] \cdot \frac{pr_{ji}^k - ct_{ji}^k}{pr_{ji}^k} \tag{32}$$

Φ is the handover result feedback factor $\Phi = NS_{tj}/(NS_{tj} + NF_{tj})$, to reflect previous handover success rate. Ω is the influence factor in game theory. We define it as:
$$\begin{cases} \Omega = 1 & \text{nash equilibrium} \\ 0 < \Omega < 1 & \text{isn't nash equilibrium} \end{cases}.$$

3.8 Mathematical Model

The objective are maximizing uu_{tj}, nu_{tj}, $\sum_{t=1}^{n} uu_{tj}$, $\sum_{t=1}^{n} nu_{tj}$ and $\sum_{t=1}^{N} \sum_{j=1}^{M} uu_{tj} + nu_{tj}$.

4 Algorithm Design

4.1 Feasible Solution and Fitness Function

The feasible solution of the problem meets $\forall t((TAS_t \subseteq NAS_{AN_{qt}}) \wedge (CS_{AN_{qt}} \cap MCS_t \neq \Phi) \wedge (MV_{AN_{qt}} \geq CV_t) \wedge (WF_t \subseteq FR_{AN_{qt}}) \wedge (TP_{AN_{qt}} \geq RS_t) \wedge (pr_{AN_{qt}i}^k \leq HP_{ti}) \wedge (AB_{AN_{qt}} - bw_{AN_{qt}i}^{kh} \geq AB_{AN_{qt}}^{\min}))$.

The fitness function is as follows:

$$Minimize\ f(x) = \begin{cases} \sum_{t=1}^{N} \left(\frac{1}{uu_{t_iAN_{qt}}} + \frac{1}{nu_{t_iAN_{qt}}} \right) & , x \text{ is a feasible solution} \\ +\infty & , x \text{ isn't a feasible solution} \end{cases}$$

4.2 Algorithm Description

The steps of our algorithm are as follows:

Step 1: Initialize relevant parameters.

Step 2: Assign random value for the antibody of the quantum coding population Y_0. Then the integer coding population X_0 is generated through mapping operation.

Step 3: Determine whether the relevant parameters of mobile terminal t and access network AN_{qt} can meet the feasible solution. If they can, handover scheme of mobile terminal t is feasible. For pre-allocating bandwidth resources, we set $AB_{AN_{qt}} = AB_{AN_{qt}} - bw_{AN_{qt}i}^{kh}$. Else, individual x_k^q is an unfeasible solution, jump to Step 5.

Step 4: Set t = t+1, if $t \leq N$ jump to Step 3; else, x_k^q is a feasible solution.

Step 5: Reset residual bandwidth values.

Step 6: Execute game analysis of the feasible solution of mobile terminal and access network for each individual, update the game factor Ω. Calculate fitness function values of individuals. Calculate affinity of antibody in population X_0, select the largest affinity antibody to store in the history optimal solution x_b, and store the corresponding quantum encode antibody in y_b.

Step 7: Clone the population Y_k in order to generate Y_k'. For each sub population $Y_{q,k}'$, select multiple antibodies according to the mutation probability p_m, then do the quantum rotation gate and quantum NOT gate operations. After the quantum sub populations' mutation operation, complete the population regeneration. Then get population Y_k'.

Step 8: Perform mapping operation on the population Y_k and Y_k'', get the corresponding integer coding population X_k and X_k''. Calculate the fitness function values of each antibody in population X_k and X_k'', and the affinity. By the clone choice operation, generate integer coding population B_k and quantum coding population B_k'.

Step 9: Compare the antibody of the largest affinity in population B_k with the antibodies affinity saved in x_b. Retain the better one and then update y_b.

Step 10: Set k = k+1. If $k \leq K$, execute quantum reorganization operation on the population B_k', treat the new generation of population as the initial population of the next iteration, jump to Step 7; else, x_b is the optimal solution.

5   Simulation and Performance Evaluation

NS2 in Linux virtual machine is used for the handover mechanism simulation.

5.1   Topology Case

We implement the simulation using the following three network topologies, which are composed of different fixed number of nodes in the full coverage of the hexagonal cellular topology. The three topologies 82, 66, 107 are named topology 1, topology 2 and topology 3. In topologies of 1–3, we set the same random function seeds. For the number of users 3, 5, 10, 20, 30, 50 cases, handover decision mechanisms 1–3 are executed 500 times, respectively. The network topologies are shown as Fig. 1.

Fig. 1. Topology 1 ~ 3

5.2 Performance Evaluation Results

This paper selects the algorithm based on VIKOR sequencing method proposed by Gallardo-Medina J.R. et al. [12] and the algorithm based on utility and game theory proposed by Chang C.J. et al. [13] as references, to evaluate the performance of the handover decision mechanism proposed in this paper. The handover decision mechanism based on QIICA with multiple users is mechanism 1, the benchmark algorithm 1 is mechanism 2 and the benchmark algorithm 2 is mechanism 3.

Figures 2 and 3 present that the mechanism 1 in both network side and user side has the highest value. This is because the other two mechanisms are lack of consideration about the user preference. With the increase in the number of users, the utility of both network side and user side slightly decrease.

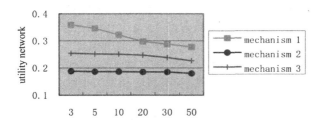

Fig. 2. Comparison of utility value of network

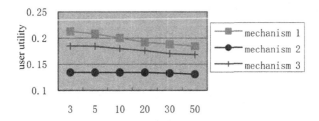

Fig. 3. Comparison of utility value of user

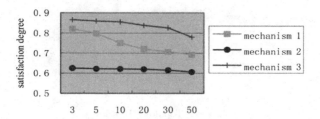

Fig. 4. Comparison of QoS satisfaction degree of user

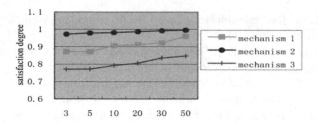

Fig. 5. Comparison of price satisfaction degree of user

Figure 4 shows that the mechanism 1 is lower than the mechanism 3 on QoS satisfaction. Because the mechanism 3 pursuits the maximum QoS utility without considering the price factor. However, mechanism 1 is to achieve the overall utility maximization.

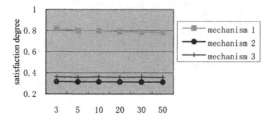

Fig. 6. Comparison of supplier preference satisfaction degree of user

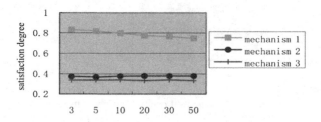

Fig. 7. Comparison of code type preference satisfaction degree of user

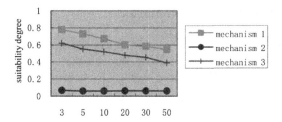

Fig. 8. Comparison of velocity suitability degree

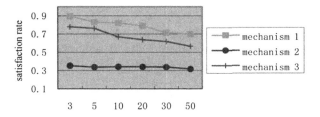

Fig. 9. Comparison of satisfaction rate of high velocity user

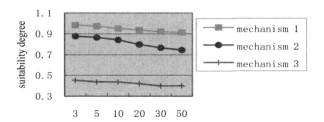

Fig. 10. Comparison of battery suitability degree

Figure 5 shows that the mechanism 2 can obtain the highest price satisfaction, and mechanism 3 has minimum price satisfaction. Because of the mechanism 2 selects the cheapest and low level service.

Figures 6, 7, 8, 9 and 10 present that the mechanism 1 is the best on satisfaction, speed fitness, the movement of user access to large coverage network and battery power fitness.

Through the above comparison, the proposed handover decision algorithm is better than the two benchmark algorithms in most criterions, so the mechanism 1 can find global optimal solution.

6   Conclusions and Future Remarks

In this paper, we propose a QIICA based handover decision mechanism, which can meet the ABC requirement better in NGI. Through the simulation based on NS2, advantages of our algorithm have been verified by both analysis and experimental results. It takes both users' preferences for suppliers, network types and costs and related interests of the network side into account, while meeting the QoS demand of users' applications.

For the next work, there are two main aspects. Firstly, we should correct and improve the concept of ABC. In addition, we are going to realize the prototype system.

Acknowledgment. This work is supported by the National Science Foundation for Distinguished Young Scholars of China under Grant No. 61225012 and No. 71325002; the Specialized Research Fund of the Doctoral Program of Higher Education for the Priority Development Areas under Grant No. 20120042130003; Liaoning BaiQianWan Talents Program under Grant No. 2013921068

References

1. Gustafsson, E., Jonsson, A.: Always best connected. IEEE Wirel. Commun. **10**(1), 49–55 (2003)
2. Wang, X., Cheng, H., Huang, M.: QoS multicast routing protocol oriented to cognitive network using competitive coevolutionary algorithm. Expert Syst. Appl. **41**(10), 4513–4528 (2014)
3. Roy, S.D., Anup, S.: Received signal strength based vertical handoff algorithm in 3G cellular network. In: IEEE International Conference on Signal Processing, Communication & Computing, pp. 326–330, Hongkong (2012)
4. Singhrova, A., Prakash, N.: Vertical handoff decision algorithm for improved quality of service in heterogeneous wireless networks. IET Commun. **6**(2), 211–223 (2012)
5. Ramirez-Perez, C., Ramos, R.V.M.: A QoS hierarchical decision scheme for vertical handoff. In: IEEE International Caribbean Conference on Devices, Circuits & Systems, pp. 1–4, San Francisco (2012)
6. Vasu, K., Maheshwari, S., Mahapatra, S., Kumar, C.S.: QoS aware fuzzy rule based vertical handoff decision algorithm for wireless heterogeneous networks. In: IEEE National Conference on Communications, pp.1–5, Washington D.C. (2011)
7. Jiang, Y., Hu, B., Chen, S.: User-network association optimization in heterogeneous wireless networks: a population game-based approach. Chin. J. Comput. **35**, 1249–1261 (2012). (in Chinese)
8. Zhu, S., Liu, F., Chai, Z., Qi, Y., Wu, J.: Simple harmonic oscillator immune optimization algorithm for solving vertical handoff decision problem in heterogeneous wireless network. Acta Physica Sinica **61**(9), 1–10 (2012). (in Chinese)
9. Çalhan, A., Çeken, C.: Case study on handoff strategies for wireless overlay networks. Comput. Stan. Interfaces **35**(1), 170–178 (2013)
10. Wang, X., Cheng, H., Huang, M.: Multi-robot navigation based QoS routing in self-organizing networks. Eng. Appl. Artif. Intell. **26**(1), 262–272 (2013)

11. Xie, X., Wang, X., Wen, Z., Huang, M.: A QoS routing protocol for cognitive networks. Chin. J. Comput. **36**(9), 1807–1815 (2013). (in Chinese)
12. Gallardo-Medina, J.R., Pineda-Rico, U., Stevens-Navarro, E.: VIKOR method for vertical handoff decision in beyond 3G wireless networks. In: IEEE 6th International Conference on Electrical Engineering, Computing Science and Automatic Control, pp. 1–5, Toluca (2009)
13. Chang, C.J., Tsai, T.L., Chen, Y.H.: Utility and game-theory based network selection scheme in heterogeneous wireless networks. In: IEEE Wireless Communications and Networking Conference, pp. 1–5, Budapest (2009)

An IEEE 802.21 Based Heterogeneous Access Network Selection Mechanism

Renzheng Wang[✉], Xingwei Wang, Fuliang Li, and Min Huang

College of Information Science and Engineering,
Northeastern University, Shenyang, China
wangrenzheng@126.com,
{wangxw, isemhuang}@mail.neu.edu.cn,
lifuliang@ise.neu.edu.cn

Abstract. With the rapid development of mobile communication technology, many heterogeneous wireless access network technologies appear, resulting in that mobile terminals often can access different networks. Therefore, an appropriate access network needs to be selected for the mobile terminal according to the context information. In this paper, we propose an IEEE 802.21 based heterogeneous access network selection mechanism. Taking the fairness among terminals and profits of network providers into account, the mechanism can meet the basic requirements of users' applications in the process of access network selection. We conduct a simulation and performance evaluation on the algorithm according to a typical network topology. Result shows that our algorithm can not only meet the performance demand but also consider network providers' profits and the fairness during network resource allocation.

Keywords: IEEE 802.21 · Selection mechanism · Heterogeneous access network

1 Introduction

With the rapid development of the mobile communication technology, many wireless access network technologies appear in people's life. Meanwhile, mobile terminals have the ability to access different networks. However, these networks usually have different characteristics in fields like coverage, bandwidth, delay, energy consumption, etc. Thus, in such heterogeneous access network environment, an appropriate access network needs to be selected for the mobile terminal, which means a reasonable allocation of network resources according to their characteristics.

However, during the selection, the context information of access networks is hard to obtain. And, the difference among them makes the seamless handover hard to achieve. Based on above considerations, the concept of Media Independent Handover was proposed and the IEEE 802.21 standard [1] was made. The standard defines three kinds of media independent handover service including the link layer event service, the network information service and the handover command service [2]. These services make the quick handover in a heterogeneous access network environment easier [3]. However, the network selection algorithm is not specified in this standard, which plays

© Springer International Publishing Switzerland 2015
D.-S. Huang and K. Han (Eds.): ICIC 2015, Part III, LNAI 9227, pp. 574–585, 2015.
DOI: 10.1007/978-3-319-22053-6_60

an important role in the process of handover [4]. Furthermore, the information service in this standard cannot provide the context of different terminals [3] at a time, which means interests of different users cannot be considered at the same time.

For the access network selection, we adopt the network-based approach, which means the network selects the appropriate network for the terminal. In the process of selection, the interest of terminals themselves as well as fairness between terminals and profits of network providers all needs to be considered. Thus, we can simplify the network selection problem in heterogeneous access network to the problem of how to allocate N terminals to M networks when the access ability and the effect of allocation result are considered. This problem is both a NP problem and a multi-objective problem [5], so that we decide to adopt the GA (Genetic Algorithm) to solve it.

The main contributions of this paper are summarized as follows. First, we make an improvement of the context information acquisition mechanism, which makes the context information of different terminals can be got at a time. In addition, we consider the application requirement and interest allocation fairness at the same time, through the reasonable design of math model and algorithm. At last, the evaluation result shows that the access network selection mechanism can reach the design goal.

The remainder of the paper is organized as follows. The related work is presented in Sect. 2. We then describe the context information acquisition mechanism, math model and access network selection algorithm in Sects. 3, 4 and 5. Section 6 presents our simulation and evaluation results. Section 7 presents our conclusions.

2 Related Work

In the field of IEEE 802.21 based handover mechanism, the work in [6] proposed a handover mechanism controlled by network itself. In this mechanism, mobile terminals send parameters of new detected access network to a specific point of service (PoS). When PoS judges that a terminal should access a new network, it selects an appropriate one and returns the result. This paper shows that the new information interaction brought by this mechanism would not have a negative effect on the performance of handover. But, this mechanism only targets at the context information of single terminal. The work in [7] proposed an improved media independent handover framework based on IEEE 802.21. This framework adopts a policy-based context-aware handover model. This model mainly includes Handoff Trigger, Connection Manager, Policy Decision Maker, local cache manager, and a Policy Repository which are placed in terminals. The context information that it needs in the process of network selection includes application requirement and preference, the context of the terminal and the parameters of access networks.

In the field of network selection algorithm, there have been some methods, which solve that problem from different angles. The work in [8] proposed an access network selection algorithm based on fuzzy TOPSIS (Technique for Order Preference by Similarity to an Ideal Solution). It can reflect the QoS requirement of different application types through the parameterized utility function and adopts different evaluation standards for applications. The work in [9] proposed a handoff decision scheme with always best connected supported, which considered the user's preference, the access network status and application QoS requirement. The work in [10] proposed a theoretical framework based on reputation system and the game theory, which can provide the service guarantee for the mobile terminal. The work in [11] proposed a linear programming based access network selection algorithm, and the optimization objective is to maximize the sum of terminals' evaluations to access networks.

Though above algorithms solved the access network selection problem from different angles, they didn't have a sufficient consideration on the requirement of different application types, the profit of network providers and the balance of interest allocation among terminals. They also did not take the fairness interest allocation among network providers or between terminals and network providers. The improved algorithm proposed in this paper can overcome the disadvantages of the above methods.

3   Context Information Acquisition Mechanism

3.1   Framework

This mechanism is based on IEEE 802.21. Figure 1 is its basic framework, which includes Terminal, PoA (Point of Attachment) and PoS (Point of Service). The PoA is classified into Serving PoA, Candidate PoA and Non-PoA (which the terminal cannot access). PoS plays a role of Information Server defined in IEEE 802.21, which is responsible for acquisition and storage of context information provided by the PoA, which includes QoS parameters, the location of access point and the coverage of access network. Rest context information is provided by terminals, which include terminal location, RSS and price. PoS is also responsible for selecting appropriate access networks for terminals through running the network selection algorithm.

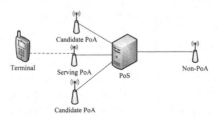

Fig. 1. Basic framework of context information acquisition mechanism

3.2   Procedure

The procedure of context information acquisition includes two parts. First, access points send the context information they can get to the PoS. Second, some terminals send the context information of their current access networks and their application types to the PoS. When handover is necessary, the PoS asks the terminal for candidate networks list and partial context information.

4 Math Model

4.1 The Depiction of Access Network

In our math model, a base station or access point represents an access network and belongs to only one network providers. In practical environment, QoS, energy consumption and price are all important [14]. For an access network i, parameters used to depict it include network type identification (NT_i), network provider identification (NP_i), coverage (CR_i, measured in meter), location of base station or access point ($APL_i = \langle APX_i, APY_i \rangle$), total bandwidth ($TB_i$, measured in Mbps), max bandwidth for each terminal ((MB_i, measured in Mbps), energy consumption (EC_i, energy consumption for unit data in unit time), local delay (DL_i) and local jitter (JT_i).

4.2 The Depiction of Mobile Terminal

Parameters used to depict the terminal j include application type (AT_j, including audio, video, data and background application), terminal location ($TL_j = \langle TX_j, TY_j \rangle$), price ($PR_{ij}$, representing the price the terminal needs to pay to the network i).

The terminal location is used to judge whether the terminal is located in the coverage area. It can also help us get the received signal strength (normalized in the scope between 0 and 1). The real RSS calculation formula is as formula (1) [12].

$$S(d) = A - 10 \times \beta \times \log d + \omega, \quad \omega \sim N(0, \sigma^2) \tag{1}$$

where $S(d)$ represents the RSS at position located d meters from the access point or base station, and the unit is dBm. A is the RSS at the access point or base station. is the error term. We can use the formula (2) to normalize $S(d)$.

$$RSS(d) = \begin{cases} 1 & 0 \leq d \leq 1 \\ 1 - \log_{CR} d & 1 < d < CR \\ 0 & d \geq CR \end{cases} \tag{2}$$

4.3 Evaluation of Access Networks

$CNS_t = \{CN_{t1}, CN_{t2}, \ldots, CN_{tN}\}$ is the candidate access networks set of terminal t, where N is the number. Evaluation parameters include RSS, coverage, bandwidth, delay, jitter, energy consumption and price. These parameters can be expressed as $\langle RSS_{tn}, CR_{tn}, BW_{tn}, DL_{tn}, JT_{tn}, EC_{tn}, PR_{tn} \rangle$ for terminal t and candidate network n ($1 \leq n \leq N$). The evaluation value to the candidate network is calculated by the GRA (Grey Relational Analysis), which mainly includes two parts, the normalization of parameters and the calculation of the grey relational grade.

For the normalization, we divide these parameters into two types. The first type parameters are the ones such as RSS_{tn}, CR_{tn} and BW_{tn}, which are normalized using the

Eq. (3). The second type parameters are the ones such as DL_{tn}, JT_{tn}, EC_{tn} and PR_{tn} which are normalized using the Eq. (4).

$$x^*_{tn}(p) = \frac{x^*_{tn} - \min_m x^p_{tm}}{\max_m x^p_{tm} - \min_m x^p_{tm}} \tag{3}$$

$$x^*_{tn}(p) = \frac{\max_m x^p_{tm} - x^*_{tn}}{\max_m x^p_{tm} - \min_m x^p_{tm}} \tag{4}$$

where $1 \leq n \leq N$, $1 \leq m \leq N$, $1 \leq t \leq T$, and x^p_{tm} is the value of parameter p.

When calculating the grey relational grade, the ideal network is the one whose parameters are all best. According to Eqs. (3) and (4), we can get that evaluation parameters of the ideal network are all 1. Therefore, we can get the grey relational grade between each candidate network and the ideal network using Eq. (5).

$$u_{tn} = \frac{1}{\sum\limits_{p=1}^{7} w_p |x^*_{tn}(p) - 1| + 1} \tag{5}$$

where $x^*_{tm}(p)$ is the normalization value of evaluation parameter p and w_p is the weight. In this paper, we regard the u_{tn} as the evaluation value and calculate weight values for different application types with Analytic Hierarchy Process.

4.4   Utility of the Terminal

The calculation method, Eqs. (6)– (9), of the utility of the terminal U_U considers both the sum of evaluation values and the fairness among terminals.

$$\bar{u} = \frac{1}{T} \sum_{t=1}^{T} u_t \tag{6}$$

$$\sigma = \sqrt{\frac{1}{T} \sum_{t=1}^{T} (u_t - \bar{u})^2} \tag{7}$$

$$\alpha = \frac{1}{2} (u_{min} + u_{max}) \tag{8}$$

$$U_U = \frac{\bar{u}}{\sigma + \alpha} \tag{9}$$

where u_t is the evaluation value of terminal t to its current selected access network, $\bar{u}$ is the average value of these evaluation values, u_{max} and u_{min} are respectively the maximum value and the minimum value in theory (which in our paper are 1 and 0), and σ is the standard deviation, which reflects the degree of variance. For verifying this calculation method, we draw the function graph when we set, $T = 2$, $u_{max} = 1$ and $u_{min} = 1$. Then we can get the Eq. (10), and the function graph is in Fig. 2.

$$U_U = \frac{u_1 + u_2}{|u_1 - u_2| + 1} \tag{10}$$

From Fig. 2 we can know that the larger and closer u_1 and u_2 are, the larger u_u is. Therefore, the utility of terminal got through above calculation method reflects not only terminals' evaluation to access networks, but also the fairness among terminals.

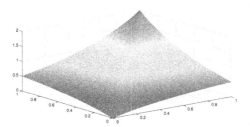

Fig. 2. Function graph of Eq. (10)

4.5 Utility of the Network Provider

The calculation method, Eqs. (11)–(16), of the utility of the network provider U_P considers both the total income and the variance of them.

$$u_p^i = \sum_{j=1}^{N_{type}} \left(v_j \cdot \sum_{k=1}^{N_p^i(j)} PR_{kp_i} \right) \tag{11}$$

$$u_p = \sum_{i=1}^{Ap(p)} u_p^i \tag{12}$$

$$\bar{u} = \frac{1}{P} \sum_{p=1}^{P} u_p \tag{13}$$

$$\sigma = \sqrt{\frac{1}{P} \sum_{p=1}^{P} (u_p - \bar{u})^2} \tag{14}$$

$$\alpha = \frac{1}{2} (u_{min} + u_{max}) \tag{15}$$

$$U_P = \frac{\bar{u}}{\sigma + \alpha} \tag{16}$$

where u_p^i is the income the network provider p get through its i th network; N_{type} is the number of application types; v_j is the representative data transmission rate of application type j; $N_p^i(j)$ is the number of terminals of application type j; PR_{kp_i} is the price for terminal k; $Ap(p)$ is the number of access networks of provider p; p is the number of network providers; u_p is total income of network provider p; u_{min} is the minimum value and u_{max} is the maximum value.

5   Access Network Selection Algorithm

5.1   Expression of Solution

as $X = (a_1, a_2, ..., a_t, ...,a_T)$, where a_t represents that the terminal t selects the a_t th network in its candidate networks list.

5.2   Design of Genetic Operators

For the selection operator, we adopt a population communication based selection operator, after considering characteristics of roulette selection and tournament selection. First, two individuals, a_1 and a_2, are selected from population A through the roulette selection, and another two individuals, b_1 and b_2, are selected from population B through the tournament selection. We conduct crossover and mutation on a_2 and b_1, and then add their offsprings, a'_1 and a'_2, to the sub-population of A. The same operation are also conducted on a_1 and b_2. Then, we conduct the roulette selection and tournament selection respectively on A and B, and the offsprings generated through crossover and mutation are added into their own sub-population. Finally we select individuals from both the sub-population and the original population as the next generation population using roulette selection or tournament selection.

For the crossover operator, we adopt the intermediate recombination, which is shown in Eqs. (17) and (18).

$$S_1(i) = P_j(i) + \alpha \times (P_k(i) - P_j(i)) \tag{17}$$

$$S_2(i) = P_k(i) + \alpha \times (P_j(i) - P_k(i)) \tag{18}$$

where $P_j(i)$ represents the ith gene in the jth chromosome, and $S_1(i)$ and $S_2(i)$ represent the ith gene of their two offspring chromosomes. α is the scale factor, which is a uniform random number in the range $[-d, 1+d]$, where $d = 0.25$ in usual.

The mutation operator we adopt is shown in Eq. (19).

$$X' = X \pm (1 + 0.5L\Delta) \tag{19}$$

where $\Delta = \sum_{i=0}^{m} \frac{a(i)}{2^i}$, $a(i)$ is set to 1 with probability $1/1m$. and to 0 with probability $1 - 1/1m$, m is usually set to 20, and L is the value range. So we can get a long step with a small probability and get a short step with a large probability.

The crossover probability and mutation probability in this paper is set with the method of adaptive probability [13], which is shown in Eqs. (20) and (21).

$$p_c = \begin{cases} k_1(f_{\max} - f')/(f_{\max} - \bar{f}) & f' \geq \bar{f} \\ k_3 & f' < \bar{f} \end{cases} \tag{20}$$

$$p_m = \begin{cases} k_2(f_{\max} - f)/(f_{\max} - \bar{f}) & f \geq \bar{f} \\ k_4 & f < \bar{f} \end{cases} \tag{21}$$

where k_1, k_2, k_3, $k_4 \leq 1.0$, according to the effect of crossover and mutation we usually set $k_1 = 1$, $k_3 = 1$, $k_2 = 0.5$ and $k_4 = 0.5$. Besides, p_c is the crossover probability, p_m is the mutation probability, $\bar{f}$ is the average value of all individuals' fitness values, $f_{\max}$ is maximum fitness value, f is the fitness value of current individual and f' is the bigger fitness value between two crossover individuals.

5.3 Fitness Function

The fitness function of the genetic algorithm designed in this paper is Eq. (22).

$$F(X) = w_U \cdot U_U(X) + w_P \cdot U_P(X) \tag{22}$$

where $U_U(X)$ and $U_P(X)$ are respectively the utility of the terminal and the network provider. w_U and w_P are weights of them. For the weights, we adopt the coefficient of variation method to calculate the w_U and w_P, which is according to the discrete degree of individuals' fitness value. The method is as Eqs. (23) and (24).

$$w_U = \frac{\sigma_U}{\bar{U}_U} \tag{23}$$

$$w_P = \frac{\sigma_P}{\bar{U}_P} \tag{24}$$

where σ_U is the standard deviation of individuals' $U_U(X)$ value, σ_P is the standard deviation of individuals' $U_P(X)$ value, $\bar{U}_U$ is the average value of individuals' $U_U(X)$ and $\bar{U}_P$ is the average value of individuals' $U_P(X)$.

5.4 Algorithm Process

Step 1: Construct the candidate access networks list for each terminal.

Step 2: Initialize population A_1 and B_1, each of which has N individuals. Initialize the maximum iteration times $N_{\max}$ and set current iteration times $N_c = 1$.

Step 3: Calculate the fitness value of each individual in population A_{N_c} and B_{N_c}. Then calculate the maximum fitness value and average fitness value in each one.

Step 4: Apply the selection operator, the crossover operator and the mutation operator introduced in last section to current two populations and get the next generation populations A_{N_c+1} and B_{N_c+1}.

Step 5: $N_c = N_c + 1$. If $N_c > N_{max}$, go on, otherwise, go to Step 3.

Step 6: Select the individual which has the largest fitness value from union of A_{N_c} and B_{N_c} as the optimal access network selection scheme.

6   Simulation and Performance Evaluation

In this section, we adopt the network topology in Fig. 3. It includes three types of networks, which are GSM, WCDMA and WLAN, and each of them belongs to different network providers. We consider three different situations which respectively have 40, 60 or 80 random distributed terminals in the topology. These terminals are respectively divided into four different application types which are voice application, video application, data application and background application.

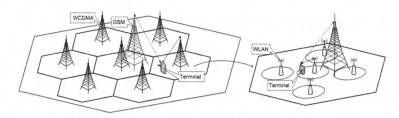

Fig. 3. Network topology

The standard algorithm we choose is a linear programming based access network selection algorithm. The objective and constraints are in Eq. (25) [11].

$$\begin{aligned} \max \quad & \sum_{i=0}^{N-1} \sum_{j=0}^{M-1} u_{ij} \cdot x_{ij} \\ s.t. \quad & \sum_{j=0}^{M-1} x_{ij} \le 1 \qquad i = 0, \ldots, N-1 \\ & \sum_{i=0}^{N-1} x_{ij} \le \phi_j \qquad j = 0, \ldots, M-1 \end{aligned} \qquad (25)$$

where N and M are respectively the number of terminals and access networks, x_{ij} represents whether terminal i selects network j, u_{ij} is the evaluation of terminal i to the network j, and ϕ_j is the limit number of terminals accessing the network j.

When making the performance evaluation, evaluation criterias include the average of evaluation to the networks, the standard deviation of the evaluation, the average of network providers' profits and the standard deviation of the profits. The compared results are in Figs. 4 ~ 7, where the algorithm proposed in this paper is Algorithm 1 and the standard algorithm is Algorithm 2.

Figure 4 shows that the Algorithm 2 has a higher average evaluation value, which is because that the sum of evaluation values to the access networks is its direct and unique optimization objective. However, for the Algorithm 1, the average evaluation value is just a part of one of its optimization objectives. Figure 5 shows that the Algorithm 1 has a lower standard deviation of evaluation values, which is because it takes the quality difference of access networks allocated to different terminals into consideration, which is included in its optimization objectives.

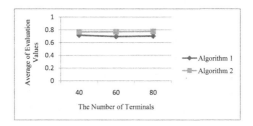

Fig. 4. Average of terminals' evaluation values to access networks

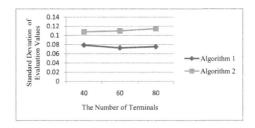

Fig. 5. Standard deviation of terminals' evaluation values to access networks

Figure 6 shows that the Algorithm 1 has a higher average value of network providers' profits, which is because it takes the profit of network providers into consideration, but the Algorithm 2 does not. Figure 7 shows that the Algorithm 1 has a lower standard deviation of network providers' profits, which makes the selection result fairer to different network providers.

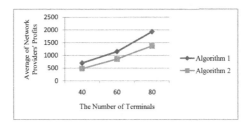

Fig. 6. Average of network providers' profits

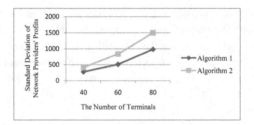

Fig. 7. Standard deviation of network providers' profits

7   Conclusions

The access network selection mechanism designed in this paper mainly includes two parts. This first part is an IEEE 802.21 based context information acquisition mechanism. It makes a good preparation for the network selection algorithm. The second part is an access network selection algorithm. The design of this algorithm considers the different requirements of different application types and profits of both terminal users and network providers. We choose the GA as the tool to solve the problem and design the genetic operators specifically for it. After the design, we conduct the simulation and performance evaluation. The result shows that the mechanism designed in this paper can achieve the design objective and has a big advantage in fairness and network providers' profits compared with the standard algorithm.

Acknowledgement.   This work is supported by the National Science Foundation for Distinguished Young Scholars of China under Grant No. 61225012 and No. 71325002; the Specialized Research Fund of the Doctoral Program of Higher Education for the Priority Development Areas under Grant No. 20120042130003; Liaoning BaiQianWan Talents Program under Grant No. 2013921068.

References

1. Oliva, A.D., Banchs, A., Soto, I.: An overview of IEEE 802.21: media-independent handover services. IEEE Wirel. Commun. **15**(4), 96–103 (2008)
2. Taniuchi, K., Ohba, Y., Fajardo, V.: IEEE 802.21: media independent handover: Features, applicability and realization. IEEE Commun. Mag. **47**(1), 112–120 (2009)
3. Xiong, M., Cao, J.: A clustering-based context-aware mechanism for IEEE 802.21 media independent handover. In: IEEE Wireless Communications and Networking Conference, pp. 1569–1574 (2013)
4. Trestian, R., Ormond, O., Muntean, G.M.: Enhanced power-friendly access network selection strategy for multimedia delivery over heterogeneous wireless networks. IEEE Trans. Broadcast. **60**(1), 85–101 (2014)
5. Jiajia, S., Wang, X., Gao, C., Huang, M.: Resource allocation scheme based on neural network and group search optimization in cloud environment. J. Softw. **25**(8), 1858–1873 (2014). (in Chinese)

6. Melia, T., Boscolo, L., Vidal, A.: IEEE 802.21 reliable event service support for network controlled handover scenarios. In: IEEE Global Telecommunications Conference, pp. 5000–5005 (2007)
7. Neves, P., Fontes, F., Sargento, S.: Enhanced media independent handover framework. In: Vehicular Technology Conference, pp. 1–5, Spain (2009)
8. Chamodrakas, I., Martakos, D.: A utility-based fuzzy TOPSIS method for energy efficient network selection in heterogeneous wireless networks. Appl. Soft Comput. 12(7), 1929–1938 (2012)
9. Wang, X., Cheng, H., Qin, P., Huang, M., Guo, L., Intan, R.: ABC supported handoff decision scheme based on population migration. In: Di Chio, C., Brabazon, A., Di Caro, G.A., Ebner, M., Farooq, M., Fink, A., Grahl, J., Greenfield, G., Machado, P., O'Neill, M., Tarantino, E., Urquhart, N. (eds.) EvoApplications 2010, Part II. LNCS, vol. 6025, pp. 111–120. Springer, Heidelberg (2010)
10. Trestian, R., Ormond, O., Muntean, G.M.: Reputation-based network selection mechanism using game theory. Phys. Commun. 4(3), 156–171 (2011)
11. Choque, J., Agüero, R., Muñoz, L.: Optimum selection of access networks within heterogeneous wireless environments based on linear programming techniques. Mob. Netw. Appl. 16(4), 412–423 (2011)
12. Herring, K.T.: Path-loss characteristics of urban wireless channels. IEEE Trans. Antennas Propag. 58(1), 171–177 (2010)
13. Tang, P.H., Tseng, M.H.: Adaptive directed mutation for real-coded genetic algorithms. Appl. Soft Comput. 13(1), 600–614 (2013)
14. Wang, X., Hou, W., Guo, L., Cao, J., Jiang, D.: Energy saving and cost reduction in multi-granularity green optical networks. Comput. Netw. 55(3), 676–688 (2011)

A Dijkstra Algorithm Based Multi-layer Satellite Network Routing Mechanism

Yinchu Sun[✉], Xingwei Wang, Fuliang Li, and Min Huang

College of Information Science and Engineering,
Northeastern University, Shenyang, China
qinqinmuji@163.com, {wangxw,isemhuang}@mail.neu.edu.cn,
lifuliang@ise.neu.edu.cn

Abstract. As the basis of satellite communication network, routing mechanism is a hot topic in satellite network. However, due to the dynamic network topology, high propagation delay and limited resource of satellite network, many great challenges have emerged in designing routing mechanism for it. In this paper, we propose a satellite network routing mechanism based on Dijkstra algorithm for multi-layer satellite network. By considering QoS requirements, service pricing, life cycle and load balance together, we first design an evaluation index, which is used to evaluate the effectiveness of the alternative route. Then we utilize the Dijkstra algorithm to select the best path for satellite network. Finally, we evaluate the proposed mechanism on the model of multi-layer satellite network and verify it from many aspects. Evaluation results show that our routing mechanism is feasible and effective.

Keywords: Multi-layer satellite network · Dijkstra algorithm · Routing mechanism

1 Introduction

With the development of wireless communication technology, satellite network has become a new type of Internet that achieves global communication seamless connection. It can supplement the lack of ground network. However, because of the increasing demand for multimedia service, satellite network's main service gradually changes from traditional voice and low-speed data service to Internet and broadband multimedia service. Therefore, building an efficient and reliable satellite network has great economic value.

Satellite network is different from the ground network. The transmission channels of inter-satellite links (ISLs) work in an open environment, which causes dynamic network topology and interference or violation to satellite network. Satellite network routing mechanism directly affects the performance of the satellite network. Dynamic topology, large transmission delay, high error rate, limited resources of nodes and unbalanced distribution of service bring big challenges to satellite network routing.

In recent years, many researchers have studied satellite network routing [1–5]. For example, a routing mechanism based on the periodicity of the satellite network is proposed in [1], and the load balance based routing mechanism is put forward in [4]. But these works just concentrate on partial characteristics of satellite network. By considering QoS requirements, service pricing, life cycle and load balance comprehensively [6], we present a satellite network routing mechanism based on Dijkstra

© Springer International Publishing Switzerland 2015
D.-S. Huang and K. Han (Eds.): ICIC 2015, Part III, LNAI 9227, pp. 586–597, 2015.
DOI: 10.1007/978-3-319-22053-6_61

algorithm. Firstly, we design an evaluation index to evaluate the effectiveness of the alternative route. Then we utilize the Dijkstra algorithm to select the best path for satellite network. Finally, we evaluate the proposed mechanism on the model of multi-layer satellite network and verify it from many aspects.

The remainder of the paper is organized as follows. The related work is presented in Sect. 2. We describe the network model and routing mechanism for multi-layer satellite network in Sect. 3. Section 4 presents our simulation and evaluation results. Our conclusions and future remarks are presented in Sect. 5.

2 Related Work

Currently, research on satellite network routing mechanism has made some progress. According to the problems, satellite routing mechanisms can be divided into five categories [8], mainly based on the constellation period, link switch, QoS requirements, load balance and path-based optimization.

The literature [1] put forward a routing mechanism based on constellation cycle. By using the slot allocation method, dynamic network topology was divided into several time slots, each of which was a static topology for routing. This mechanism demands less for the ability to process, but its adaptability is relatively poor. The literature [2] analyzed life cycle of the path to select path whose lifetime is longer. This routing mechanism meets the certain QoS requirements, but it does not take the fairness of links into account. The literature [3] proposed a routing mechanism based on the QoS requirements. According to various indicators of QoS satisfaction, they chose the best path that meets QoS requirements. This mechanism gets balance in different demands of QoS indicators, so it lacks a comprehensive strategy to optimize the QoS indicators. A routing mechanism based on load balance was studied in [4]. It uses the load as the metric to ensure the flow distribution reasonable, but it increases the expenses of ISLs signaling. The literature [5] considered the path optimization this aspect. According to the network status, they selected the path which is shorter, saves resources and meets QoS requirements. But this mechanism requires very regular network structure.

These researchers just considered partial properties of satellite network to study routing mechanism. In this paper, in order to meet the actual demands, we consider all the above-mentioned characteristics of the satellite network and present a Dijkstra algorithm based satellite network routing mechanism. Through using the QoS requirements, service pricing, life cycle and load balance as the basis for routing, it achieves the purpose of improving the QoS satisfaction, reducing link switch and saving node resource.

3 Model Design

3.1 Network Model

For multi-layer satellite network, the satellite network model reflects the relationship between nodes, determines the number of links of the node, the distance between nodes, and even affects the data transmission path and transmission reliability. According to the characteristics of multi-layer satellite network, we believe that GEO/MEO/LEO satellite network is more representative, and use it as a satellite constellation network structure of this article, as shown in Fig. 1.

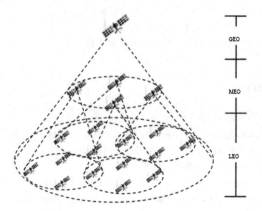

Fig. 1. Three layer satellite network structure

Through a detailed study of the multi-layer satellite network [7], the article identifies the network model by a weighted directed graph representation. As shown in Fig. 2, the weighted directed graph G represents the state of multi-layer satellite network topology; $V = \{v_0, v_1, \cdots, v_{n-1}\}$ denotes the set of satellite nodes and n is the number of satellites; $E(t) = \{e_{ij}\}$ represents the set of weight values, e_{ij} is the distance between the satellite i to j at time t, and $i, j \in [0.1]$.

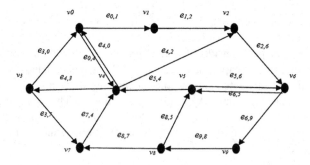

Fig. 2. Example of network model

3.2   Path Effectiveness Evaluation

Effectiveness evaluation is one of the concepts used in economics. It refers to the subjective evaluation of the user for the matching degree of services and demand functions. The article considers the characteristics of ISLs and evaluates comprehensively the QoS satisfaction, service pricing satisfaction, the life cycle fitness and load fitness as the path effectiveness evaluation of alternative paths.

QoS Satisfaction. Due to the QoS requirements of business are generally in the form of interval, this article is based on trapezoidal membership function of fuzzy mathematics to describe the changes of QoS parameters provided by user satisfaction and the service. We use PAL fuzzy enhancement algorithm to describe the user depending on the service level and the effects of parameters on QoS satisfaction.

$L_s \in [L_{Top}, L_{Low}]$ represents the service level that user sets up and the smaller L_s represents the higher service level. This article uses the PAL [10] fuzzy enhancement algorithm, calculated as shown in (1).

$$g(x) = \begin{cases} k_1 \cdot f(x)^2 & 0 \leq f(x) \leq \tau \\ 1 - k_2 \cdot (1 - f(x))^2 & \tau < f(x) \leq 1 \end{cases} \tag{1}$$

$g(x)$ is the trapezoidal membership of QoS parameters.

$$k_1 = \frac{1}{\tau} \tag{2}$$

$$k_2 = \frac{1}{1 - \tau} \tag{3}$$

$$\tau = 1 - \frac{L_s}{L_{Top} + L_{Low}} \tag{4}$$

When the value of τ is different, the function $g(x)$ is as shown in Fig. 3. Clearly, the higher the service level L_s is, the higher value of τ is, and the higher requirement for QoS parameters is.

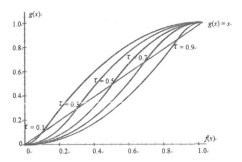

Fig. 3. Function curves of PAL algorithm

We evaluate the QoS satisfaction based on the three important parameters of bandwidth, transmission delay and error rate.

(1) *Bandwidth Satisfaction*

Bandwidth requirement belongs to the maximum and minimum constraint conditions of the QoS requirements. Alternative paths provide the business s with bandwidth which is the minimum value of available bandwidth provided by its sub-links. Assume that the bandwidth of the business s is $[er_s^l, er_s^h]$, then the bandwidth satisfaction SB is calculated as shown in (5).

$$
SB = \begin{cases}
0 & bw \leq bw_s^l \\
k_1 \cdot \left(\frac{bw - bw_s^l}{bw_s^h - bw_s^l}\right)^2 & bw_s^l < bw \leq bw_T \\
1 - k_2 \cdot \left(1 - \frac{bw - bw_s^l}{bw_s^h - bw_s^l}\right)^2 & bw_T < bw < bw_s^h \\
1 & bw \geq bw_s^h
\end{cases}
\tag{5}
$$

And,

$$
bw_T = \tau \cdot \left(bw_s^h - bw_s^l\right) + bw_s^l
\tag{6}
$$

(2) *Delay Satisfaction*

Latency requirement is QoS requirements' additive constraint. Time delay of the business s provided by alternative paths is the sum of sub-links' delay. Either in satellite network or conventional ground network, time delay is important parameter for evaluation of routing. End-to-end delay of the satellite network discussed in this paper consists of transmission delay and waiting delay.

Through analyzing ISLs time window with the method in [9], we can get the optional link to guarantee the biggest end-to-end delay dl, calculated as shown in (7).

$$
dl = p_0 + (t_n - t_1) + p_n
\tag{7}
$$

$t_n - t_1$ is the sum of transmission delay and waiting delay of the message from node 0 to node $n - 1$ on the path $R = (0, n)$. p_0 represents the time that the message wait for $hop(1)$; p_n is the transmission delay of $hop(n)$. Therefore, ideally, waiting delay is 0, then the end-to-end delay of the path is $dl = p_1 + p_2 + \cdots + p_n$.

Assuming that the demand for the delay of business s is $[dl_s^l, dl_s^h]$, and the delay satisfaction SD is calculated as shown in (8).

$$
SD = \begin{cases}
1 & dl \leq dl_s^l \\
1 - k_2 \cdot \left(1 - \frac{dl_s^h - dl}{dl_s^h - dl_s^l}\right)^2 & dl_s^l < dl \leq dl_T \\
k_1 \cdot \left(\frac{dl_s^h - dl}{dl_s^h - dl_s^l}\right)^2 & dl_T < dl < dl_s^h \\
0 & dl \geq dl_s^h
\end{cases}
\tag{8}
$$

And,

$$dl_T = dl_s^h - \tau \cdot \left(dl_s^h - dl_s^l\right) \tag{9}$$

(3) *Error Rate Satisfaction*

Error rate demand belongs to the multiplicative constraints of QoS demands. Alternative path is related to the error rate of service and sub-links provided by business. Satellite network is easy to be interfered by natural or human factors, so the importance of error rate in satellite network routing is more outstanding.

$er(h)$ represents the error rate of $hop(h)$, and the error rate of the path $R = (0, n)$ is calculated as shown in (10).

$$er = 1 - \prod_{h=1}^{n}(1 - er(h)) \tag{10}$$

Assuming that the error rate demand of business s is $[er_s^l, er_s^h]$, and the error rate satisfaction SE is calculated as shown in (11).

$$SE = \begin{cases} 1 & er \le er_s^l \\ 1 - k_2 \cdot \left(1 - \frac{er_s^h - er}{er_s^h - er_s^l}\right)^2 & er_s^l < er \le er_T \\ k_1 \cdot \left(\frac{er_s^h - er}{er_s^h - er_s^l}\right)^2 & er_s^l < er \le er_T \\ 0 & er_T < er < er_s^h \end{cases} \tag{11}$$

Service Pricing Satisfaction. Combining with limited power resource and the market pricing mechanism, we argue that the satellite network service pricing is decided by the power cost and service level. sc is the service pricing of the path $R = (0, n)$.

In (12), pb is the service pricing, P_i is the sending power and α is the adjustment coefficient, $\alpha \in (0, 1)$.

$$sc = pb \cdot (1 + \alpha \cdot \ln(\sum_{i=0}^{n-1} P_i)) \cdot \left(1 + \frac{L_{Low} - L_s}{L_{Low} - L_{Top}}\right) \tag{12}$$

We evaluate satisfaction service pricing SP through drop half parabolic distribution function, calculated as shown in (13).

$$SP = \begin{cases} \left(\frac{ap_s - sc}{ap_s}\right)^{\beta} & sc \le ap_s \\ 0 & sc > ap_s \end{cases} \tag{13}$$

ap_s is the acceptable unit price for business s and β is the adjustment coefficient.

Life Cycle Fitness. In route choice, the life cycle of the path cannot be ignored. Together with the path's available bandwidth, it determines the amount of data that can be transferred in the life cycle. The life cycle of path $R = (0, n)$ is decided by the minimum value of each link time window's period, calculated as shown in (14).

$$lf = \min(t'_h - t_h) \tag{14}$$

In the daily time, the user cannot describe all the time needed by the business. Therefore, we introduce the threshold value lf_{th} of the path's minimum life cycle, and use half of normal distribution function to evaluate life cycle satisfaction SL of the path, calculated as shown in (15).

$$SL = \begin{cases} 0 & lf \leq lf_{th} \\ 1 - e^{-\gamma \bullet (\frac{lf}{lf_{th}} - 1)^2} & lf > lf_{th} \end{cases} \tag{15}$$

γ is the adjustment coefficient and $\gamma > 0$.

Load Fitness. In this paper, we use the channel bandwidth and power utilization ratio to evaluate satellite node load. The load rate of its node i is calculated as shown in (16).

$$\eta_i = \eta_i^w + \eta_i^p - \eta_i^w \cdot \eta_i^p \tag{16}$$

η_i^p and η_i^w, respectively, represent power load rate and channel bandwidth load rate of satellite node, $\eta_i \in [0, 1], \eta_i \geq \eta_i^p, \eta_i \geq \eta_i^w$.

We consider the maximum value of load rate η_i of the path $R = (0, n)$ as the load rate η of the path. Correspondingly, its load fitness SW is calculated as shown in (17).

$$SW = \begin{cases} 1 - \eta & \forall i \in [0, n - 1], \\ & W_i \leq W_{available}^i \wedge P_i \leq P_{available}^i \\ 0 & \text{Otherwise} \end{cases} \tag{17}$$

W_i and P_i, respectively, represent the channel bandwidth and power that nodes need in the path $R = (0, n)$. $W_{available}^i$ and $P_{available}^i$, respectively, are residual channel bandwidth and the power of the nodes.

Efficiency Computing. The article introduces the weight coefficient matrix $\Lambda = [\lambda_1 \, \lambda_2 \, \lambda_3 \, \lambda_4]$, followed by the relative importance of the QoS satisfaction, service pricing satisfaction, life cycle and load fitness in the routing selection, and $\Lambda = [\lambda_1 \, \lambda_2 \, \lambda_3 \, \lambda_4]$. Path evaluation matrix is $\mathbf{E} = [SQ \, SP \, SL \, SW]^T$. Routing efficiency of alternative paths RU is as shown in (18).

$$RU = \Lambda \cdot \mathbf{E} \tag{18}$$

3.3 Algorithm Design

Mathematical Model. The article aims to maximize path effectiveness to find the best path, meanwhile meets the QoS requirements and the existence of links.

$$
\begin{aligned}
& \max(RU) \\
s.t. \quad & lf \geq lf_{th} \\
& sc \leq ap_s
\end{aligned}
\tag{19}
$$

Algorithm Description. According to the evaluation index of the path in this paper, we can get that the path length RL increases with the hops, which guarantees the condition that the length of each hop in Dijkstra algorithm is nonnegative. Therefore, we propose a Dijkstra algorithm based satellite network routing mechanism, and define the length of path RL as shown in (20).

$$
RL=
\begin{cases}
\frac{1}{RU} & \text{constraint conditions of} \\
 & \text{the mathematical model} \\
+\infty & \text{Otherwise}
\end{cases}
\tag{20}
$$

Specific steps show as below.

Input: Satellite network $G = (V, E(t))$, all the parameters of satellite nodes in the network, the QoS requirements and service level L_s of the business.

Output: The optimal path.

Step1: Initialize set $S = \emptyset$ and $T = \emptyset$, set source node V_s and destination node V_d;

Step2: Put the source node V_s into set S, then put the rest nodes into set T.

Step3: Calculate link distance between the nodes in the set T and source node V_s. If there is no link satisfied with visibility and QoS requirements, set the distance $+\infty$.

Step4: Take the node V_n from the set T, which is nearest to the node in the set V_s, into the set S. And remove V_n from the set T.

Step5: Call the path length calculation function, then using V_n as the intermediate node, calculate the distance between V_i and the source node V_s.

Step6: If the new distance is less than the original distance, update the distance from V_i to V_s and record the path, then go to step 7.

Step7: If nodes in set T have been traversed, go to step 8. Otherwise, go to step 5.

Step8: If set T is empty, go to step 9. Otherwise, go to step 4.

Step9: Ends.

4 Performance Evaluation

In this paper, the simulation example is multi-layer satellite network including 15 satellite nodes. As shown in Table 1, the constellation structure includes: 3 GEO, 3 MEO and 9 LEO.

In order to measure the performance of satellite network routing mechanism proposed, the article evaluates it mainly through satellite network capacity adopting the

Table 1. Use case of satellite constellation structure

Orbit parameters	GEO	MEO	LEO
Semi-major axis (km)	42166	27878	8378
Eccentricity ratio	0	0	0
Orbit inclination (°)	0	55	55
RAAN (°)	16E, 136E, 256E	0E, 120E, 240E	0E, 120E, 240E
Argument of perigee (°)	0	0	0
True anomaly (°)	0	0E:0, 120E:120, 240E:240	80, 200, 320
Number of satellites	3	3	9
Number of planes	1	3	3

method of the Hybrid adaptive genetic algorithm [11] based on the golden ratio. The network capacity limit results are shown in Fig. 4 and each strategy's capacity limit of variance between time slots is shown in Table 2.

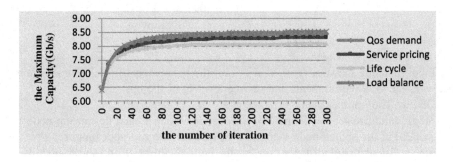

Fig. 4. The maximum capacity with different routing strategies

Table 2. The variances of maximum capacity with different routing strategies

Routing strategy	QoS demand	Service pricing	Life cycle	Load balance
Network capacity variance	0.25	0.32	0.58	0.14

We can see that putting emphasis on load balance, satellite network capacity [12] limit is higher and less volatile. While an emphasis is on the use of life cycle, the capacity limit is low and volatile. The reason is that routing mechanism focused on load balance chooses the path mainly on available resources of nodes, so the path is more diverse. Figure 5 denotes the number of nodes' ISLs and the total number of ISLs. 1–3 represent GEO satellites, 4–6 represent MEO satellites, and 7–15 represent LEO satellites. Obviously, with an emphasis on load balance, the average number of ISLs is larger. Besides, the available resources of each node are relatively balanced. So it is more rational to allocate the limited resources.

Figure 6 shows the comparison of each strategy, the corresponding strategy's variance of capacity lower limit in each time slot is shown in Table 3. Clearly, the

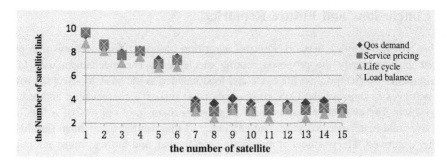

Fig. 5. The number of link in each satellite with different routing strategies

capacity lower limit is more volatile while putting an emphasis on life cycle of network, but it has a higher capacity lower limit. However, while the average number of links is larger, the routing strategy with an emphasis on load balance still gets the similar capacity lower limit with one whose key is on service pricing, because of its rationality of the resources allocation.

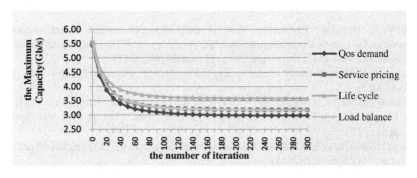

Fig. 6. The minimum capacity with different routing strategies

Table 3. The variances of minimum capacity with different routing strategies

Routing strategy	QoS demand	Service pricing	Life cycle	Load balance
Network capacity variance	0.55	1.00	1.50	0.72

In addition, the network capacity limit's difference between the minimum and maximum by using the routing strategy whose emphasis is on the life cycle. But its nodes change the moving position, which will cause large changes in resource utilization of each node, then the stability of satellite network based on this routing strategy will be worse. The satellite network routing focused on load balance can not only increase network capacity, but also help to increase the stability of the satellite net-work itself.

5   Conclusions and Future Remarks

In this paper, a satellite network Dijkstra algorithm based routing mechanism is presented and we verify the proposed routing mechanism. The mechanism uses QoS requirements, service pricing, life cycle and load balance as the basis to select the route and achieves to improve the QoS satisfaction, node resource conservation and reduce link switch. The evaluation results show that the mechanism of our design is feasible, effective, and better performance. The efficiency of this routing mechanism should be further improved. The practical of this routing mechanism and further improvement are the focus of our future research work.

Acknowledgment.   This work is supported by the National Science Foundation for Distinguished Young Scholars of China under Grant No. 61225012 and No. 71325002; the Specialized Research Fund of the Doctoral Program of Higher Education for the Priority Development Areas under Grant No. 20120042130003; Liaoning BaiQianWan Talents Program under Grant No. 2013921068.

References

1. Fdhila, R., Hamdani, T.M., Alimi, A.M.: A multi objective particles swarm optimization algorithm for solving the routing pico-satellites problem. In: IEEE International Conference on Systems, Man and Cybernetics, Seoul, pp. 1402–1407 (2012)
2. Zhang, Y., Liu, F.: Partially known routing for satellite IP networks. J. Softw. **8**(8), 1851–1858 (2013)
3. Urquizo Medina, A.N., Qiang, G.: QoS routing for LEO satellite networks. In: Zu, Q., Hu, B., Elçi, A. (eds.) ICPCA 2012 and SWS 2012. LNCS, vol. 7719, pp. 482–494. Springer, Heidelberg (2013)
4. Ma, X.: Adaptive distributed load balance routing mechanism for LEO satellite IP networks. J. Netw. **9**(4), 816–821 (2014)
5. Song, G.H., Chao, M.Y., Yang, B.W., Zhong, H., Zheng, Y.: Research on multi-path QoS routing strategy for the satellite network. In: Shen, R., Qian, W. (eds.) TT&C. LNEE, vol. 187, pp. 482–494. Springer, Heidelberg (2013)
6. Wang, X., Cheng, H., Huang, M.: QoS multicast routing protocol oriented to cognitive network using competitive coevolutionary algorithm. Expert Syst. Appl. **41**(10), 4513–4528 (2014)
7. Wang, X., Cheng, H., Li, K., Li, J., Sun, J.: A cross-layer optimization based integrated routing and grooming algorithm for green multi-granularity transport networks. J. Parallel Distrib. Comput. **73**(6), 807–822 (2013)
8. Mao, T., Zhengquan, X., Zhu, R., Hou, R.: Researching review on routing technologies of next generation satellite network. Telecommun. Sci. **28**(11), 116–120 (2012)
9. Cruz, S.H., Franck, L., Beylot, A.L.: Routing metrics for store and forward satellite constellations. IET Commun. **4**(13), 1563–1572 (2010)
10. Liu, X., Jiang, Y., Luo, X.: Improved image fuzzy-enhancement algorithm. Comput. Eng. Appl. **44**(4), 50–52 2008

11. Yang, T., Lin, P.: Spectrum allocation based on improved genetic algorithm in cognitive radio system. Comput. Simul. **31**(2), 250–254 (2014)
12. Wang, X., Wang, X., Wang, C., Huang, M.: Resource allocation in cloud environment: a model based on double multi-attribute auction mechanism. In: 6th IEEE International Conference on Cloud Computing Technology and Science (2014), Singapore, 15–18 December 2014

The Research on Optimizing Deployment Strategy for Aviation SWIM Application Servers

Haitao Zhang[1(✉)] and Zhijun Wu[2]

[1] School of Electronic Information Engineering,
Tianjin University, Tianjin Province 300072, China
13512222258@163.com
[2] School of Electronics and Information Engineering,
Civil Aviation University of China, Tianjin Province 300300, China
zjwu@cauc.edu.cn

Abstract. This paper presents a method for dynamically optimizing business application deployment for improving the overall stability of the SOA in Aviation SWIM. The solution analyses Web Service call logs with association rules analysis and deploys business applications, which are in a tight coupling relationship in the chain of synchronously interconnected service components, into one physical system. The aim of this solution is to decrease long-distance calls of Service and to enhance stability of SOA. The association rules analysis improves Apriori Algorithm based on the incidence matrix, and Hadoop cloud computing platform verifies that this algorithm can work out the result faster with better parallelism.

Keywords: SWIM · SOA · Association rules · Apriori Algorithm · Hadoop

1 Introduction

SWIM (System-Wide Information Management) is a key technology to achieve sharing and exchange of information on modern air transportation, so NGATS (Next Generation Air Transportation System) and SESAR (Single Europe Sky ATM Research) utilize SWIM based on SOA (Service-Oriented Architecture) as the major part of foundation framework of information exchange. With improvement and implementation of SWIM in recent years, flexibility, scalability and stability of SOA have drawn much more attention [1].

As the core of aviation information system, SWIM integrates and shares all types of resources in the network such as airspace management, flow management, traffic management, surveillance management, separation management and aircraft system. Also, SWIM manages data transmission from communication, navigation, surveillance data, weather information, global geographic information and all flight vehicles to provide CDT (Common Data Transmission) from civil aviation system [2] (Fig. 1). A large number of application servers are deployed in SWIM based on SOA in order to achieve the above functions. Civil Aviation Information System requires its SOA to

© Springer International Publishing Switzerland 2015
D.-S. Huang and K. Han (Eds.): ICIC 2015, Part III, LNAI 9227, pp. 598–606, 2015.
DOI: 10.1007/978-3-319-22053-6_62

ensure efficient availability and stability of service to minimize impact on the entire system from unpredictable conditions, such as emergent widespread network delay, overlong database response time and lower QoS (Quality of Service).

Thus, this paper will focus on methods of establishing resilient SOA to meet the requirements of SWIM for continuous availability and performance of service and to minimize the negative effects caused by changes in the network environment. Besides, this paper presents the deployment strategy of SWIM application services on each application server. First, analyze call relationship and coupling tightness of SWIM business application services in operation with the association rules analysis of KDD (Knowledge Discovery in Database). Then, re-deploy business application services in a tight coupling relationship in one physical system according to the computation results. This method reduces the remote procedure call to increase the service speed and throughput of distributed application servers [3].

2 Solutions

2.1 Issues and Solutions

Setting up a highly resilient SOA architecture is an important requirement of civil aviation network and SWIM. ("Resilience" means continuous availability and performance of service, and the negative modification has little impact on it.) As the main platform for the deployment of civil aviation business services, application server is an important infrastructure of SWIM, and its work focuses on implementation of the business logic. It has been proved that a key technical issue of setting up a resilient SOA architecture is to work out thread block of the application server. However, a large number of application servers of civil aviation SWIM system are decentralized and often need to call remote threads or services. Thus, the solution to thread block of the application server is important to improve the overall stability of the system, and there are usually two types of solutions [3]:

- to add the aggressive timer to the application server: the key to solve the thread block is increasing the average service speed rate of each thread by dynamically changing the probability of a timeout and timeout values, which accelerates the service speed rate of the entire server and enhance stability and availability of SOA.
- to optimize and improve the deployment of services or applications of SOA: The key of such solution is deploying business applications of a close coupling in synchronous interconnected modules chain to the same physical system as much as possible. It is responsible for re-deploying business applications that are decentralized in the network but share a large number of synchronous communications, which reduces the RMI (Remote Method Invocation) and provides a great benefit for business application programs so as to improve the overall condition and stability of SOA.

The strategy put forward for the above issue in reference [3] is to optimize the deployment of the tightly coupled services and applications of SOA in the physical system. (The tightly coupled refers to the relationship between multiple services called

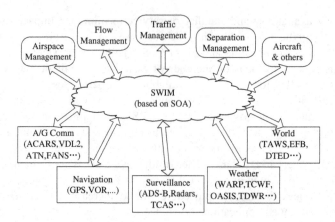

Fig. 1. The architecture of SWIM base on SOA

synchronously. Its feature is when a block occurs to the client calls service, the service would not continue until it receives a response.) If multiple tightly coupled SOA services often appear in some business, thereby synchronous interconnected modules chain will be formed in SOA. Thus, this paper presents a solution of dynamically optimizing service deployment with association rules analysis. That is to constantly analyze and regulate dynamically business applications (or services) in the synchronous interconnected modules chain together. Tightly coupled services are re-deployed in the same physical system so as to improve the overall condition and stability of SOA.

For example, such applications of synchronized dependence that are deployed in different servers typically generate HTTP or RMI communication connection. For example (Fig. 2), an external client calls application A in the server 1. Its work request will be sent to the managed task of the server 1. However, the communication between the application A and B is synchronized, so managed tasks distributing work to application A in the server 1 have to be blocked until it receives a response from the application B and when blocked, managed tasks are all in wait.

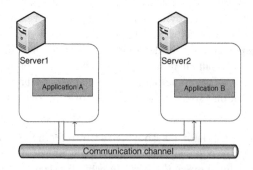

Fig. 2. RMI communications connection of a server to another one.

If the server suffers from a heavy load at this time, only a small part of managed task resources are available, so more and more managed tasks are blocked on synchronous remote service calls, and the ability of server 1 calling works would be further reduced, resulting in decline of server1 in the speed rate of overall service at a fast pace. When the service rate is lower than the arrival rate of new tasks, the server will not catch up, so the new task will be waiting in line. The worst outcome is that all managed tasks are blocked, and the server cannot perform any work, which seems to suspend. In this case, queue tasks and distributed tasks will delay for a long time and eventually fail.

The solution to this problem is to deploy tightly coupled business applications in the same server, which apparently reduces pressures from applications on resources (Fig. 3). The application A can use the local in-process protocol to call application B, which is a direct call. This avoids occurrence of the remote call and a series of problems related to the remote delay. This will reduce the blocked managed tasks, so that the server can maintain a high rate of service, effectively avoiding the queue growth to prevent timeout of the task deployment, which in the end further reduces instability in the environment of the SOA to establish a more stable and more resilient SOA system.

2.2 The Association Rules Analysis Based on Hadoop Cloud Computing Platform

To solve the above problem, this paper presents a strategy that the frequency of mutual calling between services and threads determines their deployment to ensure the stability of the SOA-based SWIM. By log analysis of the threads or services calling, services or threads often calling each other are deployed in the same physical server. Since the various types of application services in SWIM generate a large number of service calling logs, there must emerge a huge amount of computation in the association rules analysis for the calling relationships in the server. Therefore, this paper intends to present a combined solution of using cloud computing platform to increase computing speed and improving association rules analysis algorithm in the SWIM:

- To introduce the open-source Hadoop cloud computing platform [4]; The Apache Hadoop project develops open-source software for distributed computing. The

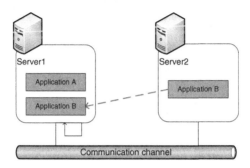

Fig. 3. Deploying tightly coupled business applications on the same server.

Apache Hadoop software library is a framework that uses simple programming models to distribute and process large data sets across clusters of computers. It is designed to extend single servers to thousands of machines that offer local computation and storage. Instead of hardware delivering high-availability, the library is to detect and handle failures at the application layer so as to offer a highly-available service on top of a cluster of computers.

- To increase the computing speed by improving the Apriori algorithm of association rules analysis and the algorithm's parallelism with Map-Reduce. Apriori algorithm and FP-growth are classical algorithms of association rules analysis. However, with a large number of candidate frequent item-sets, the FP-growth algorithm takes up a large amount of RAM, and the classic Apriori algorithm generates a large amount of computation, none of which applies to large data processing. This paper presents Apriori Algorithm based on Boolean Matrix in the reference [5] to improve the computation of inner product of vectors of Boolean matrix and enhance its parallelism. Finally this algorithm will be implemented on Map-Reduce parallel computation framework of Hadoop.

3   Using the Map-Reduce Parallel Computing Framework to Complete and Improve the Algorithm

3.1   Definitions and Features of the Apriori Algorithm Based on Boolean Matrix

Definition 1: The vector quantity of each item I_j is defined: $D_j = \begin{pmatrix} t_{1j} \\ t_{2j} \\ \vdots \\ t_{nj} \end{pmatrix}$, T_i is the

i-item, $t_{ij} = \begin{cases} 0, & I_j \notin T_i \\ 1, & I_j \in T_i \end{cases}$, $\text{support\_count}(I_j) = \sum_{i=1}^{n} t_{ij}$.

Definition 2: (1) The matrix of item set I is: $D = (D_1, D_2, \ldots, D_n) = \begin{bmatrix} d_{11} & d_{12} & \cdots & d_{1n} \\ d_{21} & d_{22} & \cdots & d_{2n} \\ \vdots & \vdots & & \vdots \\ d_{p1} & d_{p2} & \cdots & d_{pn} \end{bmatrix}$, so the value (0 or 1) of d_{ij} can be worked out according to

definition 1. (2) The vector quantity of item set $\{I_i, I_j\}$ is D_{ij} : $D_{ij} = D_i \wedge D_j = \begin{bmatrix} d_{1i} \wedge d_{1j} \\ d_{2i} \wedge d_{2j} \\ \vdots \\ d_{ni} \wedge d_{nj} \end{bmatrix}$. Thus, the support count of 2-item set $\{I_i, I_j\}$: $\text{support\_count}(I_i, I_j) = \sum_{k=1}^{n} (d_{ki} \wedge d_{kj})$.

Definition 3: The vector quantity of k-item set $\{I_1, I_2, \ldots, I_k\}$: $D_{12\ldots k} = D_1 \wedge D_2 \wedge \ldots D_k = (D_1 \wedge D_2 \wedge \ldots D_{k-1}) \wedge D_k)$; thus, support_count$\{I_1, I_2, \ldots, I_k\} = \sum_{k=1}^{n} \{(d_{p1} \wedge d_{p2} \wedge \ldots \wedge d_{p(k-1)}) \wedge D_k\}$.

Definition 4: Put any k columns vectors in Boolean matrix D into counterpoint (peer elements) "AND" operation, and the number of "1" in the result divides the total number of items, the result of which is called k- item set support counter. The minimum number of support transaction is the product of minimum support and the transaction number of transaction database.

The following features and inferences have been obtained according to the definition of item set support with Apriori features.

Feature 1: X_k is a k-item set. If the number of frequent (k-1) – item set L_{k-1} containing (k-1)–subitem sets in X_k is less than k, X_k can not be maximal frequent item set of k-dimension. I.e. the number of times that frequent k-1 item sets in frequent k-item set appears is equal to or greater than k.

The inference from feature 1 is that the number of times that each factor of frequent item set X_k appearing in frequent set of (k-1)-dimension is not less than k-1, so the number of times that some item I_i in Boolean matrix D appearing in L_{k-1} is less than k-1. Thus, columns corresponding to I_j can be deleted in Boolean matrix, and then, the number of times that I_j appearing in (k-1)-frequent item set is k in Boolean matrix, i.e. all the subsets of (k-1)-dimension that contain I_j are in (k-1)-frequent set. Thus, k combinations of items in this matrix don't need tailoring.

Feature 2: Suppose that L_k is frequent item set of k item, if T contains M and $|T| = k$, T is the transaction that can be deleted in database in following mining.

Feature 3: If $|L_k| < k+1$ (L_k is frequent k-item sets), the number of times of maximal frequent item set in mined transaction database is k. $| L_k |$ refers to the number of frequent k- item sets [6].

A large number of candidate frequent item sets are generated as Apriori Algorithm scans the database many times, so the calculation amount increases sharply. This paper makes f: D $\rightarrow$ R(1) for any given transaction database D with the method in reference [5], and R = f (D) = $(r_{ij})_{n \times m}$, n is the number of transactions, m is the number of items, transaction sets is $T_i (i \in n)$, item sets is $I_j (j \in m)$. If $I_j \in T_i$ $r_{ij} = 1$. Or $r_{ij} = 0 (i = 1, 2, \ldots, n; j = 1, 2, \ldots, m)$. Thus, after a scan, transaction database D can be mapped to the Boolean matrix with f.

After transaction database is mapped to Boolean matrix, the vector inner product computation can be applied to the row vectors in Boolean matrix to find out the possible rows in frequent item sets so as to concentrate the row vectors of Boolean matrix step by step. Thus, the frequent item sets that transaction database looks up will be inducted quickly and intuitively from the concentrated Boolean matrix.

3.2 Design of Implementation Methods Based on Map-Reduce

The equivalent Boolean matrix will be obtained after the transaction database is scanned once with Boolean matrix method (This Boolean matrix contains all data to mine frequent item sets next). Thus, the mining of association rules in transaction database has turned into the analysis of the equivalent Boolean matrix [7]. However, since the Boolean matrix generated from the large transaction database is gigantic, the conventional computation method cannot be loaded into the RAM to compute.

This paper introduces the cloud computing platform based on Hadoop, analyses and processes Boolean matrix with the Map-Reduce distributed computation frame. The basic steps are as follow:

Step 1: to divide Boolean matrix generated after scanning the transaction database once into blocks by row (to keep the transaction integrated). To decompose the huge matrix into several Splitted-Matrixes A_i (suppose the Boolean matrix corresponding to the transaction database D is R, R = $(A1,A2,...,An)^T$, T is transposition, then A_i is the m dimension row vectors of R in real number domain, i = 1,2,...,n)

Step 2: every Splitted-Matrix distributes in the node of the cluster after transposed as the data of Map function during iteration.

Step 3: Map function is the local frequency of candidate k-item sets generated from computing the inner product of vectors in each Splitted-Matrix.

Step 4: to use Reduce function to combine the local frequency of k-item sets to obtain the support of candidate k-item sets.

Step5: to combine results to obtain the corresponding global frequency of frequent k-item sets of the minimal support

Step 6: to transpose the k-item sets as candidate item sets to operate in the third iteration until the null set occurs.

To sum up, the Map-Reduce distributed computing frame based on Hadoop solves the storage issue of Boolean matrix with huge data, but also adds to computational nodes to increase the computation speed of Apriori Algorithm. This method reasonably partitions the transaction database according to the particular case of user computational node RAM during data preparation. The data sets distributed on N nodes can serve as N partitions of the integrated transaction database, during computing and generating the results. Therefore, each local node can search for the frequent item sets of the integrated transaction database.

4   Experimental Analysis

This paper puts database (from http://fimi.ua.ac.be/data/) "mushroom" and "accidents" of classic association rules analysis as a database set for verification and analysis. The above algorithm is compiled with Java and its development environment is the compiler "IntelliJ IDEA 14.02"; cloud computing platform selects Hadoop 2.50(CDH version number 5.2.0); Java JDK 1.60. The hardware environment where the experiment operates are 2 Cisco Blade Server clusters (CPU is 8 cores Xeon 2.6 GHz/RAM 16 GB DDR3), and the clusters' network connection is double 10 GB optical fibers. The experiment is carried out in a network computing environment of "VMware vSphere 5 Enterprise" virtualized to 8 computers, and each computer's operating system is Ubuntu (version 12.04).

First the experiment computes data sets "mushroom" with single node, compares computing results of classic serial Apriori Algorithm and Apriori Algorithm based on Boolean matrix with different min_supports (Fig. 4).

On account of the Boolean matrix, only one complete scan has to be taken on frequent item sets. Then, the search for frequent item sets requires more flag bits and memory space developed by the Boolean matrix R to operate. Thus, this algorithm can save much more time than the conventional algorithm with the frequent item sets increasing.

Compared with the classic algorithm, it is not obvious for the single-node Boolean matrix algorithm to increase the speed. However, the divided Boolean matrix shortens the statistical time of the candidate sets. With the increase of the matrix divisions whose corresponding compute nodes are computing concurrently, the total computation speed is much faster.

Thus, the computing results are analyzed on node 1, node 2 and node 4 with Map-Reduce computation framework (Fig. 5), which indicates this algorithm owns sound speed-up ratio with compute nodes increasing.

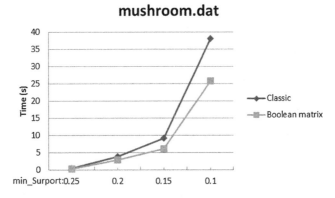

Fig. 4. data sets "mushroom" computed by single node.

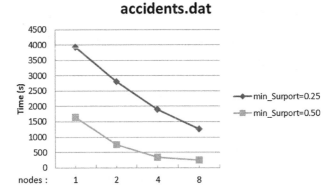

Fig. 5. data sets "accidents" computed by multi-nodes.

5 Conclusion

This paper discusses the performance and availability of SWIM of the civil aviation network, presents a deployment strategy for SWIM application and service based on the association rules analysis for the resilience issue of SOA framework in reference [3]. According to the frequent mutual calling between various types of applications and services of SWIM in the course of operating, this paper redeploys applications and services in a tight coupling relationship into the same physical system so as to increase the service speed rate of the application servers and the resilience of the SOA framework.

To solve the problem of association rules analysis of a large number service calling logs, this paper introduces Hadoop cloud computing model into the SWIM system, using the Map-Reduce parallel computation framework to complete and improve the Apriori Algorithm of Boolean Matrix in reference [5]. The experimental analysis on classical data sets act to verify the accuracy of the computation results of this algorithm and good parallelism (there is a higher speedup ratio as computational nodes increase).

Acknowledgements. The research work was supported by National Natural Science Foundation of China under Grant No. 61170328&U1333116 and Research Program of Application Foundation and Advanced Technology of Tianjin under Grant No. 12JCZDJC20900.

References

1. Luckenbaugh, G., Landriau, S.: Service oriented architecture for the next generation air transportation system. In: 2007 Integrated Communications Navigation and Surveillance Conference, pp. 1–2 (2007)
2. Harkness, D., Taylor, M.S., Jackson, G.S.: An Architecture for System-Wide Information Management. In: 25th Digital Avionics Systems Conference, pp. 1A6-1–1A6-13 (2006)
3. Antani, S., Alderman, R.G.: Build a resilient SOA infrastructure, Part 1: Why blocking application server threads can lead to a brittle SOA. http://www.ibm.com/developerworks/webservices/library/ws-soa-resilient/
4. The Apache Hadoop project. http://hadoop.apache.org/
5. Liu, Y., Yang, B.: Research of an improved Apriori algorithm in mining association rules. Comput. Appl. (Chinese) **27**(2), 418–420 (2007)
6. Miao, M., Wang, Y.: Research on improvement of Apriori algorithm based on matrix compression. Comput. Eng. Appl. (Chinese) **49**(1), 159–162 (2013)
7. Lin, C., Wu, Y., Huang, Z., Zeng, S.: Parallel research of Apriori Algorithm based on MapReduce. J. Jiangnan Univ. (Nat. Sci. Ed.) **13**(4), 411–415 (2014)

Modeling Fault Tolerated Mobile Agents by Colored Petri Nets

Shao-zhen Zhang[1], Zuo-hua Ding[1], and Jue-liang Hu[2(✉)]

[1] College of Information Science and Technology,
Zhejiang Sci-Tech University, Hangzhou 310018, Zhejiang, China
`sojen3@126.com, zuohuading@hotmail.com`
[2] College of Science, Zhejiang Sci-Tech University,
Hangzhou 310018, Zhejiang, China
`hujlhz@163.com`

Abstract. A mobile agent is a software migrating from one node to another to fulfill the task of its owner. The security problem of mobile agent is becoming a bottleneck for further development. To handle the malicious host problem in the migration path of mobile agents is one of the challenge tasks. One problem is agent blocking: the malicious host may deny the resources required by the agent and may kill the agent, thus the obtained data will loss. This paper proposes a mechanism to recover the mobile agents. We use Colored Petri Nets to model the dynamic mobile agent that owner can detect the malicious hosts and restore point before the failure, it can continue to perform the task all the components on the new (to skip the point of failure) mission path. The simulation result shows the effectiveness of the propose method.

Keywords: Mobile agents · Malicious host · Fault tolerance · Colored Petri Net · Security issues

1 Introduction

With the development of the Internet world, many network-related technologies are constantly grow and evolve. Which due to the extensive use in distributed and heterogeneous networks, mobile agent technology has been more and more people focus on software development. Broadly speaking, agent is an autonomous energy on behalf of someone or something (an organization or another agent) software components to perform a specific task. Behavior of the component is not only determined by the external event or interaction, but also determined by its own mission objectives. Agent can be divided into two categories according mobility: static agent and mobile agent. Collaboration and communication agent-based applications often involve multiple different roles between agent. In addition, a mobile agent application will involve a variety of computing environments, namely mobile agent system or server, different environments will provide different services and resources for agent to run and operate. Mobile agent system is a software component that is to receive agent and provide local resources.

© Springer International Publishing Switzerland 2015
D.-S. Huang and K. Han (Eds.): ICIC 2015, Part III, LNAI 9227, pp. 607–617, 2015.
DOI: 10.1007/978-3-319-22053-6_63

For the mobile agent system to support the agents in various application areas, the issues regarding the reliable agent execution, as well as the compatibility between two different agent systems or the secure agent migration, have been considered. Due to the nature of asynchronization, concurrence and distribution, the distributed systems become more complex. The mobile agent brings new complexity, its many factors, such as: operating in an operational environment is autonomous, open face security attacks, agent server crash, and could not find the resources [8]. In the absence of backup, mobile agent on a host failure will cause it to work on before the host, including data collection and loss of some or all of the work of the state, this failure is that we do not want to happen. In order to restore the original after a failure occurs working condition in time, the home host of mobile agent continuously detect whether the failure to issue it, if fails, the home host will sends a new agent to perform the task continue.

So we propose the model to identify the malicious host or platform to skip it from the itinerary. Here, we concentrate on agent instead of the protection of information or its itinerary. The rest of this paper is structured as follows. The definitions of the mobile agent and fault tolerance are given in Sect. 2. Section 3 presents an informal model of fault tolerated mobile agent and a solution for identifying and skipping malicious hosts in the mobile agent's itinerary. Sections 4 and 5 describes a failure tolerance model and then gives the simulation and analysis of the model. Section 6 contains a brief survey of related work. Finally, we make conclusions and propose the future work of the paper in Sect. 7.

2 Mobile Agent and Fault Tolerance

The so-called agent [12] is a program that is able to operate and migrate between the same structure and heterogeneous networks autonomously and independently, it has autonomy, intelligence, mobility, collaboration and so on. Agent is a software object [14] that is situated within an execution environment and must possess the following mandatory properties:

- **Autonomous:** On behalf of the user
- **Reactive:** Responsive to change in environment
- **Goal Driven:** Proactive acting in advance to deal with an expected situation
- **Temporally Continuous:** Continuously operating

There can be two types of operations in agent systems: *Dependent* and *Independent* [10]. *Dependent* computation means that the output of one host is input to the next host. That is to say, the agent cannot fulfill its task when at least one of the required hosts specified in route r is either not available or denies its services to the agent.

The other type is *Independent* computation. Here, the agent computation does not need the results of another host. As a consequence, the hosts contained in the pre-scribed route r can be visited in any arbitrary order. The mobile agent is free to move without the control of its owner, therefore the possibilities of its loss increases a lot. When the agent moves from h_i to h_j, the possibility of h_j being malicious is significant and thus *Blocking* or *Killed Agent* might occur, which would result in loss of the agent and the partially computed result. We can also say, in the *Blocking* agent is not transmitted further from the malicious host and the owner keeps on waiting for the agent and data. There are many techniques have been developed to add reliability and fault tolerance to mobile agent systems. These techniques include transactions, group communications and rollback recovery, and have different tradeoffs and focuses. Fault tolerance is achieved by periodically using stable storage to save the processes' states during failure free execution [15]. We provide a solution for the above problem in the next section.

3   Informal Modeling of Fault Tolerated Mobile Agent

Here we propose a model for solving the above problem. In the model, the mobile agent is able to detect malicious hosts and it can skip malicious hosts continue to perform the task. There are three types of mobile agent in the model, master agent (MA), shadow agent (SA) and shadow agent' (SA'). A variant of this model is discussed in [13, 16].

The model shows the home host h_0, which launches the agent MA, MA goes to host h_1 and then proceed further depending on the decision taken dynamically. SA is usually on the back of a host MA current position (and SA' is SA's predecessor, it produces after the MA task on a host). That is, when the SA waiting on one arbitrary host $h_{i-1}(i = 1, 2..., n)$ on the route r of task, MA is running on the next host h_i MA generated SA' after complete the calculation on h_i and moving on to the next host h_{i+1}. While SA' residing on the h_i send a messages m_1: "MA has been sent out, the target host is h_{i+1}" to SA on the h_{i-1}. After MA arrived in host h_{i+1}, a message m_2 is sent to SA and tell it: "MA has arrived h_{i+1}". Figure 1 shows an informal model assumed that h_i sends MA to h_{i+1}. The SA residing on the host h_{i-1} waits for the messages from MA and SA', when it finds that the time out has occurred it consults the MA or SA' and checks for their states. If there is no message m_1 or m_2 is received SA (or beyond a specific time window), can be regarded as h_i or h_{i+1} is malicious. For instance, the SA would understand a malicious h_i if m_1 was not received at h_{i-1} within a proper time out T. This time out should be large enough to account for the longest estimated execution time at h_i including the transmission and queuing times. Then SA create a new MA and resend it this time skipping h_i from the itinerary.

After receiving m_1 and m_2, SA start self-destructive program, and MA go to the next migration process. Such that the cycle until the task is completed to return to h_0.

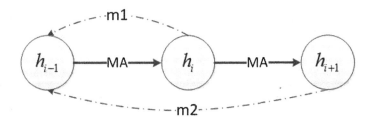

Fig. 1. Informal model

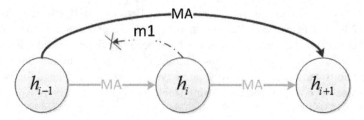

Fig. 2. Informal agent-blocking model

Note that, SA received m_1 and m_2 instructions h_i are trusted or non-malicious. Therefore SA' turn into the SA as a copy of MA, being stored on h_i and goes to sleep. Under the assumptions of the network is good, if SA is not received the message sent by h_i or h_{i+1}, h_i or h_{i+1} may be identified as malicious. If not received m_1 can be identified as problematic, located on the SA generates a new MA, skip continues a task. Figure 2 shows an informal agent-blocking model when the m_1 is not received.

There are several variations in this procedure, it may be the case that both m_1 and m_2 are not received. And sometimes the case may be that m_1 is received but m_2 is not. Both of cases shown in Figs. 2 and 3. All variety of situations shown in Table 1.

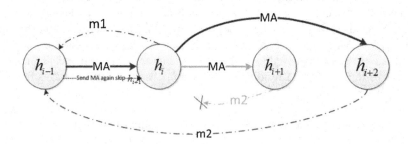

Fig. 3. Informal agent-blocking model

Table 1. All Variety of Situations

m_1	m_2	operation
√	√	OK
√	✗	Skip h_{i+1} and Continue
✗	√	Verification h_i
✗	✗	Skip h_i and Continue

4 Modeling Fault Tolerated Mobile Agents with CPN

Colored Petri Nets (CPNs) [2] is a discrete-event modeling language combining petri nets with the functional programming language Standard ML. As a language for the modeling and validation of systems, Colored Petri Nets (CPNs) play a major role in which concurrency, synchronization and communication. The primitives for definition of data types, describing data manipulation, and for creating compact and parametirsable models be provides by Standard ML. A common model shown in Fig. 4 is an agent model without fault tolerated ability.

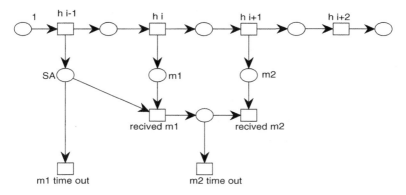

Fig. 4. Agent model without fault tolerated ability

CPN model is an executable model representing the states of the system and the events that can cause the system to change the state. Figure 5 is a model with detect the malicious host ability.

The proposed model shown in Fig. 7 has been simulated using CPNs. The CPN model has 11 places and 10 transitions. The host h_{i-1} is labeled as *t1*. There are two transitions as the timer in the model. SA is saved in places called *p7*. Place *p8* and *p9* are used to receive m_1 and m_2. The data types, constants, variables used in the model are declared in screen shot Fig. 6. The declarations are described in CPN ML. The AG represents agent types of token in the declarations.

5 Simulation and Analysis

CPN can be simulated interactively or automatically. The goal of simulation is to debug and investigate the system design and to investigate different scenarios and explore the behaviors of the system. The initial marking, agent is at h_{i-2}, there is a token $1`(1,0)@0$ at the p1 in Fig. 7. On reaching h_{i-1}(t1) a new marking M is reached. In this marking t1 keeps a SA, send a token to the place p7 and waiting for the token in the place p8 to fire the transition t5. Timing starts when SA reach the p7. If there is no token in p8, recovery_1 will be fired after a timeout. Similarly, if there is no token in p9, recovery_2 will be fired after a timeout. After recovery 1 or recovery 2 be fired, the mobile agent system will be shift to a recover state.

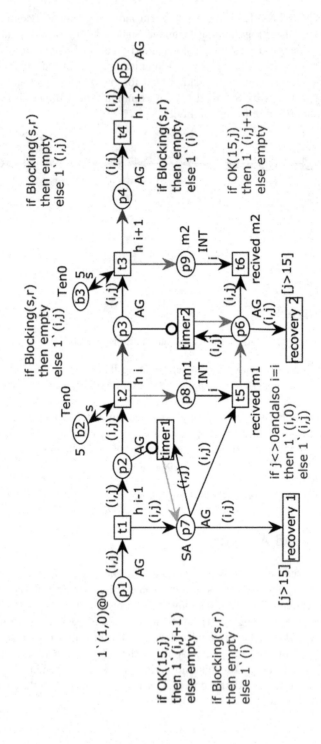

Fig. 5. Agent model with detective ability

```
▼Declarations
  ► Standard priorities
  ▼Standard declarations
    ► colset UNIT
    ► colset BOOL
    ▼ colset INT = int;
    ▼ colset INTINF = intinf;
    ▼ colset TIME = time;
    ▼ colset REAL = real;
    ▼ colset STRING = string;
    ▼ colset AG=product INT*INT  timed;
    ▼ var i,j,n:INT;
    ▼ colset Ten0 = int with 0..10;
    ▼ colset Ten1 = int with 1..10;
    ▼ var s: Ten0;
    ▼ var r: Ten1;
    ▼fun OK(15,j) = (j<=15);
    ▼fun Blocking(s,r)=(r<=s);
```

Fig. 6. Declarations

The basic idea of full state spaces is to calculate all the reachable states (markings) and all state changes (occurring binding elements) of the CPN model and represent these in a directed graph where the nodes correspond to the set of reachable marking and the arcs correspond to occurring binding elements. The statements similar to "if... then... else..." in the Fig. 7 are the condition of the corresponding arcs, each arc near the same color of the sentence is the condition for it.

Simulation can only be used to consider a finite number of executions of the model being analyzed. This makes simulations suited for detecting errors and for obtaining increased confidence in the correctness of the model. The situation of our model shows that the blocking host can be detected and agent can skipped it to continue the task. The simulation of the model proved that the blocking host could be detected.

We can take an analysis of the characteristics of agent systems using the state space tool in CPN Tools. State-space report of the model as shown in Fig. 8. According to the analysis, the agent model with fault tolerated ability shown in Fig. 7 is active, no dead transition, not exist deadlock in the system and a finite occurrence sequences. This state space analysis also shows that the mobile agent can detect the malicious host and thus prevent agent blocking in subsequent journey by skipping that malicious host.

6   Related Work

Many of the problems concerning the security of mobile agent systems, both protecting the host from malicious agents and protecting agents from malicious hosts, have been discussed in the literature. To recover the mobile agent after failure is the serious

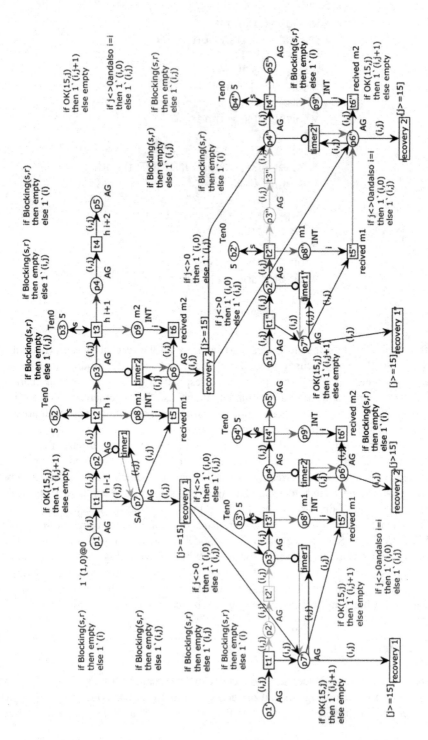

Fig. 7. Agent model with fault tolerated ability

```
Statistics
                              Liveness Properties
-----------------------       ---------------------------------------

    State Space
        Nodes:  1054          Dead Markings
        Arcs:   4827              20 [985, 983, 969, 968, 46, ... ]
        Secs:   1
        Status: Full          Dead Transition Instances
                                  None
    Scc Graph
        Nodes:  1054          Live Transition Instances
        Arcs:   4827              None
        Secs:   0

                              Fairness Properties
                              ---------------------------------------
Boundedness Properties            No infinite occurrence sequences.
-----------------------
```

Fig. 8. State-space report

research work in the growth of the agent based distributed environment. In [9], a distributed system recovery models are proposed to recover the agent. There is a number of assumptions be made about the context. They assume the existence of a close system with only controlled processes modifying the data, and then same form of rollback is possible. In [17], an approach fault tolerance is achieved by cloning the original agent and checkpointing. Similarly, Scales and Lam [1] proposed a model of check pointing and logging technique to recover a system. A focus of many of these techniques is thus how to reduce the check pointing.

The shadow model be proposed in [4–8] is significant attention within the mobile agent failure tolerance community concerning the loss of mobile agents at remote hosts that fail by crashing. Vogler et al. [3] propose that a mobile agent inject a replica into a stable storage upon arriving at an agent server. However, in the event of agent server crash, the replica remains unavailable for an unknown period. The mobile shadow scheme proposed Pears et al. [11] is a solution of a quite similar problem. The scheme is employing a pair of replica mobile agents, master and shadow, to survive remote agent server crashes.

Madkour et al. [16] proposed a protocol to facilitate identifying and skipping any blocking hosts in the mobile agent's itinerary. It is based on launching two mobile agents, primary and shadow, the shadow agent will be always lagging behind the primary and always visiting hosts that did not block the primary. Whenever a malicious host blocks the primary agent, the shadow agent will detect that event and request help from its home host.

7   Conclusion and Future Work

The paper has presented a new model for detect and recovery of failures of mobile agents. Simulation and analysis of the model shows that by two messages (m_1 and m_2) mobile agent MA can detect the malicious host and thus prevent lost of agent in itinerary by skipping that malicious host. Results and analysis show that the model offers a when fault tolerance measures are exercised.

The model discussed above is well suited for dynamic mobile agent, but there may be a situation for having the three colluded or n-colluded attacks. For this it is not be able to detect more than three malicious hosts at the same time. Such cases need to be considered.

References

1. Scales, D.J., Lam, M.: Transparent fault tolerance for parallel applications on networks of workstations. In: USENIX Annual Technical Conference, pp. 329–342 (1996)
2. CPN Tools website. http://cpntools.org/start
3. Vogler, H., Hunklemann, T., Moschgath, M.: An approach for mobile agent security and fault tolerance using distributed transactions. In: Proceedings of International Conference on Parallel and Distributed Systems (ICPADS 1997), Seoul, pp. 268–274 (1997)
4. Schneider, F.: Towards fault-tolerant añd secure agentry. In: Proceedings of 11th International Workshop on Distributed Algorithms, Saarbrucken, pp. 1–14 (1997)
5. Straßer, M., Rothermel, K., Maihöfer, C.: Providing reliable agents for electronic commerce. In: Lamersdorf, W., Merz, M. (eds.) TREC 1998. LNCS, vol. 1402, pp. 241–253. Springer, Heidelberg (1998)
6. Pleisch, S., Schiper, A.: Modeling fault-tolerant mobile agents as a sequence of agreement problems. In: Proceedings of 19th Symposium on Reliable Distributed Systems (SRDS), Nuremberg, pp. 11–20 (2000)
7. Silva, F.M., Popescu-Zeletin, R.: Mobile agent-based transactions in open environments. IEICE Trans. Commun. $E83-B$(5), 973–987 (2000)
8. Silva, L.M., Batista, V., Silva, J.G.: Fault-tolerant execution of mobile agents. In: Proceedings of International Conference on Dependable Systems and Networks, pp. 144–153, June 2000
9. (Mootaz) Elnozahy, E.N., Alvisi, L., Wang, Y., Johnson, D.: A survey of rollback recovery protocols in message-passing systems. ACM Comput. Surv. **34**(3), 375–408 (2002)
10. Cubaleska, B., Schneider, M.: A method for protecting mobile agents against denial of service attacks. In: Klusch, M., Ossowski, S., Shehory, O. (eds.) CIA 2002. LNCS (LNAI), vol. 2446, pp. 297–311. Springer, Heidelberg (2002)
11. Pears, S., Jie, X., Boldyreff, C.: Mobile agent fault tolerance for information retrieval applications: an exception handling approach. In: Proceedings of the Sixth International Symposium on Autonomous Decentralized Systems, (ISADS 2003), pp. 115–122 (2003)
12. Petrie, C.J.: Agent-based engineering, the web, and intelligence. IEEE Expert **11**(6), 24–29 (1996)
13. Venkatesan, S., Chellappan, C., Dhavachelvan, P.: Performance analysis of mobile agent failure recovery in e-service applications. Comput. Stan. Interfaces **32**, 38–43 (2010)
14. Aggarwal, M., Murgai, P.: Simulation of dynamic mobile agent model to prevent denial of service attack using CPNS. Int. J. Comput. Appl. **20**(1), 19–25 (2011)
15. Mohammad, A., Hamid, E.: A new approach for a fault tolerant mobile agent system. In: ACIS International Conference on Software Engineering, Artificial Intelligence, Networking and Parallel/Distributed Computing 12th, pp. 133–138 (2011)

16. Madkour, M.A., Eassa, F.E., Ali, A.M., Qayyum, N.U.: Securing mobile-agent-based systems against malicious hosts. World Appl. Sci. J. **29**(2), 287–297 (2014)
17. Mahajan, R., Hans, R.: A novel comparison based approach for fault tolerance in mobile agent systems. In: Mauri, J.L., Thampi, S.M., Rawat, D.B., Jin, D. (eds.) SSCC 2014. CCIS, vol. 467, pp. 221–229. Springer, Heidelberg (2014)

A Self-Adaptive Context-Aware Model
for Mobile Commerce

Munir Naveed[✉]

Al Khawarizmi International College, Abu Dhabi, UAE
`muneer.navid@khawarizmi.com`

Abstract. Use of mobile devices for the online shopping is growing ever. This paper addresses the problem of querying the contents relevant to the current context of the mobile node. We present a context-aware model that can incrementally learn the user preferences and location-based content retrieval for the purpose of one-to-one marking strategy. The model is based on Monte-Carlo sampling and tree induction method. Monte-Carlo sampling is used to construct the synopsis structure while tree induction is used to predict the user preferences in the current context. The model is evaluated using two benchmark datasets for offline testing and an application is developed to test the model online. The results show an obvious advantage of using the Monte-Carlo based tree induction method as compare to its state-of-the-art rivals.

Keywords: Context-awareness · Mobility prediction · Monte-Carlo simulations · Mobile commerce · One-to-one marketing

1 Introduction

Mobile commerce (M-commerce) is growing rapidly due to easily available mobile apps for online transactions. Two examples of such kind of apps are Boku [1] and obopay [2]. However, the mobile apps are less effective in terms of providing relevant contents to the user as compared to their web counterparts for online shopping. The main challenging issues in the development of the context-aware apps for M-Commerce are (i) the limited computational capability of the mobile devices and (ii) mobility. The user preferences can be depending on its current location. Therefore, location plays a main role in determining the current context of a mobile device.

The use of current context to display the relevant commerce contents is very crucial for the success of a corporate. It is required to reach the potential clients and establish a healthy and useful communication on one-to-one basis. The irrelevant and unfocussed contents for a client will not develop a good communication channel between the corporate and the client. In this work, the main focus is on determining the relevancy of a set of contents for a client by using the context modeling.

The paper addresses the context-awareness with the notion of identifying the user preferences based on the device's current location, user profile and the past interaction of the user with the app. The main emphasis is put on predicting the user preferences

© Springer International Publishing Switzerland 2015
D.-S. Huang and K. Han (Eds.): ICIC 2015, Part III, LNAI 9227, pp. 618–625, 2015.
DOI: 10.1007/978-3-319-22053-6_64

with respect to its current location. The contents of the M-commerce app are shown to the user based on this prediction. This kind of solution are required for various commercial and social network apps, however, we choose an advertising app to explore solutions for this problem.

In this work, we present a Monte-Carlo based tree induction approach to solve the prediction problem. The prediction of the user preferences in a mobile app is non-trivial and complex due to mobility and real-time constraints. The mobility offers the main challenges e.g. a user can move to several locations in a day and can execute a transaction at any one of them. Apparently it seems intuitive to define a context without using location. However, we cannot ignore mobility to determine the user preferences due to a strong association between the user interaction and location. For example, in our testbed, most of the users performed search for electronic devices from their home and a search for the properties while in travelling. We use a variation of decision tree for the predication preferences, however, induction tree is applied on a summarized form of the problem space— a synopsis structure is used to represent the summary form of the problem space. We use a real-time Monte-Carlo sampling [3] to generate the synopsis structure. The main contributions of this work as follow:

1. An incremental and real-time prediction technique for user-preferences
2. Evaluation of the technique using the benchmark dataset
3. Comparison with the state-of-the-art rivals technique.

2 Related Work

PhoneMonkey [4] is app that emulate the user interactions to identity the relevant contents for the current user. This contextual model is based on page number to extract information about user preferences. However, this approach does not consider location as a part of context.

RTDroid [5] is a variation of Android that provides a platform for the development of apps to run under real-time constraints. It is based on Fiji real-time VM running on real-time OS. In our case, app impose real-time constraint on search (in problem space) at application level rather than using the system calls to ensure outcomes in real-time.

Encore [6] is a platform that uses the context-awareness in establishing the secure communication between any two mobile devices. The social apps are an ideal category of mobile apps to exploit the capabilities of this platform. It uses the current context of a device to search for the nearby users and resources; the search can identify a relevant nearby user of a social app by using the profile of the current user. The profiles of two neighboring users are matched using the tags.

A context-aware model is presented by [7] where a context is entirely defined by the current location of a mobile node. It predicts the next possible location of the user. The results show a better prediction by Monte-Carlo based Markov Model, however, it is not clear what kind of mobile apps can take benefit of that approach.

Ehsan et al. [8] present a Naïve Bayes based approach to predict the user preferences. Their work uses the same dataset as we have used in the evaluation of our model.

The results in [8] shows that Gaussian Process based Naïve Bayes perform better than Full-Gaussian Process based classifier.

A Naïve-Bayes based approach is explored by [11] where Naïve-Bayes in form of utility function are used to predict the quality of experience in context of a mobile user. The context-model is a combination of quality service parameters and the current context of a mobile device. The results demonstrate the Naïve-Bayes based approach significantly perform better than its state-of-the-art rivals.

A context-aware model for mobile learning is presented by [12] where a fuzzy logic based approach is used to provide the relevant contents to a student (a learner) using the mobile device.

A novel classification method, called diversity of class probability estimation (DCPE), is presented by [13]. DCPE requires less parameter (a kind of semi-supervised approach) to tune for training purpose as compared other classification methods e.g. Naïve Bayes.

3 Problem Formulation

In this paper, we address the problem of predicting the user preferences using the user's profile and the current location. We formulate this problem as a classification function as shown in Eq. (1). S represents synopsis, c is the current context and P is the preference vector. c is a tuple $< l, a, e >$ where l is the current location, a is the current activity of the app and e is the current event at activity a.

$$f(S, c) = P \tag{1}$$

The context aware model learns to solve this problem by using a supervised learning approach. In an offline evaluation, model is trained using a large set of example data. To construct the model on the fly, S is constructed or modified periodically using the user interaction with the app. The main focus of this research work is to build S under real-time bound.

4 Real-Time Constraints

The construction of the synopsis structure is the main component of the context-aware model. The structure is modified several times after constructing its initial edifice. It can be time consuming task to build or modify a synopsis structure. If a commercial app spends more time on this task, then it lose its competitive edge as compared to other apps (that do not use this structure) due to slow response to the user actions in an activity of the app. To guarantee a realistic response time to a user action, app must impose real-time constraints on synopsis construction and its modification.

5 Synopsis Construction

Monte-Carlo (MC) sampling is a simple way to construct a synopsis and they give an upper bound on the quality of the solution. We use a variation of a real-time Monte-Carlo sampling technique given in [7]. The synopsis is construction algorithm (adapted from [7]) is given in Fig. 1. A is the set of all activities in an app, L is a location value, E is the set of all possible events in an app. The algorithm reads the current value of A, L and gets all possible events applicable i.e. E(A) in an activity (Line 1). A probability distribution is maintained in synopsis to find the most frequent interaction (i.e. a pair $< a, e >$ where $a \in A$ and $e \in E(A)$). The frequent pairs are maintained by a vector called V(a, e) as shown at Line 2, Fig. 1. The frequent pairs can grow very large when a user runs different events several times at different location. To reduce the size of V without losing the useful information, Monte-Carlo simulations are run for a fixed time T—to select half of the promising pairs from V for the current location L (as shown in Line 3). These pairs are stored in a vector Vu— which is main component of the synopsis structure. S is modified once Vu is determined for the current stream.

Algorithm MCSynopsis (A, L, E, T)
Foreach $a \in A$
$\quad\quad \Pi = E(a)$;
$\quad$ Foreach $e \in \Pi$
$\quad\quad\quad V(a, e) = Prod(L, a, e)$;
$\quad\quad\quad V^u(a, e, L)$=Simulate($V$, L, T);
$\quad\quad\quad$ Update Prod(L, a, e);
$\quad\quad$ End
$\quad$ End
$\quad\quad\quad S = S \cup V^u(a, e, L)$

Fig. 1. MCSynopsis— high level algorithm

6   Real-Time Context-Aware Tree Induction (RCTI)

RCTI is a variation of a very fast Decision Tree (VFDT) [9]. The main motivation of using a variation of VFDT is to keep the tree induction within the real-time bound and reduce the overload of modeling the context-awareness for a commercial app. Since there are several flavor of a VFDT available in the literature, we use the following Gain function for a variation of VFDT in our work. Gain of an event 'e' at an activity 'a' is computed using the probability distribution built during the synopsis construction. It is expressed in Eq. (2) where $n = |E(a)|$ where $a \in A$.

$$Gain\ (e) = \sum_{l=1}^{L} \sum_{\forall e \in E(a)} \frac{|V^u(a, e, l)|}{|V(a, e)|} \tag{2}$$

7   Experimental Setup

RCTI is evaluated using two benchmark dataset: Car Preference dataset [8] and Mobile Context-Aware (MCA) dataset [7]. These dataset have been used in some previous work.

A hold-out validation method with 70 % training examples and 30 % test examples. Performance of RCTI is measured by using precision and recall. RCTI is compared with K-means, Naïve-Bayes and Decision Tree. We use RapidMiner 5.3 [10] to run experiments for K-means, Naïve Bayes and Decision Tree. RCTI is programming in Java and attach with a Android App called 'ConAware'. The app has five activities and seven events per activity. All events are designed in the form of button click. Each button is linked to only one table in a database. For the online experiments, we enumerated the locations using four semantics {home, work, market, misc}. The online experiments are run using two android tablets and three Smartphone with Android OS for a period of two weeks. RCTI learns the user preferences incrementally and app can change the color of the buttons on an activity with respect to the location of the user. For example, most of the users explored property related contents while travelling. The travelling locations are represented by 'misc' and as soon as the user moves to a new location, the app the color of 'property' button in the search activity.

Car preference dataset [8] has four user attributes to represent a user's profile: education, age, gender and region. A car is also represented by four attributes which are body type, transmission, engine capacity and fuel consumed. A user's preference is a pair of two cars where the first element of the pair is the user's choice over the other one. The user profile and car attributes are input to the model and the preference pair is the output.

MCA dataset is based on 250 mobile app users. Each user moves to three locations: home, shop and work. For sake of data collection, each participant manually enters the name of the location. MCA takes location coordinates and the time spent on each coordinator as input and the semantics for the location are considered as output. The location semantics in MCA are home, work and shop.

8   Results

The results on Car Preference dataset are given in Table 1. The recall in Table 1 represents a ratio of correct instances to the total retrieved instances. RCTI performs better than K-Means and Naïve Bayes. However, it is not as optimal as Decision Tree. This difference is due to the difference in training time. RCTI being an anytime algorithm, constructs the tree within real-time bounds while DT is given advantage of using as much CPU cycles as it needs to converge. RCTI is better than Naïv-Bayes due to the use of probability distribution in measuring the information gain for each attribute.

Table 1. Comparison of RCTI on *Car Preference* dataset.

Model	Recall	Precision
RCTI	0.22	0.65
K-Means	0.45	0.75
Naïve-Bayes	0.30	0.59
DT	0.13	0.89

The results on Mobile Context-Aware dataset are given in Table 2. RCTI performs better than all of its rival. DT is sensitive to the distinct values where RCTI can generalize better as compared DT. Naïve-Bayes performs better than DT because the user always moved to the same next place from some certain places e.g. work place to home movement. RCTI uses a generalized i.e. a synopsis structure, therefore, it can avoid some of the problems that traditional DT suffers from.

Table 2. Results on *Mobile Context-Aware* dataset.

Model	Recall	Precision
RCTI	0.03	0.91
K-Means	0.26	0.34
Naïve-Bayes	0.13	0.79
DT	0.18	0.76

In the online experiments, RCTI takes more time to learn the user preferences as compared to Monte-Carlo Simulation model given in [7] however, RCTI adapts to the changes more quickly as compared to [7].

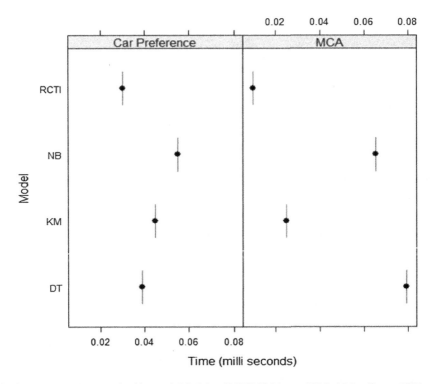

Fig. 2. Training time required by each Models—RCTI, K-Means (KM), Naïve Bayes (NB) and Decision Tree (DT).

Figure 2 demonstrates performance of models with respect to the training time required by each model for a dataset. RCTI is less sensitive to the kind of dataset and demonstrates a consistent performance to learn the user preferences within an upper bound on time. RCTI performs better than its rivals with respect to learning time because of the limits on sample size in each training episode. This is a characteristic of Monte-Carlo search that it can run simulations with an upper bound on run-time. The quality of solution by RCTI is given in Tables 1 and 2.

DT performs better than others models on car preference dataset (except RCTI) with respect to training time. The car preference dataset has less number of distinct values in each input parameter. K-Means is a better choice than DT and NB on MCA dataset. MCA dataset has three main groups—one for each semantics. In such a case, K-Means converges much earlier than other models (except RCTI).

Figure 3 demonstrate the learning profile of RCTI for car preference dataset. The accuracy represents the percentage of correctly classified instances. Since RCTI is based on a random sampling technique, therefore, the learning profile is not smooth for the whole duration of training phase. However, the learning profile gives indication of escaping the local-minimum during the training phase. For example, RCTI converges to a local better solution at 0.01 ms but then it explores the solution space in other directions and eventually discovers a globally better solution. In other words, RCTI can balance the trade-off between exploration and exploitation.

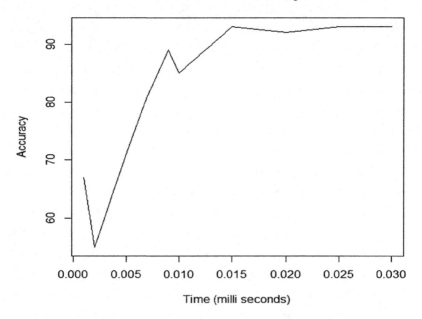

Fig. 3. RCTI's convergence profile for car preference dataset.

9   Conclusion

We present a preliminary work in the direction of the self-adaptive context-aware model which is particular suitable for the commercial apps. This model—called as RCTI—can also be used for contextual advertising on the mobile apps. RCTI exploits the capability of Monte-Carlo sampling to construct a synopsis data structure which contains rich but summarized information about the user preferences and context. RCTI then use a real-time classifier to predict a preference for a mobile user based on its current location. The model is evaluated using two datasets in an offline setup and shows a better performance on both dataset. The experimentation for the online setup is still progress, however initial results show a potential strength of RCTI in managing context-aware in real-time.

In future work, we aim to modify the RCTI classifier to reduce the time ad memory complexities of the model. We also aim to explore other classifier e.g. neural networks to speed up the classification and reduce the overload of managing the context-awareness by a mobile app.

References

1. http://www.boku.com/. Accessed 7 June 2014
2. http://www.obopay.com/corporate_website/index.php. Accessed 9 June 2014
3. Naveed, M., Crampton, A., Kitchin, D., McCluskey, T.: Real-time path planning using simulation based markovian decision process. In: AI-2011: 31st SGAI International Conference on Artificial Intelligence, Cambridge, UK (2011)
4. Nath, S., Lin, F.X., Ravindranath, L., Padhye, J.: SmartAds: bringing contextual ads to mobile apps. In: The proceedings of Mobisys 2013, Taipei, Taiwan (2013)
5. Yan, Y., Cosgrove, S., Anand, V., Kulkarni, A.: Real-time android with RTDroid. In: The proceedings of Mobisys 2014. Bretton Woods, New Hampshire, USA (2014)
6. Aditya, P., Erdelyi, V., Lentz, M., Shi, E., Bhattacharjee, B., Druschel, P.: Encore: private, context-based communication for mobile social apps. In: The proceedings of MobiSys 2014. Bretton Woods, New Hampshire, USA (2014)
7. Naveed, M.: Online learning based contextual model for mobility prediction. In: O'Grady, M.J., Vahdat-Nejad, H., Wolf, K.-H., Dragone, M., Ye, J., Röcker, C., O'Hare, G. (eds.) AmI Workshops 2013. CCIS, vol. 413, pp. 313–319. Springer, Heidelberg (2013)
8. Abbasnejad, E., Sanner, S., Bonilla, E.V., Poupart, P.: Learning community-based preferences via dirichlet process mixtures of gaussian processes. In: Proceedings of the 23rd International Joint Conference on Artificial Intelligence (IJCAI) (2013)
9. Domingos, P., Hulten, G.: Mining high-speed data streams. In: The proceedings of KDD 2000. ACM Press (2000)
10. http://rapidminer.com/products/rapidminer-studio/. Accessed 10 March 2015
11. Mitra, K., Zaslavsky, A., Ahlund, C.: Context-aware QoE modelling, measurement, and prediction in mobile computing systems. IEEE Trans. Mob. Comput. **14**(5), 920–936 (2015)
12. Al-Hmouz, A., Shen, J., Yan, J.: A machine learning based framework for adaptive mobile learning. In: Spaniol, M., Li, Q., Klamma, R., Lau, R.W. (eds.) ICWL 2009. LNCS, vol. 5686, pp. 34–43. Springer, Heidelberg (2009)
13. Xu, J., He, H., Man, H.: DCPE co-training for classification. Neurocomputing **86**, 75–85 (2012)

Research and Implementation on Autonomic Integration Technology of Smart Devices Based on DPWS

Yan-qin Mao[1(✉)], Lu Jin[2], and Su-bin Shen[1]

[1] School of Computer Science and Technology, School of Software,
Nanjing University of Posts and Telecommunications,
No. 66 Xin Mo Fan Road, Nanjing, Jiangsu, China
{yqmao,sbshen}@njupt.edu.cn
[2] Oracle Software Systems (China) Co., Ltd., Nanjing, China
njinlu@163.com

Abstract. Smart devices can take advantage of a specific protocol to provide services, but they do not have open and autonomic integration capability. After analyzing the characteristics of the existing integration middlewares, an autonomic integration technology for smart devices is proposed based on DPWS protocol. According to the requirement background of network monitoring integration system containing different types of network cameras, an autonomic integration prototype of network cameras is designed and implemented. Test results on autonomic discovery and configuration for smart devices show that DPWS is a good and open method to solve the autonomic integration problem and demonstrate that the method is realizable.

Keywords: Smart device · Devices Profile for Web Services · Autonomic integration · Web services · Service oriented architecture

1 Introduction

Today, with the rapid development of various types of smart devices, lack of unified service interfaces leads to the difficulty in developing and integrating the application system. Therefore, users tend to use a unified platform to control various smart devices in certain domain. An interoperable architecture [1] which can be seamlessly integrated with the terminal services [2] in IP network is needed urgently. It is possible for smart devices to take service-oriented architecture (SOA) [3] into account.

As a component-based technology, SOA can improve scalability and reusability [4] for distributed applications. In 1996, Gartner put forward the concept of service computing. In 1999, the practice of XML-RPC proposed Simple Object Access Protocol (SOAP) [5] and formulated the SOAP1.0 standard. In 2000, Web services are formed together with Web Services Description Language (WSDL) [6] and SOAP. Then, service-oriented architecture was shaped initially.

With the development of electronic technology, the idea of applying SOA into smart devices integration has been widely recognized in industry. Smart devices

© Springer International Publishing Switzerland 2015
D.-S. Huang and K. Han (Eds.): ICIC 2015, Part III, LNAI 9227, pp. 626–636, 2015.
DOI: 10.1007/978-3-319-22053-6_65

provide their functionalities through Web Services (WS) [7] which had been applied in networked devices gradually. Such mode is called Service-Oriented Device Architecture (SODA) [8]. Service computing researchers have been designing web services in the smart devices, focusing on low cost SOAP protocol implementation such as kSOAP, gSOAP.

Analyzing the characteristics of existing device integration technologies, for achieving autonomic integration of smart devices with limited resources, an autonomic smart devices integration solution based on DPWS is proposed. Under the background of network video surveillance applications, the device side functions and the control side functions of DPWS stack are designed and implemented respectively. Besides, an autonomic network camera integration system prototype is established. Test results about smart device discovery and configuration features of system prototype show that the autonomic smart devices integration solution based on DPWS can provide open and autonomic integration services.

2   DPWS-Based Autonomic Integration Solution of Smart Devices

Device integration refers to the process of perception and access for different types of devices. Currently, there are a variety of device integration middlewares such as Universal Plug and Play (UPnP) [9], Open Service Gateway Initiative (OSGi) [10] service platform, Java intelligent Network infrastructure (JINI) [11], Devices Profile for Web Services (DPWS) [12] and so on. As a software architecture supporting interoperability among network devices, Web services is currently the most widely used large-scale integration technology. With the standardization of DPWS specification based on web service specifications set of Organization for the Advancement of Structured Information Standards (OASIS), a new service standard for device was born.

Based on WS standards and existing Web technologies including HTTP, SOAP and XML technology, DPWS defines the message transfer format and technical details which have nothing to do with the platform. Comparing the main features with above integration middlewares, as web services-based protocol stack, DPWS can not only solve the technical problems in device integration, also has the advantages of web services. DPWS references the WS protocol suite which contains a set of network service specification standards including device discovery, description, transport, security and so on. Protocols related with device integration mainly include WS-Discovery, WS-Transfer, WS-MetadataExchange, WS-Eventing, and WS-Security.

Any device needs descript its own services information and exposes the service access points (SAPs) in an open network environment. Through device discovery mechanism a client can find the SAPs to get descriptions about the device services. Device discovery has active mode and passive mode. Interactive process of device discovery is shown in Fig. 1.

Active mode refers to that a device actively sends a multicast message named online to inform a control side (Client Application) for its existence. When the control side receives the multicast packet, the packet provides the necessary information of the device. If the control side needs to know more details about the device, metadata

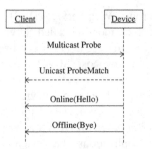

Fig. 1. Interactive process of device discovery

exchange can be used. When the device leaves the network, a multicast message named offline is sent. The control side informs notification management process to remove the device information and its associated resources when receiving the offline message.

Passive mode refers to that a control side sends a multicast message named probe which contains expected device type. All devices receive the message and match the device type. If the matching is successful, the device will send a unicast message named match to the control side. Then the control side allocates related resources about the device.

In WS-Discovering specification, Hello, Bye, Probe and ProbeMatches messages are respectively defined which represent the above messages including online, offline, probe and match. These messages are all SOAP protocol messages and follow End-point Reference model defined in WS-Addressing specification. Endpoint reference model defines the address of the message sender and message recipient in a transport protocol-independent manner and provides relative information of accessing device service for a control side.

3 Design of Network Camera Autonomic Integration System Prototype Based on DPWS

Autonomic smart device integration system should have platform-independent, robust, scalable and large-scale deployment features. In order to verify the feasibility of DPWS in device autonomic integration, under the background of network video surveillance application, a unified network video monitoring platform integrating multi-vendors and multi-types of network cameras is put forward.

Based on Browser/Server model, every type of network camera provides users with a range of common gateway interfaces (CGIs). Besides configuration function, network camera surveillance system provides online video with multiple formats, PTZ control (tilt/swivel/preset), user management and so on. Through adapter layer, original CGIs of different cameras are transformed to web services interfaces. Building blocks of network camera autonomic integration system are shown in Fig. 2.

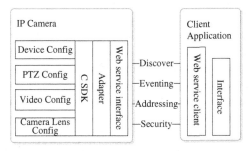

Fig. 2. Building blocks of network cameras autonomic integration system

In order to implement application interface based on web services, WSDL is used for descripting application interfaces and detail contract for web service interface is defined.

4   Implementation of Network Camera Autonomic Integration System Prototype Based on DPWS

4.1   Device Description

Service interface itself cannot provide relevant semantics information and cannot guide the control side to choose the right web service. So, ontology-based device description is adopted to solve the problem. Device is described by XML which is also known as the ontology-based device metadata. DPWS provide the basic information about the device at run time. WS4D-gSOAP tool creates the appropriate code according to the device information to send to the control side if needed. The basic description of camera defines three parts including Relationship, ThisModel and ThisDevice.

Relationship describes the relationship among boarding services in a device including hosting services and hosted services. Hosting services are responsible for describing services hosted on this host which use Types label for type classification and are identified by ServiceId tag. Hosted services use the same field Types for classification which describes the service interface of service and uses the field Port in the WSDL file to specify field name. Hosted services are identified by ServiceId also.

ThisModel specifies the factory information about Product such as the manufacturer, the manufacturer website, model name, model number and so on.

ThisDevice explains the device itself information such as aliases, firmware version, serial number information and so on.

These description data will be sent to the entity which requires it. Development Kit uses setMetadata function to read the metadata for the system and loads the description when initializing the service.

4.2   Device Discovery

1. **Control side**

The control side uses Windows Communication Foundation (WCF) to achieve WS-discovery protocol. The key issue is how to perceive the presence of the device side. In WCF, Find and Resolve operations are used for achieving basic functions of WS-discovered.

Find operation is responsible for sending Probe packets and automatically calling resolve operation. The control side can specify the type of service and other parameters to find the right device. When the device is receiving the Probe packet, ProbeMatches and other messages including metadata are sent to the control side.

Resolve operation is used to determine the specific location of services which is automatically called after receiving the response of Find operation. ProbeMatches messages received after Probe only includes limited endpoints information. However, Resolve operation can locate the address of the service itself based on endpoint information which is a conversion process from virtual address to actual URI.

In addition, WCF provides the user with an optional feature called Announcements. By default, a device sends Hello packets for online and sends Bye messages for offline. This mode allows the control side to receive online information of the devices actively, without having to make the control side to get new messages after sending Probe message. So, need to open the service Announcements to listen Hello and Bye messages.

Device probing uses Find and Resolve operations provided in WCF which consist four steps of calling process.

(a) Set version and timeout for WS-discovery protocol
```
var ep = new
UdpDiscoveryEndpoint(DiscoveryVersion.WSDiscoveryApril200
5);
ep.MaxResponseDelay = discoveryDuration;
```
(b) Set type of target device
This step hopes to find the device type of NetworkVideoTransmitter. Name space is http://www.onvif.org/ver10/network/wsdl.
```
var fc = new FindCriteria();
fc.ContractTypeNames.Add(new
XmlQualifiedName("NetworkVideoTransmitter",
@"http://www.onvif.org/ver10/network/wsdl"));
```
(c) Send Probe message and Perform Find operations through functions provided by WCF4.0 and send Probe message.
```
DiscoveryClient(ep).FindAsync(findCriteria, sync)
```
(d) Listen Hello and Bye message
Probe operation in control side uses DiscoveryClient object to find a device. At the same time control side open a bulletin service for listening Hello and Bye messages from devices.

```
var announcementEp = new
UdpAnnouncementEndpoint(DiscoveryVersion.WSDiscoveryApril
2005);
  m_host = new ServiceHost(m_announcementService);
  m_host.AddServiceEndpoint(announcementEp);
  m_host.Open();
```

The control side uses a multicast address soap.udp://239.255.255.250: 3702 to send Probe message and specify the type of device for searching in the message.

```
<s:Body>
    <Probe
xmlns="http://schemas.xmlsoap.org/ws/2005/04/discovery">
    <d:Types
xmlns:d="http://schemas.xmlsoap.org/ws/2005/04/discovery"
xmlns:dp0="http://www.onvif.org/ver10/network/wsdl">dp0:N
etworkVideoTransmitter</d:Types>
    </Probe>
</s:Body>
```

Response message of Probe message designates UUID as the endpoint reference. Resolve operation is responsible for transmitting the reference address into the actual network addresses.

```
<SOAP-ENV:Body>
    <wsd:Resolve>
      <wsa:EndpointReference>
    <wsa:Address>urn:uuid:a2bb3689-62ff-4d13-8da8-82171e091
6c9</wsa:Address>
      </wsa:EndpointReference>
    </wsd:Resolve>
</SOAP-ENV:Body>
```

2. Device side

After initializing, in addition to sending Hello message for on line, the device listens Probe multicast request. Using project code generated by WS4D-gSoap platform, Probe handler is automatically mapped once receipt of Probe message and a thread is established for recording MessageID in SOAP message and determining the type of device according to the <d: Types> field contents. If the device type matches its own type, then unicast message ProbeMatches as the response is sent to the control side. Response message specifies RelatesTo field as MessageID in ProbeMatches message. Thus, the control side identifies the packets which correspond to its requests using RelatesTo field.

For device side, operations including Hello, Bye, ProbeMatches and Resolve-Matches should be achieved.

(a) Hello message (Online) and Bye message (offline)

```
dpws_send_Hello (struct dpws_s *device, const char *types);
int dpws_send_Bye (struct dpws_s *dpws);
```

Sending Hello message includes the following processes. Firstly, generate a message ID. Secondly, generate SOAP header based on message ID. Then fill in Hello message body and send Hello messages by calling soap_send_wsd_Hello function.

Below is the definition of Hello message structure.

```
struct wsd_HelloType
{
  struct wsa__EndpointReferenceType
wsa__EndpointReference;
  char *wsd__Types;
  struct wsd__ScopesType *wsd__Scopes;
  char *wsd__XAddrs;
  unsigned int wsd__MetadataVersion;
  int __size;
};
```

Below is the Hello message sent by a device.

```
<SOAP-ENV:Body>
    <wsd:Hello>
      <wsa:EndpointReference>
      <wsa:Address>urn:uuid:a2bb3689-62ff-4d13-8da8-82171e091
6c9</wsa:Address>
      </wsa:EndpointReference>
      <wsd:Types>wsdp:Device
n1:NetworkVideoTransmitter</wsd:Types>
        <wsd:Scopes></wsd:Scopes>
      <wsd:XAddrs>
http://192.168.1.136:17612/a2bb3689-62ff-4d13-8da8-82171e
0916c9</wsd:XAddrs>
        <wsd:MetadataVersion>39139</wsd:MetadataVersion>
      </wsd:Hello>
</SOAP-ENV:Body>
```

Hello messages are sent via multicast mode. When the control side receives the SOAP message, it can make sense of the presence of this device. If using this device service, by calling the Resolve operation, EndpointReference field is parsed into service address. Meanwhile, the device sends a multicast message Bye. After receiving this message the control side releases the resources relative of the device.

Process of Bye message is similar to the Hello message.

(b) Probe response

After receiving Probe message, devices will automatically call__wsd__Probe function to process. Firstly, call wsd_process_probe function which is used for compare device type. Function wsd_gen_response_addr and dpws_header_gen_MessageId are used to handle Socket relevant details and message ID. Then, wsd_gen_ProbeMatches function is called which is responsible for filling the context structure.

```
struct wsd__ProbeMatchType
{
  struct wsa__EndpointReferenceType
wsa__EndpointReference;
  char *wsd__Types;
  struct wsd__ScopesType *wsd__Scopes;
  char *wsd__XAddrs;
  unsigned int wsd__MetadataVersion;
  int __size;
};
```

Finally, the device sent the response to the corresponding control side through wsd_send_async_ProbeMatches function. This function is a point to point transmission and the function prototype is as follows.

```
int wsd_send_async_ProbeMatches (struct soap *soap, const
char *MsgId,const char *to, struct ws4d_abs_eprlist
*matches,struct soap *req, ws4d_alloc_list * alist,
struct wsd__ProbeMatchesType *wsd__ProbeMatches)
```

A complete ProbeMatches SOAP message sent to the control side is as follows.

```
<SOAP-ENV:Header>
  <wsa:MessageID>urn:uuid:488197b8-22b6-41f4-8544-81a8292
7866a</wsa:MessageID>
  <wsa:RelatesTo>urn:uuid:fa7118b0-cfbd-11e1-bfa0-442bc17
0b0eb</wsa:RelatesTo>
  <wsa:Action>http://schemas.xmlsoap.org/ws/2005/04/disco
very/ProbeMatches</wsa:Action>
  <wsd:AppSequence MessageNumber="2"
InstanceId="5336"></wsd:AppSequence>
  </SOAP-ENV:Header>
  <SOAP-ENV:Body>
    <wsd:ProbeMatches>
      <wsd:ProbeMatch>
  <wsa:EndpointReference>
  <wsa:Address>urn:uuid:a2bb3689-62ff-4d13-8da8-82171e091
6c9</wsa:Address>
      </wsa:EndpointReference>
      <wsd:Types>wsdp:Device
n1:NetworkVideoTransmitter</wsd:Types>
      <wsd:Scopes></wsd:Scopes>
      <wsd:MetadataVersion>5336</wsd:MetadataVersion>
    </wsd:ProbeMatch>
  </wsd:ProbeMatches>
  </SOAP-ENV:Body>
```

(c) Resolve response

The control side will receive a ProbeMatches message after the Probe message. The ProbeMatches message provides the endpoint reference information as following.

```
<wsa:EndpointReference>
    <wsa:Address>
        urn:uuid:a2bb3689-62ff-4d13-8da8-82171e0916c9
    </wsa:Address>
</wsa:EndpointReference>
```

However, the endpoint address does not represent the address of the service provided. Therefore, Resolve/ResolveMatches operations are needed to achieve the conversion from endpoint address to specific services URI. After receiving corresponding Resolve messages the device will provide specific URI address and send a ResolveMatches message back to the control side. The ResolveMatches SOAP message is as follows.

```
<SOAP-ENV:Body>
    <wsd:ResolveMatches>
        <wsd:ResolveMatch>
            <wsa:EndpointReference>

    <wsa:Address>urn:uuid:a2bb3689-62ff-4d13-8da8-82171e091
6c9</wsa:Address>
            </wsa:EndpointReference>
            <wsd:XAddrs>
http://192.168.1.136:17612/a2bb3689-62ff-4d13-8da8-821
71e0916c9
            </wsd:XAddrs>
            <wsd:MetadataVersion>39139</wsd:MetadataVersion>
        </wsd:ResolveMatch>
    </wsd:ResolveMatches>
</SOAP-ENV:Body>
```

5   Test of Network Camera Autonomic Integration System Prototype Based on DPWS

5.1   Test Tools

soapUI is an excellent Web service load testing tool that directly tests the Web service interface through Web service WSDL file. It contains the complete WSDL coverage analysis including operational level and schema level and testes all paths to reach each element.

WS4D-explorer is a GUI tool provided by WS4D [13] which can analyze compatible DPWS services, also a conformable DPWS client.

5.2   Function Test

Start DPWS hosting service and sent the Hello message. The services provided by the device are waiting for multicast message Probe.

Fig. 3. Discovered device service and relevant metadata

Run software supporting WS-discovery. The software supports the WS-discovery and does not specify the type of device at the time of sending Probe message. It can find the device and add the device to the management list after finding the devices.

Run command java -jar ws4d-explorer.jar to start WS4D-Explorer and find the service provided by the device, while viewing the relevant descriptive information about the device. Figure 3 shows the service of the discovered device and relevant metadata through the ws4d explorer.

6 Conclusions and Future Works

In order to realize the SOA model in embedded intelligent devices and integrate different types of devices to solve the device autonomic discovery and configuration problems, an autonomic integration method for smart devices based on DPWS protocol stack is put forward. According to the application requirements of network video monitoring system of multi-types of network cameras, web services are implemented in embedded intelligent devices. DPWS protocol stack of device side and control side (application side) are implemented to provide a unified device automatic discovery and configuration services. Function test results show that DPWS is a good method to solve the autonomic integration problem for the devices and the method is realizable.

For all types of networked devices, developments on how to combine the device services using the Business Process Execution Language (BPEL) and provide process-based business services are the future works. In addition, researches on how to storage and deal with the data generated by large-scale integrated devices will be done later.

Acknowledgements. In this paper, the research was sponsored by the innovative research joint funding project of Jiangsu Province (Project No. BY2013095 - 108).

References

1. Zeeb, E., Moritz, G., Timmermann, D., Golatowski, F.: WS4D: toolkits for networked embedded systems based on the Devices Profile for Web Services. In: 39th International Conference on Parallel Processing Workshops, pp. 1–8. IEEE Press, New York (2010)

2. Guinard, D., Trifa, V., Karnouskos, S., Spiess, P., Savio, D.: Interacting with the SOA-based Internet of Things: discovery, query, selection, and on-demand provisioning of web services. J. IEEE Trans. Serv. Comput. **3**(3), 223–235 (2010)

3. Erl, T.: Service-Oriented Architecture: Concepts, Technology, and Design. Pearson Education, India (2006)

4. Dostal, W., Jeckle, M., Melzer, I., Zengler, B.: Service-Oriented Architectures mit Web Services. Elsevier, Munchen (2005)

5. Gudgin, M., Hadley, M., Mendelsohn, N.: SOAP Version 1.2. Technical report, W3C recommendation (2003)

6. Chinnici, R., Moreau, J., Ryman, A.: Web services description language (wsdl) version 2.0 part 1: Core language. Technical report, W3C Recommendation (2007)

7. Zeeb, E., Moritz, G., Timmermann, D,: Towards component orientation in embedded web service environments. In: 15th IEEE International Conference on Emerging Technologies and Factory Automation, pp. 1–8. IEEE Press, New York (2010)

8. SODA consortium. SODA - Technical Framework Description. Technical report, SODA consortium (2007)

9. The UPnP Forum. http://www.upnp.org

10. The OSGi Alliance. http://www.osgi.org

11. The Community Resource for Jini technology. http://www.jini.org

12. Devices Profile for Web Services (DPWS) Version 1.1, OASIS Standard

13. WS4D: Web Services for Devices. http://www.ws4d.org

Experimental Verification of the Dependence Between the Expected and Observed Visit Rate of Web Pages

Jozef Kapusta[(✉)], Michal Munk, and Martin Drlik

Constantine the Philosopher University in Nitra,
Tr. A. Hlinku 1, 949 74 Nitra, Slovakia
{jkapusta,mmunk,mdrlik}@ukf.sk

Abstract. This paper is focused on a utilization of the web usage mining and web structure mining methods. We tried to answer the question if the expected visit rate of individual web pages correlates with the observed visit rate of the same web pages. We used web server log files as a data source. We applied several log file pre-processing methods to identify the user sessions on different levels of granularity. We found out that the quality of acquired knowledge about the users' behaviour depends on the method of the session identification. We have experimentally proved a higher dependence between the observed and expected visit rates of the examined web pages in well-prepared files with identified user sessions. We found out statistically significant differences between PageRank and a real visit rate in the files with application of more advanced methods of session identification.

Keywords: Web usage mining · Web structure mining · PageRank · Support · Observed visit rate · Expected visit rate

1 Introduction

The aim of the website designers or creators is to provide information to users in a clear and understandable form. Information displayed on individual web pages is interconnected by hypertext references. The website creator can affect visitors' behaviour by defining of references between web pages. He indicates the importance of information displayed on web pages through these references. It is probably true more references head to more important web pages. These are directly accessible from the home page or are referred from other important web pages.

Web pages are mostly understood as an information resource for users. They can also provide information in an opposite direction. The website providers can collect information about their users or about users' behaviours, needs or interests.

The knowledge discovery from the web page structure is known as a web structure mining (WSM) [1]. From WSM point of view, we focused on an analysis of quality and importance of web pages based on the references (links) among web pages. Determination of the web page importance is based on the idea that the degree to which we can rely on the web page quality is transferred by the references to web pages. If the

© Springer International Publishing Switzerland 2015
D.-S. Huang and K. Han (Eds.): ICIC 2015, Part III, LNAI 9227, pp. 637–648, 2015.
DOI: 10.1007/978-3-319-22053-6_66

web page is referred to other relevant pages, the references on that web page also become important.

The second research field, closely related to the topic of the paper, is a web usage mining (WUM). In terms of WUM, a website's visitor always sends a large amount of information to the server during browsing the website. Most web servers automatically save this information in the form of records stored in log files.

The goal of this paper is to point out the relationship between the estimated importance of web pages (received by the methods of web structure mining) and visitors' actual perception of the importance of individual web pages (obtained by the methods of web usage mining).

We will prove the connection between observed and expected visit rate in the following steps:

- We will summarize results of a pre-experiment where we applied four different approaches to data pre-processing of web server's log file. We applied a sequence rule analysis (the log file collects observed visit rates of individual web pages) with the aim to assess the most suitable steps of data pre-processing for an analysis of website's visitors behaviour.
- We will calculate a PageRank of the individual pages of the examined website. The PageRank represents the probability of accesses to the web pages. In our study, it will represent an expected visit rate of the web pages.
- We will compare the *PageRank* values of the individual web pages and the value of variable *support* explaining the actual web page visit rate during the examined period (received from the log file). We will try to prove that the highest dependence of *PageRank* on variable *support* will be in the file where the most suitable steps of data pre-processing for web usage mining were applied (the most suitable steps of data pre-processing were found in the pre-experiment). Hereby we will try to prove the dependence between the estimated and observed probability of accesses to the web pages of the examined website.

We will use possible differences between the expected probability and observed probability of accesses to individual portal web pages for the purpose of identifying suspicious web pages. A suspicious web page is defined as a web page which is not ordered correctly in the hypertext structure of the website.

We will suggest an approach to identifying suspicious web pages based on the comparison of the expected and observed probability of accesses, i.e., we will suggest an approach to determine the web pages where the importance was highly assessed by website's developers but did not achieve the expected real visit rate. Conversely, we will determine the web pages which were underestimated by the developers, but which were frequently visited.

The rest of the paper is structured as follows. The second chapter deals with the related work in the WSM and WUM research area. We describe the tasks related to the initial experiment in the third chapter. The fourth chapter brings the detailed description of the experiment, in which we tried to find dependencies of the value *PageRank* on variable *support*. We provide discussion and conclusions in the last chapter.

2 Related Work

The analysis of users' behaviour represents the main objective of the web usage mining [1, 2]. Data about the accesses of website's visitors is stored in standardized text form of the log files, referred to as Common Log File (CLF).

Data pre-processing refers to the stage of processing of the web server logs for the purpose of identifying meaningful representations. Data cleaning methods are necessary because a WUM is sensitive to noise data. On the other hand, data pre-processing can represent a difficult task when the available data is incomplete or includes erroneous information. According to Cooley, Mobasher, and Srivastava [3] data pre-processing consists of data cleaning (removing irrelevant references and fields, eliminating erroneous references, adding missing references due to caching mechanisms, etc.) and data transformation (user-session identification, path completion [4, 5], etc.).

The web server log file is the primary source of anonymous data about a user (a website's visitor). Anonymous data can also cause a problem with a unique identification of a web page visitor because the visitor can visit the web page repeatedly. Therefore, the web log file can contain multiple sessions of the same visitor.

The objective of this phase of pre-processing is the user session identification [3]. The session identification method using time-window represents the most common method [6]. Using this method, each time we had found subsequent records about the web page requests where the time of the web page displaying had been higher than explicitly selected time, we divided the user visits into several sessions. Explicitly chosen time is denoted as a Standard Time Threshold (STT) and it can take different values: 5 min [7], 10 min [8], 15 min [9], or 30 min [10] or individual threshold [11]. This method is widely used because of its simplicity.

We should briefly mention the alternative methods of the user session identification, which are based on the information stored in cookies files saved on the user's computer. The cookies are tightly bound to the web browser. Even though cookies are considered the most common and the most simple method of user session identification, known issues [12, 13] limit their practical use and have to be replaced by other methods [14].

The main aim of the paper is to use the PageRank (PR) algorithm [15, 16] in WUM domain. Several authors tried to combine WSM, web content mining and WUM methods in several studies. Lorentzen [17] found quite a few studies using a combination of two sub-fields of web log mining.

Usually, the estimation of the web page quality was assured by the PR, HITS or TrustRank algorithms. However, low quality, unreliable data or spam stored in the hypertext structure caused less effective estimation of the web page quality [18, 19]. We can find a review of PR algorithms in Web Mining, their limitations and a new method for indexing web pages in [16]. An interesting approach for using web logs for improvement of website design and organization is described in [20].

Lorentzen [17] noticed that the structure mining is frequently used with other methods. For example, the Markov chain-based Site Rank and Popularity Rank combined structure and usage mining with a co-citation-based algorithm. Another approach used HITS algorithm, semantic clustering, co-link analysis and social network analysis for an automatic generation of hierarchical sitemaps for web sites, or for an automatic exploration of a topical structure of a given academic subject.

Ahmadi-Abkenari [21] introduced a web page importance metric of LogRank that worked based on analysis of different levels of clickstreams in server data set. The importance of each web page was precisely based on the observation period of log data and independent from the downloaded portion of the web.

Agichtein et al. and Meiss et al. [22, 23] used the traffic data to validate the PageRank random surfing model. Su et al. [24] proposed and experimentally evaluated a novel approach for personalized web page ranking and recommendation by integrating an association mining and PageRank.

We also found similar approaches, which combined WUM and WSM methods in other experiments [12, 16, 18, 21, 22, 24–31], but these experiments did not research quality of acquired knowledge about the users' behaviour depending on the selected method of user session identification.

3   Data Pre-Processing

We try to explain, how to link WUM and WSM methods effectively in this section. Firstly, we have to note previously realized experiments, which are not included in this paper, but their results are inevitable for the formulation of findings.

We used data stored in a standard log format in all the experiments described in this paper. We developed a crawler, which went through and analysed web pages. The crawler began on the home page and read all hyperlinks on the examined web page. If the crawler found hyperlinks to the unattended web pages, it added them to the queue. The crawler created a site map which we utilized later in the PR calculation of individual web pages.

At the same time, we used the site map as an input to the path completion algorithm. Besides the site map, the crawler collected information about the level, in which the analysed web page has been in respect to the home page. If the home page was level 1, then all web pages, which had a reference from the home page, would have been level 2. The ith level contained all web pages with references from the i-1 level. It is clear that we considered the highest level of each web page. We considered two categories of the web page levels:

- Category A included web pages of the first and second level, i.e., home page and all web pages, which were accessible from it on one click.
- Category B included all remaining levels.

When the crawler finished, we created a hypertext matrix from the site map. Consequently, we calculated PR for individual web pages according to the formula (1). The value of damping factor d was 0.85.

We used the log files of the university web site. We considered also the session identification method based on cookies in the experiment. It was necessary to change the format of the logs, change the credentials and the manner of writing and reading cookies in the web server. After that, we removed unnecessary records and accesses of crawlers from the log file. The final log file had 573020 records over a period of three weeks. We also removed records, where the information about the cookies was missing (160660 records, 28.04 %). Finally, we obtained the file log with 412360 records.

3.1 Initial Experiment

We prepared an experiment with the aim of verifying the contribution of the proposed method of user session identification using cookies. We used reputable user session identification methodology using STT for this purpose [1, 6, 11, 32, 33]. We applied the proposed method to the four different files with various levels of pre-processing.

At the same time, we intended to find out which of the pre-processed log files was the most suitable for the proposed method. We decided to identify the quality and quantity of the acquired knowledge (behavioural patterns of the users) from the individual log files for this purpose.

We followed the following methodology in the process of examining the influence of data pre-processing on the quality and quantity of extracted knowledge [33]:

1. Data acquisition – definition of observed variables in the log file (IP address, access date and time, URL).
2. Data matrices creation from the log file (information about users' accesses) and the site map (information about the web content).
3. Data pre-processing on the different levels.
4. Data analysis – user behavioural pattern finding in individual files. We used the Apriori algorithm for extraction of sequence rules implemented in the Sequence Association and Link Analysis [34] Module of STATISTICA.
5. Output data understanding – the creation of data files from the outputs of the analysis of individual files and basic characteristics calculation.
6. Comparison of obtained knowledge from the files, which were pre-processed at the different levels. We evaluated the acquired knowledge in terms of the quality and quantity of found sequence rules – user behavioural patterns. We took great care in:

- Comparison of proportion of found rules in examined files.
- Comparison of proportion of useful, trivial, or inexplicable rules in examined files.
- Comparison of variables *support* and *confidence* of found rules in examined files.

We prepared data at some levels. We obtained the final set of files:

- File **A1** – session identification using STT without the path completion,
- File **A2** – session identification using STT with the path completion,
- File **B1** – session identification using cookies without the path completion,
- File **B2** – session identification using cookies with the path completion.

We used STT = 10 min in the cases of files A1 and A2 and 10 min for cookies expiration in the case of files B1 and B2.

3.2   Results of Initial Experiment

We examined users' accesses to the web site of the university during three weeks. We obtained the sequence rules from the frequented sequences, which accomplished the minimal support (min s = 0.005) as a result of the analysis (Table 1). We obtained frequented sequences previously from the identified sequences, i.e., from the visits of individual users in the observed period.

Table 1. Discovered sequence rules in individual files.

Body	->	Head	A1	A2	B1	B2
(/), (/admissions)	->	(/admissions/admissions-results)	1	1	1	1
...	->	...	...	...	...	...
(/university-structure)	->	(/university-structure)	0	1	0	1
(/university-structure)	->	(/university-structure/faculty-of-natural-sciences)	1	1	1	1
...	->	...	...	...	...	...
(/study/accredited-study-programs)	->	(/study)	0	1	0	1
Count of derived sequence rules			51	197	43	227
Percent of derived sequence rules (Percent 1's)			21.52	83.12	18.14	95.78
Percent 0's			78.48	16.88	81.86	4.22
Cochran Q test			Q = 443.3120, df = 3, p < 0.000000			

We could see the high consistency (compliance) between the results of the sequence rule analysis in terms of the portion of the found rules in the files without the path completion (A1, B1). Simultaneously, we could see the similar compliance in the case of files with a path completion (A2, B2).

We extracted most of the rules from the file with the identified user sessions and completed paths. More rules were discovered in the files with the path completion (A2, B2).

The assessment of the quality of obtained sequence rules represented the next step in results evaluation. We assessed two characteristics - *support* and *confidence*. We found differences in quality and quantity of the found sequence rules between individual files regarding the values of variable *support*. Statistically, significant differences were found between the files without path completion (A1, B1) and between the files with path completion (A2, B2).

We should have considered files with path completion (A2, B2) as the best-pre-processed files for the extraction of user behavioural patterns.

4   Finding Dependences Between Variables *PageRank* and Variable *Support*

We merged WSM methods (PageRank) with WUM methods (sequence rule analysis) in the following experiment. We tried to answer the question if the expected visit rate of individual web pages (calculated using PR) correlates with the observed visit rate of the web pages, which were found by the WUM method in the previously described experiment and expressed by the value of variable *support*.

Variable *support* is defined as $support(X) = P(X)$. In other words, item X has a support s if s % of transactions contain X, i.e. the variable *support* means the frequency of occurrence of given set of items in the database. It represents the probability of visiting a particular web page in identified sequences (sessions). We assumed that the data reliability used in WUM would be increasing with the growth of dependence between values of *PR* and *support*.

4.1 Results

The value of *PR* and the level of the web page were added to the data obtained from the log file. It means that we calculated PR and assigned the appropriate level to each record of the log file. We made these changes to the log file examined in the initial experiment (Sect. 3.1). We analysed not only this log file, but also the log files which have been pre-processed in the same manner.

The variable *support* means the probability of individual web page visits in identified sessions. We examined the variable *support* from available statistics, and subsequently we analysed only the web pages with minimal support 0.5 %.

Table 2 shows the dependence of *PR* on the variable *support* calculated from the files with different level of data pre-processing. A directly proportional relationship was identified in all examined files. There were evident variations from normality. Therefore, we used non-parametric correlation [35] for calculation of dependence rate between *PR* and *support*.

Table 2. The dependence between *PR* and variable *support* in examined files.

Total	Valid N	Spearman R	t(N−2)	p-level
PageRank & support (A1)	47	0.4052	2.972739	0.004728
PageRank & support (A2)	42	0.4248	2.967526	0.005049
PageRank & support (B1)	46	0.4004	2.898073	0.005834
PageRank & support (B2)	39	0.4687	3.227136	0.002619

We identified the medium dependence of *PR* on variable *support* (Table 2). The dependence was greater in the files with identified paths. The correlation coefficients were statistically significant at the 1 % significance level. The greatest dependence was reached in file B2. Following these findings we could assume that the method of session identification using cookies in conjunction with the path completion had the greatest impact on the data reliability.

If we considered the web structure, i.e. the position of the web page in web site, we obtained similar results. We regarded two categories of web pages:

- Category A contains web pages of the first and second level.
- Category B contains remaining levels.

The impact of data pre-processing was not significant in category A. The coefficients of correlation (Table 3) were not significant (Spearman R < 0.4; p > 0.05).

On the opposite, we could suppose the path completion had an impact on the data reliability (Spearman R > 0.4; p < 0.05) in the case of category B (Table 4).

Table 3. The dependence between *PR* and variable *support* in category A.

Category A	Valid N	Spearman R	t(N−2)	p-level
PageRank & support (A1)	19	0.3809	1.698370	0.107664
PageRank & support (A2)	18	0.3635	1.560512	0.138198
PageRank & support (B1)	19	0.3668	1.625806	0.122384
PageRank & support (B2)	17	0.3936	1.658376	0.118001

Table 4. The dependence between *PR* and variable *support* in category B.

Category B	Valid N	Spearman R	t(N−2)	p-level
PageRank & support (A1)	28	0.3710	2.037140	0.051936
PageRank & support (A2)	24	0.4476	2.347991	0.028276
PageRank & support (B1)	27	0.3568	1.909797	0.067698
PageRank & support (B2)	22	0.4245	2.096915	0.048918

We also compared the values of *PR* and variable *support* for given categories. We considered the position of the web page in the structure of the web site. The distribution of observed variables was asymmetric. Therefore, we used non-parametric Mann-Whitney U Test for differences testing.

We found statistically significant difference of *PR* values in categories A and B at the 0.1 % significance level from the results of Mann-Whitney U Test (Table 5). The value of median was 10.3 and middle 50 % values were from the interval 9.4–10.6 in the category A. The value of median was 0.5 and middle 50 % values were from the interval 0.2–1.2 in the category B.

Table 5. Testing differences: *PR* x *category*.

	Rank sum A	Rank sum B	U	Z	p-level
PageRank	664.5	463.5	57.5	4.519811	0.000006

The statistically significant differences in variable *support* between categories A and B in examined files (Table 6) were not proven (p > 0.05).

Table 6. Testing differences: variable *support* x *category*.

	Rank sum A	Rank sum B	U	Z	p-level
support (A1)	534.0	594.0	188.0	1.690864	0.090864
support (A2)	448.5	454.5	154.5	1.563110	0.118028
support (B1)	520.0	561.0	183.0	1.639722	0.101064
support (B2)	402.0	378.0	125.0	1.755968	0.079095

The median of variable *support (A1)* was 1.6 and the middle 50 % of values were from the interval 0.6–5.2 in the category A. In contrast to these values, the median was 0.9 and the middle 50 % of values were from the interval 0.6–1.5 in the category B. This means the values of *support* were more homogeneous in category B than in category A from the variability point of view.

We also achieved similar results in files A2, B1 and B2. We noticed the differences in the variable *support* only in the variability between categories A and B.

5 Discussion and Conclusion

The authors of the original idea of PR introduced PR as a probability that the random visitor accessed a particular web page. The log file and WUM methods represented the observed visit rate of individual web pages in realized experiment. We proved experimentally that the found expected visit rate (*PR*) correlate with the observed visit rate expressed as a value of the variable *support*.

As we noted previously, we have to modify the log file in the pre-processing phase in order to obtain the real user sessions. The quality of acquired knowledge about user's behaviour depended on the selected method of the session identification and executed changes.

We proved in the experiment that there was a higher dependence between *PR* and the variable *support* in the visit rate of the examined web pages in well-prepared files with identified user sessions.

We found out statistically significant differences between *PR* and variable *support* in the files with the application of more advanced methods of session identification. We obtained the same results in the initial experiment (Sects. 3.1 and 3.2), where we tried to compare session identification methods in the four files with different levels of data pre-processing. We proved the results of the initial experiment by using a different methodology.

We verified a new proposed methodology for the comparison of methods of data pre-processing in WUM in terms of the reliability of the obtained data. We could follow these steps:

1. Data acquisition.
2. Creating of data matrices from the log file and site map.
3. Data pre-processing at different levels.
4. PR calculation for individual web pages.
5. Variable *support* calculation of the web page selected from the log files and assignment of PR to the individual web pages. In this case, we considered only web pages with value of *support* > 0.5 %.
6. Data understanding and creation of data files from the calculated characteristics.
7. Comparison of obtained characteristics in term of the dependence of *PR* on variable *support* from log files, which were pre-processed at different levels.

We omitted some steps in contrast to the method introduced in the Sect. 3.2. We did not extract the rules and did not analyse the extracted rules in terms of their quantity and quality. We verified suitability of these steps in terms of the dependence on *PR* and variable *support*. We focused on the suitability verification of proposed steps from the point of view of the dependence between *PR* and variable *support*.

The proposed methodology did not involve the extraction of sequence rules from examined files. It only claimed that well-prepared files are files which best reflect the dependence of the expected and observed visit rate (expected and the real probability).

On the other hand, the usage analysis involved rules extraction. It means that the proposed methodology is only an alternative or supplementary method. It serves mainly for experimental purposes and results verification of the original methodology (used in Sect. 3.1).

Acknowledgments. This paper is supported by the project VEGA 1/0392/13 Modelling of Stakeholders' Behaviour in Commercial Bank during the Recent Financial Crisis and Expectations of Basel Regulations under Pillar 3- Market Discipline and project KEGA 015UKF-4/2013 Modern computer science – New methods and forms for effective education.

References

1. Srivastava, J., Cooley, R., Deshpande, M., Tan, P.-N.: Web usage mining: discovery and applications of usage patterns from Web data. SIGKDD Explorations Newsletter 1, pp. 12–23 (2000)
2. Romero, C., Ventura, S., Zafra, A., Bra, P.D.: Applying web usage mining for personalizing hyperlinks in web-based adaptive educational systems. Comput. Educ. **53**, 828–840 (2009)
3. Cooley, R., Mobasher, B., Srivastava, J.: Data preparation for mining world wide web browsing patterns. Knowl. Inf. Syst. **1**, 5–32 (1999)
4. Zhang, C., Zhuang, L.: New path filling method on data preprocessing in web mining. In: Proceedings of Computer and Information Science 1, pp. 112–115 (2008)
5. Li, Y., Feng, B., Mao, Q.: Research on path completion technique in web usage mining. In: Proceedings of the 2008 International Symposium on Computer Science and Computational Technology, vol. 01, pp. 554–559. IEEE Computer Society (2008)
6. Huynh, T., Miller, J.: Empirical observations on the session timeout threshold. Inf. Process. Manag. **45**, 513–528 (2009)
7. Downey, D., Dumais, S., Horvitz, E.: Models of searching and browsing: languages, studies, and applications. In: Proceedings of the 20th international joint conference on Artifical intelligence, pp. 2740–2747. Morgan Kaufmann Publishers Inc., Hyderabad (2007)
8. Chien, S., Immorlica, N.: Semantic similarity between search engine queries using temporal correlation. In: Proceedings of the 14th International Conference on World Wide Web, pp. 2–11. ACM, Chiba, (2005)
9. He, D., Göker, A.: Detecting session boundaries from web user logs. In: Proceedings of the BCS-IRSG 22nd Annual Colloquium on Information Retrieval Research, pp. 57–66 (2000)
10. Radlinski, F., Joachims, T.: Query chains: learning to rank from implicit feedback. In: Proceedings of the Eleventh ACM SIGKDD International Conference on Knowledge discovery in Data Mining, pp. 239–248. ACM, Chicago (2005)
11. Mehrzadi, D., Feitelson, D.G.: On extracting session data from activity logs. In: Proceedings of the 5th Annual International Systems and Storage Conference, pp. 1–7. ACM, Haifa (2012)
12. Guerbas, A., Addam, O., Zaarour, O., Nagi, M., Elhajj, A., Ridley, M., Alhajj, R.: Effective web log mining and online navigational pattern prediction. Knowl. Based Syst. **49**, 50–62 (2013)

13. Cooley, R.: Web usage mining: discovery and application of interesting patterns from web data. Ph.D. thesis. University of Minnesota (2000)
14. Schmitt, E., Manning, H., Paul, Y., Tong, J.: Measuring Web Success. Forrester Report (1999)
15. Brin, S., Page, L.: The anatomy of a large-scale hypertextual web search engine. Comput. Netw. **30**, 107–117 (1998)
16. Jain, A., Sharma, R., Dixit, G., Tomar, V.: Page ranking algorithms in web mining, limitations of existing methods and a new method for indexing web pages. In: Proceedings of the 2013 International Conference on Communication Systems and Network Technologies, pp. 640–645. IEEE Computer Society (2013)
17. Lorentzen, D.G.: Webometrics benefitting from web mining? An investigation of methods and applications of two research fields. Scientometrics **99**, 409–445 (2014)
18. Lili, Y., Yingbin, W., Zhanji, G., Yizhuo, C.: Research on PageRank and hyperlink-induced topic search in web structure mining. In: Conference Research on PageRank and Hyperlink-Induced Topic Search in Web Structure Mining, pp. 1–4 (2011)
19. Wu, G., Wei, Y.: Arnoldi versus GMRES for computing PageRank: a theoretical contribution to google's PageRank problem. ACM Trans. Inf. Syst. **28**, 1–28 (2010)
20. Xu, G., Zhang, Y., Li, L.: Web Mining and Social Networking Techniques and Applications. Springer, Heidelberg (2011)
21. Ahmadi-Abkenari, F., Selamat, A.: A clickstream based web page importance metric for customized search engines. In: Nguyen, N. (ed.) Transactions on Computational Collective Intelligence XII, vol. 8240, pp. 21–41. Springer, Berlin Heidelberg (2013)
22. Agichtein, E., Brill, E., Dumais, S.: Improving web search ranking by incorporating user behavior information. In: Proceedings of the 29th Annual International ACM SIGIR Conference on Research And Development in Information Retrieval, pp. 19–26. ACM, Seattle (2006)
23. Meiss, M.R., Menczer, F., Fortunato, S., Flammini, A., Vespignani, A.: Ranking web sites with real user traffic. In: Proceedings of the 2008 International Conference on Web Search and Data Mining, pp. 65–76. ACM, Palo Alto (2008)
24. Su, J.-H., Wang, B.-W., Tseng, V.S.: Effective Ranking and Recommendation on web page retrieval by integrating association mining and PageRank. In: Proceedings of the 2008 IEEE/WIC/ACM International Conference on Web Intelligence and Intelligent Agent Technology, vol. 03, pp. 455–458. IEEE Computer Society (2008)
25. Srikant, R., Yang, Y.: Mining web logs to improve website organization. In: Proceedings of the 10th International Conference on World Wide Web, pp. 430–437. ACM, Hong Kong (2001)
26. Liu, H., Keselj, V.: Combined mining of web server logs and web contents for classi fying user navigation patterns and predicting users' future requests. Data Knowl. Eng. **61**, 304–330 (2007)
27. Das, R., Turkoglu, I.: Creating meaningful data from web logs for improving the impressiveness of a website by using path analysis method. Expert Syst. Appl. **36**, 6635–6644 (2009)
28. Yang, Q., Ling, C., Gao, J.: Mining web logs for actionable knowledge. In: Liu, J., Zhong, N. (eds.) Intelligent Technologies for Information Analysis, pp. 169–191. Springer, Heidelberg (2004)
29. Eirinaki, M., Vazirgiannis, M.: Usage-based PageRank for web personalization. In: Proceedings of the Fifth IEEE International Conference on Data Mining, pp. 130–137. IEEE Computer Society (2005)
30. Masseglia, F., Poncelet, P., Teisseire, M.: Using data mining techniques on web access logs to dynamically improve hypertext structure. SIGWEB Newsletter, 8, pp. 13–19 (1999)

31. Tripathy, A., Patra, P.K.: A Web mining architectural model of distributed crawler for internet searches using PageRank algorithm. In: Proceedings of the 2008 IEEE Asia-Pacific Services Computing Conference, pp. 513–518. IEEE Computer Society (2008)
32. Fang, Y., Huang, Z.: An improved algorithm for session identification on web log. In: Wang, F., Gong, Z., Luo, X., Lei, J. (eds.) Web Information Systems and Mining, vol. 6318, pp. 53–60. Springer, Heidelberg (2010)
33. Munk, M., Kapusta, J., Švec, P.: Data preprocessing evaluation for web log mining: reconstruction of activities of a web visitor. Procedia Comput. Sci. **1**, 2273–2280 (2010)
34. Agrawal, R., Srikant, R.: Fast algorithms for mining association rules in large databases. In: Proceedings of the 20th International Conference on Very Large Data Bases, pp. 487–499. Morgan Kaufmann Publishers Inc., (1994)
35. Pilkova, A., Volna, J., Papula, J., Holienka, M.: The influence of intellectual capital on firm performance among slovak SMEs. In: Proceedings of the 10th International Conference on Intellectual Capital, Knowledge Management and Organisational Learning (ICICKM-2013), pp. 329–338 (2013)

Bioinformatics Theory and Methods

Kernel Independent Component Analysis-Based Prediction on the Protein O-Glycosylation Sites Using Support Vectors Machine and Ensemble Classifiers

Zehao Chen[✉]

Centrale Pékin, Beihang University, Beijing, China
409275072@qq.com

Abstract. O-glycosylation means that sugar transferred to the protein. It can adjust the function of protein. To improve the prediction accuracy of O-glycosylation sites in protein, we used a new method of combining kernel independent component analysis with support vectors machine (KICA + SVM). The samples for experiment are encoded by the sparse coding with window size $w = 51$, 48 kernel independent components (feature) are extracted by kernel independent component analysis (KICA), then the prediction (classification) is done in feature space by support vector machines (SVM). The results of experiment show that the performance of KICA + SVM is better than that of KPCA + SVM, ICA + SVM, and PCA + SVM. Furthermore, we investigated the same protein sequence under various window size ($w = 5, 7, 9, 11, 21, 31, 41, 51$), and used the sum role to combine all the pre-classifiers to improve the prediction performance. The results indicate that the performance of ensembles of KICA + SVM is superior to that of pre-classifier. The prediction accuracy is about 90 %.

Keywords: Prediction · Protein · KICA · SVM · Ensemble classifier

1 Introduction

Glycosylation is the most common post-translation modification of protein in eukaryotic cells, and has important functions in secretion, antigenicity, and metabolism of glycoproteins. There are four types of glycosylation: N-linked glycosylation to the amide nitrogen of asparagines side chains, O-linked glycosylation to the hydroxyl of serine and threonine side chains (Fig. 1), C-linked glycosylation to the tryptophan side chains and GPI. Here we only focus on O-linked glycosylation protein sequence. In fact, not all serine or threonine residue are glycosylated and about 10 %–30 % protein can't be glycosylated. There are many factors which affect this process, so it is very important to predict the O-glycosylation sites.

In the engineering of biological pharmacy, the treatment effect of medicine can be improved and the toxicity of medicine can be reduced by choosing appropriate bearer protein for glycosylation modification, if the glycosylation sites is predicted. The glycosylation degree and abnormity of Glycan structure is one of the symbols of cancer, so the prediction of glycosylation sites is also important for disease survey.

© Springer International Publishing Switzerland 2015
D.-S. Huang and K. Han (Eds.): ICIC 2015, Part III, LNAI 9227, pp. 651–661, 2015.
DOI: 10.1007/978-3-319-22053-6_67

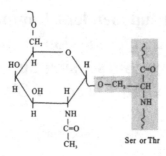

Fig. 1. The structure of o-glycosylation

Many computational methods based on artificial neural networks (ANN) and support vector machines (SVM) have been developed for prediction of O-glycosylation sites. The prediction accuracy can be achieved more than 70 % [1–3]. Yong-zi Chen [4] used a new protein bioinformatics tool, CKSAAP_OGlySite, to predict mucin-type O-glycosylation serine or threonine sites in mammalian proteins, under the composition of k-spaced amino acid pairs (CKSAAP) based encoding scheme, with the assistance of SVM. His method yielded a higher accuracy of 83.1 % and 81.4 % in predicting O-glycosylated S and T sites, respectively.

Since there are complex structure features in the protein sequence, the prediction performance is not satisfied if one classifies the original data directly without extracting the meaningful features.

Principal component analysis (PCA) is a statistical method for feature extraction, it can reduce the dimension and eliminate relativity of the original data, so it plays a key role in many research areas of science and engineering. In the reference [5], Yang XM et al. used PCA for pattern analysis and feature extracting, they first found and verified by computational methods that O-glycosylation is abundant near the C terminus for serine.

But PCA only considers about the second order statistics information of the original data, it can't eliminate the high-order relativity of each components of the data, and the most part of information of data are included in the high-order statistics feature, so PCA can't recognize the data accurately.

Independent component analysis (ICA) is a statistics method based on the high-order statistics feature of the original data, it is an linear transform which can eliminate the high-order relativity of each components of data, and make each components independent [6], so the data after ICA transform can be recognized more accurately. But similar to PCA, it only draws the linear character of the samples.

Since the nonlinear features of the original data are usually important to recognition, we need to extract the nonlinear features of the original data to improve the prediction performance.

Kernel principal component analysis (KPCA) [7] is the application of PCA in a kernel-defined feature space making use of dual representation. It can capture the nonlinear feature of the original data and make the recognition of objective more accurate. Yang XM [8] predicted the O-glycosylation sites in protein by combining KPCA with SVM or Mahalanobis distance, and obtained better performance.

In this paper, we used a new method of KICA + SVM to predict the O-glycosylation site in protein sequence. We first extracted features of original data by KICA [9], then used support vector machine (SVM) for classification in the feature space. At last, we used the sum role to ensemble the all pre-classifiers [10] to improve the prediction accuracy.

The remainder of the paper is organized as follows. Section 2 describes the algorithm of KICA + SVM for prediction. Section 3 presents protein sequence data and their coding. Prediction results are shown in Sect. 4. The conclusions are given in Sect. 5.

2 Kernel Independent Component Analysis and Support Vector Machines for Prediction

The prediction can be view as a two-class classification problem (positive and negative). For samples in various window size w = 5, 7, 9, 11, 21, 31, 41, 51, we first extract features of training and test samples with KICA, then classify the test samples by SVM. At last, we use the feature subspaces under various window sizes to construct individual classifiers, and assemble the individual SVM classifiers by the sum role to give the final prediction.

2.1 ICA

ICA is a statistics method based on the high-order statistics feature of the original data. In fact, most of the important information is included in the high-order statistics feature. Since each component of data after ICA transform is independent, we can recognize them more accurately.

The model of ICA can be showed as

$$\mathbf{x} = \mathbf{As} \tag{1}$$

where $\mathbf{s}$ is the blind source, we suppose that each component of $\mathbf{s}$ is independent, $\mathbf{A}$ is an unknown hybrid matrix, $\mathbf{x}$ is known signal. We need to find a matrix $\mathbf{W}$, such that

$$\mathbf{y} = \mathbf{Wx} \tag{2}$$

$\mathbf{y}$ should approximate $\mathbf{s}$.

There are many methods for learning the separating matrix $\mathbf{W}$. The learning algorithm for $\mathbf{W}$ based on information maximization is as the following:

$$\Delta W = (I + g(y)y^T)W \tag{3}$$

Where $g(y) = 1 - 2/(1 + e^{-y})$.

Before the learning procedure, $\mathbf{x}$ is centered and whitened by the following

$$\mathbf{x} = \mathbf{W}_0(\mathbf{x} - \mathbf{m_x}) \tag{4}$$

Where $\mathbf{m_x}$ is mean vector of $\mathbf{x}$, and $\mathbf{W_0} = \mathbf{C}^{-1/2}$, $\mathbf{C}$ is the covariance matrix of $\mathbf{x}$. Therefor the component transform is calculated as the product of the whitening matrix and the learning matrix:

$$\mathbf{W} = \mathbf{WW_0} \tag{5}$$

2.2 KICA

ICA is a linear method that can reduce the redundancy and extract the linear feature of the data, but there often exists nonlinear feature in the original data, and these nonlinear feature are often important for the pattern recognition of the data.

Kernel independent component analysis is a new algorithm which combining ICA with a kernel trick [9], named kernel ICA (KICA).

KICA solve a problem of ICA through a method to calculate a loss function by using kernel canonical correlation analysis. KICA algorithm is based on minimization of a contrast function based on the kernel idea. A contrast function measures the statistical dependence between components, so when applied to estimate components and minimize over possible demixing matrix, components that are as independent as possible are to be found.

The basic idea of KICA is to map the input data into an implicit feature space F with the kernel trick:

$$\Phi : X \in R^N \rightarrow \Phi(x) \in F \tag{6}$$

The data in the input space $X = (x_1, x_2, \ldots, x_m) \in R^N$ is mapped to a potentially much higher dimensional feature space F and the data is converted to easier form for analysis.

We run ICA in F to produce a set of nonlinear features of input data. First, the input data X is whitened in the feature space F.

The whitening matrix is expressed as follow:

$$W_w^\Phi = (\Lambda^\Phi)^{\frac{1}{2}}(V^\Phi)^T \tag{7}$$

where $(\Lambda^\Phi)^{\frac{1}{2}}$ and $(V^\Phi)^T$ are the eigen value and eigen vector matrix of covariance matrix

$$\hat{C} = \frac{1}{n}\sum_{i=1}^n \Phi(x_i)\Phi(x_i)^T, \; respectively$$

We obtain the whitened data X_W^Φ by

$$X_W^\Phi = (W_w^\Phi)^T\Phi(X) = (\Lambda^\Phi)^{-1}\alpha^T K \tag{8}$$

where K is the kernel functions, and "." denotes an inner product:

$$K_{ij} := (\Phi(x_i) \cdot \Phi(x_j)) = k(i,j) \tag{9}$$

and α denotes the eigen matrix of K. The ICA learning iteration algorithm described by Eq. (3) is executed as follows:

$$y_I^\Phi = W_I^\Phi X_W^\Phi \tag{10}$$

$$\Delta W_I^\Phi = [I + (I - \frac{2}{1 + e^{y_I^\Phi}})(y_I^\Phi)^T] W_I^\Phi \tag{11}$$

$$\hat{W}_I^\Phi = W_I^\Phi + \rho \Delta W_I^\Phi \rightarrow W_I^\Phi \tag{12}$$

until W_I^Φ converges, where the learning rate ρ is constant. Then, a new feature representation in the feature space of a test data y can be obtained:

$$s = W_I^\Phi (\Lambda^\Phi)^{-1} \alpha^T K(X, y) \tag{13}$$

where, $K(X, y) = [k(x_1, y), k(x_2, y), \ldots, k(x_n, y)], k$ is a kernel function.

The choice of kernel function must be subject to the Mercer's theorem. In this paper we use the following kernel function: (1) polynomial kernel function, $k(x_i, y) = [(y \cdot x_i) + c]^d$, (2) Gaussian radial basis kernel function (RBF), $k(x_i, y) = \exp(-\frac{\|x_i - y\|^2}{2\sigma^2})$.

We can run KICA on the vectors of the acid sequence, and obtain the kernel independent components of the data, then predict the o-glycosylation sites by using the kernel independent components.

2.3 SVM

Support Vector Machine (SVM) is a kind of supervised machine learning technology for many two classes of classification problem. The classification idea of SVM is mapping the training vectors into multidimensional space by a kernel function, and then construct a hyperplane optimally positioned between the positive and negative samples, a testing sample is then projected into the multidimensional space to determine its class affiliation based on its relative position to the hyperplane.

For a given training set $\{(\mathbf{x}_1, y_1), \cdots, (\mathbf{x}_M, y_M)\}, \mathbf{x}_i \in \mathbf{R}^d, y_i \in \{-1, 1\}$, the equation of separating hyperplane is

$$\mathbf{w} \cdot \phi(\mathbf{x}) + b = 0$$

Then the objective function of SVM is

$$\begin{aligned} &\min_{w, b, \xi} && \frac{1}{2} \|\mathbf{w}\|^2 + C \sum_{i=1}^{M} \xi_i^2 \\ &s.t. && y_i = \mathbf{w} \cdot \phi(\mathbf{x}_i) + b + \xi_i, i = 1, \ldots, M \end{aligned} \tag{14}$$

Where $\frac{2}{\|\mathbf{w}\|}$ is the margin, ξ is the margin slack vector, the parameter C controls the trade-off between the margin and the size of the slack variables.

By using Lagrange multiplier method, we can obtain the solution of (14) $\mathbf{w}^* = \sum_{i=1}^{M} \alpha_i^* \phi(\mathbf{x}_i)$, and $b^* = y_j - \sum_{i=1}^{M} \alpha_i^* K(\mathbf{x}_j, \mathbf{x}_i) - \frac{\alpha_j}{2C}$, here $\boldsymbol{\alpha}^*$ is Lagrange multiplier, $K(\mathbf{x}_i, \mathbf{x}_j)$ is kernel function (we use kernel function again). So the classifier (decision function) is shown in (15)

$$f(\mathbf{x}) = \sum_{i=1}^{M} \alpha_i^* K(\mathbf{x}_i, \mathbf{x}) + b^* \tag{15}$$

We can classify the kernel independent components of the acid sequence by using SVM.

2.4 Sum Role

The sum role [10] is a kind of ensemble method. Assume $f_i(x)(i = 1, \ldots, M)$ are the outputs of M individual SVM classifiers, the sum role calculates the average value of these outputs

$$\bar{f}(x) = \frac{1}{M} \sum_{i=1}^{M} f_i(x), \tag{16}$$

and the category of sample is decided by the sign of $\bar{f}(x)$.

For a same serine or threonine site, we can use the samples with different window size to predict whether it is o-glycosylated or not, thus we will obtain several outputs of SVM classifiers, the sum role can be used to assemble the results of these outputs and decide the category of the site.

3 Protein Sequence Data and Encoding

The protein sequence data used in this research is from glycosylation database Uniprot (v8.0) [11]. We selected 99 mammalian protein entries, each entry contains some serine and threonine residue sites which are annotated experimentally as being glycosylated, together with other serine and threonine residue sites which have no such annotations. We call the former a positive site (positive S or positive T), while the latter a negative site (negative S or negative T). Each selected protein entry (sequence) is truncated by a window (window size: w) into several subsequences with S or T residues at the center. Figure 2 shows an example of subsequences ($w = 5$).

The protein sequence (exclude S or T at the center) with a length of w-1 are used for analysis. We use the sparse coding scheme for representation of the protein sequence. In sparse coding, 21-binary sequence is used to code one site of amino acid or vacancy, for example, the site of amino acid I is coded as 100000000000000000000, the site of

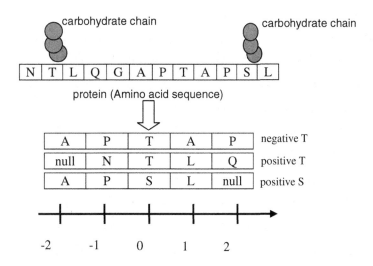

Fig. 2. Subsequences (w = 5) with serine or threonine at the center

amino acid V is coded as 010000000000000000000. Thus the total length of coded sequence or dimension of sample vector is $(w-1) *21$.

The number of samples for each class is summarized in Table 1. Since the number of negative sites is much larger than that of positive sites, we randomly chose 100 samples from each class for training, and 50 samples from each class for testing.

Table 1. Number of samples for experiments

Type	Total number	Number for training	Number for testing
Positive S	174	100	50
Positive T	292	100	50
Negative S	693	100	50
Negative T	841	100	50

4   Results and Analysis of Experiment

4.1   Position Probability Function

We experiment by using the KICA + SVM method, comparing with ICA + SVM, KPCA + SVM, and PCA + SVM. The algorithm is implemented in MATLABr2013a. Figure 3 shows the position probability function of four types of protein sequence (positive T, positive S, negative T, negative S). The color from dark blue to dark red represents a digital from 0 to 0.3074, which means the probability of one amino acid appearances at a position (−25–25). It can be seen that positive and negative have different characters. There are higher contents of proline, serine, threonine, and null in positive site.

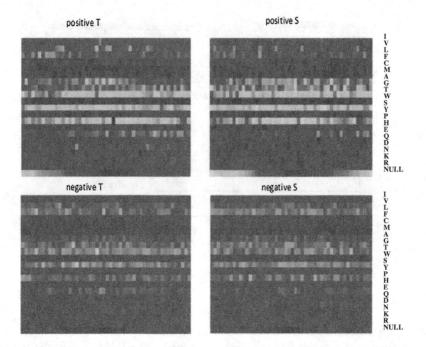

Fig. 3. Position probability function of protein sequence ($w = 51$) (Colour figure online)

4.2 Kernel Independent Components

In order to make a more quantitative analysis and comparison, we calculate the top four independent components and kernel independent components (the basis of subspace) for protein sequence. Figures 4 and 5 show these basis. It can be seen that the content of proline, serine, threonine and alanine is high in the first and the second kernel independent components, while null has a high content in the third and the forth kernel independent components. Comparing with the kernel independent components, the information included in the independent components are not obvious and abundant enough.

4.3 Prediction

We first experiment by using the samples with window size w = 51. we extracted the first 48 kernel independent components as the input data of SVM. The kernel function is polynomial kernel function and RBF, in our algorithm, we used the same kernel function in KICA and SVM. The parameter C of SVM is varied from 0.01 to 100. We tested 200 testing samples. The results of prediction are shown in Tables 2 and 3.

From Tables 2 and 3, we can see that, when using polynomial kernel function, the best performance of KICA + SVM is 83.5 %, and the best performance of KICA + SVM is 87 % when using RBF, they are better than the best performance of SVM and PCA + SVM, it's because that KPCA captures the nonlinear feature of the original data and makes the recognition of objective more accurate. We also see that the

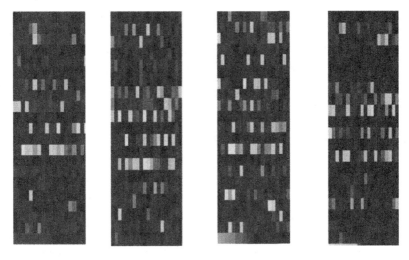

Fig. 4. The top 4 basis of ICA subspace

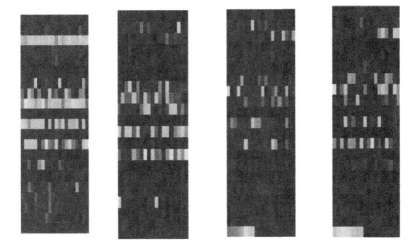

Fig. 5. The top 4 basis of KICA subspace

best parameters for polynomial kernel function and RBF are c = 5, d = 2 and σ = 5.5 when using KICA + SVM.

Furthermore, we investigated the same protein sequence with the window size w = 5, 7, 9, 11, 31, 41, 51, by using the method of KICA + SVM, the results of experiment are shown in Tables 4 and 5.

It can be seen that, (1) The best performance of prediction is obtained at the window size 21, this is because that if the window size is too small, such as 5 or 7, some input data of such short protein sequence contradict each other for the prediction, that is, some are positive and the others are negative with the same input sequence; and if the window size is too large, such as 41 or 51, the dimension of input data is higher,

Table 2. Prediction results (kernel function: polynomial c = 5)

Method										PCA + SVM		
ICA + SVM						KPCA + SVM		KICA + SVM				
d	1	2	3	1	2	3	2	3	4	2	3	4
Prediction accuracy (%)	81	**82**	80	81	82	**83.5**	**84.5**	83	80	82.5	**85.5**	83.5

Table 3. Prediction results (kernel function: RBF)

Method				PCA + SVM			ICA + SVM		KPCA + SVM		
KICA + SVM											
σ	10	12	14	10	13.5	15	14	15.5	16	13	16.5	18
Prediction accuracy (%)	83	**83**	84	81.5	**84.5**	80	84	**86**	85	85.5	**86.5**	86

Table 4. Prediction results (kernel function: RBF)

Window size	5	7	9	11	21	31	41	51
σ	5.5	5.4	6.8	8	10.5	12.9	14	16.5
Negative (%)	93	94	92	91	88	85	83	79
Positive (%)	75	73	79	84	90	92	96	94
Average (%)	84	83.5	85.5	87.5	**89**	88.5	88.5	86.5

The ensemble result with the sum role: **90**

Table 5. Prediction results (kernel function: polynomial)

Window size	5	7	9	11	21	31	41	51
c, d	c = 4, d = 4	c = 7, d = 4	c = 8, d = 3	c = 8, d = 2	c = 6, d = 2	c = 6, d = 2	c = 7, d = 2	c = 5, d = 3
Negative (%)	87	85	86	84	88.5	85	81	78
Positive (%)	77	83	85	86	85.5	88	91	93
Average (%)	82	84	85.5	85	**87**	86.5	86	85.5

The ensemble result with the sum role: **88**

so the features of samples are too complex to extract easily. So in the experiment, the window size of samples should not be too small or too large. (2) When using polynomial kernel function, the best order d of it is different for different window size, when the window size is smaller, d is larger, and when the window size is larger, d is smaller, this is because that a sample with smaller window size needs a higher order polynomial kernel function to extract its feature. (3) Since σ is a width parameter, the parameter σ is increased with the increasing of window size, when using RBF kernel function. (4) We find that, the prediction accuracy for positive is increased with the increasing of window size, and it is on the contrary for negative, this indicates that a classifier under smaller window size can predict a negative sample more accurate, but a classifier under larger window size can predict a positive sample more accurate. To improve the prediction performance, we consider to design an ensemble classifier. We use the sum role, combining all the pre-classifiers under different window size, and obtain the better performance.

5 Conclusions

We proposed a new method of KICA + SVM to realize the prediction of O-linked glycosylated sites in protein sequence. We also used the ensemble classifier to combine the advantage of every pre-classifier. The result of experiments shows that, KICA can extract the nonlinear features of data, the ensemble classifier can improve the prediction accuracy. The proposed method is effective and accurate.

References

1. Nishikawa, I., Sakamoto, H., Nouno, I., Iritani, T., Sakakibara, K., Ito, M.: Prediction of the *O*-glycosylation sites in protein by layered neural networks and support vector machines. In: Gabrys, B., Howlett, R.J., Jain, L.C. (eds.) KES 2006. LNCS (LNAI), vol. 4252, pp. 953–960. Springer, Heidelberg (2006)
2. Sasaki, K., Nagamine, N., Sakakibara, Y.: Support vector machines prediction of N- and *O*-glycosylation sites using whole sequence information and subcellular localizition. IPSJ Trans. Bioinform. **2**, 25–35 (2009)
3. Li, S., et al.: Predicting O-glycosylation sites in mammalian proteins by using SVMs. Comput. Biol. Chem. **30**, 203–208 (2006)
4. Chen, Y.: Prediction of mucin-type O-Glycosylation sites in mammaliam protein using the composition of k-spaced amino acid pairs. BMC Bioinform. **9**, 101–112 (2008)
5. Yang, X., Chen, Y.-W., Ito, M., Nishikawa, I.: Principal component analysis of O-linked glycosylation sites in protein sequence. In: IEEE Third International Conference on IIHMSP, vol. 1, pp. 121–126 (2007)
6. Wang, C.-Z., Tan, X.-F., Chen, Y.-W., Han, X.-H.: Independent component analysis-based prediction of O-linked glycosylation sites in protein using multi-layered neural networks. In: ICSP 2010 Proceedings, pp. 1761–1764 (2010)
7. Yang, X.M., Cui, X.W., Yang, X.Z.: Prediction of O-glycosylation sites in protein sequence by kernel principal component analysis. In: Proceedings of the International Conference on Computational Aspects of Social Networks, Taiyuan, China, pp. 267–270 (2010)
8. Yang, X.M.: Prediction of the protein O-glycosylation by machine learning based on kernel principal component analysis and ensemble classifiers. ICIC Express Lett. **5**(8B), 2805–2810 (2011)
9. Tateyama, T., Nakao, Z.: Brain matters emphasis in MRI by kernel independent component analysis. In: IEEE Third International Conference on IIHMSP, vol. 1, pp. 117–120 (2007)
10. Sun, S.: Ensembles of feature subspaces for object detection. In: Yu, W., He, H., Zhang, N. (eds.) ISNN 2009, Part II. LNCS, vol. 5552, pp. 996–1004. Springer, Heidelberg (2009)
11. http://www.ebi.uniprot

Semantic Role Labeling for Biomedical Corpus Using Maximum Entropy Classifier

Lei Han[✉], Dong-hong Ji, and Han Ren

School of Computer Science, Wuhan University, Wuhan, China
hattason@126.com, donghong_ji@yahoo.com,
cslotus@mail.whu.edu.cn

Abstract. Semantic role labeling (SRL) is a natural language processing (NLP) task that finds shallow semantic representations from sentences. In this paper, we construct a biomedical proposition bank and train a biomedical semantic role labeling system that can be used to facilitate relation extraction and information retrieval in biomedical domain. Firstly, we construct a proposition bank on the basis of the GENIA TreeBank following the Penn PropBank annotation. Secondly, we use GenPropBank to train a biomedical SRL system, which uses maximum entropy as a classifier. Our experimental results show that a newswire SRL system that achieves an F1 of 85.56 % in the newswire domain can only maintain an F1 of 65.43 % when ported to the biomedical domain. By using our annotated biomedical corpus, we can increase that F1 by 19.2 %.

Keywords: Semantic role labeling · Dependency parsing · Maximum entropy · Biomedical corpus

1 Introduction

The volume of biomedical literature available has experienced unprecedented growth in recent years. Automatically processing lots of literature would be an invaluable ability for large-scale experiments. For this reason, more and more information extraction (IE) systems using the technique of natural language processing (NLP) have been developed in the biomedical field. A task in the biomedical field is extraction of relations, such as protein-protein and gene-gene interactions. Currently, most biomedical relation-extraction systems fall under one of the following three approaches: cooccurrence-based, pattern-based, and machine-learning-based. However, they have the same limitation in extracting relations from complex natural language. They only extract the relations between the targets (e.g., proteins, genes) and the verbs, overlooking many adverbial and prepositional phrases describing location, manner, timing, condition, and extent. The information in such phrases may be important for precise definition and clarification of complex biological relations.

Semantic role labeling (SRL) is a shallow semantic processing task that has become increasingly popular in the NLP field over the last few years. The task is to identify many parts of a sentence that represent arguments for a given predicate and label each argument with a semantic role. The main semantic roles include

© Springer International Publishing Switzerland 2015
D.-S. Huang and K. Han (Eds.): ICIC 2015, Part III, LNAI 9227, pp. 662–668, 2015.
DOI: 10.1007/978-3-319-22053-6_68

Agent, Patient, Instrument, etc., and adjunctive semantic roles indicating Location, Time, Manner, Cause, Condition, and Extent. The input of the SRL system is a single sentence with a predicate in it and the output is the same sentence with labeled semantic roles. Consider the following example:

Input: Transcription factor GATA-3 [stimulates]$_{PRED}$ HIV-1 expression.

Output: [Transcription factor GATA-3]$_{ARG0}$ [stimulates]$_{PRED}$ [HIV-1 expression]$_{ARG1}$.

In this example, the semantic role ARG0 is the cause of stimulate and the semantic role ARG1 is the thing stimulated. This information is most valuable for IE and other tasks like question answering and automatic summarization. In the newswire domain, Morarescu [7] have demonstrated that full-parsing and SRL can improve the performance of relation extraction which increases F1 from 67 % to 82 %. This significant result leads us to conclude that SRL may have potential for relation extraction in the biomedical domain. In this paper, we aim to firstly construct a biomedical proposition corpus based on phrase structure parsing on the basis of the Genia corpus, and then build an automatic system of semantic role labeling for the biomedical corpus which used the maximum entropy classifier based on phrase structure parsing with common features and some extended features.

2 Biomedical Proposition Bank

Traditionally in open domain, most research in SRL has focused on corpus consisting of the newswire documents. SRL performs well on test sentences from the same domain, while it shows a sharp performance drop when the system tests sentences from different domains [9]. Although there have been a number of efforts to apply SRL to the biomedical domain in recent years [1, 4], the development of state-of-the-art SRL systems for the biomedical domain is hindered by the shortage of large biomedical corpora that are labeled with semantic roles. The main reason is the construction of such corpora is time consuming and expensive.

Since Penn PropBank [8] is annotated on the basis of Penn TreeBank [6], we selected a biomedical corpus which has a Penn-style treebank as the basic corpus. We choose the GENIA corpus, a collection of MEDLINE abstracts selected from the search results with the following keywords: human, blood cells, transcription factors, etc. The Penn-style treebank for GENIA currently contains 2000 abstracts which include MEDLINE UID, the title and the content annotated with part-of-speech tags and coreferences [11]. However, GENIA lacks a proposition bank in contrast with Penn TreeBank. Therefore we use the GENIA TreeBank as our basic corpus to build a biomedical proposition bank, GenPropBank, which we add the PropBank annotation into the GTB annotation. The specific steps of the GenPropBank construction as follows: First, we train an SRL system on the Penn PropBank (Wall Street Journal corpus) which can achieve an F1 of 85.56 %. Second, we use this SRL system to automatically annotate our basic corpus, and then human annotators check the system's results and correct the errors. Thus we can get the biomedical proposition bank, GenPropBank.

3 Semantic Role Labeling

Semantic role labeling (SRL) [10] is a popular semantic analysis technique of shallow semantic parsing. It can be used in a variety of natural language processing application systems in which some kind of semantic interpretation is needed, such as question and answering, information extraction, machine translation, paraphrasing, and so on. In this section, we introduce how to build a SRL system to construct the GenPropBank and test on it. The task of semantic role labeling is to find all arguments for a given predicate in a sentence and label them with semantic roles.

The first step is to parse the sentence into a syntactic parse tree. The parse tree consists of the words in the sentence, their part-of-speech tags (e.g., NN, VBZ, etc.), and nodes with syntactic categories (e.g., S, NP, VP, etc.). Figure 1 shows the syntactic parse tree for the example sentence from Sect. 1 (the semantic role labels ARG0 and ARG1 are not part of the syntactic parse tree). The gold standard SRL corpus, PropBank, was designed as an additional layer of annotation on top of the syntactic structures of the Penn TreeBank [3]. The second step is to identify all the predicates in the sentences. This can be easily accomplished by finding all instances of verbs and check their POS. The next step is the argument identification in which the SRL system has to find the boundaries for all arguments in the sentence. The annotation standard for semantic roles demands that the boundaries align with nodes in the syntactic parse tree. Thus, argument identification is to decide which nodes in the parse tree are the possible semantic roles. For example in Fig. 1, the system should find that the NP node that dominates Transcription factor GATA-3 and the NP node that dominates HIV-1 expression span arguments and all other nodes do not. Finally, the system has to determine the semantic role for all identified nodes, which is called argument classification. In our example, the first identified NP node should be labeled ARG0 and the second identified NP node should be labeled ARG1, as shown in Fig. 1.

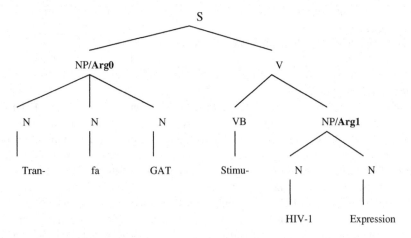

Fig. 1. A syntactic parsing tree with semantic roles added

For the predicate stimulate, ARG0 and ARG1 represent the cause of stimulate and the thing stimulated. In general, ARG0 refers to the agent and ARG1 refers to the object of the predicate. Each of the semantic roles ARG2-5 does not have a general meaning for different predicates. For example the semantic role ARG2 is the instrument for the predicate stimulate, but for the predicate increase, ARG2 is the amount increased. The semantic roles ARG0-5 are called core semantic roles, because they represent the essential arguments of a predicate. A predicate and the semantic roles of it are called a predicate-argument structure (PAS) or proposition. In addition to its core arguments, a predicate can have some adjunctive arguments. Adjunctive arguments express general attributes like time, location, manner, purpose, etc. They are labeled with ARGM and a functional tag, e.g., ARGM-LOC, ARGM-TMP, or ARGM-MNR.

In our SRL system, we select maximum entropy as a classifier to implement the semantic role labeling system. Because of the high speed and no affection in the number of classes with efficiency of maximum entropy classifier, we use one stage to label all arguments of predicates. It means that the "NULL" tag of constituents is regarded as a class like "ArgM-NULL".

4   Maximum Entropy Model and Features

By considering SRL as a machine learning problem, there are two critical problems to be solved: the choice of features and the choice of the machine learning algorithm. In this paper, we adopt the features used in other state-of-the-art SRL systems, which include the seven baseline features from the original work of Gildea and Jurafsky [2], additional features taken from Pradhan et al. [10], and feature combinations that are inspired by the system in [14, 15]. All features can be extracted from the syntactic parse tree. The features that we use in the classifier are as follows:

Baseline Features: predicate lemma; path from constituent to predicate; syntactic category; relative position to the predicate; active or passive voice; syntactic head word of the phrase; subcategory (rule expanding the predicate's parent).

Advanced Features: POS of the syntactic head word; head word and POS of the rightmost NP child if the phrase is a prepositional phrase; first/last word and POS in the constituent; syntactic category of the parent node; head word and POS of the parent; type of left and right brother; head word and POS of left and right brother; temporal key words present; partial path to predicate; projected path without directions.

Feature Combinations: predicate and phrase type; predicate and head word; predicate and path; predicate and relative position.

The machine learning algorithm in our experiments is a maximum entropy (maxent) classifier [5]. Maximum entropy classifiers do not require any independence assumptions which allow great flexibility in encoding linguistic knowledge via features. The model takes the form where y is a semantic role, x is an input vector, f_i are feature functions, λ_i are the weights that are learned during training, and Z is a normalization term. A detailed description of maximum entropy classifiers can be found in (Ratnaparkhi 1998).

$$p(y|X) = \frac{1}{Z(X)} \exp\{\sum_{i=1}^{N} \lambda_i f_i(X, y)\}$$

$$Z(X) = \sum_{y} \exp\{\sum_{i=1}^{N} \lambda_i f_i(X, y)\}$$

5 Experiments

In biomedical domain, there is one available treebank for GENIA. In contrast to WSJ, however, GENIA lacks any proposition bank. Since predicate-argument annotation is essential for training and evaluating SRL systems, we constructed GenPropBank. We adopted a semi-automatic strategy to annotate GenPropBank. Firstly, we used the Penn PropBank to train a SRL system which achieves an F1 of over 85.56 % on section 24 of the PropBank. Secondly, we used this SRL system to annotate the GENIA TreeBank automatically. Finally human annotators check the system's results and as far as possible to correct the errors. Thus we can get the biomedical proposition bank, GenPropBank.

The experiments are conducted using four fold cross validation on the biomedical data set. The 2000 abstracts in the biomedical data set are divided into four portions, and the separation is done randomly to guard against selection bias. The SRL system is trained on the three portions and tested on the remaining portion for four times.

6 Results and Discussion

The experimental results are reported in Table 1. Our experimental results show that a newswire English SRL system that achieves an F1 of 85.56 % can only maintain an F1 of 65.43 % when ported to the biomedical domain. By training the semantic role labeling system on GenPropBank, we can increase the F1 by 19.2 % on GENIA TreeBank. As can be seen from Table 1, the results of four experiments is relatively stable, indicating annotation effect of the GenPropBank is relatively steady. In the biomedical field, Dahlmeier [1] and Tsai [12, 13] have conducted similar experiments which achieve the F1 of 85.38 % and 84.25 %, compared with the performance of our systems higher 1.7 % and 0.6 %. Because the experimental data and the methods used in the experiments are completely different and there is no uniform evaluation criteria, we can not make an accurate judgment of performance differences in these similar experiment.

Table 1. The results of semantic role labeling on the biomedical test data

	Precision (%)	Recall (%)	F1 (%)
1	84.13	80.25	82.14
2	86.76	82.69	84.68
3	85.67	81.32	83.44
4	86.45	82.37	84.36
Average	85.75	81.66	83.66
Tsai [12]	87.03	81.65	84.25

7   Conclusions and Future Work

The contribution of this paper is as follows: Firstly, we construct a biomedical proposition bank, GenPropBank, on the basis of the biomedical GENIA TreeBank following the PropBank annotation. Secondly, we build a biomedical SRL system which uses GenPropBank as its training corpus. In the future, we will construct a SRL system for the nominal predicate in the biomedical corpora. In the other hand, we try to use other classifiers instead of maximum entropy classifier, and strengthen post-processing to improve system performance.

Acknowledgement. This work is supported by the Natural Science Foundation of China under Grant Nos. 61173095, 61202304.

References

1. Dahlmeier, D., Ng, H.T.: Domain adaptation for semantic role labeling in the biomedical domain. Bioinformatics **26**, 1098–1104 (2010)
2. Gildea, D., Jurafsky, D.: Automatic labeling of semantic roles. Comput. Linguist. **28**(3), 245–288 (2002)
3. Gildea, D., Palmer, M.: The necessity of syntactic parsing for predicate argument recognition. In: Proceedings of ACL 2002, pp. 239–246 (2002)
4. Gormley, M.R., Mitchell, M., Durme, B.V., et al.: Low-resource semantic role labeling. In: Proceedings of the 52nd Annual Meeting of the Association for Computational Linguistics, pp. 1177–1187 (2014)
5. Liu, T., Che, W.X., Li, S.: Semantic role labeling system using maximum entropy classifier. In: Proceedings of CoNLL 2005, pp. 189–192 (2005)
6. Marcus, M.P., Santorini, B., Marcinkiewicz, M.A.: Building a large annotated corpus of English: the Penn Treebank. Comput. Linguist. **19**, 313–330 (1993)
7. Morarescu, P., Bejan, C., Harabagiu, S.: Shallow semantics for relation extraction. In: Proceedings of IJCAI 2005 (2005)
8. Palmer, M., Gildea, D., Kingsbury, P.: The Proposition bank: an annotated corpus of semantic roles. Comput. Linguist. **31**(1), 71–106 (2005)

9. Pradhan, S., Ward, W., Martin, J.H.: Towards robust semantic role labeling. Comput. Linguist. **34**(2), 289–310 (2008)

10. Pradhan, S., Hacioglu, K., Krugler, V., et al.: Support vector learning for semantic argument classification. Mach. Learn. J. **60**(3), 11–39 (2005)

11. Tateisi, Y., Yakushiji, A., Ohta, T., Tsujii, J.: Syntax annotation for the GENIA corpus. In: Proceedings of the 2nd International Joint Conference on Natural Language Processing (IJCNLP-2005) (2005)

12. Tsai, R.T., Chou, W.C., Lin, Y.C., et al.: BIOSMILE: a semantic role labeling system for biomedical verbs using a maximum-entropy model with automatically generated template features. BMC Bioinformatics **8**, 325 (2007)

13. Tsai, R.T., Lai, P.T.: A resource-saving collective approach to biomedical semantic role labeling. BMC Bioinformatics **15**, 160 (2014)

14. Xue, N., Palmer, M.: Calibrating features for semantic role labeling. In: Proceedings of the EMNLP 2004, pp. 88–94 (2004)

15. Xue, N., Palmer, M.: Automatic semantic role labeling for chinese verbs. In: Proceedings of IJCAI2005, Edinburgh, UK, pp. 1160–1165 (2005)

Automatic Detection of Yeast and Pseudohyphal Form Cells in the Human Pathogen Candida Glabrata

Luis Frazao[1], Rui Santos[2], Miguel Cacho Teixeira[2], Nipon Theera-Umpon[1,3(✉)], and Sansanee Auephanwiriyakul[1,4]

[1] Biomedical Engineering Center, Chiang Mai University,
Chiang Mai 50200, Thailand
{luis.frazao,nipon.t}@cmu.ac.th

[2] Department of Bioengineering and IBB - Institute for Bioengineering and Biosciences,
Instituto Superior Tecnico, Universidade de Lisboa, 1049-001 Lisbon, Portugal
rui.ramos.santos@tecnico.ulisboa.pt, mnpct@ist.utl.pt

[3] Department of Electrical Engineering, Faculty of Engineering, Chiang Mai University,
Chiang Mai 50200, Thailand

[4] Department of Computer Engineering, Faculty of Engineering, Chiang Mai University,
Chiang Mai 50200, Thailand

Abstract. The morphological switching shown by Candida glabrata between oval-shaped budding yeast cells and elongated pseudohyphal growth structures has been found to be related to its ability to undergo invasive growth, and thus, predicted to affect virulence. Therefore, the morphological analysis of C. glabrata cell cultures is a procedure of clinical relevance. In the present study, a pioneering algorithm was developed to automatically detect both yeast and pseudohyphal form structures from a database of 82 phase contrast microscopy images of C. glabrata cell cultures. The algorithm produced robust results, despite some limitations in the quality of the images, having detected correctly 84.56 % and 51.94 % of the yeast form cells and pseudohyphal structures, respectively. Future work should focus on improving these results by incorporating the analysis of cell nuclei or septa positions, extracted from fluorescence images.

Keywords: Cell detection · Candida glabrata · Pseudohyphae · Yeast form cells

1 Introduction

Systemic fungal infections are a problem of increasing clinical significance, since the extensive use of antifungal drugs, both as treatment and prophylaxis, has led to a huge increase in the number of intrinsically resistant infections with fungal pathogens [1, 2]. This is particularly true for the non-*albicans Candida* species *Candida glabrata*. *C. glabrata* arose, in the past few decades, as the second most frequent pathogenic yeast, after *C. albicans*, in mucosal and invasive human fungal infections, representing 15–20 % of all infections caused by *Candida* species [3]. Unlike *C. albicans*, which is known to exhibit dimorphic growth, that is, switching from yeast to hyphal growth, *C. glabrata* is only able to switch between yeast and what is called pseudohyphal growth.

© Springer International Publishing Switzerland 2015
D.-S. Huang and K. Han (Eds.): ICIC 2015, Part III, LNAI 9227, pp. 669–678, 2015.
DOI: 10.1007/978-3-319-22053-6_69

This morphological switching has been found to underlie *C. glabrata* ability to undergo invasive growth [4], being thus an important feature to be analysed in lab strains and clinical isolates.

The purpose of the present study is to develop an algorithm that will automatically detect the cells present in phase contrast microscopy images of *C. glabrata* cell cultures. Some of such images are shown in Fig. 1. One can see here that there are essentially two kinds of structures in each image: (1) approximately circular ones, known as yeast form cells; (2) elongated filaments, known as pseudohyphae, composed of several cells with no distinct individual boundary. By detecting the elongated structures, our algorithm helps to find the regions where cells showing a pseudohyphal growth can probably be found. And by detecting the remaining oval-shaped cells, the algorithm is an indispensable aid in the calculation of the ratio between the cells showing and not showing a pseudohyphal growth, which is a measure commonly used by human analysts.

To our knowledge, this is the first time that such automatic analysis of microscopy images of *Candida* cells has been developed. Despite the inherent limitations of the quality of phase contrast microscopy images, the results obtained were impressively good, mainly for the detection of yeast form cells, since 84.56 % of these structures were detected. As for the elongated filaments, the algorithm detected approximately 51.94 % of these structures.

The remainder of this paper is organized as follows: First, the methodology that was developed to detect both yeast form and elongated form cells is described in Sect. 2. Then, the results obtained are described and discussed in Sect. 3. The final conclusions are drawn in Sect. 4.

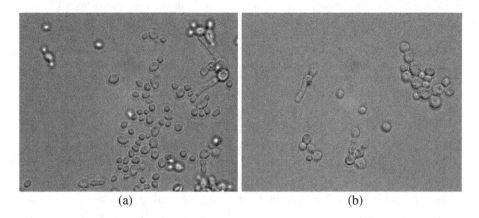

(a)   (b)

Fig. 1. Some images to be analyzed

2 Methodology

In the present work, a software was developed to automatically detect *C. glabrata* cells both in the form of budding yeast (approximately circular) and pseudohyphae (elongated structures composed of several aggregated cells, linked at the mother-to-daughter septation locus) from microscopy images of *C. glabrata* cell cultures. The algorithm was developed in MATLAB and a database of 82 images were analyzed. In order to obtain the used images, *C. glabrata* KUE100 wild-type cells were cultivated over-night in YEPD medium, containing per liter, 20 g of glucose (Merck), 20 g of peptone (Liofilmchem) and 10 g of yeast-extract (VWR Chemicals), with agitation (250 rpm). An inoculum was then prepared with an initial $OD_{600nm} = 0.05 \pm 0.0025$ in YEPD medium supplemented with 0.5 % (v/v) of isoamyl alcohol, to induce pseudohyphae formation [5], and cells were cultivated for 48 ± 1 h. After the 48 h of incubation, 5 μL of the cell suspension was observed by optical microscopy (Zeiss Axioplan Microscope, Carl Zeiss Microimaging). Images were collected with a coupled CoolSnap fx camera (Roper Scientific Photometrics).

The procedure involved in the detection of yeast form cells and pseudohyphal structures is described next.

2.1 Detection of Yeast Form Cells

Image normalization was initially performed such that the range of pixel intensities was the same for all images. One can see (e.g. in Fig. 1a) that sometimes there are high intensity blobs present in the images, which correspond to out of focus cells. Still, it is important to detect these cells along with the focused structures. A well known technique to detect bright blobs was used to detect these cells, namely the filtering of each image with a negative Laplacian of Gaussian (LoG). A scale of $\sigma = 5$ proved to produce the best results. Next, a top hat filtering was performed, using a disk structuring element with a radius of 12. The resulting image was then thresholded with a value chosen empirically. The binary components on the image boundaries were discarded from further analysis. An area opening operation was then performed, removing the components with an area below a certain value chosen empirically. Afterwards, the holes that could be present were removed with a morphological filling operation. Each object was then reduced to a point using morphological shrinking. This point was considered to be the center of the circle corresponding to the out of focus cell. The radius was then estimated using the following procedure applied to the LoG filter response image obtained before: starting from each point obtained in the last step, an array of consecutive pixels with a certain length was analyzed for each of four directions — up, down, left and right; if, for all the four directions, there is a change of sign between two consecutive pixels, namely from a positive valued to a zero or negative valued pixel, the minimum of the four distances from the center point for which such change of sign was observed was considered to be the radius of an out of focus cell; otherwise, if there is at least one direction for which no change of sign was observed, the component was discarded. Finally, to avoid estimating more than one circle for the same cell, intersections of circles above a certain amount were avoided. Namely, a circle was discarded if its center was inside a previously detected circle.

After having considered the out of focus cells, the focused yeast form cells were detected. To avoid the influence of the out of focus cells on further analysis, the binary area of their circles was dilated with a disk structuring element with radius 5, and the corresponding intensities were then replaced by the mean intensity value of the entire image. Each image was then filtered with a positive LoG with $\sigma = 2$. This method tended to highlight the cell boundary region corresponding to the cell wall. However, several discontinuities were present in these highlighted structures. Therefore, the method developed in [6] was applied in order to remove these discontinuities. This method was originally developed to detect blood vessels in medical images, using a multiscale second order structure approach. We used the same scales and parameters' values as in the original vesselness application [6] to detect bright "vessels" in the positive LoG filter response image. The results were then thresholded with a value chosen empirically. The components lying on the image boundaries were then discarded from further analysis. Afterwards, a parallel thinning algorithm [7] was applied in order to reduce each component to its skeleton, by removing pixels on the boundaries but not allowing objects to break apart. From the resulting skeletonized binary image, we applied a Circular Hough Transform (CHT) based algorithm, that uses a Phase Coding technique to compute the accumulator array [8, 9]. Circles with radius between 10 and 25 pixels were detected, and the sensitivity of the method was chosen empirically. Excessive intersection between circles was avoided by discarding a circle if its center was inside a previously detected circle.

After applying this sequence of steps, there were still a considerable amount of focused yeast form cells that were not detected. Therefore, an additional complementary approach was performed. One can easily notice from Fig. 1 that the cell wall region often appears as darker than its surroundings. Based on this characteristic, a sequence of steps similar to the one described above was used to detect focused yeast form cells, but now the vesselness technique [6] was applied to detect dark "vessels" (instead of LoG image filtering followed by bright "vessel" detection, as done before). From the set of detected circles, each circle with its center inside a previously detected circle was discarded.

2.2 Detection of Pseudohyphal Structures

After having detected the approximately circular structures, corresponding to yeast form cells, the next step in our algorithm was to detect the elongated structures, which contained the pseudohyphal cells.

We started by increasing the contrast of the images by mapping the intensity values such that 1 % of data would be saturated at low and high intensities. Next, a top hat filtering operation was performed, using a disk structuring element with radius of 10. From the resulting image, we were interested in the long structures, brighter than the background, with an arbitrary orientation. Therefore, a filtering operation was performed using a bank of the second order derivative (in x-direction) of the Gaussian filter, rotated at angles between 0° and 180° with a 15° interval. The scales that were used $\sigma_x = 5$ and $\sigma_y = 15$. Then, the response was maximized for each pixel position. The vesselness technique [6] was then used to detect bright "vessels", using the same parameters as before (Sect. 2.1). The response was thresholded with a value chosen empirically. Then, the components of the binary image were reduced to their skeletons. Since this tended to produce excessively long skeletons, linking structures that were actually not related to each other, each skeleton component was then "cut" at its branch points, unless the branch point was close enough to an endpoint (maximum 3 pixels apart). Next, an area opening operation was performed, discarding the components with an area less than a certain value chosen empirically. Pixels inside the area of previously detected circles were discarded. A second area opening operation was performed, using again a threshold chosen empirically. Then, the components that were too close to the image boundaries were discarded.

Some of the remaining lines did not correspond to cellular elongated structures. For instance, sometimes they were on the extracellular gap between clusters of yeast form cells. Thus, in order to remove these lines, the following approach was performed: when travelling along the longitudinal direction of a line, select various pixels approximately at an equal distance from each other; for the first pixel, search for the 2 pixels that lie in the transverse direction to the line (in opposite directions) and at a same distance from the line; if both pixels are inside circles previously detected as yeast form cells, the line is discarded; otherwise, search for other 2 pixels, farther apart from the line until a maximum transverse distance is reached; repeat the same procedure for the next pixel along the longitudinal direction of the line, until the line is discarded or until there are no more pixels to analyze. The distance between pixels along the longitudinal and transverse directions, as well as the maximum transverse searching distance, were chosen empirically.

Afterwards, a component (line) was discarded if any of its pixels was outside the cellular region. To find this region, the following approach was performed: calculate the edge indicator function as described in [10], using a Gaussian kernel with a standard deviation $\sigma = 1.5$; apply a morphological opening operation using a disk structuring element with a radius of 3; apply a morphological closing operation using a disk structuring element with a radius of 10; threshold the image with a value chosen empirically.

After having obtained a set of lines that were believed to be located inside and along the elongated structures of interest, a segmentation technique was applied in order to identify the area of these structures. Specifically, the distance regularized level set evolution (DRLSE) [10] segmentation technique was performed, since it can be initialized by a small area inside the elongated structures (say, the lines detected before) and evolve outwardly towards the boundary of the filaments. However, before applying this technique, an image sharpening was performed in order to increase the contrast on the edges of the structures, and thus avoiding boundary leakage during the segmentation procedure. The image was thus sharpened with an unsharp masking technique using a Gaussian kernel with a standard deviation $\sigma = 2$. Besides, there were often inhomogeneities in intensity values inside the elongated structures to be detected, which could cause the segmentation algorithm to stop before reaching the cellular boundary. Therefore, these inhomogeneities were attenuated by performing an anisotropic non-linear diffusion filtering [11, 12]. The DRLSE segmentation [10] was then performed. The segmentation algorithm was run for each detected line. The area of the plateau of the initial binary step function corresponded to the line dilated with a disk structuring element with radius of 2. The value of the plateau was set to 10. Using the double-well potential approach, the following parameters were set up: the coefficient of the weighted length term $\lambda = 5$, the parameter specifying the width of the Dirac Delta function $\varepsilon = 1.5$, the coefficient of the weighted area term $\alpha = -5$, the coefficient of the distance regularization term $\mu = 0.04$ and the time step $\Delta t = 5$. The number of iterations using this set of parameters were 500, and 10 additional iterations were run with $\alpha = 0$ in order to refine the contour (as suggested in [10]). Finally, an attempt to eliminate segmented areas not corresponding to any cellular content was performed, by discarding every final segmented area that contained a minimum amount of pixels (chosen empirically) outside the cellular region.

3   Results

The high variability of shape, resolution and contrast between structures in different microscopy images made it hard to choose parameters that produced good results globally. For instance, sometimes the cell border was not well defined (low contrast), which originated missed detections of both yeast form and elongated form cells. However, some interesting results were obtained for both types of detection.

3.1   Detection of Yeast Form Cells

The detection of oval-shaped yeast form cells produced good and robust results. From the total 4021 yeast form cells present in the database, the algorithm detected correctly 3400, i.e. 84.56 %. Some results are shown in Fig. 2. Even for cells included in dense agglomerates (Fig. 2a and b), the algorithm was often able to identify the individual structures.

Nevertheless, some limitations were observed. For instance, some isolated circular cells were missed (Fig. 2c) and groups of aggregated cells were sometimes considered as individual cells. Besides, the algorithm produced some false detections, such as in non-cellular regions or by tracing more than one circle for the same cell. However, these missed and false detections were not seen as a major problem, since the goal of the human analyst is often to calculate the percentage of cells in an image that show pseudohyphal growth in different cell populations. Thus, missing or detecting some circular cells in excess may change absolute values but will not alter significantly the relative level of pseudohyphal cells observed in different cell cultures.

A major limitation of the algorithm is, however, when it identified some portion of an elongated structure (mainly its extremities), or even the entire filament, as a group of yeast form cells (Fig. 2d). This leads the human analyst to miss a region of possible pseudohyphal growth, by assuming that only yeast form cells are present.

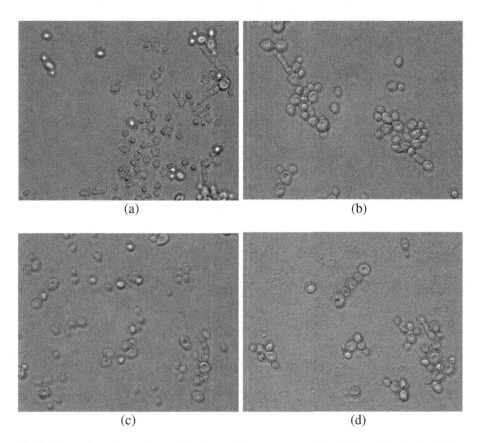

(a)   (b)

(c)   (d)

Fig. 2. Detection of yeast form cells. Detected structures are identified as blue circles (Color figure online).

3.2 Detection of Pseudohyphal Structures

Some results of the detection of elongated structures are shown in Fig. 3. The detection was very accurate for some images, where a segmentation of the entire area of interest was observed (Fig. 3a). As for several other images, only part of the area of the filaments was segmented, mainly due to intensity inhomogeneities inside these structures and a low number of iterations in the DRLSE algorithm (Fig. 3b).

Some major limitations were observed, such as missing the detection of elongated structures (Fig. 3c). This was sometimes due to excessive boundary leakage during the segmentation process, because of the low contrast in cell border regions, causing the algorithm to discard the final segmented area.

Still, from the 258 elongated structures that were observed in the database, the algorithm identified correctly 134, i.e. 51.94 %, either highlighting part of their area, the entire area or slightly more than their area.

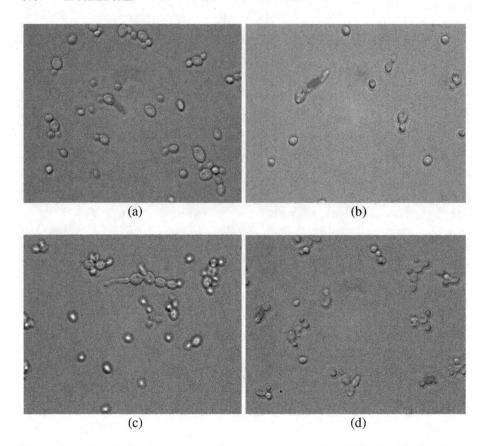

Fig. 3. Detection of elongated structures. Segmented areas are identified with red color (Color figure online).

The algorithm also showed several false detections. For instance, it tended to detect some yeast form cells that were missed in the previous steps (Fig. 3d). Besides, non-cellular content was sometimes wrongly identified as elongated structures.

These limitations in the detection of filaments could be attenuated by the comparative parallel study of fluorescence images highlighting the cell nuclei and/or cell septa positions. This study would allow to discard segmented areas with no cellular content and to count the number of pseudohyphal cells inside an elongated structure.

Finally, the analysis of the performance of any software should take into account its processing time. This was another limitation of our algorithm, mainly during the detection of elongated structures, which took in average about 7 min for each image, using a 3.30 GHz Intel Core i5-2500K processor (while only about 9 s were needed to detect circular cells). One would expect this limitation, since it takes a considerable amount of processing time to run the DRLSE algorithm, and thus, running it several times for an image proved to be computationally expensive.

4   Conclusion

The pioneering fully automatic tool that was developed in the present study produced useful and robust results, despite some limitations in the quality of the microscopy images, which we believe are representative of the average images to be analysed by the potential users of such a software.

To overcome these limitations, future work should focus on incorporating other information, such as the cell nuclei or septa positions, extracted from fluorescence images.

Despite the limitations shown by the algorithm, it can still be used to aid in tedious tasks such as identifying and counting the number of *C. glabrata* cells present in an image, and as a method for the preliminary classification of cell structures, to be followed by manual completion and correction.

Furthermore, we believe that this algorithm could be incorporated as a first stage of an automatic system that would further extract certain features from each cell and classify it as showing or not pseudohyphal differentiation.

References

1. Fidel Jr., P.L., Vazquez, J.A., Sobel, J.D.: Candida glabrata: review of epidemiology, pathogenesis, and clinical disease with comparison to C. albicans. Clin. Microbiol. Rev. **12**(1), 80–96 (1999)
2. Mishra, N., Prasad, T., Sharma, N., Payasi, A., Prasad, R., Gupta, D., Singh, R.: Pathogenicity and drug resistance in candida albicans and other yeast species. Acta Microbiol. Immunol. Hung. **54**(3), 201–235 (2007)
3. Perlroth, J., Choi, B., Spellberg, B.: Nosocomial fungal infections: epidemiology, diagnosis, and treatment. Med. Mycol. **45**(4), 321–346 (2007)
4. Csank, C., Haynes, K.: Candida glabrata displays pseudohyphal growth. FEMS Microbiol. Lett. **189**(1), 115–120 (2000)
5. Lorenz, M., Cutler, N., Heitman, J.: Characterization of alcohol-induced filamentous growth in saccharomyces cerevisiae. Mol. Biol. Cell **11**(1), 183–199 (2000)
6. Frangi, A.F., Niessen, W.J., Vincken, K.L., Viergever, M.A.: Multiscale vessel enhancement filtering. In: Wells, W.M., Colchester, A.C.F., Delp, S.L. (eds.) MICCAI 1998. LNCS, vol. 1496, pp. 130–137. Springer, Heidelberg (1998)
7. Lam, L., Lee, S.-W., Suen, C.Y.: Thinning methodologies - a comprehensive survey. IEEE Trans. Pattern Anal. Mach. Intell. **14**(9), 869–885 (1992)
8. Davies, E.R.: Machine Vision: Theory, Algorithms, Practicalities. Morgan Kaufmann Publishers Inc., San Francisco (2004)
9. Atherton, T.J., Kerbyson, D.J.: Using phase to represent radius in the coherent circle hough transform. In: IEEE Colloquium on Hough Transforms, pp. 5–10. IET (1993)
10. Li, C., Xu, C., Gui, C., Fox, M.D.: Distance regularized level set evolution and its application to image segmentation. IEEE Trans. Image Process. **19**(12), 3243–3254 (2010)

11. Kroon, D., Slump, C.H.: Coherence filtering to enhance the mandibular canal in cone-beam CT data. In: 4th Annual Symposium of the IEEE-EMBS Benelux Chapter, pp. 41–44. IEEE-EMBS Benelux Chapter, Enschede, The Netherlands (2009)
12. Kroon, D.-J., Slump, C.H., Maal, T.J.J.: Optimized anisotropic rotational invariant diffusion scheme on cone-beam CT. In: Jiang, T., Navab, N., Pluim, J.P.W., Viergever, M.A. (eds.) MICCAI 2010, Part III. LNCS, vol. 6363, pp. 221–228. Springer, Heidelberg (2010)

Semi-supervised Feature Extraction
for RNA-Seq Data Analysis

Jin-Xing Liu[1,2], Yong Xu[1(✉)], Ying-Lian Gao[3], Dong Wang[2],
Chun-Hou Zheng[4], and Jun-Liang Shang[2]

[1] Bio-Computing Research Center, Shenzhen Graduate School,
Harbin Institute of Technology, Shenzhen, China
yongxu@ymail.com
[2] School of Information Science and Engineering,
Qufu Normal University, Rizhao, China
{sdcavell,dongwshark}@126.com,
shangjunliang110@163.com
[3] Library of Qufu Normal University, Qufu Normal University, Rizhao, China
yinliangao@126.com
[4] School of Mechanical Engineering and Automation,
Anhui University, Hefei, China
zhengch99@126.com

Abstract. It is of urgency to effectively identify differentially expressed genes from RNA-Seq data. In this paper, we propose a novel method, semi-supervised feature extraction, to analyze RNA-Seq data. Our scheme is shown as follows. Firstly, we construct a graph Laplacian matrix and refine it by using labeled samples. Secondly, we find semi-supervised optimal maps by solving a generalized eigenvalue problem. Thirdly, we solve an optimal problem via joint $L_{2,1}$-norm constraint to obtain a projection matrix. Finally, we identify differentially expressed genes based on the projection matrix. The results on real RNA-Seq data sets demonstrate the feasibility and effectiveness of our method.

Keywords: Feature extraction · $L_{2,1}$-norm constraint · Spectral regression · RNA-Seq data analysis

1 Introduction

An important research topic of analyzing RNA-Seq data is to identify genes that are located in a position to differentiate across treatments/conditions. These genes are known as differentially expressed genes (DEGs).

The feature extraction methods have been proposed to identify the DEGs. Among feature extraction methods, principal component analysis (PCA) [1] is the most frequently used one. Recently, sparse methods have sprung up like mushrooms. L_1-norm (LASSO) [2] is the most well-known sparse method. Journée et al. used L_1-norm constraint to propose a Sparse PCA (SPCA) [3] which was used to identify characteristic genes by Liu et al. [4]. Witten et al. used L_1-norm constraint to propose a penalized matrix decomposition (PMD) [5] which was used to discover some molecular patterns [6].

© Springer International Publishing Switzerland 2015
D.-S. Huang and K. Han (Eds.): ICIC 2015, Part III, LNAI 9227, pp. 679–685, 2015.
DOI: 10.1007/978-3-319-22053-6_70

In machine learning and pattern recognition, semi-supervised learning methods become more and more attention. Because of its simplicity, k-nearest-neighbor (kNN) graph Laplacian method is widely used and many related methods have been put forward for dimension reduction [7]. In [8], Cai et al. proposed a spectral regression (SR) method by using kNN graph Laplacian.

In order to benefit from both SR and $L_{2,1}$-norm, we suggest a semi-supervised feature extraction (SFE) to identify DEGs. Firstly, we refine the graph construction method based on Laplacian matrix [8] to obtain semi-supervised optimal maps. Secondly, we solve an optimal problem via joint $L_{2,1}$-norm constraint to get a projection matrix. Finally, we identify DEGs based on the projection matrix.

The main contributions of our work are described as follows. Firstly, it proposes a method to refine the Laplacian graph by using labeled samples. Secondly, the $L_{2,1}$-norm constraint is used to reduce outliers and noises.

2   Methodology

In this section, we propose a semi-supervised feature extraction (SFE).

2.1   Notations and Definitions

Given a $p \times n$ matrix $\mathbf{M} = (m_{ij})$ whose i-th row and j-th column are denoted by $\mathbf{m}^i$ and $\mathbf{m}_j$, respectively. The L_2-norm and L_1-norm of matrix are defined as $\|\mathbf{M}\|_2^2 = \sum_{i=1}^{p} \|\mathbf{m}^i\|_2^2$ and $\|\mathbf{M}\|_1 = \sum_{i=1}^{p} \|\mathbf{m}^i\|_1$, respectively. The $L_{2,1}$-norm of matrix is defined as $\|\mathbf{M}\|_{2,1} = \sum_{i=1}^{p} \|\mathbf{m}^i\|_2$. Let $\mathbf{X}_l = [x_1, \cdots, x_l] \in R^{p \times l}$ and $\mathbf{X}_u = [x_1, \cdots, x_u] \in R^{p \times u}$ be labeled and unlabelled samples, respectively. Without loss of generality, let $n = l + u$ be the total number of samples and p be the number of features. These samples belong to c classes and let l_k be the number of labeled samples in k-th class ($\sum_{k=1}^{c} l_k = l$). Let $\mathbf{Y} = [y_1, \cdots, y_l]$ be the labels of samples.

2.2   The Graph Construction

We introduce the method of graph construction in this subsection as follows.

Step 1: Let $G(\mathbf{V}, \mathbf{W})$ denote a graph model with n vertices. The i-th vertex v_i corresponds to the samples x_i.
Step 2: Construct the adjacency graph. Similarly to [9], the weight matrix $\mathbf{W}$ can be defined as follows:

$$w_{ij} = \begin{cases} \exp\left(-\|x_i - x_j\|^2 / \tau\right) & x_i \text{ and } x_j \text{ are k nearest neighbors,} \\ 0 & \text{otherwise,} \end{cases} \tag{1}$$

Step 3: Refine the graph by using the sample labels. For $\mathbf{X}_l$, the $\mathbf{W}$ can be defined as follows.

$$w_{ij} = \begin{cases} 1/l_k & \text{if } x_i \text{ and } x_j \text{ have the same label,} \\ 0 & \text{if } x_i \text{ and } x_j \text{ have the different labels,} \end{cases} \tag{2}$$

where l_k is the number of labeled samples in k-th class. We compute the graph Laplacian matrix $\mathbf{L} = \mathbf{D} - \mathbf{W}$, where $\mathbf{D}$ is a diagonal matrix with the i-th diagonal element as $d_{ii} = \sum_{j=1}^n w_{ij}$.

Then, we solve the following eigenvalue problem to get $\mathbf{Z}$.

$$\mathbf{Lz} = \lambda \mathbf{Dz}, \tag{3}$$

where $\mathbf{z}$ is the eigenvector.

2.3   Solving the Optimal Problem via Joint $L_{2,1}$-norm Constraint

In [8], Cai et al. proposed a spectral regression (SR) framework for subspace learning. Find $\mathbf{A}$ which satisfies

$$\mathbf{X}^T \mathbf{A} = \mathbf{Z}. \tag{4}$$

The most popular way to solve this problem is to impose a penalty on the L_2-norm of $\mathbf{a}$:

$$\mathbf{a} = \arg\min_{\mathbf{a}} \left(\sum_{i=1}^c \left\| \mathbf{a}_i^T \mathbf{x}_i - \mathbf{z}_i \right\|_2^2 + \gamma \|\mathbf{a}_i\|_2^2 \right). \tag{5}$$

In our paper, we will replace the L_2-norm regularization in Eq. (5) with an $L_{2,1}$-norm to extract features across all data points with row sparsity. In addition, we will impose $L_{2,1}$-norm on the regression loss function and/or regularization.

Situation 1. Due to the row sparsity of $L_{2,1}$-norm, the optimal problem in Eq. (5) becomes:

$$\mathbf{A} = \arg\min_{\mathbf{A}} \left(\left\| \mathbf{X}^T \mathbf{A} - \mathbf{Z} \right\|_2^2 + \gamma \|\mathbf{A}\|_{2,1} \right). \tag{6}$$

Solving the problem in Eq. (6), we can get

$$\mathbf{A} = \mathbf{\Psi}^{-1} \mathbf{X} \left(\mathbf{X} \mathbf{\Psi}^{-1} \mathbf{X}^T + \gamma \mathbf{I} \right)^{-1} \mathbf{Z}, \tag{7}$$

where $\mathbf{\Psi}^{-1}$ is a diagonal matrix with the i-th diagonal element equal to $\psi_{ii}^{-1} = 2\|\mathbf{a}^i\|_2$.

Situation 2. If we impose $L_{2,1}$-norm on both the regression loss function and regularization simultaneously, the solution of Eq. (5) can be obtained by solving the following optimal problem:

$$\mathbf{A} = \arg\min_{\mathbf{A}} \left(\left\| \mathbf{X}^T \mathbf{A} - \mathbf{Z} \right\|_{2,1} + \gamma \|\mathbf{A}\|_{2,1} \right). \tag{8}$$

The problem in Eq. (8) is equivalent to

$$(\mathbf{\Delta}, \mathbf{A}) = \arg\min_{\mathbf{A}, \mathbf{\Delta}} \left(\|\mathbf{\Delta}\|_{2,1} + \|\mathbf{A}\|_{2,1} \right) \quad \text{s.t.} \quad \mathbf{X}^T \mathbf{A} + \gamma \mathbf{\Delta} = \mathbf{Z}. \tag{9}$$

According to the algorithm proposed by Nie et al. [10], we can obtain

$$\mathbf{U} = \mathbf{G}^{-1}\mathbf{\Theta}^T \left(\mathbf{\Theta}\mathbf{G}^{-1}\mathbf{\Theta}^T \right)^{-1} \mathbf{Z}, \tag{10}$$

where $\mathbf{\Theta} = \begin{bmatrix} \mathbf{X}^T & \gamma\mathbf{I} \end{bmatrix} \in R^{p \times k}$, $\mathbf{U} = \begin{bmatrix} \mathbf{A} \\ \mathbf{\Delta} \end{bmatrix} \in R^{k \times c}$, and the i-th element of diagonal matrix $\mathbf{G}^{-1}$ can be defined as $2\|\mathbf{u}^i\|_2$.

2.4   The Algorithm

The algorithm of our method is summarized as follows.

```
Algorithm 1. The algorithm of our method.
Input: Data matrix [X_l,X_u] ∈ R^{p×n}
    Class label Vector: Y ∈ N^{l×1}
    Sparsity-controlling parameter: γ ≥ 0
Output: Projection matrix: A ∈ R^{p×c}
Begin
    Construct the adjacency graph by Eq.(1).
    Refine the graph from the sample labels according to
Eq.(2).
    Compute the graph Laplacian matrix L = D - W.
    Solve the eigenvalue problem in Eq.(3) to get Z.
    Solve optimization problems in Eq.(6) or Eq.(8) to get
    a projection matrix A.
End
```

3   Results and Discussion

We apply our feature extraction methods (called Situation 1 and 2 as L21SFE and DL21SFE, respectively) to RNA-Seq data set to validate their performances. For performance comparisons, we use our methods to extract differentially expressed genes on real RNA-Seq data set. We compare our methods with several popularly used methods in bioinformatics, such as Sparse LDA(SLDA) [11], SPCA [3], PMD [5] and PS [12]. The publically available RNA-Seq data set, MAQC [13], is used to evaluate our method. Here, our methods use the data set of BodyMap [14] as unlabelled one to improve their performances. Table 1 lists an overview of the data sets. Here, these sets of RNA-Seq count data are downloaded from http://bowtie-bio.sf.net/recount [15]. For a comparison, 500 genes are identified by each of these methods.

Table 1. An overview of the data sets

Data	Number of samples	Number of classes	Number of reads
MAQC	14	2	71,970,164
BodyMap	19	17	2,197,622,796

The MAQC data set contains 14 samples which are derived from Stratagene's human universal reference RNA and Ambion's human brain reference RNA.

We investigate the enrichment of functional annotations by inputting the 500 genes identified by these methods into the ToppFun [16] which is publicly available at http://toppgene.cchmc.org/enrichment.jsp. The p-value of ToppFun is set to 0.01 and additional parameters are used as default. The closely related GO terms found by ToppFun are listed in Table 2.

Table 2. The GO terms of genes identified by these methods on MAQC data

Rank	Name	L21SFE	DL21SFE	SLDA	SPCA	PMD	PS
1	Genes with HCP bearing histone H3K27me3 in MEF cells	8.60E-63	2.64E-34	1.41E-54	1.41E-54	1.84E-52	1.89E-21
2	Synaptic transmission	2.95E-51	2.68E-28	1.45E-37	1.21E-37	1.33E-30	4.17E-28
3	Set 'H3K27 bound': genes in human embryonic stem cells	7.30E-42	1.52E-24	5.12E-33	2.65E-32	4.46E-28	1.64E-18
4	Cell-cell signaling	6.44E-37	5.46E-21	9.84E-31	8.07E-31	4.14E-25	2.97E-22
5	Set 'Suz12 targets': genes in human embryonic stem cells	2.95E-33	9.83E-25	1.01E-27	4.85E-27	2.12E-22	3.52E-06

As listed in this table, the term of 'Genes with HCP bearing histone H3K27me3 in MEF cells' has the lowest p-value, so it is considered as the most probable enrichment term. Moreover, L21SFE, DL21SFE, SPCA, PMD and PS can identify genes with p-value 8.60E-63, 2.64E-34, 1.41E-54, 1.41E-54, 1.84E-52 and 1.89E-21, respectively. This table also lists some other significant terms.

4   Conclusions

In this paper, we propose a semi-supervised feature extraction method to identify differentially expressed genes. Firstly, we refine the graph construction method to obtain the semi-supervised eigenmap. Then, we solve an optimal problem via joint $L_{2,1}$-norm constraint to get a projection matrix. Finally, we identify differentially expression genes based on the projection matrix. Our graph construction method can make full use of a large number of unlabelled samples. Furthermore, our methods can reduce the impact of noises and outliers by the use of $L_{2,1}$-norm constraint and produce more precise results.

Acknowledgements. This work was supported in part by the NSFC under grant Nos. 61370163 and 61272339; China Postdoctoral Science Foundation funded project, No. 2014M560264; Shandong Provincial Natural Science Foundation, under grant Nos. ZR2013FL016 and BS2014DX004; Shenzhen Municipal Science and Technology Innovation Council (Nos. JCYJ20140417172417174, CXZZ20140904154910774 and JCYJ20140904154645958).

References

1. Jolliffe, I.T.: Principal Component Analysis. Springer, New York (2002)
2. Tibshirani, R.: Regression shrinkage and selection via the lasso. J. Roy. Stat. Soc. Ser. B (Methodol.) **58**, 267–288 (1996)
3. Journée, M., Nesterov, Y., Richtarik, P., Sepulchre, R.: Generalized power method for sparse principal component analysis. J. Mach. Learn. Res. **11**, 517–553 (2010)
4. Liu, J.-X., Xu, Y., Zheng, C.-H., Wang, Y., Yang, J.-Y.: Characteristic gene selection via weighting principal components by singular values. PLoS ONE **7**, e38873 (2012)
5. Witten, D.M., Tibshirani, R., Hastie, T.: A penalized matrix decomposition, with applications to sparse principal components and canonical correlation analysis. Biostatistics **10**, 515–534 (2009)
6. Zheng, C.H., Zhang, L., Ng, V., Shiu, C.K., Huang, D.S.: Molecular pattern discovery based on penalized matrix decomposition. IEEE/ACM Trans. Comput. Biol. Bioinf. **8**, 1592–1603 (2011)
7. France, S.L., Douglas Carroll, J., Xiong, H.: Distance metrics for high dimensional nearest neighborhood recovery: compression and normalization. Inf. Sci. **184**, 92–110 (2012)
8. Cai, D., He, X., Han, J.: Spectral regression for efficient regularized subspace learning. In: IEEE 11th International Conference on Computer Vision, 2007. ICCV 2007, pp. 1–8 (2007)
9. Belkin, M., Niyogi, P.: Laplacian eigenmaps and spectral techniques for embedding and clustering. In: NIPS, pp. 585–591 (2001)
10. Nie, F., Huang, H., Cai, X., Ding, C.: Efficient and robust feature selection via joint l2, 1-norms minimization. Adv. Neural Inf. Process. Syst. **23**, 1813–1821 (2010)
11. Cai, D., He, X., Han, J.: SRDA: an efficient algorithm for large-scale discriminant analysis. IEEE Trans. Knowl. Data Eng. **20**, 1–12 (2008)
12. Li, J., Witten, D.M., Johnstone, I.M., Tibshirani, R.: Normalization, testing, and false discovery rate estimation for RNA-sequencing data. Biostatistics **13**, 523–538 (2012)

13. Bullard, J.H., Purdom, E., Hansen, K.D., Dudoit, S.: Evaluation of statistical methods for normalization and differential expression in mRNA-Seq experiments. BMC Bioinf. **11**, 94 (2010)
14. Tonner, P., Srinivasasainagendra, V., Zhang, S., Zhi, D.: Detecting transcription of ribosomal protein pseudogenes in diverse human tissues from RNA-seq data. BMC Genomics **13**, 412 (2012)
15. Frazee, A., Langmead, B., Leek, J.: ReCount: a multi-experiment resource of analysis-ready RNA-seq gene count datasets. BMC Bioinf. **12**, 449 (2011)
16. Chen, J., Bardes, E.E., Aronow, B.J., Jegga, A.G.: ToppGene suite for gene list enrichment analysis and candidate gene prioritization. Nucleic Acids Res. **37**, W305–W311 (2009)

Compound Identification Using Random Projection for Gas Chromatography-Mass Spectrometry Data

Li-Li Cao, Zhi-Shui Zhang, Peng Chen, and Jun Zhang[✉]

School of Electronic Engineering and Automation, Anhui University,
Hefei 230601, Anhui, China
wwwzhangjun@163.com

Abstract. In general, compound identification through library searching is performed on original mass spectral space by using some developed similarity measure. In this paper, the original mass spectral space was transformed into binary space by random projection. The hamming distance between query and reference the vector of binary space are calculated. The Mass Spectral Library 2005 (NIST05) main library is used as reference database and the replicate library is used as query data. With the number of binary digits increasing, the accuracy of compound identification is also increased. When the number set as 2076 bits, random projection achieve better identification performance than corresponding three similarity measures.

Keywords: Random projection · Mass spectrometry · Compound identification

1 Introduction

Gas chromatography coupled to mass spectrometry (GC-MS) is one of the most widely employed analytical technique for analyzing chemical or biological samples in many fields [1]. Many mass spectral similarity measures have been developed for the spectrum matching-based compound identification [2–5]. In recent year, Koo et al. introduced wavelet and Fourier transform based composite measures and showed that the proposed similarity scores perform better than the original dot product version [6]. Kim et al. use some statistical approach to find the optimal weight factors through a reference library for compound identification and declare that the widely used original weights is not the optimal one with the mass spectral library updated [7]. Kim et al. also proposed a composite similarity measure based on partial and semi-partial correlations [8]. Koo et al. compared the performance of several spectral similarity measures and found that the composite semi-partial correlation measure is the best one, but it is also the most time-consuming similarity measure [9].

During the past decades, in one side, the size of some commercial mass spectral libraries have been increased remarkably. If the original mass spectra data can be transformed into binary vector and keep their relative space distance, the library search will be speeded up. Local sensitive hashing (LSH) [10, 11] is a technique which use different hash function to map the original real vectors into binary vector and keep their

© Springer International Publishing Switzerland 2015
D.-S. Huang and K. Han (Eds.): ICIC 2015, Part III, LNAI 9227, pp. 686–692, 2015.
DOI: 10.1007/978-3-319-22053-6_71

relative distance with each other. Among all kinds LSH algorithm, random projection hashing algorithm [12] is designed to approximate the cosine distance between vectors. It is a very useful method known for its simplicity.

The objective of this work was to use random projection hashing method to map original mass spectra into binary vector. After conversion, by calculating hamming distance between two molecular binary vector, the computational time of library search will be reduced greatly. Furthermore, the mass spectra can be mapped to any number of binary digits according to requirement since random projection hashing is very flexible.

2   Materials and Methods

2.1   NIST EI Mass Spectral and Repetitive Library

Currently, two mass spectral libraries are available in existing commercial NIST/EPA/NIH Mass Spectral Library: the main and replicate library. In this study, Mass Spectral Library 2005 (NIST05) main library containing 163,195 mass spectra is used as reference database. The replicate library containing 23,290 mass spectra is used as query data. The chemical compounds are identified by Chemical Abstracts Service (CAS) registry number.

2.2   Spectral Similarity Measures

The mass spectra can be treated as vectors. Let $X = (x_1, x_2, \ldots, x_n)$ and $Y = (y_1, y_2, \ldots, y_n)$ be the query and reference mass spectra, respectively. Stein use the weighted spectra to calculate similarity [13], which are defined as follows:

$$X^w = (x_1^w, \ldots \ldots, x_n^w) \tag{1}$$

$$Y^w = (y_1^w, \ldots \ldots, y_n^w) \tag{2}$$

where $x_i^w = x_i^a \cdot m_i^b$ and $y_i^w = y_i^a \cdot m_i^b$, i = 1,...,n, m_i, is m/z value of the ith fragment ion, n is the number of mass-to-charge ratios considered for computation, a and b are the weight factors for peak intensity and m/z value, respectively. In this work, the weight factors are set as (a, b) = (0.53, 1.3).

2.2.1   Stein and Scott's Composite Similarity Measure (SS)
The similarity measure is derived from a weighted average of two items. First item is the cosine similarity. The second item is defined a ratio of peak pairs as follows:

$$S_R(X, Y) = \frac{1}{N_c} \sum_{i=2}^{N_c} (\frac{y_i}{y_{i-1}} \cdot \frac{x_{i-1}}{x_i})^n \tag{3}$$

where n = 1 if the first intensity ratio is less than the second, otherwise n = −1. xi, yi are non-zero intensities having common m/z value and N_c is the number of non-zero peaks in both the reference and the query spectra. The composite similarity is calculated as:

$$S_{ss}(X^w, Y^w) = \frac{N_x \cdot S_c(X^w, Y^w) + N_c \cdot S_R(X, Y)}{N_x + N_c} \tag{4}$$

where N_x is the number of non-zero peak intensities in the query spectrum.

2.2.2   Discrete Fourier and Wavelet Transform Composite Similarity Measure

The original spectral signal $X = (x_1, x_2, \ldots, x_n)$ is converted by discrete Fourier transform (DFT) into a new signal $X_f = (x_1^f, x_2^f, \ldots, x_n^f)$ as follows:

$$x_k^f = \sum_{d=1}^{n} x_d \exp(-\frac{2\pi i}{n} kd), \quad k = 1, \ldots, n \tag{5}$$

where i is imaginary unit and $\exp(-(2\pi i/n)kd)$ is a primitive nth root of unity. The original Eq. (5) can also be converted into the following formula:

$$x_k^f = \sum_{d=1}^{n} x_d \cos(-\frac{2\pi i}{n} kd) + i \sum_{d=1}^{n} x_d \sin(-\frac{2\pi i}{n} kd), \quad k = 1, \ldots, n \tag{6}$$

The converted signal consists of real and imaginary part. The real part of a signal is $X^{FR} = (x_1^{FR}, x_2^{FR}, \ldots, x_n^{FR})$, which can be calculated by the following equations:

$$x_k^{fr} = \sum_{d=1}^{n} x_d \cos(-\frac{2\pi i}{n} kd) \quad k = 1, \ldots, n \tag{7}$$

The discrete wavelet transform (DWT) convert a discrete time domain signal into a time-frequency domain signal. A signal pass through a low-pass filter and a high-pass filter, two subsets of signals are formed: approximations and details. The approximation and details are defined as follows:

$$\text{Approximation DWT: } x_k^A = \sum_{d=1}^{n} x_d g[2k - d - 1] \quad k = 1, \ldots, n \tag{8}$$

$$\text{detail DWT: } x_k^D = \sum_{d=1}^{n} x_d h[2k - d - 1] \quad k = 1, \ldots, n \tag{9}$$

where g and h are low-pass and the high-pass filter respectively. The detail DWT is converted into $Y^{FR} = (y_1^{FR}, y_2^{FR}, \ldots, y_n^{FR})$. In literature [14] only use real part in DFT and detail part in DWT to replace the second item of composite similarity measure defined in Eq. (4).

$$S_{DFT.F}(X^w, Y^w) = \frac{N_x \cdot S_c(X^w, Y^w) + N_c \cdot S_R(X^{FR}, Y^{FR})}{N_x + N_c} \tag{10}$$

$$S_{DWT.D}(X^w, Y^w) = \frac{N_x \cdot S_c(X^w, Y^w) + N_c \cdot S_R(X^D, Y^D)}{N_x + N_c} \tag{11}$$

2.3 Random Projection

The random projections algorithm implements the Johnson and Lindenstrauss lemma. In order to reduce the dimensionality of a given mass spectral vector X defined before (with n variables), a set of random vectors $\gamma = \{\rho_i\}_{i=1}^{n}$ is generated where $\rho_i \in R^m$ are column vectors and $\|\rho_i\|_{l_2} = 1$. In this work, the vector $\|\rho_i\|_{l_2} = 1$ follow a normal distribution N(0,1) over the m dimensional unit sphere. The vectors in γ are used to form the columns of a $m \times n$ conversion matrix:

$$W = (\rho_1|\rho_2|\ldots|\rho_n) \tag{12}$$

The lower-dimensional (m-dimensional) subspace is obtained by:

$$\tilde{X}_i = \text{sgn}(X_i * W') \tag{13}$$

where the Sgn function is added to make the conversion into binary vector, if the digit of X*W > 0, the corresponding bit set as 1, otherwise set as 0. $\tilde{X}$ is the m bits binary vector.

2.4 Performance Measurement

The identification accuracy is used to evaluate the performance of three spectral similarity measures and our converted binary vector. In this work, the compound is considered as correct identification through identical Chemical Abstract Service (CAS) registry number. Therefore, the accuracy of identification is calculated by:

$$accuracy = \frac{number\ of\ spectra\ matched\ correctly}{number\ of\ spectra\ queried} \tag{14}$$

3 Results and Discussion

The NIST replicate library including 23290 mass spectra is used as query data and the main spectral library is used as reference database. Random projection are used to convert the original mass spectral data into different numbers of binary digits for library searching. The number of binary digits is set as 64, 128, 256, 512, 768, 1024 and 2176 bits respectively. Hamming distance between the binary vector of query and reference spectrum is calculated to measure the similarity.

Figure 1 and Table 1 show the identification accuracy of random projection and three corresponding similarity measures. Since the randomness of random projection, five experiments are performed for each number of binary digits setup. With the increase of the number of binary digits, the accuracy of compound identification is increased. Even the number of binary digits setup as 64 bits, the identification accuracy of rank 1 can reach to 65.5 %. When the number of binary digits setup as 256 bits, random projection easily overcome SS measure, as to 768 bits, it achieve the same identification performance with DFT.R and DWT.D measures. With the number of binary digits set as 1024, the identification accuracy of rank 1 reach 82.5 %, random projection even excels DFT.R and DWT.D in identification performance. Since the growth rate is slow when the number of binary digits over 768 bits, a large number 2176 bits are set up for comparison. At this setup, the identification accuracy of rank 1 even reach to 83.1 %.

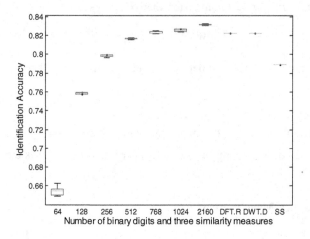

Fig. 1. Compound identification results of random projection and three similarity measures in rank 1.

Table 1. Compound identification performance

Random projection and similarity measures	Accuracy at rank (%)					
	1	1–2	1–3	1–4	1–5	1–10
SS	0.787	0.897	0.928	0.947	0.957	0.973
DFT.R	0.821	0.915	0.949	0.964	0.972	0.986
DWT.D	0.823	0.914	0.947	0.962	0.97	0.984
64 bits	0.655 ± 0.0051	0.779 ± 0.0044	0.831 ± 0.0048	0.86 ± 0.005	0.879 ± 0.005	0.922 ± 0.004
128 bits	0.758 ± 0.0011	0.87 ± 0.0008	0.912 ± 0.0006	0.932 ± 0.0008	0.944 ± 0.0004	0.969 ± 0.0007
256 bits	0.798 ± 0.0013	0.9 ± 0.0018	0.937 ± 0.0011	0.953 ± 0.0021	0.963 ± 0.0018	0.985 ± 0.0009
512 bits	0.816 ± 0.0006	0.913 ± 0.001	0.947 ± 0.0011	0.962 ± 0.0009	9.97 ± 0.0007	0.984 ± 0.0004
768 bits	0.823 ± 0.0011	0.917 ± 0.0017	0.95 ± 0.0009	0.964 ± 0.0009	0.973 ± 0.0004	0.986 ± 0.0006
1024 bits	0.825 ± 0.0013	0.919 ± 0.0009	0.952 ± 0.0009	0.966 ± 0.0003	0.974 ± 0.0003	0.987 ± 0.0002
2176 bits	0.831 ± 0.0007	0.923 ± 0.0004	0.954 ± 0.0008	0.968 ± 0.0007	0.975 ± 0.0002	0.987 ± 0.0002

Table 1 depicts the relationship between identification accuracy and the number of top ranks. Only top ten ranked compound are listed in Table 1. The identification accuracy can be increased to 98.7 ± 0.0002 % by the random projection (1024 bits to 2176 bits) if the top 10 compounds were considered (Table 1).

4 Conclusions

Mass spectral library searching through spectrum matching is widely used method for compound identification in analysis of GC-MS data. In order to reduced computational time, the original mass spectral space was transformed into binary space by random projection. The library searching was performed based on hamming distance between query and reference binary vector. The experiments show that the random projection achieved high identification accuracy than three corresponding similarity measures (SS, DFT.R and DWT.D).

Acknowledgments. This work was supported by National Natural Science Foundation of China under grant nos. 61271098 and 61032007, and Provincial Natural Science Research Program of Higher Education Institutions of Anhui Province under grant no. KJ2012A005.

References

1. Denkert, C., et al.: Mass spectrometry-based metabolic profiling reveals different metabolite patterns in invasive ovarian carcinomas and ovarian borderline tumors. Cancer Res. **66**(22), 10795–10804 (2006)
2. Stein, S.E., Scott, D.R.: Optimization and testing of mass spectral library search algorithms for compound identification. J. Am. Soc. Mass Spectrom. **5**(9), 859–866 (1994)
3. McLafferty, F.W., et al.: Comparison of algorithms and databases for matching unknown mass spectra. J. Am. Soc. Mass Spectrom. **9**(1), 92–95 (1998)
4. Hertz, H.S., Hites, R.A., Biemann, K.: Identification of mass spectra by computer-searching a file of known spectra. Anal. Chem. **43**(6), 681–691 (1971)
5. Visvanathan, A.: Information-Theoretic Mass Spectral Library Search for Comprehensive Two-Dimensional Gas Chromatography with Mass Spectrometry. ProQuest, Ann Arbor (2008)
6. Koo, I., Zhang, X., Kim, S.: Wavelet- and Fourier-transform-based spectrum similarity approaches to compound identification in gas chromatography/mass spectrometry. Anal. Chem. **83**(14), 5631–5638 (2011)
7. Kim, S., et al.: A method of finding optimal weight factors for compound identification in gas chromatography-mass spectrometry. Bioinformatics **28**(8), 1158–1163 (2012)
8. Kim, S., et al.: Compound identification using partial and semipartial correlations for gas chromatography-mass spectrometry data. Anal. Chem. **84**(15), 6477–6487 (2012)
9. Koo, I., Kim, S., Zhang, X.: Comparative analysis of mass spectral matching-based compound identification in gas chromatography-mass spectrometry. J. Chromatogr. A **1298**, 132–138 (2013)
10. Indyk, P., Motwani, R.: Approximate nearest neighbors: towards removing the curse of dimensionality. In: Proceedings of the Thirtieth Annual ACM Symposium on Theory of Computing. ACM (1998)

11. Gionis, A., Indyk, P., Motwani, R.: Similarity search in high dimensions via hashing. In: VLDB (1999)
12. Charikar, M.S.: Similarity estimation techniques from rounding algorithms. In: Proceedings of the Thiry-Fourth Annual ACM Symposium on Theory of Computing. ACM (2002)
13. Stein, S.E., Scott, D.R.: Optimization and testing of mass spectral library search algorithms for compound identification. J. Am. Soc. Mass Spectrom. **5**(9), 859–866 (1994)
14. Koo, I., Zhang, X., Kim, S.: Wavelet- and Fourier-transform-based spectrum similarity approaches to compound identification in gas chromatography/mass spectrometry. Anal. Chem. **83**(14), 5631–5638 (2011)

A Random Projection Ensemble Approach to Drug-Target Interaction Prediction

Peng Chen[1($\boxtimes$)], ShanShan Hu[1], Bing Wang[2], and Jun Zhang[3]

[1] Institute of Health Sciences, Anhui University, Hefei 230601, Anhui, China
bigeagle@mail.ustc.edu.cn
[2] School of Electronics and Information Engineering,
Tongji University, Shanghai 804201, China
[3] College of Electrical Engineering and Automation,
Anhui University, Hefei 230601, Anhui, China

Abstract. Drug-target interaction prediction is very important in drug development. Since determining drug-target interactions is costly and time-consuming by experiments, it is a complement to determine the interactions by computational method. To address the issue, a random projection ensemble approach is proposed and drug-compounds are encoded with feature descriptors by software "PaDEL-Descriptor", while target proteins are encoded with physicochemical properties of amino acids. From 544 properties in AAindex1, 34 relatively independent physicochemical properties are extracted. Random projection on the vector of drug-target pair with different dimensions can map the original space onto a reduced one and thus yield a transformed vector with fixed dimension. Several random projections build an ensemble REPTree system. Experimental results showed that our method significantly outperformed and ran faster than other state-of-the-art drug-target predictors.

Keywords: Random projection · Drug-target interaction · REPTree · Ensemble system

1 Introduction

Drug-target interaction is to identify whether a pair of drug and target can be interacted or not. It is a key in the drug discovery for specific disease [1]. Before a drug candidate was synthesized [2], several difficulties need to be overcome. The first difficulty is how to find out the drug effects to different people [3] and the second one is to trace and elucidate the drug effects along the biological interaction pathways in human beings. Moreover, since drug discovery is costly and time-consuming and the number of new drug approvals is quite low per year, computational methods are complement to the drug discovery. Computational methods can be used to identify the sensitivity and toxicity before a drug candidate was approved [2], and they can save time and money to a great extent.

Many works have developed different computational methods for analyzing and identifying drug-target interactions. They can be divided into various classes: docking simulations [4], literature text mining [5], methods combining chemical structure, genomic sequence, and 3D structure information [6], and so on.

© Springer International Publishing Switzerland 2015
D.-S. Huang and K. Han (Eds.): ICIC 2015, Part III, LNAI 9227, pp. 693–699, 2015.
DOI: 10.1007/978-3-319-22053-6_72

Here we propose a random projection ensemble approach for drug-target interactions based on the REPTree algorithm [7] by using random projection [8] to map original data onto a rather smaller space. To encode the input to the classifier ensemble, drug-compounds are encoded with feature descriptors by software "PaDEL-Descriptor", while target proteins are encoded with physicochemical properties of amino acids. From AAindex1, 34 relatively independent physicochemical properties are extracted. Random projection on the vector of drug-target pair with different dimensions can map the original space into a reduced one and thus yield a transformed vector with fixed dimension. Several random projections build an ensemble REPTree system. Experimental results showed that our method significantly outperformed and ran faster than other state-of-the-art drug-target predictors, on the commonly used drug-target benchmark sets.

2   Methods

2.1   Feature Vector Representing a Target Protein

To encode target protein, AAindex1 database is used which contains 544 amino acid properties [9]. Most of them are relevant, so like as our previous work [10], irrelevant ones with a correlation coefficient (CC) of 0.5 are extracted. The CC of each two properties is computed and the number of relevant properties is counted. Ranking the relevant number in descend, a list of properties is obtained. For the top one property, we remove all of the next properties related to the top one. Step by step, each property related to the previous one is removed from the list. Finally, 34 properties are retained, where each two properties have a CC less than 0.5 [10].

For the ith target protein chain, the whole chain is considered in this work. In order to investigate the evolution of protein residue in terms of physicochemical property, an encoding schema integrating amino acid properties and sequence profile is used to represent the residue. The sequence profile for one residue created by PSI-Blast with default parameters is then multiplied by each amino acid property, where the property for one amino acid is multiplied by the score of the sequence profile for the same amino acid. Therefore, the profile SP^k for residue k and one amino acid property scale, Aap, are both vectors with 1×20 dimensions. Thereafter, $MSK^k = SP^k \times Aap$ for residue k represents the multiplication of the corresponding sequence profile by the scale, whose jth element $MSK^{k,j} = SP^{k,j} \times Aap^j, j = 1, \ldots, 20$. The standard deviation of MSK^k, TP^k, is used to represent the kth residue. As a result, the th target protein is vectorized as $TP = [TP^1, \ldots, TP^k, \ldots, TP^{lenSeq}]^T$, where $lenSeq$ is the length of the target sequence. A similar vector representation can be found in our previous work [10–12].

2.2   Feature Vector Representing a Drug Candidate

In order to encode drug candidate, PaDEL-Descriptor software is used. PaDEL-Descriptor is a software for calculating molecular descriptors and fingerprints which currently calculates different descriptors (1D, 2D descriptors, and 3D descriptors) and

10 types of fingerprints [13]. A molecular descriptor is the final result of a logic and mathematical procedure which transforms chemical information encoded within a symbolic representation of a molecule into a useful number or the result of some standardized experiment [14]. In this work, 1D and 2D descriptors are used, meanwhile salt is removed from a molecule which assumes that the largest fragment is the desired molecule. In addition, aromaticity information is removed and aromaticity is automatically detected in the molecule before calculation of descriptors. As a result, 1444 descriptors are used to encode one drug molecule. So the th drug candidate can be formulated as $D^i = [D^i_1, D^i_2, \ldots, D^i_{1444}]^T$. These 1D and 2D descriptors and fingerprints are calculated mainly using The Chemistry Development Kit [13]. These descriptors include atom type electrotopological state descriptors, McGowan volume, molecular linear free energy relation descriptors, ring counts, and count of chemical substructures identified by Klekota and Roth [15].

For the pair of drug-target, DT^i, whose target is encoded by the AAindex1 property Aap, it can be formulated as a *(1444 + lenSeq)*-D vector given by

$$V^{i,Aap} = [D^i, TP^i_{Aap}]^T = [D^i_1, D^i_2, \ldots, D^i_{1444}, TP^{i,Aap}_1, TP^{i,Aap}_2, \ldots, TP^{i,Aap}_{lenSeq}]^T, \quad (1)$$

where *lenSeq* is the length of the target sequence.

The corresponding target value T^i is 1 or 0, denoting whether the drug-target pair is in interaction or not. Actually, our method expects to learn the relationship between input vectors V^{Aap} and the corresponding target arrays T and try to make its output as close to the target T as possible, where Aap denotes that the target is encoded in terms of the irrelevant AAindex1 property Aap.

2.3   Random Projection on REPTree

Random projection is a data reduction technique that projects a high dimensional data onto an low-dimensional subspace [16, 17]. Given the original data vector, $X \in \mathbb{R}^{N \times L_1}$, the linear random projection is to multiply the original vector by a random matrix $R = \mathbb{R}^{L_1 \times L_2}$. The projection

$$X^R = XR = \sum_i x_i r_i \quad (2)$$

yields a dimensionality reduced vector $X^R \in \mathbb{R}^{N \times L_2}$, where x_i is the ith sample of the original data, r_i is the ith column of the random matrix, and $L_2 \ll L_1$. The matrix R consists of random values and each column has been normalized to unity. In the Eq. 1, each original data sample with dimension L_1 has been replaced by a random, non-orthogonal direction L_2 in the reduced-dimensional space [17]. Therefore, the dimensionality of original data is reduced from (1444 + *lenSeq*) to a rather small value.

REPTree is a fast tree learner that uses reduced-error pruning [7], based on information gain/variation reduction as the splitting principle, and optimize for speed by sorting values for numeric attributes once. This work adopts the default *numFolds* parameter of the REPTree (default 3 in WEKA software) that determines the size of the pruning set: the data is divided equally into that number of parts and the last one used as an independent test set to estimate the error at each node.

Previous results showed that the generalization error caused by one classifier can be compensated by other classifiers, therefore using tree ensembles can lead to significant improvement in prediction accuracy [18]. For our drug-target interaction prediction problem, the ensemble of simple trees votes for the most popular class of drug-target interaction. Given the set of training data $V_{tr}^{k,Aap} = \{(X_i^{R^k,Aap}, Y_i)\}|_{i=1}^{N}$ in terms of AAindex1 property Aap, after multiplied by the random projection R^k, let the number of training instances be N, the number of features in the classifier be L_2. Then the data $V^{k,Aap}$ is generated as an input to a REPTree and thus it forms a classifier $CF^{k,Aap}(x)$, where x is a training instance.

After all of REPTree classifiers with random projection are generated, they vote for the most popular class and thus the prediction of the ensemble is

$$R(X) = majority\,vote\,\{CF^{k,Aap}(x)\}_{Aap=1}^{34}, \tag{3}$$

where x is a query instance.

2.4   Data Sets and Prediction Evaluation

We used the drug-target datasets in literature [6] for our study. It excluded drug-target pairs that lack experimental information and finally contains a total of 4797 pairs, of which 2,719 for enzymes, 1,372 for ion channels, 630 for GPCRs, and 82 for nuclear receptors. The lists of the pairs can be found in reference [6]. All these datasets were regarded as the positive ones and a total of 9588 negative ones were extracted as the same as in the literature [6].

In this work we adopted four evaluation measures to show the ability of our model objectively, criteria of Recall (Rec), precision (Prec), F-measure (F1), and Matthews correlation coefficient (MCC) [11, 19].

3   Results

3.1   Performance of Drug-Target Interaction Prediction

In this work, each dataset of drug-target interactions is divided into training data set $\aleph_{tr}$ and test one $\aleph_{ts}$ by 10-fold cross-validation. That is to say, the dataset is divided into 10 subsets with roughly the same number of instances and one subset is regarded as the test set while the others are grouped as training set. The test subset is selected one-by-one and finally all of the instances are tested. Then different random projections are used to map the original dataset onto a rather lower space, in this work 5 dimensionality space mapped. For achieving better random projection, the training data set $\aleph_{tr}$ is divided into training subset $\aleph_{tr}^{sub}$ and test subset $\aleph_{ts}^{sub}$ by 10-fold cross-validation. Running the REPTree classifier by the random projection technique, predictions on the test subset $\aleph_{ts}^{sub}$ by the training subset $\aleph_{tr}^{sub}$ are obtained. Only random projections yielding top performance are retained. Running REPTree classifier, given the top random projection, on the training data set $\aleph_{tr}$ and test one $\aleph_{ts}$ yields the final predictions.

In details, there are 34 random projections R^k on the original data matrix for each of 34 independent AAindex1 properties Aap. The ensemble of the 34 classifiers by random projections yields prediction for the training data subset $\aleph_{tr}^{sub}$ and the prediction for the test data subset $\aleph_{ts}^{sub}$. The 34 random projections are retained if the prediction accuracy is larger than 0.75 for $\aleph_{tr}$. Repeating the classifier ensemble by random projections R^k, several top predictions are obtained by random projections R^k $(k = 1 \sim K)$, on $\aleph_{tr}$ and $\aleph_{ts}$. Combining the K predictions yields the final prediction. Table 1 shows the performance comparison of the ensemble ones for the four protein target classes. Here the dimensionality of the original data is reduced from $(1444 + lenSeq)$ to 5. For drugs, they are encoded as vectors with fixed length, 1444; while for protein targets with different sequence lengths, they will be encoded as vectors with different sequence lengths. The longest sequence length is obtained for the original space dimensionality $maxLenSeq$ of random projections. Target sequence with shorter length $lenSeq$ is encoded into a subspace Ψ^{lenSeq} in the space $\Psi^{maxLenSeq}$, i.e., $\Psi^{lenSeq} \in \Psi^{maxLenSeq}$. From the Table 1, it can be seen that the ensemble system tested on nuclear receptors class performs better than that on other classes. It yields an accuracy of 0.911 as well as a precision of 0.889 at a recall of 0.837.

Table 1. Prediction performance of the REPTree classifier ensemble with majority vote technique, i.e., the ensemble system predicts a drug-target pair to be interacting if all of REPTree classifiers in the ensemble predict it to be interacting.

Dataset	Class	Rec	Acc	Prec	F1
Training[a]	Enzymes	0.970	0.944	0.876	0.921
	Ion-channels	0.986	0.886	0.751	0.853
	GPCRs	0.994	0.892	0.758	0.0860
	Nuclear-receptors	0.709	0.812	0.722	0.716
Test[b]	Enzymes	0.972	0.900	0.782	0.867
	Ion-channels	0.993	0.89	0.755	0.858
	GPCRs	1.002	0.85	0.693	0.818
	Nuclear receptors	0.8371	0.91	0.889	0.862

[a]Prediction on the training dataset $\mathbb{N}_{tr}$.
[b]Prediction on the test dataset $\mathbb{N}_{ts}$.

3.2 Comparison with Other Methods

We compare our method with other two methods: the work in reference [6] and the random predictor on the same datasets. Table 2 shows the performance comparison in accuracy of our method with other two methods. The random predictions implemented here and ran 100 times. The average performance is appended at the bottom of the table. Our method yields accuracies of 0.900, 0.89, 0.852, and 0.911 for classes of enzymes, ion channels, GPCRs, and nuclear receptors, respectively, and achieves improvements of 4.5 % to 8.2 % than the work [6]. Also results showed that our method outperforms the random predictor by 2 times of score.

Table 2. Accuracy comparison of our method with two methods on the same datasets.

Method	Type	Enzymes	Ion channels	GPCRs	Nuclear receptors
Our method	REPTree	0.900	0.89	0.852	0.911
Ref [6]	kNN	0.855	0.808	0.785	0.857
Random predictor		0.489	0.489	0.488	0.488

4 Conclusions

This paper proposes an ensemble of REPTree classifiers by random projection to identify drug-target interactions. For each independent AAindex1 property, the original encoders for drug-target interaction transformed by different random projections are input into a REPTree classifier. There are 34 REPTree classifiers with respect to AAindex1 property. The ensemble of these REPTree classifiers can yield good prediction on drug-target interactions. Therefore, our method is simple for only statistical amino acid properties are applied. Moreover, the dimensionality reduction of random projection is adopted here to reduce the original encoder space. More importantly, the random projection technique can handle protein chains with different numbers of amino acids and get unified encoder space. Actually, the random projection technique provides a useful mechanism such that it reduces the high dimensional original data and makes the data more diverse and thus, the method yields a good prediction on drug-target interactions. Results show that our method outperforms other state-of-the-art methods of drug-target interaction prediction.

Acknowledgment. This work was supported by the National Natural Science Foundation of China (Nos. 61300058, 61271098 and 61472282).

References

1. Knowles, J., Gromo, G.: A guide to drug discovery: target selection in drug discovery. Nat. Rev. Drug Discov. 2(1), 63–69 (2003)
2. Johnson, W.: Predicting human safety: screening and computational approaches. Drug Discov. Today 5(10), 445–454 (2000)
3. Wood, A.J., Evans, W.E., McLeod, H.L.: Pharmacogenomics drug disposition, drug targets, and side effects. N. Engl. J. Med. 348(6), 538–549 (2003)
4. Cheng, A.C., Coleman, R.G., Smyth, K.T., Cao, Q., Soulard, P., Carey, D.R., Salzberg, A.C., Huang, E.S.: Structure-based maximal affinity model predicts small-molecule druggability. Nat. Biotechnol. 25(1), 71–75 (2007)
5. Zhu, S., Okuno, Y., Tsujimoto, G., Mamitsuka, H.: A probabilistic model for mining implicit chemical compound-gene relations from literature. Bioinformatics 21(Suppl 2), ii245–ii251 (2005)

6. He, Z., Zhang, J., Shi, X.H., Hu, L.L., Kong, X., Cai, Y.D., Chou, K.C.: Predicting drug-target interaction networks based on functional groups and biological features. PLoS ONE **5**(3), e9603 (2010)
7. Esposito, F., Malerba, D., Semeraro, G., Tamma, V.: The effects of pruning methods on the predictive accuracy of induced decision trees. Appl. Stoch. Models Bus. Ind. **15**, 277–299 (1999)
8. Schclar, A., Rokach, L.: Random projection ensemble classifiers. In: Filipe, J., Cordeiro, J. (eds.) ICEIS 2009. LNBIP, vol. 24, pp. 309–316. Springer, Heidelberg (2009)
9. Kawashima, S., Pokarowski, P., Pokarowska, M., Kolinski, A., Katayama, T., Kanehisa, M.: Aaindex: amino acid index database, progress report 2008. Nucleic Acids Res. **36**(Database issue), D202–D205 (2008)
10. Chen, P., Li, J., Wong, L., Kuwahara, H., Huang, J.Z., Gao, X.: Accurate prediction of hot spot residues through physicochemical characteristics of amino acid sequences. Proteins **81** (8), 1351–1362 (2013)
11. Chen, P., Li, J.: Sequence-based identification of interface residues by an integrative profile combining hydrophobic and evolutionary information. BMC Bioinf. **11**, 402 (2010)
12. Chen, P., Wong, L., Li, J.: Detection of outlier residues for improving interface prediction in protein heterocomplexes. IEEE/ACM Trans. Comput. Biol. Bioinf. **9**(4), 1155–1165 (2012)
13. Yap, C.W.: Padel-descriptor: an open source software to calculate molecular descriptors and fingerprints. J. Comput. Chem. **32**(7), 1466–1474 (2012)
14. Todeschini, R., Consonni, V.: Handbook of Molecular Descriptors, vol. 11. Wiley, New York (2008)
15. Klekota, J., Roth, F.P.: Chemical substructures that enrich for biological activity. Bioinformatics **24**(21), 2518–2525 (2008)
16. Papadimitriou, C.H., Raghavan, P., Tamaki, H., Vempala, S.: Latent semantic indexing: a probabilistic analysis (1998)
17. Kaski, S.: Dimensionality reduction by random mapping: fast similarity computation for clustering. In: Proceedings of the 1998 IEEE International Joint Conference on Neural Networks. IEEE World Congress on Computational Intelligence, vol. 1, pp. 413–418 (1998)
18. Chen, P., Huang, J.Z., Gao, X.: LigandRFs: random forest ensemble to identify ligand-binding residues from sequence information alone. BMC Bioinf. **15**(Suppl 15), S4 (2014)
19. Wang, B., Chen, P., Huang, D.S., Li, J.J., Lok, T.M., Lyu, M.R.: Predicting protein interaction sites from residue spatial sequence profile and evolution rate. FEBS Lett. **580**(2), 380–384 (2006)

Kernel Local Fisher Discriminant Analysis-Based Prediction on Protein O-Glycosylation Sites Using SVM

Xuemei Yang[1] and Shiliang Sun[2(✉)]

[1] School of Mathematics and Information Science,
Xianyang Normal University, Xianyang, China
yangxuemei691226@163.com
[2] Department of Computer Science and Technology,
East China Normal University, Shanghai, China
slsun@cs.ecnu.edu.cn

Abstract. O-glycosylation means that sugar transferred to the protein. It can adjust the function of protein. To obtain a higher prediction accuracy of O-glycosylation sites, we used a method of Kernel Local Fisher Discriminant Analysis (KLFDA). The original data are projected into the subspace constructed by KLFDA, and the local feature vectors are extracted. Then the prediction (classification) is done in feature subspace by support vector machines (SVM). The results of experiments show that compared with LFDA, FDA, KPCA, PCA and ICA, the prediction accuracy of KLFDA is the best.

Keywords: KLFDA · Glycosylation · Prediction · Protein · SVM

1 Introduction

Glycosylation means that sugar transferred to the protein under glycosyltransferase, it is an important post-translation modification of protein. The position of O-linked glycosylation is the hydroxyl of serine(S) and threonine (T) side chains. In fact, not all S or T residues are glycosylated, so we need to predict the O-glycosylation sites [1].

In recent years, some computational prediction approaches such as artificial neural network (ANN) and support vector machine (SVM) were applied to carry out the prediction [2–4].

Since there are complex structure features in the protein sequence, the prediction performance is not satisfied if one classifies the original data directly without extracting the meaningful features.

Principal component analysis (PCA) [5] is a statistical method for feature extraction. In our previous work, we used PCA for pattern analysis, and verified some properties of glycosylated protein.

The characters drawn by PCA are the best characters for description, but not the best characters for classification. Fisher discriminant analysis (FDA) [6] searches for a linear transformation which aims at the best characters for classification. However, it can not obtain a satisfied result if the data are multimodal. Local Fisher discriminant

© Springer International Publishing Switzerland 2015
D.-S. Huang and K. Han (Eds.): ICIC 2015, Part III, LNAI 9227, pp. 700–705, 2015.
DOI: 10.1007/978-3-319-22053-6_73

analysis (LFDA) [7] can deal with effectively the multimodal data, but it's still a linear method. Kernel Local Fisher discriminant analysis (KLFDA) [7] overcomes this fault, which can extract the nonlinear characters of the sample. Nobody used this method to predict the O-glycosylation sites up to now.

Since the glycosylated acid sequence(positive) includes positive T and positive S, and the non-glycosylated acid sequence(negative) includes negative T and negative S, they have different characters, namely, though positive T and positive S are both positive, they also have different characters. So the data are multimodal data.

In this paper, we tried to use KLFDA for extracting the nonlinear characters of protein sequence, and predicted the O-glycosylation sites by using SVM [8].

The rest of the paper is organized as follows. In Sect. 2, KLFDA and SVM is briefly reviewed. In Sect. 3, we describe protein sequence data and the prediction results. Finally, we give conclusions in Sect. 4.

2 Kernel Local Fisher Discriminant Analysis and Support Vector Machine

2.1 Kernel Local Fisher Discriminant Analysis

KLFDA is a method that extends LFDA to non-linear circumstance by projecting the samples into a kernel feature space.

Consider the problem of classification. Let $x_i \in \mathbf{R}^d (i = 1, 2, \cdots, n)$ denote the d-dimensional samples with corresponding class labels $y_i \in \{1, 2, \cdots, c\}$. Let $\phi(x)$ be a non-linear projection from the original space $\mathbf{R}^d$ to a reproducing kernel Hilbert space H, we define an affinity matrix A which element $A_{i,j}$ denotes the affinity between x_i and x_j as the follow

$$A_{i,j} = \exp(-\frac{\left\|x_i - x_j\right\|^2}{s^2}), \tag{1}$$

where $s(> 0)$ is a tuning parameter. Let

$$\tilde{W}_{i,j}^{(w)} \equiv \begin{cases} A_{i,j}/n_l & \text{if } y_i = y_j = l \\ 0 & \text{if } y_i \neq y_j \end{cases}, \tag{2}$$

$$\tilde{W}_{i,j}^{(b)} \equiv \begin{cases} A_{i,j}(1/n - 1/n_l) & \text{if } y_i = y_j = l \\ 1/n & \text{if } y_i \neq y_j \end{cases}. \tag{3}$$

where n_l is the sample number of class l. Then we have

$$\tilde{L}^{(w)} \equiv \tilde{D}^{(w)} - \tilde{W}^{(w)} \tag{4}$$

where $\tilde{D}^{(w)}$ is the n-dimensional diagonal matrix with the i-th diagonal element being $\tilde{D}_{i,j}^{(w)} \equiv \sum_{j=1}^{n} \tilde{W}_{i,j}^{(w)}$. We also have

$$\tilde{L}^{(b)} \equiv \tilde{D}^{(b)} - \tilde{W}^{(b)} \tag{5}$$

where $\tilde{D}^{(b)}$ is the n-dimensional diagonal matrix with the i-th diagonal element being $\tilde{D}_{i,j}^{(b)} \equiv \sum_{j=1}^{n} \tilde{W}_{i,j}^{(b)}$.

Defining kernel function $\tilde{K}_{i,j} = \langle \phi(x_i), \phi(x_j) \rangle = K(x_i, x_j)$, solving the following equation,

$$\tilde{K}\tilde{L}^{(b)}\tilde{K}\tilde{\alpha} = \tilde{\lambda}\tilde{K}\tilde{L}^{(w)}\tilde{K}\tilde{\alpha}. \tag{6}$$

We obtain the generalized eigenvectors $\{\tilde{\alpha}_k\}_{k=1}^{n}$ and the corresponding generalized eigenvalues $\tilde{\lambda}_1 \geq \tilde{\lambda}_2 \geq \ldots \geq \tilde{\lambda}_n$. Then the projection of the sample x by KLFDA is

$$\left(\sqrt{\tilde{\lambda}_1}\tilde{\alpha}_1 | \sqrt{\tilde{\lambda}_2}\tilde{\alpha}_2 | \cdots | \sqrt{\tilde{\lambda}_n}\tilde{\alpha}_n\right)^T \begin{pmatrix} K(x_1, x) \\ K(x_2, x) \\ \vdots \\ K(x_n, x) \end{pmatrix}. \tag{7}$$

2.2 Support Vector Machine

SVM is a statistics learning method, which aims to seek a hyperplane $\mathbf{w} \cdot \phi(\mathbf{x}) + b = 0$, which can separate optimally the positive and negative samples. The objective function of SVM is

$$\begin{array}{cc} \min_{w,b,\xi} & \frac{1}{2}\|w\|^2 + C\sum_{i=1}^{n}\xi_i^2 \\ s.t. & y_i = w \cdot \phi(x_i) + b + \xi_i \end{array}, \tag{8}$$

where $\frac{2}{\|\mathbf{w}\|}$ is the margin, ξ is the margin slack vector, and the parameter C is the penalty factor. The decision function of SVM is

$$f(\mathbf{x}) = \sum_{i=1}^{M} \alpha_i^* K(\mathbf{x}_i, \mathbf{x}) + b^*, \tag{9}$$

where $\boldsymbol{\alpha}^*$ is Lagrange multiplier. The category of test sample is decided by the sign of $f(\mathbf{x})$.

3 Prediction on Protein Sequence Data

The training and the test samples of acid sequences used in the experiments are the same in the reference [5], and the sparse coding scheme is employed to code one site of acid or vacancy.

The prediction can be view as a two-class classification problem (positive and negative). We first extract features of training and test samples by KLFDA, comparing with FDA, LFDA, PCA, ICA and KPCA, then classify the test samples by SVM.

The algorithm is implemented in MATLAB7.8.0. We take the window size w = 21. In this paper, we use the Gaussian kernel function, the parameter C in SVM is $1 \sim 100$.

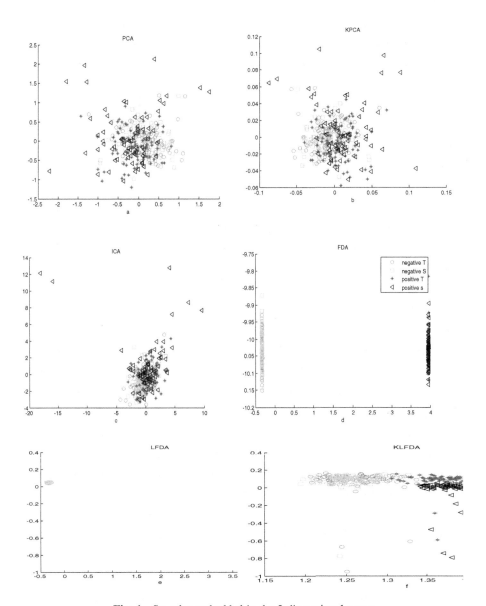

Fig. 1. Samples embedded in the 2-dimensional space

3.1 Projection

Figure 1 depicts the samples projected into the two-dimensional subspace founded by each method. The horizontal axis and the vertical axis represent respectively the first feature and the second feature. For FDA, the second feature is selected randomly.

From Fig. 1, we see that both KLFDA and LFDA can well separate the positive samples from the negative samples, and clearly preserves within-class multimodality simultaneously, but KLFDA extracts more meaningful information due to its nonlinear property. FDA separates the samples in different classes well, but losses the within-class multimodality, namely, '*' and 'Δ' are mixed. For PCA, ICA and KPCA, samples in different classes are all mixed in one cluster.

3.2 Prediction

Table 1 shows the prediction accuracy rate by different methods. It's shown that the results of FDA is better than PCA. This is because that the characters drawn by PCA is the best characters for description but not for classification, and FDA aims at the best characters for classification. ICA is a method of high-order statistics, so its performance is better than that of PCA and FDA. Since there are many complex nonlinear features in the protein sequence, and KPCA can extract the nonlinear features of original data, KPCA classifies more accurately than PCA, ICA and FDA. KLFDA works the best, owing to its capability to capture the nonlinearity and multimodality. LFDA also works well, since it can deal with multimodality, but it does not compare favorably with KLFDA.

Table 1. Recognition performance of different methods (window size:21)

Method	PCA	FDA	ICA	KPCA ($\sigma = 10$)	LFDA (s = 3)	KLFDA ($\sigma = 4$)
Feature number	66	1	64	66	2	2
Accuracy(%)	81.5	82.5	83.5	84.5	85	87

4 Conclusion

We predict protein o-glycosylation sites by using KLFDA. KLFDA projects the samples onto a nonlinear space H, and LFDA is used in H for dimensionality reduction and feature extraction. The experimental results show that KLFDA is superior to other methods for the classification of multimodal data.

Acknowledgment. This work is partially supported by the Scientific Research Project of Education Department of Shaanxi Province (No. 2013JK1125), the Nature Science Fund Project of Shaanxi Province (No. 2014JM1032), and the Science and Technology Project of National Bureau of Quality Inspection (No. 2013QK152).

References

1. Hansen, E.L., Tolstrup, N., Gooley, A.A., Williams, K.L., Brunak, S.: NetOglyc: prediction of mucin type O-glycosylation sites based on sequence context and surface accessibility. Glycoconj J. **15**, 115–116 (1998)
2. Nishikawa, I., Sakamoto, H., Nouno, I., Iritani, T., Sakakibara, K., Ito, M.: Prediction of the *O*-glycosylation sites in protein by layered neural networks and support vector machines. In: Gabrys, B., Howlett, R.J., Jain, L.C. (eds.) KES 2006. LNCS, vol. 4252, pp. 953–960. Springer, Heidelberg (2006)
3. Chen, Y.Z., Tang, Y.R., Sheng, Z.Y., Zhang, Z.D.: Prediction of mucin-type O-glycosylation sites in mammalian protein using the composition of k-spaced amino acid pairs. BMC Bioinf. **9**, 101–112 (2008)
4. Sakamoto, H., Nakajima, Y., Sakakibara, K., Ito, M., Nishikawa, I.: Prediction of the *O*-glycosylation by support vector machines and semi-supervised learning. In: Köppen, M., Kasabov, N., Coghill, G. (eds.) ICONIP 2008, Part I. LNCS, vol. 5506, pp. 986–994. Springer, Heidelberg (2009)
5. Yang, X.M., Chen, Y.W., Masahiro, I., Ikuko, N.: Principal component analysis of O-linked glycosylation sites in protein sequence. In: Proceedings of IEEE 3rd International Conference on Intelligent Information Hiding and Multimedia Signal Processing, Taiwan, pp. 121–126 (2007)
6. Zhou, K., Ai, C.Z., Dong, P.P., Fan, X.R., Yang, L.: A novel model to predict O-glycosylatin sites using highly unbalanced dataset. Glycoconj J. **29**, 551–564 (2012)
7. Masashi, S.: Dimensionality reduction of multimodal labeled data by local fisher discriminant analysis. J. Mach. Learn. Res. **8**, 1027–1051 (2007)
8. J, Shawe Taylor, Sun, S.L.: A review of optimization methodologies in support vector machines. Neurocomputing **74**(17), 3609–3618 (2011)

Identification of Colorectal Cancer Candidate Genes Based on Subnetwork Extraction Algorithm

Ran Wei[1], Hai-Tao Li[2], Yanjun Wang[1], Chun-Hou Zheng[2], and Junfeng Xia[1(✉)]

[1] Key Laboratory of Intelligent Computing and Signal Processing, Ministry of Education,
and Institute of Health Sciences, School of Computer Science and Technology,
Anhui University, Hefei 230039, Anhui, China
jfxia@ahu.edu.cn
[2] College of Electrical Engineering and Automation, Anhui University,
Hefei 230601, Anhui, China

Abstract. Colorectal cancer (CRC) is one of the most common malignancies that could threaten human health. As the molecular mechanism of CRC has not yet been completely uncovered, identifying related genes of this disease is an important area of CRC research that could provide new insights into gene function as well as potential targets for CRC treatment. Here we used a subnetwork extraction algorithm (Limited K-walks algorithm) to discover CRC related genes based on protein-protein interaction network. In particular, we computationally predicted two genes (*UBC* and *SMAD4*) as putative key genes of CRC. Therapy targeting on the functions of these two key genes may provide a promising therapeutic strategy for CRC treatment.

Keywords: Colorectal cancer · Pathway analysis · Protein-protein interaction network · Subnetwork extraction algorithm

1 Introduction

Colorectal cancer (CRC) is one of the digestive tract tumors that threaten human health seriously. As the second leading cause of malignant tumor mortality in the developed countries [1], the incidence of CRC is also showing a rising trend in developing countries [2]. A larger number of genetic and environmental factors have been established for the CRC. The molecular mechanisms in CRC are not only involved in the cell proliferation, but also related to the regulation of cell apoptosis [3]. Many patients with CRC are diagnosed at a late stage, leading to high mortality rates and low survivals. So the early diagnosis and detection are very important for CRC patients. It can significantly improve the treatment and reduce the risk of recurrence. Therefore, it is very crucial to identify candidate genes associated with CRC, which can be acted as biomarkers for further clinical application.

Many efforts have been made on discovering the genomic changes in CRC [4] and recently the Cancer Genome Atlas Network has cataloged recurrent genomic abnormalities in CRC [5], which provides a unique resource for comprehensive discovery of mutations and genes that can be further mined. In addition, with the

© Springer International Publishing Switzerland 2015
D.-S. Huang and K. Han (Eds.): ICIC 2015, Part III, LNAI 9227, pp. 706–712, 2015.
DOI: 10.1007/978-3-319-22053-6_74

available of high-quality protein-protein interaction (PPI) data, network analysis has been used to detect cancer genes, pathways and biomarkers for precision medicine. Lots of PPI network-based algorithms have been developed to discover cancer associated genes, including gastric cancer [6] and hepatocellular carcinoma [7].

In this paper, we develop a method to discover candidate genes related to CRC by constructing a functional protein interaction network. The Limited K-walks algorithm [8] in GenRev [9] was utilized to extract the significant network modules and detect new candidate genes in CRC. Further analysis suggests that some of the identified candidate genes are of great importance to the development of CRC. The result may help us to find the mechanism of CRC, and then design an effective way to treat CRC.

2   Materials and Methods

2.1   Collection of Known CRC Related Genes

CRC related genes are collected from the following three datasets: (1) 13 genes were picked up from Gene Associated with a High Susceptibility of Colorectal Cancer (http://www.cancer.gov/cancertopics/pdq/genetics/colorectal/HealthProfessional/page2) by setting the keywords as human colorectal cancer; (2) 67 colorectal cancer related genes were found from [5]; (3) 38 genes were obtained from Atlas of Genetics and Cytogenetics in Oncology and Haematology database (http://atlasgeneticsoncology.org/Tumors/colonID5006.html). After combing these genes, we obtained 74 known colorectal cancer related genes.

2.2   Protein-Protein Interaction (PPI) Network

PPI networks provide a lot of valuable information in terms of understanding of cellular function and biological processes. In the PPI network, the nodes represent genes and edges represent interaction. Innumerable studies have shown that proteins in the same subnetwork are most likely to share similar or common biological functions [10, 11]. In this study, the PPI network was constructed based on the protein interaction information retrieved from BioGRID (the Biological General Repository for Interaction Datasets, http://thebiogrid.org/) (version 3.2) [12], which is an online interaction repository with data compiled through comprehensive curation efforts.

2.3   Methods for Selection of New Candidate Genes

The Limited K-walks algorithm [8] in GenRev [9] was used to explore the functional relevance genes, which is freely available at http://bioinfo.Mc.vanderbilt.edu/GenRev.html. It can map the genes to the network and extract the subnetworks that are interacted with each other by giving an interconnected network or a set of seed genes. For detailed description, please refer to [9].

2.4   Gene Set Enrichment Analysis

DAVID [13], a functional annotation tool, was used to analyze gene lists in order to understand the biological meaning. In this study, the Gene Ontology (GO) enrichment analysis tool in DAVID was use to functionally analyze the gene set. We selected the GO terms with the enrichment P-value smaller than 0.05 followed by Benjamin multiple testing correction method. For details, please see [13].

3   Result and Discussion

3.1   Identified Candidate Genes

In our method, the known CRC related genes were denoted as terminal genes, and the linker genes were the identified CRC candidate genes in the functional subnetwork. Overall, 114 candidate genes were obtained. We obtained the degree and betweenness of these 114 candidate genes and 74 known CRC related gene, where the top 20 of these genes were listed in Table 1. The degree and betweenness are two significant topological properties in PPI network. The degree is the number of edges adjacent to a node (gene). The betweenness of a node is defined as the ratio of the total number of shortest paths going through a node to the number of paths that pass through the nodes. As detailed information were listed in Table 1, we can see that the *UBC* has a highest degree of 121 and highest betweenness of 0.7502.

3.2   Result of Gene Set Enrichment Analysis

As a functional annotation tool, DAVID can be performed GO and KEGG enrichment analysis. In our research, we used GO to analyze the 114 significant candidate genes. There were 193 terms enriched by the 114 candidate genes. The "Count" (the number of genes associated with the gene set) items in the output of DAVID for these 13 GO terms with the FDR smaller than 0.05 were shown in the Table 2. We can find that some biological processes are involved in cancer pathways, including regulation of phosphorylation, signal transduction, signal pathway and so on. Overall, the GO enrichment analysis elucidates that the identified candidate genes may implicate in tumorigenesis.

3.3   Analysis of the Relationship of Two Candidate Genes

In Table 1(a), we can find seven linker genes with a high degree of top 20 genes, including *UBC*, *SMAD4*, *SMAD7*, *HDAC2*, *ERBB2IP*, *AR*, and *TSC22D1*. Meanwhile, the betweenness of top 20 genes contained eight linker genes in Table 1(b), including *UBC*, *SMAD4*, *FBCN1*, *HPCA*, *PEX5*, *NOV*, *ADAM12*, *HDAC2*. In the following section we will discuss two linker genes (*UBC* and *SMAD4*) with higher degree and betweenness.

UBC (ubiquitin C) is the gene encodes ubiquitin C protein in mammals and maintains cellular ubiquitin (Ub) levels [14]. Conjugation of ubiquitin monomers or polymers can lead to various effects within a cell. These ubiquitin proteins participate in many cellular

Table 1. List the top 20 Genes ranked according to their degree and betweenness in the subnetwork, respectively.

(a)		(b)	
Gene	Degree	Gene	betweenness
*UBC	121	*UBC	0.7502
SMAD2	33	CTNNB1	0.0727
SMAD3	31	SMAD3	0.0495
TP53	31	SMAD2	0.0449
MYC	28	*SMAD4	0.0407
CTNNB1	28	PIK3R1	0.0377
*SMAD4	22	MYC	0.0367
TGFBR1	20	TP53	0.03578
*SMAD7	18	*FBLN1	0.0291
*HDAC2	15	APC	0.0274
*ERBB2IP	14	BMPR1A	0.0274
MSH2	14	*HPCA	0.0212
*AR	14	*PEX5	0.0212
PIK3R1	13	LRP5	0.0194
ATM	12	ACVR1B	0.0186
AXIN1	12	*NOV	0.0175
MLH1	12	TGFBR1	0.0167
PIK3CA	10	*ADAM12	0.0167
*TSC22D1	10	PTEN	0.0162
APC	10	*HDAC2	0.0148

*represents the linker gene (the identified CRC candidate gene)

processes in a variety of eukaryotic cells, including the degradation of abnormal and short-lived proteins, chromatin structure, cell cycle regulation, DNA repair kinase modification, endocytosis, and regulation of other cell signaling pathways. It is known that many proteins investigated by clinical cancer researchers are involved in ubiquitin pathways. Some researchers suggested that the *UBC* is the best selection for future expression profiling of liver tissues [15]. In addition, Xiang et al. revealed that the *UBC* gene network is helpful in the prognosis of multiple subtypes of breast cancers [16]. Based on the aforementioned discoveries [17], we may conclude that the *UBC* gene is also a promising target for CRC therapy.

SMAD4 (SMAD family member 4) is a protein coding gene that codes a member of the Smad family of signal transduction protein. The Smad family can be activated by transmembrane receptor kinases, and are involved in numerous physiological and pathological processes. SMAD4 can accumulate in the nucleus and regulate the transcription

Table 2. List the top 13 GO terms with the FDR small than 0.05

Term	Count	P-value	Benjamini	FDR
GO:0010605: negative regulation of macromolecule metabolic process	19	4.62E-06	7.92E-04	0.0076
GO:0019220: regulation of phosphate metabolic process	17	3.67E-07	5.02E-04	6.03E-04
GO:0051174: regulation of phosphorus metabolic process	17	3.67E-07	5.02E-04	6.03E-04
GO:0010558: negative regulation of macromolecule biosynthetic process	17	1.80E-06	6.16E-04	0.0030
GO:0031327: negative regulation of cellular biosynthetic process	17	2.49E-06	6.84E-04	0.0041
GO:0009890: negative regulation of biosynthetic process	17	3.28E-06	7.49E-04	0.0054
GO:0042325: regulation of phosphorylation	16	1.21E-06	5.51E-04	0.0020
GO:0007167: enzyme linked receptor protein signaling pathway	14	1.06E-06	7.29E-04	0.0017
GO:0009968: negative regulation of signal transduction	11	4.41E-06	8.64E-04	0.0073
GO:0010648: negative regulation of cell communication	11	1.22E-05	0.0017	0.0200
GO:0007507: heart development	10	2.53E-05	0.0029	0.0416
GO:0007178: transmembrane receptor protein serine/threonine kinase signaling pathway	8	9.47E-06	0.0014	0.0156
GO:0030509: BMP signaling pathway	6	1.56E-05	0.0019	0.0257

of target genes with other activated Smad proteins. It is noteworthy that the mutation of *SMAD4* gene can lead to human CRC [18]. In addition, previous study has revealed that the *SMAD4* link to pancreatic cancer [19]. Further work will be needed to explore and confirm the function of *SMAD4* in CRC.

4   Conclusion

As one of the most common malignancies and a major cause of cancer-related death, the CRC related genes need to be found urgently. These genes may help to reveal the

CRC mechanism and then can design an effective way to take care of the cancer patients. Our study utilized the Limited K-walks algorithm to discover novel genes that are linked with known CRC genes. Some genes, which were participated in signal transduction or pathway, have a direct relationship with CRC. However, others, which took part in the metabolic process or biosynthetic process, may play an indirect role in CRC. Our result may give a new insight to understand CRC.

Acknowledgments. This work was supported by National Natural Science Foundation of China (31301101 and 61272339), the Anhui Provincial Natural Science Foundation (1408085QF106), the Specialized Research Fund for the Doctoral Program of Higher Education (20133401120011), and the Technology Foundation for Selected Overseas Chinese Scholars from Department of Human Resources and Social Security of Anhui Province (No. [2014]-243).

References

1. Kamangar, F., Dores, G.M., Anderson, W.F.: Patterns of cancer incidence, mortality, and prevalence across five continents: defining priorities to reduce cancer disparities in different geographic regions of the world. J. Clin. Oncol. **24**(14), 2137–2150 (2006)
2. Eilstein, D., Hedelin, G., Schaffer, P.: Incidence of colorectal cancer in Bas-Rhin, trend and prediction in 2009. Bull. Cancer **87**(7–8), 595–599 (2000)
3. Hickman, E.S., Moroni, M.C., Helin, K.: The role of p53 and pRB in apoptosis and cancer. Curr. Opin. Genet. Dev. **12**(1), 60–66 (2002)
4. Fearon, E.R.: Molecular genetics of colorectal cancer. Annu. Rev. Pathol. **6**, 479–507 (2011)
5. Cancer Genome Atlas Network: Comprehensive molecular characterization of human colon and rectal cancer. Nature **487**(7407), 330–337 (2012)
6. Jiang, Y., Shu, Y., Shi, Y., et al.: Identifying gastric cancer related genes using the shortest path algorithm and protein-protein interaction network. BioMed Res. Int. **2014** (2014). Article ID: 371397
7. Jiang, M., Chen, Y., Zhang, Y., et al.: Identification of hepatocellular carcinoma related genes with k-th shortest paths in a protein–protein interaction network. Mol. BioSyst. **9**(11), 2720–2728 (2013)
8. Dupont, P., Callut, J., Dooms, G., et al.: Relevant subgraph extraction from random walks in a graph. Universite catholique de Louvain, UCL/INGI, Number RR, 7 (2006)
9. Zheng, S., Zhao, Z.: GenRev: exploring functional relevance of genes in molecular networks. Genomics **99**(3), 183–188 (2012)
10. Kourmpetis, Y.A.I., Van Dijk, A.D.J., Bink, M.C.A.M., et al.: Bayesian Markov random field analysis for protein function prediction based on network data. PloS One **5**(2), e9293 (2010)
11. Ng, K.L., Ciou, J.S., Huang, C.H.: Prediction of protein functions based on function–function correlation relations. Comput. Biol. Med. **40**(3), 300–305 (2010)
12. Stark, C., Breitkreutz, B.J., Reguly, T., Boucher, L., Breitkreutz, A., Tyers, M.: Biogrid: a general repository for interaction datasets. Nucleic Acids Res. **34**, D535–D5359 (2006)
13. Huang, D.W., Sherman, B.T., Lempicki, R.A.: Systematic and integrative analysis of large gene lists using DAVID bioinformatics resources. Nat. Protoc. **4**(1), 44–57 (2008)
14. Ryu, K.Y., Maehr, R., Gilchrist, C.A., et al.: The mouse polyubiquitin gene UbC is essential for fetal liver development, cell-cycle progression and stress tolerance. EMBO J. **26**(11), 2693–2706 (2007)

15. Verhelst, G., Lauwers, S., Zissis, G., et al.: Selection of optimal internal controls for gene expression profiling of liver disease. Biotechniques **35**(3), 456–460 (2003)
16. Xiang, Y., Zhang, J., Huang, K.: Mining the tissue-tissue gene co-expression network for tumor microenvironment study and biomarker prediction. BMC Genomics **14**(Suppl. 5), S4 (2013)
17. Mani, A., Gelmann, E.P.: The ubiquitin-proteasome pathway and its role in cancer. J. Clin. Oncol. **23**(21), 4776–4789 (2005)
18. Miyaki, M., Iijima, T., Konishi, M., et al.: Higher frequency of Smad4 gene mutation in human colorectal cancer with distant metastasis. Oncogene **18**(20), 3098–3103 (1999)
19. Blackford, A., Serrano, O.K., Wolfgang, C.L., et al.: SMAD4 gene mutations are associated with poor prognosis in pancreatic cancer. Clin. Cancer Res. **15**(14), 4674–4679 (2009)

Detection of Protein-Protein Interactions from Amino Acid Sequences Using a Rotation Forest Model with a Novel PR-LPQ Descriptor

Leon Wong[(✉)], Zhu-Hong You, Shuai Li, Yu-An Huang,
and Gang Liu

College of Computer Science and Software Engineering,
Shenzhen University, Shenzhen 518060, Guangdong, China
lg_wong@foxmail.com

Abstract. Protein-protein interactions (PPIs) play an essential role in almost all cellular processes. In this article, a sequence-based method is proposed to detect PPIs by combining Rotation Forest (RF) model with a novel feature representation. In the procedure of the feature representation, we first adopt the Physicochemical Property Response Matrix (PR) method to transform the amino acids sequence into a matrix and then employ the Local Phase Quantization (LPQ)-based texture descriptor to extract the local phrase information in the matrix. When performed on the PPIs dataset of *Saccharomyces cerevisiae*, the proposed method achieves the high prediction accuracy of 93.92 % with 91.10 % sensitivity at 96.45 % precision. Compared with the existing sequence-based method, the results of the proposed method demonstrate that it is a meaningful tool for future proteomics research.

Keywords: Protein-Protein interactions · Rotation forest · Local phase quantization · Physicochemical property response matrix (PR)

1 Introduction

Since the Human Genome Project (HGP) has made great progress in biotechnology, the proteomics, a cutting-edge research direction, springs up. However, PPIs play extremely important roles in nearly all cellular processes. In the last decades, many researchers have proposed innovative techniques for detecting PPIs. And PPIs data for various species have been amassed, which is ascribable to the advancement of large-scale experimental technologies such as yeast two-hybrid (Y2H) screens [2, 4], tandem affinity purification (TAP) [1] and other high-throughput biological techniques. Nevertheless, PPI pairs from experiments are just a small part of the whole PPI networks [3, 5, 6], and the experimental methods also consume too much time and money with low rates of true positive and true negative predictions [7–11].

Great deals of computational methods emerge in response to the needs of the PPIs prediction. The methods are based on different data types such as gene neighborhood, sequence conservation among interacting proteins, phylogenetic profiles, gene fusion, literature mining knowledge and combining interaction information with various data

© Springer International Publishing Switzerland 2015
D.-S. Huang and K. Han (Eds.): ICIC 2015, Part III, LNAI 9227, pp. 713–720, 2015.
DOI: 10.1007/978-3-319-22053-6_75

sources. But these methods are hardly employed with such unavailable pre-knowledge of the proteins. Lately, many particular methods that gain information indirectly from transforming the amino acid sequence to the image and directly from the amino acid sequence [19, 20]. Large numbers of computational experiments proved that PPIs can be predicted by using the information of amino acid sequences alone [23].

Among them, *Shen*'s work based on SVM is one of the well-performance works [20], in which it clusters the 20 amino acids into seven classes corresponding to their dipoles and volumes of the side chains and applies the conjoint triad method to extract the feature from the protein pair. But it is unable to express the effect of neighboring in which the PPIs are more likely to be responded in the non-continuous segments of the sequence. And *Guo*'s work used the auto-covariance method that is to discover the information in the segments of the non-continuous amino acids sequence with accuracy rate of 86.55 % [14], while we got well performance by employing correlation coefficient and auto-correlation descriptors in our previous works, respectively [12, 15, 16].

In this study, we report a particular sequence-based method that combines Rotation Forest and the feature extraction of the image processing methods. In detail, we adopt the Physicochemical Property Response Matrix (PR) method and the Local Phase Quantization (LPQ) method for feature extraction. Specifically, PR method is to generate a matrix from the protein sequence. And the LPQ method is to extract the feature vectors that represent the complex and essential coefficients. Then we employ the RF model for predicting PPIs. To evaluate our proposed method, *Saccharomyces cerevisiae* dataset is applied. The accuracy, precision and sensitivity of the experiment results are 93.92 %, 96.45 % and 91.10 % respectively.

2   Materials and Methodology

2.1   Generation of the Data Set

In the proposed method we evaluated, the data are derived from yeast used in the study of Guo et al. [14]. The *Saccharomyces cerevisiae* core subset of Database of Interacting Proteins (DIP) offers the PPI dataset we need. The tautological protein pairs that hold a protein that the residues are fewer than 50 or the sequence identity is more than 40 % are filtered. The whole dataset is assembled by protein pairs of the number of 11,188, in which half of the dataset is positive and the rest is negative. It must note that the non-tautological dataset we have used is same as *Guo*'s.

2.2   Feature Vector Extraction

For more effective prediction of PPIs, it is inevitable to extract feature vectors from the protein sequences. What's more, feature vectors represent the essence of information of the encoded proteins. In this section, we adopt a protein representation model and an effective texture descriptor. Before feature extraction, it is necessary to undertake the preprocessing referred to a protein representation—Physicochemical Property Response

Matrix (PR) [13]. First the physicochemical property response matrix $PRM(i,j) \in R_{N \times N}$ is generated from the protein sequence $P = (p_1, p_2, ..., p_N)$ corresponding to a specific physicochemical property such as Hydrophobicity index, alpha-CH chemical shifts, Signal sequence helical potential and so on, and setting the value of the sum of the two values by indexing the position i and j in the physicochemical property for $PRM(i,j)$. Consider

$$PRM(i,j) = index(p_i) + index(p_j) \ i,j = 1, ..., N, \tag{1}$$

where $index(a)$ returns the value of the specific property for amino acid a.

In our method, we apply the Hydrophobicity index as the physicochemical property and the values are 0.61, 0.60, 0.06, 0.46, 1.07, 0, 0.47, 0.07, 0.61, 2.22, 1.53, 1.15, 1.18, 2.02, 1.95, 0.05, 0.05, 2.65, 1.88, and 1.32 corresponding to the amino acids 'A' 'R' 'N' 'D' 'C' 'Q' 'E' 'G' 'H' 'I' 'L' 'K' 'M' 'F' 'P' 'S' 'T' 'W' 'Y' and 'V'. For example, if the protein sequence P = 'ARN', the PRM is as follow:

$$PRM = \begin{bmatrix} 0.61 + 0.61 & 0.61 + 0.60 & 0.61 + 0.06 \\ 0.60 + 0.61 & 0.60 + 0.60 & 0.60 + 0.06 \\ 0.06 + 0.61 & 0.06 + 0.60 & 0.06 + 0.06 \end{bmatrix}.$$

Then the PRM matrix is treated as an image and compressed to 250 × 250 if larger. And Local Phase Quantization (LPQ) is an effective texture descriptor. The foundation of the LPQ is the blur invariance property of the Fourier phase spectrum. And the operation is as follow:

$$g(x) = f(x) \times h(x), \tag{2}$$

where $g(x)$ is denoted as the observed image, $f(x)$ is as the original image, and $h(x)$ is as the blur function. And the Fourier is as follow:

$$G(x) = F(x) \times H(x), \tag{3}$$

where the function $G(x)$, $F(x)$ and $H(x)$ are the Fourier transforming of $g(x)$, $f(x)$ and $h(x)$ respectively.

In LPQ, it operates the Fourier Transform on the local image to reflect the local information effectively. That is, the operation of the Fourier Transform is as follow:

$$F(u,x) = \sum_{y \in N_{m \times n}} f(x-y)e^{-j^2 \pi u^T y = w_u^T f_x}. \tag{4}$$

The local phase information is extracted from the 2-D short-term Fourier Transform (STFT) that is worked out a rectangular neighborhood transformed from each pixel position. After STFT, it could retain four complex coefficients that match four fied 2-D frequencies. And that are divided into real and imaginary parts and then quantized as integers between 0–255 using a binary coding scheme. A normalized histogram of such coefficients generated is as the final feature vector we need.

2.3   Rotation Forest Classifier

Rotation Forest is an excellent classifier without processing the trade-off between the accuracy and diversity in the design of multiple classifier system [22]. Due to its high classification accuracy and non-coincident errors, multiple classifier system is an active research field in machine learning and pattern recognition.

Assuming that the size of a training sample set matrix X is $N \times n$. The matrix X represents for N training samples and n features. Set a label of figure 1 or -1 for each sample, in which figure 1 means PPI and figure -1 means non-PPI, and the labels are denoted as $Y = [y_1, y_2, ..., y_N]^T$. Set K for the amount of the feature subset and L for the amount of the decision trees in RF. Denote the D_i as a decision trees. Note that the parameters L and K must be set before training. The process of the training for an individual classifier D_i is as following steps:

Step 1: Split the feature set F into K subsets at random. Arranging that each feature subset holds $M = n/K$ features.

Step 2: Put F_{ij} to be the *jth* subset of features for training classifier D_i, and X_{ij} to be the dataset X for the features in F_{ij}. In each nonempty subset, it is picked out from $X_{ij}^{'}$ at random. Then to organize a new training set, three fifth of the dataset describe a bootstrap subset of targets that is denoted as the new set $X_{ij}^{'}$. Subsequently, apply PCA on $X_{ij}^{'}$ to generate the coefficients that are stored in a matrix C_{ij} denoted by the coefficients of principal components, $a_{i1}^{(1)}, ..., a_{ij}^{M_j}$, the size of each is $M \times 1$.

Step 3: Build a sparse rotation matrix Ri by organizing the obtained vectors Cij with the coefficients of principal component, $a_{i1}^{(1)}, ..., a_{ij}^{M_j}$, as follows:

$$R_i = \begin{bmatrix} a_{i1}^{(1)}, ..., a_{ij}^{M_j} & \{0\} & \cdots & \{0\} \\ \{0\} & a_{i2}^{(1)}, ..., a_{i2}^{M_2} & \{0\} & \cdots \\ \vdots & \{0\} & \cdots & \cdots \\ \vdots & \cdots & \cdots & \cdots \\ \{0\} & \{0\} & \cdots & a_{iK}^{(1)}, ..., a_{iK}^{M_K} \end{bmatrix}. \quad (5)$$

For matching the order of the features set F, it is necessary to rearranging the columns of R_i and then build $R_i^a (sizeN \times n)$. (Y, XR_i^a) is set as the training set for building classifier D_i. After the training phase, for a given test sample x, the probability $d_{i,j}(xR_i^a)$ is assigned by the classifier D_i, and the classifier assumes the sample x has correlation with the decision trees. Then, calculate an average $\mu_j(x)$ of all the probabilities as follow:

$$\mu_j(x) = \frac{1}{L} \sum_{i=1}^{L} d_{i,j}(xR_i^a), \ j = 1, ..., c. \quad (6)$$

Finally, assign x to the class with the largest confidence.

3   Experiments and Results

3.1   Evaluation Measures

In order to measure the prediction performance of the proposed method, *Sensitivity*, *Precision*, Matthews's correlation coefficient (*MCC*), and overall *Accuracy* were calculated. The definitions of these measures are defined as follows:

$$Accuracy = \frac{TP + TN}{TP + FP + TN + FN}, \tag{7}$$

$$Sensitivity = \frac{TP}{TP + FN}, \tag{8}$$

$$Precision = \frac{TP}{TP + FP}, \tag{9}$$

$$MCC = \frac{TP \times TN - FP \times FN}{\sqrt{(TP + FN) \times (TN + FP) \times (TP + FP) \times (TN + FN)}}, \tag{10}$$

where true positive (*TP*) stands for the numeral of true PPIs that are of correct prediction; false negative (*FN*) stands for the numeral of true PPIs that are predicted incorrectly for its non-interacting; false positive (*FP*) represents the numeral of the true non-interacting pairs that are predicted to the opposite side, and true negative (*TN*) represents the numeral of true non-interacting pairs that are of correct prediction; Mathew's correlation coefficient is abbreviated to *MCC*.

3.2   Parameter Selection

In our method, we need to set the two vital parameters that the parameter *L* represents the amount of the decision trees in the RF and the parameter *K* is for the amount of the split features. In Fig. 1, we set the *K* = 8 and let *L* be from 3 to 30. After finishing the

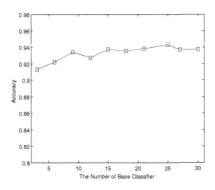

Fig. 1. Overall accuracy rate with an increased *L* of decision trees' amount

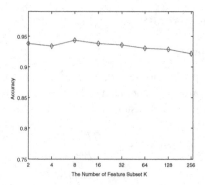

Fig. 2. Overall accuracy rate with an increased K of features subset amount

training with the increased L by the interval of 3, the accuracy achieves at 94.01 % with L of 24. For the best result, we train the RF at $K = 8$ and L from 21 to 27. Finally, 25 of the L make a best result with accuracy at 94.32 %. The following is to find the best value of K that is set to the nth power of 2 because the amount of the features is 512 and we split it evenly. Figure 2 shows that the best parameter of K can be set to 8.

3.3   Prediction Performance of Proposed Model

For the stability of calculation result, a 5-times random selecting is operated. For each, the four fifth of the processed dataset are as the training set and the rest are for testing.

Table 1 shows the results of the proposed method. From Table 1, it is obvious that the precisions are more than 95.9 %, and the sensitivities are more than 90.72 %. What's more, employing the proposed method the accuracy of the PPI prediction

Table 1. The prediction result of the test dataset using proposed method.

Model	Testing set	Sensitivity (%)	Precision (%)	Accuracy (%)	MCC (%)
Proposed method	1	91.35	97.03	94.32	89.27
	2	90.72	95.90	93.47	87.78
	3	91.35	96.76	94.19	89.03
	4	91.26	96.38	93.97	88.64
	5	90.81	96.18	93.65	88.09
	Average	**91.10 ± 0.31**	**96.45 ± 0.45**	**93.92 ± 0.36**	**88.56 ± 0.63**
Zhou's work	SVM + LD	87.37 ± 0.22	89.50 ± 0.60	88.56 ± 0.33	77.15 ± 0.68
Guo's work	ACC	89.93 ± 3.68	88.87 ± 6.16	89.33 ± 2.67	*N/A*
	AC	87.30 ± 0.22	87.82 ± 4.33	87.36 ± 1.38	*N/A*
Yangs' work	Cod1	75.81 ± 1.20	74.75 ± 1.23	75.08 ± 1.13	*N/A*
	Cod2	76.77 ± 0.69	82.17 ± 1.35	80.04 ± 1.06	*N/A*
	Cod3	78.14 ± 0.90	81.86 ± 0.99	80.41 ± 0.47	*N/A*
	Cod4	81.03 ± 1.74	90.24 ± 1.34	86.15 ± 1.17	*N/A*

model is 93.92 % ± 0.36. From the statistics of the table, it shows that an average MCC value of 88.09 % represents that the proposed method perform well. Moreover, the standard deviation values of the sensitivity, precision, accuracy and MCC are as low as 0.31 %, 0.45 %, 0.36 %, and 0.63 % respectively.

Besides, many other researchers who are keen on investigations of predicting PPIs come up with their own sequence-based methods. We compare our method with those of Guo et al. [14], Zhou et al. [18], and Yang et al. [19] so as to evaluate the prediction performances of the RF model. It can be observed from Table 1 that the proposed method can achieve better performance than all of these methods which highlights the superiority of the proposed method. Moreover, From Table 1 we can see that the proposed method is robust and of high accuracy.

4 Discussion and Conclusions

In this study, we propose a novel approach for detecting PPIs by combining the RF classifier with a LPQ method. The RF classifier is employed to build up the prediction model, and the LPQ method is applied to extract the sequence information of proteins. As a popular feature extraction approach widely used in the image processing, LPQ-based texture descriptor can extract the local phrase information from a matrix. Experiments on *S. cerevisiae* dataset demonstrate the superiority of the proposed method over existing predictors. We believe that the proposed method will be a bright and meaningful method for future proteomics research.

Acknowledgments. This work is supported in part by the National Science Foundation of China, under Grants 61373086, 61202347, and 61401385, in part by the Guangdong Natural Science Foundation under Grant 2014A030313555, and in part by the Shenzhen Scientific Research and Development Funding Program under grants JCYJ20140418095735569.

References

1. Gavin, A.C., Bosche, M., Krause, R., Grandi, P.: Functional organization of the yeast proteome by systematic analysis of protein complexes. Nature **415**(6868), 141–147 (2002)
2. Ito, T., Chiba, T., Ozawa, R., Yoshida, M., Hattori, M., Sakaki, Y.: A comprehensive two-hybrid analysis to explore the yeast protein interactome. Proc. Natl. Acad. Sci. USA **98** (8), 4569–4574 (2001)
3. Zhu, L., You, Z.H., Wang, B., Huang, D.S.: t-LSE: a novel robust geometric approach for modeling protein-protein interaction networks. PLOS One, **8**(4), Article e58368 (2013)
4. Krogan, N.J., Cagney, G., Yu, H.Y., Zhong, G.Q.: Global landscape of protein complexes in the yeast Saccharomyces cerevisiae. Nature **440**(7084), 637–643 (2006)
5. Guo, Y., Yu, L., Wen, Z., Li, M.: Using support vector machine combined with auto covariance to predict proteinprotein interactions from protein sequences. Nucleic Acids Res. **36**(9), 3025–3030 (2008)
6. You, Z.-H., Yin, Z., Han, K., Huang, D.-S., Zhou, X.: A semisupervised learning approach to predict synthetic genetic interactions by combining functional and topological properties of functional gene network. BMC Bioinform. **11**, 343 (2010)

7. You, Z.-H., Lei, Y.-K., Gui, J., Huang, D.S., Zhou, X.: Using manifold embedding for assessing and predicting protein interactions from high-throughput experimental data. Bioinformatics 26, 2744–2751 (2010)
8. Xia, J.-F., You, Z.-H., Wu, M., Wang, S.-L., Zhao, X.-M.: Improved method for predicting pi-turns in proteins using a twostage classifier. Protein Pept. Lett. 17, 1117–1122 (2010)
9. Lei, Y.K., You, Z.H., Ji, Z., Zhu, L., Huang, D.S.: Assessing and predicting protein interactions by combining manifold embedding with multiple information integration. BMC Bioinformatics, 13, Article S3 (2012)
10. You, Z.-H., Li, L., Yu, H., Chen, S., Wang, S.-L.: Increasing reliability of protein interactome by combining heterogeneous data sources with weighted network topological metrics. In: Huang, D.-S., Zhao, Z., Bevilacqua, V., Figueroa, J.C. (eds.) ICIC 2010. LNCS, vol. 6215, pp. 657–663. Springer, Heidelberg (2010)
11. Lei, Y.-K., You, Z.-H., Dong, T., Jiang, Y.-X., Yang, J.-A.: Increasing reliability of protein interaction by fast manifold embedding. Pattern Recogn. Lett. 34, 372–379 (2013)
12. You, Z.H., Yu, J.Z., Zhu, L., Li, S., Wen, Z.K.: A MapReduce based parallel SVM for large-scale predicting protein-protein interactions. Neurocomputing 145(5), 37–43 (2014)
13. You, Z.H., Yin, Z., Han, K., Huang, D.S., Zhou, X.B.: A semi-supervised learning approach to predict synthetic genetic interactions by combining functional and topological properties of functional gene network. BMC Bioinform. 11, 343 (2010)
14. Guo, Y., Yu, L., Wen, Z., Li, M.: Using support vector machine combined with auto covariance to predict protein-protein interactions from protein sequences. Nucleic Acids Res. 36(9), 3025–3030 (2008)
15. Xia, J.F., Han, K., Huang, D.S.: Sequence-based prediction of protein-protein interactions by means of rotation forest and autocorrelation descriptor. Protein Pept. Lett. 17(1), 137–145 (2010)
16. Shi, M.G., Xia, J.F., Li, X.L., Huang, D.S.: Predicting protein-protein interactions from sequence using correlation coefficient and high-quality interaction dataset. Amino Acids 38 (3), 891–899 (2010)
17. You, Z.H., Li, S., Gao, X., Luo, X., Ji, Z.: Large-scale protein-protein interactions detection by integrating big biosensing data with computational model. BioMed Res. Int. 2014, 9, Article 598129 (2014)
18. Zhou, Y.Z., Gao, Y., Zheng, Y.Y.: Prediction of protein-protein interactions using local description of amino acid sequence. In: Zhou, M., Tan, H. (eds.) CSE 2011, Part II. CCIS, vol. 202, pp. 254–262. Springer, Heidelberg (2011)
19. Yang, L., Xia, J.F., Gui, J.: Prediction of protein-protein interactions from protein sequence using local descriptors. Protein Pept. Lett. 17(9), 1085–1090 (2010)
20. Shen, J., Zhang, J., Luo, X., Zhu, W., Yu, K., Chen, K., Li, Y., Jiang, H.: Predicting protein-protein interactions based only on sequences information. Proc. Natl. Acad. Sci. USA 104(11), 4337–4341 (2007)
21. You, Z.H., Zhu, L., Zheng, C.H., Yu, H.J., Deng, S.P., Ji, Z.: Prediction of protein-protein interactions from amino acid sequences using a novel multi-scale continuous and discontinuous feature set. BMC Bioinformatics 15(S15), Article S9 (2014)
22. You, Z.H., Lei, Y.K., Zhu, L., Xia, J.F., Wang, B.: Prediction of protein-protein interactions from amino acid sequences with ensemble extreme learning machines and principal component analysis. BMC Bioinform. 14(8), 10 (2013)
23. Luo, X., You, Z.H., Zhou, M.C., Li, S., Leung, H.: A highly efficient approach to protein interactome mapping based on collaborative filtering framework. Scientific reports, Report 5, Article 7702 (2015)

Healthcare and Medical Methods

Age-Related Alterations in the Sign Series Entropy of Short-Term Pulse Rate Variability

Yongxin Chou[1,2] and Aihua Zhang[1,2(✉)]

[1] College of Electrical and Information Engineering, Lanzhou University of Technology,
No. 287 LanGongPing Street, Qilihe District, Lanzhou 730050, China
lutzhangah@163.com
[2] Key Laboratory of Gansu Advanced Control for Industrial Processes,
Lanzhou 730050, China

Abstract. So far, the pulse rate variability (PRV) analysis methods cannot effectively extract the nonlinear changes of heart beat and need long time data. So a non-linear approach, sign series entropy analysis (SSEA), is employed to derive age-related alterations from short-term PRV, and a probabilistic neural network (PNN) is designed to classify subjects according to their ages. Continuous non-invasive blood pressure signals are chosen to generate short-term PRV signals as the experimental data, and their time domain and frequency domain parameters are also extracted for comparison. The experimental results show that the sign series entropy has a significant difference between young and old subjects, even if the PRV is corrupted by heavy noises; and PNN can accurately classify subjects. SSEA is more suitable for analyzing short-term PRV signals.

Keywords: Short-term pulse rate variability · Sign series entropy · Probabilistic neural network · Age classification

1 Introduction

Heart rate variability (HRV) is the result of balance between sympathetic and parasympathetic systems. Various studies have shown that HRV contains abundant physiological and pathological information on cardiovascular neural systems [1]. HRV is derived from ECG signals. However, during recording ECG signals, multiple electrode attachments and cable connections is not convenient for data sampling. Compared with HRV, Pulse rate variability (PRV) can be derived from pulse signals, which are easily collected [2]. In addition, a wide range of studies have shown that PRV can be a surrogate of HRV [3, 4].

At present, HRV is analyzed either in the time domain or in the frequency domain. In addition, there are geometric and nonlinear approaches [5] to deal with it. Some nonlinear approaches have shown that aging has a profound impact on HRV [6]. However, most methods need long-term HRV or PRV signals, some even over 24 h. In clinical applications, methods suitable for short-term PRV (<10 min) are more useful.

© Springer International Publishing Switzerland 2015
D.-S. Huang and K. Han (Eds.): ICIC 2015, Part III, LNAI 9227, pp. 723–729, 2015.
DOI: 10.1007/978-3-319-22053-6_76

The nonlinear method, sign series entropy analysis (SSEA), was employed to analyze short-term HRV signals [7], and the experimental results showed that the sign series entropy (SSE) evidently changed along with aging. Up to present, it is unknown whether SSEA is suitable for analyzing short-term PRV signals.

In this study, the principle of SSEA is first introduced. Then, the SSEA method is employed to analyze short-term PRV signals from young and old subjects. Moreover, the parameters of PRV signals in the time and the frequency domains are extracted for comparison. Finally, a probabilistic neural network is utilized to classify subjects based on SSE, the time and the frequency domain parameters, respectively.

2   Materials and Methods

2.1   Data

The PRV signals are extracted from the continuous non-invasive blood pressure signals from the MIMIC/Fantasia database [8]. Two groups of healthy subjects, 20 young and 20 old subjects undergo 120 min of supine resting while continuous ECG and respiration signals are collected. All subjects watch the movie, Fantasia (Disney, 1940), to maintain wakefulness, and all signals are digitized at 250 Hz. Only a half of subjects of each group are investigated to record continuous non-invasive blood pressure signals with the duration of 66 min.

Compared to ECG signals, the pulse beats in blood pressure signals are uncalibrated. So the dynamic difference threshold detection algorithm [9] is used to calibrate the pulse beats, and each beat annotation is verified by visual inspection. Then the pulse rate variability is obtained from continuous pulse beat intervals.

2.2   Sign Series Entropy Analysis of Short-Term Pulse Rate Variability

Denote PRV signals as $\{PP(i)\}$, $i = 1, 2, ..., N$, where $PP(i)$ is the interval between the i-th and the $(i + 1)$-th pulse beat, and N is the number of pulse beat intervals. Here, three symbols are employed to represent the variation directions of PRV signals:

$$s(i) = \begin{cases} 0, & PP(i+1) < PP(i) \\ 1, & PP(i+1) = PP(i) \quad i = 1, 2, \cdots, N-1 \\ 2, & PP(i+1) > PP(i) \end{cases} \tag{1}$$

Where, $s(i) = 0, 1, 2$ indicates the pulse beat interval is descendant, invariant, and ascendent, respectively. From Eq. (1), the magnitude of PRV signals is coarse-grained, and only the change direction is remained.

To seek for the rule of signal $s(i)$'s change, a vector series with the size of $(N - m) \times m$ is defined as:

$$S(i) = [s(i), s(i + 1), ..., s(i + (m - 1))], \quad i = 1, 2, ..., N - m \tag{2}$$

Equation (2) show the process of generating $S(i)$. Each vector replaces a kind of changing mode of PRV signals. When the length of $S(i)$ is m, there will have $M = 3^m$ continuous modes. The probability of each mode is:

$$p_j = \frac{N_j}{N - m}, \quad j = 1, 2, \ldots, M \tag{3}$$

Where N_j is the number of the j-th mode in $S(i)$. Then the information entropy of each mode is:

$$SSE(m) = -\sum_{j=1}^{M} p_j \log_2 p_j \tag{4}$$

Here, we rename the above information entropy as the sign series entropy (SSE), and its value indicates the average irregularity of $S(i)$.

2.3 Probabilistic Neural Network

Probabilistic neural network (PNN) proposed by Donald F. Specht has been widely applied for its capabilities in learning and generalization. It consists in three layers, i.e., input, hidden and output layers, where the input layer just receives input signals. The hidden layer has a number of neurons for classifying these input signals by multidimensional kernels. Here, we choose the Gaussian function as the kernels of PNN. The output layer outputs the final results of PNN [10].

In this study, the PRV signals are derived from 20 subjects. Each PRV signal is divided into 6 successive segments as the short-term PRV signals.

Then, the SSE of these short-term PRV signals is computed and input into PNN. Meanwhile, some parameters in the time and the frequency domains are extracted for comparison to determine whether SSE evidently changes along with aging. In the time domain, the parameters are the mean value (MEANPP), the standard deviation (SDPP), the root mean square of successive differences of pulse beat to beat intervals (RMSSD), and the proportion of successive intervals differences over 50 ms (pNN50). In the frequency domain, we compute the power spectrum density (here we use the Welch PSD) after resampling the short-term PRV signals at 2 Hz with cubic spline interpolation. Then, the normalized low frequency (LF) power (0.04–0.15 Hz), high frequency (HF) power (0.15–0.4) and the location of their peaks (peakLF and peakHF), as well as the LF/HF ratio are extracted from the Welch PSD.

We extract these parameters from the short-term PRV signals. For each parameter, we can obtain 120 samples. Then, t-test is used to test whether the parameter significantly changes between the young and the old. Those parameters that significantly change are chosen and combined as new models (here, model just represents the different types inputs consist of the parameters). For each parameter with 120 samples, 100 of them are randomly chosen as the training data, and the rest are as the testing data when training a PNN. Further, we compare the accuracy in classification between SSE and the other types of inputs.

3 Results

3.1 The Results of Sign Series Entropy Analysis

The Result of SSE. Figure 1 shows the SSE distribution of the young and the old subjects, where the first 60 samples are the SSE values of the old subjects, and the rest are the SSE values of the young subjects. From this figure, the SSE value increases as the increase of aging, and there is a significantly difference between these two groups ($P = 6.4885e-23$).

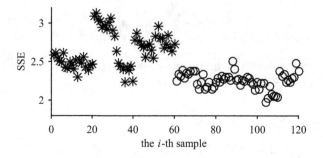

Fig. 1. The comparison of SSE ($m = 2$) between the young and the old group where '*' represents the old subjects, and 'o' means the young subjects

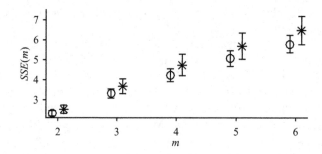

Fig. 2. The SSE results with different length of construct vector when the length of short-term PRV is 10 min (mean ± standard deviation), where '*' represents the old subjects, and 'o' is the young subjects

The Influence of the Length of Construct Vector to SSE. According to the theory of SSEA, the longer the construct vector, the more kinds the variation modes of PRV, the more computation time of SSEA. So it is important to choose an appropriate value of m. The result with different values of m is shown as Fig. 2. From this figure, the SSE value increases as the increase of m. The t-test results corresponding to $m = 2$ to 6 are $P = 0.0016$ to 0.0019. The computation time increases from 0.6936 ms to 1.0301 ms (The computation time is obtained under the implementation environment of Intel (R) Core (TM) i7-4510U CPU). From the t-test results, the difference of SSE values between the young and the old subjects is not significant along with m. However, the computation time of SSEA evidently increases. Considering both the accuracy and real-time performance of SSEA, $m = 2$ is a good choice in SSEA.

The Noise Immunity of SSEA. The PRV signals, which are corrupted by random noise to imitate the effect of inaccuracy of beat intervals detection, are analyzed by SSEA. The result is shown as Fig. 3, as the signal to noise rate decreases from 34 dB to 16 dB, the SSE values of the old subjects decrease, whereas those of the young subjects do not obviously change. There are significantly difference in the SSE values between the young and the old subjects, even if the PRV signals have heavy noise.

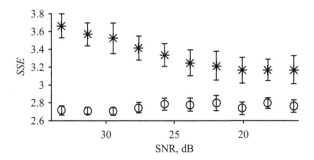

Fig. 3. The noise immunity of SSEA when the length of short-term PRV is 10 min and $m = 2$ (mean ± standard deviation), where '*' represents the old subjects, and 'o' is the young subjects

3.2 The Results of Probabilistic Neural Network

The Results of t-test. For each parameter, we extract 120 samples from the short-term PRV signals, and the t-test is employed to determine whether there is a significant distinction between the young and the older subjects for this parameter. The results are as follows: MEANPP: $P = 0.9159$. SDPP: $P = 1.6251e\text{-}06$. RMSSD: $P = 0.0013$. pNN50: $P = 1.0564e\text{-}19$. LF: $P = 3.5892e\text{-}08$. peakLF: $P = 0.1678$. HF: $P = 1.7986e\text{-}05$. peakHF: $P = 0.1678$. LF/HF: $P = 0.8992$. SSE: $P = 3.8432e\text{-}22$.

The Results of Model Building. From the t-test results, SDPP, pNN50, LF, HF and SSE, which have more significantly difference than other parameters, are chosen to build models for classifying the subjects based on aging. There are 15 models with 5 parameters, as listed in Table 1, where '*' means that the parameter in this row is chosen, and 1–15 in the first row refer to the number of models.

The Results of PNN. In this study, PNN is trained based on the samples of 15 models to classify the young and the old subjects. The accuracy of all models are over 70 %; model 1 is the most accurate, with its accuracy of 93.3 %; the accuracy of model 10 is 90 %; and the accuracy of models 2, 3, 6, 7 are above 85 %. Model 1 has only one parameter, SSE, so SSE is the most related with aging among all these parameters.

Table 1. The models consist of SSE, SDPP, pNN50, LF and HF.

	1	2	3	4	5	6	7	8	9	10	11	12	13	14	15
SSE	*	*	*	*	*										
SDPP		*	*	*	*	*	*	*	*						
pNN50			*	*	*		*	*	*	*	*	*			
LF				*	*			*	*		*	*	*	*	
HF				*					*			*		*	*

4 Discussion

In this work, SSEA is employed to extract the temporal and structural information of the short-term PRV signals of the young and the old subjects. Compared with the SSEA results of the short-term HRV signals in [7], the same conclusion we get about the SSE values of HRV and PRV, they all increase as the increase of aging, but the SSE values of PRV change significantly compared with those of HRV (HRV: $P = 3.5686e\text{-}12$, PRV: $P = 6.4885e\text{-}23$). Moreover, the result of PNN for model 1 (only contains SSE) shows that the SSE value is valid to classify the young and the old subjects (the accuracy >93 %).

In addition, to compare the parameters of the time and the frequency domains with SSE, the 15 models are employed as the inputs of PNN to classify the subjects. The accuracy of models with one parameter (as list in Table 1, model 1, 6, 10, 13 and 15) are 93.3 %, 86.7 %, 90 %, 76.7 % and 80 %, respectively; the model that consists of SSE is the most accurate. In conclusion, the SSE value of short-term pulse rate variability is evidently related with the variation of age.

5 Conclusion

In this study, the sign series entropy analysis method is employed to derive the alterations of aging from the young and the old subjects, and the probabilistic neural network is designed to classify the subjects by their ages. The experimental results show that, compared with the commonly used parameters, SSE is significantly changed as the increase of age; and the PNN trained with SSE is efficient to classify the subjects. Therefore, because of the simplicity and the computational efficiency of SSEA method, it can be employed to analyze the short-term PRV signals in the portable medical devices.

Acknowledgments. This work was supported by the open project program of the national laboratory of pattern recognition (grant 201407347), the natural science foundation of Gansu province (grant 1308RJZA225, 145RJ2A065) and the national natural science foundation of China (grant 81360229).

References

1. Thayer, J.F., Yamamoto, S.S., Brosschot, J.F.: The relationship of autonomic imbalance, heart rate variability and cardiovascular disease risk factors. Int. J. Cardiol. **141**, 122–131 (2010)
2. Jang, D.G., Farooq, U., Park, S.H., Hahn, M.: A robust method for pulse peak determination in a digital volume pulse waveform with a wandering baseline. IEEE Trans. Biomed. Circuits Syst. **8**(5), 729–737 (2014)
3. Heathers, J.A.: Smartphone-enabled pulse rate variability: an alternative methodology for the collection of heart rate variability in psychophysiological research. Int. J. Psychophysiol. **89**(3), 297–304 (2013)
4. Schäfer, A., Vagedes, J.: How accurate is pulse rate variability as an estimate of heart rate variability? A review on studies comparing photoplethysmographic technology with an electrocardiogram. Int. J. Cardiol. **166**(1), 15–29 (2013)
5. Mohan, A., James, F., Fazil, S., Joseph, P.K.: Design and development of a heart rate variability analyzer. J. Med. Syst. **36**(3), 1365–1371 (2012)
6. Takahashi, A.C., Porta, A., Melo, R.C., Quitério, R.J., Silva, E.D., Borghi, S.A., Tobaldini, E., Montano, N., Catai, A.M.: Aging reduces complexity of heart rate variability assessed by conditional entropy and symbolic analysis. Intern. Emerg. Med. **7**(3), 229–235 (2012)
7. Bian, C., Ma, Q., Si, J., Wu, X., Shao, J., Ning, X., Wang, D.: Sign series entropy analysis of short-term heart rate variability. Chin. Sci. Bull. **54**(24), 4610–4615 (2009)
8. The MIMIC Database. http://www.physionet.org/cgi-bin/atm/ATM
9. Chou, Y.X., Zhang, A.H., Wang, P., Gu, J.: Pulse rate variability estimation method based on sliding window iterative DFT and Hilbert transform. J. Med. Biol. Eng. **34**(4), 347–355 (2014)
10. Wang, J.S., Chiang, W.C., Hsu, Y.L., Yang, Y.T.: ECG arrhythmia classification using a probabilistic neural network with a feature reduction method. Neurocomputing **116**, 38–45 (2013)

An Assessment Method of Tongue Image Quality Based on Random Forest in Traditional Chinese Medicine

Xinfeng Zhang, Yazhen Wang[✉], Guangqin Hu, and Jing Zhang

Signal and Information Processing Lab, Beijing University of Technology,
No. 100, Pingleyuan, Chaoyang District, Beijing 100124, China
zxf@bjut.edu.cn, {wyzsnowing,yiyijiuzai}@126.com,
hdmh@163.com.cn

Abstract. In the study and practice of the tongue characterization, experienced doctors found that a large number of the tongue images collected by tongue image instrument don't meet the clinical requirement, which will directly affects the final result of tongue image analysis. In this paper, the automatic quality evaluation of tongue image is designed for the first time through the following steps. First, the original tongue images are processed. Second, statistics of local normalized luminance based on natural scene statistics (NSS) model, color, geometric and texture features of tongue images are extracted respectively. Finally, the Random Forest classifier is used to classify. Experimental results show that the method we proposed can get a better evaluation of tongue image quality. This approach can provide reliably reference data for assisted tongue image analysis.

Keywords: Tongue image · Quality assessment · NSS · Random forest

1 Introduction

Traditional tongue diagnosis mainly depends on the doctor's subjective analysis and manual record, which can't collect and save the features of the tongue images automatically. In Signal and Information Processing Lab of Beijing University of Technology, tongue image analysis instrument (TIAI) has been built to acquire and analyze the tongue image [1, 2]. The emergence of TIAI can solve the manual record problem of tongue images characteristics and can give a relatively accurate diagnosis result. But there exist some problems. Tongue image analysis by TIAI includes four steps: image acquisition, pre-processing, feature extraction and recognition. But in the process of acquiring an image, there is not a system that can evaluate the quality of tongue image. The judgment whether the tongue image is qualified or not belongs to the domain of image quality evaluation in essence.

In this paper, we propose an assessment method of tongue image quality based on the Random Forest in TCM in order to pick out the qualified images. In Diagnostics of Chinese Medicine [3], the requirements of tongue inspection are sticking out tongue, fully exposing tongue, to be natural, relaxed, and the tongue should be flat diastole. But in the study of tongue characterization, experienced doctors found that a larger number of tongue images collected by TIAI don't meet the clinical diagnostic requirement.

© Springer International Publishing Switzerland 2015
D.-S. Huang and K. Han (Eds.): ICIC 2015, Part III, LNAI 9227, pp. 730–737, 2015.
DOI: 10.1007/978-3-319-22053-6_77

Those images are shown by Fig. 1(b) and (d). So, we put forward a method to distinguish the qualified and unqualified tongue images. So far, we haven't found the papers on TCM tongue image quality evaluation at home and abroad.

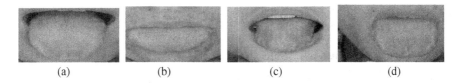

(a)   (b)   (c)   (d)

Fig. 1. (a) Qualified image. (b) Unexposed fully tongue image. (c) No relaxed tongue image. (d) Blur tongue image

2   Pre-processing Image

In this paper, the tongue images (all labelled by medical experts) we used are collected from the people who undergo a medical examination by tongue image acquisition device (see Fig. 2(a)) in hospital of Beijng University of Technology. Size of the original image is 4271×2848 as shown in Fig. 2(b).

(a)   (b)

Fig. 2. (a) Tongue image acquisition device. (b) Original image.

The original tongue images includes tongue (target area), and background regions. For the whole image, there are a lot of backgrounds having nothing to do with the study object. So target area will be found through pre-processing.

3   Feature Extraction

In the process of acquisition, there will be a variety of unqualified tongue images due to the shaking of human or human tongue. In order to determine whether it is qualified or not, feature based on NSS model is extracted. Color and geometric features are extracted to determine whether it is the fully exposed tongue image, and texture features are extracted to determine whether the tongue image is relaxed or not.

3.1   Feature Extraction Based on NSS

In [4, 5] we know that natural images are not necessarily images of natural environments such as trees or skies. Any natural light image that is captured by an optical camera and is not subjected to artificial processing on a computer is regarded as a natural image. Tongue images also are the natural images, and also possess certain regular statistical properties [6, 7]. We use locally normalized luminance coefficients to quantify possible losses of "naturalness" in the image. So in this paper, we have the same features with recently introduced NSS based Blind/Referenceless Image Spatial Quality Evaluator (BRISQUE) proposed in [4, 5].

GGD (Generalized Gaussian Model) to Fit the MSCN Coefficients. Ruderman [8] found that normalized luminance values strongly tend to a unit normal Gaussian characteristic for natural image. We utilize the pre-processing model to transform luminances as mean subtracted contrast normalized (MSCN) coefficients. The MSCN coefficients have characteristic statistical properties that are changed by the presence of distortion. Figure 3 shows that MSCN coefficient distributions are symmetric. In this article, we use GGD distribution where the GGD with zero mean is given by Eq. (1).

$$f\left(x;\alpha,\delta^2\right) = \left(\alpha/\left(2\beta\Gamma\left(1/\alpha\right)\right)\right)\exp\left(-\left(|x|/\beta\right)^\alpha\right) \tag{1}$$

$$\beta = \sigma\left(\Gamma\left(1/\alpha\right)/\Gamma\left(3/\alpha\right)\right)^{1/2} \tag{2}$$

Where, $\Gamma\left(\alpha\right)$ is the gamma function, α controls the shape of the distribution while δ^2 control the variance.

As illustrated in Fig. 3, we also can see that unqualified tongue images have a heavy tail, which can distinguish two kinds of tongue images.

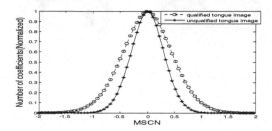

Fig. 3. Distribution of MSCN coefficients of qualified and unqualified tongue images

AGGD (Asymmetric Generalized Gaussian Distribution) to Fit the Paired Products of Neighboring MSCN Coefficients. We model the statistical relationships between neighboring pixels using the empirical distributions of paired products of adjacent MSCN coefficients in four orientations–horizontal, vertical, main-diagonal and secondary-diagonal. We utilize AGGD model to fit the paired products of adjacent MSCN coefficients, as illustrated in Eq. (3):

$$f\left(x;\alpha,\delta_l^2,\delta_r^2\right) = \begin{cases} \left(\alpha/\left(\left(\beta_l+\beta_r\right)\Gamma\left(1/\alpha\right)\right)\right)\exp\left(-\left(\left(-x\right)/\beta_l\right)^\alpha\right) & x < 0 \\ \left(\alpha/\left(\left(\beta_l+\beta_r\right)\Gamma\left(1/\alpha\right)\right)\right)\exp\left(-\left(\left(-x\right)/\beta_r\right)^\alpha\right) & x \geq 0 \end{cases} \tag{3}$$

$$\beta_l = \sigma_l\left(\Gamma\left(1/\alpha\right)/\Gamma\left(3/\alpha\right)\right)^{1/2} \tag{4}$$

$$\beta_r = \sigma_r \left(\Gamma \left(1/\alpha \right) / \Gamma \left(3/\alpha \right) \right)^{1/2} \tag{5}$$

α controls the shape of the distribution. δ_l^2 and δ_r^2 are control the left and right variances of the model, respectively.

Feature Selection. For each tongue image, the parameters (α, δ^2) of the best GGD fit and parameters $(\alpha, \eta, \delta_l^2, \delta_r^2)$ of the best AGGD fit are extracted. For each paired product, 16 parameters (4 parameters $\times$ 4 orientations) are computed as 16 features. Thus, a total of 18 features of each tongue image. Images are naturally multiscale, and distortions affect image structure across scales. Research [5] has demonstrated that increasing the number of scales beyond 2 did not improve the performance. Thus, the features are computed at two scales.

3.2 Feature Extraction of Color

In the image, the main colors are tongue color and complexion. According to statistics on the probability of all colors, we choose four most frequently occurring colors as the color feature. Images we used are in RGB color space. So we convert it into HSV [9] color space, because the HSV is closer to the way of human vision. Quantify hue (H), saturation (S) and value (V) respectively with unequal interval. Then H, S, V can be divided into 16 portions, 4 portions, 4 portions respectively, and integrated them into a one-dimensional histogram according to L = 16 h + 4 s + v.

3.3 Feature Extraction of Geometric

As shown in Fig. 1(a) and (b), the area of qualified tongue image is relatively large, and the aspect ratio is relatively small. But the unqualified tongue image is just on the contrary. Before we extract the geometric features, we should find the tongue body. In [10], we know that binary H component can display the contours of tongue and both sides of tongue and binary V component can present the contours of tongue base in HSV color space. So we use HSV color space. Convert RGB color space into HSV color space, and extract the H, S, V component respectively. Then binarize the image of H and V component, as shown by Fig. 4(a) and (b). We implement closing and opening operation of mathematical morphology on binary H component (Fig. 4(c)). Binary H component and V component are fused (Fig. 4(d)), and there are some lips, so we remove small area (Fig. 4(e)). The area and aspect ratio are extracted as the geometric features.

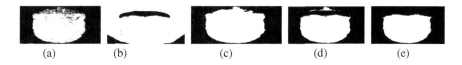

(a)   (b)   (c)   (d)   (e)

Fig. 4. (a) Binary tongue image of H. (b) Binary tongue image of V. (c) Morphological operations of H. (d) Fusion of tongue image. (e) Remove small area.

3.4   Feature Extraction of Texture

As shown in Fig. 1(c), unqualified tongue images have a big coarseness in tongue body edge. So, the tongue body edge should be found, and then the texture features are extracted. These features are more complex, and they don't have the regularity or periodicity and so on characteristics. So, statistical-based method can be used for extracting the features of the texture. In statistical-based method, Tamura [11] texture is the method that based on the human visual property. Thus, we utilize this method to extract the texture features. Consider the coarseness, contrast, roughness as the texture features. We can get the tongue body though the method of 3.3. Then we set the pixel value of the tongue image's middle region to zero automatically and then get the tongue body edge. Texture features of target tongue images are extracted.

4   Random Forest

Random Forest [12, 13] grows many classification trees. To classify a new object from an input vector, put the input vector down each of the trees in the forest. Each tree gives a classification, the forest chooses the classification having the most votes. It gives estimates of what variables are important in the classification. Therefore, we used Random Forest in this paper. Each tree is grown in the following way: If the number of cases in the training set is N, sample N cases at random – but with replacement, from the original data. This sample will be the training set for growing the tree. If there are M input variables, a number $m \ll M$ is specified such that at each node, m variables are selected at random out of the M and the best split on these m is used to split the node. The value of m is held constant during the forest growing. Each tree is grown to the largest extent possible. There is no pruning.

5   Experimental Results

We verify the generalization ability by oob error estimate and cross-validation (CV) using same data set.

Table 1. Time efficiency and generalization error comparison of CV and oob error rate

Method	Cross-validation	OOB error rate
Time(s)	64.1467	3.8604
Generalization error	0.1975	0.1848

From Table 1, we know that oob and CV have the similar generalization error. But the time complexity of CV is much bigger than oob. So Random Forest has a better performance. RF also can give estimates of what variables are important in the classification (see Fig. 5). Gini index and accuracy decreases for each feature over all trees in the forest gives a fast feature importance. From Fig. 5, we can know that the features that we choose are useful.

The performance of classifier is based on the scale of the RF. In order to study the identification of the RF under different number of decision tree, we use the same date set, same candidate attributes and different RF size. As illustrated in Table 2, different number of RF trees has different identification accuracy. We set the number of RF trees between 20 and 300, and get the best number automatically. The highest classification accuracy is 87.65 %.

In our experiment, we select 324 qualified and 324 unqualified tongue images. A quarter of samples are randomly selected as testing samples, the remaining 3/4 as training. We extract 36 dimensions of normalized luminance features, 4 dimensions of color features, 2 dimensions of geometric features and 3 dimensions of texture features,

Table 2. The number of decision trees influence on the identification accuracy

Number of trees Accuracy (%)	Testing set	Training set
10	77.78%	98.77%
30	87.04%	100%
50	84.57%	100%
100	82.10%	100%
200	84.57%	100%
500	85.80%	100%

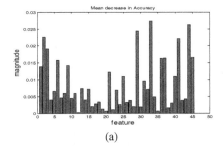 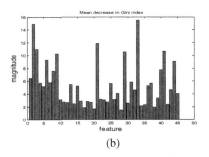

(a)   (b)

Fig. 5. (a) Mean decrease in accuracy (b) mean decrease in Gini index

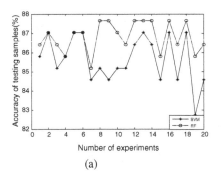 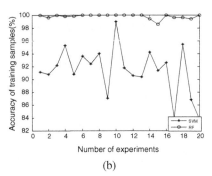

(a)   (b)

Fig. 6. Distribution of accuracy. (a) Testing samples (b) training samples

a total of 45 dimensions, then take the characteristics of data into normalization, using the RF and SVM to classify. Different testing and training samples are randomly selected respectively. Take 20 groups of experiments, the results distribution of testing samples and training samples are illustrated in Fig. 6. As shown by Fig. 6, for the same date, the classification accuracy rate of RF is higher than SVM on testing and training samples. Through the above experiment, we can know that RF is benefit to picking the qualified tongue images automatically.

6   Conclusions

Through the method we proposed, we can automatically pick out the qualified and unqualified tongue image in the clinical medicine and development of tongue characterization. Experiment results show this method can evaluate the quality of tongue image and have a better classification results. This approach is expected to be applied to the next generation of TIAI. We expect that the tongue image quality can be evaluated when it is acquired, which is our further work.

Acknowledgements.   The work is supports by the National Natural Science Foundation of China (No. 61201360), the Importation and Development of High-Caliber Talents Project of Beijing Municipal Institutions (CIT&TCD201504018) and General projects of Beijing Municipal Education Commission (No. JE334001201201). China.

References

1. Shen, L.S., Wei, B.G., Cai, Y.H., et al.: Image Analysis for Tongue Characterization. Chin. J. Electron. **12**(3), 317–323 (2005). (in Chinese)
2. Cao, M.L, Cai, Y.H., et al.: Recent process in new portable device for tongue image analysis. In: IEEE International Conference on Neural Networks and Signal Processing, pp. 492–499 (2008). (in Chinese)
3. Zhu, W.F.: Diagnostics of Chinese Medicine. People's Medical Publishing House (2011). (in Chinese)
4. Mittal, A., Moorthy, A.K., Bovik, A.C.: Blind/reference image spatial quality evaluation. In: Forty Fifth Asilomar Conference on Signals, Systems and Computer (ASILOMAR), pp. 723–727 (2012)
5. Mittal, A., Moorthy, A.K., Bovik, A.C.: No-reference image quality assessment in the spatial domain. IEEE Trans. Image Proc. **21**(12), 4695–4708 (2012)
6. Sheiki, H.R., Bovik, A.C., Cormack, L.: No-reference quality assessment using natural scene statistics: JPEG2000. IEEE Trans. Image Process. **14**(11), 1918–1927 (2005)
7. Moorthy, A.K., Bovik, A.C.: A two-step framework for constructing blind image quality indices. IEEE Signal Process. Lett. **14**(11), 1918–1927 (2005)
8. Ruderman, D.L.: The statistics of natural images. Netw. Comput. Neural Syst. **5**(4), 517–548 (1994)
9. Liu, F., Liu, X.Y., Chen, Y.: An efficient detection method for rare colored capsule based on RGB and HSV color space. In: IEEE International Conference on Granular Computing, pp. 175–178 (2014)

10. Sun, X.L., Pang, C.Y.: An improved snake model method on tongue segmentation. J. Changchun Univ. Sci. Technol. **36**(5), 154–156 (2013). (in Chinese)
11. Zhang, X.D., Shen, P.Y., Gao, J.R., et al.: A license plate recognition system based on tamura texture in complex conditions. In: International Conference on Information and Automation, pp. 1947–1952 (2010)
12. Brieman, L.: Random forests. Mach. Learn. **45**, 5–32 (2001)
13. Vrushali, Y.K., Pradeep, K.S.: Pruning of random forest classifier: a survey and future directions. In: International Conference on Data Science and Engineering, pp. 64–68 (2012)

Detection of Epileptic Seizures in EEG Signals with Rule-Based Interpretation by Random Forest Approach

Guanjin Wang[1,3(✉)], Zhaohong Deng[2], and Kup-Sze Choi[1,3]

[1] School of Nursing, Hong Kong Polytechnic University, Hong Kong, China
guanjin.br.wang@connect.polyu.hk
[2] School of Digital Media, Jiangnan University, Wuxi, Jiangsu, China
dzh666828@aliyun.com
[3] Centre for Smart Health, School of Nursing, Hong Kong Polytechnic University,
Hong Kong, China
thomasks.choi@polyu.edu.hk

Abstract. Epilepsy is a common neurological disorder and characterized by recurrent seizures. Although many classification methods have been applied to classify EEG signals for detection of epilepsy, little attention is paid on accurate epileptic seizure detection methods with comprehensible and transparent interpretation. This study develops a detection framework and focuses on doing a comparative study by applying the four rule-based classifiers, i.e., the decision tree algorithm C4.5, the random forest algorithm (RF), the support vector machine (SVM) based decision tree algorithm (SVM + C4.5) and the SVM based RF algorithm (SVM + RF), to two-group and three-group classification and the most challenging five-group classification on epileptic seizures in EEG signals. The experimental results justify that in addition to high interpretability, RF has the competitive advantage for two-group and three-group classification with the average accuracy of 0.9896 and 0.9600. More importantly, its performance is highlighted in five-group classification with the highest average accuracy of 0.8260 in contrast to other three rule-based classifiers.

Keywords: Seizure detection · EEG · Random forest · SVM · Ensemble learning approach

1 Introduction

Epilepsy is a common brain disorder, characterized by recurrent seizures [1]. EEG (electroencephalogram) signals are widely in use to detect the epilepsy by directly recording the brain's electrical activity. However, they still encounter the clinical difficulties. Generally, as stated in [2], there are two techniques involved in the detection system which are feature extraction techniques on the EEG input signals and classification techniques on extracted features. In this study, in order to achieve an accurate epileptic seizure detection with comprehensible and transparent interpretation, we develop a detection framework in which the often-used short time Fourier transform

© Springer International Publishing Switzerland 2015
D.-S. Huang and K. Han (Eds.): ICIC 2015, Part III, LNAI 9227, pp. 738–744, 2015.
DOI: 10.1007/978-3-319-22053-6_78

(STFT) method [3] is used to extract features of EEG signals, and four rule-based classifiers - decision tree algorithm C4.5 [4], RF [5] and two ensemble learning approaches – SVM + C4.5 and SVM + RF [6] - are taken to do comparison for the detection of epileptic seizures in EEG signals in two-, three- and the most challenging five- group classification.

This paper is organized as follows. Section 2 describes the proposed detection framework. Section 3 discusses the EEG dataset and the STFT algorithm for feature extraction. The experiment results and conclusion are given in Sects. 4 and 5.

2   The Proposed Detection Framework

The proposed detection framework for two-, three- and five-group classification in EEG signals can be described as three stages.

In the first stage, a feature extraction method – STFT is run on EEG signals to generate the training and testing datasets with the extracted features.

In the second stage, four rule-based classifiers – the decision tree algorithm C4.5 [4], the RF algorithm [5], two ensemble learning approaches called SVM + C4.5 and SVM + RF are utilized on the training dataset to construct comprehensive and transparent rules for classification on extracted features. As two comprehensible decision tree classifiers, C4.5 learns rules by splitting the training dataset into subsets based on an attribute value test, and RF is an ensemble learning method for classification, which consists of many decision tree classifiers and aggregates the results. As we may know well, SVM is the most typical kernel-based classifier in supervised learning [6–8]. In order to save the space of the paper, here we do not review these three classifiers, and their details can be seen in [6, 7]. SVM + C4.5 and SVM + RF here begin with the training dataset which is used to construct the SVM model by tuning the parameters through cross-validation (CV). And then they extract the support vectors (SVs) from the model constructed by the best fold of cross validation. After that, the SVs are put into the built SVM model to get the predicted labels, and these predicted labels will take place of the original actual labels of the SVs to form an artificial dataset. The purpose of the replacement of labels is to maximize the simulation of the prediction by the SVM model [7]. The reasons we perform this rule extraction technique from SVM is that it is a way to have an insight into black-box model and eliminate the noise which is class overlapping in the data [8]. The SVs are used to construct two rule sets separately by C4.5 and RF, which are fixed by tuning the parameters through cross validation.

In the third stage, the rule sets generated from four classifiers are evaluated on the testing dataset and the corresponding results are compared.

Let us keep in mind that SVM has been applied in the analysis of EEG signals and C4.5 has been used only for two-group and three-group seizure classification. However, RF and the ensemble learning approaches SVM + RF and SVM + C4.5 have never been applied to classification of EEG signals before, especially for multi-class classification of EEG signals. Although the ensemble learning approach SVM + RF exhibits the superiority over other three rule-based classifiers in [7] for diagnosis of diabetes, we indeed need evidence to support its feasibility in detection of epileptic seizures in EEG signals.

3   Datasets and Feature Extraction

3.1   Datasets

The dataset used in this study is from the University of Bonn, Germany [9]. This dataset has five subsets (A, B, C, D, E) and each contains 100 single-channel EEG segments captured in 23.6 s. The sampling rate of all subsets is 173.6 Hz. Subsets A and B consist of EEG signals taken from five normal volunteers using standardized electrode placement scheme. The volunteers were relaxed in awake state with eyes open (subset A) and eyes closed (subset B) respectively. Subsets C, D and E are EEG signals carried out on epileptic subjects of presurgical diagnosis. Segments in subset C were recorded from the hippocampal formation of the opposite hemisphere of the brain and those in subset D were recorded from the epileptogenic zone. Both subsets C and D contain signals measured only during seizure free intervals, while subset E contains data recorded only during seizure activity. Figure 1 shows the EEG signals of the five groups.

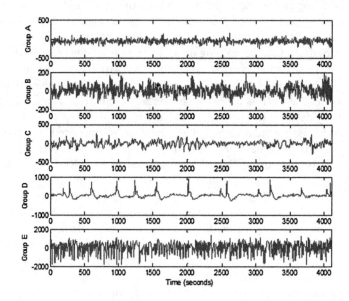

Fig. 1.   EEG signals from Group A to Group E

Three experiments (Expt 1, 2 and 3) are carried out to evaluate the performance of classifiers in classifying the EEG signals into two, three and five groups respectively.

Expt 1:   Two-group classification. The goal is to find the distinctive features extracted from sub-band frequency analysis and to accurately classify healthy and epileptic subjects. The EEG signals of data sets A and B are used to represent a group of 200 healthy subjects, whereas the epileptic seizure EEG signals in set E are used to represent a group of 100 epileptic subjects. Here we use the SVM model [6].

Expt 2: Three-group classification. The goal is to accurately classify three groups of subjects, namely Group 1, the healthy subjects; Group 2, the epileptic subjects during a seizure-free interval; and Group 3, epileptic subjects during a seizure. The data used in this experiment are arranged such that data sets A and B are combined to represent Group 1; sets C and D are merged to represent Group 2, and finally, set E represents Group 3. Here we use the multi-class SVM model [10].

Expt 3: Five-group classification. As stated in [11], differences between all five groups of subjects in the original dataset should be recognized. Therefore the goal of the third experiment is to classify the EGG segments from different extracranial and intracranial recording regions, and particularly from different physiological states brain activities. Here we use the multi-class SVM model [10]. This is the most challenging task and so far no previous research has been conducted for five-group classification by using the complete EEG dataset from Bonn University of Bonn.

3.2 Feature Extraction

In this study, short time Fourier transform (STFT) [3] is utilized for feature extraction. To perform STFT, a small sliding window is used for the Fourier transform. The spectrogram is computed with a one second hamming window for every half second, which is a widely adopted approach in EEG signal processing systems. For a given continuous EEG signal $x(t)$, a function of limited width window $g(t)$ and the centre of a small window u, STFT can be computed by

$$F_{STFT}(t,f) = \int_{-\infty}^{\infty} x(t)g * (t - u) e^{-j2\pi ft} dt \tag{1}$$

where F is a transformation function, mapping the EEG signals into the time-frequency plane.

Firstly, the STFT method distributes EEG signals into different local stationary signal segments. A group of spectra of local signals is then obtained through Fourier transform. Also, the time-varying characteristics of the signals with discrepancy in local spectrum at different times can be seen. Finally the energy of the EEG signals is separated into five frequency bands which are listed in Table 1. Figure 2 illustrates the extracted EEG signals of group A.

Table 1. Five frequency sub-bands

Band	Delta	Theta	Alpha	Beta	Gamma
Frequency range (Hz)	0–4	4–8	8–15	15–30	30–60

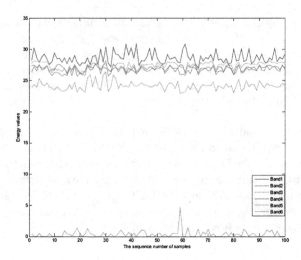

Fig. 2. Extracted features of EEG signal in group A by STFT

4 Results and Discussion

In this experimental study, STFT technique is first applied on the original EEG signals to get the dataset with extracted features. After that, we utilize the rule based classifiers – RF, C4.5, SVM + RF and SVM + C4.5 – on the dataset with extracted features on detection of epileptic seizures.

For SVM + C4.5 and SVM + RF, we run on the dataset ten times in which each time we separate the dataset into training and test dataset with a ratio 9:1 randomly. In the first running time, 90 % of the dataset is used for training SVMs by 10-fold Cross Validation (CV). 10-fold CV is to search for the optimal parameter C and kernel parameter delta of SVM by the grid-search. Also the SVM model is constructed by the best fold which obtains the best classification rate on that fold's test set, and finally the built SVM model is tested on the remaining 10 % dataset. For the remaining nine times, in order to ensure the fair performance of the trained model, the nine shuffled datasets are trained

Table 2. Average accuracy results in experiments of 10-Fold CV for 10 runs

	Expt 1 (mean ± SD)	Expt 2 (mean ± SD)	Expt 3 (mean ± SD)
SVM	0.9963 ± 0.0117	0.9660 ± 0.0325	0.6704 ± 0.0501
RF	0.9930 ± 0.0044	0.9569 ± 0.0369	0.8311 ± 0.0515
C4.5	0.9852 ± 0.0259	0.9311 ± 0.0286	0.7178 ± 0.0555
RF on SVs	0.9900 ± 0.0161	0.9432 ± 0.0467	0.9216 ± 0.0348
C4.5 on SVs	0.9917 ± 0.0118	0.9218 ± 0.0708	0.9102 ± 0.0362

by SVM with the same chosen parameters. Additionally, RF and C4.5 are implemented on the same dataset for ten runs. The average accuracy results of SVM, RF and C4.5 in three experiments of 10-fold CV for 10 runs are shown in Table 2. After SVM, in terms of SVM + RF and SVM + C4.5, RF and C4.5 run with the obtained SVs with the predicted labels. Their results are also listed in Table 2.

In terms of Table 2, both RF and C4.5 work well on the obtained SVs, but SVM + RF and SVM + C4.5 are heavily dependent on SVM's performance. Therefore, these five classifiers actually perform well in Expt 1 and Expt 2. However, RF has obvious advantage over other four classifiers for five-group classification in Expt 3.

In order to evaluate the proposed detection framework, SVM + RF, SVM + C4.5, RF and C4.5 are tested on the corresponding testing dataset for 10 runs. The average accuracy results for the testing set for 10 runs were shown in Table 3.

Table 3. Average accuracy results for the testing sets in experiments for ten runs

	Expt 1 (mean ± SD)	Expt 2 (mean ± SD)	Expt 3 (mean ± SD)
SVM + RF	0.9967 ± 0.0105	0.9720 ± 0.0253	0.6980 ± 0.0797
SVM + C4.5	0.9933 ± 0.0141	0.9433 ± 0.0312	0.6820 ± 0.0649
SVM	0.9940 ± 0.0123	0.9400 ± 0.0351	0.6600 ± 0.0736
RF	0.9896 ± 0.0157	0.9600 ± 0.0311	0.8260 ± 0.0525
C4.5	0.9920 ± 0.0144	0.9367 ± 0.0416	0.7220 ± 0.0537

Table 3 demonstrates that in Expt 1, two ensemble learning approaches SVM + RF and SVM + C4.5 show the best average classification results on two-group classification which are 0.9967 and 0.9933, respectively. RF obtains the worst accuracy result which still achieves at 0.9896. Thus, for two-group classification, all the classifiers perform very well without significant performance difference among them. In Expt 2, SVM + RF and RF yield better classification results, i.e., 0.9720 and 0.9600, respectively, on three-group classification. The remaining classifiers' performances are below 0.9500. In Expt 3, RF stands out in five-group classification with the highest accuracy of 0.8260 in four classifiers. C4.5 obtains the second highest accuracy of 0.7220 and the other two classifiers are with the accuracies below 0.7000. SVM gets the average accuracies of 0.9940, 0.9400 and 0.6600, respectively, in two-, three-, and five-group classification, which means that SVM has the general performance but no explainable ability at all.

As the result, we can conclude that among all four rule-based classifiers, due to its strong decision-tree-based ensemble learning capability, RF is at least comparable to and even outperforms C4.5, SVM + RF and SVM + C4.5 in detection of EEG signals in two-, three-and five-group classification. In the most challenging five-group classification, it has significant advantage over C4.5, SVM + RF and SVM + C4.5. It might be due to the SVM's poor performance on five-group classification (0.6704) in Table 2, which directly affects the rule extraction from SVM by the ensemble learning approaches. On the other hand, during the experiments, RF does not need to tune a bunch of parameters like SVM which also makes it easy to implement and scale up.

5 Conclusion

This study utilizes the comprehensive and transparent rule-based approaches – RF, C4.5, SVM + RF and SVM + C4.5, to detect epileptic seizures in EEG signals for two-, three- and five- group classification. STFT is for feature selection at the data preparation stage. Experimental results indicate that RF has the competitive advantage in two- and three-group classification of EEG signals. Moreover, it has the obvious advantage in the most challenging five-group classification with the highest average accuracy of 0.8260 in contrast to other three rule-based classifiers. These rule sets generated by RF can be regarded as a second choice for diagnosis of epilepsy and the results from the model are easy to understand by users after a brief explanation. Further research is to prune the rule set of RF such that the obtained rule set is much less without degrading the classi-fication accuracy of EEG signals a lot.

Acknowledgement. This work was supported in part by the Research Grants Council of the Hong Kong SAR (PolyU5134/12E), the Hong Kong Polytechnic University (G-UC93).

References

1. Benbadis, S.R., Hauser, W.A.: An estimate of the prevalence of psychogenic non-epileptic seizures. Seizure **9**(4), 280–281 (2000)
2. Acharya, U.R., et al.: Automated EEG analysis of epilepsy: a review. Knowl.-Based Syst. **45**, 147–165 (2013)
3. Griffin, D., Lim, J.S.: Signal estimation from modified short-time Fourier transform. IEEE Trans. Acoust. Speech Signal Process. **32**(2), 236–243 (1984)
4. Quinlan, J.R.: C4.5: Programs for Machine Learning. Elsevier, Amsterdam (2014)
5. Liaw, A., Wiener, M.: Classification and regression by randomForest. R News **2**(3), 18–22 (2002)
6. Barakat, N., Bradley, A.P.: Rule extraction from support vector machines: a review. Neurocomputing **74**(1), 178–190 (2010)
7. Han, L., et al.: Rule extraction from support vector machines using ensemble learning approach: an application for diagnosis of diabetes. IEEE J. Biomed. Health Inform. (2014)
8. Martens, D., Baesens, B., Van Gestel, T.: Decompositional rule extraction from support vector machines by active learning. IEEE Trans. Knowl. Data Eng. **21**(2), 178–191 (2009)
9. Andrzejak, R.G., et al.: Indications of nonlinear deterministic and finite-dimensional structures in time series of brain electrical activity: Dependence on recording region and brain state. Phys. Rev. E **64**(6), 061907 (2001)
10. Weston, J. Watkins, C.: Multi-class support vector machines. Technical report CSD-TR-98-04, Royal Holloway, University of London (1998)
11. Barry, R.J., et al.: EEG differences between eyes-closed and eyes-open resting conditions. Clin. Neurophysiol. **118**(12), 2765–2773 (2007)

Sparse-View X-ray Computed Tomography Reconstruction via Mumford-Shah Total Variation Regularization

Bo Chen[1,2], Chen Zhang[1], Zhao-Ying Bian[3], Wen-Sheng Chen[1,2(✉)],
Jian-Hua Ma[3(✉)], Qing-Hua Zou[1], and Xiao-Hui Zhou[1]

[1] College of Mathematics and Computational Science, Shenzhen University,
Shenzhen 518060, People's Republic of China
chenws@szu.edu.cn
[2] Shenzhen Key Laboratory of Media Security,
Shenzhen 518060, People's Republic of China
[3] Department of Biomedical Engineering, Southern Medical University,
Guangzhou 510515, People's Republic of China
jhma@smu.edu.cn

Abstract. The regularization plays an important role in the sparse-view x-ray computer tomography (CT) reconstruction. Based on the piecewise constant assumption, total variation (TV) regularization has been widely discussed for the sparse-view CT reconstruction. However, TV minimization often leads to some loss of the image edge information during reducing the image noise and artifacts. To overcome the drawback of TV regularization, this paper proposes to introduce a novel Mumford-Shah total variation (MSTV) regularization by integrating TV minimization and Mumford–Shah segmentation. Subsequently, a penalized weighted least-squares (PWLS) scheme with MSTV is presented for the sparse-view CT reconstruction. To evaluate the performance of our PWLS-MSTV algorithm, both qualitative and quantitative analyses are executed via phantom experiments. Experimental results show that the proposed PWLS-MSTV algorithm can attain notable gains in terms of accuracy and resolution properties over the TV regularization based algorithm.

Keywords: Statistical image reconstruction · Regularization · Image segmentation · Mumford-Shah total variation

1 Introduction

With the increasing concerns on the x-ray radiation exposure to patients, minimizing the exposure risk has been one of the major endeavors in current computed tomography (CT) examinations [1, 2]. In order to reduce radiation dose, two classes of strategies have been extensively discussed: (1) lower the x-ray flux towards each detector bins by lower x-ray tube current-measured by milliampere-seconds (mAs) or lower x-ray tube voltage-measured by kilovoltage-peak (kVp); and (2) lower the required number of projection views during the inspection. Although the two strategies can reduce the radiation exposure risk considerably, noisy and artifacts would

© Springer International Publishing Switzerland 2015
D.-S. Huang and K. Han (Eds.): ICIC 2015, Part III, LNAI 9227, pp. 745–751, 2015.
DOI: 10.1007/978-3-319-22053-6_79

be inevitably introduced into the resulting image. Acquiring a high-diagnostic CT images from low-dose or sparse-view acquisitions would naturally be an important and interesting research topic in the CT field. In this work, we focus on the sparse-view CT image reconstruction.

Due to the data inconsistency in the sparse-view acquisitions, the CT image reconstructed by the conventional filtered back-projection (FBP) algorithm usually suffers serious streak artifacts and noise. To address this problem, various image reconstruction methods by incorporating adequate prior information about the desired image have been widely studied [3–11]. Based on the piecewise constant assumption of the desired image, the total variation (TV) based projection onto convex sets (POCS) reconstruction strategy has shown its effectiveness for dealing with the data insufficiency from sparse-view sampling [3, 4]. Furthermore, to address the limitations of the original TV constrain with isotropic edge property, different weighted-TVs, as extensions of the original one, were proposed recently [9–13].

In this work, aiming to improve the TV regularization, we introduce a useful variation of TV called Mumford-Shah TV, namely MSTV [14], which not only considers the TV norm of to-be-estimated image, but also considers the corresponding edge of to-be-estimated image. With considering the measure of corresponding edge, a higher resolution and a better quality of the reconstructed image can be achieved. To compare the MSTV regularization with the TV regularization fairly, the MSTV regularization was adapted under the penalized weighted least-squares (PWLS) criteria for CT image reconstruction from sparse-view projection measurements. For simplicity, the present algorithm is termed as "PWLS-MSTV". Compared with the TV reconstruction algorithm (namely, termed as "PWLS-TV"), qualitative and quantitative evaluations were carried out on the digital phantoms in terms of accuracy and resolution properties.

2 Description of the Proposed Method

2.1 Mumford-Shah Total Variation

The MSTV was first proposed by Shah [14] in the image segmentation and extensively studied in image segmentation and image restoration [12, 15]. Actually, the widely used MSTV is an approximation proposed by Alicandro et al. [16]:

$$MSTV_\varepsilon(u, v) = \int_\Omega v^2 |\nabla u| dx + \alpha \int_\Omega \left(\varepsilon |v|^2 + \frac{(v-1)^2}{4\varepsilon} \right) dx \qquad (1)$$

where Ω is a bounded domain, ∇u is the gradient of image u, and v is the edge function of image u, which is approximate to zero in the edge of image u while it is approximate to one in other region of image 300th, ε is a small positive constant and α is a positive weight which needs to be tuned manually. Alicandro et al. [16] also proved the Γ-convergence of this function to

$$MSTV(u) = \int_{\Omega\setminus K} |\nabla u| dx + \alpha \int_K \frac{|u^+ - u^-|}{1 + |u^+ - u^-|} dH^1 + |D^c u|(\Omega) \tag{2}$$

where u^+ and u^- denote the image values on two sides of the edge set K, H^1 is the one-dimensional Hausdorff measure and $D^c u$ is the Cantor part of the measure-valued derivative D_u. Through the definition of MSTV as shown in (2), it is obvious that MSTV not only considered the TV norm of image u in the image domain except for the edge, but also considered the measure of edge set K. Therefore, MSTV regularization brings more powerful regularity of solution than TV regularization.

2.2 PWLS-MSTV Minimization

Inspired by the studies of MSTV in image restoration [12], we propose the cost function for CT image reconstruction as follows:

$$\min_{u \geq 0, v} (y - Hu)^T G^{-1}(y - Hu) + \beta_2 MSTV_\varepsilon(u, v) \tag{3}$$

where $G = \frac{1}{\beta_1} HH^T + \Sigma$, β_1 and β_2 are two hyperparameters to balance these two terms, namely, the fidelity term and the regularization term. And u is the vector of attenuation coefficients to be reconstructed, symbol T denotes the matrix transpose. The operator H represents the system or projection matrix with the size of $M \times N$. The element of h_{ij} is the length of the intersection of projection ray i with pixel j. Σ is a diagonal matrix with the ith element of σ_i^2 which is the variance of sinogram data y_i. Additionally, by introducing a new vector f, we have

$$\min_f (y - Hf)^T \Sigma^{-1}(y - Hf) + \beta_1 \|f - u\|^2 = (y - Hu)^T G^{-1}(y - Hu). \tag{4}$$

Hence, solving formula (4) is equal to solve the below formula:

$$\min_{u \geq 0, f, v} (y - Hf)^T \Sigma^{-1}(y - Hf) + \beta_1 \|f - u\|^2 + \beta_2 MSTV_\varepsilon(u, v). \tag{5}$$

Namely,

$$\min_{u \geq 0, f, v} (y - Hf)^T \Sigma^{-1}(y - Hf) + \beta_1 \|f - u\|^2 + \beta_2 \int_\Omega v^2 |\nabla u| dx$$
$$+ \gamma \int_\Omega \frac{(v - 1)^2}{4\varepsilon} + \varepsilon |\nabla v|^2 dx \tag{6}$$

Here, the parameter $\gamma = \beta_2 \alpha$ for simplifying the redundant parameters. To solve the cost function in Eq. (6), an alternating optimization method referring to [7] was adopetd in this work.

3 Experimental Results

Figure 1 is a slice of the XCAT phantom, which contains head anatomy structures with a tumor lesion. We find a geometry which was representative of a monoenergetic fan-beam CT scanner setup. The imaging parameters of the CT scanner were as follows: (1) each rotation includes 1160 projection views that are evenly spaced on a circular orbit; (2) the number of channels per view was 672; (3) the distance from the detector arrays to the X-ray source is 1040 mm; (4) the distance from the rotation center to the X-ray source is 570 mm; and (5) the space of each detector bin is 1.407 mm. All the reconstructed images were composed 512 × 512 square pixels and the size of pixel was 0.625 × 0.625 mm. Each projection datum along an X-ray through the sectional image was calculated based on the known densities and intersection areas of the ray with the geometric shapes of the objects in the sectional image.

As previous studies mentioned in [11], we first simulated the noise-free sinogram data $\hat{y}$, then generated the noisy transmission measurement I according to the statistical model of the prelogarithm projection data, namely,

$$I = Poisson(I_0 \exp(-\hat{y})) + Normal(0, \sigma_e^2) \tag{7}$$

where I_0 is the incident X-ray intensity and σ_e^2 is the background electronic noise variance. In the simulation, I_0 and σ_e^2 were set to 1.0×10^6 and 11.0, respectively. Finally, the noisy sinogram data y were calculated by performing the logarithm transformation on the transmission data I.

Figure 2 represents the visualization-based evaluation among recontruced results. The reconstructed images are corresponding to the FBP, PWLS-TV and PWLS-MSTV methods in the column. From top to bottom corresponds results reconstructed from 30-, 40-, 60- view projections, respectively. It can be seen that the reconstructed images by PWLS-MSTV are superior to other methods from each projection view, especially from 30-view, with more structure details preserving. The result also illustrates that the PWLS-MSTV method can achieve more close results to the true phantom image than the PWLS-TV method.

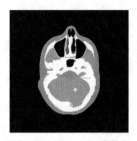

Fig. 1. A slice of digital XCAT phantom that contains head anatomy structures.

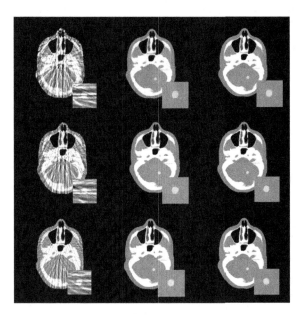

Fig. 2. Images reconstructed by the FBP with ramp filter, PWLS-TV andPWLS-MSTV methods from 30-, 40-, 60- view projections, respectively. The first, second and third row correspond results from 30-, 40-, 60- view projections, respectively. The first, second and third column correspond FBP, PWLS-TV and PWLS-MSTV.

For comparing the ability of noise reduction with both the PWLS-MSTV and the PWLS-TV methods, the peak signal to noise ratio (PSNR) is used. Table 1 lists the PSNRs of the resulting images by both the PWLS-MSTV and the PWLS-TV methods from 30-, 40-, and 60- view projections, respectively. The proposed PWLS-MSTV method achieves higher PSNRs than the PWLS-TV method, which means more strong ability of noise reduction.

For comparing the ability of keeping the similarity between the reconstructed and ideal images with PWLS-TV, the universal quality index (UQI) [18] is studied in the region of interest (ROI)-based analysis. UQI is employed for measuring the similarity between the reconstructed and true images. The higher value UQI has, the higher similarity to the true image the reconstructed image have. Figure 3 shows the UQI comparison among the FBP, PWLS-TV and the proposed PWLS-MSTV methods. The result demonstrates that the reconstructed image with PWLS-MSTV has higher similarity than the reconstructed images with both FBP and PWLS-TV.

Table 1. The PSNRs of results by both the PWLS-MSTV and the PWLS-TV methods.

Projection views	PWLS-TV	PWLS-MSTV
30 views	34.82 dB	36.56 dB
40 views	37.62 dB	38.90 dB
60 views	39.13 dB	40.59 dB

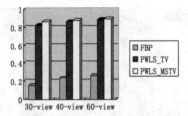

Fig. 3. The UQIs of FBP, PWLS-TV and PWLS-MSTV with different projection views.

4   Conclusion

To improve the performance of TV regularization for the sparse-view CT reconstruction, we introduced an improved TV-based regularization, namely MSTV regularization, which integrates TV minimization and Mumford–Shah segmentation. The proposed PWLS-MSTV algorithm jointly solves the CT image reconstruction and segmentation simultaneously. The Mumford–Shah segmentation information could make good the loss resulted by TV regularization. And these reconstruction and segmentation results mutually promote. Experimental results demonstrated that the proposed PWLS-MSTV algorithm can attain notable gains in terms of accuracy and resolution properties over the PWLS-TV algorithm.

Acknowledgements. This work is partially supported by the National Natural Science Foundation of China (61272252, 81371544), the National Science Technology Major Project of the Ministry of Science and Technology of China (2014BAI17B02) and Science and Technology Planning Project of Shenzhen City (JCYJ20140828163633997, JCYJ20130326111024546). The authors also acknowledge the Key Laboratory of Medical Image Processing in Southern Medical University for providing the experimental data. Additionally, the authors also acknowledge Bar Leach for sharing their work.

References

1. Brenner, D., Hall, E.: Computed tomography—an increasing source of radiation exposure. New Engl. J. Med. **357**, 2277–2284 (2007)
2. Einstein, A., Henzlova, M., Rajagopalan, S.: Estimating risk of cancer associated with radiation exposure from 64-slice computed tomography coronary angiography. J. Am. Med. Assoc. **298**(3), 317–323 (2007)
3. Li, X., Orchard, M.: Spatially adaptive image denoising under over complete expansion. In: Proceedings of 2000 International Conference on Image Processing, vol. 3, pp. 300–303 (2000)
4. Sidky, E., Pan, X.: Image reconstruction in circular cone-beam computed tomography by constrained, total-variation minimization. Phys. Med. Biol. **53**(17), 4777–4807 (2008)
5. Li, T., Li, X., Wang, J., Wen, J., Lu, H., Hsieh, J., Liang, Z.: Nonlinear sinogram smoothing for low-dose X-ray CT. IEEE Trans. Nucl. Sci. **51**(5), 2505–2513 (2004)
6. Elbakri, I., Fessler, J.: Statistical image reconstruction for polyenergetic X-ray computed tomography. IEEE Trans. Med. Imaging **21**(2), 89–99 (2002)

7. Niu, S., Gao, Y., Bian, Z., Huang, J., Chen, W., Yu, G., Liang, Z., Ma, J.: Sparse-view X-ray CT reconstruction via total generalized variation regularization. Phys. Med. Biol. **59**(12), 2997–3017 (2014)
8. Gao, Y., Bian, Z., Huang, J., Zhang, Y., Niu, S., Feng, Q., Chen, W., Liang, Z., Ma, J.: Low-dose X-ray computed tomography image reconstruction with a combined low-mAs and sparse-view protocol. Opt. Express **22**(12), 15190–15210 (2014)
9. Liu, Y., Ma, J., Fan, Y., Liang, Z.: Adaptive-weighted total variation minimization for sparse data toward low-dose X-ray computed tomography image reconstruction. Phys. Med. Biol. **57**(23), 7923–7956 (2012)
10. Liu, Y., Liang, Z., Ma, J., Lu, H., Wang, K., Zhang, H., Moore, W.: Total variation stokes strategy for sparse-view X-ray CT image reconstruction. IEEE Trans. Med. Imaging **33**(3), 749–763 (2014)
11. Wang, J., Li, T., Lu, H., Liang, Z.: Penalized weighted least-squares approach to sinogram noise reduction and image reconstruction for low-dose X-ray computed tomography. IEEE Trans. Med. Imaging **25**(10), 1272–1283 (2006)
12. Bar, L., Sochen, N., Kiryati, N.: Semi-blind image restoration via Mumford-Shah regularization. IEEE Trans. Image Process. **15**(2), 483–493 (2006)
13. Rudin, L., Osher, S., Fatemi, E.: Nonlinear total variation based noise removal algorithms. Physica D **60**(1–4), 259–268 (1992)
14. Shah, J.: A common framework for curve evolution, segmentation and anisotropic diffusion. In: Proceedings of IEEE Conference on Computer Vision and Pattern Recognition, pp. 136–142 (1996)
15. Bar, L., Brook, A., Sochen, N., Kiryati, N.: Deblurring of color images corrupted by impulsive noise. IEEE Trans. Image Process. **16**(4), 1101–1111 (2007)
16. Alicandro, R., Braides, A., Shah, J.: Free discontinuity problems via functional involving the norm of the gradient and their approximation. Interfaces Free Bound **1**, 17–37 (1999)
17. Weisstein, E.: Minimal residual method. MathWorld—A Wolfram Web Resource. http://mathworld.wolfram.com/MinimalResidualMethod.html
18. Wang, Z., Bovik, A.: A universal image quality index. IEEE Signal Process. Lett. **9**(3), 81–84 (2002)

The Utilisation of Dynamic Neural Networks for Medical Data Classifications- Survey with Case Study

Abir Jaafar Hussain[1], Paul Fergus[1(✉)], Dhiya Al-Jumeily[1], Haya Alaskar[2], and Naeem Radi[3]

[1] Liverpool John Moores University, Byroom Street, Liverpool L3 3AF, UK
{a.hussain,P.fergus,d.aljumeily}@ljmu.ac.uk
[2] Department of Computer Science, Salman Bin Abdulaziz University,
Al-Kharj, Saudi Arabia
[3] Al-Khawarizmi International College, Abu Dhabi, UAE
n.radi@khawarizmi.com

Abstract. Various recurrent neural networks have been utilised for medical data analysis and classifications. In this paper, the ability of using dynamic neural network to medicine related problems has been examined. Furthermore, a survey on the use of recurrent neural network architectures in medical applications will be discussed. A case study using the Elman, the Jordan and Layer recurrent networks for the classifications of Uterine Electrohysterography signals for the prediction of term and preterm delivery for pregnant women are presented.

Keywords: Uterine EMG signals · Dynamic neural network · Features extraction

1 Introduction

The development of medical information systems has played an important role in medical societies. The aim of these developments is to improve the utilisation of technology in medical applications. Expert systems and different Artificial Intelligence methods and techniques have been used and developed to improve decision support tools for medical purposes. One of the most widely-used classification tools for medical application is Artificial Neural Network (ANN). ANNs have the ability to identify differences between groups of signals, which were utilised to identify different types of diseases and illnesses. This is related to their characteristics of self-learning, self-organization, non-linearity, and parallel processing compared with linear traditional classifiers [16]. However, the feed-forward neural networks suffer from some limitation specially when dealing with temporal pattern. Recurrent neural networks (RNNs) have advantages over feedforward neural networks. They have the ability to discover the hidden structure of the medical time signal. Existing studies have indicated that RNN has the ability to perform pattern recognition in medical time-series data and has obtained high accuracy in the classification of medical signals [5, 11, 19]. In addition, it has been shown that RNN has the ability to provide an insight into the feature used to represent biological signals. Therefore, the employment of a dynamic tool to deal with time-series data classification is highly

© Springer International Publishing Switzerland 2015
D.-S. Huang and K. Han (Eds.): ICIC 2015, Part III, LNAI 9227, pp. 752–758, 2015.
DOI: 10.1007/978-3-319-22053-6_80

recommended. This type of neural network has a memory that is capable of storing information from past behaviours [10]. One of the most important applications of RNN is to model or identify temporal patterns, as Chung et al. have stated in his work [4]. Different studies have indicated that RNN can be applied to non-linear decision boundaries [9]. In addition, the main advantages of recurrent neural networks is their ability to deal with static and dynamical situations.

In this paper, the applications of recurrent neural networks for medical data classification will be discussed. Furthermore, there is a strong body of evidence emerging that suggests the analysis of uterine electrical signals, from the abdominal surface (Electrohysterography – EHG), could provide a viable way of diagnosing true labour and even predict preterm deliveries. Hence, the performance of three types of recurrent neural network architectures including the Elman, the Jordan and the Layer recurrent networks for the classifications of Uterine EHG signals for the prediction of term and preterm delivery for pregnant women will be presented and discussed as a case study to validate our research for the applicability of recurrent neural networks as classification algorithms for medical data.

2 Recurrent Neural Networks for Classification

In the last couple of years, various medical applications based on RNN have been developed. One of the most prominent applications of RNN is pattern recognition, such as automated diagnostic systems. The RNN can utilize nonlinear decision boundaries and process memory of the state, which is crucial for the classification task [9, 19]. Numbers of studies have confirmed that RNN has the ability to distinguish linear and nonlinear relations in the signals. In addition, they have proven that RNN enjoys signal recognition abilities [19]. The researchers are attempted to investigate the ability of RNN to classify biological signals (e.g. EEG, ECG and EMG). The procedure for signals classification is performed in two stages. The first step is extracting the features. These features will be used as an input to RNN classifier. This will follow by preforming the classifier techniques.

Currently, most research work is based on using recurrent neural network for EEG signals classification. Koskela et al. [12] have been addressed the utilising of recurrent self-organising map (RSOM) to EEG signals for epileptic. It has been applied to detect the activity of epileptic on EEG signals. The EEG sample was 200 Hz. The features that have been used on this experiment are spectral features and they were extracted with 256 size of window. They used wavelet transform to extract signals from each window, and sixteen energy features from the wavelet domain have been computed for each window. The data divided into training and testing set, the training set contains *150987 × 16* dimension vectors, and the epileptic activity was *5430* patterns on the training set. The RSOM network has been run to classify the EEG signal to normal or epileptic activity. Their results show that RSOM achieved better clustering result than self-organising map (SOM). They conclude that using context memory for detecting the EEG epileptic activity has enhanced the classification performance on SOM [12].

Another study was presented using the Elman network to classify the mental diseases on EEG signals combined with wavelet pre-processing. Petrosian et al. [19] investigated the ability of RNN employed with wavelet pre-processing methods for diagnosis of epileptic seizures in EEG signals and for the early detection of Alzheimer's Disease (AD) in EEG signals. For diagnosis of epileptic seizures in EEG signals analysis task, the authors examined the ability of recurrent neural networks (RNN) combined wavelet transformation methods to predict the onset of epileptic seizures. The signals were collected from EEG channel. The recurrent neural network was trained based on decoupled extended kalmen filter (DEKF) algorithm. In Alzheimer's disease in EEG signals detection task, the RNN has been used to distinguish between AD and healthy groups. In that study, the authors have used network training algorithm based on Extended Kalman Filter (EKF). The signals were obtained from ten healthy persons and ten early AD patients. The EEG signals were recorded using 9 channels with 2 min length and with 512 Hz sampling rate. EEG has been recorded to monitor the subject during the eyes closed resting state. The Fourier power spectra methods have been used to analysis the row EEG signals. The band pass FIR filter has been used to filter each EEG signals into four sub groups (delta, theta, alpha and beta). Furthermore, the fourth levels wavelets filter has been used on raw EEG signals. In the study, the inputs of the RNN were the original channel signals and the derived delta, theta, alpha and beta for each signal as well as their wavelet filtered subbands at levels 1–6. From their experiments, the best RNN result was achieved using parietal channel P3 raw signals as well as wavelet decomposed sub-bands at levels 4 as inputs. RNN achieved high performance to classify AD with 80 % sensitivity and 100 % specificity. Petrosian et al. in their work [19] have been shown that the combination between RNN and wavelet approach has the ability to analysis EEG signals of early AD detections.

In addition, their study attempted to classify different type of conditions related to human muscles. For example, Ilbay et al. [11] used the Elman recurrent neural network (RNN) for automated diagnosis of Carpal Tunnel Syndrome (CTS). It has been applied on patients suffering from various Carpal Tunnel Syndrome symptoms such as right CTS, left CTS and bilateral CTS. In this experiment, the study has collected EMG signals from 350 patients who suffer from CTS (left, right and bilateral) symptoms and signs. Nerve conduction study (NCS) was applied by using surface electrode to record the EMG signals on both hands for each patient. NCS measures how fast electrical signals can be sent through nerves. Therefore, they are able to diagnosis the Carpal Tunnel Syndrome and the result of this test are used to evaluate the degree of any nerve damage. During NCS test, surface electrodes are located on patient's hand and wrist, and then electrical signals are created to stimulate the nerves in the wrist, forearm and fingers. Sensory responses are collected from the index finger (median nerve) or little finger (ulnar nerve), with ring electrodes. The features were extracted from these signals. They were right median motor latency, left median motor latency, right median sensory latency, left median sensory latency which has been used as RNN inputs. RNNs are trained with the Levenberg-Marquardt algorithm. The result of this research has shown that RNN obtains 94 % classification accuracy, which is higher than MLPNN with 88 %.

In the field of biomedical, the analysis of EHG signals with powerful and advanced methodologies is becoming required. EHG is a technique for measuring electrical activity of the uterus muscle during pregnancy, through uterine contractions [8, 18]. EHG is one form of electromyography (EMG), the measurement of activity in muscular tissue. Electromyography (EMG) technique is considered as helpful and effective method to detect the preterm labour. EHG is very sufficient measurements of recording electrical activity, because it measures the contraction directly, rather than the physical respond of contractions, which may get lost amongst other physical noise and disturbance [14]. In this section, the ability of recurrent neural networks to forecast EHG signals will be investigated. The analysis and characterization of Uterine EHG signals is very challenging and this is related to their low signal to noise ratio (SNR). The ability of RNNs to forecast EHG signal can be used for pre-processing EHG signal. The signal pre-processing aim to improve signal to noise ratio [3]. The main objective of this experiment is to explore the possibility of applying RNNs network as filtering method to increase uterine EHG signal to noise ratio value.

The data used in this research were recorded at the Department of Obstetrics and Gynaecology, Medical Centre, Ljubljana between 1997 and 2006 [20]. In the Term-Preterm ElectroHysteroGram (TPEHG) database, there are 300 records of patient. These records are openly available, via the TPEHG dataset, in Physionet website. The signals in this study were already collected by [6]. Each record was collected by regular examinations at the 22nd week of gestation or around the 32nd week of gestation. The signal in records was 30 min long, had a sampling frequency (fs) of 20 Hz, and had a 16-bit resolution over a range of ±2.5 mV.

Prior to sampling, the signals were sent through an analogue three-pole Butterworth filter, in the range of 1–5 Hz. The recording time shows the gestational age. Each recording was classified as a full-term or preterm delivery, after birth. Figures 2 and 3 show two examples of EHG signals from different record. The recordings were categorised as four types as follows:

1. Early – Term: Recordings made early, signed as a term delivery
2. Early – Preterm: Recordings made early, signed as a preterm delivery
3. Late - Term: Recordings made late, signed as a term delivery
4. Late – Preterm: Recordings made late, signed as a preterm delivery

Two experiments have taken place on 76 EHG signals with 38 preterm and 38 term values. The model was trained over channel 3 following the recommended of Fele-Žorž et al. [6]. The first experiment used the RNNs to model EHG signals before filtering. While the second experiment modelled the EMG signals with the RNNs after using a band-pass filter configured between 0.3 Hz and 4 Hz.

3 Modelling RNN for Forecasting

In this section, the steps that have been used to build the RNNs to model the EHG signals are presented. The maximization of the quality of uterine EHG signal that produced by RNN networks can be achieved by evaluating signal-to-noise ratio. These measurements have been designed to hold the highest amount of information from EHG signal as possible and smallest amount of noise.

An experiment was taken place to examine the performance of the network. The performance was evaluated using the Mean Squared Error (MSE), and Signal-to-Noise

Table 1. Comparison of different types of recurrent neural networks

	SNR	r	MSE
Elman (before filtering)	7.9256	0.4506	0.0033
Elman (after filtering 0.3 Hz–4 Hz)	13.7702	0.2363	1.4504e-04
Jordan (Before filtering)	16.138	0.856	0.0011
Jordan (after filtering 0.3 Hz–4 Hz)	16.7627	0.8550	4.0464e-04
layrecnet(Before filtering)	21.1003	0.8445	0.0066
layrecnet(after filtering 0.3 Hz–4 Hz)	33.0693	0.9642	5.1178e-05

Ratio (SNR). Table 1 shows the average results for the mean squared error, correlation coefficient (r), and signal-noise-ratio (SNR) using 76 Uterine EHG signals. The best forecasting performance is measured by the SNR, which is a key measure of predictability with higher values for SNR indicating better predictability. Table 1 shows the performance comparison of different type of recurrent neural networks for EHG noise reduction. These different recurrent neural network models were used to reduce the noises in the uterine EHG signals.

Among these models, it compared: Elman, Jordan network, and Layer recurrent neural networks with each layer has a recurrent connection with a tap delay associated with it (layrecnet). The results show that RNNs are able to model the nonlinear relation on the EHG signals. The layrecnet model provides the highest SNR measurement. Furthermore, the MSE and correlation coefficient values on this result indicated that layrecnet neural network is better predictor than other recurrent neural networks. Therefore, layrecnet is considered the best model among the benchmarks recurrent neural network to remove noise from EHG signals.

In this experiment, the ability of using recurrent neural network architectures to forecast EHG signals to obtain high SNR has been presented. In the experiments, the result demonstrated that recurrent neural networks are capable to filtering the Uterine EHG signals, achieving very high signal to noise ratio. In order to assess the performance of the various neural networks for processing EHG signals, the means error (MSE), and the correlation coefficient (r), have also been calculated. The result demonstrated that recurrent models are able to capture temporal behavior of the signals.

To further analyse the results and to compare with more standards types of machine learning algorithms, an additional 11 items of clinical information representing the pregnancy duration at the time of recording, maternal age, number of previous deliveries (parity), previous abortions, weight at the time of recording, hypertension, diabetes, bleeding first trimester, bleeding second trimester, funnelling, smoker are added to the original TPEHG feature set. The data set consists of 150 preterm data samples and 108 term data samples. The data have been split up as follows: 40 % of the data has been selected randomly as training data, with 20 % for validation and 40 % as testing data. The performance of the machine learning algorithms was evaluated using the mean error and the percentage accuracy.

4   Conclusion

This paper has introduced different applications of recurrent neural network for the purpose of analysing medical time series. Previous studies have demonstrated that RNN had considerable achievement in discriminating the biological signals. Some of these studies had compared the RNN result with MLP and their results were confirmed that RNN obtained better classification accuracy than MLP. RNN have the higher ability to analysis and classify the different types of biomedical signals. This paper has also presented the application of recurrent neural network for filtering EHG signals, which is one of the diagnosing approaches to detect labour. Various recurrent neural networks are applied for the prediction of EHG and filtering. Results showed that the RNNs can successfully filtering EHG signals with high SNR.

References

1. Arvind, R., Karthik, B., Sriraam, N., Kannan, J.K.: Automated detection of PD resting tremor using PSD with recurrent neural network classifier. In: 2010 International Conference on Advances in Recent Technologies in Communication and Computing, pp. 414–417 (2010). doi:10.1109/ARTCom.2010.33
2. Baghamoradi, S., Naji, M., Aryadoost, H.: Evaluation of cepstral analysis of EHG signals to prediction of preterm labor. In: 18th Iranian Conference on Biomedical Engineering, pp. 1–3, Tehran, Iran (2011)
3. Chendeb, M., Khalil, M., Hewson, D., Duchêne, J.: Classification of non stationary signals using multiscale decomposition. J. Biomed. Sci. Eng. 03(02), 193–199 (2010). doi:10.4236/jbise.2010.32025
4. Chung, J.R., Kwon, J., Choe, Y.: Evolution of recollection and prediction in neural networks. In: 2009 International Joint Conference on Neural Networks, pp. 571–577 (2009). doi:10.1109/IJCNN.2009.5179065
5. Übeyli, E.D.: Recurrent neural networks employing Lyapunov exponents for analysis of ECG signals. Expert Syst. Appl. 37(2), 1192–1199 (2010). doi:10.1016/j.eswa.2009.06.022
6. Fele-Žorž, G., Kavšek, G., Novak-Antolič, Z., Jager, F.: A comparison of various linear and non-linear signal processing techniques to separate uterine EMG records of term and pre-term delivery groups. Med. Biol. Eng. Compu. 46(9), 911–922 (2008). doi:10.1007/s11517-008-0350-y
7. Forney, E.M., Anderson, C.W.: Classification of EEG during imagined mental tasks by forecasting with Elman recurrent neural networks. In: The 2011 International Joint Conference on Neural Networks, pp. 2749–2755 (2011). doi:10.1109/IJCNN.2011.6033579
8. Garfield, R.E., Maner, W.L., MacKay, L.B., Schlembach, D., Saade, G.R.: Comparing uterine electromyography activity of antepartum patients versus term labor patients. Am. J. Obstet. Gynecol. 193(1), 23–29 (2005). doi:10.1016/j.ajog.2005.01.050
9. Guler, N., Ubeyli, E., Guler, I.: Recurrent neural networks employing Lyapunov exponents for EEG signals classification. Expert Syst. Appl. 29(3), 506–514 (2005). doi:10.1016/j.eswa.2005.04.011
10. Haykin, S.: Neural Networks A comprehensive Foundation. Prentice Hall PTR, New Jersey (1998)

11. Ilbay, K., Übeyli, E. D., Ilbay, G., Budak, F.: A new application of recurrent neural networks for EMG-based diagnosis of carpal tunnel syndrome. In: Cardot, H. (ed.), Recurrent Neural Network for Temporal Data Processing. InTech (2011). doi:10.5772/631

12. Koskela, T., Varsta, M., Heikkonen, J., Kaski, K.: Temporal sequence processing using recurrent SOM. In: Proceedings of the Second International Conference Knowledge-Based Intelligent Electronic Systems, pp. 21–23. IEEE, Adelaide. http://ieeexplore.ieee.org/xpls/abs_all.jsp?arnumber=725861&tag=1. Accessed 1998

13. Kumar, S.P., Sriraam, N., Benakop, P.G.: Automated detection of epileptic seizures using wavelet entropy feature with recurrent neural network classifier. In: TENCON 2008–2008 IEEE Region 10 Conference, pp. 1–5. IEEE, Hyderabad (2008). doi:10.1109/TENCON.2008.4766836

14. Leman, H., Marque, C., Gondry, J.: Use of the electrohysterogram signal for characterization of contractions during pregnancy. IEEE Trans. Bio-med. Eng. **46**(10), 1222–1229 (1999)

15. Ling, S.H., Leung, F.H.F., Leung, K.F., Lam, H.K., Iu, H.H.C.: An improved GA based modified dynamic neural network for cantonese-digit speech recognition. In: Grimm, M., Korschel, K. (eds.) Robust Speech Recognition and Understanding, p. 460. I-Tech, Vienna (2007)

16. Liu, B., Wang, M., Yu, H., Yu, L., Liu, Z.: Study of feature classification methods in BCI based on neural networks. In: Annual International Conference of the IEEE Engineering in Medicine and Biology Society. IEEE Engineering in Medicine and Biology Society. Conference, vol. 3, pp. 2932–2935. China (2005). doi:10.1109/IEMBS.2005.1617088

17. Makarov, V.A., Song, Y., Velarde, M.G., Hübner, D., Cruse, H.: Elements for a general memory structure: properties of recurrent neural networks used to form situation models. Biol. Cybern. **98**(5), 371–395 (2008). doi:10.1007/s00422-008-0221-5

18. Marshall, J.: Regulation of activity in uterine smooth muscle. Physiol. Rev. Suppl. (1962)

19. Petrosian, A.A., Prokhorov, D.V., Schiffer, R.B.: Recurrent neural network based approach for early recognition of alzheimer ' s disease in EEG. Clin. Neurophysiol. **112**(8), 1378–1387 (2001)

20. Physionet: Physionet (2011). http://physionet.org/

Neural Network Classification of Blood Vessels and Tubules Based on Haralick Features Evaluated in Histological Images of Kidney Biopsy

Vitoantonio Bevilacqua[1(✉)], Nicola Pietroleonardo[1,2], Vito Triggiani[1,2],
Loreto Gesualdo[2], Anna Maria Di Palma[2], Michele Rossini[2],
Giuseppe Dalfino[2], and Nico Mastrofilippo[2]

[1] Dipartimento di Ingegneria Elettrica e dell'Informazione, Politecnico di Bari, Bari, Italy
vitoantonio.bevilacqua@poliba.it
[2] Department of Emergency and Organ Transplantation, Nephrology Unit, University of Bari
Aldo Moro, Bari, Italy

Abstract. In this paper, we present a Computer Aided Diagnosis that implements a supervised approach to discriminate vessels versus tubules that are two different types of structural elements in images of biopsy tissue. In particular, in this work we formerly describe an innovative preliminary step to segment region of interest, then the procedure to extract from them significant features and finally present and discuss the Back Propagation Neural Network binary classifier performance that shows Precision 91 % and Recall 91 %.

Keywords: Computer aided diagnosis · Neural network · Image segmentation · Vessels · Histological image · Haralick features

1 Introduction

Computer-aided diagnosis (CAD), are procedures in medicine that assist physicians in the interpretation data such as medical images. These systems analyze the image to detect and characterize regions of interest, with the aim to help the physicians to improve diagnostic accuracy, paying his attention to the most suspicious areas of the image for the presence of pathology. CAD systems have been applied to several methods of diagnostic imaging [1, 2], all of which are generally based on radiological images. Within a few years, this field of research is also expanding in other medical fields, such as histopathology [3], in which the type of images is completely different than the radiological images. In this paper we present the design and the implementation of CAD system for segmentation and discrimination of blood vessels versus tubules from biopsies in kidney tissue through the elaboration of histological images. This is an important step for the evaluation of the suitability of a kidney transplant by the Karpinski score [4]. The final Karpinski score is in turn composed of 4 scores (glomerulosclerosis, interstitial fibrosis, tubular atrophy and vascular disease). The vascular score which is calculated in correspondence of the vessels, is important because donor vessel score of 3/3 was associated with a 100 % incidence of delayed graft function. The objective of the segmentation of the vessels is to identify not completely closed vessels,

© Springer International Publishing Switzerland 2015
D.-S. Huang and K. Han (Eds.): ICIC 2015, Part III, LNAI 9227, pp. 759–765, 2015.
DOI: 10.1007/978-3-319-22053-6_81

on which it is possible to determine the vascular score. The work presented in this paper is a single task of a wider project that we are developing for the determination of all four score of Karpinski.

2 Materials and Methods

In this section, we describe in details the image acquisition of the kidney biopsies, the vessel segmentation, the feature extraction and the vessel classification on a data set composed of 221 regions of interest consisting of: 71 vessels and 150 tubules.

2.1 Histological Slide Preparation and Acquisition

Kidney biopsy slide (KBS) preparation and digital acquisition have been conducted at the Department of Emergency and Organ Transplantation (DETO), University of Bari Aldo Moro (Bari, Italy). All KBSs have been prepared by expert lab technicians of DETO according to Renal Biopsy Guidelines of the Ad Hoc Committee appointed by the Renal Pathology Society [5]. Once the KBS is ready, it can be analysed at the microscope by nephropathologist of DETO and digitally acquired. In details, each KBS was digitally acquired using the microscope Aperio ScanScope CS featuring a 20× optical zoom. The acquired RGB images have the following characteristics: Resolution: 0.50 μm/pixel; Compression Standard: JPEG. For evaluation of vessel we used PAS (Periodic Acid Schiff) staining slide.

2.2 Vessel Segmentation

This task consists of three main steps: detection the lumen position; feature extraction of the membrane; classification.

Lumen Position Detection. The first step of the segmentation is to detect the position of the lumen in the acquired image. In order to identify the position of each lumen, the RGB image was converted in gray-scale image. Lumens are detected using a combination of a thresholding process and a morphological evaluation [6]. In particular, all pixels with a value greater than or equal to a threshold (a value of 0.6 guarantees good results) are labelled as lumen components. A new black and white image was created. Each pixel of the mask was stained white if the corresponding pixel in the RGB image is labelled as lumen pixel, black otherwise. In order to have a final lumen mask with a number of connected components equal to the number of lumens, some morphological operations are executed successively:

1. Labelling of connected components;
2. Removal of blobs with area lower than 6000 pixel and greater 180000 pixel(which represent noise);
3. Evaluation of each connected component on the basis of its shape factors: solidity and Area/Filled Area ratio.

The result of these steps showed in Fig. 1.

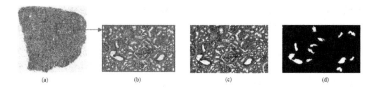

Fig. 1. (a, b) RGB image. (c) Gray-scale image. (d) Lumen mask obtained after the step 3.

This mask is processed for remove false positive lumen. In general this false positive consist in lumen of tubules.

A metric M that considering 3 factors has been developed: (1) Number of the closer connected components; (2) Distance among the closer connected component; (3) Area ratios with the closer connected components.

The metric M is based on a specific feature of the kidney tissue: tubules are very close to each other while the vessels are distant to each other, so with this metric we find this feature to remove tubules from the mask preserving the near vessels.

As follows:

$$M = \frac{N_{DC} + N_{CCSA} - N_{VCCSA}}{N_{CN}}, M \in [0, 1] \tag{1}$$

Where:

- N_{DC}: *Distant Components Number* - Is the number of components that have a distance ≥ 65 pixels. The bigger is the number of distant components, the more likely the membrane is thicker.
- N_{CCSA}: *Number of Close to each other Components with a Small Area.* - Is the number of components that have a distance in the range [30, 65] pixels and that have a small area compared to one of component in analysis, ≤ 20 %. It is most likely that tubules with small lumen are near the vessels.
- N_{VCCSA}: *Number of Very Close to each other Components with a Similar Area.* - Is the number of components that have a distance <30 pixels and which have an area comparable to the component in analysis, ≥ 60 %. This parameter is used to compensate the rare situations in which two vessels are close and have a lumen with similar area.
- N_{CN}: *Number of Components in the Neighbourhood* - Is the number of components in the neighbourhood of the component in analysis.

The result of the application of this metric is shown in Fig. 2.

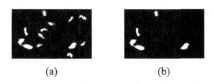

(a)   (b)

Fig. 2. (a) Mask of Fig. 1. (b) Lumen Mask obtained after processing with defined metric.

Extraction Features. This subsection describes differences between the two classes (vessels/tubules) and the corresponding extracted features used for classification.

Vessel Vs Tubule. Vessels and tubules are composed of two basic elements: the lumen and the membrane. The first is positioned at the centre of the element and is distinguishable for its white colour. In the membrane are present the cell nuclei. Differences between the membrane of the vessels and tubules are in the texture and colouring. As regards tubules, the membrane has an irregular texture and a reddish colour with a high saturation level. The PAS staining causes a high degree of saturation of this component. Furthermore, this staining causes at the membrane a saturation that is reduced in intensity going to the outside.

Extracted Features. This subsection describes the membrane features that were extracted for the classification step and the adopted methodology to extract the membrane itself.

The small sizes of ROIs obtained by preliminary segmentation step, ensure that processing algorithms of ROIs are not time-consuming.

In this work, a Region Growing [6] procedure was developed for the extraction of the membrane. Results of region growing algorithm is shown in Fig. 3.

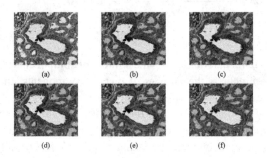

Fig. 3. (a) ROI Vessel. (b) Lumen. (c) Pixel seed for start procedure. (d) First iteration (e) i-th iteration. (f) End procedure.

In literature, there are 3 kinds of histological features: geometric, chromatic and texture-based. In this work was chosen a texture-based feature, in particular 5 Haralick texture features [7] to describe the texture of membrane: Contrast, Correlation, Energy, Homogeneity and Entropy. This 5 descriptors are calculated on the values extracted from each of the 4 asymmetric co-occurrence matrices associated with each of the following 5 images:

1. a RGB image normalized by using the method explained in [8] on the original RGB image;
2. a RGB image representing the Haematoxylin component [8];
3. a RGB representing the Eosin component [8];
4. a gray-scale image normalized by using the stretching contrasted of the relative section;
5. a gray-scale image representing the saturation channel of HSV image. For every image is calculated the 256×256 gray-scale co-occurrence matrix in to 4 directions showed in the Fig. 4, with fixed distance $D = 1$.

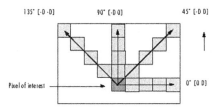

Fig. 4. Co-occurrence matrix: 4 directions with D = 1.

So for each of the 5 images we obtained 20 features corresponding to 4 co-occurrence matrices multiplied by the 5 Haralick descriptors. One more most important feature is evaluated considering an index of the saturation of membrane and finally we used a 101 features pattern.

Classification. The binary classifier (Vessel vs Tubule) classification step was performed with Back Propagation Neural Network (BPNN) [2, 9]. The classifier was trained using all the 101 features presented previously extracted from a data set randomly divided in three parts: Training Set (50 %), Validation Set (30 %) and Test Set (20 %). The classifier has been trained using the Early Stopping algorithm in order to monitor the validation error during training phases. We implemented a 101 inputs BPNN. Since the NN is used as a classifier, the output layer is composed of 1 logsig neuron. Since output neuron values can range [0, 1], the result of the classification is the class corresponding to the output neuron. The optimized topology is composed of only one hidden layer with 101 neurons. The activation function is the Hyperbolic Tangent Sigmoid function for the hidden layer and Log-Sigmoid function for the output layer.

The BPNN was trained with the Neural Network toolbox of Matlab.

3 Experimental Results

This section reports the experimental results obtained during the training and test phase.

Training Results. Table 1 shows the confusion matrix of the NN obtained for the validation set.

Table 1. NN confusion matrix - validation set.

		Prediction outcome		
		Vessel	Tubule	
Actual value	Vessel	23	3	88,5 %
	Tubule	1	39	97,5 %
		95,8 %	92,9 %	93,9 %

The classification performance are evaluated with three metrics [10]: Accuracy, Recall, Precision and Specificity, where:

$$Accuracy = \frac{TP + TN}{TP + FP + FN + TN} \qquad (2)$$

$$Recall = \frac{TP}{TP + FN} \qquad (3)$$

$$Precision = \frac{TP}{TP + FP} \qquad (4)$$

$$Specificity = \frac{TN}{TN + FP} \qquad (5)$$

Test Results. Table 2 shows the confusion matrix and the performance metrics obtained for the test set.

Table 2. NN confusion matrix - test set.

		Prediction outcome		
		Vessel	Tubule	
Actual value	Vessel	11	1	91,7 %
	Tubule	1	31	96,9 %
		91,7 %	96,9 %	95,5 %

4   Discussion

The classification performance of the validation set shows that the NN is able to classify Vessels and Tubules with a Precision greater than 88 %, Accuracy greater than 93 % and Recall greater than 95 %. Analysing the performance on the Test Set, the Neural Network assures good performance in terms of Precision (91 %) and Recall (91 %). Concluding the classifier performed the Test Set classification with good results. It is worth to note that the precision and recall level have been reached with poor number of example for training and validation.

5   Conclusion and Future Work

Good results on discrimination between vessels and tubules shows how the phase discussed and implemented in this work, taking into account only the determination of the angiosclerosis score, can validly be used to design and implement, in the future, a new CAD to support the evaluation of the overall index of Karpinski, depending from other scores like glomerulosclerosis, interstitial fibrosis and tubular atrophy. At present, the performance, in terms of recall index of the proposed classifier, shows that despite the reduced number of samples of vessels used to build the training set, the features, discussing in this paper, perform to generalize the detection of vessels extracted from the same section.

Acknowledgments. Smart Health 2.0 project (granted by Italian Ministry of University and Research) aims at developing ICT solutions for e-Health in the field of primary, secondary (early diagnosis), and tertiary prevention of diseases along life span.

The author would like to thank Ivan di Bari and Vincenzo Gesualdo for their valuable support.

References

1. Doi, K.: Computer-aided diagnosis in medical imaging: historical review, current status and future potential. Comput. Med. Imaging Graph. **31–4**, 198–211 (2007)
2. Bevilacqua, V.: Three-dimensional virtual colonoscopy for automatic polyps detection by artificial neural network approach: new tests on an enlarged cohort of polyps. Neurocomputing (2013). ISSN: 0925-2312–. doi:10.1016/j.neucom.2012.03.026
3. He, L., Long, L.R., Antani, S., Thoma, G.: Computer Assisted Diagnosis in Histopathology, pp. 271–287. iConcept Press, Kowloon (2010)
4. Karpinski, J., Lajoie, G., Cattran, D., Fenton, S., Zaltzman, J., Cardella, C., Cole, E.: Outcome of kidney transplantation from high-risk donors is determined by both structure and function. Transplantation **67**(8), 1162–1167 (1999)
5. Walker, P.D., Cavallo, T., Bonsib, S.M.: Practice guidelines for the renal biopsy. Mod. Pathol. Nat. **17**(22), 1555–1563 (2004)
6. Gonzalez, R.C., Woods, R.E.: Digital Image Processing, 3rd edn. Pearson, London (2007)
7. Haralick, R.M.: Statistical and structural approaches to texture. IEEE **67**(5), 786–804 (1979)
8. Macenko, M., Niethammer, M., Marron, J.S., Borland, D., Woosley, J.T., Guan, X., Schmitt, C., Thomas, N.E.: A method for normalizing histology slides for quantitative analysis. ISBI **9**, 1107–1110 (2009)
9. Kotsiantis, S.B., Zaharakis, I.D., Pintelas, P.E.: Machine learning: a review of classification and combining techniques. Artif. Intell. Rev. **26**(3), 159–190 (2006). Springer
10. Sokolova, M., Lapalme, G.: A systematic analysis of performance measures for classification tasks. Inf. Process. Manag. **45–4**, 427–437 (2009). Elsevier

Dependable Healthcare Service Automation: A Holistic Approach

Kaiyu Wan[1,2] and Vangalur Alagar[1,2(✉)]

[1] Xi'an Jiaotong-Liverpool University, Suzhou, People's Republic of China
[2] Concordia University, Montreal, Canada
`Kaiyu.Wan@xjtlu.edu.cn`, `alagar@cse.concordia.ca`

Abstract. In order to minimize healthcare cost and maximize the utility of the system, healthcare service must be automated in a patient-centric open distributed system that will provide dependable services for all people, in particular the elderly, physically challenged, and those who live in remote areas, at all times. Healthcare providers are often driven by economic factors, whereas patients look for dependable services. In order to promote dependability the creation of healthcare services must be founded on accurate medical knowledge and delivered through certified medical devices and professionals. Hence, it is necessary to integrate Trust determinants of Service Requester with Economic aspects of Service Provider and Medical Knowledge Science aspects of healthcare experts in automating a healthcare system. Motivated by this goal we discuss a holistic approach for developing a dependable service automation system in which the inherent relationships of Trust, Economic principles, and Knowledge are harmoniously integrated.

Keywords: Healthcare service automation · Personal health model · Economic theory · Trust theory · Knowledge science

1 Introduction

In most of the countries in the world the demand for dependable healthcare far exceeds the capabilities of existing healthcare systems. An open patient-centric distributed service system in which patients, physicians, medical devices, and clinical units are networked, and are empowered to share knowledge in a trustworthy manner seems to be the best way to maximize the health service utilization and minimize service cost. Such a network can offer not only universal health coverage but also universal health awareness. Some of the challenges to overcome in developing such a system include getting regulatory approvals, developing cyber medical devices and software-based

The research work reported in this paper was supported by research grants from National Natural Science Foundation of China (Project Number 61103029), Research Development Funding of Xi'an Jiaotong-Liverpool University (Project Number RDF-13-02-06), and Discovery Grants Program, Natural Sciences and Engineering Research Council of Canada (NSERC).

D.-S. Huang and K. Han (Eds.): ICIC 2015, Part III, LNAI 9227, pp. 766–777, 2015.
DOI: 10.1007/978-3-319-22053-6_82

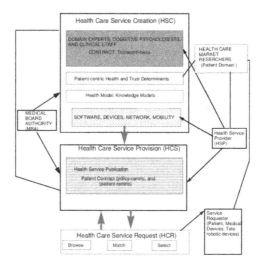

Fig. 1. Conceptual model of healthcare service automation

solutions for their interoperability, and automated knowledge-sharing among experts. In this paper we address the impact of knowledge-sharing and consumer trust on health service creation. From the many unofficial reports available on the Internet we understand that health care in most of the countries, developed or under-developed, are not available for the poor, those living in remote areas, and the elderly. Given the huge size of this population and the might of modern day mobile technology, achieving an acceptable level of healthcare for all should certainly be a priority. It is certainly not beyond the means of health care industry, because cost of communication and knowledge sharing is much lower than in current monolithic models. Patients can communicate with remote health care facility through wearable devices attached to them and receive timely medical help, if not medication. Of course, communication across borders should be allowed for timely care of patients. Physicians who are willing to be counseled online for free should be allowed by health care providers. The framework that we put forth in this paper supports these features and may be adapted and modified by health care providers to suit their policies in creating a dependable knowledge sharing medium for health care.

1.1 Contributions

Motivated by the economic, social, and knowledge science aspects the paper puts forth a conceptual model of Healthcare Service Automation. The distinguished features of the model and its merits are discussed in Sect. 2. In Sect. 3 we argue that the primary goal of on-line health care giver must be to earn the trust of consumers. We explain the consumer-side expectations in using on-line healthcare and how these should be met by the health service provider. In Sect. 4 we explain different dimensions of intelligence and creativity that are fundamental to the creation of intelligent systems, such as healthcare, and suggest the importance of wisdom, ethics and trust in automating healthcare services. We conclude the paper in Sect. 5 with remarks on implementation issues, and a summary of our ongoing work in Healthcare Cyber Physical System.

2   Conceptual Model of Healthcare Service Automation

We reckon that a total automation of a healthcare service system may not be possible and desirable. Healthcare systems that integrate patients, medical devices, physicians, and clinical staff will always need a "supervisory control" to manage crisis situations and offer expert opinions. The "knowledge and wisdom" of medical experts are required for a dependable diagnosis. Often a team of experts may need to share their knowledge and experience in deciding the treatment process. Such a collaboration among experts uses tacit knowledge, whereas the phase preceding that might use cognitive knowledge. In order to share and use these two types of knowledge, trust is required [13, 18].

The essence of trust is that the confidence it builds does not happen instantaneously, rather it is built gradually over a series of stages. Initially, its effect is to make patients comfortable in believing the quality of advertised procedures and services. Once the initial trust is formed, patients ultimately purchase the service through the healthcare provider whom they trust. Following this stage, patients might share personal health information (their medical records) with total confidence. Next, they share their experience among social groups, acting as agents in recommending the healthcare providers they trust and the quality of services they received. These trusting stages have been studied from clinical side in [11], and from service automation side in [10, 15]. An important conclusion in these studies is that social trust adds economic value to the service provider. Motivated by this study [12] the three trust dimensions purpose, process, and performance may be associated with patient behavior. Because some patient groups might be more knowledgeable than other patient groups about the healthcare domain and the technology through which services can be obtained, we must add cognitive and affective trust to behavioral trust and generate additional trust dimensions for patient groups.

The above trust considerations are factored into the generic conceptual healthcare service model shown in Fig. 1. This model absorbs the features from the service automation model introduced by us [2], specializes it for healthcare services and enriches it by adding user-centric health model and knowledge-centric service creation platform. It is also generic in the sense that it is general to fit a family of healthcare providers. A Healthcare Service Provider (HSP) in a region or in a specific health domain can adapt this model to meet the requirements mandated by regulatory and health policies. A HSP has a team of Healthcare Market Researchers (HMR) who survey the health scenario of the population for whom the HSP intents to serve. Clearly the HSP and its HMR should agree on a Health Model. Among the many research reports from governments and other agencies on healthcare the three most comprehensive reports are World Health Organization Report [7], World Health Organization Report [14], and Canadian Report [23]. These reports solely target Social Determinants of Health, but do not discuss a wide range of health determinants arising from individual behavior, environmental impact, and health care availability. The model that we have created [20], shown in Fig. 2, is much richer and more comprehensive. In our model health determinants are grouped into three spiral layers. The determinants within each layer are tightly coupled, in the sense that they strongly influence one another. The determinants in a layer have a relationship with each determinant in the next inner layer. Thus, the determinants in layer INDIVIDUAL are most important to be included in the "Personal Mobile Device" of a client. That is, each client (patient) must collect her information on "genetics", "child development", "biological information", "lifestyle that impacts health", and "the health goals" to be achieved. To this set of determinants an individual should add social determinants and environmental determinants. This personal model of a patient is the Personal Domain Knowledge (PDM) of an individual. A collection of PDMs characterize the health profile of a region that is of interest for a HSP. Healthcare Market Researchers (HMR) assess patient groups, their health status, health goals, required medical treatments, and quality of service expectations and communicate it to HSP and HSC. A patient can simply upload it's PDM to the HMR site, with privacy restrictions and constraints. That is, patients voluntarily provide their health determinants, their health goals, privacy and safety requirements to HMR. Thus our model promotes voluntary trusting of patients in their HSP.

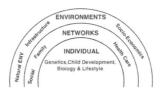

Fig. 2. Personal health care model

Domain experts, and cognitive psychologists in Health Service Creation (HSC) layer receive the collection of patient profiles, and create a Healthcare Service Model in HSC layer. The health experts are aware of the domain knowledge requirements, knowledge to be shared by them for collaborative decision making and interpreting information streams from medical devices. The software engineers are guided by health experts and turn the health goals received from patients into "functional requirements" of the system. From the safety and privacy requirements of patient profiles they create "patient-centric trust determinants". A software can automatically transform it into "trustworthiness specifications" which is a contract to be met by the HSP in producing and delivering healthcare services. MBA is a Trusted Authority for Healthcare Domain who needs to certify the publication of services in HCS. Such a certification is to ensure that published healthcare information and quality attributes of service are both accurate, and trustworthy. Most importantly the MBA team will initially validate the contract and will allow publication of services that fulfills the contract. Since the contract is formulated from patient profiles and the MBA validates it prior to service publication, at the instant of browsing and selecting services patients are guaranteed to get only trustworthy services. Without certification of healthcare services and information associated with every step of their application, no healthcare provider can publish it in layer HCS. Patients and other devices (agents) acting on their behalf (HCR layer) can browse, match and select the services they need from those services published in HCS layer. These two layers can be implemented following IoT principles.

3   Integrating Consumer Trust and Economic Goals of HSP

In this section we justify the conceptual model with respect to the trust relationship between HCR (consumer) and (HSC, HSP). A consumer's trust of the automated system is a combination of trusting the HSP who provides health services, and trusting the automation which is the medium for providing heath services. The former is a combination of cognitive trust (CT), which is consumer's confidence and willingness to rely on the competence and reliability of the HSP, and affective trust (AT), which is characterized by the confidence placed on HSP from "emotions" generated by the level of concern demonstrated by the HSP.

3.1   Consumer Determinants for Trusting HSP

The important determinants to be included in determining CT are (1) Expertise of HSP: which will convince the consumer on the competence, understanding, faith, and motives of the HSP, (2) Service Performance: which informs the consumer in advance the set of quality attributes of healthcare service, (3) Similarity: which will affirm that the goals of consumer (getting services) and HSP (delivering services) have many commonalities, and (4) Reputation: which lists the rating of HSP by the MBA and the social reputation of the HAP. Affective trust (AT) is built from Reputation. An on-line healthcare service system might also advertise "promotional items". Consumers who participate in this "socialization" process might be persuaded to get into trusting relationship with HSP.

3.2   Consumer Determinants for Trusting the Automation

In this section we list the consumer trust determinants for trusting the automation itself. According to Lee and See [15] trust in automation is "the belief that the system will help consumers achieve their goals, even in situations characterized by uncertainties and vulnerabilities". Based on this definition, trust of consumer in automation was classified [15] into one of three dimensions Performance, Process, and Purpose (PPP). We can specialize them for healthcare service automation. Performance is to be regarded as a combination of Competence, Information Accuracy, Reliability, and Accountability. Competence is patient's estimate of the suitability of the healthcare system to achieve her goal. Information accuracy is patient's judgment on the level of precision of the information given by the system. Reliability over time is patient's prediction on how long (and how far) in the future the information given by the system will remain valid. Responsibility reflects patient's assessment on the functional completeness of the system. Process is to be regarded as a composition of Dependability, Understandability, Control, and Predictability. Dependability is patient's assessment on the consistency and reliability of system response to her interactions. Understandability reflects the patient's ability to grasp the functionality of the system. Control stands for the patient's emotion "am I in control of my personal information or not?" Predictability is patient's assessment of "whether or not her expectations are met". Purpose is defined to include Motives, Benevolence, and Faith. Motives assess "whether or not the intended use of the system is truthfully communicated". Benevolence is patient's view on "whether or not the system has her interests and goals while delivering services". Faith is the belief that the system can be relied upon in future.

3.3 Consumer Trust-Based Economic Goals of HSP

In this section we invoke the economic theory classification from AgriFood Sector [9] and adapt it to healthcare industry. The economic theory [9] has been classified into three broad areas, respectively called Transaction Cost Economics (TCE), Information Economics (INE), and Socio-Economics (SOE). We explain how this may be adapted for healthcare sector. According to Johnson [10] earning trust adds economic value. As such we explore the kinds of trust that add economic value under each of the classified areas, and suggest service creation activities that will enhance system exhibit dependability properties [3] in all contexts.

The theory of TCE [9] states that every service transaction has a quantifiable direct and indirect cost, as well as an unquantifiable cost. In healthcare sector, the direct cost is the cost involved in patient monitoring, diagnosis, clinical testing, and healthcare service implementation. The indirect cost may include cost of healthcare information dissemination through healthcare network and other media, and monitoring the different transaction phases. The unquantifiable cost is usually the "cost of disadvantage" in the event of posting inaccurate or misleading or incomplete information, contract violation, and not meeting service provision rules within certain time constraints. This TCE theory may be adapted by HSP in the following manner, especially to trust building by offsetting the cost of disadvantage.

– Information should be disseminated to patients without interruption in a proactive manner. Human interaction and processing associated information in a timely fashion are central to healthcare. So, information processing and dissemination must be optimized. The goal of HSP must be to increase understanding, and improve protocols for patient-physician interaction. All collected information in every interaction scenario must automatically be recorded. This requires the inclusion of system availability and system reliability aspects in trustworthy contract of HSC layer.
– HSP should explain to patients how complexity will be contained, because complexity causes errors. The patients should understand the medical treatment, be convinced that the treatment is suitable, become aware of diagnostic process, and steps and available choices. The patients should be educated on drug doses, times of administration, correct methods for using medical devices, and different methods of surgical dressings.
– The information communicated by HSP to HCR must be correct and current. That is, HSP must ensure integrity of information in HCS layer. Patients must be informed on security and privacy aspects for sharing their medical information in the healthcare network, and told the conditions for delivering services in total confidence.

– A service whose quality violates the ethical and social values in an environment should not be provided by the HSP. That is, safety, respect for social and cultural norms, and ethical principles (especially for servicing elderly and physically challenged) must be explicitly included in the set of non-functional requirements of HSC layer.

In summary, the trust attributes related to Health TCE are safety, security (privacy and integrity included), availability, accountability (transaction monitoring), and reliability. The HSP and its HMR must gather these trust attributes and communicate to HSC layer.

The INE theory [9] can be adapted for healthcare because the relationship between HCR and any HSP is asymmetric. In general, HSP is expected to be fully knowledgeable about its services, whereas HCR may have little or no knowledge on what services to expect from a HSP. Since the patient is most likely to accept only services that have potential benefits for her and high economic value (low cost), the HSPs should provide knowledge-intensive tutoring to the HCR before delivering health services. HSPs should employ intelligent yet simple to use interfaces in HCS layer to convey the knowledge to consumers in order that they may be able to evaluate the value potential of offered services. An important aspect of INE theory is that "feeding copious information to patients may not help to sell health services". The theory suggests that in order to overcome information asymmetry, accurate health knowledge must be filtered, screened, and delivered directly to patients. Another important implication is that the learning curve required by patients must be made "less steep" through "long term" contacts and tutorials. In essence, through INE theory the HSP should learn to "trust building through direct interaction with patients". Since the PMDs of individuals are directly uploaded to HMR site, it becomes easier for HSC and HMR to collaborate in building trust with patients. This must be done at HCS (interface design). Both HSP and HMR should collaborate in fulfilling INE trust at HCS layer (drafting patient contract description), and HCF layer (delivering services to satisfy user preferences).

We adapt the SOE theory [9] to healthcare in suggesting that the HSP pay attention to "trust recommendations" and publish them in their service. Social networks, family and cultural links, and environmental determinants influence patient behavior. That is the reason why we included them in our spiral health model (Fig. 2). Patients may seek recommendations from their families, and social groups. They may consult Environmental Agents and other trusted healthcare authorities (such as government reports) to guide them in choosing best healthcare vendors. Hence the HSP might receive rankings from independent authorities and by patient groups. We suggest that the HSP include these rankings in HCS layer.

4   Healthcare Knowledge Sharing

In this section we explore the trust factors that are necessary for sharing knowledge in healthcare system automation. Healthcare professionals (physicians, experts in specific medical domains) need to share domain knowledge, knowledge arising out of experience, evidence-based knowledge, and knowledge gained from mining patient and clinical data. In addition, medical experts and software engineers need to share their knowledge in automating healthcare services, and HMR and HSP should share knowledge identified from the data gathered from patient domain. Without knowledge sharing, time-critical and patient-centric decision making in an automated environment is not viable.

4.1 Essential Knowledge Types for Healthcare

In a survey of healthcare knowledge types, their modalities, and mechanisms for managing knowledge Abidi [1] has enumerated a set of knowledge types that have an impact on clinical decision-making process. These include Patient Knowledge, Practitioner Knowledge, Healthcare Domain Knowledge, Resource Knowledge, Process Knowledge, and Organizational Knowledge. Clinical decision support systems (CDSS) have been hailed for their potential to reduce medical errors [4], and improve healthcare quality [24]. Evidence plays a crucial role in CDSS. Evidence can be categorized as Literature-based Evidence (LBE), Practice-based Evidence (PBE), Patient-directed Evidence (PDE). Hence to support CDSS in the healthcare network, three knowledge-bases, one for each evidence type, should also exist in the healthcare network. We summarize below these knowledge-bases, and explain how each will be shared.

– Domain Knowledge: This is the core medical domain knowledge. Typically, domain knowledge is an aggregation of knowledge from different sub-domains, and from related domains.
– Patient Knowledge: This knowledge-base includes patient medical records, history of administered medicines, medical observations and clinical readings, physician's remarks and inferences, and history of experts and treatments given by them.
– Practitioner Knowledge: Both cognitive and tacit knowledge will be part of this knowledge-base. Knowledge representation methods [1] are suitable for recording cognitive knowledge, whereas tacit knowledge can only be annotated with pointers to the experts who should be consulted off-line. The accumulated wisdom with reference to specific patient populations is part of tacit knowledge. This knowledge may undergo changes as more patient medical history becomes available. It may even be refined after a collection of experts share their experiences.
– Organizational and Process Knowledge: Organizational policies and case-based work-flows are recorded in this knowledge-base. This knowledge will be shared by the entire healthcare staff and patients will be given access to this knowledge.
– CDSS Knowledge-bases: This contains three collections: research literature (for LBE), practice-based evidences (for PBE), and patient-directed medical information (for PDE). The first two are intended for the use of clinical healthcare personnel, and the third one is targeted to patient community.
– Design Knowledge: Because design is "creative" and comes out of "wisdom, intelligence, and accumulated experience" design knowledge is both cognitive and tacit. All design decisions, design artifacts, domain-specific design patterns, and illustrations on their use are part of this knowledge-base. There is a need for "mutual trust" in sharing the design wisdom in order that the software engineers in HSC layer

develop healthcare service automation that can be used with minimal learning effort by every segment of society in other layers.

- Market Research Knowledge: The purpose of this knowledge-base is towards developing the system components that meet consumer expectations, as explained in Sect. 3.2. This knowledge-base will contain the PMDs of consumers, their environmental constraints, their trust expectations, and goals.
- Social Knowledge: This knowledge-base, collected by Social Psychologists, includes individual and social profiles of patients, physicians, and other actors who will interact with the healthcare system. This knowledge needs to be shared with HMR and HSC in order to synthesize it with the knowledge gathered by them.

4.2 Trust for Sharing Knowledge in Healthcare Automation

Regardless of specific representations, knowledge management study [8, 21] places Data (D), Information (I), Knowledge (K), Wisdom (W) (called DIKW hierarchy) in a pyramid structure, with D at the bottom of the pyramid and W at the peak of the pyramid. To this pyramid we can add Ethics (E) on top of W and get the hierarchy DIKWE. The lower part of the pyramid (DIK) can be called cognitive or explicit knowledge, because all of data, information, and some parts of knowledge (for information processing) can be explicitly written down and hence formally represented. The higher part of the pyramid (KWE) can be called tacit knowledge because not all aspects of this knowledge may be explicitly written down. It is possible to map the knowledge types enumerated in Sect. 4.1 with the levels in this hierarchy.

Many publications on knowledge hierarchy remark that wisdom is to be used for the well-being of others [8, 21]. For healthcare, wisdom involves a balanced coordination of (medical) mind and ethics. It is agreed by many researchers that "intelligence is a prerequisite for creativity since creative people generate lots of new ideas, analyze them, and choose the better ones". In essence, creativity leads to making head ways in a given field, generating new knowledge and wisdom. In particular, Intelligence "for intelligent healthcare" can be injected in system development by a collection of creative system developers. This "group effect" in creating large service systems will be reflected in the "intelligent behavior of the system" created by the experts. So, we may regard intelligence and wisdom with ethics as essential prerequisites for creativity in healthcare. The creativity arising from this combination will have six desirable effects [22]. These are

- ability to recognize similarities and differences between medical ideas, things (devices), and case studies (knowledge, wisdom, ethics),
- ability to put together known/old information (literature research and clinical experience) and theories in a novel manner (wisdom, ethics),
- aesthetic taste and imagination (for discovering new design paradigms) (usability, social welfare),
- ability to make decisions (at critical moments), change directions after weighing the pros and cons (sharing knowledge, ethical collaboration),
- drive for accomplishment of chosen goal (completing goal-oriented task), and
- intuition and inquisitiveness (necessary ingredients for discovery).

In sharing knowledge there is both risk and advantage. The risk is that unauthorized access to knowledge may lead to loss of integrity. The integrity of explicit knowledge can be preserved by applying security mechanisms based on trusted cryptography, authentication, and information flow controls [19]. However, to reduce the level of uncertainty associated with tacit knowledge transfer trusting relationship is essential. Tacit knowledge transfer requires willingness to share and capacity to communicate. The main barriers for tacit knowledge transfer are (1) difficulty in identifying the persons who have tacit knowledge on a specific topic (domain), (2) unwillingness of a group member to share and his incapacity to communicate, and (3) a feeling of competitive disadvantage (risk) in sharing. To overcome these barriers trust relationship must be promoted among the members of the groups who will share a knowledge-base. The two kinds of trust studied towards improving trust are (1) affect-based trust, and (2) cognition-based trust. Affect-based trust is grounded in mutual care and concern for members of the group, and in the ability to reach a compromise on the goal-oriented creative process of the system. Cognition-based trust is grounded on the competence of individuals in the group, and their reliable cooperation among themselves in using explicit knowledge. Based on the statistical evidence [13] we may say that affect-based trust has a positive effect on the willingness to share tacit knowledge, while cognition based trust has a positive effect on the willingness to use tacit knowledge. When trusted collaboration among the experts and system developers are based on affective and cognitive trust the automated system can be expected to provide dependable medical services.

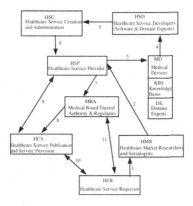

Fig. 3. Trust flow in healthcare model

5   Conclusion

The conceptual healthcare model in Fig. 1 can be expanded to include the Integrated Clinical Environment (ICE) [17] and trusted interoperable medical devices and regulatory issues (TIMDR) [11, 16]. The expanded model is intended as a blueprint for healthcare service automation. The rationale to develop our healthcare model independent of ICE and TIMDR is that their trust domains are different and mechanisms for enforcing them are different. In the case of ICE and TIMDR, it is necessary to verify that medical devices and their manufacturers are trustworthy, and protocols for their communication are free of errors. Whereas the nature of trust in our model permeates through "humans and their agents" (not only through objects) that interact in the healthcare network.

Consequently, independently investigating the trust determinants do not affect the overall trustworthiness of the system. In reality, such a "divide and conquer" approach can only strengthen the overall trustworthiness factor. The study [5] proposes an automatic framework to integrate ISO/IWA1 practices to reduce excessive efforts of patient image quantity examination in a nuclear medicine department. They claim to have validated this approach with a pilot study and prove its correctness vis-a-vis traditional processes. Our healthcare model has several stakeholders and a validation of our model would require a huge amount of their combined time and effort. Moreover we want to emphasize that "a model is neither correct nor incorrect, it should only be useful". We need to build prototypes based on our model and field test them. Based on that experience the model will evolve. However, a theoretical analysis of trust propagation scenario shown in Fig. 3 is rather straightforward. In this diagram, each edge is labeled by a number, and reading them in sequence will convince us that trust determinants flow across appropriate stakeholders. On this basis we conclude that our approach leads to integrating consumer-centric trust with system-centric dependability criteria. We believe that this trust propagation model is consistent with the theoretical issues raised on trust and human interaction of automated systems [18]. A rigorous comparison between different dependable healthcare models is part of our ongoing work.

References

1. Abidi, S.S.R.: Healthcare knowledge management: the art of the possible. In: Riaño, D. (ed.) K4CARE 2007. LNCS (LNAI), vol. 4924, pp. 1–20. Springer, Heidelberg (2008)
2. Alagar, V., Wan, K.: Integrating trust and economic theories with knowledge science for dependable service automation. In: Mauw, S., Jensen, C.D. (eds.) STM 2014. LNCS, vol. 8743, pp. 1–16. Springer, Heidelberg (2014)
3. Avizienis, A., Laprie, J.C., Randell, B., Landwehr, C.: Basic concepts and taxonomy of dependable and secure computing. IEEE Trans. Dependable Sec. Comput. 1(1), 11–33 (2004)
4. Bates, D.W., Cohen, M., Leape, L.L., et al.: Reducing the frequency of errors in medicine using information technology. J. Am. Med. Inf. Assoc. 8, 299–308 (2001)
5. Chen, J.J., Su, W.C.: Integrating ISO/IWA1 practices with an Automatic Patient Image Quantity Examination Approach to Achieve the Patient Safety Goal in a Nuclear Medicine Department. In: 2014 IEEE International Conference on Internet of Things (iThings 2014), Green Computing and Communications (GreenCom 2014), and Cyber-Physical-Social Computing (CPSCom 2014), pp. 335–338 (2014)
6. Cross, R., Baird, L.: Technology is not enough: improving performance by building organizational memory. Sloan Manag. Rev. 41(3), 69–78 (2000)
7. Euro Report. Social Determinants of Health: The Solid Facts. http://www.euro.who.int/en/publications/abstracts/social-determinants-of-health.-thesolid-facts
8. Fricke, M.: The Knowledge Pyramid: a critique of the DIKW hierarchy. J. Inf. Sci. 35(2), 131–142 (2009)

9. Fritz, M., Hausen, T., Schefer, G., Canavari, M.: Trust and electronic commerce in the agrifood sector: a trust model and experimental experience. Presented at the XIth International Congress of the EAAE (European Association of Agricultural Economists), "The Future of Rural Europe in the Global Agri-Food System", Copenhagen, Denmark, 24–27 August 2005

10. Johnson, D., Grayson, K.: Cognitive and affective trust in service relationship. J. Bus. Res. **58**, 500–507 (2005)

11. Hatcliff, J., Vasserman, E., Weininger, S., Goldman, J.: An overview of regulatory and trust issues for the integrated clinical environment. In: Proceedings of HCMDSS 2011 (2011)

12. Hoffmann, H., Söllner, M.: Incorporating behavioral trust theory into system development for ubiquitous applications. Pers. Ubiquit. Comput. **18**, 117–128 (2014)

13. Holste, J.S., Fields, D.: Trust and tacit knowledge: sharing and use. J. Knowl. Manag. **14**(1), 128–140 (2010)

14. WHO Report. Policies and strategies to promote social equity in health. http://whqlibdoc.who.int/euro/-1993/EURICPRPD414(2).pdf

15. Lee, J.D., See, K.A.: Trust in automation: designing for appropriate reliance. Hum. Factors **46**(1), 50–80 (2004)

16. Logan, M., Patel, B. (eds): Medical Device Interoperability: A Safer Path Forward Priority Issues from the 2012 AAMIFDA Interoperability Summit (2012)

17. Moorman, B.: Medical device interoperability: standards overview. ITWorld, 132–138, March-April 2010

18. Muir, B.M.: Trust in automation: Part I: theoretical issues in the study of trust and human intervention in automated systems. Ergonomics **37**(11), 1905–1922 (1994)

19. Myers, A., Liskov, B.: A decentralized model for information flow control. In: Proceedings of the 16th ACM Symposium on Operating System Principles, Saint Malo, France, October 1977

20. Nguyen, N., Alagar, V.: Context aware mobile health care framework. Research Report, Department of Computer Science, Concordia University, May 2014

21. Rowley, J.: The wisdom hierarchy: representation of the DIKW hierarchy. J. Inf. Sci. **33**(2), 163–180 (2007)

22. Sternberg, R.J.: What is the common thread of creativity? Its dialectical relation to intelligence and wisdom. Am. Psychol. **56**(4), 360–363 (2001)

23. Canadian Report. Social Determinants of Health: The Canadian Facts. http://www.thecanadianfacts.org/

24. Teich, J.M., Wrinn, M.M.: Clinical decision support system come of age. MD Comput. **17**(1), 43–46 (2000)

Information Security

Improved Concurrent Signature on Conic Curve

Zhanhu Li and Xuemei Yang[(✉)]

School of Mathematics and Information Science,
Xianyang Normal University, Xianyang, China
lizhanhu75@126.com, yangxuemei691226@163.com

Abstract. As the proposing of concurrent signatures, the bottleneck of the third party is completely gotten rid of in the electronic payment system, which greatly improves the safety and fairness of electronic payment systems. In order to further improve the operation efficiency of it, a concurrent signature based on conic curve was put forward, which greatly improved operational efficiency of signature. However, it can't fulfill un-forge ability. In order to compensate for this defect, an improved concurrent signature protocol on conic curve is proposed. By analyzing this scheme, it can not only fulfill the un-forge ability, but also has higher efficiency comparing with the one proposed by Chen L as well. Furthermore, the scheme is also based on discrete logarithm in conic curves, so it has the same security.

Keywords: Concurrent signature · Conic curve · Discrete logarithm

1 Introduction

In 2004, Chen [1] proposed the Concurrent Signatures (CS) the first time in Cryptology – EUROCRYPT, which changes the traditional mode of fair exchange message. The core idea of it is using ring signature technology to generate an ambiguous agreement between each signature, while the signature information is bound with the corresponding signer identity. Its main purpose includes two aspects: first, though the either party knows the identity of signer, none of them can prove it to the third party. Second, the receiver can ensure the true signature information source, the signer can ensure that only his designated receiver can verify the validity of the signature as well.

In the initial stage, the initiator randomly selects a private key and generates a key-stone through a single mapping function, and hides it in the signature. Finally, the initiator sends the signature to recipient. After the recipient receives the message, he uses the same method to generate an equally ambiguous signature structure, One thing to note is that the key-stone used by the recipient must be the one sent by initiator. Before opening the private key, even though both sides know the identity of the signature, neither can prove it to the third part. Once opening the private key, the third part can distinguish the identity of signer by using the signature and private. So the ambiguity of signature is removed, both signatures have true validity. Therefore, the application of concurrent signature have aroused great interest of many researchers and scholars [2–5].

© Springer International Publishing Switzerland 2015
D.-S. Huang and K. Han (Eds.): ICIC 2015, Part III, LNAI 9227, pp. 781–786, 2015.
DOI: 10.1007/978-3-319-22053-6_83

In 90's of last century, the studying of applications of conic curve over finite field in the large number factorization and cryptography aroused great interest. In 1996, for a conic curve over finite field, Zhang Mingzhi produced the addition operation on conic curve for the first time, and proved that was a additive Abel group [6]. In 1998, Cao Zhenfu [7, 8] proposed a public key cryptosystem based on a conic curve over finite fields. As the advantages of embedding plaintext conveniently, easy points operating, and fast in speed etc., especially the easy calculating of inverse on conic curve, the cryptographic algorithm design has a strong attraction to more and more peoples.

In 2002, the authors of reference [9–11] denoted that there existed an algorithms of high speed of calculating the integral multiple operation of pointing in expression integer by standard binary system, which could reduce about 1/4 calculation by using it on the operation between points of conic curve. In 2005, Sun Qi and Wang Biao [12, 13] expanded the conic curve over finite field onto ring Z_n, and denoted that it had the advantages of embedding plaintext conveniently, easy points operating, and fast in speed etc. The author of reference [14] further studied the characters of $C_n(a, b)$ over ring Z_n. They pointed out that there was always a basic point which order was N_n in $C_n(a, b)$. they also pointed out that the protocol they produced had the advantage of the easier calculating under some condition compared with the one on $E_n(a, b)$. In 2010, Li Guojing [15] further studied the add operations between point on conic curve, and designed a new algorithm to calculate multiplication. The results show that it makes a great improvement in multiplication compared with the one defined in reference [14]. In reference [16], an anonymous proxy signature based on the conic curves over Z_n was produced, and in reference [17], a digital signature based on conic curve over Eisenstein ring was put forward.

In 2011, the author put forward a CS scheme by using the cryptology on conic curve [18]. Though it achieved a great efficiency, it can not fulfill un-forgeable. In this paper, we use the advantages of conic curve to propose an improved CS scheme on conic curve over ring Z_n, which not only fulfills the un-forge ability, but also has higher efficiency comparing with the one proposed by Chen L as well. Furthermore, the scheme is also based on discrete logarithm in conic curves, so it has the same security.

2   The Conic Curve C_n(a, b) Over Ring Z_n

Let Z_n be a module ring of n. In reference [12], the conic curve $C_n(a, b)$ over ring Z_n is defined as the set of all solutions of congruence equation

$$y^2 = ax^2 - bx(\bmod\ n)$$

Where $n = pq$, p, q are two large different odd prime numbers. $(a, n) = (b, n) = 1$.

Theory [12]. Let $A \in C_n(a, b)$, we define the lest integer number k as the order of A satisfied $kA = 0$, and denote it as $O(A)$. $\forall A \in C_n(a, b)$, there is a only point (A_p, A_q) correspondence to A in $C_p(a, b) \times C_q(a, b)$, and it's order is equal to $lcm[O(A_p), O(A_q)]$.

Deduction [12]. Let $n = pq$, where p, q are two large different odd prime numbers, and $\left(\frac{a}{p}\right) = \left(\frac{a}{q}\right) = -1$, $p + 1 = 2r$, $q + 1 = 2s$, r, s are also prime numbers, then there exist a point G in conic curve $C_n(a, b)$, which order is $N_n = 2rs$. G is a basic point of $C_n(a, b)$, and set $S = \{0, G, 2G, \ldots (N_n - 1)G\}$ is a subgroup of $C_n(a, b)$.

The hardness problem on conic curve—discrete algorithm problem in S: Given two different points $M, N \in S$, it is hard to compute $e \in Z, e > 0$ such that $M = eN$.

About the details of the operation between points of conic curve and public key encryption protocols over conic curve we can see them from references [6–15].

3 Concurrent Signature on Conic Curve

The concurrent signature on conic curve mainly is composed following parts:

(1) Setup of parameters

PKG chose a conic curve with cryptography $C_n(a, b)$ as defined in the above part. Let G be a basic point of $C_n(a, b)$, which order is $N_n = 2rs$, $H : \{0, 1\}^* \to Z_{N_n}^*$ is a secure *hash* function. Let $d \in Z_{N_n}^*$ be private key, and $Q = dG (\bmod n)$ the public key corresponding to d.

Open n, a, b, G, Q, N_n as public key, take d as the private key.

(2) The algorithm of CS

Let A be the signature sponsor and B match signer. Their key-pairs is (d_A, Q_A) and (d_B, Q_B) respectively. The algorithm of CS is following two steps.

(1) A randomly selects two integer numbers k, t such that $(k, N_n) = 1$, and computes
$P_0 = kG = (x_0, y_0)(\bmod n)$, $\gamma = x_0 (\bmod N_n)$,
(2) $u_A = tG - \gamma Q_B (\bmod n)$, $v_A = t + \gamma d_A H(m)(\bmod N_n)$,

If $\gamma = 0$ or $u_A = 0$, reselects the integer numbers k, t. Else the A's signature for a message m is $S_A = (\gamma, m_A, u_A, v_A)$.

(3) Verification

After receiving the S_A, B first examines if $v_A G = 0$ is holds, output error, else, B verifies if following equation holds:

$$v_A G = u_A + \gamma Q_B + \gamma H(m) Q_A (\bmod n)$$

if not, output error, else, B generates his signature of m_B, which method is similar to A, and not generate γ by himself, but use the one sent by A. Finally B generates his signature: $S_B = (\gamma, m_B, u_B, v_B)$, and send it to A. After A receives the S_B, she also verifies the correctness of S_B, and she must ensure that the γ B used is the one she sent, else she corrupt the protocol. If A verifies that the B's signature is true, then she opens the k.

4 Analysis of Scheme

(1) Correctness

Let $u_A = tG - \gamma Q_B(\mathrm{mod}\, n)$, $v_A = t + \gamma d_A H(m)(\mathrm{mod}\, N_n)$, then equation

$$v_A G = u_A + \gamma Q_B + \gamma H(m)Q_A(\mathrm{mod}\, n)$$

is established.

Proof: $v_A G = tG + \gamma Q_A(\mathrm{mod}\, n) = u_A + \gamma Q_B + \gamma H(m)Q_A(\mathrm{mod}\, n)$

(2) Ambiguity: For any outside party, either A or B is able to generate this signature. Before opening the keystone k, both A and B know who is the initial signer, but neither of them can prove it. While opening the keystone k, anyone can ensure who is the initial signer at once.

(3) Verifiability. When any authenticator obtains the k and (γ, u, v), he can work out the γ by k, then he examines $vG = u + \gamma Q_i + \gamma H(m)Q_i(\mathrm{mod}\, n)$, $i \in \{A, B\}$. If γ is the coefficient of Q_B, he can deduce that A is the initial signer and B is receiver. Else γ is the coefficient of Q_A, he can deduce that B is the initial signer and A is receiver. If neither of them is established, then he output error.

(4) Un-forge ability. Suppose *Eve* poses as A send a legitimate signature to B. As the keystone k is operated in the course of signature through integral multiple of point, it is hard to solve the k through P_0 and G based on the discrete logarithm on conic curve. As a result, he can't work out the private key of A. On the other hand, if *Eve* picks a appropriate k' by himself, and works out corresponding γ' and u'. However, to calculate v', he can wok out the true v' just only under knowing the private key of A. So the protocol proposed in this paper is un-forge ability.

(5) Analysis of operation efficiency: In reference [1], there are four times modular index operation and five times multiple operation altogether in the course of signature and verify. In this paper, there are four times integer multiple point and one time additive between points on $C_n(a, b)$ altogether in the course of signature and verify. Because the computation order of complexity of $x \times y(\mathrm{mod}\, n)$ is $O(q^2)$ [19], and the one of $x^y(\mathrm{mod}\, n)$ is $O(q^3)$ [20], where q is the bit length of x, y, n expressed by standard binary, then comparing the times of operating and computation order of complexity, where the times of operating denotes the *hash* operation as *Inv* on group, integer multiple point as *Pm*, additive between points as *Ad*, modular index operation as *Em*, integer multiple as *P*, we can deduce that the protocol in this paper has an improvement of efficient of signature and verify under the same secure condition from Table 1.

Table 1. Comparison of different methods

Concurrent signation	Computation order of complexity	Computation load
Reference [1]	$4\ Em + 5\ P$	$O(q^3)$
This paper	$3\ Em + 4\ Ad$	$O(q^2)$

where q is the bit length of q and N_n in reference [1] and this paper.

5 Conclusion

As the exist CS scheme on conic curve over ring Z_n can't resist un-forge ability, an improved scheme is put forward in this paper. Compared with the finite field, conic curves have advantages of embedding plaintext conveniently, easy points operating, and fast in speed etc. So it has higher efficiency than the one proposed in reference [1] as well. Furthermore, the scheme has the same security based on the discrete logarithm in conic curves.

Acknowledgment. This work was supported by Natural Science Foundation of Xian Yang Normal University (14XSYK007) and Natural Science Foundation of Shaanxi Province (2014JM1002) and Natural Science Foundation of Xian Yang Normal University (14XSYK004).

References

1. Chen, L., Kudla, C., Paterson, K.G.: Concurrent signatures. In: Cachin, C., Camenisch, J.L. (eds.) EUROCRYPT 2004. LNCS, vol. 3027, pp. 287–305. Springer, Heidelberg (2004)
2. Xiao, H.Y., Zhang, M.Q., Yang, X.Y., Zhou, X.W.: Fair exchange protocol based on concurrent signature. Comput. Eng. Appl. **45**(23), 206–207 (2009)
3. Liu, J.: Fairness improvement of pay word protocol based on concurrent signature. J. Comput. Appl. **30**(6), 1493–1494 (2010)
4. Susilo, W., Au, M.H., Wang, Y., Wong, D.S.: Fairness in concurrent signatures revisited. In: Boyd, C., Simpson, L. (eds.) ACISP 2013. LNCS, vol. 7959, pp. 318–329. Springer, Heidelberg (2013)
5. Xin, L.: Concurrent-secure blind signature scheme without random oracles. In: Proceedings of the 2012 International Conference on Information Technology and Management Science (ICITMS 2012), 2013, pp. 535–544 (2012)
6. Mingzhi, Z.: Factoring integers with conics. J. Sichuan Univ. (Nat. Sci. Ed.) **33**(4), 356–359 (1996)
7. Cao, Z.: A public key cryptosystem based on a conic over finite fields Fp. In: Advances in Cryptology-Chinacrypt 1998, pp. 45–49. Science Press, Beijing (1998)
8. Cao, Z.: Conic analog of RSA cryptosystem and some improved RSA cryptosystems. J. Nat. Sci. Heilongjiang Univ. **4**, 15–18 (1994)
9. Dai, Z., Pei, D., Yang, J. et al.: Cryptanalysis of a public key cryptosystem based on conic curves. In: The International Workshop on Cryptographic Techniques and E-Commerce, Hong Kong (2000)
10. Sun, Q., Zhang, Q.F., Peng, G.H.: A new algorithm on the Dickson poynomial $g_e(x, 1)$ public key cryptosystem. J. Sichuan Univ. (Nat. Sci. Ed.) **1**, 18–23 (2002)

11. Sun, Q., Zhang, Q.F., Peng, G.H.: A algorithm of calculating an integer multiple of group and it's application in public key cryptosystem. In: Advances in Cryptology–Chinacrypt 2002, pp 117–124. Science Press, Electronic Industry, Beijing (2002)
12. Sun, Q., Zhu, W.Y., Wang, B.: The conic curves over Z_n and public-key cryptosystem protocol. J Sichuan Univ. (Nat. Sci. Ed.) **42**(3), 471–478 (2005)
13. Wang, B., Zhu, W.Y., Sun, Q.: Public-key cryptosystem based on the conic curves over ring Zn. J. Sichuan Univ. (Eng. Sci. Ed.) **37**(5), 112–117 (2005)
14. Wang, B., Fang, Y.J., Lin, H.: QV signature protocol on conic curve over ring Zn. Sci. China **39**(2), 212–217 (2009)
15. Li, G., Li, Y., Wen, T.: Improved addition of conic curve over the ring Zn. J. Northeast. Univ. (Nat. Sci.) **31**(1), 28–30 (2013)
16. Jia, N., Li, Z.C.: Anonymous proxy signature based on the conic curves over ring Zn. Microcomput. Inf. **26**(6–3), 78–79 (2010)
17. Pan, R., Wang, L.J., Li, X., Li, D.D.: Digital signature based on conic curve over Eisenstein ring. Comput. Eng. **36**(6), 169–172 (2010)
18. Li, Z., Fan, K., Li, H.: Concurrent signature protocol on conic curve over ring Zn. In: 2011 International Conference on Electronics, Communications and Control, vol. 1, pp. 194–197 (2011)
19. Lu, K.D.: Computer Cryptography, pp. 225–226. Science Press, Qing Hua University, Beijing (1998)
20. Stinson, D.R.: Cryptography Theory and Practice, 2nd edn. Science Press, Electronic Industry, Beijing (2003)

Double Compression Detection in MPEG-4 Videos Based on Block Artifact Measurement with Variation of Prediction Footprint

Peisong He[1,2(✉)], Tanfeng Sun[1,2], Xinghao Jiang[1,2(✉)], and Shilin Wang[1,2]

[1] School of Electronic Information and Electrical Engineering,
Shanghai Jiao Tong University, Shanghai, China
{gokeyhps,tfsun,xhjiang,wsl}@sjtu.edu.cn
[2] National Engineering Lab on Information Content Analysis Techniques,
GT036001, Shanghai, China

Abstract. In this paper, we propose a novel scheme to detect double MPEG-4 compression with block artifact analysis. An adaptive measurement of block artifact in decompressed frames is proposed and then combined with the Variation of Prediction Footprint (VPF) in an effective way. Based on such measurement, periodic analysis is used to detect double compression. The proposed scheme is verified on several publically available standard videos and compared with the state-of-the-art method. Experimental results demonstrate that it has more robust detection capability.

Keywords: Video forensics · MPEG-4 · Double compression · Block artifact

1 Introduction

In recent years, MPEG-4 based codec has been applied for a considerable proportion of surveillance systems and video cameras. There are imperative needs to detect the integrity and authenticity of MPEG-4 videos when digital videos are proposed as the evidence in front of a court of law. Double compression detection method is one of the important authentication methods in video forensics [1], since almost all the tampered videos must go through at least twice encoding processes.

According to whether the structure of GOP between the primary and the secondary compression are the same or not, video double compression can be categorized into two classes. For the double compression detection with the same GOP structure, some successful methods have been proposed. In [2–4], distributions of quantized DCT coefficients are the main clue to expose double compression. The first digit statistics is used as another powerful tool to detect double compression in [5, 6]. In [7], Subramanyam and Emmanuel used principles of estimation theory to detect double quantization. In [8], Wang et al. apply the Markov based features to detect double compression in MPEG-4 videos. For double compression detection with different GOP structures, most of the above-mentioned methods fail to achieve satisfactory performance. Luo et al. [9] consider this issue by analyzing the feature curve of recompression videos after removing

© Springer International Publishing Switzerland 2015
D.-S. Huang and K. Han (Eds.): ICIC 2015, Part III, LNAI 9227, pp. 787–793, 2015.
DOI: 10.1007/978-3-319-22053-6_84

different frames for a given sequence. In [10], Bestagini et al. proposed a method to identify the type of codec used in the first coding step of doubly encoded videos. The main drawbacks of [9, 10] is that the performance drops rapidly when the strength of last compression increases. In [11], the technique based on the variation of the macroblock prediction types in the re-encoded P-frames is proposed. However, the method in [11] still has some limitations: this method may become unreliable when the content of videos contains complex texture.

To overcome these drawbacks, we propose a double MPEG-4 compression detection scheme based on block artifact measurement with VPF. The proposed scheme first obtains the strength of block artifact from differences between decompressed frames of the input video and their deblocked versions using an adaptive deblocking algorithm. Then, the strength of block artifact is combined with the VPF extracted from video stream to generate the measurement sequence. Finally, periodic analysis is applied to the measurement sequence to detect double compression. Experimental results demonstrate the proposed approach has more robust double compression detection capability compared with the state-of-the-art method.

The rest of this paper is organized as follows. Section 2 introduces the properties of block artifact. Section 3 detailedly presents the proposed scheme. Section 4 reports the results of experimental tests and Sect. 5 draws the conclusions.

2   Analysis of Block Artifact

In MPEG-4 encoded video sequences, there are three types of frames: intra-coded frames (I-frames), predictive-coded frames (P-frames) and bi-directionally predictive-coded frames (B-frames) in each GOP. For simplicity, B frames are not considered in this work. Since the transform coding and quantization is independent from one block to another in MPEG-4 compression, the discontinuity often exists in the boundaries of 8×8 blocks. This kind of distortion is called as block artifact. Block artifact may exist in both I-frames and P-frames. However, the block artifact of P-frames is different from that of I-frames, since the texture coding step is carried out in prediction residues in inter-coded macroblocks (P-MBs) instead of content in current blocks.

Different from single compression, the block artifact of recompressed frames is affected by compression degradations in the first compression. If a MPEG-4 video is double compressed with different GOP structure (referred to as double compression hereinafter), there will be four kinds of re-encoded frames: I-P frames (the P-frames previously encoded as I-frames in the first compression), I-I frames, P-P frames and P-I frames. In first compression, the quantization error introduced by intra-coding can weaken the correlation between the I-frame and its previous P-frame. Because of it, the difference between the current block and the corresponding reference block increases. When quantiser scale is relative small, the degradation due to recompression is slight. The block artifact of original I-frames will be retained in the second inter-coding process. Thus, in decompression domain, the I-P frames always perform more severe block artifact on boundaries between on-grid blocks than adjacent P-P frames. Please note this kind of block artifacts in P-I and I-I frames (referred to as double compressed I frames) are also distinct. However, block artifact in double compressed I frames cannot be used as the clue of double compression directly, since it is caused by intra-coding process in second compression.

3   The Proposed Approach

In this section, a novel double compression detection scheme based on block artifact analysis is described.

3.1   The Measurement of Block Artifact

As discussed in Sect. 2, block artifact between on-grid blocks is more distinct in I-P frames than adjacent P-P frames. It is used as the clue of double compression in this work. In order to measure the strength of block artifact, differences between the decompression frames and their deblocked versions are computed. Inspired by [12], an adaptive postfiltering method based on both spatial and frequency information is applied to deblock decompressed frames. Denote a decompressed video sequence of the input video as $\mathbf{X} = \{\mathbf{X}_1, \mathbf{X}_2, \ldots, \mathbf{X}_N\}$, where N is the length of the video sequence and $\mathbf{X}_n$ represents the luminance component of n-th decompressed frame. The deblocking method for each decompressed frame runs as follows:

Step 1. Setting Block Semaphores: The decompressed frame is first divided into non-overlapping blocks. Then, the 8×8 transform coefficients of each block are computed using discrete cosine transform. A block having only DC component can cause both horizontal and vertical blocking artifact. In this case, both the horizontal blocking semaphore (HBS) and the vertical blocking semaphore (VBS) of the corresponding block are set to 1. When only the coefficients in the top row of 8×8 transform coefficients are nonzero, the block may have vertical block artifact. In this case, the VBS is set to 1. When only the coefficients in the far left column have nonzero values, the HBS is set to 1. Do such operation for all the non-overlapping blocks in each decompressed frame.

Step 2. Horizontal Deblocking: If the HBS of the block and its horizontally adjacent block are both equal to 1, a seven-tap low pass filter $\mathbf{f}_1$ is applied to smooth the boundary between these blocks. The seven-tap is of the form: $\mathbf{f}_1 = \left[\frac{1}{8}, \frac{1}{8}, \frac{1}{8}, \frac{1}{4}, \frac{1}{8}, \frac{1}{8}, \frac{1}{8}\right]$. If this condition is not satisfied, the absolute value of difference between pixels near the boundary is calculated. When the difference is smaller than $2 \times QP$ (quantizer scaler), the weak low pass filter $\mathbf{f}_2$ is applied to smooth the block boundary, where the weak filter is $\left[\frac{1}{4}, \frac{1}{2}, \frac{1}{4}\right]$ in this work.

Step 3. Vertical Deblocking: The vertical filtering is performed in the same way as horizontal filtering.

After deblocking, the deblocked version $\tilde{\mathbf{X}}_n$ of $\mathbf{X}_n$ is generated. This measurement of n-th frame can be calculated as follows:

$$E(n) = \sum_{n=1}^{L} \sum_{m=1}^{M} |\tilde{\mathbf{X}}_n(l,m) - \mathbf{X}_n(l,m)| \tag{1}$$

Then, apply this operation for each decompressed frame to get measurement sequence $E(n)$ of the input video, where $n = 0, 1, \ldots, N - 1$. Since I-frames have strong block artifact both in single encoded videos and double encoded videos, $E(n)$ of such frames cannot be used directly. The elements located at frames kG_2, where G_2 is GOP size in second compression and $k = 0, \ldots, \lfloor N/G_2 \rfloor$, are replaced to the average value of its previous element and following element, formally: $E(kG_2) = (E(kG_2 - 1) + E(kG_2 + 1))/2$. Finally, the ratio between the local strength and the sequence's average strength is calculated to get the measurement sequence of block artifact $\tilde{\mathbf{E}}(n)$ as follows: $\tilde{\mathbf{E}}(n) = E(n)/E_{ave}$, where $E_{ave} = \frac{1}{N} \sum_{n=0}^{N-1} E(n)$.

3.2   Combination Between Block Artifact and VPF

To further improve robustness of the proposed method, the measurement of block artifact is combined with VPF. In [11], the abnormal fluctuation of marcoblocks is applied to locate the VPF and the strength of VPF is measured as the function of numbers for different macroblock types between the current frame and its adjacent frames. In this work, we still use the condition in [11] to locate VPF and get the set Ω containing VPFs of the input video. Different from [11], the strength of block artifact is used as the measurement for VPF. Besides, the block artifact of frames not in Ω is also considered with the adjustable parameter. Formally, the modified measurement sequence $V(n)$ is defined as follows:

$$V(n) = \begin{cases} \tilde{\mathbf{E}}(n), & if \ n \in \Omega \\ \alpha \tilde{\mathbf{E}}(n), & \text{otherwise} \end{cases} \tag{2}$$

where $\alpha \in [0, 1]$ is the parameter to adjust the strength of block artifact in frames not in Ω and $n = 0, 1, \ldots, N - 1$. By selecting $\alpha \in [0, 1]$, the detector can obtain the tradeoff between block artifact measurement and VPF to get more effective double compression detection ability.

3.3   Periodic Analysis

Periodic analysis is applied to the measurement sequence $V(n)$ to expose double compression. Since coding error propagation can severely degrade the quality of the MPEG-4 compression video, it is reasonable to assume that the number of candidate GOP sizes is finite. We empirically set that the maximum value of GOP as 300 and regard videos whose length are less than 1500 frames as short clips in this work. The set of candidate GOP size $\mathbf{C} = \{2, 3, \ldots C_{max}\}$ is defined. C_{max} is set as one-fifth of the video length when the input video is the short clip, otherwise C_{max} is set as 300.

Then, a function $\psi(c)$ is defined to measure how well the choice of $c \in \mathbf{C}$ models the periodicity of signal $V(n)$. The function $\psi(c)$ maps each candidate value c to a fitness value ψ. Different from [11], the composite fitness function $\psi(c) = \psi_1(c) - \psi_2(c)$ is defined as follows: $\psi_1(c)$ calculates the average value of elements in integer multiples of the candidate GOP size c; $\psi_2(c)$ penalizes the presence of periodic caused by other GOP candidates as follows: $\psi_2(c) = \max\limits_{z \in [1, c-1]} \psi_1(z)$.

Finally, the value of $\psi(c)$ is calculated, iterating over each of the candidate GOP size in $\mathbf{C}$. The highest value is compared to the threshold T_ψ to assign the class to the input video where T_ψ is a preset threshold. If the highest value is above the threshold T_ψ, the input video is classified as double compression.

4 Experiments

In this section, the performance of our algorithm in double compression detection is evaluated. Fourteen widely known YUV sequences[1] with 352×288 resolution (CIF) are selected as source sequences. In order to test the reliability of the proposed scheme in presence of short clips, only the first 250 frames for each YUV sequence are considered. According to the definition of $C_{\max}$, it is easy to extend the proposed scheme to longer clips. For all the experiments, the codec built in MPEG-4 of libavcodec library (through FFmpeg) is used to encode and decode all the videos. In our tests, the first bitrate B_1 and second bitrate B_2 were both selected from the set $\{100, 300, 500, 700\}$(kbps) with constant bit rate (CBR) mode. The values of G_1 were selected from the set $\{10, 15, 30, 40\}$ while G_2 from the set $\{9, 16, 33, 50\}$. We create a set $\mathbf{S}$ containing 224 singly encoded videos with all combinations of G_2 and B_2 for all YUV sequences and another set $\mathbf{D}$ containing 3584 doubly encoded videos with all combinations of G_1, B_1 in first compression and G_2, B_2 in second compression for all YUV sequences. In order to provide a fair comparison, we replace the GOP candidate selection method in [11] to our proposed method in all experiments.

The performance of double compression detection is investigated as a function of first and second bitrate. The area under the Receiver Operating Characteristic (ROC) curve is adopted to depict the performance of detector. The area under the ROC curve (AUC) is averaged over 20 times of experiments by randomly selecting double compressed videos with the corresponding combination (B_1, B_2) in set $\mathbf{D}$. The values for AUC achieved by the proposed methods and [11] are shown in Table 1. The values of α is experimentally set as 0.8. The best results for each combination (B_1, B_2) are highlighted.

As shown in Table 1, it can be observed that our method always performs better. The proposed method gets distinct improvement when videos have relatively good quality (e.g. $B_1 > 100$ kbps and $B_2 > 100$ kbps). Concretely, the average AUC of the proposed method is 0.947 while the method in [11] is 0.921 in these cases. And we

---

[1] Freely available at this website: http://trace.eas.asu.edu/yuv/. Chosen sequences are: akiyo, bridge-close, bridge-far, coastguard, container, foreman, hall, highway, mobile, news, paris, silent, tempete, waterfall.

Table 1. AUC of double compression detection

(B_1, B_2)		100	300	500	700
100	Proposed [11]	0.958	**0.998**	**0.996**	**0.997**
		0.971	0.998	0.993	0.996
300	Proposed [11]	0.849	**0.956**	**0.987**	**0.992**
		0.866	0.933	0.970	0.977
500	Proposed [11]	**0.766**	**0.923**	**0.954**	**0.979**
		0.747	0.873	0.929	0.949
700	Proposed [11]	**0.710**	**0.870**	**0.903**	**0.956**
		0.707	0.858	0.883	0.915

observed that the fingerprint of VPF is inefficient in videos containing moving complex texture while the proposed method still performs well. In such videos, magnitudes of high-frequency components in motion residual are relatively large, (e.g. "waterfall" sequence). On the other hand, the improvement of the proposed method compared with [11] is marginal when perceptual quality of the recompressed video is very low (e.g. $B_1 = 100$ kbps or $B_2 = 100$ kbps). It is because the severe quantization error may be propagated to following P-frames in the same GOP and such quantization error may degrade the block artifact of I-P frames measured by the post-filtering method. However, all the values of the AUC still arrive 0.95 when $B_1 \leq B_2$ for our method.

5 Conclusion

In this paper, a novel double MPEG-4 compression detection scheme with analysis of block artifact is proposed. The block artifact measurement is combined with VPF using the adjustable parameter α to get more robust detection ability. Experimental results show that the proposed scheme provides better discriminative performances compared with [11]. The future works focus on finding another deblocking algorithm to reduce the influence of low compression bitrate and extending this scheme to other video compression standards.

Acknowledgments. This work was supported by the National Natural Science Foundation of China (No. 61272249, 61272439), and the Specialized Research Fund for the Doctoral Program of Higher Education (No. 20120073110053).

References

1. Milani, S., Fontani, M., Bestagini, P., Barni, M., Piva, A., Tagliasacchi, M., Tubaro, S.: An overview on video forensics. APSIPA Trans. Signal Inf. Process. **1**, e2 (2012)
2. Wang, W., Farid, H.: Exposing digital forgeries in video by detecting double mpeg compression. In: 8th Workshop on Multimedia and Security, pp. 37–47. ACM (2006)

3. Wang, W., Farid, H.: Exposing digital forgeries in video by detecting double quantization. In: Proceedings of the 11th ACM Workshop on Multimedia and Security, pp. 39–48. ACM (2009)

4. Xu, J., Su, Y., Liu, Q.: Detection of double MPEG-2 compression based on distributions of DCT coefficients. Int. J. Pattern Recogn. Artif. Intell. **27**(01), 1354001 (2013)

5. Chen, W., Shi, Y.Q.: Detection of double MPEG compression based on first digit statistics. In: Kim, H.-J., Katzenbeisser, S., Ho, A.T.S. (eds.) IWDW 2008. LNCS, vol. 5450, pp. 16–30. Springer, Heidelberg (2009)

6. Sun, T., Wang, W., Jiang, X.: Exposing video forgeries by detecting MPEG double compression. In: 2012 IEEE International Conference on Acoustics, Speech and Signal Processing (ICASSP), pp. 1389–1392. IEEE (2012)

7. Subramanyam, A., Emmanuel, S.: Pixel estimation based video forgery detection. In: 2013 IEEE International Conference on Acoustics, Speech and Signal Processing (ICASSP), pp. 3038–3042. IEEE (2013)

8. Jiang, X., Wang, W., Sun, T., Shi, Y.Q., Wang, S.: Detection of double compression in MPEG-4 videos based on markov statistics. Signal Process. Lett. IEEE **20**(5), 447–450 (2013)

9. Luo, W., Wu, M., Huang, J.: MPEG recompression detection based on block artifacts. In: Electronic Imaging 2008, pp. 68190X–68190X. International Society for Optics and Photonics (2008)

10. Bestagini, P., Allam, A., Milani, S., Tagliasacchi, M., Tubaro, S.: Video codec identification. In: 2012 IEEE International Conference on Acoustics, Speech and Signal Processing (ICASSP), pp. 2257–2260. IEEE (2012)

11. Vázquez-Padin, D., Fontani, M., Bianchi, T., Comesana, P., Piva, A., Barni, M.: Detection of video double encoding with gop size estimation. In: IEEE International Workshop on Information Forensics and Security (WIFS) (2012)

12. Park, H., Lee, Y.L.: A postprocessing method for reducing quantization effects in low bit-rate moving picture coding. IEEE Trans. Circ. Syst. Video Technol. **9**(1), 161–171 (1999)

Author Index

Alagar, Vangalur 766
Alaskar, Haya 752
Aljaaf, Ahmed J. 101
Al-Jumaily, Mohammed 101
Al-Jumeily, Dhiya 101, 752
Auephanwiriyakul, Sansanee 669

Bevilacqua, Vitoantonio 264, 759
Bian, Zhao-Ying 745
Biasi, Luigi 264

Calvo, Hiram 491
Cao, Li-Li 686
Caporusso, Nicholas 264
Carnimeo, Leonarda 115
Castro-Cabrera, Carlos Eduardo 187
Cervantes, Jair 72, 79
Chen, Bo 745
Chen, Fang 363
Chen, Jie-Min 375, 453
Chen, Liang 34
Chen, Peng 323, 686, 693
Chen, Qiao-ling 34
Chen, Qiumei 354
Chen, Wen-Sheng 745
Chen, YanFeng 251
Chen, Zehao 419, 651
Chen, Zhong 510
Chen, Ziyi 13
Chiroma, Haruna 387
Choi, Kup-Sze 738
Chou, Yongxin 723

Dalfino, Giuseppe 759
Deng, Zhaohong 738
Di Palma, Anna Maria 759
Ding, Zuo-hua 607
Drlik, Martin 637
Du, Ji-Xiang 233

Encheva, Sylvia 179
Ese, Torleiv 179

Fan, Guolong 243
Fang, Jianwen 363
Fergus, Paul 101, 752
Figueroa-García, Juan Carlos 187
Frazao, Luis 669
Fu, JinLan 471

Gao, Ying-Lian 679
García-Lamont, Farid 72, 79
Gelbukh, Alexander 491
Gesualdo, Loreto 759
Ghazali, Rozaida 387
Gong, Dunwei 87
Gong, Shu 151

Han, Fei 94
Han, Lei 662
He, Peisong 787
Herawan, Tutut 387
Hong, SanLiang 251
Hou, Lei 275
Hu, Guangqin 208, 730
Hu, Jue-liang 607
Hu, ShanShan 693
Huang, Huajuan 43
Huang, Min 561, 574, 586
Huang, Yu-An 713
Hussain, Abir Jaafar 101, 752

Islam, M.M. Manjurul 538

Ji, Dong-hong 662
Ji, Zhiwei 3
Jia, Zhi-juan 34
Jiang, Chao 441
Jiang, Xinghao 787
Jin, Lu 626
Jing, Yun 318
Jo, Kang-Hyun 289

Kang, Hee-Jun 551
Kapusta, Jozef 637

Khan, Sheraz A.   526, 538
Kim, Jong-Myon   526, 538
Kleschev, Alexander   519
Kobayashi, Yoshinori   304
Kuno, Yoshinori   304

Lam, Antony   304
Lasisi, Ayodele   387
Lau, Raymond Y.K.   109
Lee, Hong-Hee   139
Li, Ce   297
Li, Chengdong   167
Li, Fuliang   561, 574, 586
Li, Hai-Tao   706
Li, Hongyu   282
Li, Jin   354
Li, Jing-Jing   453
Li, Li   471
Li, Ming   297
Li, Shuai   713
Li, Taisong   463
Li, Xiao-Ping   323
Li, Xue   251
Li, Yan   407
Li, Zhanhu   781
Liang, Qingjun   282
Liu, Feng   407
Liu, Gang   713
Liu, Guodong   199
Liu, Haitao   341
Liu, Jin-Xing   679
Liu, Tian-Lan   397
Liu, Tingting   561
López, Asdrúbal   79
López-Chau, Asdrúbal   72
Lu, Jianwei   282
Luo, Ming   453
Luo, Qifang   65
Lv, Gang   331

Ma, Chenglong   463
Ma, Jian-Hua   745
Ma, Jinwen   13
Ma, Qiang   297
Mannan, Md. Abdul   304
Mao, Yan-qin   626
Mastrofilippo, Nico   759
Mastronardi, Giuseppe   264
Medo, Michal   429

Meng, Guanmin   3
Munk, Michal   429, 481, 500, 637
Munková, Daša   481, 500

Naveed, Munir   618
Nitti, Rosamaria   115
Niu, Lu   55

Pepe, Antonio   264
Pietroleonardo, Nicola   759
Pilkova, Anna   429

Qian, Yang   318
Qiao, Shilei   65
Qin, Jun   397
Qiu, Jie   471

Radi, Naeem   101, 752
Rashedul Islam, Md.   526
Ren, Han   662
Ren, Weina   167
Rodriguez, Lisbeth   79
Rossini, Michele   759
Ruan, Jiechang   151
Ruiz Castilla, José S.   79

Santos, Rui   669
Shalfeeva, Elena   519
Shang, Jun-Liang   679
Shao, Wenkai   151
Shen, Su-bin   626
Shi, Yubo   311
Stranovská, Eva   500
Sun, Fenglin   87
Sun, Jing   87
Sun, QiYan   275
Sun, Shiliang   700
Sun, Tanfeng   787
Sun, Xiaoyan   87
Sun, Yinchu   586
Sun, Zhan-Li   318
Sun, Zhi-Qiang   34
Svec, Peter   429

Tang, Wen-Sheng   397
Teixeira, Miguel Cacho   669
Theera-Umpon, Nipon   669
Tran, Minh-Duc   551
Tran, Quoc-Hoan   139

Triggiani, Vito   759
Trueba, Adrián   79

Wahyono,   289
Wan, Kaiyu   766
Wan, Xin   463
Wang, Bing   323, 693
Wang, Dong   679
Wang, Guanjin   738
Wang, Huidong   167
Wang, Jing   233, 471
Wang, Renzheng   574
Wang, Rui   65
Wang, Shaokai   109
Wang, Sheng-Chun   397
Wang, Shilin   787
Wang, Shulin   363
Wang, Xiao-Feng   331
Wang, Xingwei   561, 574, 586
Wang, Xuan   275
Wang, Yajie   127, 159
Wang, Yanjun   706
Wang, Ya-Ping   318
Wang, Yazhen   208, 730
Wang, Yongxiong   311
Wang, Zhengzi   199
Wang, Zhenhua   25
Wei, Ran   706
Wei, Xiuxi   43
Wong, Leon   713
Wu, Bu-Xiao   375
Wu, QingXiang   251, 275
Wu, Ran   341
Wu, Zhijun   598

Xia, Junfeng   706
Xia, Qibiao   3
Xia, Sen   323
Xiao, Jing   375, 453
Xiao, Limei   297
Xie, HaiHui   251
Xie, Zhihua   199

Xiong, Shengwu   510
Xu, Yong   679

Yang, Liu   341
Yang, Xuemei   25, 700, 781
Yang, Zijiang   441
Ye, Yunming   109, 407
Yi, Jianqiang   167
You, Zhu-Hong   713

Zhai, Chuan-Min   233
Zhang, Aihua   297, 723
Zhang, Chen   745
Zhang, De-Xiang   318
Zhang, Haitao   598
Zhang, Hong   221
Zhang, Jian-Ming   94
Zhang, Jing   208, 730
Zhang, Jun   323, 686, 693
Zhang, Lin   282
Zhang, Ruiling   510
Zhang, Shao-zhen   607
Zhang, Weiyuan   127, 159
Zhang, Wenping   221
Zhang, Xiaofeng   407
Zhang, Xinfeng   208, 730
Zhang, Yan   463
Zhang, Yanbang   243
Zhang, Ying   55
Zhang, Yu-Hui   233
Zhang, Zhen   463
Zhang, Zhi-Shui   686
Zhao, Junlong   55
Zhao, Min-Ru   94
Zhao, Weidong   341
Zhao, Yahong   407
Zheng, Chun-Hou   679, 706
Zhou, Xiao-Hui   745
Zhou, Yongquan   65
Zhou, Yuxiang   65
Zou, Le   331
Zou, Qing-Hua   745

Printed in the United States
By Bookmasters